Identification of Unusual Pathogenic Gram-Negative Aerobic and Facultatively Anaerobic Bacteria

second edition

Identification of Unusual Pathogenic Gram-Negative Aerobic and Facultatively Anaerobic Bacteria

second edition

ROBBIN S. WEYANT, Ph.D.
C. WAYNE MOSS, Ph.D.
ROBERT E. WEAVER, M.D., Ph.D.
DANNIE G. HOLLIS, M.S.
JEAN G. JORDAN, B.S.
ELLEN C. COOK, M.S.
MARYAM I. DANESHVAR, Ph.D.

Emerging Bacterial and Mycotic Diseases Branch
Division of Bacterial and Mycotic Diseases
National Center for Infectious Diseases
Centers for Disease Control and Prevention
U.S. Public Health Service
U.S. Department of Health and Human Services
Atlanta, Georgia 30333

Photographs by James D. Howard and William A. Clark

Williams & Wilkins
A WAVERLY COMPANY

BALTIMORE • PHILADELPHIA • LONDON • PARIS • BANGKOK
BUENOS AIRES • HONG KONG • MUNICH • SYDNEY • TOKYO • WROCLAW

Editor: William R. Hensyl
Managing Editor: Kim Jones
Production Coordinator: Raymond E. Reter
Copy Editor: Caral Shields Nolley, Sherri Wolf
Illustration Planner: Ray Lowman
Typesetter: Graphic Sciences Corporation, Cedar Rapids, Iowa
Printer & Binder: Edwards Brothers, Inc., Ann Arbor, Michigan
Digitized Illustrations: Trinity Graphics, Cherry Hill, New Jersey

Cover design and text design Copyright © 1996 Williams & Wilkins

351 West Camden Street
Baltimore, Maryland 21201-2436 USA

Rose Tree Corporate Center
1400 North Providence Road
Building II, Suite 5025
Media, Pennsylvania 19063-2043 USA

All rights reserved. This book is protected by copyright. No part of this book may be reproduced in any form or by any means, including photocopying, or utilized by any information storage and retrieval system without written permission from the copyright owner.

Accurate indications, adverse reactions and dosage schedules for drugs are provided in this book, but it is possible that they may change. The reader is urged to review the package information data of the manufacturers of the medications mentioned.

Printed in the United States of America

First Edition, 1984

Library of Congress Cataloging-in-Publication Data

Identification of unusual pathogenic gram-negative aerobic and facultatively
  anaerobic bacteria / Robbin S. Weyant . . . [et al.] :
  with photographs by James D. Howard and William A. Clark. — 2nd ed.
    p. cm.
  Includes bibliographical references and index.
  ISBN 0-683-00615-0
    1. Anaerobic bacteria—Laboratory manuals.   2. Gram-negative
  bacteria—Laboratory manuals.   I. Weyant, Robbin S.
    [DNLM: 1. Gram-Negative Bacteria—laboratory manuals.   2. Gram-
  Negative Aerobic Bacteria—classification.   3. Gram-Negative
  Facultatively Anaerobic Rods—classification.   QW 25 I192 1995]
  QR89.5.I34   1995
  616'.0145—dc20
  DNLM/DLC
  for Library of Congress   95-12416
                                                                             CIP

The publishers have made every effort to trace the copyright holders for borrowed material. If they have inadvertently overlooked any, they will be pleased to make the necessary arrangements at the first opportunity.

To purchase additional copies of this book, call our customer service department at **(800) 638-0672** or fax orders to **(800) 447-8438.** For other book services, including chapter reprints and large quantity sales, ask for the Special Sales department.

Canadian customers should call **(800) 268-4178,** or fax **(905) 470-6780.** For all other calls originating outside of the United States, please call **(410) 528-4223** or fax us at **(410) 528-8550.**

Visit Williams & Wilkins on the Internet: **http://www.wwilkins.com** or contact our customer service department at **custserv@wwilkins.com.** Williams & Wilkins customer service representatives are available from 8:30 am to 6:00 pm, EST, Monday through Friday, for telephone access.

96 97 98 99 00
1 2 3 4 5 6 7 8 9 10

Dedication—Elizabeth O. King

The research of bacteriologist Elizabeth O. King (1912–1966) forms the nucleus of this manual, which we dedicate to her.

Ms. King was born in Atlanta in 1912. She received a Bachelor of Science in Zoology degree from the University of Georgia and a Master of Science in Medical Technology degree from Emory University. During World War II she served as a commissioned officer (bacteriology) in the Army at Fort Detrick, Frederick, Maryland. From 1946–1948 she worked in the Emory University Hospital. In 1948, Ms. King joined the staff of the Communicable Disease Center, now the Centers for Disease Control and Prevention (CDC) in Atlanta, where she worked until her untimely death from cancer in April 1966.

Ms. King's first assignment at CDC was in the Diphtheria Laboratory, where she worked for three years with Dr. Elizabeth Parsons. In 1951, she transferred to the newly established General Bacteriology Laboratory, where she worked under the Director, Dr. Martha K. Ward. There she became intensely interested in the Gram-negative rods that are associated with disease, but which do not belong to the family *Enterobacteriaceae*. Many of them were poorly classified. Having worked in two different hospital laboratories, Ms. King was keenly aware of the diagnostic dilemma these bacteria presented. With characteristic concentration and determination, she set herself to the task of bringing some order into the classification of these organisms. This she did by means of a systematic approach, a keen power of observation, and a zeal for her task.

Ms. King was internationally known as an authority on a variety of unusual bacteria. Her rational and systematic approach led to the characterization and establishment of *Flavobacterium meningosepticum* as a new species of bacteria and the proof that this organism was an etiological agent of nosocomial epidemics of meningitis in infants.

The same approach was used to simplify the identification of such organisms as *Acinetobacter* species (then *Herellea vaginicola* and *Mima polymorpha*), *Pasteurella* species, and *Campylobacter* (then *Vibrio*) *fetus*. Ms. King was also among the first workers in the United States to perceive the importance of *Listeria* in human infections and to alert the medical profession to the prevalence of listeriosis.

Her observations were generously transmitted to others who sought her advice through correspondence, bench training, and formal courses at CDC.

Ms. King's contributions and devotion to her profession were an inspiration to those who had the good fortune of knowing her. This manual brings together in a single volume the legacy of information from Ms. King's own work and from that of her colleagues.

Foreword

A Note from the Editors

It was a distinct pleasure for those of us at the Editorial Office of the Bergey's Manual Trust to help bring this second edition of an important volume to print. We believe this is a valuable contribution from one of the leading medical laboratories in the world, which has compiled immense amounts of data on unusual pathogens and herein present clear and useful devices for their identification. The book should find wide use by workers in hospital and public health laboratories who frequently encounter bacteria, not normally thought of as pathogenic, causing human and animal infections.

We are indebted to Dr. Weyant and his colleagues at the Centers for Disease Control and Prevention for the excellence of the manuscript and the cooperative and cordial relationship they have maintained with our office. We would also like to acknowledge the fine help of Beverly J. Weber with a portion of the preliminary word processing.

John G. Holt, *Editor-in-Chief*
Betty J. Caldwell, *Editorial Assistant*
BERGEY'S MANUAL TRUST

Preface to the Second Edition

The first edition of this manual, published in 1984, presented an approach to identification of unusual Gram-negative pathogenic bacteria based on morphological and biochemical data from more than 60,000 strains studied by Elizabeth O. King and her successors in the Special Bacteriology Reference Laboratory (SBRL) beginning in the 1950s. It was an expansion and an update of Ms. King's original "Round-Table" and the revisions that had been made in the 1970s. For the first time, written descriptions, photographs depicting colonial and cellular morphology, and biochemical test methods for the identification of unusual species and unknown "groups" were included. The purpose of publishing this information was to make available to clinical bacteriologists the standard approach used by this laboratory in the identification of rarely encountered unusual pathogens.

In response to many requests from clinical and public health microbiologists, we have revised and updated the manual. For many organisms described in the first edition, additional data collected over the last decade have been added to the biochemical tabulations. For other organisms, the 1984 tables have been retained. In some cases, the tables were not updated because the number of strains previously tabulated was considered sufficient to demonstrate the biochemical variability within the species or group. In other cases, the tables were not updated because few, if any, additional cultures were received. As clinical laboratories have gained experience in identifying the more commonly encountered species [e.g., *Pseudomonas cepacia* (recently proposed as *Burkholderia cepacia*), *Eikenella corrodens*, *Xanthomonas maltophilia* (recently proposed as *Stenotrophomonas maltophilia*), and *Shewanella putrefaciens*], fewer of these isolates have been submitted to the SBRL.

The most significant change in the second edition is the addition of a section on the use of gas-liquid chromatographic analysis of whole-cell fatty acids for bacterial identification. This section is the result of more than 25 years of systematic study of unusual Gram-negative pathogens by the Analytical Chemistry Laboratory, Division of Bacterial & Mycotic Diseases, Centers for Disease Control and Prevention. The inclusion of both biochemical and cellular fatty acid information in this edition is representative of the long-standing collaboration between these two laboratories.

This manual is designed so that the biochemical and chromatographical sections can be used separately or in combination for the identification of an unknown isolate. In instances where fatty acid profile analysis is especially useful in differentiating biochemically similar species, references in the biochemical identification keys are made to the appropriate chromatographic descriptions. For laboratories with both biochemical and chromatographic capabilities, these methods can be used in combination to confirm the identification of an isolate.

The scope of the second edition has evolved somewhat, in keeping with the changing responsibilities of the SBRL. To the best of our knowledge, all proposed taxonomic changes through April 1994 have been noted. Recently emerging pathogens, such as *Afipia*, *Bartonella*, and *Roseomonas* species have been added, along with additional unnamed "groups." Conversely, biochemical and morphological descriptions of selected *Enterobacteriaceae*, *Vibrionaceae*, and *Legionellaceae* species that were included in the previous edition have been deleted from the second edition since these analyses are now performed in other CDC reference laboratories. The fatty acid profiles for some of these organisms, however, have been included so that they may be differentiated from the other groups and genera covered more completely in this edition. A few of the more frequently encountered pathogens, such as *Neisseria gonorrhoeae*, *N. meningitidis*, *Haemophilus influenzae*, *Bordetella pertussis*, and *Francisella tularensis* have been retained in this edition, even though the identification of these species is performed in other CDC reference laboratories. These organisms have been retained so that the reader can compare them with the less frequently encountered species presented in this manual.

The information presented in this manual represents our experience with known reference strains that we have obtained and patient isolates that have been submitted to us. This information should not be considered all-inclusive with respect to the identification of unusual Gram-negative bacteria. With the increasing incidence of immunosuppression in humans, due to either disease processes or as a by-product of therapy, the probability of encountering an opportunistic agent that we have not studied has increased. In fact, many clinical isolates that are submitted to us cannot be identified to a specific species at the time we study them. We place these isolates into various CDC "groups," which are classified taxonomically at a later date. The possibility of a new or previously undescribed organism should always be considered when studying an isolate that does not conform to the species or groups presented herein.

Preface to the First Edition

The Special Bacteriology Reference Laboratory, Division of Bacterial Diseases, Center for Infectious Diseases, Centers for Disease Control (CDC), Atlanta, Georgia, has on file the recorded characteristics of more than 60,000 unusual pathogenic bacteria that have been submitted for identification over the past 30 years. Until now, only a small portion of these data has been available in published or summary form. Bacteriologists in hospitals and clinical laboratories need ready access to these data to help them identify these strange and unusual organisms. To meet this need, this manual presents descriptions of species and as yet unnamed "groups," the compiled detailed characteristics of strains examined, laboratory media and methods used in identification, photographs of colonies and photomicrographs of stained preparations of selected strains of the various species and groups, sources of strains submitted for identification, and pertinent references.

In 1948, Ms. Elizabeth O. King came to CDC when the organization, established to study malaria and the exotic diseases brought back by servicemen and women in World War II, was in its infancy. CDC needed a laboratory to study organisms that the State health departments were having difficulty identifying. So CDC established the General Bacteriology Laboratory under Dr. Martha K. Ward, and in 1951, Ms. King was appointed to the laboratory staff because of her broad background in bacteriology.

Ms. King spent much time in research, trying to match up these unidentified organisms with known cultures obtained from culture collections and from specialists throughout the world.

Each culture that was submitted for identification by the State laboratories was propagated, frozen, and stored at below 42° C, thus preserving it for future study and comparison. The large collection thus amassed is still extant.

Each culture (as submitted and as a subculture) was stained for microscopic examination and inoculated into a battery of test media. The results were recorded over a 7-day period (some were observed for 2–3 weeks for certain characteristics helpful in identification). Known cultures for many named species were obtained and studied with the same battery of tests, so that valid comparisons could be made between known and unknown cultures. (Ms. King would never send out a report identifying an organism with a named species until she had studied a known culture of that species. If she could not be quite sure, she would merely report "resembles" a certain organism.) Results were recorded on data cards illustrated in Figure 1.

The culture data being collected were compiled first on McBee punchcards and later on cards designed by Dr. Ward and Ms. King (Fig. 2). Data presented in the charts in this manual were compiled from the second set of cards.

In 1964, at a round table meeting entitled "Current Trends in Diagnostic Microbiology," Ms. King first presented her "Identification of Unusual Gram-Negative Pathogenic Bacteria." There, in Washington, DC, at the annual meeting of the American Society for Microbiology, Ms. King gave the results of research she had done since 1951 in the form of the "Round-Table." From that time until the present, the General Bacteriology Laboratory (now the Special Bacteriology Reference Laboratory) has collected more information and data; its staff, under Dr. Robert E. Weaver, has revised and enlarged Ms. King's original "Round-Table" several times (1967, 1972, 1974).

Ms. King used the "Round-Table" in teaching the Laboratory Methods in Special Medical Bacteriology Course at CDC, and the Bacteriology Training Branch, with the help of Dr. Weaver's staff, has continued this practice. Thus, since 1964, CDC has published and distributed this document to hundreds of clinical bacteriologists. Now, using the "Round-Table" as a basis, we present this manual, an expanded and up-to-date volume, to clinical bacteriologists, as a further aid in identifying "strange and unusual bacteria" of clinical significance.

The user of this manual should keep in mind that many bacteriologists, working over a period of 25 years, recorded the culture results from which these data have been compiled. During that time, there were changes in the formulations of some of the media and in the procedures for some of the tests. These circumstances have undoubtedly produced an unknown degree of error in the data.

CULTURE NO.	SENDER'S NO	DATE REC'D.	IDENTIFIED AS:
SENDER:	STATE H.D.		
SOURCE			PATIENT
MORPHOLOGY OF ORIGINAL CULT.: GRAM			
PURE CULT.: H.I. SLANT (24 HRS.): GRAM			
H.I. BROTH (24 HRS.)			FLAGELLA
TYPE OF GROWTH: BLOOD AGAR (H/R) (24 HR. CO_2)			
			A.F.: SPORE:

H.I. SLANT: H.I. BROTH:

GROWTH TEMP.: 5° 25° 35° 42° 52° O_2 REQUIREMENTS ODOR

BASE	24	48	1 wk		BASE	24	48	1 wk		BASE	24	48	1 wk		24	
GLUCOSE					CATALASE					TSI SI / BUTT					LITMUS	48
D-XYLOSE					OXIDASE K/ROUT.					H_2S PAPER BUTT					MILK	
MANNITOL					MAC CONKEYS					MR /VP ROUT.						1 wk
LACTOSE					S.S.					MR /VP SPEC.					PIGMENT; GROWTH	
SUCROSE					CITRATE					MOTILITY WP DIFCO						
MALTOSE					CETRIMIDE					GELATIN					WATER SOLUBLE:	
FRUCTOSE					UREA SERUM PLAIN					ESCULIN HYD.					FLO.	
CONTROL					NITRATE INF. ROUT.					% 0					TECH.	
10% SLANT LACTOSE					INDOLE INF. ROUT.					N_aCl 6					HIT.	

CDC 52.55 (Formerly 56.16)
Rev. 3/93

LABORATORY DATA CARD

Figure 1A

BASE	24	48	1 wk		BASE	24	48	1 wk		PAI:
GLUCOSE					GLYCOGEN					SEROLOGY, NOTES
D-XYLOSE					ERYTHRITOL					
MANNITOL					MELIBIOSE					
LACTOSE					MELEZITOSE					
SUCROSE					STARCH					
MALTOSE					BLANK					
GLYCEROL					FAT HYD.					
SALICIN					NUT. AGAR 25/35					
L-ARABINOSE					TOMATO JUICE					
ADONITOL					KETOGLUCONATE					
DULCITOL					ACETAMIDE					
D-GALACTOSE					SERINE					
FRUCTOSE					TARTRATE					
MANNOSE										
RHAMNOSE										
TREHALOSE										
RAFFINOSE					PPA.					
SORBITOL					LYSINE					
INOSITOL					ARGININE					
CELLOBIOSE					ORNITHINE					
INULIN					BLANK					
DEXTRIN					KCN					

Figure 1B

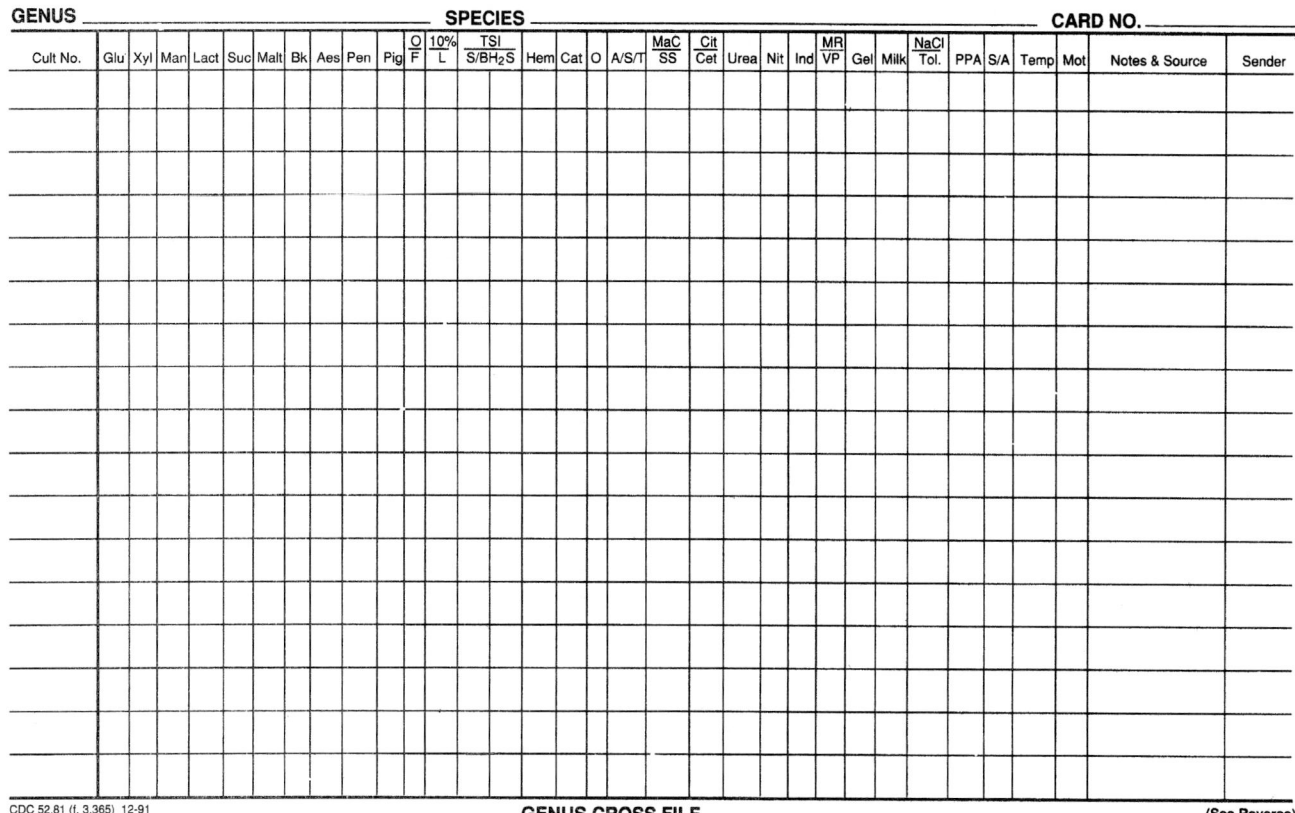

Figure 2A

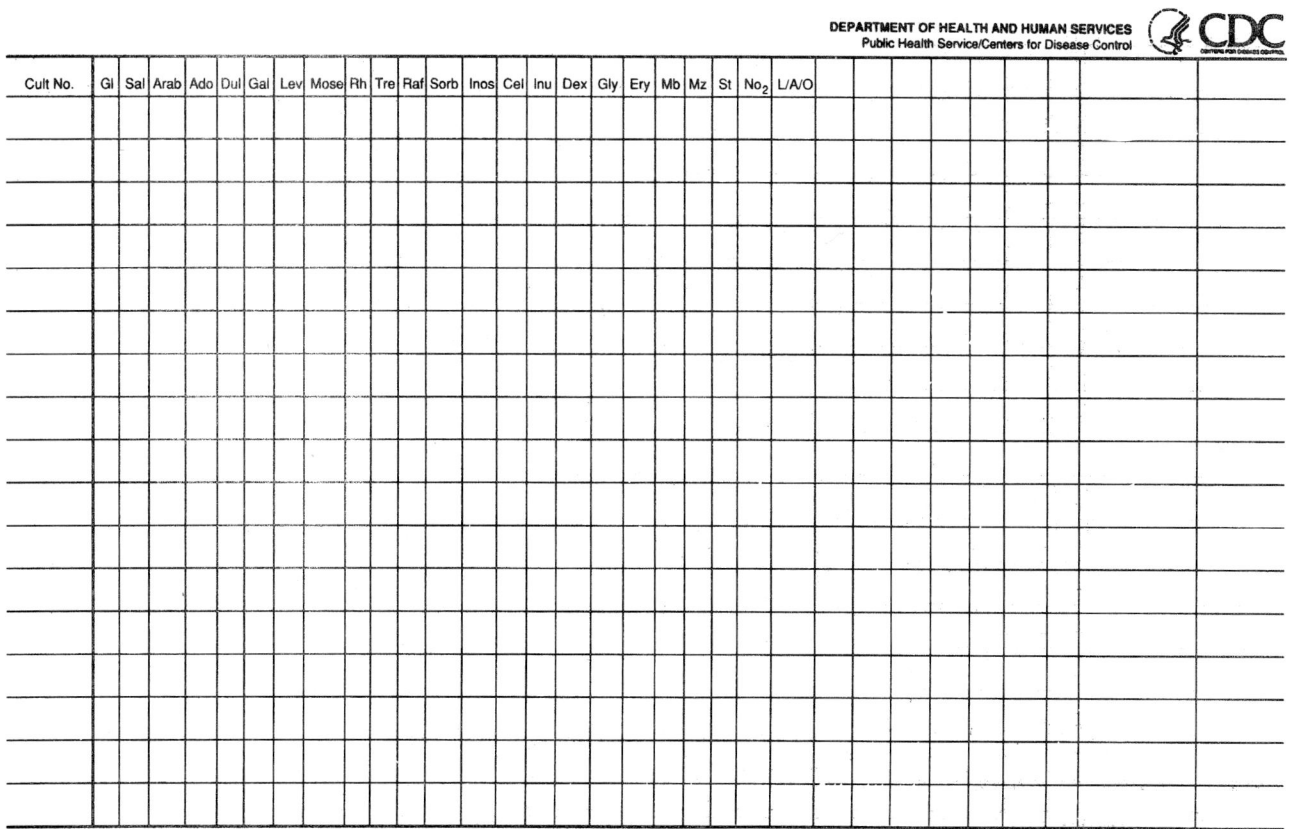

Figure 2B

Acknowledgments

We acknowledge the support that Dr. Albert Balows, then Assistant Director for Science, Center for Infectious Diseases, Centers for Disease Control, gave us in writing for the first edition of this manual. Mr. Earl Long, Director (1961–1982) of the Laboratory Section, Georgia Department of Human Resources (GDHR), generously contributed facilities and resources for part of the work.

Mr. Harvey Tatum, through his close association as a laboratory investigator with Ms. King and his meticulous work with the organisms dealt with in the manual, contributed greatly to the reliability of the data. He also provided valuable information for writing the historical background alluded to in the Preface and first section (Introduction) of the manual. The chapter on "Miscellaneous Gram-Negative Bacteria," written by Mr. Tatum (with W. H. Ewing and R. E. Weaver) in the second edition of the Manual of Clinical Microbiology (American Society for Microbiology, Washington, DC, 1974) was a valuable source of information.

In addition, we thank Dr. John Holt and Ms. Bets Caldwell of the Bergey's Manual Trust for supplying clerical and editorial assistance in the production of the second edition of this manual; and Dr. Don J. Brenner, Chief, Investigation and Surveillance Laboratory Section, Centers for Disease Control and Prevention for his generous support and review of the second edition.

We also thank present and former members of the Special Bacteriology Reference Laboratory staff and former members of the Analytical Chemistry Laboratory for helping to develop these data, Sara R. Cote for labeling the chromatograms, and Mr. Ralph Ramsey (GDHR) for his help with the photography.

We are also grateful for the patience, support, and understanding of our families during the preparation and writing of this manual.

Table of Contents

Dedication—Elizabeth O. King v
Foreword vii
Preface to the Second Edition ix
Preface to the First Edition xi
Acknowledgments xv
Recent Nomenclature Changes xix
Symbols and Abbreviations xxi
Symbols for CDC "Groups" with Published Names xxv
Evaluation of Clinical Bacterial Specimens Submitted for Characterization and Identification ...1
Media and Methods 4
Introduction to King's Identification Key 24

1 Gram-Negative Organisms: An Approach to Identification (Guide to Presumptive Identification) ...25
   Media Used for Presumptive Identification 27
   Presumptive Identification Key 30
   Presumptive Identification Tables 33
   Guide to Presumptive Identification: Code Book 66

2 Bacterial Identification Using King's Key and "Round-Table" Charts 109
   Key for Identification of Unusual Pathogenic Gram-Negative Aerobic and Facultatively Anaerobic Bacteria ..110
   Index to "Round-Table" Charts 128
   Charts 1–46 130

3 Descriptions of Species 223

4 Bacterial Identification by Cellular Fatty Acid Analysis 565
   Introduction 566
   List of Figures 570
   List of Tables 573
   Protocols 576
   Gas Chromatograms and Cellular Fatty Acid Composition Tables 581
   Names of Fatty Acids Found in Bacteria 583

Index 723

Recent Nomenclature Changes
November 1994

The following nomenclature changes recently have been proposed. Listed below are the names used in this manual and the proposed name changes. Unless otherwise indicated, the reference for the name change is cited in the individual species description.

Name used in this manual	Proposed name (reference)
Flavobacterium breve	*Empedobacter brevis* (2)
Flavobacterium gleum	*Chryseobacterium gleum* (2)
Flavobacterium indologenes	*Chryseobacterium indologenes* (2)
Flavobacterium meningosepticum	*Chryseobacterium meningosepticum* (2)
Flavobacterium mizutaii	*Sphingobacterium mizutaii*
Flavobacterium multivorum	*Sphingobacterium multivorum*
Flavobacterium spiritivorum	*Sphingobacterium spiritivorum*
Flavobacterium thalpophilum	*Sphingobacterium thalpophilum*
Flavobacterium yabuuchiae	*Sphingobacterium yabuuchiae*
Pseudomonas cepacia	*Burkholderia cepacia*
Pseudomonas diminuta	*Brevundimonas diminuta* (1)
Pseudomonas gladioli	*Burkholderia gladioli*
Pseudomonas mallei	*Burkholderia mallei*
Pseudomonas pickettii	*Burkholderia pickettii*
Pseudomonas pseudomallei	*Burkholderia pseudomallei*
Pseudomonas vesicularis	*Brevundimonas vesicularis* (1)
Weeksella zoohelcum	*Bergeyella zoohelcum* (2)
Xanthomonas maltophilia	*Stenotrophomonas maltophilia*

1. Segers, P., M. Vancanneyt, B. Pot, U. Torck, B. Hoste, D. Dewettinck, E. Falsen, K. Kersters, and P. De Vos. 1994. Classification of *Pseudomonas diminuta* Leifson and Hugh 1954 and *Pseudomonas vesicularis* Büsing, Döll, and Freytag 1953 in *Brevundimonas* gen. nov. as *Brevundimonas diminuta* comb. nov. and *Brevundimonas vesicularis* comb. nov., respectively. Int. J. Syst. Bacteriol. 44: 499–510.

2. VanDamme, P., J.-F. Bernardet, P. Segers, K. Kersters, and B. Holmes. 1994. New perspectives in the classification of the flavobacteria: description of *Chryseobacterium* gen. nov., *Bergeyella* gen. nov., and *Empedobacter* nom. rev. Int. J. Syst. Bacteriol. 44: 827–831.

Symbols and Abbreviations

Symbols

[ ]	key tests
( )	late reaction, 3 to 7 days
+	positive reaction (90% or more strains tested were positive) within 48 h, except with gelatin, within 14 days
+, gas	with nitrate test, indicates nitrates reduced to nitrites, incomplete reduction of nitrites with gas formation
−	negative reaction (10% or less strains tested were positive); with action on blood, indicates no reaction
+ or (+)	positives and late positives together total 90% or more
>	more than
<	less than
←	from

Abbreviations

A	acid
AFB	Air Force Base (Brooks)
al or α	α-hemolysis
amb	amber
ATCC	American Type Culture Collection (Rockville, MD)
BAP	rabbit blood agar plate
BBL	"Baltimore Biological Laboratory," a Division of Bioquest (Becton, Dickinson, and Company, Cockeysville, MD)
β	β-hemolysis around isolated colonies
B-G	Bordet-Gengou
br	brown (pigment); browning of blood; broad rod
bsrs	bipolar staining short straight rod
C	Celsius
c	coagulation
CA	cysteine agar
cc	coccobacillus
cc,pr	paired coccobacilli
cd	diplococcus
CDC	Centers for Disease Control & Prevention (Atlanta, GA)
ch	chains
char	chartreuse
chs	short chains
cl	clot
co	coral
CTA	cystine trypticase agar (BBL)
CYE	cysteine yeast extract (agar)
DF	dysgonic (fastidious) fermenter
dskfil	disk-like "roll of coins" morphology

EF	eugonic (nonfastidious) fermenter
EO	eugonic oxidizer
F	fermentative (glucose); Enteric fermentation base
FA	fluorescent antibody
FB	flagella broth
fc	filaments, curved
G	gas
gl	gliding
gr	greening of blood, usually accompanied by lysis; growth
h	hour(s) (incubation)
HB	initials of a patient
HIA	Heart infusion agar (Difco)
HIB	Heart infusion broth (Difco)
H_2S	hydrogen sulfide
inc	incomplete
ins	insoluble
ins gr	insufficient growth
II	"II-forms," pleomorphism where rod-shaped cells often appear thin to very thin in the central region, with thicker ends
IR	indicator reduced
k	alkaline
l	long
L	lateral
lav	lavender
lg	light growth
LG	lavender-green coloration under heavy growth, indication of proteolysis
lr	long rod
lrch	long rods in chains
lrs	long, straight rods
ltrt	long, thin, tapered rods
ltspcr	long, thin, spindle-shaped, curved rods
ly	diffuse lysis of red cells extending out from heavy growth; no action by individual colonies
m	medium; motile
M	*Moraxella*-like
mbr	medium length broad rod
MM	motility medium
mod	moderate
MR	methyl red, methyl red test
mrc	medium curved rod
mrs	medium straight rod
msr	medium length slender rod
NaCl	sodium chloride
nc	no change
NCTC	National Collection of Type Cultures (London)
neg	negative
ngr	no growth
nm	non-motile
n-o	nonoxidative (glucose)

NO	nonoxidizer
NT	test not done
nut br	nutrient broth
O	oxidative (glucose), oxidizer
occ	occasional
OF	oxidation-fermentation
ONPG	O-nitrophenyl-β-D-galactopyranoside (test)
or	orange
p	polar
p,1–2	1–2 polar (flagella)
p,1–4	1–4 polar (flagella)
p,>2	more than two polar (flagella)
pb ac	lead acetate
pep	peptonization
pe	peritrichous (flagella)
pk	pink
p,L	polar and lateral (flagella), only one or two flagella on most cells
plr	pleomorphic rod
pr	paired
plrs	pleomorphic straight rod
PPA	phenylpyruvic acid
pt	polar tuft
pu	purple
r	rod
rc	curved rod
rs	straight rod
RST	rapid sugar test, also rapid fermentation test (RFT)
RTD	routine test dilution
rv	very short rod
SBRL	Special Bacteriology Reference Laboratory (CDC)
sbr	short broad rod
sc	comma-shaped, spiral
sl	slight, slightly
s-lr	short to long rods
sol	soluble
sp.	species (singular)
spp.	species (plural)
spr	short pleomorphic rods
srs	short straight rod
SS	*Salmonella-Shigella* (agar)
subsp	subspecies
szf	serum zone formation (clear top in milk)
t	thin
T	total
ta	tan
TGY	tryptone glucose yeast extract medium
Thio	Fluid thioglycollate medium (BBL)
TM	Thayer-Martin

tr	trace
trt	thin tapered rods
TSI	triple sugar iron (agar)
tsrs	thin, short, straight rods
v	variable (11–89% tested positive)
vbr	variable length broad rod
V-P	Voges-Proskauer (test)
vps-mr	vacuolated, pale-staining short to medium rod
w	weak, weakly
wh	white
WHO	World Health Organization (Geneva)
WO	Weak oxidizer
WRAMIR	Walter Reed Army Medical Institute of Research (Washington, DC)
yel	yellow

Symbols for CDC "Groups" with Published Names

DF-1	*Capnocytophaga* spp.
DF-2	*Capnocytophaga canimorsus*
DF-2-like	*Capnocytophaga cynodegmi*
EF-6	*Vibrio fluvialis*
EF-9	*Tatumella ptyseos*
EO-1	*Pseudomonas cepacia*
HB-1	*Eikenella corrodens*
HB-5	*Pasteurella bettyae*
M-1	*Kingella kingae*
M-3	*Moraxella atlantae*
M-4	*Oligella urethralis*
M-4f (also IVf)	*Flavobacterium odoratum*
M-5	*Neisseria weaveri*
M-6	*Neisseria elongata* subsp. *nitroreducens*
NO-2	*Bordetella holmesii*
TM-1	*Kingella denitrificans*
Pink coccoid (I through IV)	*Roseomonas* spp.
I	*Pseudomonas maltophilia*
Ia	*Pseudomonas diminuta*
Ib	*Pseudomonas putrefaciens*
IIa	*Flavobacterium meningosepticum*
IId	*Cardiobacterium hominis*
IIf	*Weeksella virosa*
IIj	*Weeksella zoohelcum*
IIk-1	*Pseudomonas paucimobilis*
IIk-2	*Flavobacterium multivorum*
IIIa, IIIb	*Alcaligenes xylosoxidans* subsp. *xylosoxidans*
IVa	*Bordetella bronchiseptica*
IVb	*Bordetella parapertussis*
IVe	*Oligella ureolytica*
IVf (also M-4f)	*Flavobacterium odoratum*
Va-1	*Pseudomonas pickettii*
Va-2	*Pseudomonas pickettii*
Vb-1	*Pseudomonas stutzeri*
Vd	*Ochrobactrum anthropi*
Vd-3	*Agrobacterium radiobacter* biovar 1
Ve-1	*Chryseomonas luteola*
Ve-2	*Flavimonas oryzihabitans*
VI	*Alcaligenes faecalis*

Evaluation of Clinical Bacterial Specimens Submitted for Characterization and Identification

In this section, we describe how clinical bacterial specimens have been evaluated through the years in the Special Bacteriology Reference Laboratory, Division of Bacterial & Mycotic Diseases, National Center for Infectious Diseases, CDC.

Bacterial growth from the submitted specimen is Gram-stained and streaked (quadrant method) onto a 5% defibrinated rabbit blood agar plate (when sheep blood is employed, the hemolytic reactions may vary from those obtained on rabbit blood). The agar is stabbed several times with the inoculating loop in the primary quadrant streak to create an area of reduced oxygen tension for enhancement of hemolysis. Then the plate is inverted and incubated overnight in a candle jar at 35° C. Reactions on blood agar are described below.

After overnight incubation (20–24 h), the blood plate is observed for the amount of growth, colonial morphology, and hemolytic reaction. An isolated colony of the organism is picked to a heart infusion broth (HIB), a heart infusion agar (HIA) slant, and a triple sugar iron (TSI) agar slant (streaked and stabbed). When colonies are very small, several are picked. All of the cultures are incubated for 18–24 h at 35° C in room atmosphere. If the organism appears to grow poorly on the blood agar plate, a blood agar slant is also inoculated; the four tubes of media are incubated in a candle jar at 35° C. After the colony is picked, the oxidase test is performed. After incubation, Gram stains are made from the HIA and HIB cultures, and the cellular morphology is noted. The TSI slant is examined to determine presumptively the culture's mode of carbohydrate utilization (see below, "Oxidation, Fermentation, or Nonutilization of Carbohydrates").

The bacterial isolate, now of known cellular morphology and with a presumed mode of carbohydrate metabolism, is characterized biochemically by using a basic set of differential media: carbohydrates glucose, xylose, mannitol, lactose, sucrose, and maltose in appropriate base [enteric fermentation base is used for fermenters, and oxidation-fermentation (OF) base semisolid agar is used for nonfermenters of glucose]; MacConkey agar slant; SS agar slant; Simmons citrate agar; motility medium; urea agar (Christensen's); nitrate broths (routine and special); nitrite broths (0.1 and 0.01%); tryptone broth for indole; litmus milk; nutrient broth; nutrient broth with 6% sodium chloride; gelatin; HIA slant; esculin agar slant; three tryptose glucose yeast (TGY) agar slants for temperature range testing (25°, 35°, and 42° C); acetamide, serine, and tartrate alkalinization slants; rapid gelatin slant; and TSI agar (a strip of lead acetate paper is suspended above the slant).

Additional media for fermenters are MR-VP broth for the methyl red test, and an additional HIA slant to grow a culture for inoculating an MR-VP broth for the Voges-Proskauer test (Coblentz method).

Additional media for the oxidizers and nonutilizers of carbohydrates are an OF-base control containing no carbohydrates, a cetrimide agar slant, "Flo" and "Tech" agar slants, and a heart infusion tyrosine agar slant.

Special media for *Neisseria* species include the carbohydrates glucose, lactose, sucrose, maltose, and fructose in CTA base.

The decarboxylase tests for lysine and ornithine, the dihydrolase test for arginine, and other tests are useful in characterizing certain microorganisms, as indicated in the tables of biochemical data.

INOCULATION OF DIFFERENTIAL MEDIA

With a straight needle and inoculum from the HIA slant, the Simmons citrate is inoculated lightly on the slant, the motility medium is inoculated by stabbing once about 5 mm under the surface, and the six OF base carbohydrates and the control tube and/or CTA carbohydrates are inoculated by stabbing four times approximately 2.5 cm under the surface (for the *Neisseria* cultures, a heavy inoculum is introduced with a loop and mixed throughout the entire top 5–10 mm of the CTA medium).

With a capillary pipette and the HIB culture, the litmus milk and the gelatin are inoculated with 4 drops of inoculum, and the remaining media with 1 drop of inoculum each. When the agar slants are inoculated, they are rotated to ensure that the inoculum is evenly distributed over the surface of the agar.

When growth of fastidious microorganisms needs to be enhanced, 1–2 drops of sterile rabbit serum are added to each of the media.

After the media are inoculated, they are incubated at 35° C in room atmosphere and maintained under those conditions, unless the growth temperature tests indicate that either 25° or 42° C is closer to optimal temperature.

Fastidious organisms are incubated in the candle jar or CO_2 incubator, unless they are shown not to require such an atmosphere.

All differential media are examined after 1, 2, and 7 days of incubation, with the following exceptions: 1) The TGY slants are examined after 18–24 h of incubation for comparative growth. The tube with the best growth is used for the catalase test. Negative catalase reactions are confirmed by testing all of the TGY slants having growth. If no growth appears at any temperature, the slants are reincubated, or HIA slants are inoculated and incubated at the prescribed temperatures. 2) The nitrate reduction and indole production tests are performed on 48-h cultures. 3) The methyl red test is performed on 48-h cultures (poorly growing cultures are assayed after 4–5 days of incubation). 4) Gelatin liquefaction is tested at 1, 2, 5, 7, and 14 days before a negative reaction is reported.

REACTIONS ON RABBIT BLOOD AGAR

β-hemolysis is the same as that described for the streptococci, except that the total destruction of all the red blood cells within the hemolytic area is not as important as it is for the streptococci. Some microorganisms show little hemolysis surrounding individual colonies, but show marked hemolysis in the stabs. For example, *Listeria monocytogenes* shows very weak hemolytic action under the streaks and usually none at all under isolated colonies. In the stabs, this organism shows a very distinct, narrow, clear zone of hemolysis at 24 h. Other microorganisms show a well-defined zone of hemolysis extending out from the heavy portion of the streak. β-hemolysis usually has a very sharp and well defined edge. The zones in this case seem to be devoid of hemoglobin and appear colorless.

Incomplete β-hemolysis is exemplified by *Gardnerella vaginalis*, which, after 3–4 days of incubation in a candle jar, shows wide but incomplete zones of hemolysis. Even upon prolonged incubation, the hemolytic action never becomes completely clear, nor will the edge of the zone become well defined.

Lysis is the term used when the red blood cells are lysed but the hemoglobin is not decolorized (destroyed) as it is in β-hemolysis. In this case, the edge of the zone is not well defined. Some organisms lyse the whole area of the plate; some produce lysis under the heavy portion of the streak; well-isolated colonies never show any action on the blood.

Greening is the same general action as lysis, except that the lysed area has a greenish tinge.

α-hemolysis is pronounced greening under heavy growth and around well isolated colonies, usually without a lysed area. Some microorganisms show this only on prolonged incubation. These usually show α-hemolysis around stabs and a faint reaction under the heavy growth within 24 h. Generally, these microorganisms are poor growers. There is no lysis of red blood cells.

Browning is a reaction whereby the blood turns dark brown and completely opaque under the area of heaviest bacterial growth. Well isolated colonies sometimes show α-hemolysis.

Lavender-green is a reaction in which lavender and green appear under the areas of heavy growth. Very proteolytic organisms, such as *Pseudomonas aeruginosa, Xanthomonas maltophilia, Aeromonas hydrophila,* and *Flavobacterium meningosepticum* produce this reaction.

Soluble brown pigments are produced by some bacteria, such as melanin-producing strains of *P. aeruginosa*. These pigments diffuse from the heavy portions of growth.

Lysis-inhibition zones occur with some organisms; the blood immediately surrounding an isolated colony remains unchanged, but an area of lysis occurs outside the zone of intact cells.

Compacted red blood cells are noted with some organisms. A well-isolated colony has a very narrow zone of lysis immediately surrounding the colony, and beyond this a "compacted" and narrow dense band of red blood cells occurs.

OXIDATION, FERMENTATION, OR NONUTILIZATION OF CARBOHYDRATES

TSI agar is generally used to differentiate fermentative from oxidative and noncarbohydrate-utilizing microorganisms. An organism that produces in 24 h an acid butt with or without an acid slant is considered fermentative. An oxidizer or nonutilizer of sugars produces no change in the butt and slant, or produces alkalinization of the slant. Occasionally, an oxidizer produces slight acidity on the slant. However, a poorly growing organism that produces a degree of acidity of the slant and no change in the butt is tentatively designated a fermentative organism. In addition, some fermentative organisms (usually fastidious) do not acidify either the butt or the slant of TSI agar.

In cases when there is no change in the butt of TSI agar or the interpretation of the reaction is doubtful, an oxidation-fermentation (OF) test is used to establish whether an organism ferments, oxidizes, or does not produce acid from glucose.

Enteric fermentation base medium with carbohydrates is used to detect acid production by fermentative microorganisms. Oxidative organisms usually do not produce a detectable amount of acid in the enteric base carbohydrate medium for one of two reasons: 1) the accumulation of acid products produced by oxidation generally is less than that produced by fermentation; or 2) sufficient alkaline products frequently are produced from the peptone in the medium to neutralize the acid produced by oxidative metabolism.

An OF-base medium is used to determine the pattern of carbohydrate metabolism of nonfermentative bacteria [the Special OF Medium (King) is the medium upon which carbohydrate reactions given in this manual are based]. OF-base medium contains 0.3% agar and less peptone than does the fermentation base medium. The agar inhibits diffusion of the acid produced and concentrates it in the area of bacterial growth.

After the biochemical characteristics of an unknown organism are determined as described above, and with media and methods as detailed in the following chapter, the reactions are carefully recorded (see Fig. 1 on p. xii). The keys then are used to select a group of species, the characteristics of which can be compared to those of the unknown to attempt to arrive at the correct identification.

Media[a] and Methods

List of Titles

- Acetate agar
- Alkalinization of acetamide, serine, and tartrate
- Amylosucrase test with 5% sucrose agar
- Alkaline phosphatase
- Benedict's reagent (double strength)
- Blood agar
- Bordet-Gengou agar without peptone (Difco)
- Catalase reaction
- Cetrimide agar
- Chocolate agar
- Citrate agar (Simmons) (Difco)
- Coccobacilli and cocci, differentiation of
- Cysteine agar (carbohydrate base for *Francisella*)
- CTA medium (BBL)
- "Decarboxylase tests" (Moeller's method) (decarboxylases for lysine and ornithine, dihydrolase for arginine)
- Dye tolerance test for *Brucella*
- Enteric fermentation base (carbohydrate fermentation) medium (Difco)
- Esculin medium
- Flagella broth
- Flagella stains
- Fluid thioglycollate medium (BBL)
- Gas production from fluid thioglycollate medium (BBL)
- Gel formation (*Brucella canis*)
- Gelatin liquefaction
- Glucose cysteine peptone agar
- Gram stain
- Growth factor test for *Haemophilus*
- Heart infusion agar (Difco)
- Heart infusion broth (Difco)
- Heart infusion tyrosine (HIT) agar
- Indole spot test
- Indole test
- 2-Ketogluconate test
- 3-Ketolactonate medium for *Agrobacterium*
- Litmus milk
- MacConkey agar (Difco)
- Maintenance of bacteria in motility medium
- Methyl red (MR) test
- Motility test
- Nitrate broth, semiaerobic (Stanier)
- Nitrate reduction test
- Nitrite reduction test
- Nutrient agar (Difco) growth test for pathogenic *Neisseria*
- OF (Oxidation/Fermentation) medium
- ONPG (*O*-Nitrophenyl-β-D-Galactopyranoside) test
- Oxidase test
- Peptic digest of blood
- Phenylalanine agar
- Pigment production by pseudomonads
- Porphyrin test for *Haemophilus*
- Potassium hydroxide test for verifying the Gram reaction of bacteria
- Preservation of bacteria by freezing
- Rapid gelatin slant test
- Rapid sugar test for *Neisseria*
- Saline, buffered (pH 7.0)
- SS agar (Difco)
- Salt tolerance test
- Selective medium for pathogenic *Neisseria* (Thayer-Martin medium modified)
- Transformation assay for *Acinetobacter* and *Moraxella osloensis*
- Transformation assay for *Psychrobacter*
- Triple sugar iron agar (TSI agar)
- Tryptone glucose yeast extract (TGY) agar
- Urea agar (Christensen's)
- Voges-Proskauer (VP) test (Coblentz method)

---

[a]The composition of each commercially-prepared medium cited in this manual is stated with quantities of the ingredients presented in one liter of the finished medium.

ACETATE AGAR

This medium is used to determine whether the tested organism can use acetate as a sole source of carbon; breakdown of sodium acetate causes the pH of the medium to shift toward the alkaline range, turning the indicator blue.

The medium consists of sodium chloride, 5 g; sodium acetate, 2 g; magnesium sulfate, 0.2 g; monoammonium phosphate, 1 g; dipotassium phosphate (anhydrous), 1 g; bromthymol blue, 0.08 g; agar, 20 g; and distilled water, 1 l. The pH is 6.7 ± 0.2.

The medium is tubed in 3-ml portions into 13- × 100-mm screwcapped tubes, autoclaved at 121° C for 15 min, and slanted for cooling. Each slant is inoculated with 1 drop of an 18- to 24-h heart infusion broth culture. These are incubated at 35° C and observed for up to 7 days. Positive reactions are indicated by the medium becoming alkalinized and blue.

Reference: Trabulsi, L.R., and W.H. Ewing. 1962. Sodium acetate medium for differentiation of *Shigella* and *Escherichia* cultures. Public Health Lab. *20*: 137–140.

ALKALINIZATION OF ACETAMIDE, SERINE, AND TARTRATE

These media are used to determine whether the tested organism can produce an alkaline reaction in the presence of acetamide, serine, or tartrate as the sole source of carbon.

The medium consists of 22.2 mg Simmons agar base that is dissolved in 1 l of distilled water and supplemented with either acetamide (10 g), L-serine (5 g), or D-tartrate (5 g). The pH is adjusted to 6.8 ± 0.2 and tubed in 4-ml portions into 13- × 100-mm screwcapped tubes, autoclaved at 121° C for 15 min, and slanted for cooling.

To perform the alkalinization test, the slant is inoculated with 1 drop of an 18- to 24-h heart infusion broth culture. These are incubated at the optimal growth temperature for the organism and observed for up to 7 days. Positive reactions are indicated by the medium becoming alkalinized and changing color from green to blue.

ALKALINE PHOSPHATASE

The purpose of the test is to detect alkaline phosphatase activity, which is useful in the differentiation of some *Haemophilus* species.

The test reagent consists of 100 mg of *p*-nitrophenyl phosphate disodium hexahydrate suspended in 25 ml of an alkaline buffer solution (0.01 M glycine, 1 mM $MgCl_2$). The reagent is dispensed in 0.3-ml amounts into 13- × 100-mm screwcapped tubes and stored at −20° C until used.

The test is performed by making a heavy suspension of 18- to 24-h growth from a blood agar plate in 0.3 ml of 0.85% NaCl, then transferring this suspension to a thawed tube of alkaline phosphate test reagent. The tube is incubated at 35° C in air for 6 h. The development of a distinct yellow color indicates alkaline phosphatase activity.

AMYLOSUCRASE TEST WITH 5% SUCROSE AGAR

The test is used to detect an amylosucrase enzyme produced by some *Neisseria* strains. The enzyme synthesizes an iodine-reacting polysaccharide from sucrose. *N. meningitidis*, *N. gonorrhoeae*, and *N. lactamica* do not produce the enzyme, but many other neisseriae do.

The medium consists of heart infusion agar (Difco), 20 g (for composition, see p. 12); sucrose, 25 g; and distilled water, 500 ml. Six-milliliter amounts are dispensed into 16- × 125-mm screwcapped tubes and autoclaved at 121° C for 15 min. The medium is melted as needed, cooled to 50° C, and poured into 60- × 15-mm Petri dishes.

The culture to be tested is inoculated onto the agar plate and incubated at 35° C for 48 h. To the growth is applied 1–2 drops of a solution containing 0.2% iodine and 0.4% potassium iodide (Burke's modification of Gram's iodine solution) freshly diluted 1:5 with distilled water.

A positive test is indicated by a rapid darkening of the colonies (a dark reddish brown or blue-black develops, then fades; the color can be restored by reapplying the iodine solution). Corresponding colonies on sucrose-free medium do not react in this manner when tested with iodine.

Reference: Lennette, E.H., E.H. Spaulding, J.P. Truant (Eds.). 1974. Manual of Clinical Microbiology, 2nd ed., Washington, American Society for Microbiology, pp. 119–120.

BENEDICT'S REAGENT (DOUBLE STRENGTH)

The double-strength reagent is used for the 2-ketogluconate test; it may be diluted 1:1 with distilled water for use in the 3-ketolactonate test.

The ingredients are copper sulfate $CuSO_4 \cdot 5H_2O$, 17.3 g; sodium citrate, 173 g; sodium carbonate (anhydrous), 100 g; and distilled water, to a final volume of 500 ml.

The citrate and carbonate are dissolved by heating in about 300 ml of water; the solution is then filtered; the copper sulfate is dissolved by heating in 50 ml of water; it is then poured slowly into the first solution, with constant stirring; the solution is cooled, made up to a final volume of 500 ml, and dispensed into a 500-ml ground-glass stoppered bottle. (This reagent keeps indefinitely.)

BLOOD AGAR

The medium consists of heart infusion agar (Difco), 40 g (for composition, see p. 12); and distilled water, 1

L 250-ml amounts of the melted medium are added to each of four 500-ml Erlenmeyer flasks and autoclaved at 121° C for 15 min. After the mixture has cooled to 50° C, 12.5-ml amounts of sterile defibrinated rabbit blood are added aseptically to each flask. This is mixed by swirling with care, to keep bubbles from forming. Then, 20-ml portions are poured into Petri dishes.

Reference: King, E.O. 1964. The Identification of Unusual Pathogenic Gram-Negative Bacteria, Atlanta, Communicable Disease Center.

- **BORDET-GENGOU AGAR WITHOUT PEPTONE (DIFCO)**

This medium is especially useful for growing *Bordetella pertussis*. *B. pertussis* cultures inoculated onto the medium produce colonies surrounded by a β-hemolytic zone. The hemolytic zone is more prominent around stabs made in the agar.

The medium is prepared according to the manufacturer's instructions (the dehydrated medium is composed of potato infusion, 125 g; sodium chloride, 5.5 g; and Bacto-agar 20 g); in addition, 5 g of Bacto-agar/liter is added. The medium is tubed in 5-ml amounts into 16- × 125-mm screwcapped tubes for slants, and in 15-ml amounts into 18- × 150-mm screwcapped tubes for plates. The medium is autoclaved at 121° C for 15 min and cooled to approximately 50° C. Seventeen percent blood (i.e., 1 ml of blood to 5 ml of medium) is added to the tubes, the contents of which are then mixed well and either slanted or poured. Alternatively, the tubed medium can be stored in the refrigerator, melted and blood added as needed.

The slants and plates are inoculated and then incubated at 35° C in a candle jar for 24–48 h.

- **CATALASE REACTION**

Purpose of the test: to detect the presence of the enzyme catalase that promotes breakdown of hydrogen peroxide with the evolution of oxygen gas.

The reaction is determined on an 18- to 24-h culture grown on a TGY extract agar slant. Several drops of a 3% hydrogen peroxide solution are deposited on the agar slant growth. A positive test is indicated by the active liberation of gas bubbles from the bacterial growth exposed to the hydrogen peroxide. The reagent lasts indefinitely when refrigerated. Care should be taken not to use 30% hydrogen peroxide because the reaction can be more pronounced and might produce a contamination hazard. Growth media containing blood are avoided, because the catalase in blood sometimes gives false-positive results.

Reference: Levin, M., and D.Q. Anderson. 1932. Two new species of bacteria causing mustiness in eggs. J. Bacteriol. *23:* 337–347.

- **CETRIMIDE AGAR**

This medium contains cetrimide, a toxic substance that inhibits the growth of many bacteria but permits *Pseudomonas aeruginosa* and certain other bacteria to grow.

The medium consists of heart infusion agar (Difco), 40 g (for composition, see p. 12); cetrimide (hexadecyltrimethylammonium bromide) (Eastman Organic Chemicals, Rochester, NY), 4 ml of a 22.5% solution in distilled water; and distilled water, 1 l. The medium is tubed in 5-ml amounts into 15- × 125-mm tubes, autoclaved at 121° C for 15 min, and slanted for cooling.

The slants are inoculated with 1 drop of an 18- to 24-h HIB culture and incubated at 35° C for up to 7 days. The slants are examined for bacterial growth (a positive reaction).

Reference: Lowburg E.J.L. 1955. The use of a new cetrimide product in a selective medium for *Pseudomonas pyocyanea*. J. Clin. Pathol. *8:* 47–48.

- **CHOCOLATE AGAR**

This medium, used to grow fastidious organisms like *Neisseria gonorrhoeae*, is prepared with heart infusion agar (Difco), 40 g (for composition see p. 12), Bacto-supplement B (a sterile yeast concentrate), 10 ml; rabbit blood, 40 ml; and distilled water, 1 l.

The heart infusion agar and water are mixed together and sterilized at 121° C for 20 min. This is cooled to 70–75° C, and the rabbit blood is added. The mixture is kept at that temperature until the blood turns chocolate-brown. Then supplement B is added, the mixture is cooled to approximately 55° C, and poured into Petri dishes.

Reference: McLeod, J.W., J.C. Coates, F.C. Happold, D.P. Priestley, and B. Wheatley. 1934. Cultivation of the gonococcus as a method in the diagnosis of gonorrhoea with special reference to the oxydase reaction and to the value of air reinforced in its carbon dioxide content. J. Pathol. Bacteriol. *39:* 221–231.

- **CITRATE AGAR (SIMMONS) (DIFCO)**

This medium is used to determine whether an organism uses citrate when it is incorporated as the sole source of carbon.

The medium (composed of magnesium sulfate, 0.2 g; monoammonium phosphate, 1 g; dipotassium phosphate, 1 g; sodium citrate, 2 g; sodium chloride, 5 g; Bacto-agar, 15 g; and Bacto-bromthymol blue, 0.08 g) is prepared according to the manufacturer's instructions and tubed in 5-ml amounts into 12- × 125-mm tubes, autoclaved at 121° C for 15 min, and slanted for cooling.

The agar slant is inoculated lightly with a needle, by using growth from an 18- to 24-h heart infusion agar slant, and incubated at 35° C for up to 7 days. A positive reaction, denoting alkalinization, is indicated by the development of a dark blue color (growth without color change in the medium is not considered a positive test).

Reference: Simmons, J.S. 1926. A culture medium for differentiating organisms of typhoid-colon-*aerogenes* group and for isolation of certain fungi. J. Infect. Dis. *39*: 201–214.

- **COCCOBACILLI AND COCCI, DIFFERENTIATION OF**

This procedure involves the microscopic (wet mount) examination of cells taken from rabbit blood agar or susceptibility testing plates that have grown peripherally to disks of the antibiotics penicillin G, ampicillin, or sulfadiazine. True *Neisseria* species retain the coccoid form, but *Acinetobacter* and *Moraxella* cultures produce long rods and filaments. Low concentrations of the antibiotics block septum formation more strongly than synthesis of lateral cell-wall components. Therefore, bacteria that normally divide and produce short rods do not divide but continue to grow and produce filaments.

The cells to be examined are removed from the outer margins of the zones of growth inhibition surrounding the antibiotic disks on plates incubated 18–24 h. The cells are suspended in a small loopful of sterile broth and used to make a wet mount preparation (alternately, the smear is dried and stained for 30 sec with crystal violet). The preparation is examined by using the oil immersion objective and (for the wet mount) reduced illumination. The slide is scanned for filaments.

Reference: Catlin, B.W. 1975. Cellular elongation under the influence of antibacterial agents: a way to differentiate coccobacilli from cocci. J. Clin. Microbiol. *1*: 102–105.

- **CYSTEINE AGAR (CARBOHYDRATE BASE FOR *FRANCISELLA*)**

A special base with cysteine is required to demonstrate acid production from carbohydrates by *Francisella*.

The medium consists of two solutions: solution A, consisting of Bacto-peptone, 10 g; beef extract, 1.5 g; NaCl, 5 g; and cysteine-HCl, 1 g; dissolved in 500 ml of distilled water, pH 7.2; and solution B, consisting of agar, 15 g; melted in 500 ml of distilled water. Solutions A and B are mixed, and 2 ml of a 1.5% aqueous solution of phenol red (dissolve 1.5 g of phenol red in 5 ml of 0.1 N NaOH and add 95 ml of water) is added. The medium is sterilized at 121° C for 20 min, and filter-sterilized carbohydrate solution is added to a final concentration of 1%. The medium is tubed, 5 ml/16- × 125-mm tube, and slanted.

Each slant is inoculated heavily with a loop of growth from a young glucose cysteine peptone agar culture of *Francisella tularensis*; this is incubated with caps loose in a 35° C incubator without CO_2.

- **CTA MEDIUM (BBL)**

This medium is used to determine carbohydrate utilization by *Neisseria* cultures.

The medium consists of cystine, 0.5 g; trypticase peptone, 20 g; sodium chloride, 5 g; sodium sulfite, 0.5 g; phenol red, 0.017 g; agar, 2.5 g; and distilled water, 1 L (pH 7.3). The medium, in 5-ml amounts, is dispensed into 16- × 125-mm screwcapped tubes and autoclaved at 121° C for 15 min. The medium is cooled to approximately 50° C, and filter-sterilized carbohydrates are added aseptically to a final concentration of 1%.

A heavy inoculum is introduced into the agar deeps with a loop and mixed throughout the entire top 5–10 mm of the medium. The inoculum is obtained from bacterial growth on an 18- to 24-h HIA slant. The tubes, with caps tightened, are incubated at 35° C for up to 7 days. Acid production is indicated by a yellow color.

Reference: Vera, H.D. 1948. A simple medium for identification and maintenance of the gonococcus and other bacteria. J. Bacteriol. *55*: 531–536.

- **"DECARBOXYLASE TESTS" (MOELLER'S METHOD) (DECARBOXYLASES FOR LYSINE AND ORNITHINE, DIHYDROLASE FOR ARGININE)**

The purpose of the test is to measure the enzymatic ability of an organism to decarboxylate (or hydrolyze) an amino acid, forming an amine, with resulting alkalinity.

The basal medium for this test is prepared with peptone (Orthana) (marketed as "Thiotone"), 5 g; beef extract, 5 g; bromcresol purple, 0.01 g; cresol red, 0.005 g; glucose, 0.5 g; pyridoxal, 0.005 g; and distilled water, 1 L; the pH is 6.0.

The basal medium is divided into four equal parts: one part is tubed as a control; to each of the other three parts are added, respectively, 1% L-arginine monohydrochloride, 1% lysine dihydrochloride, and 1% L-ornithine dihydrochloride (if DL-amino acids are used, the concentration is doubled). The pH of the ornithine portion is readjusted before the medium is sterilized. The media are tubed in 13- × 100-mm screwcapped tubes, 1.5 ml/tube, and autoclaved at 121° C for 10 min.

For nonfermenting organisms, very heavy suspensions (≥McFarland #8 turbidity standard) prepared with young bacteria, 18–24 h old, from blood agar cultures are made in each of the three decarboxylase media and the control medium; a 4-mm layer of melted petrolatum is added to each tube. The cultures are incubated at 35–37° C for 5–6 days. Alkaline reactions are recorded as positive.

With fermentative organisms, the tubes are inoculated with a drop of an 18- to 24-h HIB culture.

Reference: Moeller, V. 1955. Simplified tests for some amino acid decarboxylases and for the arginine dihydrolase system. Acta Pathol. Microbiol. Scand. *36:* 158–172.

• DYE TOLERANCE TEST FOR BRUCELLA

The purpose of the test is to determine whether cultures of *Brucella* species are inhibited by basic fuchsin and/or thionin dyes at concentrations of 1:50,000 and 1:100,000.

The basal medium, heart infusion agar (Difco) (for composition, see p. 12), is prepared according to the manufacturer's directions. Twenty-milliliter amounts of the medium are dispensed into 18- × 150-mm screw-capped tubes and autoclaved at 121° C for 15 min; they are cooled to 50° C, and the prescribed concentration of dye is added. The warm agar is mixed and poured into Petri dishes. The solidified agar in plates is held in the 35° C incubator for 1–2 h before being used.

Dyes. Stock solutions, 0.1% in distilled water of both dyes, basic fuchsin and thionin, are prepared. Each dye solution is dispensed into an 8-oz screwcapped bottle and autoclaved at 121° C for 15 min (the dye solutions are routinely discarded after 30 days). The final dye concentrations routinely employed in the media are 1:50,000 (0.4 ml of 0.1% dye solution) and 1:100,000 (0.2 ml of 0.1% dye solution).

The inoculum used is prepared by suspending a loopful of bacterial growth from a 24- to 48-h agar slant into 1 ml of sterile physiological saline. A sterile cotton swab is immersed into the bacterial suspension and used to inoculate the dye plates and one control plate containing no dye. A single streak is made across each of the dye plates and, last, the control plate, without reloading the swab. The plates are inverted and incubated at 35° C for up to 4 days (if the cultures require an increased carbon dioxide tension, the plates are incubated in a candle jar). The plates are then examined for bacterial growth.

Reference: Alton, G.G., L.M. Jones, and D.E. Pietz. 1975. Laboratory Techniques in Brucellosis, 2nd ed., Geneva, World Health Organization, pp. 42–44.

• ENTERIC FERMENTATION BASE (CARBOHYDRATE FERMENTATION) MEDIUM

This medium is used to determine the ability of an organism to ferment a specific carbohydrate that is incorporated in a basal medium, thereby producing acid or acid with visible gas.

The medium consists of peptone, 10 g; meat extract, 3 g; sodium chloride, 5 g; Andrade's indicator, 10 ml; and distilled water, 1 L. The pH is adjusted to 7.2. The medium is tubed in 3-ml amounts into 15- × 125-mm tubes containing inverted Durham vials and autoclaved at 121° C for 15 min. After cooling, 0.3 ml of a 10% filter-sterilized carbohydrate solution is added aseptically to each tube (to give a final concentration of 1% carbohydrate).

Each tube is inoculated with 1 drop of an 18- to 24-h heart infusion broth culture and incubated at 35° C for up to 7 days. The tubes are examined for acid (indicated by a red color) and gas production.

Andrade's indicator consists of acid fuchsin, 2 g; sodium hydroxide (1 N), 160 ml; and distilled water, 1 L; autoclaved at 121° C for 20 min.

The fuchsin is dissolved in the distilled water, and the sodium hydroxide is added. If, after overnight standing, the fuchsin is not sufficiently decolorized (not darker than faint straw-color when brought up into a 10-ml pipette), an additional 1 or 2 ml of alkali is added. Andrade's indicator improves somewhat on aging and should be prepared in a sufficiently large amount to last for several years. It should be prepared and aged for approximately 6 months before being used.

References: Ewing, W.B., and B.R. Davis. 1970. Media and Tests for Differentiation of *Enterobacteriaceae*. Atlanta, Center for Disease Control.

Andrade, E. 1905–06. Influence of glycerine in differentiating certain bacteria. J. Med. Res. *14:* 551–556.

• ESCULIN MEDIUM

This medium is used to determine whether an organism will hydrolyze the glycoside esculin.

The medium consists of esculin, 1 g; ferric citrate, 0.5 g; heart infusion agar (Difco), 40 g (for composition see p. 12); and distilled water, 1 L. The ingredients are heated to dissolve, then cooled to 55° C. The pH is adjusted to 7.0. The medium is tubed in 5-ml amounts into 15- × 125-mm tubes, autoclaved at 121° C for 15 min, and slanted for cooling.

The medium is inoculated with 1 drop of a 24-h heart infusion broth culture and incubated at 35° C for up to 7 days. The slants are examined for esculin hydrolysis (esculin shows fluoresence under the ultraviolet rays of a Wood's lamp; when esculin is hydrolyzed, the medium turns reddish black and the esculin no longer shows fluoresence). An inoculated control tube of the medium without esculin may be useful in interpreting results when there are problems.

Reference: Sneath, P.H.A. 1956. Cultural and biochemical characteristics of the genus *Chromobacterium*. J. Gen. Microbiol. *15:* 70–98.

• FLAGELLA BROTH

This liquid medium is high in phosphate that, according to Leifson, enhances flagella production by certain bacteria.

The medium consists of Bacto-tryptose (Difco), 10 g; dipotassium phosphate (anhydrous), 1 g; sodium chloride, 2.5 g; and distilled water, 1 L. The pH is 7.0. The medium is tubed in 5-ml amounts into 15- × 125-mm tubes and autoclaved at 121° C for 15 min.

The medium is inoculated with 1 drop of an actively motile 18- to 24-h broth culture. Or, to enhance motility, bacteria are picked into flagella broth from the leading edge of young culture growth in a semisolid motility medium (Gard plate). The broth subculture is incubated at 25° C for 18–24 h. The organisms are stained with flagella stain to determine flagellar arrangement and morphology.

Reference: Leifson, E. 1951. Staining, shape, and arrangement of bacterial flagella. J. Bacteriol. *62:* 377–389.

• FLAGELLA STAINS

Flagella stains are used to demonstrate the presence, the site of attachment, and the morphology of flagella. Both the Leifson stain and the silver stain have been used successfully in the Special Bacteriology Reference Laboratory. Recently, we have replaced these methods with a commercial preparation of the Ryu stain for routine flagella detection.

Leifson stain. The flagella stain is prepared as follows. Solution A consists of basic fuchsin (certified for flagella staining), 0.6 g, and 95% ethanol, 50 ml. Alternatively, solution A can be prepared by adding 0.45 g of pararosanilin acetate and 0.15 g of pararosanilin hydrochloride to 50 ml of 95% ethyl alcohol. The mixture is shaken, and the dye is allowed to dissolve overnight.

Solution B consists of sodium chloride, 0.75 g; tannic acid (chemically pure grade), 1.5 g; and distilled water, 100 ml. Solutions A and B are combined and mixed thoroughly. The pH is checked and, if necessary, adjusted to 4.8–5.0 with 1 N sodium hydroxide. The stain is stored at 4° C for 2–3 days to improve its effectiveness. Thereafter, 50-ml portions of the stain are chilled to −20° C or below in screwcapped tubes for future use (the ultra-cold stain is stable indefinitely).

For use, the stain is warmed to room temperature (20–25° C) and remixed. Immediately after use, it is stored at 4° C (stain is stable for about 1 month at 4° C).

Culture. With most motile cultures, smears for flagella staining are made directly from young (usually 18–24 h) cultures on blood agar or a 25° C tryptone glucose yeast extract agar slant. With a needle, and avoiding the agar, a small amount of growth, equivalent to a 1-mm colony, is transferred to a test tube containing 3 ml of distilled water. The growth is emulsified into a droplet of water against the side of the tube, then mixed with the water in the tube to produce a faint opalescence. From this, a smear is prepared.

If the culture is not actively motile, or if it is poorly flagellated, the center of a tube or plate of motility medium (Gard plate) is inoculated with a small amount of the culture on a needle and incubated at the optimum temperature for motility. If the culture is motile, it begins to spread well away from the point of inoculum; a bit of growth from the actively growing edge of the culture is picked and transferred to flagella broth. The broth culture is incubated at the optimum temperature for motility for 18–24 h or until growth appears. Twenty-five hundredths of a milliliter of 37% formaldehyde is added to the broth culture (5 ml), which is allowed to stand for 15 min and centrifuged about 10 min in a table-top centrifuge. The supernatant is decanted, the packed cells are mixed in the drop or two of liquid left in the bottom of the tube, and distilled water is added to three-fourths the volume of the tube. After the mixture is centrifuged again, the water is decanted, and the cells are resuspended in sufficient water to attain a faint opalescence.

Slides. The flagella stain will not work unless the slides are scrupulously clean. New "precleaned" slides should be soaked for 4 days at ambient temperature (20–25° C) in acid dichromate solution. For soaking, 10 slides are placed in a glass rack, then immersed in dichromate in a covered staining jar. Slides are rinsed in 10 changes of tap water, then 2 changes of distilled water, and then air-dried. Cleaned slides are handled with forceps and stored in a covered container.

Smear. Before the smear is prepared, the slide is passed slowly several times through the tip of a blue Bunsen flame to burn off any insoluble residue (a yellow flame will deposit carbon on the slide, rendering it unusable). The slide is cooled, flamed side up, on a paper towel. With a wax pencil, a thick line is drawn across the width of the slide to contain the stain in an area approximately 2.5 × 4.5 cm.

A large loopful of culture suspension is deposited onto the center of the slide adjacent to the wax line and allowed to run down the length of the tilted slide (slides that do not permit the droplet to glide down easily are discarded as unclean). The slides are air-dried on a level surface.

Staining. Without being heat-fixed, the slide is placed on a level staining rack. One milliliter of stain at ambient temperature is added, and a timer is set for the required time. (The optimum staining time is determined for each new batch of stain by staining three or more slides of a known flagellated organism for varying times [5–15 min]). The slide is washed with tap water, air-dried in an upright position, and examined with the oil immersion lens.

References: Leifson, E. 1951. Staining, shape, and arrangement of bacterial flagella. J. Bacteriol. *62:* 377–389.

Clark, W.A. 1976. A simplified Leifson flagella stain. J. Clin. Microbiol. *3:* 632–634.

Silver stain. The flagella stain is prepared as follows. The mordant consists of 25 ml of saturated aluminum potassium sulfate (kept as a stock), 50 ml of a fresh 10% tannic acid solution, and 5 ml of 5% ferric chloride solution (kept as a stock); these ingredients are mixed together and stored in a dark bottle at 20–26° C. This mordant is stable several weeks.

The silver stain consists of a 5% silver nitrate solution (90 ml) to which is added, drop by drop, 2–5 ml of concentrated ammonium hydroxide until the precipitate that forms dissolves. Then, additional silver nitrate solution is added until the solution becomes faintly cloudy. This solution, when stored in a brown bottle at 20–26° C, is stable for several months.

The culture, slides, and smears are prepared as described above for the Leifson stain. However, the requirement for scrupulously clean slides does not seem to be as stringent as it is in the Leifson procedure.

The slides are placed on a staining rack and flooded with the mordant, which is allowed to act for 4 min. The slide is rinsed with distilled water, then flooded with the silver stain. The stain is heated by passing a flame under the slide until steam begins to rise. Then the flame is removed and the stain is allowed to act for 2–4 min (background staining is lighter with the shorter time). The slide is rinsed with distilled water and air-dried.

Reference: West, M., N.M. Burdash, and F. Friemath. 1977. Simplified silver-plating stain for flagella. J. Clin. Microbiol. *6:* 414–419.

Ryu stain. The commercial preparation (Carr-Scarborough Microbiologicals Cat. No. 75–950) contains 0.6% crystal violet, 2% tannic acid, 2.5% phenol, and 5.7% aluminum potassium sulfate. Upon receipt of a new lot, the optimum staining time should be determined by staining control slides at intervals between 1 and 5 min.

The culture, slides, and smears are prepared as described above for the Leifson and silver stains. As with the silver stain, the requirement for scrupulously clean slides does not seem to be as stringent as it is in the Leifson procedure.

After the slide has air-dried, flood it with the staining reagent and allow the reagent to remain on the smear for the optimum staining time. Then rinse with tap water and allow to air-dry.

References: Ryu, E. 1937. A simple method of staining bacterial flagella. Kitasato Arch. Exp. Med. *14:* 218–219.

Kodaka, H., A.Y. Armfield, G.L. Lombard, and V.R. Dowell, Jr. 1982. Practical procedure for demonstrating bacterial flagella. J. Clin. Microbiol. *16:* 948–952.

- **FLUID THIOGLYCOLLATE MEDIUM (BBL)**

This medium is used for a variety of reasons: for demonstrating growth characteristics, cellular morphology, and cellular arrangement, especially with streptococci, lactobacilli, *Gardnerella vaginalis,* and *Streptobacillus,* and for demonstrating gas production (see "Gas Production from Fluid Thioglycollate Medium"); and for promoting growth of fastidious organisms (serum is added when required).

The medium consists of pancreatic digest of casein (trypticase-BBL), 15 g; L-cystine, 0.5 g; dextrose (anhydrous), 5 g; yeast extract, 5 g; sodium chloride, 2.5 g; sodium thioglycollate, 0.5 g; resazurin, 0.001 g; agar, 0.75 g; and distilled water, 1 L; final pH was 7.1 ± 0.1.

The medium is prepared according to the manufacturer's instructions, tubed while hot, 9 ml/ tube in 16- $\times$ 125-mm screwcapped tubes, and autoclaved at 121° C for 15 min. It is stored in the dark at room temperature with caps tightly closed. It is discarded after a third of the medium in the tube has become oxidized (indicated by the pink color of the indicator).

- **GAS PRODUCTION FROM FLUID THIOGLYCOLLATE MEDIUM (BBL)**

This is a sensitive method for demonstrating gas production from a fermentable carbohydrate (glucose) when gas production is not obvious by other tests. In particular, the test is useful for detecting gas production by cultures of *Actinobacillus actinomycetemcomitans* and *Haemophilus aphrophilus.*

A 48- to 72-h fluid thioglycollate medium culture of the organism is used. A red-hot inoculating needle is plunged into the broth near the side of the glass tube. In a positive test, gas bubbles appear along the line of insertion. (This test is performed in a biological safety cabinet.)

Reference: King, E.O., and H.W. Tatum. 1962. *Actinobacillus actinomycetemcomitans* and *Haemophilus aphrophilus.* J. Infect. Dis. *111:* 85–94.

- **GEL FORMATION (*BRUCELLA CANIS*)**

Brucella canis usually forms a stringy mass or "gel" of increased viscosity when suspended in phenolized saline.

A heavy suspension is prepared with 24- to 48-h bacterial growth in phenolized saline (0.5% phenol in buffered saline) in a 13- $\times$ 100-mm screwcapped test tube and is allowed to stand for 15–30 min. (If the test is negative after this period, and *B. canis* is still suspected, the tube is incubated overnight.)

In testing for gel formation, a bacteriological loop is inserted into the suspension. If a gel has been formed, a short string is formed as the loop is withdrawn. The gel also can be detected by tilting the tube and comparing the viscosity of the suspension in the tube with that of buffered saline in a control tube.

Reference: Weaver, R.E., H.W. Tatum, and D.G. Hollis. 1972. The Identification of Unusual Pathogenic Gram Negative Bacteria (E.O. King). Preliminary Revision. Atlanta, Center for Disease Control.

• GELATIN LIQUEFACTION

The purpose of the test is to determine the ability of an organism to produce proteolytic enzymes (gelatinases) that liquefy gelatin.

The test medium consists of 12% Bacto-gelatin in heart infusion broth (HIB) (Difco) (for composition see p. 12); pH 7.4. The medium is tubed in 5-ml amounts into 15- $\times$ 125-mm screwcapped tubes and autoclaved at 121° C for 15 min.

The medium is inoculated with 4–5 drops of a 24-h HIB culture and incubated at 35° C for up to 14 days. The medium is incubated at 25° C if the organism grows better at 25° C than at 35° C.

The culture is checked daily for liquefaction by placing the tube into an ice water bath or into the refrigerator, along with an uninoculated control. Liquefaction is assayed only after the control has hardened (gelled). If liquefaction occurs within 14 days, the culture is considered to be positive.

Reference: Ewing, W.H. 1962. *Enterobacteriaceae*, Biochemical Methods for Group Differentiation. Atlanta, US Public Health Service. Publication 734 (revised).

• GLUCOSE CYSTEINE PEPTONE AGAR

This medium is used for the cultivation of *Francisella tularensis*.

The medium consists of two solutions: solution A: Bacto-peptone, 20 g; meat extract, 3 g; sodium chloride, 10 g; glucose, 50 g; cysteine-HCl, 2 g; and distilled water, 1 L (the pH is adjusted to 7.2); and solution B: agar, 30 g; distilled water, 1 L. A and B solutions are combined, mixed thoroughly, dispensed in 500-ml amounts into 1000-ml Erlenmeyer flasks, and autoclaved at 121° C for 20 min. The medium is cooled to 50° C, and 25 ml of defibrinated rabbit blood is added to each flask. Five-milliliter quantities are dispensed into 15- $\times$ 125-mm screwcapped tubes, and 20-ml quantities are placed into Petri dishes. The tubes are slanted.

• GRAM STAIN[b]

The Gram stain enables us to differentiate bacteria into two major groups, depending upon the chemical nature of the cell wall: Gram-positive organisms, which retain the crystal violet-iodine complex (blue) in the cell wall during decolorization, and Gram-negative organisms, which do not; the latter take up safranin (red).

The solutions are prepared as follows:

Crystal violet consists of solutions A and B. Solution A is crystal violet powder (99% dye content), 13.87 g (the amount of dye used is adjusted in proportion to the dye content of the powder, stated on the label) and 200 ml of 95% ethanol. Solution B is ammonium oxalate, 8 g in 800 ml distilled water. Solutions A and B are mixed and allowed to stand overnight or until the dye dissolves. The solution is filtered through coarse filter paper.

Gram's iodine consists of iodine crystals, 1 g; potassium iodide, 2 g; and distilled water, 300 ml.

The **decolorizer** is 95% ethanol.

The **counterstain** is safranin-0, 3.41 g; 95% ethanol, 100 ml; and distilled water, 900 ml.

The smears are prepared by depositing a thin suspension of bacteria on a clean glass slide and allowing it to air dry; smears are heat-fixed gently.

Staining procedure. The smear is covered with crystal violet for 30 sec, then rinsed with tap water; the smear is covered with Gram's iodine for 30 sec, then rinsed with tap water; the smear is decolorized with ethanol for about 30 sec, or until the wash is clear, then is counterstained about 30 sec with safranin.

Reference: Bartholomew, J.W. 1962. Variables influencing results, and the precise definition of steps in Gram staining as a means of standardizing the results obtained. Stain Technol. *37:* 139–155.

• GROWTH FACTOR TEST FOR *HAEMOPHILUS*

The purpose of the test is to determine the requirement for different growth factors (X = heme, V = nicotinamide adenine dinucleotide) by various species of *Haemophilus*.

The medium used in the test is heart infusion agar (Difco) (for composition see p. 12); it is prepared according to the manufacturer's directions. The melted sterile medium is dispensed in 20-ml amounts into Petri dishes.

The inoculum is prepared by making a heavy suspension in sterile distilled water with the growth from an overnight (18- to 20-h) chocolate agar culture. The test plate is inoculated with three widely distributed 25-mm-diameter smears, with a loopful of suspension for each smear. A growth-factor strip (one of each, X, V, and XV) is laid in the center of each smear. (Strips are cut into quarters; one quarter is used for each assay.) The plates are incubated at 35° C for 18–24 h. If no growth occurs around any of the factors, then a newly inoculated factor plate is incubated in a candle jar for 18–24 h at 35° C.

Factor strips. The strips, X, V, and XV (Taxo, BBL), are obtained from Bioquest, Cockeysville, MD.

---

[b]We have also found a commercially prepared set of reagents (Carr-Scarborough Microbiologicals) to be satisfactory.

- **HEART INFUSION AGAR (DIFCO)**

The medium consists of heart infusion agar (Difco), 40 g (composed of beef heart infusion, 500 g; Bacto-tryptose, 10 g; sodium chloride, 5 g; and Bacto-agar, 15 g); and distilled water, 1 l. The mixture is heated to dissolve the agar, then dispensed in 5-ml portions into 15- × 125-mm tubes. These are autoclaved at 121° C for 15 min, then cooled in a slanted position to obtain a 2.5-cm butt and a 4-cm slant.

- **HEART INFUSION BROTH (DIFCO)**

The medium is prepared as follows: heart infusion broth (Difco), 25 g (composed of beef heart infusion, 500 g; Bacto-tryptose, 10 g; and sodium chloride, 5 g); and distilled water, 1 L. Four-milliliter amounts are dispensed into 15- × 125-mm test tubes and autoclaved at 121° C for 15 min.

- **HEART INFUSION TYROSINE (HIT) AGAR**

This medium is used to enhance the production of brown soluble pigments by certain bacteria.

The medium consists of heart infusion agar (Difco), 40 g (for composition see above); L-tyrosine, 1 g; distilled water, 1 L.

The medium is tubed in 7-ml amounts in 15- × 125-mm tubes, sterilized at 121° C for 15 min, and slanted.

Reference: King, E.O. 1964. The Identification of Unusual Pathogenic Gram-Negative Bacteria. Atlanta, Communicable Disease Center.

- **INDOLE SPOT TEST**

This is a rapid test in which a minimal amount of reagent is used for determining the ability of a microorganism to produce indole on media containing tryptophan. In this laboratory, the test is used specifically on suspected *Haemophilus* strains; generally it has not been used for other genera.

The reagent is 1% paradimethylaminocinnamaldehyde in 10% HCl (90 ml of distilled water with 10 ml of concentrated HCl added).

The culture is grown on a chocolate agar plate at 35° C for 18–24 h (we use a candle-jar atmosphere). No. 1 Whatman filter paper is placed in a Petri dish and moistened with 1 drop of the test reagent. The colonies to be tested are taken with a loop and spread over a small area of the moist filter paper. The development of a blue to blue-green color indicates the presence of indole. Development of any other color indicates a negative test.

References: Vracko, M.D., and J.C. Sherris. 1963. Indole-spot test in bacteriology. Am. J. Clin. Pathol. *39:* 429–432.

Sutter, V.L., and W.T. Carter. 1972. Evaluation of media and reagents for indole-spot tests in anaerobic bacteriology. Am. J. Clin. Pathol. *58:* 335–338.

- **INDOLE TEST**

The purpose of the test is to determine the ability of an organism to split the tryptophan molecule to form the compound indole.

The medium for culture growth is 2% Bacto-tryptone broth (rich in tryptophan) (pH 7.2–7.4). The broth is tubed in 4-ml amounts into 15- × 125-mm tubes and autoclaved at 121° C for 15 min. Heart infusion broth is used to test organisms that do not grow well in Bacto-tryptone broth.

The broth is inoculated with 1 drop of a 24-h heart infusion broth culture and incubated at 35° C for 48 h. Thereafter, 1 ml of xylene is added to the broth culture. The mixture is shaken vigorously to extract the indole and allowed to stand until the xylene forms a layer on top of the aqueous phase. Then, 0.5 ml of Ehrlich's reagent (see below) is added down the side of the tube. The presence of indole is indicated by the development of a pink ring below the xylene layer.

Ehrlich's reagent consists of 95% ethanol, 95 ml; paradimethylaminobenzaldehyde, 1 g; and hydrochloric acid (concentrated), 20 ml. The aldehyde is dissolved in the alcohol, and then the acid is slowly added. The reagent is prepared in small quantities and stored at 4° C.

Reference: Bohme, A. 1905. Die anwendung der Ehrlichsen Indolreaktion für bakteriologische Zwecke. Centr. Bakt. I Abt. Orig. *40:* 129–133.

- **2-KETOGLUCONATE TEST**

The purpose of the test is to determine the ability of an organism to oxidize gluconic acid as a sole carbon source to the reducing compound 2-ketogluconate.

The medium consists of monopotassium phosphate, 5.4 g; potassium nitrate, 2 g; potassium gluconate, 20 g; and distilled water, 1 L. The pH is 6.5. The mixture is filter-sterilized and dispensed aseptically in 1-ml amounts into sterile 13- × 100-mm screwcapped tubes.

The medium is inoculated with one loopful of a 24-h heart infusion broth culture and incubated at 35° C for 3 days.

2-Ketogluconate reaction. Three-tenths of a milliliter of double-strength Benedict's reagent is added to the culture broth. The tube is placed in a boiling water bath for 10 min. In place of the Benedict's reagent, a Clinitest (Ames) tablet can be added to the culture broth (heating is not required). The presence of 2-ketogluconate is indicated by the appearance of a copper-colored precipitate.

Reference: Moore, H.B., and M.J. Pickett. 1960. The *Pseudomonas-Achromobacter* group. Can. J. Microbiol. *6:* 35–42.

- **3-KETOLACTONATE MEDIUM FOR *AGROBACTERIUM***

This medium is used to determine whether an organism has the property, probably limited to members of the genus *Agrobacterium*, of producing 3-ketoglycosides from the corresponding disaccharides and bionic acids.

The medium consists of lactose, 10 g; yeast extract (Difco), 1 g; agar, 20 g; and distilled water, 1 L. The medium is tubed in 20-ml amounts into 20- × 150-mm screwcapped tubes and autoclaved at 121° C for 15 min. The contents of one tube are poured into a Petri dish.

A TGY agar slant is inoculated with the test culture and incubated at 25° C for 2–3 days. Then two 3-ketolactonate medium plates are inoculated by smearing some of the TGY slant growth onto the agar plate. The area of inoculation is never larger than a circle with a diameter of 1 cm. A positive control (i.e., a strain of *Agrobacterium radiobacter* biovar 1) is also inoculated onto the test plates. The plates are incubated at 25° C for 2–7 days. After 2 days of incubation, one of the plates is flooded with single-strength Benedict's solution. A positive reaction is indicated by a large yellow zone occurring, within 30 min, around the bacterial growth. If no zone appears, the other plate, after 7 days' incubation, is tested in the same manner.

Reference: Bernaerts, M.J., and J. De Ley. 1963. A biochemical test for crown gall bacteria. *Nature* 197: 406–407.

• LITMUS MILK

This medium is used to detect acid production from lactose fermentation, production of caseolytic enzymes, casein coagulation, and reduction of litmus indicator by bacteria.

Litmus milk (Difco), 105 g (composed of Bacto-skim milk, 100 g and Bacto-litmus, 5 g), is mixed into 1 L of distilled water to obtain a homogeneous suspension. The suspension is dispensed in 6-ml amounts into 15- × 125-mm tubes and autoclaved at 121° C for 10 min.

The medium is inoculated with 4–5 drops of a 24-h heart infusion broth culture and incubated at 35° C for up to 7 days. The culture is observed daily for changes in composition: a blue color indicates a pH shift to the alkaline range; pink indicates acid production; and loss of color, a reduction of the indicator. Also acid clot, rennet clot, or peptonization are noted.

Reference: King, E.O. 1964. *The Identification of Unusual Pathogenic Gram-Negative Bacteria*. Atlanta, Communicable Disease Center.

• MACCONKEY AGAR (DIFCO)

This medium is used to detect an organism's ability to grow in a moderate concentration of bile salts, as in enteric environments.

The medium [composed of Bacto-peptone, 17 g; proteose peptone (Difco)], 3 g; Bacto-lactose, 10 g; Bacto-bile salts #3, 1.5 g; sodium chloride, 5 g; Bacto-agar, 13.5 g; Bacto-neutral red, 0.03 g; and Bacto-crystal violet, 0.001 g) is prepared according to the manufacturer's directions, tubed in 5-ml amounts into 15- × 125-mm tubes, autoclaved at 121° C for 15 min, and slanted.

Slants are inoculated with a drop of a 24-h heart infusion broth culture and incubated at 35° C for up to 7 days. The slants are observed for growth, indicating a positive reaction.

• MAINTENANCE OF BACTERIA IN MOTILITY MEDIUM

Many strains of bacteria encountered in the Special Bacteriology Reference Laboratory are maintained satisfactorily for extended periods (at least 1 yr) at "room" temperature (around 25° C) in motility medium (see below). The medium in screwcapped tubes is inoculated with a needle to 2 mm below the agar surface and incubated at optimal temperature until growth appears. Then the tube is closed to prevent evaporation and stored at room temperature.

• METHYL RED (MR) TEST

The purpose of the test is to determine an organism's ability to produce and maintain stable acid endproducts from glucose fermentation.

The medium for growing the organism is MR-VP medium (Difco) (composed of buffered peptone, 7 g; Bacto-dextrose, 5 g; and dipotassium phosphate, 5 g). It is prepared according to the manufacturer's instructions, tubed in 4-ml amounts into 15- × 125-mm tubes, and autoclaved at 121° C for 10 min.

The broth medium is inoculated with 1–2 drops of a 24-h HIB culture and incubated at 35° C for 48 h (4–5 days for poorly growing cultures).

Five to six drops of methyl red reagent (see below) are added to the broth culture. A positive test is a bright red (acid); a weakly positive test is a red-orange; and a negative reaction is yellow or orange.

The methyl red reagent is prepared by dissolving 0.1 g of methyl red in 300 ml of 95% ethanol. Distilled water is added to make a final volume of 500 ml.

Reference: Clark, W.M., and H.A. Lubs. 1915. The differentiation of bacteria of the colon-aerogenes family by the use of indicators. *J. Infect. Dis.* 17: 161–173.

• MOTILITY TEST

Motility is a useful characteristic for classifying bacteria. The simplest test for motility is to observe the bacteria in a culture droplet under the microscope (see "Culture droplet," below). If the organism appears nonmotile, it should be inoculated into motility medium to determine whether it spreads throughout the soft agar (motile) or remains at the site of inoculum (nonmotile) (see "Motility medium," below).

Culture droplet. A line is drawn on a clean slide with a glass-marking pencil, and a loopful of culture suspension is placed on the slide adjacent to the line. The microscope with the 10× objective in place is focused on the line. The light is decreased by closing down the substage diaphragm, and the slide is moved to focus on the edge of the droplet.

The 40× objective is swung into place, and the organisms are observed. Motile organisms change position with respect to one another; in Brownian movement, the organisms, although quite active because of molecular bombardment, remain in the same relative position with respect to other organisms in the field.

Motility medium. The medium consists of motility test medium (Difco), 16 g (composed of Bacto-tryptose, 10 g; sodium chloride, 5 g; and Bacto-agar, 5 g); nutrient broth (Difco), 4 g (composed of Bacto-beef extract, 3 g; and Bacto-peptone, 5 g); sodium chloride, 1 g; and distilled water, 1 L. (This medium contains a 0.4% final concentration of agar.) The medium is heated to melt the agar, 8-ml amounts are dispensed into 15- × 125-mm tubes, and the tubes are autoclaved at 121° C for 15 min. The inoculum, obtained from an 18- to 24-h HIA slant culture, is stabbed into the center of the tube to a depth of 5 mm. The tubes are incubated at 35° C (or 25° C if the organism did not exhibit motility at 35° C) and examined for up to 7 days. Motile organisms grow away from the inoculation site into the surrounding medium. Nonmotile organisms remain localized at the inoculation site.

If an organism does not appear motile in motility medium, one can check OF medium for motility, if previously inoculated. (Motility medium has 0.4% agar, and OF media have 0.2–0.3% agar; this lesser amount of agar further enhances expression of motility.)

The Gard plate also is useful for determining motility. A Petri dish containing motility medium is inoculated at the center of the plate with the test organism and incubated at 25° C for up to 7 days. If motile, the organisms spread away from the point of inoculation; a subculture is picked from the edge of the spreading growth (center of increased concentration of motile cells) into flagella broth and incubated at 25° C until growth appears. A flagella stain is then prepared.

Reference: Gard, S. 1938. Das Schwarmphanomen in der Salmonella-Gruppe und seine praktische Ausnutzung. Zeitschr. Hyg. Infektionskr. *120*: 615–619.

- **NITRATE BROTH, SEMIAEROBIC (STANIER)**

This medium is used for testing *Pseudomonas* strains for denitrification. It is used selectively for organisms one would expect to produce gas from nitrate but that do not do so in ordinary nitrate broths.

The broth medium consists of disodium phosphate and monopotassium phosphate, 1 M, pH 6.8, 40 ml; vitamin-free mineral solution (Hutner) (see below), 20 ml; ammonium sulfate, 1 g; glycerol, 10 g; potassium nitrate, 10 g; yeast extract, 5 g; distilled water, to an 800-ml volume.

The ingredients are combined and heated slightly in a double water bath to effect solution; they are then filter-sterilized.

One gram of Ionagar #2 (Oxoid) or its equivalent is dissolved in 200 ml of glass-distilled water and autoclaved at 121° C for 15 min.

All ingredients are combined aseptically and dispensed in 5-ml volumes in sterile 16- × 125-mm screw-capped tubes.

Hutner's vitamin-free mineral solution contains nitrilo-triacetic acid, 10 g; magnesium sulfate, 14.45 g; calcium chloride ($2H_2O$), 3.335 g; ferrous sulfate ($7H_2O$), 0.099 g; ammonium molybdate ($4H_2O$), 0.0093 mg; and concentrated metals solution (see below), 50 ml.

The nitrilotriacetic acid is dissolved and neutralized with potassium hydroxide, about 7.3 g. The salts and 50 ml of concentrated metals solution are added, the pH is adjusted to 6.8, and glass-distilled water is added to make a final volume of 1 L.

Concentrated metals solution consists of ethylenediaminetetraacetic acid, 2.5 g; zinc sulfate ($7H_2O$), 10.95 g; ferrous sulfate ($7H_2O$), 5 g; manganese sulfate (H_2O), 1.54 g; copper sulfate ($5H_2O$), 0.392 g; cobalt nitrate ($6H_2O$), 0.248 g; sodium borate ($10H_2O$), 0.177 g; sulfuric acid, a few drops; and distilled water to make 1 L.

A tube of nitrate broth is inoculated with 5–7 drops of a 24-h heart infusion broth culture using a Pasteur pipette. The inoculation is initiated in the bottom of the tube and brought upward carefully to the medium surface, in order not to introduce air bubbles into the medium with the pipette. After a 24-h incubation, the organism is subcultured in the same way into the same medium, melted, and cooled to around 40° C. A 1-cm cap of sterile petroleum jelly is applied over the surface of the inoculated medium. The tubes are incubated at 35° C for up to 7 days. The tubes are examined daily for 1) gas production, resulting in the formation of a space between the petroleum jelly cap and the medium, and 2) cellular growth out from the area of the line of inoculation.

References: Stanier, R.Y., N.J. Palleroni, and M. Doudoroff. 1966. The aerobic pseudomonads: a taxonomic study. J. Gen. Microbiol. *43*: 159–271.

Cohen-Bazire, G., W.R. Sistrom, and R.Y. Stanier. 1957. Kinetic studies of pigment synthesis by non-sulfur purple bacteria. J. Cell Comp. Physiol. *49*: 25–68.

- **NITRATE REDUCTION TEST**

The purpose of the test is to determine an organism's ability to reduce nitrate to either nitrite, free nitrogen gas, or other reduced nitrogen compounds.

The broth media for assaying nitrate reduction are: 1) routine nitrate broth: Bacto-peptone, 20 g; potassium nitrate, 2 g; and distilled water, 1 L; and 2) special nitrate broth (for the more fastidious organisms): heart infusion broth (Difco), 25 g; potassium nitrate, 2 g; and distilled water, 1 L. In each case the pH is adjusted to 7.0.

Each of the broth media is dispensed in 4-ml amounts into 15- × 125-mm tubes containing inverted Durham vials and autoclaved at 121° C for 15 min.

The reagents for detecting the presence of nitrite are solution #1: sulfanilic acid, 2.8 g; distilled water, 250 ml; and glacial acetic acid, 100 ml; and solution #2: dimethyl-α-naphthylamine, 2.1 ml; distilled water, 250 ml; and glacial acetic acid, 100 ml.

An alternative solution #2 is made up of distilled water, 120 ml; glacial acetic acid, 30 ml; and 1,6-Cleve's acid (5-amino-2-naphthalenesulfonic acid), 0.2 g. Water is added to the Cleve's acid, and the mixture is warmed and shaken frequently until most of the compound is dissolved; do not boil, because boiling will cause almost all of the compound to dissolve. The solution is filtered and cooled, and acetic acid is added to the filtrate. The reagent is stored in the refrigerator.

The broth tubes are inoculated with 1–2 drops of a 24-h HIB culture and incubated at 35° C for 48 h. After incubation, the Durham tube is observed for gas formation and the broth is tested for the presence of nitrite.

Test for Presence of Nitrite

To the nitrate broth culture are added 0.25 ml of nitrite test solution #1, then 0.25 ml of nitrite test solution #2. The presence of nitrite is evidenced by the development of a red color within 1–2 min. (Color produced with alternative solution #2 fades rapidly but can be restored by adding more of the reagent.)

If there is gas in the Durham vial and the mixture turns red after solutions #1 and #2 are added, nitrate has been reduced to nitrite, and part of the nitrite has been reduced to a gaseous substance.

If red is not observed, either 1) the nitrate has not been reduced, or 2) the nitrate has been reduced beyond the oxidation level of nitrite (if gas is produced, it is detected in the Durham vial). To determine which has taken place, a little powdered zinc is added to the tube. If nitrates are still present, they will be reduced by the zinc and a red color will develop in 5–10 min.

• NITRITE REDUCTION TEST

The purpose of the test is to determine whether an organism can reduce nitrites to gaseous nitrogen or to other compounds containing nitrogen. The test is performed in media containing two different concentrations of nitrite (0.1% and 0.01%). With some organisms, nitrite reduction is more easily demonstrated in 0.01% nitrite.

The broth medium for assaying nitrite reduction consists of heart infusion broth (Difco), 25 g; potassium nitrite, 1 g for 0.1% and 0.1 g for 0.01%; and distilled water, 1 L. The pH is adjusted to 7.0.

The test for the presence of nitrite (see Nitrate Reduction Test above) is performed on a 2- to 5-day-old nitrite broth culture. If the broth does not become red within 1 min, the nitrite has been reduced. Gas production from nitrite reduction is observed in the Durham vial. If the broth does not become red and no gas production is observed, zinc dust is added to ensure that the nitrite has not been oxidized to nitrate. If it has, the mixture becomes red.

• NUTRIENT AGAR (DIFCO) GROWTH TEST FOR PATHOGENIC NEISSERIA

Cultures of *Neisseria gonorrhoeae* and *N. meningitidis* will not grow on nutrient agar at temperatures of 25° or 35° C (a few strains of *N. meningitidis*, usually isolated from human carriers, will grow at 35° C). Most nonpathogenic *Neisseria* cultures grow at 35° C, but variable results have been noted at 25° C.

The medium, which does not contain NaCl, is prepared according to the manufacturer's directions, dispensed in 5-ml amounts into 15- × 125-mm tubes, and slanted for cooling.

One drop of an 18- to 24-h heart infusion broth culture or a light suspension of bacteria is used to inoculate each slant. The cultures are incubated in a candle jar at 35° C for up to 7 days. A blood agar slant, inoculated at the same time, serves as a positive control for culture viability.

Reference: Hollis, D.G., G.L. Wiggins, and R.E. Weaver. 1969. *Neisseria lactamicus* sp. n., a lactose-fermenting species resembling *Neisseria meningitidis*. Appl. Microbiol. 17: 71–77.

• OF (OXIDATION/FERMENTATION) MEDIUM

This medium is used to determine whether an organism produces acid from carbohydrates. The basal media for detecting carbohydrate utilization and mode of utilization are:

Special OF Medium (King). (This is the medium upon which carbohydrate utilization is based in this manual): Bacto-casitone, 2 g; phenol red, 2 ml of a 1.5% aqueous solution (dissolve 1.5 g of phenol red in 5 ml of 0.1 N NaOH and add 95 ml of water); agar, 3 g; and distilled water, 1 L (the pH is 7.3).

Reference: Riley, P.S., H.W. Tatum, and R.E. Weaver. 1972. *Pseudomonas putrefaciens* isolates from clinical specimens. Appl. Microbiol. 24: 798–800.

OF Basal Medium (Difco). (This medium also has been used successfully to determine carbohydrate utilization): 9.4 g of the Difco product [Bacto-tryptone, 2 g; sodium chloride, 5 g; dipotassium phosphate (anhydrous), 0.3 g; bromthymol blue, 0.08 g; and agar, 2 g] and 1 L of distilled water, made according to the manufacturer's instructions (the pH is 6.8–6.9; the color of the medium is green).

Each basal medium is tubed in 6-ml amounts into 15- × 125-mm test tubes and autoclaved at 121° C for 15 min. The basal media are cooled to approximately 50° C, and the filter-sterilized carbohydrates are added aseptically to a final concentration of 1%.

Production of Acid from Carbohydrates

For determining whether acid is produced from carbohydrates, the agar deeps are inoculated with bacterial growth from a 24-h heart infusion agar slant by stabbing a needle 4–5 times into the medium to a depth of 2.5 cm. The tubes are incubated at 35° C for up to 7 days. Acid production is indicated by the color indicator changing to yellow.

Nonfermentative bacteria are routinely tested for their ability to produce acid from six carbohydrates (glucose, xylose, mannitol, lactose, sucrose, and maltose) in Special OF Medium (King). In addition to the six tubes containing carbohydrates, a control tube containing the OF base without carbohydrate also is inoculated.

Weak acid formation can be detected by comparing the tube containing the medium with carbohydrate with the inoculated tube containing medium with no carbohydrate. Most bacteria that can grow in the OF base produce an alkaline reaction in the control tube. If the color of the medium in a tube containing carbohydrate remains about the same color as it was before the medium was inoculated and if the inoculated medium in the control tube becomes a deeper red (i.e., becomes alkaline), the culture being tested is considered weakly positive—i.e., assuming the amount of growth is about the same in both tubes.

If screwcapped tubes are used to contain OF media, the caps are loosened during incubation to allow for air exchange. Otherwise, the control tube and tubes containing carbohydrates that are not oxidized might not become alkaline.

Mode of Acid Production from Glucose

For determining whether the culture oxidized or fermented glucose, duplicate tubes of 1% glucose OF medium are inoculated.[c] The medium in one of the tubes is then overlaid with 1–2 cm of sterile petroleum jelly; both tubes are incubated at 35° C for up to 7 days. Acid formation in both tubes indicates a carbohydrate fermenter. Acid only in the tube that is not overlaid with petroleum jelly indicates an oxidizer. The absence of acid in either tube indicates that either the organism did not use glucose, or that during glucose metabolism, the organism did not produce detectable amounts of organic acids.

Reference: Hugh, R., and E. Leifson. 1953. The taxonomic significance of fermentative versus oxidative metabolism of carbohydrates by various Gram-negative bacteria. J. Bacteriol. 66: 24–26.

For a discussion of the relative merits of the use of bromthymol blue and phenol red indicators in OF media, see Webster, J.A. and R. Hugh. 1979. Int. J. Syst. Bacteriol. 29: 336–337.

- **ONPG (O-NITROPHENYL-β-D-GALACTOPYRANOSIDE) TEST**

The purpose of the test is to detect "late" lactose fermenters that, although possessing the enzyme β-galactosidase for cleaving the lactose molecule, do not have the enzyme galactosidase-permease for transporting lactose inside the cell, where β-galactosidase can split the lactose. ONPG is hydrolyzed, if β-galactosidase is present, to produce O-nitrophenol, a yellow compound.

Medium assay. The medium consists of Bacto-tryptose (Difco), 4 g; Bacto-yeast extract (Difco), 4 g; sodium chloride, 5 g; agar, 3 g; and distilled water, 800 ml. The medium is autoclaved at 121° C for 15 min. Then it is cooled to approximately 50° C, and to it is added aseptically 200 ml of a 1% ONPG solution (filter-sterilized). The medium is dispensed in 3-ml amounts into 13- × 100-mm screwcapped tubes.

The agar deeps are inoculated with a 24-h heart infusion agar slant culture by stabbing a needle to a depth of about 1.5 cm directly into the center of the tube. The tubes are incubated at 35° C and observed at 24 and 48 h to determine if the area around the bacterial growth has become yellow.

Tablet assay. One ONPG tablet (Key Scientific Products, Los Angeles) is added to 1 ml of distilled water. The suspension is inoculated with a heavy suspension of bacterial growth from a 24-h heart infusion agar slant and incubated at 35° C for 6 h. A positive reaction for β-galactosidase activity is indicated by a yellow color.

Reference (medium assay): Negut, M., and G. Hermann. 1975. A comparison of two methods for detecting β-D-galactosidase. Public Health Lab. 33: 190–193.

- **OXIDASE TEST**

The purpose of the test is to detect in organisms the presence of oxidase enzymes associated with the cyto-

---

[c]When the Special OF Medium (King) is used for determining whether an organism oxidizes or ferments glucose, the amount of casitone in the medium is increased to 5 g/liter. If less casitone is used, some organisms that oxidize glucose may produce an acid reaction under the petroleum jelly seal. *Reference:* King, E.O. 1964. The Identification of Unusual Pathogenic Gram-Negative Bacteria. Atlanta, Communicable Disease Center.

chrome respiratory system. The reagent, a dye, acts as an electron acceptor and changes color in the presence of the oxidase enzymes.

The reagent for the test is a 0.5% aqueous solution of tetramethyl-*p*-phenylenediamine hydrochloride. The solution is stable 1 wk at 4–10° C, or indefinitely when frozen. It is convenient to prepare the oxidase reagent in small amounts: 50-mg amounts of the powder are transferred to 10 or 12, 15- × 125-mm cotton-plugged tubes; these are stored in a brown, screwcapped bottle with some calcium sulfate as a drying agent in the bottom. For use, 10-ml of distilled water is added to one tube, the reagent is dissolved, and the solution is transferred to a small dropper bottle.

To test a culture, 1 drop of oxidase reagent is added to colonies on a plate. The colonies of oxidase-positive organisms become light blue (weak reaction) to blue-black (strong reaction). Because the reagent eventually kills the organisms, colonies must be subcultured as soon as possible after the color begins to change.

Kovacs' modification is a more sensitive procedure for detecting oxidase, and it is used to confirm all negative results obtained by the routine colony procedure. One to two drops of oxidase reagent are placed on a piece of filter paper; a loopful of the organisms from the agar is mixed into the reagent on the paper; the development of a light to dark blue color within 10 sec indicates a positive reaction (a late reaction within 10 sec also is considered weak). The oxidase test performed with commercially prepared oxidase reagent stabilized with citric acid may give negative results with some *Pasteurella* species.

CAUTION: some metals in bacteriological loops produce false-positive reactions. Platinum-iridium wire (15%) loops are satisfactory. If another kind of loop is used, it should be tested first by exposing it to the oxidase reagent on filter paper.

References: King, E.O., and H.W. Tatum. 1962. *Actinobacillus actinomycetemcomitans* and *Haemophilus aphrophilus*. J. Infect. Dis. *111:* 85–94.

Kovacs, N. 1956. Identification of *Pseudomonas pyocyanea* by the oxidase reaction. Nature *178:* 703.

- **PEPTIC DIGEST OF BLOOD**

Peptic digest is used as a medium supplement for the cultivation of *Haemophilus* cultures. It is prepared as follows: 150 ml of 0.9% aqueous sodium chloride is autoclaved (121° C for 15 min) in a bottle having a ground glass stopper, but with the stopper separately wrapped in paper and the bottle cotton-plugged. The bottle is placed in a 55° C water bath, and the following are added to it: hydrochloric acid, concentrated, 6 ml; sterile sheep blood, defibrinated, 50 ml; and pepsin, 1 g. The mixture is kept in the water bath at 55° C overnight, with occasional shaking during the first 2 h. Twelve milliliters of 20% sodium hydroxide is added, then more is added slowly, with intermittent shaking, until the pH reaches 7.6. Thereafter, the pH is returned to the range of 7.0–7.2 with 1 N hydrochloric acid. Five-tenths of a milliliter of chloroform is added to maintain sterility, and the glass-stoppered bottle is stored in the refrigerator. For the cultivation of *Haemophilus* species, the peptic digest is added to broth and agar media at a concentration of 2 to 5%.

Reference: Fildes, P. 1920. A new medium for the growth of *B. influenzae*. Br. J. Exp. Pathol. *1:* 129–130.

- **PHENYLALANINE AGAR**

This medium is used to determine the ability of an organism to deaminate the amino acid phenylalanine to phenylpyruvic acid.

The medium consists of yeast extract (Difco), 3 g; DL-phenylalanine, 2 g (or L-phenylalanine, 1 g); disodium phosphate (anhydrous), 1 g; sodium chloride, 5 g; agar, 12 g; and distilled water, 1 L. The medium is tubed in 3-ml amounts into 13- × 100-mm screw-capped tubes, autoclaved at 121° C for 10 min, and slanted for cooling.

The medium is inoculated heavily with bacterial growth from an HIA slant or a blood agar plate and incubated up to 24 h at 35° C. Four or five drops of 10% ferric chloride solution are dropped onto the growth on the slant. The formation of phenylpyruvate is indicated by the slant becoming green.

For testing members of the *Enterobacteriaceae*, the ferric chloride sometimes has to be added after only a 4- to 6-h incubation period. Continued incubation results in a false-negative reaction.

Reference: Ewing, W.H., B.R. Davis, and R.W. Reaves. 1957. Phenylalanine and malonate media and their use in enteric bacteriology. Public Health Lab. *15:* 153–160.

- **PIGMENT PRODUCTION BY PSEUDOMONADS**

Frequently, the pseudomonads encountered in the diagnostic laboratory do not produce pigment readily, if at all, in commonly employed laboratory media.

Studies by E. O. King and her associates indicated that the kind of peptone used in the basal medium markedly affects pigment production, so they developed two simple pigment-enhancing media for routine use: medium A ("Tech") for enhancing the elaboration of pyocyanin and pyorubrin, and medium B ("Flo") for stimulating the production of pyoverdin (fluorescein).

Medium A ("Tech") consists of Bacto-peptone (Difco), 20 g; glycerol, 10 ml; magnesium chloride, 1.4 g; potassium sulfate, 10 g; agar, 15 g; and distilled water, 1 L. Medium B ("Flo") consists of Proteose peptone #3

(Difco), 20 g; glycerol, 10 ml; dipostassium phosphate (anhydrous), 1.5 g; magnesium sulfate (hydrated), 1.5 g; agar, 15 g; and distilled water, 1 l. The pH of each medium is adjusted to 7.2. The media are dispensed in 7-ml amounts into 15- × 125-mm tubes, sterilized at 121° C for 15 min, and slanted for cooling.

Each slant is inoculated with 1 drop of a 24-h heart infusion broth culture and incubated at 35° C for up to 7 days. The pigments usually become evident within 3 days.

The pyocyanin-producing cultures produce pigments ranging in color from a faint aqua to a dark blue or green on medium A ("Tech") and bright green on medium B ("Flo"). *Pseudomonas aeruginosa* is the only known species of bacteria that produces pyocyanin. Pyocyanin is soluble in chloroform and changes from blue in the alkaline pH range to pink in the acid pH range. To confirm the presence of the pigment, 1–2 ml of chloroform is added to a slant culture with the suspected pyocyanin, and the agar is mashed to extract the pigment. The chloroform layer becomes blue as it absorbs the pigment.

Some of the blue chloroform is pipetted off into a small test tube, and 1 or 2 drops of 1 N HCl are added. Pyocyanin turns pink. **CAUTION: chloroform may be carcinogenic.**

The pyorubrin-producing cultures produce pigments ranging from light pink to dark maroon on the "Tech" medium. The pyoverdin-producing cultures produce a pigment on the "Flo" medium that fluoresces when excited by ultraviolet light (i.e., when a Wood's lamp is used in the dark). Pyoverdin production often is enhanced by incubation at 25° C. Some *Pseudomonas aeruginosa* cultures produce a dark brown pigment on the "Flo" medium.

Reference: King, E.O., M.K. Ward, and D.E. Raney. 1954. Two simple media for the demonstration of pyocyanin and fluorescein. Lab. Clin. Med. *44:* 301–307.

- **PORPHYRIN TEST FOR *HAEMOPHILUS***

This is a test for rapidly demonstrating porphyrin synthesis in *Haemophilus* strains. It detects the ability of bacteria to use δ-aminolaevulinic acid in the biosynthesis of porphobilinogen and porphyrins. The method permits detection of some of the enzyme activities involved in the hemin biosynthetic pathway, and offers a more reproducible alternative to the traditional tests for hemin requirement in *Haemophilus* strains.

The culture is grown on a chocolate agar plate under increased carbon dioxide tension (candle jar) at 35° C for 24–48 h. A heavy suspension of the bacterial growth is made in 0.5 ml of the substrate solution (see below) contained in a 12- × 75-mm tube. The reaction mixture is incubated at 35° C for 4 h. Then the mixture is checked with a Wood's lamp. The mixture fluoresces if the δ-aminolaevulinic acid has been converted to porphyrins (a positive test). If the reaction is negative, the tube is incubated overnight, 18–24 h, and retested. If there is no fluorescence, one volume of Kovacs' modification of Ehrlich's reagent (*p*-dimethylaminobenzaldehyde, 5 g; amyl alcohol, 75 ml; and concentrated hydrochloric acid, 25 ml) is added. The mixture is shaken vigorously and allowed to stand to permit separation of the water and alcohol phases. A red color in the **lower aqueous** phase indicates the conversion of δ-aminolaevulinic acid to porphobilinogens (a positive test). (A red color develops in the alcohol phase when indole is present.) Either reaction, fluorescence with the Wood's lamp or a red reaction in the aqueous phase with Kovacs' reagent, indicates that the organism does not require hemin (X-factor) for growth.

Substrate solution. δ-Aminolaevulinic acid hydrochloride (Sigma), 2 mM (33.5 mg), and magnesium sulfate (hydrated), 0.8 mM (19.72 mg), in 100 ml of 0.1 M phosphate buffer, pH 6.9 [Sorensen: 6.3 g $NaH_2PO_4 \cdot H_2O$ and 14.8 g $Na_2HPO_4 \cdot 7H_2O$ (or 19.7 g $Na_2HPO_4 \cdot 12H_2O$) in 1 L H_2O].

Reference: Kilian, M. 1974. A rapid method for the differentiation of *Haemophilus* strains—the porphyrin test. Acta Pathol. Microbiol. Scand. Sect. B *82:* 835–842.

- **POTASSIUM HYDROXIDE TEST FOR VERIFYING THE GRAM REACTION OF BACTERIA**

The purpose of this test is to enable the user to distinguish between Gram-positive and Gram-negative bacteria when the Gram stain results are dubious.

One to two drops of 3% KOH are placed on a glass slide. A loopful of young bacterial growth from agar is mixed with the reagent. If, on raising the loop, a thread of slime follows it, the organism is considered to be Gram-negative; Gram-positive organisms do not form the slime thread.

It has been found that some *Bacillus* strains form the slime thread, particularly older cultures that have lost their Gram-positivity.

References: Ryu, E. 1938. On the Gram-differentiation of bacteria by the simplest method. J. Jpn. Soc. Vet. Sci. *17:* 31.

Gregersen, T. 1978. A rapid method for distinction of gram-negative from gram-positive bacteria. Eur. J. Appl. Microbiol. Biotechnol. *5:* 123–127.

- **PRESERVATION OF BACTERIA BY FREEZING**

Stock cultures ranging from eugonic enteric organisms to the more fastidious organisms, such as *Campylobacter fetus*, *Haemophilus* species, and *Bordetella pertussis*, can be preserved indefinitely by freezing them in defibrinated blood and storing at low temperatures (< –40° C). Their survival over many years is excellent, and they can be recultivated with relative ease.

Freezing tubes are made from polypropylene with stoppers attached (#701, Walter Sarstedt Inc., Princeton, NJ). These particular kinds of tubes, which withstand the sudden changes in temperature to which they are subjected, must be used. The tubes are autoclaved without rinsing.

Cultures are labeled by using 1/2- × 3/4-in strips of nonwaterproof adhesive tape (we use "White Cross," American White Cross Laboratories, New Rochelle, NY) on which the name or number of the culture has been typewritten. The tape is affixed near the top of the tube.

Storage boxes are constructed of 3/8-in marine waterproof plywood, with inside dimensions of 15 × 6 × 3 in. A 1- × 6- × 15-in block of styrofoam, prepunched with 3/8-in holes in line each way, is placed in the bottom of the box. (The size of the box and styrofoam insert depends on the inside dimensions of the freezer.) The diameters of the holes are slightly less than the diameters of the tubes to allow a snug fit. A 20-penny nail is used satisfactorily as a punch. Before a new box is used, it is prechilled in a deep freeze. Alternatively, tubes can be stored in 5 × 5 × 2 in. liquid nitrogen freezer boxes. When fitted with a 9 × 9 grid, one of these boxes can hold as many as 324 cultures.

Cultures to be frozen are transferred to a heart infusion agar slant and incubated 18–24 h, or until growth appears. (Any carbohydrate-free agar medium that supports good growth can be used.) Bacteria are suspended in 7–8 drops of sterile defibrinated rabbit blood added to each slant with a capillary pipette. (Other kinds of antimicrobial-free blood probably can be used.) The suspension is aspirated and delivered to a freezing tube to about two-thirds volume.

The tubes with their contents are cooled to freezing by placing them in the deep freeze.

To recultivate the frozen organism, the chilled tube is held at the meniscus of the blood and rotated, with thumb and forefinger, until a small portion of the culture thaws. A droplet of the suspension is removed from the tube with a Pasteur pipette or other suitable means and inoculated to an appropriate medium for growth. The tube is returned immediately to the freezer to prevent further thawing.

• **RAPID GELATIN SLANT TEST**

This test, which uses a pH indicator, may occasionally demonstrate gelatin liquifaction earlier than the standard method. The medium consists of 1 g ammonium phosphate dibasic; 0.2 g potassium chloride; 0.2 g heptahydrous magnesium sulfate; 0.5 g yeast extract (Difco); 0.5 g casitone (Difco); 10 g gelatin (Difco); 15 g agar (Difco); and 0.02 g phenol red. The ingredients are dissolved in 1 L of distilled water and the pH is adjusted to 6.5. The medium is dispensed into 13- × 100-mm screwcapped tubes and autoclaved at 121° C for 10 min. Allow the tubes to cool in a tilted position so that a slant is produced.

The slant is inoculated heavily with a loopful of growth from a 24- to 72-h heart infusion agar or rabbit blood agar culture and incubated in air at the optimum growth temperature. Gelatin hydrolysis is indicated by the development of a pink to red color in the medium. The test is read at 1, 2, and 7 days.

Reference: Pickett, M.J., J.R. Greenwood, and S.M. Harvey. 1991. Tests for detecting degradation of gelatin: comparison of five methods. *J. Clin. Microbiol.* 29: 2322–2325.

• **RAPID SUGAR TEST FOR NEISSERIA**

Neisseria gonorrhoeae is presumptively identified in the clinical laboratory by growth and colonial morphology on selective media, cell morphology, and appearance with the Gram reaction and oxidase reaction. It is confirmed either by the fluorescent antibody test or by its reaction on various carbohydrate media. The rapid sugar test is a time-saving modification of the standard procedure for the latter. Although it was developed for testing *Neisseriae*, we have found it useful for other fastidious organisms, such as *Haemophilus* species, *Capnocytophaga canimorsus*, and *Kingella denitrificans*.

Reagents. The buffer-salt solution consists of dipotassium phosphate (anhydrous), 0.04 g; monopotassium phosphate, 0.01 g; potassium chloride, 0.8 g; 1% phenol red solution (aqueous), 0.4 ml; and distilled water, 100 ml. The pH is 7.0. The solution is filter-sterilized and stored at 4° C in a sterile, screwcapped bottle. The desired carbohydrate solutions are made up to 20% concentrations in distilled water or in broth (peptone, 10 g; meat extract, 3 g; sodium chloride, 5 g; distilled water, 1 L), and the pH is adjusted to 7.0. These solutions are filter-sterilized.

Test. The 0.1 ml of buffer-salt solution is placed in a 10- × 75-mm sterile tube. One drop of the appropriate carbohydrate solution is added to the tube. For a control, 1 drop of distilled water or broth is added to the tube. Then 1 drop of a very heavy 18- to 24-h culture suspension (prepared in 0.35 ml of the buffer-salt solution) is added. (Alternatively, a loopful of the suspected organism is added directly into a tube of the buffer-salt-carbohydrate solution.) The tube is incubated in a 35° C water bath for 4 h and read at 30-min intervals. A positive test is indicated by acid formation (yellow color).

References: Kellog, D.S., and E.M. Turner. 1973. Rapid fermentation confirmation of *Neisseria gonorrhoeae*. *Appl. Microbiol.* 25: 550–552.

Brown, W.J. 1976. Stability of working reagents for the modified rapid fermentation test (MRFT). Health. Lab. Sci. *14:* 172–176.

Hollis, D.G., F.O. Sottnek, W.J. Brown, and R.E. Weaver. 1980. Use of the rapid fermentation test in determining reactions of fastidious bacteria in clinical laboratories. J. Clin. Microbiol. *12:* 620–623.

- **SALINE, BUFFERED (pH 7.0)**

Stock solutions: Na_2HPO_4, 9.5 g/l; KH_2PO_4, 9.1 g/L. Working solution: Na_2HPO_4 stock solution, 61.1 ml; KH_2PO_4 stock solution, 38.9 ml; NaCl, 8.5 g; distilled water, 900 ml.

- **SS AGAR (DIFCO)**

This is a differentially selective medium for the isolation of pathogenic enteric bacilli, especially those belonging to the genera *Shigella* and *Salmonella*.

The medium consists of Bacto-beef extract, 5 g; proteose peptone (Difco), 5 g; Bacto-lactose, 10 g; Bacto-bile salts No. 3, 8.5 g; sodium citrate, 8.5 g; sodium thiosulfate, 8.5 g; ferric citrate, 1 g; Bacto-agar, 13.5 g; Bacto-brilliant green, 0.00033 g; and Bacto-neutral red, 0.025 g. It is prepared according to the manufacturer's instructions. It is dispensed while hot in 5-ml amounts into 15- × 125-mm tubes and slanted (it is not autoclaved).

Slants are inoculated with 1 drop of a 24-h heart infusion broth culture and incubated at 35° C for up to 7 days.

- **SALT TOLERANCE TEST**

The purpose of the test is to determine the ability of organisms to grow in different concentrations of sodium ion.

The medium consists of Bacto-nutrient broth (Difco), 8 g; distilled water, 1 L; and various concentrations (0, 6, 7, 8, 9, 10, 11, and 12%) of sodium chloride. Each quantity of sodium chloride is dissolved in two-thirds of the required amount of nutrient broth in a volumetric flask. The flasks are autoclaved at 121° C for 15 min. The broths are cooled to room temperature, brought to volume with additional sterile nutrient broth, and mixed well. The media are dispensed aseptically in 4-ml amounts into 16- × 125-mm screwcapped tubes.

Each medium is inoculated with 1 drop of an 18- to 24-h heart infusion broth culture and incubated at 35° C for 24 h. The cultures are examined for growth or growth inhibition in the different concentrations of sodium chloride.

Reference: King, E.O. 1964. The Identification of Unusual Pathogenic Gram-Negative Bacteria. Atlanta, Communicable Disease Center.

- **SELECTIVE MEDIUM FOR PATHOGENIC NEISSERIA (THAYER-MARTIN MEDIUM-MODIFIED)**

The medium consists of Mueller-Hinton agar (BBL), 38 g (composed of beef infusion, 300 g; Acidicase peptone (BBL), 17.5 g; starch, 1.5 g; and agar, 17 g) and distilled water, 1 L. Twenty-five-milliliter amounts are dispensed into 2-oz screwcapped bottles, autoclaved at 121° C for 15 min, and cooled to 50° C. Twenty-five hundredths of a milliliter of V-C-N Inhibitor (BBL) (vancomycin, colistin, nystatin)[d] antibiotic mixture is added. (V-C-N in 10-ml vials is reconstituted, then dispensed in 0.25-ml amounts into 13- × 100-mm screwcapped tubes, and frozen.) The contents are mixed well and poured in 20-ml amounts into Petri dishes.

NOTE: *Neisseria lactamica* and an occasional strain of other *Neisseria* species, as well as some species of Gram-negative rods, for example, *Kingella denitrificans* and *Weeksella virosa*, can grow on this medium.

Reference: Thayer, J.D., and J.E. Martin. 1966. Improved medium for cultivation of *Neisseria gonorrhoeae* and *Neisseria meningitidis*. Public Health Rep. *81:* 559–562.

- **TRANSFORMATION ASSAY FOR *ACINETOBACTER* AND *MORAXELLA OSLOENSIS***

The identification of these coccobacillary bacteria by biochemical techniques may be inconclusive. Transformation provides a more definitive and critical assessment of the taxonomic placement of these organisms. Juni (1972, 1974) has described a transformation procedure for determining genetic relatedness for *Acinetobacter* species and *Moraxella osloensis* cultures. The procedure involves mixing crude DNA extracts of test strains with the cellular mass of an amino acid auxotroph. Genetically related test strains will convert the auxotroph to prototrophy. This conversion is evidenced by the subsequent growth of colonies from the auxotroph-crude DNA test mixture upon a basal mineral salts medium.

Lactate mineral medium (solution A) is prepared as follows: to 800 ml of distilled water the following chemicals are added, one at a time (each chemical is allowed to dissolve before the next one is added): monopotassium phosphate (anhydrous), 1.5 g; disodium phosphate (anhydrous), 13.5 g; magnesium sulfate, 0.1 g; ammonium chloride, 2 g in 100 ml of distilled water, added slowly with stirring; calcium chloride (anhydrous), 10 ml of a 0.1% aqueous solution; ferric sulfate (hydrated), 5 ml of a 0.01% solution; lactic acid (reagent grade), 5 ml; and distilled water, added to make 1 L. The solution is dispensed in 150-ml portions into 8-oz screwcapped bottles and autoclaved at 121° C for 15 min.

Basal Medium (solution B) is prepared as follows: 30 g of agar is heated in 1 L of distilled water to dissolve; this is dispensed in 50-ml amounts into 8-oz screwcapped bottles and autoclaved at 121° C for 15 min.

[d]Or the equivalent.

One bottle of solution A is added to the contents of one bottle of solution B (melted and cooled to 50° C). Twenty-milliliter amounts of the medium are poured into 15- × 100-mm Petri plates. If the plates are not used immediately after drying in a 35° C incubator for about 15 h, they are stored in airtight plastic bags and placed in the refrigerator (4° C). Stored in this manner, they retain their effectiveness for at least 3 wk.

Preparing crude transforming DNA. The culture to be used as the DNA donor is grown on the surface of any suitable medium, e.g., HIA, blood agar, or nutrient agar. The age of the culture is not critical. A loopful of cell growth is suspended in approximately 0.5 ml of a sterile solution of 0.05% sodium dodecyl sulfate in standard saline citrate (0.15 M NaCl, 0.015 M trisodium citrate) contained in a tightly capped test tube. The cell paste is suspended to immerse all of the visible cellular material. The suspension is then heated in a 60° C water bath for 15 to 60 min to lyse the cells and produce a sterile DNA-transforming preparation. This kind of DNA solution may be stored at 3–5° C for months.

Transformation assay. A small portion of bacterial cell mass from a 24-h culture of the recipient strain to be transformed is applied to a section of an HIA plate (the complete medium).

A loopful of the crude DNA preparation is used to suspend and spread the cell paste in an area about 1 cm in diameter. A portion of the cell mass of the recipient strain is also spread onto another section of the plate to serve as a non-DNA-treated control. A third section of the same plate is streaked with a loopful of the DNA preparation to test its sterility. The plate is then inverted and incubated at 35° C for 4–6 h for *Acinetobacter* assays or 8 h to overnight for *Moraxella osloensis* assays. The incubation period may be prolonged if necessary.

After the incubation period on the complete medium, assuming that the DNA preparation is sterile, cellular growth from each of the growth areas is streaked heavily onto a section of the lactate mineral agar plate (minimal medium), upon which only the prototrophic transformants will grow. If the DNA preparation is not sterile, it is heated for an additional 30 min at 60° C and retested. The minimal medium plates are incubated inverted at 35° C for 24 h.

Transformant colonies are clearly visible, but the non-DNA-treated control cells do not grow on this selective medium. The *M. osloensis* strains are incubated for an additional 24 h if no growth is noted during the initial 24-h incubation period. A small percentage of strains that produce no visible transformant colonies at 24 h are positive at 48 h.

NOTE: Transformation procedures have been described for *Moraxella urethralis* and *Neisseria gonorrhoeae*: Janik, A., E. Juni, and G.A. Heym. 1976. Genetic transformation as a tool for detection of *Neisseria gonorrhoeae*. J. Clin. Microbiol. *4:* 71–81.; Juni, E. 1976. Genetic transformation assays for identification of strains of *Moraxella urethralis*. J. Clin. Microbiol. *5:* 227–235.

References: Juni, E. 1972. Interspecies transformation of *Acinetobacter*: genetic evidence for a ubiquitous genus. J. Bacteriol. *112:* 917–931.

Juni, E. 1974. Simple genetic transformation assay for rapid diagnosis of *Moraxella osloensis*. Appl. Microbiol. *27:* 16–24.

• TRANSFORMATION ASSAY FOR *PSYCHROBACTER*

This method for confirming the identification of *Psychrobacter* species is based on the ability of DNA from an unknown test isolate to transform an auxotrophic strain of *P. immobilis*. The assay method is similar to the transformation assay described above for *Acinetobacter* and *Moraxella osloensis*.

The assay is performed on M9A medium, which supports the growth of wild type *Psychrobacter* strains but not the *P. immobilis* autotroph CDC KC1838. M9A medium consists of the following ingredients:

Casein-salt solution. Combine 0.45 g of magnesium sulfate heptahydrate, 3 g of sodium chloride, 1 g of potassium phosphate, 2.8 g of sodium phosphate dibasic, and 8 g casein hydrolysate (vitamin and salt free) in 500 ml of double-distilled water. Add and dissolve the magnesium sulfate before adding the other ingredients. Autoclave at 121° C for 20 min and cool to 100° C.

Agar. Combine 15 g Bacto agar with double-distilled water to a final volume of 500 ml in a 2 L flask. Autoclave at 121° C for 20 min and cool to 100° C.

Glucose 50% solution. Combine 25 g of glucose with double-distilled water to a final volume of 50 ml. Sterilize by passing through a 0.22-μm filter.

Lactate 60% solution. Combine 30 g of sodium lactate with double-distilled water to a final volume of 50 ml. Autoclave at 121° C for 15 min.

To make M9A medium, aseptically add 500 ml of the casein-salt solution to the agar solution, mix well, and cool to 55° C. Then aseptically add 8 ml of the glucose solution and 5 ml of the lactate solution, mix well, and dispense the final mixture into 100-mm Petri plates at 20 ml/plate.

Preparing the crude transforming DNA. Suspend a loopful of the culture to be tested and a loopful of a "known" culture of *Psychrobacter immobilis* in 0.5 ml of the lysis buffer described in the *Acinetobacter* transformation protocol above, and incubate at 60° C for 1 h. If the DNA is not tested on the same day that it is prepared, refrigerate (4–5° C) until it is tested.

Transformation assay. Perform the assay using the method for the *Acinetobacter* transformation assay described above. Substitute the M9A medium for the lactate mineral agar to transfer the DNA-treated cultures after incubation on heart infusion agar. For the test to be valid, the following conditions must be met: 1) there must be no growth on the area of the M9A medium inoculated with the untreated auxotroph, and 2) there must be growth on the area inoculated with the auxotroph treated with the DNA from the positive control strain. The test is positive if there is growth in the area of the M9A medium inoculated with the auxotroph that had been overlaid with the crude DNA preparation from the test culture.

References: Juni, E., and G.A. Heym. 1980. Transformation assay for identification of psychrotrophic achromobacters. Appl. Environ. Microbiol. 40: 1106–1114.

Hudson, M.J., D.G. Hollis, R.E. Weaver, and C.G. Galvis. 1987. Relationship of CDC group EO-2 and *Psychrobacter immobilis*. J. Clin. Microbiol. 25: 1907–1910.

• TRIPLE SUGAR IRON AGAR (TSI AGAR)

This medium is used to determine an organism's ability to use fermentatively specific carbohydrates incorporated in a basal growth medium, with or without the production of gas, and to test for the production of hydrogen sulfide.

The medium contains beef extract (Difco), 3 g; yeast extract (Difco), 3 g; Bacto-peptone (Difco), 15 g; proteose peptone (Difco), 5 g; lactose, 10 g; sucrose, 10 g; glucose, 1 g; ferrous sulfate, 0.2 g; sodium chloride, 5 g; sodium thiosulfate, 0.3 g; agar, 15 g; phenol red, 0.024 g; and distilled water, 1 l. After being heated to dissolve the agar, the medium is dispensed in 5-ml portions into 15- × 125-mm tubes and autoclaved at 121° C for 15 min. Tubes are slanted to obtain approximately a 2.5-cm butt and a 4-cm slant.

Growth from an 18- to 24-h 35° C blood agar plate or HIB culture is used to inoculate the slant and stab the butt of the medium. A strip of lead acetate paper is inserted in the tube and held in place with the cotton plug so that it extends about 2 cm below the plug, close to, but not touching the agar. The culture is incubated at 35° C for 7 days. Blackening of the paper is recorded as positive for H_2S production. Hydrogen sulfide formed in the agar butt is also recorded. Acid, alkaline, or neutral reactions, occurring both in the agar butt and on the slant, and gas formation in the butt are noted as evidence of carbohydrate fermentation or lack of it.

Reference: Sulkin, S.E., and J.C. Willett. 1940. A triple sugar-ferrous sulfate medium for use in identification of enteric organisms. J. Lab. Clin. Med. 25: 649–653.

• TRYPTONE GLUCOSE YEAST EXTRACT (TGY) AGAR

This medium is used to determine the range of temperature in which an organism grows.

The medium, tryptone glucose yeast extract (TGY) agar consists of tryptone, 5 g; glucose, 1 g; yeast extract (Difco), 5 g; dipotassium phosphate, 1 g; agar, 15 g; and distilled water, 1 L. The pH is adjusted to 6.8–7.0. The medium is tubed in 5-ml amounts into 15- × 125-mm tubes, autoclaved at 121° C for 15 min, and slanted for cooling.

For each culture tested, three slants are inoculated with 1 drop of a 24-h HIB culture. The tubes are incubated at 25°, 35°, and 42° C, respectively, for 18–24 h. The relative amount of growth at each temperature is recorded.

Reference: Haynes, W.C. 1951. *Pseudomonas aeruginosa*—its characterization and identification. J. Gen. Microbiol. 5: 939–950.

• UREA AGAR (CHRISTENSEN'S)

This medium is used to determine an organism's ability to hydrolyze urea, forming ammonia by action of the enzyme urease.

The medium is prepared with Bacto-urea agar base (Difco), 29 g [composed of Bacto-peptone, 1 g; Bacto-dextrose, 1 g; sodium chloride, 5 g; monopotassium phosphate, 2 g; urea (Difco), 20 g; and Bacto-phenol red, 0.012 g] and distilled water, 100 ml; this solution is Seitz-filtered. Bacto-agar, 15 g, and distilled water are autoclaved at 121° C for 15 min. The agar is cooled to 55° C, and the urea solution is added. The medium is tubed in 5-ml amounts into 15- × 125-mm screwcapped tubes and slanted. The slants are inoculated with 1 drop of a 24-h HIB culture and incubated at 35° C for up to 7 days. A positive test is indicated by the development of a pronounced pink color in the agar. If the organism does not grow on the slant, a second slant is inoculated heavily with growth from an 18- to 24-h blood agar plate or a heart infusion agar slant.

Reference: Christensen, W.B. 1946. Urea decomposition as a means of differentiating *Proteus* and paracolon cultures from each other and from *Salmonella* and *Shigella* types. J. Bacteriol. 52: 461–466.

• VOGES-PROSKAUER (VP) TEST (COBLENTZ METHOD)

The purpose of the test is to determine the ability of some organisms to produce a neutral end product, acetylmethylcarbinol (acetoin), from glucose fermentation.

The medium [MR-VP (Difco)] is composed of buffered peptone, 7 g; glucose, 5 g; dipotassium phosphate (anhydrous), 5 g; distilled water, 1 L. The pH is adjusted to 6.9. The medium is tubed in 2-ml amounts into 18- × 150-mm screwcapped tubes and autoclaved at 121° C for 10 min.

The tubes are heavily inoculated with the growth from an 18- to 24-h HIA slant. The broth culture is incubated at 35° C for 6 h.

Test for acetylmethylcarbinol. The reagents are A) 5% alpha naphthol in 95% ethanol (deteriorates after

30 days at room temperature), and B) 0.3% creatinine in 40% potassium hydroxide (keeps at least 2 wk at room temperature).

The test is initiated by adding 0.6 ml of reagent A to the culture broth and shaking gently to disperse the reagent. Then, 0.2 ml of reagent B is added, and the mixture is shaken vigorously and observed for 5–10 min, with occasional shaking. A positive reaction is indicated by the development of a pink to cherry-red color.

Reference: Coblentz, L.M. 1943. Rapid detection of the production of acetylmethyl-carbinol. Am. J. Publ. Health 33: 815–817.

Introduction To King's Identification Key

In Miss King's original edition of the "Identification of Unusual Gram Negative Bacteria," a key for identifying aerobes was included. In the Key, the various species and unnamed groups of bacteria were placed in categories based on three characteristics: 1) the mechanism by which the organism produced acid from glucose, fermentation, or oxidation (organisms not producing acid from glucose were designated as glucose nonoxidizers); 2) the ability of the organism to grow on MacConkey agar; and 3) the oxidase reaction of the organism. An organism was to be identified by comparing its biochemical profile to the profiles of known organisms having the same "key" characteristics (referring to the glucose, MacConkey, and oxidase reactions) as those of the unknown organism.

In this edition of the manual, the revised Key appears in two places. First, the Key is part of the "Gram-Negative Organisms: An Approach to Identification (Guide to Presumptive Identification)" (p. 25). Second, it is a guide to the "Round-Table" Charts, (p. 109).

In the Key, a species or unnamed group has been considered positive for a characteristic if 90% or more of the strains have been positive for the characteristic. If 11% through 89% of the strains have been positive, the organism has been considered variable for the characteristic and has been included in both a positive and negative category. If less than 11% of the strains have been positive, the organism has been considered negative for the characteristic.

The Key is only a guide for identification. If an organism cannot be identified as one of the species listed in its category on the Key, one should consider the possibility that it belongs to a species or group for which less than 11% of the strains have one or more of the key characteristics exhibited by the unidentified culture. This is more likely to occur with the oxidase or MacConkey reaction than with the glucose reaction. Such a culture, therefore, should also be compared with species that differ from it in its oxidase or MacConkey reaction, or in both.

If an organism with a negative Gram stain cannot be identified in the schema, one should consider the possibility that it is a Gram-positive organism, particularly if it does not grow on MacConkey agar.

X- and V-factor-requiring *Haemophilus* species are not included in the Key. These organisms are tabulated in "Round-Table" Chart 45 and are described individually in Chapter 3, "Description of Species." Organisms that fail to grow on heart infusion agar without rabbit blood should be tested for X- and V-factor requirements.

There is also the possibility that an organism that cannot be identified with this Key belongs to a species that we have neither studied nor encountered.

Gram-Negative Organisms: An Approach to Identification (Guide to Presumptive Identification)

This Guide was developed originally by R. E. Weaver and D. G. Hollis in response to the question, "How can we identify these organisms in the clinical laboratory without doing the total number of tests that are done in a reference laboratory?" The Guide is based on results obtained with conventional procedures described in this manual.

The Guide consists of three parts: 1) a Key for the identification of Gram-negative aerobes, 2) 12 tables, and 3) a code book. The Key contains 13 categories. Twelve of the categories are based on the method by which an organism produces acid from glucose, its ability to grow on MacConkey agar, and the oxidase reaction. There is a corresponding table for each of these categories. The 13th category of the Key is for an organism that acidifies the OF basal medium in the absence of glucose.

In the Key and the Tables, a species or unnamed group is considered positive for a key characteristic if 90% or more of the strains have been positive for that characteristic. If 11% through 89% of the strains have been positive for that characteristic, the organism is considered to be variable for the characteristic. If only 10% or less of the strains have been positive, the organism is considered to be negative for the characteristic. If a species or unnamed group is variable for one or more of the three key characteristics, the species or group is listed in more than one category of the Key. For example, if a species ferments glucose 95% of the time, grows on MacConkey agar 65% of the time, and is oxidase-positive 2% of the time, it is listed, because of the variable MacConkey reaction, in two categories of the Key (and on the two corresponding Tables): 1) glucose fermenters, MacConkey-positive, oxidase-negative, and 2) glucose fermenters, MacConkey-negative, oxidase-negative.

The first step in arriving at a presumptive identification is to determine the three key characteristics of the culture being studied. After this has been done, the organism is tested for the additional reactions or characteristics included in the Table listing the various species and unnamed groups of bacteria that have the same three key characteristics. For example, if an unknown culture ferments glucose, grows on MacConkey agar, and has a positive oxidase reaction, it should be tested for the additional reactions listed in Table 1.2: lysine and ornithine decarboxylase, arginine dihydrolase, indole, urease, and fermentation of mannitol, lactose, and sucrose. A presumptive identification is reached by comparing the results obtained in these tests with the reactions listed for the various organisms in the Table.

An alternative method for reaching the presumptive identification is to translate the results into a number (the number of digits in the number varies for each Table). As shown at the top of each Table, each reaction has been assigned a numerical value of 1, 2, or 4. The reactions to be used for calculating the value of each digit are separated by the bold lines. For Table 1.2, the first digit is determined by the lysine, arginine, and ornithine reactions; the second digit by the indole, urea, and mannitol reactions; and the third digit by the lactose and sucrose reactions. The calculation of the three-digit number for an organism that produces positive reactions for ornithine decarboxylase, indole, urease, and fermentation of sucrose is illustrated below. The number is 432.

To find which organism(s) may produce these reactions, the number 432 is found in the Code Book on the pages titled "Glucose fermenters, MacConkey-positive/oxidase-positive." The presumptive identification for the organism in the illustration is *Pasteurella pneumotropica*.

	Lysine	Arginine	Ornithine	Indole	Urea	D-Mannitol	Lactose	Sucrose
Numerical value of positive result	1	2	4	1	2	4	1	2
Test result	−	−	+	+	+	−	−	+
Score of result	0	0	4	1	2	0	0	2
Sum of scores (code number)	4			3			2	

Some of the code numbers may be generated from the reactions produced by more than one species. In these instances, the organisms that would produce the number are listed along with additional characteristics that can be used to differentiate those organisms. For example, the reactions of both *Actinobacillus lignieresii* and *A. suis* may generate the code number 062. The Code Book indicates that the esculin hydrolysis test can be used to differentiate these two species of bacteria.

For each of the Tables in the Guide, there is a corresponding set of code numbers. In the Presumptive Identification Key (p. 30), the table number and the pages of the Code Book which correspond to each category are indicated.

Media Used for Presumptive Identification

The media that can be used to determine the key reactions are:

For Mode of Glucose utilization:
    Triple sugar iron agar
    OF D-glucose
    OF base, control
MacConkey agar (growth)
Blood agar (oxidase test)

The media used for the presumptive identification of Gram-negative aerobic bacteria are listed below for each of the 12 tables.

Table 1.1. Glucose fermenters: MacConkey-positive/oxidase-negative

Urea agar	Lactose broth
Tryptone broth (indole)	Sucrose broth
Lysine, Moeller base	Motility agar (incubate at 35° C)
Arginine, Moeller base	Motility agar (incubate at 25° C)
Ornithine, Moeller base	SS agar slant
Moeller base, control	

Table 1.2. Glucose fermenters: MacConkey-positive/oxidase-positive

Lysine, Moeller base	D-Mannitol broth
Arginine, Moeller base	Lactose broth
Ornithine, Moeller base	Sucrose broth
Moeller base, control	20% NaCl solution, sterile
Tryptone broth (indole)	MR broth (optional test)
Urea agar	MR broth with 3% NaCl (optional test)

One drop of the 20% NaCl solution is added to each 1 ml of the Moeller base media and tryptone broth to enhance the growth of sodium-requiring organisms. The use of MR broth (that is essentially void of NaCl) and MR broth with 3% NaCl is to aid in the identification of the sodium-requiring organisms.

Table 1.3. Glucose fermenters: MacConkey-negative/oxidase-negative

Heart infusion agar slant (catalase)	Maltose broth
Nitrate broth, with insert tube	D-Xylose broth
Tryptone broth (indole)	Esculin agar slant
Lactose broth	

Table 1.4. Glucose fermenters: MacConkey-negative/oxidase-positive

Heart infusion agar slant (catalase)
Tryptone broth (indole)
Urea agar
Nitrate broth, with insert tube
Ornithine, Moeller base
Moeller base, control

D-Xylose broth
D-Mannitol broth
Sucrose broth
Maltose broth
Blood agar (hemolysis, if not previously inoculated)

Table 1.5. Glucose oxidizers: MacConkey-positive/oxidase-negative

OF D-mannitol
OF lactose
OF sucrose
OF base, control
Nitrate broth, with insert tube

Lysine, Moeller base
Arginine, Moeller base
Moeller base, control
Esculin agar slant
Motility agar

Table 1.6. Glucose oxidizers: MacConkey-positive/oxidase-positive

Tech agar slant
Flo agar slant
Tryptone broth (indole)
TSI slant (if not previously inoculated)
Lysine, Moeller base
Arginine, Moeller base
Moeller base, control
Nitrate broth, with insert tube

Urea agar
OF D-xylose
OF D-mannitol
OF lactose
OF sucrose
OF maltose
OF base, control

To aid in the identification of apyocyanogenic strains of *Pseudomonas aeruginosa*, it is recommended that cultures also be tested for the ability to grow at 42° C. *P. aeruginosa* grows at 42° C while *P. fluorescens* and *P. putida* do not.

Table 1.7. Glucose oxidizers: MacConkey-negative/oxidase-negative

Urea agar
Tryptone broth (indole)
Nitrate broth, with insert tube
OF D-xylose
OF sucrose

OF maltose
OF base, control
Arginine, Moeller base
Moeller base, control

Table 1.8. Glucose oxidizers: MacConkey-negative/oxidase-positive

Tryptone broth (indole)
Urea agar
OF D-xylose
OF D-mannitol
OF lactose

OF sucrose
OF maltose
OF base, control
Nitrate broth, with insert tube

Table 1.9. Nonoxidizers: MacConkey-positive/oxidase-negative

OF maltose	Urea agar
OF D-xylose	Nitrate broth, with insert tube
OF base, control	Blood agar (hemolysis, if not previously
Motility agar	inoculated)

Table 1.10. Nonoxidizers: MacConkey-positive/oxidase-positive

OF D-mannitol	Urea agar
OF D-xylose	SS agar slant
OF base, control	Motility agar
Nitrate broth, with insert tube	Heart infusion agar slant (incubate at 25° C
TSI slant (if not previously inoculated)	for flagella stain)
	Nitrite broth, with insert tube (optional test)

Table 1.11. Nonoxidizers: MacConkey-negative/oxidase-negative

Urea agar
Nitrate broth, with insert tube
Motility agar

Table 1.12. Nonoxidizers: MacConkey-negative/oxidase-positive

Tryptone broth (indole)	Nitrate broth, with insert tube
Urea agar	Motility agar
OF D-xylose	Heart infusion agar slant (catalase)
OF base, control	

NOTES

All strains were not tested in every test.

Symbols:

− = 0–10% +
v = 11–89% +
+ = 90–100% + within 48 h
+ or (+) = <90% + in 48 h, >90% + in 3–7 days

Test results are recorded as above; numbers are percentage-positive reactions.

Some *Bacillus* species may give a negative Gram stain reaction. The possibility that an organism, particularly if it does not grow on MacConkey agar, may be a *Bacillus* species should be considered if it can not be identified as a species of Gram-negative bacteria.

Microscopic morphology is important for the recognition of some genera such as *Neisseria* and *Capnocytophaga*.

Presumptive Identification Key

Glucose Fermenters

Table No.		Code Book Page(s)

MacConkey-positive

1.1 Oxidase-negative 33

Chromobacterium violaceum, Enterobacteriaceae, Haemophilus aphrophilus, Pasteurella bettyae.

1.2 Oxidase-positive 34

Aeromonas species, *Actinobacillus equuli, A. hominis, A. lignieresii, A. suis,* Bisgaard's taxon 16, *Chromobacterium violaceum,* EF-4a, *Haemophilus aphrophilus, Neisseria mucosa, N. sicca, N. subflava, Pasteurella aerogenes, P. bettyae, P. gallinarum, P. haemolytica, P. pneumotropica, P. trehalosi, Plesiomonas shigelloides, Suttonella indologenes, Vibrio* species.

MacConkey-negative

1.3 Oxidase-negative 36

Actinobacillus actinomycetemcomitans, A. ureae[a], Bisgaard's taxon 16[a], Capnocytophaga species (DF-1), DF-3, DF-3-like, *Gardnerella vaginalis, Haemophilus aphrophilus, Leptotrichia buccalis, Pasteurella bettyae, P. canis[a], P. dagmatis[a], P. multocida[a], P. pneumotropica[a], P. stomatis[a], Streptobacillus moniliformis.*

1.4 Oxidase-positive 38

Actinobacillus actinomycetemcomitans, A. equuli, A. lignieresii, A. suis, A. ureae, Bisgaard's taxon 16, *Capnocytophaga canimorsus, C. cynodegmi, Cardiobacterium hominis,* EF-4a, *Haemophilus aphrophilus, Kingella denitrificans, K. kingae, Neisseria lactamica, N. mucosa, N. sicca, N. subflava, Pasteurella bettyae, P. canis, P. dagmatis, P. gallinarum, P. haemolytica, P. multocida, P. pneumotropica, P. stomatis, Riemerella anatipestifer, Simonsiella* species, *Suttonella indologenes, Vibrio* species.

Glucose Oxidizers

MacConkey-positive

1.5 Oxidase-negative 42

Saccharolytic *Acinetobacter* species, *Brucella canis, Chryseomonas luteola, Flavimonas oryzihabitans,* O-1, *Pseudomonas cepacia, P. gladioli, P. mallei, Roseomonas gilardii, Sphingomonas paucimobilis* and *S. parapaucimobilis, Xanthomonas maltophilia.*

1.6 Oxidase-positive 44

"*Achromobacter*" group B, "*Achromobacter*" group E, *Acidovorax delafieldii, A. temperans, Agrobacterium radiobacter* biovar 1, "*Agrobacterium* yellow group," *Alcaligenes xylosoxidans* subsp. *xylosoxidans, Brucella* species (*B. abortus, B. canis, B. melitensis, B. suis*), EF-4b, EO-2, EO-3, *Flavobacterium breve, F. meningosepticum, F. multivorum, Flavobacterium* species (IIb), *F. spiritivorum, F. thalpophilum, F. yabuuchiae, Francisella philomira-*

---

[a] Included in this section because oxidase reaction may vary with the reagent used.

gia, Methylobacterium species, *Neisseria canis, N. elongata* subsp. *nitroreducens, Ochrobactrum anthropi,* O-1, *Pseudomonas aeruginosa, P. cepacia, "P. denitrificans," P. diminuta, P. fluorescens, P. gladioli, Pseudomonas*-like group 2, *P. mallei, P. mendocina, P. pertucinogena, P. pickettii* (biovars 1 and 2), *P. pseudomallei, P. putida, P. stutzeri, "P. thomasii", P. vesicularis,* Saccharolytic *Psychrobacter immobilis, Roseomonas fauriae, R. gilardii, Shewanella putrefaciens* biotype 1, *Sphingomonas paucimobilis/parapaucimobilis,* Ic, Vb-3.

MacConkey-negative

1.7 Oxidase-negative 48

Brucella canis, Francisella tularensis, O-1, *Pseudomonas mallei, Roseomonas gilardii, Sphingomonas paucimobilis/parapaucimobilis.*

1.8 Oxidase-positive 50

Acidovorax delafieldii, A. facilis, A. temperans, "Agrobacterium yellow group", *Balneatrix alpica, Brucella* species (*B. abortus, B. canis, B. melitensis, B. suis*), EF-4b, E0-2, *Flavobacterium meningosepticum, F. mizutaii, Flavobacterium* species (IIb), *F. spiritivorum, F. yabuuchiae, Francisella philomiragia, Methylobacterium* species, *Neisseria canis, N. elongata* subsp. *nitroreducens, N. gonorrhoeae, N. lactamica, N. meningitidis, N. polysaccharea,* O-1, O-2, *Pseudomonas mallei, P. vesicularis, Riemerella anatipestifer, Roseomonas fauriae, R. gilardii, Sphingomonas paucimobilis/parapaucimobilis,* IIe, IIh, Ili.

Nonoxidizers

MacConkey-positive

1.9 Oxidase-negative 54

Asaccharolytic *Acinetobacter* species, *Bordetella holmesii, B. parapertussis, Brucella canis,* NO-1, *Roseomonas gilardii, Xanthomonas maltophilia.*

1.10 Oxidase-positive 56

Acidovorax delafieldii; Afipia felis; "Agrobacterium yellow group"; *Alcaligenes eutrophus; A. faecalis; Alcaligenes*-like group 1; *A. piechaudii; A. xylosoxidans* subsp. *denitrificans; A. xylosoxidans* subsp. *xylosoxidans; Bordetella avium; B. bronchiseptica; Brucella abortus; B. canis; B. melitensis; Campylobacter* species; *Comamonas acidovorans; Comamonas* species (includes *C. testosteroni* and *C. terrigena*); *Flavobacterium odoratum;* Gilardi rod group 1; *Methylobacterium* species; *Moraxella atlantae; M. osloensis; M. phenylpyruvica; Neisseria canis; N. elongata* subsp. *elongata; N. elongata* subsp. *glycolytica; N. elongata* subsp. *nitroreducens; N. flavescens; N. mucosa; N. sicca; N. subflava; N. weaveri; Ochrobactrum anthropi; Oligella ureolytica; O. urethralis; Pseudomonas alcaligenes; P. diminuta; P. pseudoalcaligenes; Pseudomonas* species CDC Group 1; *P. vesicularis;* Asaccharolytic *Psychrobacter immobilis; Roseomonas cervicalis; R. fauriae; Roseomonas* genomospecies 4, 5, and 6; *R. gilardii; Shewanella putrefaciens* biotypes 1 and 2; *Taylorella equigenitalis;* IIg; IVc-2.

MacConkey-negative

1.11 Oxidase-negative 62

Acinetobacter species (nonhemolytic/asaccharolytic strains), *Bartonella bacilliformis, Bartonella* (formerly *Rochalimaea*) species, *Bordetella holmesii, Brucella canis, Francisella tularensis,* NO-1, *Roseomonas gilardii.*

1.12 Oxidase-positive 63

Acidovorax delafieldii; A. facilis; Afipia broomeae; A. clevelandensis; A. felis; Afipia genomospecies 1, 2, and 3; "*Agrobacterium* yellow group"; *Bartonella* (formerly *Rochalimaea*) species; *Bordetella pertussis; Brucella* species (*B. abortus, B. canis, B. melitensis*); *Campylobacter* species; *Eikenella corrodens; Kingella denitrificans; Methylobacterium* species; *Moraxella atlantae; M. bovis; Moraxella (Branhamella) catarrhalis; M. lacunata; M. lincolnii; M. nonliquefaciens; M. osloensis; M. phenylpyruvica; Neisseria canis; N. cinerea; N. elongata* subsp. *elongata; N. elongata* subsp. *glycolytica; N. elongata* subsp. *nitroreducens; N. flavescens; N. mucosa; N. sicca; N. subflava; N. weaveri; Oligella ureolytica;* O-2; *Pseudomonas vesicularis; Psychrobacter immobilis* (asaccharolytic strains); *Roseomonas fauriae; Roseomonas* genomospecies 5 and 6; *R. gilardii; Taylorella equigenitalis; Weeksella virosa; W. zoohelcum.*

Acid Producers in OF Basal Medium Without Carbohydrate

MacConkey-positive

Oxidase-positive 127

OFBA-1: see page 127

Table 1.1. Glucose fermenters: MacConkey-positive/oxidase-negative*[a]

Tests	Number of Strains	D-Glucose	Growth on MacConkey	Oxidase	Urea	Indole	Lysine	Arginine	Ornithine	Lactose	Sucrose	Motility 35°C	Motility 25°C	Growth on SS
Numerical value of positive result					1	2	4	1	2	4	1	2	4	1
Enterobacteriaceae		+	+	−										
Chromobacterium violaceum	37	+ 100	+ 100	v 67	v 5 (14)	v 21	− 0	+ 100	− 0	− 0	v 20 (6)	+	+	+ or (+) 71 (23)
Haemophilus aphrophilus	203	+ 98 (2)	v 19	v 28	− 0	− 0	− 0	− 0	− 0	+ 96 (4)	+ 99 (1)	−	−	− 0
Pasteurella bettyae	88	+ 100	v 20 (24)	v 61	− 0	+ 100	− 0	− 0	− 0	− 0	− 0	−	−	− 0

*See Presumptive Identification Code p. 66.
[a]Where the blank spaces occur, either the percentages were not calculated or they were not available.

SECTION 1: AN APPROACH TO IDENTIFICATION

Table 1.2. Glucose fermenters: MacConkey-positive/oxidase-positive*a

Tests	Number of Strains	D-Glucose	Growth on MacConkey	Oxidase	Lysine	Arginine	Ornithine	Indole	Urea	D-Mannitol	Lactose	Sucrose
Numerical value of positive result					1	2	4	1	2	4	1	2
Aeromonas species, *Vibrio* species, or *Plesiomonas shigelloides*		+	v	+								
Actinobacillus equuli	19	+ 100	v 84 (5)	+ 100	- 0	- 0	- 0	- 0	+ 100	+ 100	+ 95	+ 100
Actinobacillus hominis	1	+	+	+	-	-	-	-	+	+	+	+
Actinobacillus lignieresii	30	+ 96 (4)	v 67 (8)	+ 100	- 0	- 0	- 0	- 0	+ 100	+ 91 (9)	v 17 (61)	+ 96 (4)
Actinobacillus suis	33	+ 94 (6)	+ or (+) 82 (12)	+ 100	- 0	- 0	- 0	- 0	+ 97 (3)	v 54	+ or (+) 79 (18)	+ 94 (6)
Bisgaard's taxon 16	30	+ 100	v 7 (10)	+ 100	- 0	- 0	- 0	+ 100	- 0	- 0	- 0	+ 100
Chromobacterium violaceum	37	+ 100	+ 100	v 67	- 0	+ 100	- 0	v 21	v 5 (14)	- 0	- 0	v 20 (6)
EF-4a	97	+ 100	v 42 (8)	+ 100	- 0	v 77 (2)	- 0	- 0	- 0	- 0	- 0	- 0
Haemophilus aphrophilus	203	+ 98 (2)	v 19	v 28	- 0	- 0	- 0	- 0	- 0	- 0	+ 96 (4)	+ 99 (1)
Neisseria mucosa	30	+ or (+) 83 (10)	v 57 (3)	+ 100	-	-	-	-	-	-	- 0	+ 90 (10)
Neisseria sicca	43	+ or (+) 78 (14)	v 68	+ 100	-	-	-	-	-	-	- 0	+ 92 (8)
Neisseria subflava	153	v 68 (6)	v 47 (2)	+ 100	-	-	-	-	-	-	- 0	v 56 (5)
Pasteurella aerogenes	16	+ 100	+ 100	+ 100	- 0	- 0	v 88	- 0	+ 100	- 6	- 0	+ 94
Pasteurella bettyae	88	+ 100	v 20 (24)	v 61	- 0	- 0	- 0	+ 100	- 0	- 0	- 0	- 0
Pasteurella gallinarum	10	v 90 (10)	v 20 (10)	+ 90	- 0	- 0	v 25	- 0	- 0	- 0	- 0	+ 100
Pasteurella haemolytica: Reference strains	2	+ 100	+ 100	+ 100	- 0	- 0	- 0	- 0	- 0	+ 100	v (50)	+ 100

SECTION 1: AN APPROACH TO IDENTIFICATION

Organism	n											
Pasteurella haemolytica: phenotypically similar strains	28	+ 96 (4)	v 64 (14)	+ 96	− 0	− 0	− 0	− 0	− 0	+ 96 (4)	v 7 (43)	+ 96 (4)
Pasteurella pneumotropica	107	+ 97 (3)	v 36 (17)	+ 99	v 33	− 0	+ 100	+ 90	+ 95 (1)	− 2 (1)	v 14 (39)	+ 97 (3)
Pasteurella trehalosi	11	+ 100	+ 100	+ 100	− 0	− 0	− 0	− 0	− 0	+ 82 (18)	− 0	+ 100
Suttonella indologenes	6	+ 100	v 17	+ 100	− 0	− 0	− 0	+ 100	− 0	− 0	− 0	+ 100

*See Presumptive Identification Code p. 67.
[a]Where blank spaces occur, either the percentages were not calculated or they were not available.

Table 1.3. Glucose fermenters: MacConkey-negative/oxidase-negative*

Tests	Number of Strains	D-Glucose	Growth on MacConkey	Oxidase	Catalase	Nitrate Reduction	Indole	Lactose	Maltose	D-Xylose	Esculin hydrolysis
Numerical value of positive result					1	2	4	1	2	4	1
Actinobacillus actinomycetemcomitans	120	+ or (+) 83 (16)	− 4 (1)	v 19	+ 99	+ 100	− 0	− 0	+ or (+) 80 (15)	v 33 (9)	− 0
Actinobacillus ureae	97	+ 100	− 0	+[a] 99	v 63	+ 99	− 0	− 0	+ 91 (5)	− 0	− 0
Bisgaard's taxon 16	30	+ 100	v 7 (10)	+[a] 100	+ 100	+ 100	+ 100	− 0	+ 100	− 0	− 0
Capnocytophaga species (DF-1)	155	+ 90 (10)	− 0	− 7	− 7	v 63	− 0	v 75 (11)	+ or (+) 86 (14)	− 0	v 81 (2)
DF-3	21	+ or (+) 86 (14)	− 0	− 0	− 0	− 0	v 71 (14)	+ or (+) 52 (43)	+ or (+) 81 (19)	+ or (+) 86 (14)	+ 100
DF-3-like	7	+ or (+) 43 (57)	− 0	− 0	v 29	− 0	+ 100	+ or (+) 57 (43)	+ or (+) 57 (43)	− 0	v 57 (14)
Gardnerella vaginalis	126	+ or (+) 81 (13)	− 5	− 6	− 2	− 0	− 0	− (1)	+ or (+) 87 (13)	v 36 (9)	− 0
Haemophilus aphrophilus	203	+ 98 (2)	v 19	v 28	− 5	+ 100	− 0	+ 96 (4)	+ 99 (1)	− 0	− 0
Leptotrichia buccalis	9	+ or (+) 88 (12)	− 0	− 0	− 0	− 0	− 0	v 44 (11)	+ or (+) 67 (33)	− 0	v 89
Pasteurella bettyae	88	+ 100	v 20 (24)	v 61	− 1	+ 100	+,w 100	− 0	− 0	− 0	− 0
Pasteurella canis	31	+ 100	− 0	+[a] 92	+ 96	+ 100	+ 96	− 0	− 0	− 0	− 0
Pasteurella dagmatis	129	+ 100	− 1 (2)	+[a] 98	+ 99	+ 100	+ 100	− 3	+ 100	− 0	− 0
Pasteurella multocida[b]	225	+ 98 (2)	− 1 (1)	+[a] 96	+ 98	+ 99	+ 98	− 8 (<1)	− <1	v 81 (4)	− 0
Pasteurella pneumotropica	107	+ 97 (3)	v 36 (17)	+[a] 99	+ 100	+ 100	+ 90	v 14 (39)	+ 97 (3)	+ or (+) 76 (19)	− 0

Pasteurella stomatis	8	+ 100	− 0	+^{a} 100	+ 100	+ 100	+ 100	− 0	− 0	− 0	− 0
Streptobacillus moniliformis	48	+ or (+) 79 (21)	− 0	− 3	− 2	− 0	− 0	− 0	+ or (+) 81 (19)	− (4)	v 47 (9)

*See Presumptive Identification Code p. 69.
^{a}Oxidase reaction for this organism can vary depending upon the method used.
^{b}For differentiation of the subspecies of *P. multocida* see pages 462.

Table 1.4. Glucose fermenters: MacConkey-negative/oxidase-positive*[a]

Tests	Number of Strains	D-Glucose	Growth on MacConkey	Oxidase	Catalase	Indole	Urea	Nitrate reduction
Numerical value of positive result					1	2	4	1
Vibrio species		+	v	+				
Actinobacillus actinomycetemcomitans	120	+ or (+) 83 (16)	– 4 (1)	v 19	+ 99	– 0	– 0	+ 100
Actinobacillus equuli	19	+ 100	v 84 (5)	+ 100	v 68 (5)	– 0	+ 100	+ 100
Actinobacillus lignieresii	30	+ 96 (4)	v 67 (8)	+ 100	v 89	– 0	+ 100	+ 100
Actinobacillus suis	33	+ 94 (6)	+ or (+) 82 (12)	+ 100	v 85	– 0	+ 97 (3)	+ 100
Actinobacillus ureae	97	+ 100	– 0	+ 99	v 63	– 0	+ 100	+ 99
Bisgaard's taxon 16	30	+ 100	v 7 (10)	+ 100	+ 100	+ 100	– 0	+ 100
Capnocytophaga canimorsus	90	v 48 (32)	– 0	+ 100	+ 100	– 0	– 0	– 0
Capnocytophaga cynodegmi	13	+ or (+) 69 (31)	– 0	+ 100	+ 100	– 0	– 0	– 0
Cardiobacterium hominis	65	+ or (+) 80 (20)	– 0	+ 100	– 1	+ 100	– 0	– 0
EF-4a	97	+ 100	v 42 (8)	+ 100	+ 100	– 0	– 0	+ 97
Haemophilus aphrophilus	203	+ 98 (2)	v 19	v 28	– 5	– 0	– 0	+ 100
Kingella denitrificans	60	+ or (+) 30 (62)	– 0	+ 100	– 10	– 0	– 0	+ 93
Kingella kingae	137	+ or (+) 41 (54)	– 9 (1)	+ 100	– 0	– 0	– 0	– 3 (1)
Neisseria lactamica	287	+ 96 (3)	– 0	+ 100	+ 100	–	– 0	– 0

(continued on page 40)

Nitrate to gas	Ornithine Decarboxylase	D-Xylose	D-Mannitol	Sucrose	Maltose	β-like Hemolysis
2	4	1	2	4	1	2
− 0	−	v 33 (9)	v 66 (16)	− 0	+ or (+) 80 (15)	− 0
− 0	− 0	+ 100	+ 100	+ 100	+ 95	− 0
− 0	− 0	+ or (+) 87 (13)	+ 91 (9)	+ 96 (4)	+ or (+) 83 (17)	− 4
− 0	− 0	+ 94 (6)	v 54	+ 94 (6)	+ 94 (6)	v 76
− 0	− 0	− 0	+ 99 (1)	+ 99 (1)	+ 91 (5)	− 0
− 0	− 0	− 0	− 0	+ 100	+ 100	− 0
− 0	− 0	− 0	− 0	− 0	+ or (+) 53 (39)	− 4
− 0	− 0	− 0	− 0	v 30 (46)	+ or (+) 69 (31)	v 67
−	−	− 0	+ or (+) 52 (43)	+ or (+) 63 (35)	+ or (+) 69 (31)	− 0
v 62	− 0	− 0	− 0	− 0	− 0	− 1
− 0	− 0	− 0	− 0	+ 99 (1)	+ 99 (1)	− 0
v 88	−	− 0	− 0	− 0	− 0	− 0
− 0	− 0	− 0	− 0	− 0	+ or (+) 71 (29)	v 81
−	−	−	− 0	− 0	+ 98 (2)	− 0

(continued on page 41)

Table 1.4. Glucose fermenters: MacConkey-negative/oxidase-positive*a *(continued from page 38)*

Tests	Number of Strains	D-Glucose	Growth on MacConkey	Oxidase	Catalase	Indole	Urea	Nitrate reduction
Neisseria mucosa	30	+ or (+) 83 (10)	v 57 (3)	+ 100	+ 100	-	- 0	+ 100
Neisseria sicca	43	+ or (+) 78 (14)	v 68	+ 100	v 80	-	- 0	- 0
Neisseria subflava	153	v 68 (6)	v 47 (2)	+ 100	v 80	-	- 0	- 0
Pasteurella bettyae	88	+ 100	v 20 (24)	v 61	- 1	+ 100	- 0	+ 100
Pasteurella canis	31	+ 100	- 0	+ 92	+ 96	+ 96	- 0	+ 100
Pasteurella dagmatis	129	+ 100	- 1 (2)	+ 98	+ 99	+ 100	+ 95 (5)	+ 100
Pasteurella gallinarum	10	+ 90 (10)	v 20 (10)	+ 90	+ 100	- 0	- 0	+ 100
Pasteurella haemolytica: reference strains	2	+ 100	+ 100	+ 100	+ 100	- 0	- 0	+ 100
Pasteurella haemolytica: phenotypically similar strains	28	+ 96 (4)	v 64 (14)	+ 96	+ 96	- 0	- 0	+ 100
Pasteurella multocida[b]	225	+ 98 (2)	- 1 (1)	+ 96	+ 98	+ 98	- 0	+ 99
Pasteurella pneumotropica	107	+ 97 (3)	v 36 (17)	+ 99	+ 100	+ 90	+ 95 (1)	+ 100
Pasteurella stomatis	6	+ 100	- 0	+ 100	+ 100	+ 100	- 0	+ 100
Riemerella anatipestifer	5	+ or (+) 60 (40)	- 0	+ 100	+ 100	- 0	v 40 (20)	- 0
Simonsiella crassa: type strain	1	+, w	-	+	-	-	-	+
Simonsiella-like strains	3	+ 100	- 0	+ 100	+ 100	- 0	- 0	v 67
Suttonella indologenes	6	+ 100	v 17	+ 100	v 17	+ 100	- 0	- 0

*See Presumptive Identification Code p. 72.
[a]Where the blank spaces occur, either the percentages were not calculated or they were not available.
[b]For differentiation of the subspecies of *P. multocida* see pages 462.

(continued from page 39)

Nitrate to gas	Ornithine Decarboxylase	D-Xylose	D-Mannitol	Sucrose	Maltose	β-like Hemolysis
+ 100	−	−	− 0	+ 90(10)	+ 90 (10)	− 0
−	−	−	− 0	+ 92 (8)	+ 95 (5)	− 7
−	−	−	− 0	v 56 (5)	+ 99 (1)	− 9
− 2	− 0	− 0	− 0	− 0	− 0	− 0
− 4	+ 100	− 0	− 0	+ 100	− 0	− 0
− 0	− 0	− 0	− 0	+ 100	+ 100	− 0
− 0	v 25	v 33	− 0	+ 100	+ 100	− 0
− 0	− 0	+ 100	+ 100	+ 100	+ 100	v 50
− 0	− 0	+ 96 (4)	+ 96 (4)	+ 96 (4)	+ 93 (7)	v 75
− 0	+ 93	v 81 (4)	+ 97 (3)	+ 96 (3)	− <1	− 2
− 0	+ 100	+ or (+) 76 (19)	− 2 (1)	+ 97 (3)	+ 97 (3)	− 0
− 0	− 0	− 0	− 0	+ 100	− 0	− 0
− 0	−	− 0	− 0	− 0	+ or (+) 40 (60)	− 0
+	−	−	−	+,w	−	−
v 67	−	− 0	− 0	− 0	+ 100	v 67
− 0	−	− 0	− 0	+ 100	+ or (+) 17 (83)	− 0

Table 1.5. Glucose oxidizers: MacConkey-positive/oxidase-negative*a

Tests	Number of Strains	D-Glucose	Growth on MacConkey	Oxidase	OF D-Mannitol
Numerical value of positive result					1
Saccharolytic *Acinetobacter* species (both hemolytic and nonhemolytic strains)	77	+ 95 (5)	+ 96	– 1	– 0
Brucella canis	28	+ or (+) 81 (14)	v 12 (29)	v 72	– 0
Chryseomonas luteola	34	+ 100	+ 100	– 0	+ or (+) 76 (18)
Flavimonas oryzihabitans	36	+ 100	+ 100	– 0	+ 100
O-1	62	+ or (+) 69 (31)	v 6 (40)	v 77	– 0
Pseudomonas cepacia	159	+ 100	+ 100	v 86	+ 100
Pseudomonas gladioli	58	+ 98 (2)	+ 97 (3)	v 47	+ 91 (9)
Pseudomonas mallei	8	+ 100	v 88	v 25	v 62 (14)
Roseomonas gilardii	21	v (43)	+ or (+) 43 (52)	v 52	v 14 (38)
Sphingomonas paucimobilis/ parapaucimobilis	134	+ 93 (7)	v 10 (13)	v 75	– 0
Xanthomonas maltophilia	228	+ or (+) 85 (5)	+ 100	– 0	– 0

*See Presumptive Identification Code p. 75.
aWhere the blank spaces occur, either the percentages were not calculated or they were not available.
bOne strain tested.

OF Lactose	OF Sucrose	Nitrate Reduction	Lysine	Arginine	Esculin Hydrolysis	Motility
2	4	1	2	4	1	2
v 69 (13)	– 0	– 8	– 5 (5)	v 15 (5)	– 0	– 0
– 0	– 0	+ 100	–	–	– 0	– 0
v 3 (24)	v 12	v 62	– 0	+ 100	+ 100	+
v 14 (22)	v 25	– 6	– 7	v 14	– 0	+
– 0	– 0	– 0	– 0	– 4	+ 93 (2)	v
+ 99 (1)	v 86 (1)	v 57	v 80	– 0	v 63 (6)	+
v 9 (28)	– 0	v 43	– 0	– 2	– 0	+
v 12 (62)	– 0	+ 100	– 0	+ 100	– 0	– 0
– 0	– 0	– 5	–[b]	–[b]	– 0	v 33
+ 93 (7)	+ 93 (7)	– 3	– 0	– 8	+ 91	v
v 60 (1)	v 63 (1)	v 39	+ 93	– 0	v 39	+

SECTION 1: AN APPROACH TO IDENTIFICATION

Table 1.6. Glucose oxidizers: MacConkey-positive/oxidase-positive*

Tests	Number of Strains	D-Glucose	Growth on MacConkey	Oxidase	Pyocyanin	Pyoverdin (Fluorescin)	Indole	H₂S TSI	Lysine
Numerical value of positive result					1	2	4	1	2
"Achromobacter" group B	3	+ 100	+ 100	+ 100	- 0	- 0	- 0	- 0	- 0
"Achromobacter" group E	2	+ 100	+ 100	+ 100	- 0	- 0	- 0	v (50)	- 0
Acidovorax delafieldii	2	(+) (100)	(+) (100)	+ 100	- 0	- 0	- 0	- 0	- 0
Acidovorax temperans	2	+ 100	+ or (+) 50 (50)	+ 100	- 0	- 0	- 0	- 0	- 0
Agrobacterium radiobacter biovar 1	66	+ 94 (6)	+ 100	+ 100	- 0	- 0	- 0	v 11 (3)	- 0
"Agrobacterium yellow group"	2	(+) (100)	v (50)	+ 100	- 0	- 0	- 0	- 0	- 0
Alcaligenes xylosoxidans subsp. xylosoxidans	135	v 78	+ 100	+ 100	- 0	- 0	- 0	- 0	- 0
Brucella species	347	+ or (+) 80 (10)	v 23 (27)	+ 92	- 0	- 0	- 0	- 0	
EF-4b	34	+ or (+) 70 (26)	v 65 (6)	+ 100	- 0	- 0	- 0	- 0	- 0
EO-2	11	+ 91 (9)	v 64 (18)	+ 100	- 0	- 0	- 0	- 0	-
EO-3	7	+ 100	+ 100	+ 100	- 0	- 0	- 0	- 0	-
Flavobacterium breve: type strain and 6 similar strains	7	+ or (+) 86 (14)	+ 100	+ 100	- 0	- 0	+ 100	- 0	- 0
Flavobacterium meningosepticum	148	+ 95 (4)	v 89 (3)	+ 99	- 0	- 0	+ 100	- 3	-
Flavobacterium multivorum	22	+ 100	+ 100	+ 100	- 0	- 0	- 0	- 0	-
Flavobacterium species (IIb)	155	+ 92 (6)	v 54 (9)	+ 96	- 0	- 0	+ 98	- 1	-
Flavobacterium spiritivorum	13	+ 100	v (46)	+ 100	- 0	- 0	- 0	- 0	- 0
Flavobacterium thalpophilum	10	+ 100	+ 100	+ 100	- 0	- 0	- 0	- 0	- 0

(continued on page 46)

Arginine	Nitrate Reduction	Nitrate to gas	Urea	OF D-Xylose	OF D-Mannitol	OF Lactose	OF Sucrose	OF Maltose
4	1	2	4	1	2	4	1	2
+ 100	+ 100	+ 100	+ 100	+ 100	v (67)	v (33)	(+) (100)	+ or (+) 33 (67)
+ 100	+ 100	+ 100	+ 100	+ 100	- 0	(+) (100)	(+) (100)	+ 100
+ 100	+ 100	- 0	+ 100	(+) (100)	+ or (+) 50 (50)	- 0	- 0	- 0
- 0	+ 100	+* 100	v 50	- 0	v 50	- 0	- 0	- 0
- 8	v 83	- 5	+ or (+) 88 (9)	+ 97 (3)	+ 94 (6)	+ or (+) 79 (21)	+ 95 (5)	+ 97 (3)
- 0	- 0	- 0	(+) (100)	+ or (+) 50 (50)	- 0	(+) (100)	(+) (100)	+ 100
v 13	+ 100	v 60	- 0	+ 99	- 0	- 0	- 0	- 0
-	+ 100	v 44	+ 99	+ 90 (10)	- 0	- 0	- 0	- 0
- 0	+ 97	- 0	- 0	- 0	- 0	- 0	- 0	- 0
-	+ 100	v 18	+ or (+) 36 (55)	+ 91 (9)	- 0	+ or (+) 45 (55)	- 0	- (9)
-	- 0	- 0	+ or (+) 14 (86)	+ 100	+ or (+) 57 (43)	+ or (+) 71 (29)	- 0	v 14 (14)
- 0	- 0	- 0	- 0	- 0	- 0	- 0	- 0	+ or (+) 86 (14)
-	- 0	-	- 3 (5)	- 2 (1)	+ 91 (8)	v 42 (15)	- 0	+ 93 (7)
- 0	- 0	- 0	+ 95	+ 100	- 0	+ 100	+ 100	+ 100
-	v 22	- 0	v 14 (28)	v 30 (1)	- 10	- 0	v 13 (1)	+ 92 (6)
v 25	- 0	- 0	+ or (+) 62 (38)	+ 92 (8)	+ 100	+ 92 (8)	+ 100	+ 92 (8)
- 0	+ 100	- 0	+ 90 (10)	+ 100	- 0	+ 100	+ 100	+ 100

(continued on page 47)

Table 1.6. Glucose oxidizers: MacConkey-positive/oxidase-positive* *(continued from page 44)*

Tests	Number of Strains	D-Glucose	Growth on MacConkey	Oxidase	Pyocyanin	Pyoverdin (Fluorescin)	Indole	H₂S TSI	Lysine
Flavobacterium yabuuchiae	2	+ 100	v (50)	+ 100	- 0	- 0	- 0	- 0	- 0
Francisella philomiragia	16	+ or (+) 63 (37)	v 19 (13)	+[a] 100	- 0	- 0	- 0	+ or (+) 56 (44)	- 0
Methylobacterium species	90	v 40	v 15	+ 96	- 0	- 0	- 0	- 0	
Neisseria canis	1	(+)[b]	(+)	+	-	-	-	-	-
Neisseria elongata subsp. *nitroreducens*	26	v 23w	v 19 (35)	+ 100	- 0	- 0	- 0	- 0	- 0
Ochrobactrum anthropi	71	+ or (+) 86 (13)	+ 100	+ 100	- 0	- 0	- 0	v 49	- 0
O-1	62	+ or (+) 69 (31)	v 6 (40)	v 77	- 0	- 0	- 0	- 0	- 0
Pseudomonas aeruginosa	201	+ 97 (1)	+ 100	+ 99	v 46	v 65	- 0	- 0	- 0
Pseudomonas cepacia	159	+ 100	+ 100	v 86	- 0	- 0	- 0	- 0	v 80
"*Pseudomonas denitrificans*" type strain	1	+	+	+	-	-	-	-	-
"*Pseudomonas denitrificans*" similar strains	28	+ 100	+ 100	+ 100	- 0	- 0	- 0	- 7	- 0
Pseudomonas diminuta	68	v 21 (9)	+ 97 (3)	+ 100	- 0	- 0	- 0	- 0	- 0
Pseudomonas fluorescens	155	+ 100	+ 100	+ 97	- 0	+ 96	- 0	- 0	- 0
Pseudomonas gladioli	58	+ 98 (2)	+ 97 (3)	v 47	- 0	- 0	- 0	- 0	- 0
Pseudomonas-like group 2	11	+ 100	+ 100	+ 100	- 0	- 0	- 0	- 0	- 0
Pseudomonas mallei	8	+ 100	v 88	v 25	- 0	- 0	- 0	- 0	- 0
Pseudomonas mendocina	4	+ 100	+ 100	+ 100	- 0	- 0	- 0	- 0	- 0
Pseudomonas pertucinogena	2	+ or (+) 50 (50)	+ 100	+ 100	- 0	- 0	- 0	- 0	- 0
Pseudomonas pickettii biovar 1	70	+ 100	+ 99 (1)	+ 100	- 0	- 0	- 0	- 0	- 0
Pseudomonas pickettii biovar 2	54	+ 100	+ 100	+ 100	- 0	- 0	- 0	- 0	- 0

(continued on page 48)

(continued from page 45)

Arginine	Nitrate Reduction	Nitrate to gas	Urea	OF D-Xylose	OF D-Mannitol	OF Lactose	OF Sucrose	OF Maltose
v 50	− 0	− 0	v 50	+ 100	+ 100	+ 100	+ 100	+ 100
− 0	− 0	− 0	− 0	− 0	− 0	− 0	+ or (+) 63 (37)	+ or (+) 63 (37)
−	v 25	− 0	v 29 (26)	+ 94	− 2	− 0	− 0	− 2
−	+	−	−	−b	−b	−b	−b	−b
− 0	+ 100	− 0	− 0	− 0	− 0	− 0	− 0	− 0
v 68	+ 100	+ 99	+ 92 (8)	+ 96 (4)	v 46 (34)	− 0	v 28 (25)	v 32 (25)
− 4	− 0	− 0	− 2	− 0	− 0	− 0	− 0	− 0
+ 100	+ 98	+ 93	v 48 (9)	+ 90 (1)	v 70 (3)	− (1)	− 0	− (<1)
− 0	v 57	− 0	v 60 (18)	+ 100	+ 100	+ 99 (1)	v 86 (1)	+ 99 (1)
+	+	+	−	(+)	−	+	−	+
+ 100	+ 100	v 71	v 25 (18)	+ or (+) 43 (57)	− 0	+ 96 (4)	− 0	+ 100
− 0	− 3	− 0	v (13)	− 0	− 0	− 0	− 0	− 0
+ 97	v 19	− 3	v 21 (31)	+ 100	v 53 (2)	v 24 (3)	v 48	− 2
− 2	v 43	− 0	v 30 (28)	+ 98 (2)	+ 91 (9)	v 9 (28)	− 0	− 0
v 30	v 18	− 0	+ 91 (9)	+ 100	+ 100	+ 100	− 0	− 0
+ 100	+ 100	− 0	v 12	v 12 (50)	v 62 (14)	v 12 (62)	− 0	v (75)
+ 100	+ 100	+ 100	v 50	+ or (+) 75 (25)	− 0	− 0	− 0	− 0
− 0	− 0	− 0	− 0	+ or (+) 50 (50)	− 0	− 0	− 0	− 0
− 6	+ 100	v 86	+ 100	+ 100	− 0	+ 100	− 0	+ 100
− 0	+ 100	+ 100*	+ 100	+ 100	− 0	− 0	− 0	− 0

(continued on page 49)

Table 1.6. Glucose oxidizers: MacConkey-positive/oxidase-positive* *(continued from page 46)*

Tests	Number of Strains	D-Glucose	Growth on MacConkey	Oxidase	Pyocyanin	Pyoverdin (Fluorescin)	Indole	H₂S TSI	Lysine
Pseudomonas pseudomallei	70	+ 100	+ 100	+ 100	- 0	- 0	- 0	- 0	- 0
Pseudomonas putida	16	100	+ 100	+ 100	- 0	+ 93	- 0	- 0	- 0
Pseudomonas stutzeri	28	96 (4) +	+ 100	+ 100	- 0	- 0	- 0	- 0	- 0
"*Pseudomonas thomasii*"	31	+ 100	+ 100	+ 100	- 0	- 0	- 0	- 0	- 0
Pseudomonas vesicularis	94	+ or (+) 87 (12)	v 43 (23)	+ 98	- 0	- 0	- 0	- 0	- 0
Psychrobacter immobilis: saccharolytic strains	7	+ or (+) 57 (43)	+ 100	+ 100	-		- 0	- 0	- 0
Roseomonas fauriae	5	v 20	+ or (+) 60 (40)	+ 100	- 0	- 0	- 0	- 0	-
Roseomonas gilardii	21	v (43)	+ or (+) 43 (52)	v 52	- 0	- 0	- 0	- 0	-
Shewanella putrefaciens, biotype 1	24	v 17 (33)	+ 100	+ 100	- 0	- 0	- 0	+ 96	- 0
Sphingomonas paucimobilis/ parapaucimobilis	134	+ 93 (7)	v 10 (13)	v 75	- 0	- 0	- 0	- 0	- 0
Ic	34	+ 97	+ 100	+ 100	- 0	- 0	- 0	- 3	-
Vb-3	65	+ 100	+ 100	+ 100	- 0	- 0	- 0	- 0	- 0

*See Presumptive Identification Code p. 78; where the blank spaces occur, either the percentages were not calculated or they were not available.
[a]Kovacs' modification.
[b]CTA base.
[c]The volume of gas may be small. Gas may not be detected unless incubated at 25° C or in semiaerobic nitrate broth; therefore, in arriving at code numbers, the possibility that this reaction may appear negative was allowed for.

Table 1.7. Glucose oxidizers: MacConkey-negative/oxidase-negative*

Tests	Number of Strains	D-Glucose	Growth on MacConkey	Oxidase
Numerical value of positive test				
Brucella canis	28	+ or (+) 81 (14)	v 12 (29)	v 72
Francisella tularensis[a] (all biogroups)	115	+ or (+) 84 (16)	- 3 (2)	- 0
O-1	62	+ or (+) 69 (31)	v 6 (40)	v 77
Pseudomonas mallei	8	+ 100	v 88	v 25
Roseomonas gilardii	21	v (43)	+ or (+) 43 (52)	v 52
Sphingomonas paucimobilis/ parapaucimobilis	134	+ 93 (7)	v 10 (13)	v 75

*See Presumptive Identification Code p. 90; where the blank spaces occur, either the percentages were not calculated or they were not available.
[a]Carbohydrate medium containing cysteine or cystine must be used to demonstrate acid production from carbohydrates. Identification on basis of enhancement of growth by, or requirement for, cystine or cysteine in media and by serology.

(continued from page 47)

Arginine	Nitrate Reduction	Nitrate to gas	Urea	OF D-Xylose	OF D-Mannitol	OF Lactose	OF Sucrose	OF Maltose
+ 100	+ 100	+ 100[c]	v 13 (8)	+ or (+) 86 (14)	+ 94 (6)	+ 99 (1)	v 66 (4)	+ 99 (1)
+ 100	- 0	- 0	v 13 (44)	+ 100	v 25	v 25 (13)	- 0	v 31
- 0	+ 100	+ 100	v 33 (22)	+ 93 (7)	+ or (+) 89 (4)	- 0	- 0	+ 100
- 3	v 13	- 0	+ or (+) 81 (19)	+ 100	+ 100	+ 100	- 0	+ 100
- 7	- 5	- 0	2 (5)	v 27 (9)	- 0	- 0	- 0	+ 94 (6)
v 29	v 86	- 0	v (43)	+ or (+) 57 (43)	- 0	+ or (+) 57 (43)	- 0	- 0
-	+ 100	v 20	+ 100	+ or (+) 80 (20)	- 0	- 0	- 0	- 0
-	- 5	- 0	+ or (+) 71 (29)	v 19 (57)	v 14 (38)	- 0	- 0	- 0
- 0	+ 100	- 0	v 4 (8)	- 0	- 0	- 0	+ 96 (4)	+ 92 (8)
- 8	- 3	- 0	6 (3)	96 (4)	- 0	+ 93 (7)	+ 93 (7)	+ 97 (3)
+ 100	+ 100	- 0	v 18 (15)	- 0	- 0	- 0	- 0	+ 100
+ 100	+ 100	+ 100	v 12 (45)	+ 97 (3)	v 65 (9)	- 0	- 0	+ 95 (5)

Urea	Indole	Nitrate Reduction	OF D-Xylose	OF Sucrose	OF Maltose	Arginine Dihydrolase
1	2	4	1	2	4	1
+ 100	- 0	+ 100	+ or (+) 81 (19)	- 0	- 0	-
- 0	- (<1)	- 0	-	3	2	-
- 2	- 0	- 0	- 0	- 0	- 0	- 4
v 12	- 0	+ 100	v 12 (50)	- 0	v (75)	+ 100
+ or (+) 71 (29)	- 0	- 5	v 19 (57)	- 0	- 0	-
- 6 (3)	- 0	- 3	+ 96 (4)	+ 93 (7)	+ 97 (3)	- 8

Table 1.8. Glucose oxidizers: MacConkey-negative/oxidase-positive*a

Tests	Number of Strains	D-Glucose	Growth on MacConkey	Oxidase	Indole
Numerical value of positive result					1
Acidovorax delafieldii	2	(+) (100)	(+) (100)	+ 100	- 0
Acidovorax facilis	2	+ or (+) 50 (50)	- 0	+ 100	- 0
Acidovorax temperans	2	+ 100	+ or (+) 50 (50)	+ 100	- 0
"*Agrobacterium* yellow group"	2	(+) (100)	v (50)	+ 100	- 0
Balneatrix alpica	1	+	-	+	+
Brucella species	347	+ or (+) 80 (10)	v 23 (27)	+ 92	- 0
EF-4b	34	+ or (+) 70 (26)	v 65 (6)	+ 100	- 0
EO-2	11	+ 91 (9)	v 64 (18)	+ 100	- 0
Flavobacterium meningosepticum	148	+ 95 (4)	+ or (+) 89 (3)	+ 99	+ 100
Flavobacterium mizutaii	6	+ or (+) 67 (33)	- 0	+ 100	- 0
Flavobacterium species (IIb)	155	+ 92 (6)	v 54 (9)	+ 96	+ 98
Flavobacterium spiritivorum	13	+ 100	v (46)	+ 100	- 0
Flavobacterium yabuuchiae	2	+ 100	v (50)	+ 100	- 0
Francisella philomiragia	16	+ or (+) 63 (37)	v 19 (13)	+ 100	- 0
Methylobacterium species	90	v 40	v 15	+ 96	- 0

(continued on page 52)

Urea	OF D-Xylose	OF D-Mannitol	OF Lactose	OF Sucrose	OF Maltose	Nitrate
2	4	1	2	4	1	2
+ 100	(+) (100)	+ or (+) 50 (50)	- 0	- 0	- 0	+ 100
(+) (100)	(+) (100)	(+) (100)	- 0	- 0	- 0	+ 100
v 50	- 0	v 50	- 0	- 0	- 0	+ 100
(+) (100)	+ or (+) 50 (50)	- 0	(+) (100)	(+) (100)	+ 100	- 0
-	-	(+)	-	-	+	+
+ 99	+ 90 (10)	- 0	- 0	- 0	- 0	+ 100
- 0	- 0	- 0	- 0	- 0	- 0	+ 97
+ or (+) 36 (55)	+ 91 (9)	- 0	+ or (+) 45 (55)	- 0	- (9)	+ 100
- 3 (5)	- 2 (1)	+ 91 (8)	v 42 (15)	- 0	+ 93 (7)	- 0
- 0	(+) (100)	- 0	+ 100	+ or (+) 50 (50)	+ or (+) 50 (50)	- 0
v 14 (28)	v 30 (1)	- 10	- 0	v 13 (1)	+ 92 (6)	v 22
+ or (+) 62 (38)	+ 92 (8)	+ 100	+ 92 (8)	+ 100	+ 92 (8)	- 0
v 50	+ 100	+ 100	+ 100	+ 100	+ 100	- 0
- 0	- 0	- 0	- 0	+ or (+) 63 (37)	+ or (+) 63 (37)	- 0
v 29 (26)	+ 94	- 2	- 0	- 0	- 2	v 25

(continued on page 53)

Table 1.8. Glucose oxidizers: MacConkey-negative/oxidase-positive*ª (continued from page 50)

Tests	Number of Strains	D-Glucose	Growth on MacConkey	Oxidase	Indole
Neisseria canis type strain	1	(+)	(+)	+	−
Neisseria elongata subsp. *nitroreducens*	26	v 23	v 19 (35)	+ 100	− 0
Neisseria gonorrhoeae[b]	197	+ 90 (3)	− 0	+ 100	−
Neisseria lactamica[b]	287	+ 96 (3)	− 0	+ 100	−
Neisseria meningitidis[b]	375	+ 99	− 0	+ 100	−
Neisseria polysaccharea	1	+	−	+	−
O-1	62	+ or (+) 69 (31)	v 6 (40)	v 77	− 0
O-2	66	v 73 (11)	− 9	+ 97	− 0
Pseudomonas mallei	8	+ 100	v 88	v 25	− 0
Pseudomonas vesicularis	94	+ or (+) 87 (12)	v 43 (23)	+ 98	− 0
Riemerella anatipestifer	5	+ or (+) 60 (40)	− 0	+ 100	− 0
Roseomonas fauriae	5	v 20	+ or (+) 60 (40)	+ 100	− 0
Roseomonas gilardii	21	v (43)	+ or (+) 43 (52)	v 52	− 0
Sphingomonas paucimobilis/parapaucimobilis	134	+ 93 (7)	v 10 (13)	v 75	− 0
IIe	30	+ or (+) 83 (17)	− 3	+ 100	+ 100
IIh	21	+ or (+) 85 (15)	− 0	+ 100	+ 100
IIi	23	+ 91 (9)	− 0	+ 100	+ 100

*See Presumptive Identification Code p. 91.
[a]Where the blank spaces occur, either the percentages were not calculated or they were not available.
[b]Carbohydrate reactions obtained in CTA medium; *Neisseria* species also recognized by morphology.

(continued from page 51)

Urea	OF D-Xylose	OF D-Mannitol	OF Lactose	OF Sucrose	OF Maltose	Nitrate
−	−	−	−	−	−	+
− 0	− 0	− 0	− 0	− 0	− 0	+ 100
−	−	− 0	− 0	− 0	− 0	− 0
− 0	−	− 0	+ 95 (5)	− 0	+ 98 (2)	− 0
−	−	− 0	− 0	− 0	+ 99	− 0
−	−	−	−	+w	+	−
− 2	− 0	− 0	− 0	− 0	− 0	− 0
v 12	− 2	− 2	− 2	+ or (+) 64 (36)	+ or (+) 71 (27)	v 15
v 12	v 12 (50)	v 62 (14)	v 12 (62)	− 0	v (75)	+ 100
− 2 (5)	v 27 (9)	− 0	− 0	− 0	+ 94 (6)	− 5
v 40 (20)	− 0	− 0	− 0	− 0	+ or (+) 40 (60)	− 0
+ 100	+ or (+) 80 (20)	− 0	− 0	− 0	− 0	+ 100
+ or (+) 71 (29)	v 19 (57)	v 14 (38)	− 0	− 0	− 0	− 5
− 6 (3)	+ 96 (4)	− 0	+ 93 (7)	+ 93 (7)	+ 97 (3)	− 3
− 0	− 0	− 0	− 0	− 0	+ 97 (3)	− 0
− 0	− 5	− 0	− 0	− 0	+ 95	− 0
v 14 (18)	+ or (+) 87 (13)	− 0	+ 91 (9)	+ 91 (9)	+ 91 (9)	− 0

Table 1.9. Nonoxidizers: MacConkey-positive/oxidase-negative*[a]

Tests	Number of Strains	D-Glucose	Growth on MacConkey	Oxidase
Numerical value of positive result				
Acinetobacter species (asaccharolytic/nonhemolytic strains)	249	− 0	v 73 (4)	− <1
Acinetobacter species (asaccharolytic/β-hemolytic strains)	21	− 0	+ 95	− 0
Bordetella parapertussis	12	− 0	+ 100	− 0
Brucella canis	28	+ or (+) 81 (14)	v 12 (29)	v 72
NO-1	22	− 0	v 5 (15)	− 5
Bordetella holmesii (NO-2)	13	− 0	+ or (+) 77 (23)	− 0
Roseomonas gilardii	21	v (43)	+ or (+) 43 (52)	v 52
Xanthomonas maltophilia	228	+ or (+) 85 (5)	+ 100	− 0

* See Presumptive Identification Code p. 97.
[a] Where the blank spaces occur, either the percentages were not calculated or they were not available.

OF Maltose	Motility	Urea	Nitrate Reduction	β-Hemolysis	OF D-Xylose
1	2	4	1	2	4
- 0	- 0	v 5 (14)	- 5	- 0	- 0
- 0	- 0	v 21 (10)	- 10	+ 100	- 5
- 0	- 0	+ 100	- 0	+ 100	- 0
- 0	- 0	+ 100	+ 100	- 0	+ or (+) 81 (19)
- 0	- 0	- 5	+ 100	- 0	- 0
- 0	- 0	- 0	- 0	- 0	- 0
- 0	v 33	+ or (+) 71 (29)	- 5	- 0	v 19 (57)
+ 100	+	v 3 (12)	v 39	- 2	v 35 (1)

Table 1.10. Nonoxidizers: MacConkey-positive/oxidase-positive*

Tests	Number of Strains	D-Glucose	Growth on MacConkey	Oxidase	OF D-Mannitol	OF D-Xylose	Nitrate to Gas
Numerical value of positive result					1	2	4
Acidovorax delafieldii	2	(+) (100)	(+) (100)	+ 100	+ or (+) 50 (50)	(+) (100)	- 0
Afipia felis	16	- 0	v (44)	+ 100	- 0	(+) (93)	- 0
"*Agrobacterium* yellow group"	2	(+) (100)	v (50)	+ 100	- 0	+ or (+) 50 (50)	- 0
Alcaligenes eutrophus	1	-	+	+	-	-	-
Alcaligenes faecalis	49	- 0	+ 100	+ 100	- 0	- 0	- 0
Alcaligenes-like group 1	8	- 0	+ 100	+ 100	- 0	- 0	+ 100
Alcaligenes piechaudii	4	- 0	+ 100	+ 100	- 0	- 0	- 0
Alcaligenes xylosoxidans subsp. *denitrificans*	4	- 0	+ 100	+ 100	- 0	- 0	+ 100
Alcaligenes xylosoxidans subsp. *xylosoxidans*	135	v 78	+ 100	+ 100	- 0	+ 99	v 60
Bordetella avium	1	-	+	+	-	-	-
Bordetella bronchiseptica	85	- 0	+ 100	+ 100	- 1	- 7	- 0
Brucella species	347	+ or (+) 80 (10)	v 23 (27)	+ 92	- 0	+ 90 (10)	v 44
Campylobacter species		-	v	+	-	-	-
Comamonas acidovorans	69	- 0	+ 100	+ 100	+ 100	- 0	- 0
Comamonas species[a]	28	- 0	+ 96 (4)	+ 100	- 0	- 0	- 0

(continued on page 58)

Nitrate Reduction	H$_2$S TSI	Urea	Growth on SS	Motility	1-2 Polar Flagella	>2 Polar Flagella	Peritrichous (lateral & polar)	Pink Insoluble Pigment
1	2	4	1	2	4	1	2	4
+ 100	- 0	+ 100	- 0	v	v	-	-	- 0
+ 100	- 0	+ or (+) 88 (12)	- 0	+	+	-	-	-
- 0	- 0	(+) (100)	- 0	v	v	-	-	- 0
+	-	(+)	-	+	-	-	+	-
- 0	- 0	- 2	+ 100	+	-	-	+	-
+ 100	- 0	v 75	v 13	+	-	-	+	-
+ 100	- 0	- 0	+ 100	+	-	-	+	-
+ 100	- 0	- 0	+ 100	+	-	-	+	-
+ 100	- 0	- 0	+ 98	+	-	-	+	-
-	-	-	+	+	-	-	+	-
+ 92	- 0	+ 99	+ 99	+	-	-	+	-
+ 100	- 0	+ 99	- 0	-	-	-	-	-
+	-	-	-	+	+	-	-	-
+ 99	- 0	- (4)	v 63 (20)	+	-	+	-	-
+ 96	- 0	v 7 (14)	v 21 (11)	+	-	+	-	-

(continued on page 59)

Table 1.10. Nonoxidizers: MacConkey-positive/oxidase-positive* *(continued from page 56)*

Tests	Number of Strains	D-Glucose	Growth on MacConkey	Oxidase	OF D-Mannitol	OF D-Xylose	Nitrate to Gas
Flavobacterium odoratum	74	- 0	+ 91 (5)	+ 99	- 0	- 0	- 0
Gilardi rod group 1	15	- 0	+ 93	+ 100	- 0	- 0	- 0
Methylobacterium species	90	v 40	v 15	+ 96	- 2	+ 94	- 0
Moraxella atlantae	73	- 0	+ or (+) 80 (20)	+ 100	- 0	- 0	- 0
Moraxella osloensis	163	- 0	v 70	+ 100	- 0	- 0	- 0
Moraxella phenylpyruvica	50	- 0	v 80 (6)	+ 100	- 0	- 0	- 0
Neisseria canis type strain[c]	1	(+)	(+)	+	-	-	-
Neisseria elongata subsp. *elongata*	15	- 0	v 13 (53)	+ 100	- 0	- 0	- 0
Neisseria elongata subsp. *glycolytica*	2	- 0	+ or (+) 50 (50)	+ 100	- 0	- 0	- 0
Neisseria elongata subsp. *nitroreducens*	26	v 23	v 19 (35)	+ 100	- 0	- 0	- 0
Neisseria flavescens[c]	10	- 0	v 67	+ 100	- 0	-	- 0
Neisseria mucosa[c]	30	+ or (+) 83 (10)	v 57 (3)	+ 100	- 0	-	+ 100
Neisseria sicca[c]	43	+ or (+) 78 (14)	v 68	+ 100	- 0	-	- 0
Neisseria subflava[c]	153	v 68 (6)	v 47 (2)	+ 100	- 0	-	- 0
Neisseria weaveri	132	- 0	v 27 (18)	+ 100	- 0	- 0	- 0
Ochrobactrum anthropi	71	+ or (+) 86 (13)	+ 100	+ 100	v 46 (34)	+ 96 (4)	+ 99
Oligella ureolytica	37	- 0	v 62 (27)	+ 100	- 0	- 0	v 60

(continued on page 60)

(continued from page 57)

Nitrate Reduction	H₂S TSI	Urea	Growth on SS	Motility	1-2 Polar Flagella	>2 Polar Flagella	Peritrichous (lateral & polar)	Pink Insoluble Pigment
- 0	- 0	+ 100	v 30 (11)	-	-	-	-	-
- 0	- 0	- 0	v 80	-	-	-	-	-
v 25	- 0	v 29 (26)	- 0	+[b]	+	-	-	+ 99
- 0	- 0	- 0	- 0	-	-	-	-	-
v 24	- 0	- 0	- 0	-	-	-	-	-
v 68	- 0	+ 100	- 0	-	-	-	-	-
+	-	-	-	-	-	-	-	-
- 0	- 0	- 0	- 0	-	-	-	-	-
- 0	- 0	- 0	- 0	-	-	-	-	-
+ 100	- 0	- 0	- 0	-	-	-	-	-
- 0	- 0	- 0	- 0	-	-	-	-	-
+ 100	- 0	- 0	- 0	-	-	-	-	-
- 0	- 0	- 0	- 0	-	-	-	-	-
- 0	- 0	- 0	- 0	-	-	-	-	-
- 0	- 0	- 0	- 0	-	-	-	-	-
+ 100	v 49	+ 92 (8)	+ 96 (1)	+	-	-	+	-
+ 100	- 0	+ 97	- 5	v	-	-	v[b]	-

(continued on page 61)

Table 1.10. Nonoxidizers: MacConkey-positive/oxidase-positive* *(continued from page 58)*

Tests	Number of Strains	D-Glucose	Growth on MacConkey	Oxidase	OF D-Mannitol	OF D-Xylose	Nitrate to Gas
Oligella urethralis	22	- 0	+ 96	+ 100	- 0	- 0	- 0
Pseudomonas alcaligenes	26	- 0	+ 96	+ 96	- 0	- 0	- 0
Pseudomonas diminuta	68	v 21 (9)	+ 97 (3)	+ 100	- 0	- 0	- 0
Pseudomonas pseudoalcaligenes	34	- 9	+ 100	+ 100	- 0	v 18 (12)	- 0
Pseudomonas species CDC Group 1	31	- 0	+ 97 (3)	+ 100	- 0	- 0	+ 100
Pseudomonas vesicularis	94	+ or (+) 87 (12)	v 43 (23)	+ 98	- 0	v 27 (9)	- 0
Psychrobacter immobilis, asaccharolytic strains	5	- 0	v 40	+ 100	- 0	- 0	- 0
Roseomonas cervicalis	7	- 0	+ 100	+ 100	- 0	v 43	- 0
Roseomonas fauriae	5	v 20	+ or (+) 60 (40)	+ 100	- 0	+ or (+) 80 (20)	v 20
Roseomonas genomospecies 4	3	- 0	+ 100	+ 100	- 0	+ 100	- 0
Roseomonas genomospecies 5	3	- 0	+ or (+) 67 (33)	+ 100	- 0	v 67	- 0
Roseomonas genomospecies 6	1	-	(+w)	+	-	-	-
Roseomonas gilardii	21	v (43)	+ or (+) 43 (52)	v 52	v 14 (38)	v 19 (57)	- 0
Shewanella putrefaciens, biotype 1	24	v 17 (33)	+ 100	+ 100	- 0	- 0	- 0
Shewanella putrefaciens, biotype 2	26	- 0	+ 100	+ 100	- 0	- 0	- 0
Taylorella equigenitalis	1	-	(+)	+	-	-	-
IIg	12	- 0	+ 100	+ 100	- 0	- 0	- 0
IVc-2	36	- 0	+ 94 (6)	+ 100	- 0	- 0	- 0

*See Presumptive Identification Code p. 98; where the blank spaces occur, either the percentages were not calculated or they were not available.
[a]Strains similar to the type strains of *C. testosteroni* and *C. terrigena*.
[b]Motility may be delayed or difficult to demonstrate.
[c]Carbohydrate reactions obtained in CTA medium.

(continued from page 59)

Nitrate Reduction	H$_2$S TSI	Urea	Growth on SS	Motility	1–2 Polar Flagella	>2 Polar Flagella	Peritrichous (lateral & polar)	Pink Insoluble Pigment
- 0	- 0	- 0	- 9	-	-	-	-	-
v 54	- 0	- 0	v 38 (8)	+	+	-	-	-
- 3	- 0	v (13)	- 1 (1)	+	+	-	-	-
+ 100	- 0	- 3 (6)	+ 90 (3)	+	+	-	-	-
+ 100	- 0	- 3 (7)	v 30 (6)	+	+	-	-	-
- 5	- 0	- 2 (5)	- 1	+	+	-	-	-
v 40 (20)	- 0	v (20)	- 0	-	-	-	-	-
- 0	- 0	+ or (+) 86 (14)	- 0	+	+	-	-	+ 100
+ 100	- 0	+ 100	v 20	+	+	-	-	+ 100
+ 100	- 0	+ or (+) 67 (33)	- 0	v	v	-	-	+ 100
- 0	- 0	+ 100	- 0	-	-	-	-	+ 100
+	-	+	-	+	+	-	-	+
- 5	- 0	+ or (+) 71 (29)	- 0	v 33	v 33	- 0	- 0	+ 100
+ 100	+ 96	v 4 (8)	- (8)	+	+	-	-	-
+ 100	+ 100	v 42	+ 92 (4)	+	+	-	-	-
-	-	-	-	-	-	-	-	-
- 0	- 0	- 0	- 0	-	-	-	-	-
v 11	- 0	+ 100	- 3 (6)	+	-	-	+	-

Table 1.11. Nonoxidizers: MacConkey-negative/oxidase-negative*

Tests	Number of Strains	D-Glucose	Growth on MacConkey	Oxidase	Urea	Nitrate Reduction	Motility
Numerical value of positive result					1	2	4
Acinetobacter species (nonhemolytic/asaccharolytic strains)	249	– 0	v 73 (4)	– <1	v 5 (14)	– 5	– 0
Bartonella bacilliformis	2	– 0	– 0	– 0	–	–^{a}	+ 100
Bartonella (formerly *Rochalimaea*) species	18	– 0	–	v 13 w	– 0	–^{b}	– 0
Brucella canis	28	+ or (+) 81 (14)	v 12 (29)	v 72	+ 100	+ 100	–
Francisella tularensis, all biogroups	115	+ or (+)^b 84 (16)	– 3(2)	– 0	– 0	– 0	–
NO-1	22	– 0	v 5 (15)	– 5	– 5	+ 100	–
Bordetella holmesii (NO-2)	13	– 0	+ or (+) 77 (23)	– 0	– 0	– 0	– 0
Roseomonas gilardii	21	v (43)	+ or (+) 43 (52)	v 52	+ or (+) 71 (29)	– 5	v 19 (57)

*See Presumptive Identification Code p. 103; where the blank spaces occur, either the percentages were not calculated or they were not available.
^{a} Fails to grow in nitrate test media.
^{b} D-Glucose reaction determined in cysteine agar with phenol red indicator.

Table 1.12. Nonoxidizers: MacConkey-negative/oxidase-positive*

Tests	Number of Strains	D-Glucose	Growth on MacConkey	Oxidase	Indole	Urea	OF D-Xylose	Nitrate Reduction	Motility	Catalase
Numerical value of positive result					1	2	4	1	2	4
Acidovorax delafieldii	2	(+) 100	(+) (100)	+ 100	− 0	+ 100	(+) (100)	+ 100	v	+ 100
Acidovorax facilis	2	+ or (+) 50 (50)	− 0	+ 100	− 0	(+) (100)	(+) (100)	+ 100	v	+ 100
Afipia broomeae	3	− 0	− 0	+ 100	− 0	+ or (+) 33 (67)[a]	(+) (100)	− 0	+	+ 100
Afipia clevelandensis	1	−	−	+	−	(+)[a]	−	−	+	−
Afipia felis	16	− 0	v (44)	+ 100	− 0	+ or (+) 88 (12)[a]	(+) (93)	+ 100	+	v 37
Afipia genomospecies 1	2	− 0	− 0	+ 100	− 0	+ 100[a]	(+) (100)	− 0	+	v 50
Afipia genomospecies 2	1	−	−	+	−	+[a]	(+)	−	+	+
Afipia genomospecies 3	1	−	−	+	−	(+)[a]	(+)	−	+	−
"*Agrobacterium* yellow group"	2	(+) (100)	v 50	+ 100	− 0	(+) (100)	+ or (+) 50 (50)	− 0	+	+ 100
Bartonella (formerly *Rochalimaea*) species	18	− 0	− 0	v 13 w	− 0	− 0[a]	− 0	−[b]	−	−[c]
Bordetella pertussis	51	−	−	+ 94	−	− 0[a]	−	−	−	+ 100
Brucella species	347	+ or (+) 80 (10)	v 23 (27)	+ 92	− 0	+ 99	+ 90 (10)	+ 100	−	+ 100
Campylobacter species		−	v	+	−	−	−	+	+	+
Eikenella corrodens	506	− 0	− 0	+ 100	− 0	− 0	− 0	+ 99	−	− 8
Kingella denitrificans	60	+ or (+) 30 (62)	− 0	+ 100	− 0	− 0	− 0	+ 93	− 0	− 10

(continued on page 64)

Table 1.12. Nonoxidizers: MacConkey-negative/oxidase-positive* (continued from page 63)

Tests	Number of Strains	D-Glucose	Growth on MacConkey	Oxidase	Indole	Urea	OF D-Xylose	Nitrate Reduction	Motility	Catalase
Methylobacterium species	90	v 40	v 15	+ 96	- 0	v 29 (26)	+ 94	v 25	v	+ 100
Moraxella atlantae	73	- 0	+ or (+) 80 (20)	+ 100	- 0	- 0	- 0	- 0	-	+ 95
Moraxella bovis	7	- 0	- 0	+ 100	- 0	- 0	- 0	v 14	- 0	v 14
Moraxella (Branhamella) catarrhalis	74	- 0	- 5	+ 100	- 0	- 0	- 0	+ 92	-	+ 100
Moraxella lacunata	66	- 0	- 2	+ 100	- 0	- 0	- 0	+ 98	-	+ 100
Moraxella lincolnii	1	-	-	+	-	-	-	-	-	+
Moraxella nonliquifaciens	243	- 0	- 8 (2)	+ 100	- 0	- 0	- 0	+ 95	-	+ 95
Moraxella osloensis	163	- 0	v 70	+ 100	- 0	- 0	- 0	v 24	-	+ 95
Moraxella phenylpyruvica	50	- 0	v 80 (6)	+ 100	- 0	+ 100	- 0	v 68	-	+ 90
Neisseria canis	1	(+)	(+)	+	-	-	-	+	-	+
Neisseria cinerea[d]	58	- 0	- 0	+ 100	- 0	- 0	- 0	- 0	-	+ 100
Neisseria elongata subsp. elongata	15	- 0	v 13 (53)	+ 100	- 0	- 0	- 0	- 0	-	- 0
Neisseria elongata subsp. glycolytica	2	- 0	+ or (+) 50 (50)	+ 100	- 0	- 0	- 0	- 0	-	+ 100
Neisseria elongata subsp. nitroreducens	26	v 23	v 19 (35)	+ 100	- 0	- 0	- 0	+ 100	-	- 0
Neisseria flavescens[d]	10	- 0	v 67	+ 100	-	- 0	-	- 0	-	+ 100
Neisseria mucosa[d]	30	+ or (+) 83 (10)	v 57 (3)	+ 100	-	- 0	-	+ 100	-	+ 100
Neisseria sicca[d]	43	+ or (+) 78 (14)	v 68	+ 100	-	- 0	-	- 0	-	v 80

SECTION 1: AN APPROACH TO IDENTIFICATION

Neisseria subflava[d]	153	v 68 (6)	v 47 (2)	+ 100	–	–	–	–	v 80
Neisseria weaveri	132	– 0	v 27 (18)	+ 100	–	– 0	– 0	– 0	+ 100
Oligella ureolytica	37	– 0	v 62 (27)	+ 100	–	– 0	+ 97	+ 100	+ 100
O-2	66	v 73 (11)	– 9	+ 97	–	– 2	v 12	v 15	v 91
Pseudomonas vesicularis	94	+ or (+) 87 (12)	v 43 (23)	+ 98	–	v 27 (9)	– 2 (5)	– 5	v 83
Psychrobacter immobilis, asaccharolytic strains	5	– 0	v 40	+ 100	–	– 0	v (20)	v 40 (20)	+ 100
Roseomonas fauriae	5	v 20	+ or (+) 60 (40)	+ 100	–	+ or (+) 80 (20)	+ 100	+ 100	+ 100
Roseomonas genomospecies 5	3	– 0	+ or (+) 67 (33)	+ 100	–	v 67	+ 100	– 0	+ 100
Roseomonas genomospecies 6	1	–	(+)	+	–	–	+	+	+
Roseomonas gilardii	21	v (43)	+ or (+) 43 (52)	v 52	– 0	v 19 (57)	+ or (+) 71 (29)	– 5	+ or (+) 33
Taylorella equigenitalis	1	–	(+)	+	+ 100	–	–	–	+
Weeksella virosa	87	– 0	– (10)	+ 100	+ 98	– 0	– 0	– 0	+ 98
Weeksella zoohelcum	41	– 0	– 2	+ 100	+ 100	– 0	+ 100	– 0	+ 100

*See Presumptive Identification Code p. 104; where the blank spaces occur, either the percentages were not calculated or they were not available.
[a]Slant was inoculated heavily.
[b]Fails to grow in nitrate test media.
[c]Test performed by scraping growth from 4-day rabbit-blood cultures onto a glass slide and applying the reagent.
[d]Carbohydrates were tested in CTA base.

Guide To Presumptive Identification: Code Book

Table 1.1. Glucose fermenters: MacConkey-positive/oxidase-negative

These organisms must be differentiated from the species of *Enterobacteriaceae*, which have not been included on the chart.

Numerical Code	Presumptive Identification
0160	*Chromobacterium violaceum*
0161	*Chromobacterium violaceum*
0170	*Chromobacterium violaceum*
0171	*Chromobacterium violaceum*
0410	*Haemophilus aphrophilus*
1160	*Chromobacterium violaceum*
1161	*Chromobacterium violaceum*
1170	*Chromobacterium violaceum*
1171	*Chromobacterium violaceum*
2000	*Pasteurella bettyae*
2160	*Chromobacterium violaceum*
2161	*Chromobacterium violaceum*
2170	*Chromobacterium violaceum*
2171	*Chromobacterium violaceum*
3160	*Chromobacterium violaceum*
3161	*Chromobacterium violaceum*
3170	*Chromobacterium violaceum*
3171	*Chromobacterium violaceum*

Table 1.2. Glucose fermenters: MacConkey-positive/oxidase-positive

These organisms must be differentiated from the species of *Vibrio, Aeromonas,* and *Plesiomonas shigelloides,* which have not been included on the chart.

Numerical Code	Presumptive Identification
000	EF-4a: catalase +, 100%; rod; maltose neg, 0%
	Neisseria subflava: catalase v, 80%; coccus; determine reactions for differentiation of *Neisseria* species; maltose +, 99 (1)%
002	*Pasteurella gallinarum:* D-mannitol neg, 0%; β-like hemolysis neg, 0%; rod
	Pasteurella trehalosi: D-mannitol + or (+), 82 (18)%; rod
	Neisseria subflava: cocci; yellow pigmentation of Loeffler slant +, 99%; nitrate neg, 0%
	Neisseria sicca: cocci; yellow pigmentation of Loeffler slant neg, 0%; nitrate neg, 0%
	Neisseria mucosa: cocci; slight yellow pigmentation of Loeffler slant v, 50%; nitrate + with gas, 100%
003	*Haemophilus aphrophilus*
010	*Pasteurella bettyae*
012	Bisgaard's taxon 16: nitrate +, 100%
	Suttonella indologenes: nitrate neg, 0%
022	*Actinobacillus suis:* gas from D-glucose neg, 0%; esculin hydrolysis +, 100%
	Pasteurella aerogenes: gas from D-glucose +, 100%; esculin hydrolysis neg, 0%
023	*Actinobacillus suis:* see 022
	Pasteurella aerogenes: see 022
042	*Pasteurella haemolytica:* D-xylose +, 96 (4)%
	Pasteurella trehalosi: D-xylose neg, 0%
043	*Pasteurella haemolytica*
062	*Actinobacillus lignieresii:* esculin hydrolysis neg, 0%
	Actinobacillus suis: esculin hydrolysis +, 100%
063	*Actinobacillus equuli:* esculin hydrolysis neg, 0%; trehalose +, 100%
	Actinobacillus hominis (one strain studied): esculin hydrolysis (+); mannose neg
	Actinobacillus lignieresii: esculin hydrolysis neg, 0%; trehalose neg, 0%
	Actinobacillus suis: esculin hydrolysis +, 100%; mannose +, 91 (9)%
200	*Chromobacterium violaceum:* growth on SS + or (+), 71 (23)%; citrate v, 68 (9)%; usually purple pigmented
	EF-4a: growth on SS neg, 1%; citrate neg, 3 (1)%
202	*Chromobacterium violaceum*
210	*Chromobacterium violaceum*
212	*Chromobacterium violaceum*
220	*Chromobacterium violaceum*
222	*Chromobacterium violaceum*
230	*Chromobacterium violaceum*

(continued)

Table 1.2. Glucose fermenters: MacConkey-positive/oxidase-positive *(continued)*

Numerical Code	Presumptive Identification
232	*Chromobacterium violaceum*
402	*Pasteurella gallinarum*
422	*Pasteurella aerogenes*
423	*Pasteurella aerogenes*
432	*Pasteurella pneumotropica*
433	*Pasteurella pneumotropica*
532	*Pasteurella pneumotropica*
533	*Pasteurella pneumotropica*

Table 1.3. Glucose fermenters: MacConkey-negative/oxidase-negative

Numerical Code	Presumptive Identification
000	*Capnocytophaga* species (DF-1): long to filamentous rods; lactose v, 86%; succinic and acetic acids are major endproducts of D-glucose fermentation
	Gardnerella vaginalis: short to medium length rods; delayed, incomplete hemolysis on rabbit and human blood agar; lactose neg, 1%
	Leptotrichia buccalis: lactic acid is the major endproduct of D-glucose fermentation
	Streptobacillus moniliformis: long pleomorphic filamentous rods with "monilia"-like swellings; growth usually requires serum enhancement
001	*Capnocytophaga* species (DF-1): continue incubation of D-xylose for 7 days, neg, 0%; lactose v, 86%; long to filamentous rods
	DF-3: continue incubation of D-xylose for 7 days, + or (+), 86 (14)%; predominantly coccoid and short rods; propionic, lactic, and succinic acids are major endproducts of D-glucose fermentation
	Leptotrichia buccalis: lactic acid is the major endproduct of D-glucose fermentation
	Streptobacillus moniliformis: long pleomorphic filamentous rods with "monilia"-like swellings; growth usually requires serum enhancement
010	*Capnocytophaga* species (DF-1): see 000
	Leptotrichia buccalis: see 000
011	*Capnocytophaga* species (DF-1): see 001
	DF-3: see 001
	Leptotrichia buccalis: see 001
020	*Capnocytophaga* species (DF-1): see 000
	Gardnerella vaginalis: see 000
	Leptotrichia buccalis: see 000
	Streptobacillus moniliformis: see 000
021	*Capnocytophaga* species (DF-1): see 001
	DF-3: see 001
	Leptotrichia buccalis: see 001
	Streptobacillus moniliformis: see 001
030	*Capnocytophaga* species (DF-1): see 000
	Lepotrichia buccalis: see 000
031	*Capnocytophaga* species (DF-1): see 001
	DF-3: see 001
	Leptotrichia buccalis: see 001
040	*Gardnerella vaginalis*
041	DF-3
046	*Gardnerella vaginalis*
051	DF-3
061	DF-3

(continued)

Table 1.3. Glucose fermenters: MacConkey-negative/oxidase-negative *(continued)*

Numerical Code	Presumptive Identification
071	DF-3
200	*Capnocytophaga* species (DF-1)
201	*Capnocytophaga* species (DF-1)
210	*Capnocytophaga* species (DF-1)
211	*Capnocytophaga* species (DF-1)
220	*Actinobacillus ureae:* urea +, 100%; D-mannitol +, 99 (1)%
	Capnocytophaga species (DF-1): urea neg, 0%; D-mannitol neg, 0%
221	*Capnocytophaga* species (DF-1)
230	*Capnocytophaga* species (DF-1): long to filamentous rods; esculin hydrolysis v, 83%
	Haemophilus aphrophilus: coccoid to short rods; esculin hydrolysis neg, 0%
231	*Capnocytophaga* species (DF-1)
300	*Actinobacillus actinomycetemcomitans*
320	*Actinobacillus actinomycetemcomitans:* urea neg, 0%
	Actinobacillus ureae: urea +, 100%
340	*Actinobacillus actinomycetemcomitans*
360	*Actinobacillus actinomycetemcomitans*
400	DF-3-like
401	DF-3: D-xylose + or (+), 86 (14)%; sucrose + or (+), 62 (33)%
	DF-3-like: D-xylose neg, 0%; sucrose neg, 0%
410	DF-3-like
411	DF-3: see 401
	DF-3-like: see 401
420	DF-3-like
421	DF-3: see 401
	DF-3-like: see 401
430	DF-3-like
431	DF-3: see 401
	DF-3-like: see 401
441	DF-3
451	DF-3
461	DF-3
471	DF-3
500	DF-3-like
501	DF-3-like
510	DF-3-like
511	DF-3-like

(continued)

Table 1.3. Glucose fermenters: MacConkey-negative/oxidase-negative *(continued)*

Numerical Code	Presumptive Identification
520	DF-3-like
521	DF-3-like
530	DF-3-like
531	DF-3-like
600	*Pasteurella bettyae*
700	*Pasteurella canis:* D-mannitol neg, 0%; ornithine decarboxylase +, 100%
	Pasteurella multocida[a]: D-mannitol +, 97 (3)%
	Pasteurella stomatis: D-mannitol neg, 0%; ornithine decarboxylase neg, 0%
720	Bisgaard's taxon 16: ornithine neg, 0%; urea neg, 0%
	Pasteurella dagmatis: ornithine neg, 0%; urea +, 95 (5)%
	Pasteurella pneumotropica: ornithine +, 100%; urea +, 95 (1)%
730	*Pasteurella pneumotropica*
740	*Pasteurella multocida*[a]
760	*Pasteurella pneumotropica*
770	*Pasteurella pneumotropica*

[a]For differentiation of the subspecies of *P. multicida,* see page 462.

Table 1.4. Glucose fermenters: MacConkey-negative/oxidase-positive

These organisms must be differentiated from the species of *Vibrionaceae*, which have not been included on the chart.

Numerical Code	Presumptive Identification
0000	*Kingella kingae*
0001	*Kingella kingae:* rod
	Neisseria subflava: coccus
0002	*Kingella kingae*
0003	*Kingella kingae*
0041	*Neisseria subflava:* yellow pigmentation on Loeffler slant +, 99%
	Neisseria sicca: pigmentation on Loeffler slant neg, 0%
0100	*Kingella denitrificans*
0141	*Haemophilus aphrophilus*
0300	*Kingella denitrificans*
0340	*Simonsiella crassa,* type strain
1000	*Capnocytophaga canimorsus* (if maltose reaction is delayed): inulin and raffinose neg, 0%; lactose + or (+), 56 (39)%
	Capnocytophaga cynodegmi (if maltose reaction is delayed): inulin v, 77 (11)%; raffinose v, 70 (10)%; lactose + or (+), 69 (31)%
	Riemerella anatipestifer: lactose neg, 0%
1001	*Capnocytophaga canimorsus:* long thin rod, see 1000
	Capnocytophaga cynodegmi: long thin rod, see 1000
	Neisseria lactamica: coccus: ONPG +, 100%
	Neisseria subflava: coccus: ONPG neg, 0%
	Riemerella anatipestifer: rod, see 1000
	Simonsiella-like species: "roll of coins" morphology
1002	*Capnocytophaga cynodegmi*
1003	*Capnocytophaga cynodegmi:* long thin rod
	Simonsiella-like species: "roll of coins" morphology
1040	*Capnocytophaga cynodegmi*
1041	*Capnocytophaga cynodegmi:* long thin rod
	Neisseria sicca: coccus, see 0041
	Neisseria subflava: coccus, see 0041
1042	*Capnocytophaga cynodegmi*
1043	*Capnocytophaga cynodegmi*
1100	*Actinobacillus actinomycetemcomitans:* no strains have been encountered that were neg in D-xylose, D-mannitol, and maltose; TSI slant acid, +, 100%
	EF-4a: TSI slant acid, neg, 3%

(continued)

Table 1.4. Glucose fermenters: MacConkey-negative/oxidase-positive *(continued)*

Numerical Code	Presumptive Identification
1101	*Actinobacillus actinomycetemcomitans*: small rod
	Simonsiella-like species: "roll of coins" morphology
1103	*Simonsiella*-like species
1110	*Actinobacillus actinomycetemcomitans*
1111	*Actinobacillus actinomycetemcomitans*
1120	*Actinobacillus actinomycetemcomitans*
1121	*Actinobacillus actinomycetemcomitans*
1130	*Actinobacillus actinomycetemcomitans*
1131	*Actinobacillus actinomycetemcomitans*
1141	*Pasteurella gallinarum*
1151	*Pasteurella gallinarum*
1171	*Pasteurella haemolytica*
1173	*Pasteurella haemolytica*
1201	*Simonsiella*-like species
1203	*Simonsiella*-like species
1300	EF-4a
1301	*Simonsiella*-like species
1303	*Simonsiella*-like species
1341	*Neisseria mucosa*
1541	*Pasteurella gallinarum*
1551	*Pasteurella gallinarum*
2000	*Cardiobacterum hominis*
2001	*Cardiobacterium hominis*
2020	*Cardiobacterium hominis*
2021	*Cardiobacterium hominis*
2040	*Cardiobacterium hominis*: continue incubation of D-mannitol, + or (+), 52 (43)%; alkaline phosphatase neg; casein neg; Tween 40 hydrolysis neg
	Suttonella indologenes: D-mannitol neg at 7 days, 0%; alkaline phosphatase +, 100%; casein +; Tween 20 and 40 hydrolysis +
2041	*Cardiobacterium hominis*: see 2040
	Suttonella indologenes: see 2040
2060	*Cardiobacterium hominis*
2061	*Cardiobacterium hominis*
2100	*Pasteurella bettyae*
3040	*Kingella indologenes*
3041	*Kingella indologenes*
3140	*Pasteurella stomatis*

(continued)

Table 1.4. Glucose fermenters: MacConkey-negative/oxidase-positive *(continued)*

Numerical Code	Presumptive Identification
3141	Bisgaard's taxon 16
3540	*Pasteurella canis*
3560	*Pasteurella multocida*
3570	*Pasteurella multocida*
4151	*Actinobacillus suis*
4153	*Actinobacillus suis*
4160	*Actinobacillus lignieresii*
4161	*Actinobacillus lignieresii:* continue incubation of D-xylose for 7 days, + or (+), 87 (13)%; usually animal source or history of animal contact; galactose +, 90 (5)%
	Actinobacillus ureae: human sources, mainly respiratory; D-xylose neg. at 7 days; galactose neg, 0%
4170	*Actinobacillus lignieresii*
4171	*Actinobacillus equuli:* esculin neg, 0%; trehalose +, 100%
	Actinobacillus lignieresii: esculin neg, 0%; trehalose neg, 0%
	Actinobacillus suis: esculin +, 100%; trehalose +, 100%
4173	*Actinobacillus suis*
5000	*Riemerella anatipestifer*
5001	*Riemerella anatipestifer*
5151	*Actinobacillus suis*
5153	*Actinobacillus suis*
5160	*Actinobacillus lignieresii*
5161	*Actinobacillus lignieresii:* see 4161
	Actinobacillus ureae: see 4161
5170	*Actinobacillus lignieresii*
5171	*Actinobacillus equuli:* see 4171
	Actinobacillus lignieresii: see 4171
	Actinobacillus suis: see 4171
5173	*Actinobacillus suis*
7141	*Pasteurella dagmatis*
7541	*Pasteurella pneumotropica*
7551	*Pasteurella pneumotropica*

[a]For differentiation of the subspecies of *P. multicida,* see page 462.

Table 1.5. Glucose oxidizers: MacConkey-positive/oxidase-negative

Numerical Code	Presumptive Identification
000	Saccharolytic *Acinetobacter* species: pink growth pigment neg, 0%
	Roseomonas gilardii: pink insoluble pigment +, 100%
001	O-1
002	*Roseomonas gilardii*
003	O-1
010	Saccharolytic *Acinetobacter* species (8% nitrate positive): *Acinetobacter* transformation +, 100%; can also be confirmed by cellular fatty acid analysis
	Brucella species, probably *B. canis:* confirm by serology and dye inhibition tests; can also be confirmed by cellular fatty acid analysis
022	*Xanthomonas maltophilia*
023	*Xanthomonas maltophilia*
032	*Xanthomonas maltophilia*
033	*Xanthomonas maltophilia*
040	Saccharolytic *Acinetobacter* species
043	*Chryseomonas luteola*
050	Saccharolytic *Acinetobacter* species (8% nitrate positive); can also be differentiated by cellular fatty acid analysis
	Pseudomonas mallei: nitrate +, 100%; can also be differentiated by cellular fatty acid analysis
053	*Chryseomonas luteola*
100	*Roseomonas gilardii*
102	*Flavimonas oryzihabitans:* OF maltose +, 97 (3)%; yellow insoluble pigment +, 100%
	Pseudomonas gladioli: OF maltose neg, 0%; nonpigmented
	Roseomonas gilardii: OF maltose neg, 0%; pink insoluble pigment +, 100%
112	*Pseudomonas gladioli*
142	*Flavimonas oryzihabitans*
143	*Chryseomonas luteola*
150	*Pseudomonas mallei*
153	*Chryseomonas luteola*
200	Saccharolytic *Acinetobacter* species
210	Saccharolytic *Acinetobacter* species (8% nitrate positive)
222	*Xanthomonas maltophilia*
223	*Xanthomonas maltophilia*
232	*Xanthomonas maltophilia*
233	*Xanthomonas maltophilia*
240	Saccharolytic *Acinetobacter* species
243	*Chryseomonas luteola*

(continued)

Table 1.5. Glucose oxidizers: MacConkey-positive/oxidase-negative *(continued)*

Numerical Code	Presumptive Identification
250	Saccharolytic *Acinetobacter* species (8% nitrate positive): see 050
	Pseudomonas mallei: see 050
253	*Chryseomonas luteola*
302	*Flavimonas oryzihabitans:* yellow insoluble pigment +, 100%; OF maltose +, 97 (3)%; H_2S on lead acetate paper over TSI +, 97%; 1 to 2 polar flagella
	Pseudomonas cepacia: variable pigmentation; OF maltose +, 99 (1)%; H_2S on lead acetate paper over TSI neg., 0%; >2 polar flagella
	Pseudomonas gladioli: OF maltose neg, 0%; H_2S on lead acetate paper over TSI v, 31 (5)%; yellow insoluble pigment neg, 0%
303	*Pseudomonas cepacia*
312	*Pseudomonas cepacia:* see 302
	Pseudomonas gladioli: see 302
313	*Pseudomonas cepacia*
322	*Pseudomonas cepacia*
323	*Pseudomonas cepacia*
332	*Pseudomonas cepacia*
333	*Pseudomonas cepacia*
342	*Flavimonas oryzihabitans*
343	*Chryseomonas luteola*
350	*Pseudomonas mallei*
353	*Chryseomonas luteola*
422	*Xanthomonas maltophilia*
423	*Xanthomomas maltophilia*
432	*Xanthomonas maltophilia*
433	*Xanthomonas maltophilia*
443	*Chryseomonas luteola*
453	*Chryseomonas luteola*
502	*Flavimonas oryzihabitans*
542	*Flavimonas oryzihabitans*
543	*Chryseomonas luteola*
553	*Chryseomonas luteola*
601	*Sphingomonas parapaucimobilis:* OF glycerol and OF rhamnose pos; results from two reference strains
	Sphingomonas paucimobilis: OF glycerol and OF rhamnose neg; results from type strain only
603	*Sphingomonas parapaucimobilis:* see 601
	Sphingomonas paucimobilis: see 601
622	*Xanthomonas maltophilia*
623	*Xanthomonas maltophilia*

(continued)

Table 1.5. Glucose oxidizers: MacConkey-positive/oxidase-negative *(continued)*

Numerical Code	Presumptive Identification
632	*Xanthomonas maltophilia*
633	*Xanthomonas maltophilia*
643	*Chryseomonas luteola*
653	*Chryseomonas luteola*
702	*Flavimonas oryzihabitans:* see 302
	Pseudomonas cepacia: see 302
703	*Pseudomonas cepacia*
712	*Pseudomonas cepacia*
713	*Pseudomonas cepacia*
722	*Pseudomonas cepacia*
723	*Pseudomonas cepacia*
732	*Pseudomonas cepacia*
733	*Pseudomonas cepacia*
742	*Flavimonas oryzihabitans*
743	*Chryseomonas luteola*
753	*Chryseomonas luteola*

Table 1.6. Glucose oxidizers: MacConkey-positive/oxidase-positive

Numerical Code	Presumptive Identification
00000	*Francisella philomiragia*: nonmotile; growth in nutrient broth with 0% NaCl neg, 0%; H$_2$S in TSI + or (+), 56 (44)%
	O-1: motile; yellow insoluble pigment +, 100%
	Pseudomonas diminuta: motile; insoluble pigment neg, 0%; OF D-xylose neg, 0%
	Pseudomonas pertucinogena: motile; insoluble pigment neg, 0%; OF D-xylose + or (+), 50 (50)%
	Saccharolytic *Psychrobacter immobilis*: nonmotile; growth in nutrient broth with 0% NaCl + or (+), 43 (57)%; H$_2$S in TSI neg, 0%
	Roseomonas gilardii: motility v, 33%; pink insoluble pigment +, 100%; growth in nutrient broth with 0% NaCl +, 100%
00001	*Francisella philomiragia*
00002	"*Agrobacterium* yellow group": motile with 1–2 polar flagella; growth in nutrient broth with 0% NaCl + or (+), 50 (50)%; continue incubation: OF lactose and OF sucrose are (+), (100)%
	Francisella philomiragia: nonmotile; growth in nutrient broth with 0% NaCl-neg, 0%; continue incubation: OF lactose-neg, 0%; OF sucrose + or (+), 63 (37)%
	Pseudomonas vesicularis: motile with 1–2 polar flagella; growth in nutrient broth with 0% NaCl +, 95%; OF lactose neg, 0%; OF sucrose neg, 0%
00003	"*Agrobacterium* yellow group": see 00002
	Francisella philomiragia: see 00002
00010	EO-3: continue incubation: OF lactose + or (+), 71 (29)%; Simmons citrate + or (+), 29(71)%
	Methylobacterium species: OF lactose, neg, 0%; Simmons citrate neg, 2 (3)%; pink to coral insoluble pigment +, 99%
	Pseudomonas pertucinogena: OF lactose-neg, 0%; Simmons citrate neg, 0%; insoluble pink pigment neg, 0%
	Saccharolytic *Psychrobacter immobilis*: continue incubation, OF lactose, + or (+), 57 (43)%; Simmons citrate neg, 0%
	Roseomonas gilardii: OF lactose neg, 0%; pink insoluble pigment +, 100%; Simmons citrate +, 100%
00012	"*Agrobacterium* yellow group": motile, 1–2 polar flagella; continue incubation, OF lactose (+), (100)%; OF D-mannitol neg, 0%
	EO-3: nonmotile; continue incubation, OF lactose + or (+), 71 (29)%; OF D-mannitol + or (+), 57 (43)%
	Pseudomonas vesicularis: motile, 1–2 polar (short wavelength); OF lactose neg, 0%; OF D-mannitol neg, 0%
00013	"*Agrobacterium* yellow group"
00020	*Roseomonas gilardii*
00030	EO-3: nonmotile; pink insoluble pigment neg, 0%; oxidase + by direct application of reagent to colonies
	Pseudomonas gladioli: motile; pink insoluble pigment neg, 0%; oxidase + only by the Kovacs' method
	Roseomonas gilardii: pink insoluble pigment +, 100%; motility v, 33%
00032	EO-3
00033	*Agrobacterium radiobacter* biovar 1
00040	Saccharolytic *Psychrobacter immobilis*
00042	"*Agrobacterium* yellow group"

(continued)

Table 1.6. Glucose oxidizers: MacConkey-positive/oxidase-positive *(continued)*

Numerical Code	Presumptive Identification
00043	"*Agrobacterium* yellow group"
00050	EO-3: see 00010
	Saccharolytic *Psychrobacter immobilis:* see 00010
00052	"*Agrobacterium* yellow group": motile with 1–2 polar flagella; continue incubation: OF D-mannitol neg, 0%
	EO-3: nonmotile; continue incubation: OF D-mannitol + or (+), 57 (43)%
00053	*Sphingomonas paucimobilis/parapaucimobilis:* 3-ketolactonate neg, 0%
	"*Agrobacterium* yellow group": 3-ketolactonate +, 100%
00070	EO-3: nonmotile; yellow insoluble pigment +, 100%
	Pseudomonas gladioli: motile; yellow insoluble pigment neg, 0%
00072	EO-3: nonmotile; lysine decarboxylase neg, 0%; esculin neg, 0%
	Pseudomonas cepacia: motile, >2 polar flagella; lysine decarboxylase v, 80%; esculin v, 63 (6)%
	"*Pseudomonas thomasii*": motile, <3 polar flagella; lysine decarboxylase neg, 0%; esculin neg., 0%
00073	*Agrobacterium radiobacter* biovar 1: see 00473
	Flavobacterium spiritivorum: see 00473
	Flavobacterium yabuuchiae: see 00473
	Pseudomonas cepacia: see 00473
00100	*Acidovorax temperans:* rod; motile with 1–2 polar flagella; OF lactose neg, 0%
	EF-4b: rod; nonmotile; OF lactose neg, 0%; catalase +, 100%
	Neisseria canis (one strain): coccus; nonmotile
	Neisseria elongata subsp. *nitroreducens:* rod; nonmotile; OF lactose neg, 0%; catalase neg, 0%
	Saccharolytic *Psychrobacter immobilis:* rod; nonmotile; continue incubation: OF lactose + or (+), 57 (43)%; usually prefers 25°C
00110	*Alcaligenes xylosoxidans* subsp. *xylosoxidans:* Simmons citrate +, 95%; peritrichous flagella; growth on SS +, 98%; OF lactose neg, 0%; no pigment
	EO-2: nonmotile; Simmons citrate + or (+), 64 (36)%; growth on SS neg, 0%; OF lactose + or (+), 45 (55)%
	Methylobacterium species: Simmons citrate neg, 2 (3)%; polar flagella; growth on SS neg, 0%; OF lactose neg, 0%; pink to coral insoluble pigment +, 99%
	Saccharolytic *Psychrobacter immobilis:* nonmotile; continue incubation: OF lactose + or (+), 57 (43)%; Simmons citrate neg, 0%; growth on SS neg, 0%
00120	*Acidovorax temperans*
00130	*Pseudomonas gladioli*
00133	*Agrobacterium radiobacter* biovar 1
00140	Saccharolytic *Psychrobacter immobilis*
00150	EO-2: see 00110
	Saccharolytic *Psychrobacter immobilis:* see 00110
00170	*Pseudomonas gladioli*

(continued)

Table 1.6. Glucose oxidizers: MacConkey-positive/oxidase-positive *(continued)*

Numerical Code	Presumptive Identification
00172	*Pseudomonas cepacia:* see 00072
	"Pseudomonas thomasii": see 00072
00173	*Agrobacterium radiobacter* biovar 1: see 00073
	Pseudomonas cepacia see 00073
00300	*Acidovorax temperans*
00310	*Alcaligenes xylosoxidans* subsp. *xylosoxidans:* see 00110
	EO-2: see 00110
00312	*Pseudomonas stutzeri*
00320	*Acidovorax temperans*
00332	*Pseudomonas stutzeri*
00350	EO-2
00400	*Pseudomonas diminuta:* motile (1–2 polar flagella); continue incubation: OF D-xylose neg, 0%; OF lactose neg, 0%; Simmons citrate neg, 1%; pink insoluble pigment neg, 0%
	Saccharolytic *Psychrobacter immobilis:* nonmotile; continue incubation: OF D-xylose + or (+) 57 (43)%; OF lactose + or (+), 57 (43)%; Simmons citrate neg, 0%; pink insoluble pigment neg, 0%
	Roseomonas gilardii: Simmons citrate +, 100%; OF D-xylose v, 19 (57)%; OF lactose neg, 0%; pink insoluble pigment +, 100%
00402	*"Agrobacterium* yellow group"
00403	*"Agrobacterium* yellow group"
00410	EO-3: see 00010
	Methylobacterium species: see 00010
	Saccharolytic *Psychrobacter immobilis:* see 00010
	Roseomonas gilardii: see 00010
00412	*"Agrobacterium* yellow group": see 00052
	EO-3: see 00052
00413	*"Agrobacterium* yellow group"
00420	*Roseomonas gilardii*
00430	EO-3: see 00030
	Pseudomonas gladioli: see 00030
	Roseomonas gilardii: see 00030
00432	EO-3
00433	*Agrobacterium radiobacter* biovar 1
00440	Saccharolytic *Psychrobacter immobilis*
00442	*"Agrobacterium* yellow group"
00443	*"Agrobacterium* yellow group"
00450	EO-3: see 00010
	Saccharolytic *Psychrobacter immobilis:* see 00010

(continued)

Table 1.6. Glucose oxidizers: MacConkey-positive/oxidase-positive (continued)

Numerical Code	Presumptive Identification
00452	"*Agrobacterium* yellow group": see 00052
	EO-3: see 00052
00453	"*Agrobacterium* yellow group": motile, but may be hard to demonstrate; frank yellow insoluble pigment +, 100%
	Flavobacterium multivorum: nonmotile; slightly yellow insoluble pigment v, 57%
00470	EO-3: nonmotile; yellow insoluble pigment +, 100%
	Pseudomonas gladioli: motile; OF dulcitol + or (+), 78 (22)%; OF inositol +, 100%
	Pseudomonas-like group 2: motile; OF dulcitol neg, 0%; OF inositol neg., 0%
00472	EO-3: see 00072
	Pseudomonas cepacia: see 00072
	"*Pseudomonas thomasii*": see 00072
00473	*Agrobacterium radiobacter* biovar 1: motile (polar and lateral flagella); H_2S (lead acetate) +, 100%; Simmons citrate + or (+), 97 (3)%; 3-ketolactonate +, 100%
	Flavobacterium spiritivorum: nonmotile; H_2S (lead acetate) v, 56%; Simmons citrate neg, 0%; 3-ketolactonate neg, 0%; acid phosphatase +
	Flavobacterium yabuuchiae: nonmotile; H_2S (lead acetate) neg, 0%; Simmons citrate neg, 0%; 3-ketolactonate neg, 0%; acid phosphatase neg
	Pseudomonas cepacia: motile (polar flagella); H_2S (lead acetate) neg, 0%; Simmons citrate + or (+), 94 (5)%; 3-ketolactonate neg, 0%
00500	*Acidovorax temperans*: pink insoluble pigment neg, 0%; motile; esculin neg, 0%
	Saccharolytic *Psychrobacter immobilis*: pink insoluble pigment neg, 0%; nonmotile; esculin neg, 0%
	Roseomonas fauriae: pink insoluble pigment +, 100%; motile; esculin +, 100%
00510	*Brucella*: nonmotile; nonpigmented; tiny, coccoid; Simmons citrate neg, 0%; esculin neg, 0%; OF lactose neg, 0%
	EO-2: nonmotile; Simmons citrate + or (+), 64 (36)%; esculin neg, 0%; OF lactose + or (+), 45 (55)%
	Methylobacterium species: motile; pink to coral insoluble pigment +, 99%; pleomorphic rods, vacuolated; Simmons citrate neg, 2 (3)%; esculin neg, 0%; OF lactose neg, 0%
	Pseudomonas pickettii biovar 2 (if gas from nitrate is not detected): weakly motile; nonpigmented; rods (not pleomorphic); Simmons citrate +, 100%; esculin neg, 0%; OF lactose neg, 0%
	Saccharolytic *Psychrobacter immobilis*: nonmotile; Simmons citrate neg, 0%; esculin neg, 0%; OF lactose + or (+), 57 (43)%
	Roseomonas fauriae: motile; pink insoluble pigment +, 100%; esculin +, 100%; OF lactose neg, 0%; Simmons citrate v, 60 (20)%
00520	*Acidovorax temperans*
00530	*Pseudomonas gladioli*
00533	*Agrobacterium radiobacter* biovar 1
00540	Saccharolytic *Psychrobacter immobilis*
00550	EO-2: see 00110
	Saccharolytic *Psychrobacter immobilis*: see 00110
00552	*Pseudomonas pickettii* biovar 1
00553	*Flavobacterium thalpophilum*

(continued)

Table 1.6. Glucose oxidizers: MacConkey-positive/oxidase-positive *(continued)*

Numerical Code	Presumptive Identification
00570	*Pseudomonas gladioli:* see 00470
	Pseudomonas-like group 2: see 00470
00572	*Pseudomonas cepacia:* see 00072
	"Pseudomonas thomasii": see 00072
00573	*Agrobacterium radiobacter* biovar 1: see 00473
	Pseudomonas cepacia: see 00473
00700	*Acidovorax temperans:* see 00500
	Roseomonas fauriae: see 00500
00710	*Brucella:* nonmotile; growth on SS neg, 0%; Simmons citrate neg, 0%; tiny, coccoid
	EO-2: nonmotile; Simmons citrate + or (+), 64 (36)%; esculin neg, 0%
	Ochrobactrum anthropi: motile, peritrichous flagella, predominantly polar and lateral; growth on SS +, 96 (1)%; esculin v, 38 (11)%
	Pseudomonas pickettii biovar 2: motile, polar flagella; growth on SS neg, 0%; esculin neg., 0%; rod
	Roseomonas fauriae: motile with 1–2 polar flagella; pink insoluble pigment +, 100%; esculin +, 100%
00711	*Ochrobactrum anthropi*
00712	*Ochrobactrum anthropi:* motile, peritichous flagella, predominately polar and lateral; H$_2$S on lead acetate paper +, 100%; arginine v, 68%
	Pseudomonas stutzeri: motile, polar flagella; H$_2$S on lead acetate paper v, 36%; arginine neg., 0%
00713	*Ochrobactrum anthropi*
00720	*Acidovorax temperans*
00730	*Ochrobactrum anthropi*
00731	*Ochrobactrum anthropi*
00732	*Ochrobactrum anthropi:* see 00712
	Pseudomonas stutzeri: see 00712
00733	*Ochrobactrum anthropi*
00750	EO-2
00752	*Pseudomonas pickettii* biovar 1
01000	*Francisella philomiragia*
01001	*Francisella philomiragia*
01002	*Francisella philomiragia*
01003	*Francisella philomiragia*
01033	*Agrobacterium radiobacter* biovar 1
01073	*Agrobacterium radiobacter* biovar 1
01103	*Shewanella putrefaciens* biotype 1
01133	*Agrobacterium radiobacter* biovar 1
01173	*Agrobacterium radiobacter* biovar 1

(continued)

Table 1.6. Glucose oxidizers: MacConkey-positive/oxidase-positive *(continued)*

Numerical Code	Presumptive Identification
01433	*Agrobacterium radiobacter* biovar 1
01473	*Agrobacterium radiobacter* biovar 1
01503	*Shewanella putrefaciens* biotype 1
01533	*Agrobacterium radiobacter* biovar 1
01573	*Agrobacterium radiobacter* biovar 1
01710	*Ochrobactrum anthropi*
01711	*Ochrobactrum anthropi*
01713	*Ochrobactrum anthropi*
01730	*Ochrobactrum anthropi*
01731	*Ochrobactrum anthropi*
01732	*Ochrobactrum anthropi*
01733	*Ochrobactrum anthropi*
02072	*Pseudomonas cepacia*
02073	*Pseudomonas cepacia*
02172	*Pseudomonas cepacia*
02173	*Pseudomonas cepacia*
02472	*Pseudomonas cepacia*
02473	*Pseudomonas cepacia*
02572	*Pseudomonas cepacia*
02573	*Pseudomonas cepacia*
04000	Saccharolytic *Psychrobacter immobilis*
04010	*Pseudomonas putida:* motile; Simmons citrate +, 94 (6)%
	Saccharolytic *Psychrobacter immobilis:* nonmotile; Simmons citrate neg, 0%
04030	*Pseudomonas putida*
04040	Saccharolytic *Psychrobacter immobilis*
04050	Saccharolytic *Psychrobacter immobilis*
04073	*Flavobacterium spiritivorum:* acid phosphatase +
	Flavobacterium yabuuchiae: acid phosphatase neg
04100	*Pseudomonas mallei:* growth in nutrient broth with 6% NaCl neg, 0%; OF D-mannitol v, 62 (14)%; OF maltose v, (75)%
	Saccharolytic *Psychrobacter immobilis:* growth in nutrient broth with 6% NaCl +, 100%; OF D-mannitol neg, 0%; OF maltose neg, 0%
04102	*Pseudomonas mallei:* nonmotile; growth on SS agar neg, 0%
	Ic: motile; growth on SS agar +, 100%

(continued)

Table 1.6. Glucose oxidizers: MacConkey-positive/oxidase-positive *(continued)*

Numerical Code	Presumptive Identification
04110	*Alcaligenes xylosoxidans* subsp. *xylosoxidans:* motile; growth on SS +, 98%; 42°C v, 84%; growth in nutrient broth with 6% NaCl v, 69%
	Pseudomonas mallei: nonmotile; growth on SS neg., 0%; 42°C neg, 0%; growth in nutrient broth with 6% NaCl neg, 0%
	Saccharolytic *Psychrobacter immobilis:* nonmotile; growth on SS agar neg, 0%; growth at 42°C neg, 0%; growth in nutrient broth with 6% NaCl +, 100%
04112	*Pseudomonas mallei*
04120	*Pseudomonas mallei*
04122	*Pseudomonas mallei*
04130	*Pseudomonas mallei*
04132	*Pseudomonas mallei*
04140	*Pseudomonas mallei:* see 04100
	Saccharolytic *Psychrobacter immobilis:* see 04100
04142	*Pseudomonas mallei:* motility neg, 0%; growth on SS agar neg, 0%; Simmons citrate neg, 0%
	"*Pseudomonas denitrificans*"-similar strains: motility +, 1–2 polar flagella; growth on SS agar + or (+), 89 (4)%; Simmons citrate +, 96%
04150	*Pseudomonas mallei:* see 04100
	Saccharolytic *Psychrobacter immobilis:* see 04100
04152	*Pseudomonas mallei:* see 04142
	"*Pseudomonas denitrificans*"-similar strains: see 04142
04160	*Pseudomonas mallei*
04162	*Pseudomonas mallei:* nonmotile
	Pseudomonas pseudomallei: motile
04163	*Pseudomonas pseudomallei*
04170	*Pseudomonas mallei*
04172	*Pseudomonas mallei:* nonmotile
	Pseudomonas pseudomallei: motile
04173	*Pseudomonas pseudomallei*
04300	*Pseudomonas mendocina*
04310	*Alcaligenes xylosoxidans* subsp. *xylosoxidans:* OF D-xylose reaction stronger than OF D-glucose reaction; peritrichous flagella
	Pseudomonas aeruginosa: OF D-glucose reaction as strong or stronger than OF D-xylose reaction; polar flagella; acetamide +, 100%
	Pseudomonas mendocina: OF D-glucose reaction as strong or stronger than OF D-xylose reaction, polar flagella; acetamide neg, 0%
04312	Vb-3
04330	*Pseudomonas aeruginosa*
04332	Vb-3
04342	"*Pseudomonas denitrificans*" type strain

(continued)

Table 1.6. Glucose oxidizers: MacConkey-positive/oxidase-positive *(continued)*

Numerical Code	Presumptive Identification
04352	*"Pseudomonas denitrificans"* type strain
04362	*Pseudomonas pseudomallei*
04363	*Pseudomonas pseudomallei*
04372	*Pseudomonas pseudomallei*
04373	*Pseudomonas pseudomallei*
04400	Saccharolytic *Psychrobacter immobilis*
04410	*Pseudomonas putida*
04430	*Pseudomonas putida*
04440	Saccharolytic *Psychrobacter immobilis*
04450	Saccharolytic *Psychrobacter immobilis*
04470	*Pseudomonas*-like group 2
04473	*Flavobacterium spiritivorum:* acid phosphatase +
	Flavobacterium yabuuchiae: acid phosphatase neg
04500	*Acidovorax delafieldii:* motile; growth in nutrient broth with 6% NaCl neg, 0%; OF D-mannitol + or (+), 50 (50)%; OF maltose neg, 0%
	Pseudomonas mallei: nonmotile; growth in nutrient broth with 6% NaCl neg, 0%; OF D-mannitol v, 62 (14)%; OF maltose v, (75)%
	Saccharolytic *Psychrobacter immobilis:* nonmotile; growth in nutrient broth with 6% NaCl +, 100%; OF D-mannitol neg, 0%; OF maltose neg, 0%
04502	*Pseudomonas mallei:* see 4102
	Ic: see 04102
04510	*Acidovorax delafieldii:* see 04500
	Pseudomonas mallei: see 04500
	Saccharolytic *Psychrobacter immobilis:* see 04500
04512	*Pseudomonas mallei*
04520	*Acidovorax delafieldii:* see 04500
	Pseudomonas mallei: see 04500
04522	*Pseudomonas mallei*
04530	*Acidovorax delafieldii:* see 04500
	Pseudomonas mallei: see 04500
04532	*Pseudomonas mallei*
04540	*Pseudomonas mallei:* see 04100
	Saccharolytic *Psychrobacter immobilis:* see 04100
04542	*"Pseudomonas denitrificans"*-similar strains: see 04142
	Pseudomonas mallei: see 04142
04550	*Pseudomonas mallei:* see 04100
	Saccharolytic *Psychrobacter immobilis:* see 04100

(continued)

Table 1.6. Glucose oxidizers: MacConkey-positive/oxidase-positive *(continued)*

Numerical Code	Presumptive Identification
04552	*"Pseudomonas denitrificans"*-similar strains: see 04142
	Pseudomonas mallei: see 04142
04560	*Pseudomonas mallei*
04562	*Pseudomonas mallei*
04570	*Pseudomonas*-like group 2: motile; Simmons citrate +, 100%
	Pseudomonas mallei: nonmotile; Simmons citrate neg, 0%
04572	*Pseudomonas mallei*
04700	*Pseudomonas mendocina*
04710	*"Achromobacter"* group B: acetamide neg., 0%; polar and lateral flagella (average wavelength); OF lactose v, (33)%; can be differentiated from the other species with this code by cellular fatty acid analysis
	Ochrobactrum anthropi: acetamide neg, 4%; peritrichous flagella, predominantly polar and lateral (fairly short wavelength); OF lactose neg, 0%; can be differentiated from the other species with this code by cellular fatty acid analysis
	Pseudomonas aeruginosa: acetamide +, 100%; polar flagella; H_2S (PbAc) neg, 4%; can be differentiated from the other species with this code by cellular fatty acid analysis
	Pseudomonas mendocina: acetamide neg, 0%; polar flagella; H_2S (PbAc)+, 100%; can be differentiated from the other species with this code by cellular fatty acid analysis
04711	*"Achromobacter"* group B: motile with polar and lateral flagella (average wavelength); esculin +, 100%; OF lactose v, (33)%; can be differentiated from *Ochrobactrum anthropi* by cellular fatty acid analysis
	Ochrobactrum anthropi: motile with peritrichous flagella, predominantly single polar and lateral (fairly short wavelength); esculin variable, 38 (11)%; OF lactose neg, 0%; can be differentiated from *"Achromobacter"* group B by cellular fatty acid analysis
04712	*"Achromobacter"* group B: polar and lateral flagella; growth at 42°C +, 100%; OF raffinose neg, 0%; can be differentiated from *O. anthropi* by cellular fatty acid analysis
	"Achromobacter" group E: polar and lateral flagella; growth at 42°C +, 100%; OF raffinose +, 100%; can be differentiated from *O. anthropi* by cellular fatty acid analysis
	Ochrobactrum anthropi: peritrichous flagella, predominantly single polar and lateral (fairly short wavelength); growth at 42°C v, 56%; can be differentiated from *"Achromobacter"* groups B and E by cellular fatty acid analysis
	Vb-3: 1-2 polar flagella; growth at 42°C neg, 0%; can be differentiated from the other species with this code by cellular fatty acid analysis
04713	*"Achromobacter"* group B: see 04712
	"Achromobacter" group E: see 04712
	Ochrobactrum anthropi: see 04712
04730	*"Achromobacter"* group B: see 04710
	Ochrobactrum anthropi: see 04710
	Pseudomonas aeruginosa: see 04710
04731	*"Achromobacter"* group B: see 04711
	Ochrobactrum anthropi: see 04711
04732	*"Achromobacter"* group B: see 04712
	Ochrobactrum anthropi: see 04712
	Vb-3: see 04712

(continued)

Table 1.6. Glucose oxidizers: MacConkey-positive/oxidase-positive *(continued)*

Numerical Code	Presumptive Identification
04733	*"Achromobacter"* group B: see 04711
	Ochrobactrum anthropi: see 04711
04742	*"Pseudomonas denitrificans"*-similar strains
04750	*"Achromobacter"* group B
04751	*"Achromobacter"* group B
04752	*"Achromobacter"* group B: OF D-mannitol v, (67)%; OF sorbitol + or (+), 67(33)%; esculin hydrolysis +, 100%
	"Achromobacter" group E: OF D-mannitol neg, 0%; OF sorbitol neg., 0%; esculin hydrolysis +, 100%
	"Pseudomonas denitrificans,"-similar strains: OF D-mannitol neg, 0%; esculin hydrolysis neg, 0%
04753	*"Achromobacter"* group B: see 04752
	"Achromobacter" group E: see 04752
04762	*Pseudomonas pseudomallei*
04763	*Pseudomonas pseudomallei*
04770	*"Achromobacter"* group B
04771	*"Achromobacter"* group B
04772	*"Achromobacter"* group B: motile with polar and lateral flagella; urea +, 100%; growth on SS agar +, 100%; gelatin hydrolysis neg, 0%; peptonization of litmus milk neg, 0%
	Pseudomonas pseudomallei: motile with polar tufts of three or more flagella; urea variable, 13 (8)%; growth on SS agar variable, 8 (31)%; gelatin hydrolysis v, 79%; peptonization of litmus milk +, 96%
04773	*"Achromobacter"* group B: see 04772
	Pseudomonas pseudomallei: see 04772
05710	*Ochrobactrum anthropi*
05711	*Ochrobactrum anthropi*
05712	*"Achromobacter"* group E: see 04712
	Ochrobactrum anthropi: see 04712
05713	*Ochrobactrum anthropi:* see 04712
	"Achromobacter" group E: see 04712
05730	*Ochrobactrum anthropi*
05731	*Ochrobactrum anthropi*
05732	*Ochrobactrum anthropi*
05733	*Ochrobactrum anthropi*
05752	*"Achromobacter"* group E
05753	*"Achromobacter"* group E
14310	*Pseudomonas aeruginosa*
14330	*Pseudomonas aeruginosa*
14710	*Pseudomonas aeruginosa*
14730	*Pseudomonas aeruginosa*

(continued)

Table 1.6. Glucose oxidizers: MacConkey-positive/oxidase-positive *(continued)*

Numerical Code	Presumptive Identification
24010	*Pseudomonas fluorescens:* gelatin + at 25°C
	Pseudomonas putida: gelatin neg. at 25°C
24011	*Pseudomonas fluorescens*
24030	*Pseudomonas fluorescens:* see 24010
	Pseudomonas putida: see 24010
24031	*Pseudomonas fluorescens*
24050	*Pseudomonas fluorescens*
24051	*Pseudomonas fluorescens*
24070	*Pseudomonas fluorescens*
24071	*Pseudomonas fluorescens*
24110	*Pseudomonas fluorescens*
24111	*Pseudomonas fluorescens*
24130	*Pseudomonas fluorescens*
24131	*Pseudomonas fluorescens*
24150	*Pseudomonas fluorescens*
24151	*Pseudomonas fluorescens*
24170	*Pseudomonas fluorescens*
24171	*Pseudomonas fluorescens*
24310	*Pseudomonas aeruginosa*
24330	*Pseudomonas aeruginosa*
24410	*Pseudomonas fluorescens:* see 24010
	Pseudomonas putida: see 24010
24411	*Pseudomonas fluorescens*
24430	*Pseudomonas fluorescens:* see 24010
	Pseudomonas putida: see 24010
24431	*Pseudomonas fluorescens*
24450	*Pseudomonas fluorescens*
24451	*Pseudomonas fluorescens*
24470	*Pseudomonas fluorescens*
24471	*Pseudomonas fluorescens*
24510	*Pseudomonas fluorescens*
24511	*Pseudomonas fluorescens*
24530	*Pseudomonas fluorescens*
24531	*Pseudomonas fluorescens*
24550	*Pseudomonas fluorescens*
24551	*Pseudomonas fluorescens*

(continued)

Table 1.6. Glucose oxidizers: MacConkey-positive/oxidase-positive *(continued)*

Numerical Code	Presumptive Identification
24570	*Pseudomonas fluorescens*
24571	*Pseudomonas fluorescens*
24710	*Pseudomonas aeruginosa*
24730	*Pseudomonas aeruginosa*
34310	*Pseudomonas aeruginosa*
34330	*Pseudomonas aeruginosa*
34710	*Pseudomonas aeruginosa*
34730	*Pseudomonas aeruginosa*
40000	*Flavobacterium breve*
40002	*Flavobacterium breve:* OF trehalose neg, 0% (two strains); ONPG neg, 0% (three strains); esculin hydrolysis neg, 0% (three strains); *slight* yellow insoluble pigment +, 100% (three strains) *Flavobacterium* species (IIb): OF trehalose +, 100% (14 strains); ONPG v, 57% (14 strains); *frank* yellow insoluble pigment +, 99% (155 strains); esculin hydrolysis v, 70% (149 strains)
40003	*Flavobacterium* species (IIb)
40012	*Flavobacterium* species (IIb)
40013	*Flavobacterium* species (IIb)
40022	*Flavobacterium meningosepticum*
40062	*Flavobacterium meningosepticum*
40102	*Flavobacterium* species (IIb)
40103	*Flavobacterium* species (IIb)
40112	*Flavobacterium* species (IIb)
40113	*Flavobacterium* species (IIb)
40402	*Flavobacterium* species (IIb)
40403	*Flavobacterium* species (IIb)
40412	*Flavobacterium* species (IIb)
40413	*Flavobacterium* species (IIb)
40502	*Flavobacterium* species (IIb)
40503	*Flavobacterium* species (IIb)
40512	*Flavobacterium* species (IIb)
40513	*Flavobacterium* species (IIb)

Table 1.7. Glucose oxidizers: MacConkey-negative/oxidase-negative

Numerical Code	Presumptive Identification
000	*Francisella tularensis*: insoluble pigment neg, 0%
	O-1: yellow insoluble pigment +, 100%
	Roseomonas gilardii: pink insoluble pigment +, 100%
010	*Roseomonas gilardii*
070	*Sphingomonas paucimobilis/parapaucimobilis*
100	*Roseomonas gilardii*
110	*Roseomonas gilardii*
401	*Pseudomonas mallei*
411	*Pseudomonas mallei*
441	*Pseudomonas mallei*
451	*Pseudomonas mallei*
500	*Brucella* species, probably *B. canis*
501	*Pseudomonas mallei*
510	*Brucella* species, probably *B. canis*
511	*Pseudomonas mallei*
541	*Pseudomonas mallei*
551	*Pseudomonas mallei*

Table 1.8. Glucose oxidizers: MacConkey-negative/oxidase-positive

Numerical Code	Presumptive Identification
000	*Francisella philomiragia:* rod; OF sucrose (continued incubation) + or (+), 63 (37)%; OF maltose (continued incubation) + or (+), 63 (37)%; insoluble yellow to orange pigment neg, 0%
	Neisseria gonorrhoeae: coccus
	O-1: rod; OF sucrose neg, 0%; OF maltose neg., 0%; insoluble yellow to orange pigment +, 100%
	O-2: rod; OF sucrose (continued incubation) + or (+), 64 (36)%; OF maltose (continued incubation) + or (+), 71 (27)%; yellow to orange insoluble pigment +, 100%
	Riemerella anatipestifer: rod; OF sucrose neg, 0%; OF maltose (continued incubation) + or (+), 40 (60)%; pink, yellow, to orange insoluble pigment v, 14%
	Roseomonas gilardii: rod; OF sucrose neg, 0%; OF maltose neg, 0%; insoluble pink pigment +, 100%
001	"*Agrobacterium* yellow group": rod; OF lactose (continued incubation) (+), (100)%; OF sucrose (continued incubation) (+), (100)%; esculin hydrolysis + or (+), 50 (50)%; yellow or yellow to orange insoluble pigment +, 100%
	Francisella philomiragia: rod; OF lactose neg, 0%; OF sucrose (continued incubation) + or (+), 63 (37)%; esculin hydrolysis neg, 0%; yellow or yellow to orange insoluble pigment neg, 0%
	Neisseria meningitidis: coccus; amylosucrose neg, 0%
	Neisseria polysaccharea: coccus; amylosucrose +
	O-2: rod; OF lactose neg, 2%; OF sucrose (continued incubation) + or (+), 64 (36)%; esculin hydrolysis v, 64%; yellow or yellow to orange insoluble pigment +, 100%
	Pseudomonas vesicularis: rod; OF lactose neg, 0%; OF sucrose neg, 0%; esculin hydrolysis + or (+), 88 (6)%; yellow or yellow to orange insoluble pigment v, 52%
	Riemerella anatipestifer: rod; OF lactose neg, 0%; OF sucrose neg, 0%; esculin hydrolysis neg, 0%; yellow or yellow to orange insoluble pigment (v, 14%) neg, 0%
002	*Acidovorax facilis:* rod; OF D-mannitol (continued incubation) (+), (100)%; OF sucrose neg, 0%; arginine dihydrolase +, 100%; catalase +, 100%; OF D-xylose (+), (100)%; motility + (may be difficult to demonstrate). Cellular fatty acid analysis will differentiate *A. facilis* from *P. mallei*
	Acidovorax temperans: rod; OF D-mannitol v, 50%; OF sucrose neg, 0%; arginine dihydrolase neg, 0%; catalase +, 100%; OF D-xylose-neg, 0%; motility + (may be difficult to demonstrate). Cellular fatty acid analysis will differentiate *A. temperans* from EF-4b
	EF-4b: rod; OF D-mannitol neg, 0%; OF sucrose neg, 0%; arginine dihydrolase neg, 0%; catalase +, 100%; motility neg. Cellular fatty acid analysis will differentiate Ef-4b from *A. temperans*
	Neisseria canis: cocci
	Neisseria elongata subsp. *nitroreducens:* rod; OF D-mannitol neg, 0%; OF sucrose neg, 0%; arginine dihydrolase neg., 0%; catalase neg, 0%; motility neg
	O-2: rod; OF D-mannitol neg, 2%; OF sucrose (continued incubation) + or (+), 64 (36)%; arginine dihydrolase v, 22%; catalase +, 91%; motility neg
	Pseudomonas mallei: rod; OF D-mannitol v, 62 (14)%; OF sucrose neg, 0%; arginine dihydrolase +, 100%; catalase +, 100%; motility neg; cellular fatty acid analysis will differentiate *P. mallei* from *Acidovorax facilis*
003	O-2: OF sucrose (continued incubation) + or (+), 64 (36)%; arginine dihydrolase v, 22%; yellow to orange insoluble pigment +, 100
	Pseudomonas mallei: OF sucrose neg, 0%; arginine dihydrolase +, 100%; yellow to orange insoluble pigment neg, 0%

(continued)

Table 1.8. Glucose oxidizers: MacConkey-negative/oxidase-positive *(continued)*

Numerical Code	Presumptive Identification
010	*Roseomonas gilardii*
012	*Acidovorax facilis:* see 002
	Acidovorax temperans: see 002
	Pseudomonas mallei: see 002
013	*Pseudomonas mallei*
020	*Flavobacterium mizutaii*
021	"*Agrobacterium* yellow group": rod; nitrite (0.01%) neg, 0%; motility + (may be difficult to demonstrate)
	Flavobacterium mizutaii: rod; nitrite (0.01%) +, 100%; motility neg
	Neisseria lactamica: coccus; nitrite (0.01%) +, 100%; motility neg
022	*Pseudomonas mallei*
023	*Pseudomonas mallei*
032	*Pseudomonas mallei*
033	*Pseudomonas mallei*
040	*Francisella philomiragia:* yellow to orange insoluble pigment neg, 0%; esculin hydrolysis neg, 0%; motility neg, 0%
	O-2: yellow to orange insoluble pigment +, 100%; esculin hydrolysis v, 64%; motility v, 20%
041	"*Agrobacterium* yellow group": see 001
	Francisella philomiragia: see 001
	Neisseria polysaccharea: see 001
	O-2: see 001
042	O-2
043	O-2
060	*Flavobacterium mizutaii*
061	"*Agrobacterium* yellow group": see 021
	Flavobacterium mizutaii: see 021
101	*Flavobacterium* species (IIb): esculin hydrolysis v, 70%; yellow insoluble pigment +, 99%
	IIe: esculin hydrolysis neg, 0%; yellow insoluble pigment neg, 7% weak
	IIh: esculin hydrolysis +, 100%; yellow insoluble pigment neg, 0%
103	*Balneatrix alpica* (single strain): motility + (may be difficult to demonstrate); yellow insoluble pigment neg; esculin hydrolysis neg
	Flavobacterium species (IIb): motility neg, 0%; yellow insoluble pigment +, 99%; esculin hydrolysis v, 70%
111	*Flavobacterium meningosepticum*
113	*Balneatrix alpica*
131	*Flavobacterium meningosepticum*
141	*Flavobacterium* species (IIb)

(continued)

Table 1.8. Glucose oxidizers: MacConkey-negative/oxidase-positive *(continued)*

Numerical Code	Presumptive Identification
143	*Flavobacterium* species (IIb)
161	IIi
200	O-2: OF sucrose (continued incubation) + or (+), 64 (36)%; yellow or yellow to orange insoluble pigment +, 100%; pink insoluble pigment neg, 0%
	Riemerella anatipestifer: OF sucrose neg, 0%; yellow or yellow to orange insoluble pigment neg, 0%; pink insoluble pigment neg, 0%
	Roseomonas gilardii: OF sucrose neg, 0%; yellow or yellow to orange insoluble pigment neg, 0%; pink insoluble pigment +, 100%
201	"*Agrobacterium* yellow group": see 001
	O-2: see 001
	Riemerella anatipestifer: see 001
202	*Acidovorax delafieldii:* motility +, (may be difficult to demonstrate); arginine dihydrolase +, 100%; catalase neg, 0%; OF D-xylose (+), (100)%
	Acidovorax facilis: motility +, (may be difficult to demonstrate); arginine dihydrolase +, 100%; catalase +, 100%; OF D-xylose (+), (100)%
	Acidovorax temperans: motility +, (may be difficult to demonstrate); arginine dihydrolase neg, 0%; catalase +, 100%; OF D-xylose neg, 0%
	O-2: this is the only species or group with this code that produces a yellow to orange insoluble pigment
	Pseudomonas mallei: motility neg, 0%; arginine dihydrolase +, 100%; catalase +, 100%; of OF D-xylose v, 12 (50)%
	Roseomonas fauriae: this is the only species or group with this code that produces a pink insoluble pigment
203	O-2: see 003
	Pseudomonas mallei: see 003
210	*Roseomonas gilardii*
212	*Acidovorax delafieldii:* see 202
	Acidovorax facilis: see 202
	Acidovorax temperans: see 202
	Pseudomonas mallei: see 202
213	*Pseudomonas mallei*
221	"*Agrobacterium* yellow group"
222	*Pseudomonas mallei*
223	*Pseudomonas mallei*
232	*Pseudomonas mallei*
233	*Pseudomonas mallei*
240	O-2
241	"*Agrobacterium* yellow group": see 001
	O-2: see 001

(continued)

Table 1.8. Glucose oxidizers: MacConkey-negative/oxidase-positive *(continued)*

Numerical Code	Presumptive Identification
242	O-2
243	O-2
261	"*Agrobacterium* yellow group"
301	*Flavobacterium* species (IIb)
303	*Flavobacterium* species (IIb)
341	*Flavobacterium* species (IIb)
343	*Flavobacterium* species (IIb)
361	IIi
400	*Methylobacterium* species: vacuolated rod; Simmons citrate neg, 2 (3)%; OF D-mannitol neg, 2%
	Roseomonas gilardii: rod, not vacuolated; Simmons citrate +, 100%; OF D-mannitol v, 14 (38)%
401	"*Agrobacterium* yellow group": OF lactose and OF sucrose (continued incubation) (+), (100)%
	Pseudomonas vesicularis: OF lactose and OF sucrose neg, 0%
402	*Acidovorax facilis*: motility +, (may be difficult to demonstrate); OF D-mannitol (+), (100)%; OF lactose neg, 0%; arginine dihydrolase +, 100%; Simmons citrate neg, 0%; gelatin +, (7–14 day incubation); pink to coral insoluble pigment neg, 0%
	EO-2: OF D-mannitol neg, 0%; OF lactose (continued incubation) + or (+), 45 (55)%; Simmons citrate + or (+), 64 (36)%; pink to coral insoluble pigment neg, 0%
	Methylobacterium species: OF lactose neg, 0%; pink to coral insoluble pigment +, 99%
	Pseudomonas mallei: nonmotile; OF D-mannitol v, 62 (14)%; OF lactose (continued incubation) v, 12 (62)%; Simmons citrate neg, 0%; gelatin neg, 0%; arginine dihydrolase +, 100%; pink to coral insoluble pigment neg, 0%
403	*Pseudomonas mallei*
410	*Roseomonas gilardii*
412	*Acidovorax facilis*: see 012
	Pseudomonas mallei: see 012
413	*Pseudomonas mallei*
420	*Flavobacterium mizutaii*
421	"*Agrobacterium* yellow group": see 021
	Flavobacterium mizutaii: see 021
422	EO-2: see 003
	Pseudomonas mallei: see 003
423	*Pseudomonas mallei*
432	*Pseudomonas mallei*
433	*Pseudomonas mallei*
441	"*Agrobacterium* yellow group"
460	*Flavobacterium mizutaii*

(continued)

Table 1.8. Glucose oxidizers: MacConkey-negative/oxidase-positive (continued)

Numerical Code	Presumptive Identification
461	"*Agrobacterium* yellow group": nitrite (0.01%) neg, 0%; motility +, 100%; urea (continued incubation) (+), (100)% *Flavobacterium mizutaii:* nitrite (0.01%) +, 100%; motility neg, 0%; urea neg, 0% *Sphingomonas paucimobilis/parapaucimobilis:* motility + (weak); urea neg, 6 (3)%
471	*Flavobacterium spiritivorum:* acid phosphatase + *Flavobacterium yabuuchiae:* acid phosphatase neg
501	*Flavobacterium* species (IIb)
503	*Flavobacterium* species (IIb)
541	*Flavobacterium* species (IIb)
543	*Flavobacterium* species (IIb)
561	IIi
600	*Methylobacterium* species: see 400 *Roseomonas gilardii:* see 400
601	"*Agrobacterium* yellow group"
602	*Acidovorax delafieldii:* pink to coral insoluble pigment neg; catalase neg, 0%; arginine dihydrolase +, 100%; citrate +, 100%; OF D-mannitol (continued incubation) + or (+), 50 (50)%; OF lactose neg, 0% *Acidovorax facilis:* motility +, (may be difficult to demonstrate); pink to coral insoluble pigment neg; catalase +, 100%; arginine dihydrolase +, 100%; citrate neg, 0%; OF D-mannitol (continued incubation) (+), (100)%; OF lactose neg, 0% *Brucella* species: pink to coral insoluble pigment neg; catalase +, 100%; arginine dihydrolase (not determined); citrate neg, 0%; OF D-mannitol neg, 0%; OF lactose neg, 0% EO-2: pink to coral insoluble pigment neg; catalase v, 82%; arginine dihydrolase (not determined); citrate + or (+), 64 (36)%; OF D-mannitol neg, 0%; OF lactose (continued incubation) + or(+), 45 (55)% *Methylobacterium* species: vacuolated rod; pink to coral insoluble pigment +, 99%; catalase +, 100%; argininedihydrolase (not determined); citrate neg, 2 (3)%; OF D-mannitol neg, 2%; OF lactose neg, 0% *Pseudomonas mallei:* nonmotile; pink to coral insoluble pigment neg; catalase +, 100%; arginine dihydrolase +, 100%; citrate neg,0%; OF D-mannitol (continued incubation) v, 62 (14)%; OF lactose (continued incubation) v, 12 (62)% *Roseomonas fauriae:* rod, not vacuolated; pink insoluble pigment +, 100%; catalase +, 100%; arginine dihydrolase (not determined); citrate v, 60 (20)%; OF D-mannitol neg, 0%; OF lactose neg, 0%
603	*Pseudomonas mallei*
610	*Roseomonas gilardii*
612	*Acidovorax delafieldii:* see 202 *Acidovorax facilis:* see 202 *Pseudomonas mallei:* see 202
613	*Pseudomonas mallei*
621	"*Agrobacterium* yellow group"

(continued)

Table 1.8. Glucose oxidizers: MacConkey-negative/oxidase-positive *(continued)*

Numerical Code	Presumptive Identification
622	EO-2: see 003
	Pseudomonas mallei: see 003
623	*Pseudomonas mallei*
632	*Pseudomonas mallei*
633	*Pseudomonas mallei*
641	"*Agrobacterium* yellow group"
661	"*Agrobacterium* yellow group"
671	*Flavobacterium spiritivorum:* see 471
	Flavobacterium yabuuchiae: see 471
701	*Flavobacterium* species (IIb)
703	*Flavobacterium* species (IIb)
741	*Flavobacterium* species (IIb)
743	*Flavobacterium* species (IIb)
761	IIi

Table 1.9. Nonoxidizers: MacConkey-positive/oxidase-negative

Numerical Code	Presumptive Identification
00	Nonhemolytic asaccharolytic *Acinetobacter* species: *Acinetobacter* transformation +, 100%; pink insoluble pigment neg, 0%; Simmons citrate v, 27 (3)%; brown soluble pigment v, 10%; the species and groups that share this code can be differentiated by cellular fatty acid analysis
	Bordetella holmesii (NO-2): *Acinetobacter* transformation neg, 0%; pink insoluble pigment neg, 0%; Simmons citrate neg, 0%; brown soluble pigment +, 100%; the species and groups that share this code can be differentiated by cellular fatty acid analysis
	Roseomonas gilardii: pink insoluble pigment +, 100%; Simmons citrate +, 100%; brown soluble pigment neg, 0%; the species and groups that share this code can be differentiated by cellular fatty acid analysis
01	NO-1: can be differentiated by cellular fatty acid analysis from a rare nitrate + *Acinetobacter* species
02	β-hemolytic asaccharolytic *Acinetobacter* species
04	*Roseomonas gilardii*
20	*Roseomonas gilardii*
24	*Roseomonas gilardii*
30	*Xanthomonas maltophilia*
31	*Xanthomonas maltophilia*
34	*Xanthomonas maltophilia*
35	*Xanthomonas maltophilia*
40	*Roseomonas gilardii*: see 00
	Nonhemolytic asaccharolytic *Acinetobacter* species: see 00
41	*Brucella* species, possibly *B. canis*: can be differentiated by cellular fatty acid analysis and cellular morphology from a rare nitrate + *Acinetobacter* species
42	β-hemolytic asaccharolytic *Acinetobacter* species: brown soluble pigment neg, 0% (a pale yellow or tan pigment may be detected); the species and groups that share this code can be differentiated by cellular fatty acid analysis
	Bordetella parapertussis: brown soluble pigment +, 100% (demonstrated best on a medium containing tyrosine); the species and groups that share this code can be differentiated by cellular fatty acid analysis
44	*Roseomonas gilardii*
45	*Brucella* species, possibly *B. canis*
60	*Roseomonas gilardii*
64	*Roseomonas gilardii*
70	*Xanthomonas maltophilia*
71	*Xanthomonas maltophilia*
74	*Xanthomonas maltophilia*
75	*Xanthomonas maltophilia*

Table 1.10. Nonoxidizers: MacConkey-positive/oxidase-positive

Numerical Code	Presumptive Identification
0000	Cocci: *Neisseria flavescens:* CTA maltose neg, 0% *Neisseria sicca:* CTA maltose +, 95 (5)%; no pigment on Loeffler slant, 0% *Neisseria subflava:* CTA maltose +, 99 (1)%; yellow pigment on Loeffler slant +, 99% Rods: "*Agrobacterium* yellow group": OF D-glucose (+), (100)%; OF maltose +, 100%; nitrite reduction neg, 0%; catalase +, 100% *Moraxella atlantae:* phenylalanine neg, 0%; nitrite reduction neg, 3%; growth on SS agar neg, 0%; catalase +, 95% *Moraxella osloensis:* phenylalanine v, 14%; nitrite reduction neg, 0%; growth on SS agar neg, 0%; catalase +, 95% *Neisseria elongata* subsp. *elongata:* catalase neg, 0%; nitrite reduction +, 92%; D-glucose neg in rapid sugar test *Neisseria elongata* subsp. *glycolytica:* catalase +, 100%; D-glucose + in rapid sugar test; nitrite reduction +, 100% *Neisseria weaveri:* catalase +, 100%; D-glucose neg in rapid sugar test; associated with dog-bite wounds; 0.01% nitrite reduction +, 100%; phenylalanine v, 71% *Oligella urethralis:* phenylalanine +, 100%; nitrite reduction +, 100%; growth on SS agar neg, 9% Gilardi rod group 1: nitrite reduction neg, 0%; phenylalanine +, 100%; growth on SS agar v, 80% Asaccharolytic *Psychrobacter immobilis: Psychrobacter* transformation +, 100%; usually prefers 25°C growth temperature; this species can be differentiated from the other species that share this code by cellular fatty acid analysis *Taylorella equigenitalis:* not associated with humans; does not require X factor but growth is stimulated by it IIg: indole +, 100% (all other species or groups with this code are indole neg.).
0004	*Roseomonas gilardii*
0010	Gilardi rod group 1
0032	*Alcaligenes faecalis:* nitrite reduction +, 100% *Bordetella avium:* nitrite reduction neg, 0%; associated with turkeys and other birds
0060	"*Agrobacterium* yellow group": OF maltose +, 100%; OF sucrose +, 100% *Pseudomonas alcaligenes:* OF maltose neg, 0%; flagella with "normal" wavelength and amplitude *Pseudomonas diminuta:* OF maltose neg, 0%; flagella with short wavelength and low amplitude *Pseudomonas vesicularis:* OF maltose +, 94 (6)%; OF sucrose neg, 0%; flagella similar to those of *P. diminuta* but with slightly longer wavelength
0064	*Roseomonas cervicalis:* OF glycerol neg, 0%; OF D-mannitol neg, 0% *Roseomonas gilardii:* OF glycerol +, 100%; OF D-mannitol v, 14 (38)%
0070	*Pseudomonas alcaligenes*

(continued)

Table 1.10. Nonoxidizers: MacConkey-positive/oxidase-positive *(continued)*

Numerical Code	Presumptive Identification
0100	*Moraxella osloensis:* rod; catalase +, 95%; nitrite reduction neg, 0%; this species can be differentiated from the other species that share this code by cellular fatty acid analysis *Neisseria canis:* coccus *Neisseria elongata* subsp. *nitroreducens:* rod; catalase neg, 8%; nitrite reduction +, 100% (may require incubation for 48 h, gas not produced); this species can be differentiated from the other species that share this code by cellular fatty acid analysis Asaccharolytic *Psychrobacter immobilis:* rod; catalase +, 100%; *Psychrobacter* transformation +, 100%; optimal growth temperature usually 25°C; this species can be differentiated from the other species that share this code by cellular fatty acid analysis
0121	*Comamonas* species
0122	*Alcaligenes eutrophus*
0131	*Comamonas* species
0132	*Alcaligenes piechaudii*
0160	*Afipia felis:* straight rods; OF D-xylose (+), (93)%; urea + or (+), 88 (12)%; growth at 42°C neg., 0%; Simmons citrate neg, 0% *Campylobacter* species: thin, curved, and wavy rods *Pseudomonas alcaligenes:* straight or slightly curved rods; growth at 42°C neg, 0%; aerobic; OF D-xylose neg, 0%; urea neg, 0%; Simmons citrate v, 57 (8)%
0170	*Pseudomonas alcaligenes:* OF fructose neg, 0%; growth at 42°C neg., 0% *Pseudomonas pseudoalcaligenes:* OF fructose + or (+), 79 (21)%; growth at 42°C +, 94%
0360	*Shewanella putrefaciens* biotype 1
0370	*Shewanella putrefaciens* biotype 2
0400	"*Agrobacterium* yellow group": OF maltose +, 100% *Flavobacterium odoratum:* nitrite reduction v, 83%; gelatin hydrolysis +, 96%; yellow insoluble pigment v, 85%; usually produces a fruity odor; OF maltose neg, 0% *Moraxella phenylpyruvica:* nitrite reduction-neg, 0%; gelatin hydrolysis-neg, 0%; yellow insoluble pigment neg, 0%; OF maltose neg, 0% Asaccharolytic *Psychrobacter immobilis:* yellow insoluble pigment-neg, 0%; *Psychrobacter* transformation +, 100%; gelatin hydrolysis neg, 0%; optimal growth temperature usually 25°C; OF maltose neg, 0%
0404	*Roseomonas* genomospecies 5: OF glycerol v, 33%; Simmons citrate v, 33%; motility neg, 0% *Roseomonas gilardii:* OF glycerol +, 100%; Simmons citrate +, 100%; motility v, 33%
0410	*Flavobacterium odoratum*
0422	IVc-2
0460	"*Agrobacterium* yellow group": see 0060 *Pseudomonas diminuta:* see 0060
0464	*Roseomonas cervicalis:* see 0064 *Roseomonas gilardii:* see 0064

(continued)

Table 1.10. Nonoxidizers: MacConkey-positive/oxidase-positive *(continued)*

Numerical Code	Presumptive Identification
0500	*Acidovorax delafieldii*: OF D-glucose (+), (100)%; this species can be differentiated from other species that share this code by cellular fatty acid analysis
	Moraxella phenylpyruvica: OF D-glucose neg, 0%; coccoid to filamentous thick rods; nonmotile; this species can be differentiated from the other species that share this code by cellular fatty acid analysis
	Oligella ureolytica: OF D-glucose neg, 0%; small coccoid to short rods, some filaments; motile, but may be delayed or difficult to detect; this species can be differentiated from the other species that share this code by cellular fatty acid analysis
	Asaccharolytic *Psychrobacter immobilis*: OF D-glucose neg, 0%; coccobacilli; *Psychrobacter* transformation +, 100%; nonmotile; optimal growth temperature usually 25°C; this species can be differentiated from the other species and groups that share this code by cellular fatty acid analysis
0502	*Oligella ureolytica*
0520	*Oligella ureolytica*
0521	*Pseudomonas testosteroni*
0522	*Alcaligenes eutrophus* (one strain studied): citrate +; nitrate +; gas from nitrate neg; cellular fatty acid profile is identical to IVc-2 except that 19:10 cyc is not present
	Oligella ureolytica: citrate v, 14 (16)%; small coccoid to short rods, may have filamentous forms; nitrate +, 100%; gas from nitrate v, 60%; urine is major source, 92%; predominantly from male patients
	IVc-2: citrate +, 100%; medium rods; nitrate v, 11%; gas from nitrate neg, 0%; rarely from urine, 5%
0531	*Comamonas* species
0532	*Bordetella bronchiseptica*
0560	*Acidovorax delafieldii*: OF D-glucose (+), (100)%; OF D-mannitol + or (+), 50 (50)%
	Afipia felis: OF D-glucose neg, 0%; OF D-mannitol neg, 0%
0564	*Roseomonas fauriae*: OF D-xylose + or (+), 80 (20)%
	Roseomonas genomospecies 6 (one strain studied): OF D-xylose neg
0574	*Roseomonas fauriae*
0760	*Shewanella putrefaciens* biotype 1
0770	*Shewanella putrefaciens* biotype 2
1004	*Roseomonas gilardii*
1064	*Roseomonas gilardii*
1121	*Comamonas acidovorans*
1131	*Comamonas acidovorans*
1404	*Roseomonas gilardii*
1464	*Roseomonas gilardii*
1500	*Acidovorax delafieldii*
1560	*Acidovorax delafieldii*
2000	"*Agrobacterium* yellow group"
2004	*Roseomonas gilardii*
2060	"*Agrobacterium* yellow group": OF lactose (+), (100)%; flagella have "normal" wavelength and amplitude
	Pseudomonas vesicularis: OF lactose neg, 0%; flagella have short wavelength and amplitude

(continued)

Table 1.10. Nonoxidizers: MacConkey-positive/oxidase-positive *(continued)*

Numerical Code	Presumptive Identification
2064	*Methylobacterium* species: pleomorphic vacuolated rods; Simmons citrate neg, 2 (3)%
	Roseomonas cervicalis: rods, not vacuolated; Simmons citrate + or (+), 86 (14)%; OF glycerol neg, 0%
	Roseomonas gilardii: rods, not vacuolated; Simmons citrate +, 100%; OF glycerol +, 100%
2104	*Roseomonas* genomospecies 4
2132	*Alcaligenes xylosoxidans* subsp. *xylosoxidans*
2160	*Afipia felis*
2164	*Methylobacterium* species: pleomorphic vacuolated rods
	Roseomonas genomospecies 4: rods, not vacuolated
2170	*Pseudomonas pseudoalcaligenes*
2400	"*Agrobacterium* yellow group"
2404	*Roseomonas* genomospecies 5: see 0404
	Roseomonas gilardii: see 0404
2460	"*Agrobacterium* yellow group"
2464	*Methylobacterium* species: see 2064
	Roseomonas cervicalis: see 2064
	Roseomonas gilardii: see 2064
2500	*Acidovorax delafieldii*: Simmons citrate +, 100%; OF D-mannitol + or (+), 50 (50)%
	Brucella species: Simmons citrate neg, 0%; OF D-mannitol neg, 0%
2504	*Roseomonas* genomospecies 4
2560	*Acidovorax delafieldii*: see 0560
	Afipia felis: see 0560
2564	*Methylobacterium* species: pleomorphic vacuolated rods; esculin neg, 0%
	Roseomonas fauriae: rods, not vacuolated; esculin hydrolysis +, 100%
	Roseomonas genomospecies 4: rods, not vacuolated; esculin hydrolysis neg, 0%
2574	*Roseomonas fauriae*
3004	*Roseomonas gilardii*
3064	*Roseomonas gilardii*
3404	*Roseomonas gilardii*
3464	*Roseomonas gilardii*
3500	*Acidovorax delafieldii*
3560	*Acidovorax delafieldii*
4100	*Neisseria mucosa*
4122	*Alcaligenes*-like group 1

(continued)

Table 1.10. Nonoxidizers: MacConkey-positive/oxidase-positive *(continued)*

Numerical Code	Presumptive Identification
4132	*Alcaligenes*-like group 1: when observed, growth on SS agar is usually weak and delayed. Can be differentiated from *A. xylosoxidans* subsp. *denitrificans* by cellular fatty acid analysis
	Alcaligenes xylosoxidans subsp. *denitrificans:* moderate to heavy growth on SS agar within 48 h. Can be differentiated from *Alcaligenes*-like group 1 by cellular fatty acid analysis
4160	*Pseudomonas* species CDC group 1
4170	*Pseudomonas* species CDC group 1
4500	*Oligella ureolytica*
4502	*Oligella ureolytica*
4520	*Oligella ureolytica*
4522	*Alcaligenes*-like group 1: urea not rapid, "average" length flagella
	Oligella ureolytica: urea rapid, flagella relatively long
4532	*Alcaligenes*-like group 1
4564	*Roseomonas fauriae*
4574	*Roseomonas fauriae*
6132	*Alcaligenes xylosoxidans* subsp. *xylosoxidans*
6500	*Brucella*
6532	*Ochrobactrum anthropi*
6564	*Roseomonas fauriae*
6574	*Roseomonas fauriae*
6732	*Ochrobactrum anthropi*
7532	*Ochrobactrum anthropi*
7732	*Ochrobactrum anthropi*

Table 1.11. Nonoxidizers: MacConkey-negative/oxidase-negative

Numerical Code	Presumptive Identification
0	*Acinetobacter* species (nonhemolytic/asaccharolytic strains): short broad rods; OF D-glucose neg, 0%; pink insoluble pigment neg, 0%; this group can be differentiated from the other species and groups that share this code by cellular fatty acid analysis
	Bartonella (formerly *Rochalimaea*) species: thin rod forms; RST D-glucose neg, 0%; pink insoluble pigment neg, 0%; this species can be differentiated from the other species and groups that share this code by cellular fatty acid analysis
	Francisella tularensis: small coccoid cells; D-glucose + or (+), 84 (16)% in cysteine-supplemented base; pink insoluble pigment neg, 0%; this species can be differentiated from the other species and groups that share this code by cellular fatty acid analysis
	Bordetella holmesii (NO-2): continue incubation: growth on MacConkey agar + or (+), 77 (23)%; pink insoluble pigment neg, 0%; this group can be differentiated from the other species and groups that share this code by the production of a brown soluble pigment and cellular fatty acid analysis
	Roseomonas gilardii: pink insoluble pigment +, 100%; this species can be differentiated from the other species and groups that share this code by cellular fatty acid analysis
1	*Acinetobacter* species (nonhemolytic/asaccharolytic strains): pink insoluble pigment neg, 0%
	Roseomonas gilardii: pink insoluble pigment +, 100%
2	NO-1
3	*Brucella* species, possibly *B. canis*
4	*Bartonella bacilliformis:* fastidious; pink insoluble pigment neg, 0%
	Roseomonas gilardii: nonfastidious; pink insoluble pigment +, 100%
5	*Roseomonas gilardii*

Table 1.12. Nonoxidizers: MacConkey-negative/oxidase-positive

Numerical Code	Presumptive Identification
00	*Bartonella* (formerly *Rochalimaea*) species: rod; fastidious, usually requires 5–7 days' incubation for primary growth
	Moraxella bovis: rod; gelatin hydrolysis +, 100%; β-hemolysis +, 100%; probably an animal source
	Neisseria elongata subsp. *elongata:* rod; gelatin not hydrolyzed; β-hemolysis neg, 0%
	Neisseria sicca: coccus; yellow pigment on Loeffler slant neg, 0%
	Neisseria subflava: coccus; yellow pigment on Loeffler slant +, 99%
01	*Eikenella corrodens:* gelatin hydrolysis neg, 0%; nitrite reduction neg, 0%; gas from nitrate neg, 0%; lysine v, 82%; ornithine +, 98%
	Kingella denitrificans: in fermentation base D-glucose + or (+), 30 (62)%; gelatin hydrolysis neg, 0%; gas from nitrate v, 88%
	Moraxella bovis: gelatin hydrolysis +, 100%; β-hemolysis +, 100%
	Neisseria elongata subsp. *nitroreducens:* gelatin hydrolysis neg, 0%; nitrite reduction +, 100%; gas from nitrate neg, 0%
02	*Afipia clevelandensis* (one strain studied): esculin neg; H_2S (lead acetate paper) neg; phenylalanine neg
	Afipia genomospecies 3 (one strain studied): esculin neg; H_2S (lead acetate paper) (+); phenylalanine +
	Pseudomonas vesicularis: esculin + or (+), 88 (6)%; OF D-glucose + or (+), 87 (12)%
03	*Afipia felis*

(continued)

Table 1.12. Nonoxidizers: MacConkey-negative/oxidase-positive (continued)

Numerical Code	Presumptive Identification
04	Cocci:
	Neisseria cinerea: CTA maltose neg, 0%; amylosucrase neg, 0%
	Neisseria flavescens: CTA maltose neg, 0%; amylosucrase +, 100%
	Neisseria sicca: CTA maltose +, 95 (5)%; yellow pigment on Loeffler slant neg, 0%
	Neisseria subflava: CTA maltose +, 99 (1)%; yellow pigment on Loeffler-slant +, 99%
	Rod:
	Bordetella pertussis: special media required for growth or enhancement of growth; confirmation by agglutination or direct fluorescent antibody reactions; hemolytic on Bordet-Gengou medium
	Moraxella atlantae: gelatin hydrolysis neg, 0%; growth in nutrient broth with 0% NaCl neg, 0%
	Moraxella bovis: gelatin hydrolysis +, 100%
	Moraxella lincolnii (one strain studied): coccus-like to plump rods, coccal forms often appear in pairs or short chains of cells with "flattened" sides; gelatin hydrolysis neg; growth in nutrient broth with 0% NaCl neg
	Moraxella osloensis: gelatin hydrolysis neg, 0%; sodium acetate +, 100%; nitrite reduction neg, 0%; growth in nutrient broth with 0% NaCl +, 98%
	Neisseria elongata subsp. *glycolytica:* nitrite reduction +, 100%; growth in nutrient broth with 0% NaCl +, 100%; weak acid production from D-glucose may be detected using the rapid sugar test
	Neisseria weaveri: gelatin hydrolysis neg, 0%; sodium acetate neg, 4%; growth in nutrient broth with 0% NaCl v, 85%; associated with dog-bite wounds
	O-2: OF maltose + or (+), 71 (27)%; yellow to orange insoluble pigment +, 100%
	Psychrobacter immobilis (asaccharolytic strains): optimal growth temperature usually 25°C.
	Roseomonas gilardii: pink insoluble pigment +, 100%.
	Taylorella equigenitalis (one strain studied): growth in nutrient broth with 0% NaCl neg; growth in nutrient broth with 6% NaCl +; usually from nonhuman sources.
05	*Acidovorax facilis:* rod; OF sucrose neg, 0%; OF D-mannitol (+), (100)%
	Moraxella bovis: rod; β-like hemolysis +, 100%
	Moraxella (Branhamella) catarrhalis: coccus; gas from nitrate neg, 0%
	Moraxella lacunata: rod; sodium acetate neg, 0%; digestion of Loeffler slant +, 100%; not hemolytic
	Moraxella nonliquefaciens: rod; sodium acetate neg, 0%; digestion of Loeffler slant neg, 0%; not hemolytic
	Moraxella osloensis: rod; sodium acetate +, 100%; digestion of Loeffler slant neg, 0%; not hemolytic
	Neisseria canis: coccus; CTA D-glucose (+); gas from nitrate neg, 0%
	Neisseria mucosa: coccus; CTA D-glucose + or (+), 83 (10)%; gas from nitrate +, 100%
	O-2: rod; OF sucrose + or (+), 64 (36)%; OF D-mannitol neg, 0%; yellow to orange insoluble pigment +, 100%
	Psychrobacter immobilis (asaccharolytic strains): coccobacillus; *Psychrobacter* transformation +, 100%; optimal growth temperature usually 25°C; cellular fatty acid analysis will differentiate this species from the other species and groups that share this code

(continued)

Table 1.12. Nonoxidizers: MacConkey-negative/oxidase-positive *(continued)*

Numerical Code	Presumptive Identification
06	*Afipia broomeae:* urea + or (+), 33 (67)%; esculin neg, 0%; OF sucrose neg, 0%; OF maltose neg, 0%; OF D-xylose (+), (100)%
	"*Agrobacterium* yellow group": yellow insoluble pigment +, 100%; OF sucrose (+), (100)%; OF maltose +, 100%; OF D-xylose + or (+), 50 (50)%
	O-2: yellow to orange insoluble pigment +, 100%; OF sucrose + or (+), 64 (36)%; esculin v, 64%; OF maltose + or (+), 71 (27)%; OF D-xylose neg, 2%
	Pseudomonas vesicularis: urea neg, 2 (5)%; esculin + or (+), 88 (6)%; OF sucrose neg., 0%; OF maltose +, 94 (6)%; OF D-xylose v, 27 (9)%
	Roseomonas gilardii: pink insoluble pigment +, 100%
07	*Acidovorax facilis:* straight rods; OF sucrose neg, 0%; OF D-glucose + or (+), 50 (50)%
	Afipia felis: straight and pleomorphic rods; OF D-glucose neg, 0%; OF sucrose neg, 0%
	Campylobacter species: curved or spiral rods
	O-2: straight rods; OF sucrose + or (+), 64 (36)%
14	*Weeksella virosa*
22	*Afipia clevelandensis:* Simmons citrate neg, 0%; H_2S (lead acetate paper) neg, 0%
	Afipia genomospecies 1: Simmons citrate +, 100%
	Afipia genomospecies 3: Simmons citrate neg, 0%; H_2S (lead acetate paper) +, 100%
23	*Afipia felis*
24	*Moraxella phenylpyruvica:* prefers growth at 35°C; OF sucrose neg, 0%
	O-2: OF sucrose + or (+), 64(36)%
	Asaccharolytic *Psychrobacter immobilis:* prefers growth at <35°C; *Psychrobacter* transformation +, 100%; OF sucrose neg, 0%
	Roseomonas gilardii: pink insoluble pigment +, 100%; Simmons citrate +, 100%; OF D-mannitol v, 14 (38)%; motility v, 33%
	Roseomonas genomospecies 5: pink insoluble pigment +, 100%; Simmons citrate v, 33%; OF D-mannitol neg, 0%; motility neg, 0%
25	*Acidovorax delafieldii:* catalase neg, 0%; OF D-xylose (+), (100)%
	Acidovorax facilis: catalase +, 100%; OF D-xylose (+), (100)%
	Moraxella phenylpyruvica: catalase +, 90%; OF D-xylose neg, 0%; OF sucrose neg, 0%; gas from nitrate neg, 0%; nonmotile; prefers growth at 35°C
	Oligella ureolytica: catalase +, 100%; OF D-xylose neg, 0%; OF sucrose neg, 0%; gas from nitrate v, 60%; recheck for motility and/or flagella
	O-2: catalase +, 91%; OF D-xylose neg, 2%; OF sucrose + or (+), 64 (36)%
	Asaccharolytic *Psychrobacter immobilis:* catalase +, 100%; OF D-xylose neg, 0%; OF sucrose neg, 0%; gas from nitrate neg, 0%; nonmotile; prefers growth at <35°C

(continued)

Table 1.12. Nonoxidizers: MacConkey-negative/oxidase-positive *(continued)*

Numerical Code	Presumptive Identification
26	*Afipia broomeae:* Simmons citrate neg, 0%; OF sucrose neg, 0%; cannot be differentiated from *Afipia* genomospecies 2 by biochemical or cellular fatty acid analysis
	Afipia genomospecies 1: Simmons citrate +, 100%; OF sucrose neg, 0%
	Afipia genomospecies 2: Simmons citrate neg, 0%; OF sucrose neg, 0%; cannot be differentiated from *Afipia broomeae* by biochemical or cellular fatty acid analysis
	"*Agrobacterium* yellow group": yellow insoluble pigment +, 100%; OF sucrose (+), (100)%; OF D-xylose + or (+), 50 (50)%
	O-2: yellow to orange insoluble pigment +, 100%; OF D-xylose neg, 2%; OF sucrose + or (+), 64 (36)%
	Roseomonas gilardii: the only species with this code that produces a pink insoluble pigment
27	*Acidovorax delafieldii:* OF D-xylose (+), (100)%; OF sucrose neg, 0%; OF D-mannitol + or (+), 50 (50)%; motile with polar flagella; catalase neg, 0%
	Acidovorax facilis: OF D-xylose (+), (100)%; OF sucrose neg, 0%; OF D-mannitol (+), (100)%; motile with polar flagella; catalase +, 100%
	Afipia felis: OF D-xylose (+), (93)%; OF sucrose neg, 0%; OF D-mannitol neg, 0%; catalase v, 37%; motile with polar flagella
	Oligella ureolytica: OF D-xylose neg, 0%; OF sucrose neg, 0%; OF D-mannitol neg, 0%; catalase +, 100%; motile with peritrichous flagella
	O-2: OF sucrose + or (+), 64 (36)%; OF D-mannitol neg, 0%; motile with polar or polar and lateral flagella
	Roseomonas fauriae: pink insoluble pigment +, 100%; OF D-xylose + or (+), 80 (20)%
	Roseomonas genomospecies 6: pink insoluble pigment +; OF D-xylose neg, 0%
34	*Weeksella zoohelcum*
42	*Afipia* genomospecies 3: esculin neg, 0%
	Pseudomonas vesicularis: esculin +, 88 (6)%
43	*Afipia felis*
44	*Methylobacterium* species: vacuolated rod; Simmons citrate neg, 2 (3)%
	Roseomonas gilardii: rod, not vacuolated; Simmons citrate +, 100%
45	*Acidovorax facilis:* pink to coral insoluble pigment neg, 0%
	Methylobacterium species: pink to coral insoluble pigment +, 99%
46	*Afipia broomeae:* pink insoluble pigment-neg, 0%; OF maltose neg, 0%
	"*Agrobacterium* yellow group": pink insoluble pigment neg, 0%; yellow insoluble pigment +, 100%; OF maltose +, 100%; OF sucrose (+), (100)%
	Methylobacterium species: pink to coral insoluble pigment + 99%; vacuolated rod; Simmons citrate neg, 2 (3)%; OF maltose neg, 0%
	Pseudomonas vesicularis: pink insoluble pigment neg, 0%; yellow to orange insoluble pigment v, 52%; OF maltose +, 94 (6)%; OF sucrose neg, 0%
	Roseomonas gilardii: pink insoluble pigment +, 100%; rod, not vacuolated; Simmons citrate +, 100%; OF maltose neg, 0%
47	*Acidovorax facilis:* nitrate +, 100%; OF D-mannitol (+), (100)%
	Afipia felis: nitrate +, 100%; OF D-mannitol neg, 0%
	Pseudomonas vesicularis: nitrate neg, 0%; OF D-mannitol neg, 0%

(continued)

Table 1.12. Nonoxidizers: MacConkey-negative/oxidase-positive *(continued)*

Numerical Code	Presumptive Identification
62	*Afipia* genomospecies 1: Simmons citrate +, 100%
	Afipia genomospecies 3: Simmons citrate neg, 0%
63	*Afipia felis*
64	*Methylobacterium* species: vacuolated rod; pink to coral insoluble pigment +, 99%
	Roseomonas genomospecies 5: rod, not vacuolated; pink insoluble pigment +, 100%; Simmons citrate v, 33%; OF D-mannitol neg, 0%; motility neg, 0%
	Roseomonas gilardii: rod, not vacuolated; pink insoluble pigment +, 100%; Simmons citrate +, 100%; OF D-mannitol v, 14 (38)%; motility v, 33%
65	*Acidovorax delafieldii:* rods; OF D-glucose (+), (100)%; no insoluble pigment; catalase neg, 0%; recheck motility (+)
	Acidovorax facilis: rods; OF D-glucose + or (+), 50 (50)%; no insoluble pigment; catalase +, 100%; recheck motility (+)
	Brucella species: tiny coccoid forms; recheck motility (-); no insoluble pigment
	Methylobacterium species: pleomorphic rod; recheck motility (+); pink to coral insoluble pigment +, 99%
66	*Afipia broomeae:* pink insoluble pigment neg, 0%; Simmons citrate neg, 0%; cannot be differentiated from *Afipia* genomospecies 2 by biochemical or cellular fatty acid analysis
	Afipia genomospecies 1: pink insoluble pigment neg, 0%; Simmons citrate +, 100%
	Afipia genomospecies 2: pink insoluble pigment neg, 0%; Simmons citrate neg, 0%; cannot be differentiated from *Afipia broomeae* by biochemical or cellular fatty acid analysis
	"*Agrobacterium* yellow group": the only species with this code that has a yellow insoluble pigment +, 100%
	Methylobacterium species: pink to coral insoluble pigment +, 99%; vacuolated rod; Simmons citrate neg., 2 (3)%
	Roseomonas gilardii: pink insoluble pigment +, 100%; rod, not vacuolated; Simmons citrate +, 100%
67	*Acidovorax delafieldii:* pink insoluble pigment neg, 0%; OF D-mannitol + or (+), 50 (50)%; catalase neg, 0%
	Acidovorax facilis: pink insoluble pigment neg, 0%; OF D-mannitol (+), (100)%; catalase +, 100%
	Afipia felis: pink insoluble pigment neg, 0%; OF D-mannitol neg, 0%
	Methylobacterium species: pink to coral insoluble pigment +, 99%; vacuolated rod
	Roseomonas fauriae: pink insoluble pigment +, 100%; rod, not vacuolated

Bacterial Identification Using King's Key and "Round-Table" Charts

Key for the Identification of Unusual Pathogenic Gram-Negative Aerobic and Facultatively Anaerobic Bacteria

Glucose Fermenters

Table 2.1. MacConkey-positive/oxidase-negative

Organism	MacConkey % +	Oxidase % +	"Round-Table" Chart #	Comments
Chromobacterium violaceum	100	67	1	
Enterobacteriaceae				Usually (>90%) grow on MacConkey agar and are oxidase-negative
Haemophilus aphrophilus	19	28	6	
Pasteurella bettyae	20 (24)	61	2	

Glucose Fermenters
(continued)

Table 2.2. MacConkey-positive/oxidase-positive

Organism	MacConkey %+	Oxidase %+	"Round-Table" Chart #	Comments
Actinobacillus equuli	84 (5)	100	4	
Actinobacillus hominis	100	100	4	Data is presented for the reference strain (NCTC 11529) only
Actinobacillus lignieresii	67 (8)	100	4	
Actinobacillus suis	82 (12)	100	4	
Bisgaard's taxon 16	7 (10)	100	2	
Chromobacterium violaceum	100	67	1	
EF-4a	42 (8)	100	3	
Haemophilus aphrophilus	19	28	6	
Neisseria mucosa	57 (3)	100	9, 46	
Neisseria sicca	68	100	9, 46	
Neisseria subflava	47 (2)	100	9, 46	
Pasteurella aerogenes	100	100	2	
Pasteurella bettyae	20 (24)	61	2	
Pasteurella gallinarum	20 (10)	90	3	
Pasteurella haemolytica	64 (14)	96	3	
Pasteurella pneumotropica	36 (17)	99	2	
Pasteurella trehalosi	100	100	3	
Suttonella indologenes	17		12	
Vibrio species, *Aeromonas* species, and *Plesiomonas shigelloides*				Most *Vibrio* species grow on MacConkey agar and are oxidase-positive

Glucose Fermenters
(continued)

Table 2.3. MacConkey-negative/oxidase-negative[a]

Organism	MacConkey %+	Oxidase %+	"Round-Table" Chart #	Comments
Actinobacillus actinomycetemcomitans	4 (1)	19	6	
Actinobacillus ureae	0	99	4	
Bisgaard's taxon 16	7 (10)	100	2	
Capnocytophaga species (DF-1)	0	7	7	This group represents three species: *C. ochracea, C. gingivalis,* and *C. sputigena*
DF-3	0	0	5	
DF-3-like	0	0	5	
Gardnerella vaginalis	5	6	5	
Haemophilus aphrophilus	19	28	6	
Leptotrichia buccalis	0	0	7	
Pasteurella bettyae	20 (24)	61	2	
Pasteurella canis	0	92	1	
Pasteurella dagmatis	1 (2)	98	1	
Pasteurella multocida	1 (1)	96	1	
Pasteurella pneumotropica	36 (17)	99	2	
Pasteurella stomatis	0	100	1	
Streptobacillus moniliformis	0	3	6	

[a]Depending upon the method used, some *Pasteurella* and *Actinobacillus* isolates, and also Bisgaard's taxon 16, may test oxidase-negative.

Glucose Fermenters *(continued)*

Table 2.4. MacConkey-negative/oxidase-positive

Organism	MacConkey %+	Oxidase %+	"Round-Table" Chart #	Comments
Actinobacillus actinomycetemcomitans	4 (1)	19	6	
Actinobacillus equuli	84 (5)	100	4	
Actinobacillus lignieresii	67 (8)	100	4	
Actinobacillus suis	82 (12)	100	4	
Actinobacillus ureae	0	99	4	
Bisgaard's taxon 16	7 (10)	100	2	
Capnocytophaga canimorsus	0	100	7	
Capnocytophaga cynodegmi	0	100	7	
Cardiobacterium hominis	0	100	5	
EF-4a	42 (8)	100	3	
Haemophilus aphrophilus	19	28	6	
Kingella denitrificans	0	100	12	
Kingella kingae	9 (1)	100	12	
Neisseria lactamica	0	100	9, 46	
Neisseria mucosa	57 (3)	100	9, 46	
Neisseria sicca	68	100	9, 46	
Neisseria subflava	47 (2)	100	9, 46	
Pasteurella bettyae	20 (24)	61	2	
Pasteurella canis	0	92	1	
Pasteurella dagmatis	1 (2)	98	1	
Pasteurella gallinarum	20 (10)	90	3	
Pasteurella haemolytica	64 (14)	96	3	
Pasteurella multocida	1 (1)	96	1	
Pasteurella pneumotropica	36 (17)	99	2	

(continued)

Glucose Fermenters
(continued)

Table 2.4. MacConkey-negative/oxidase-positive *(continued)*

Organism	MacConkey %+	Oxidase %+	"Round-Table" Chart #	Comments
Pasteurella stomatis	0	100	1	
Riemerella anatipestifer	0	100	12	
Simonsiella species	0	100	11	
Suttonella indologenes	17	100	12	
Vibrio species	-	+		

SECTION 2: KING'S KEY AND "ROUND-TABLE" CHARTS

Glucose Oxidizers

Table 2.5. MacConkey-positive/oxidase-negative

Organism	MacConkey %+	Oxidase %+	"Round-Table" Chart #	Comments
Nonhemolytic/saccharolytic *Acinetobacter* species	95	2	22	Species that may show these characteristics include: *A. calcoaceticus*, *A. baumannii* (D-glucose 95%), *A. radioresistans* (D-glucose 33%), genomospecies 3, 13, 10, and 15 (D-glucose 50%)
β-hemolytic/saccharolytic *Acinetobacter* species	100	0	22	Species that may show these characteristics include: Tjernberg's genomospecies 14, Bouvet's genomospecies 14, *A. haemolyticus* (D-glucose 52%), and genomospecies 6 (D-glucose 66%)
Brucella canis	12 (29)	72	40, 44	
Chryseomonas luteola	100	0	29	Using the Kovacs' modification of the oxidase test, a weak reaction was observed after 10 sec with 4 of 34 strains tested
Flavimonas oryzihabitans	100	0	29	Using the Kovacs' modification of the oxidase test, a weak reaction was observed after 10 sec with a few strains
O-1	6 (40)	77	28	
Pseudomonas cepacia	100	86	33	
Pseudomonas gladioli	97 (3)	47	33	Most strains were only oxidase-positive with the Kovacs' method; reactions were frequently weak
Pseudomonas mallei	88	25	34	Weak growth on MacConkey agar was observed with 7 of 8 strains studied
Roseomonas gilardii	43 (52)	52	30	
Sphingomonas paucimobilis/parapaucimobilis	10 (13)	75	28	Neither the type strain of *S. paucimobilis* nor two reference strains of *S. parapaucimobilis* grew on MacConkey agar; however, growth was observed with 31 of 134 phenotypically similar strains
Xanthomonas maltophilia	100	0	29	Using the Kovacs' method, 14 of 73 strains tested produced a weak oxidase reaction after 10 sec; D-glucose 85 (5)%, frequently weak

Glucose Oxidizers *(continued)*

Table 2.6. MacConkey-positive/oxidase-positive

Organism	MacConkey %+	Oxidase %+	"Round-Table" Chart #	Comments
"*Achromobacter*" group B	100	100	27	3 strains studied
"*Achromobacter*" group E	100	100	27	2 strains studied
Acidovorax delafieldii	(100)	100	25	2 strains studied. Growth on MacConkey agar was weak and late for both strains studied
Acidovorax temperans	50 (50)	100	25	2 strains studied
Agrobacterium radiobacter biovar 1	100	100	27	
"*Agrobacterium* yellow group"	(50)	100	28	2 strains studied
Alcaligenes xylosoxidans subsp. *xylosoxidans*	100	100	24	D-glucose 78%
Brucella abortus	50 (20)	96	40, 44	D-glucose 50 (25)%
Brucella canis	12 (29)	72	40, 44	D-glucose 81 (14)%
Brucella melitensis	62	100	40, 44	D-glucose 33%
Brucella suis	23 (15)	95	40, 44	
EF-4b	65 (6)	100	10	D-glucose 70(26)%
EO-2	64 (18)	100	22	
EO-3	100	100	22	7 strains studied
Flavobacterium breve	100	100	14	7 strains studied
Flavobacterium meningosepticum	89 (3)	99	13	*F. meningosepticum* is a genetically heterogeneous species
Flavobacterium multivorum	100	100	17	
Flavobacterium species (IIb)	54 (9)	96	14	This group may contain strains of *F. indologenes*, *F. gleum*, or other unnamed species
Flavobacterium spiritivorum	(46)	100	17	
Flavobacterium thalpophilum	100	100	17	
Flavobacterium yabuuchiae	(50)	100	17	2 strains studied
Francisella philomiragia	19 (13)	100	41	D-glucose reaction is weak and occasionally delayed

Glucose Oxidizers
(continued)

Methylobacterium species	15	30	
Neisseria canis	(100)	96	1 strain studied
Neisseria elongata subsp. nitroreducens	19 (35)	+	8, 46	D-glucose 23%; when observed, the reaction is weak
Ochrobactrum anthropi	100	100	10, 46	
O-1	100	100	27	
Pseudomonas aeruginosa	6 (40)	77	28	
Pseudomonas cepacia	100	99	32	
"Pseudomonas denitrificans"	100	86	33	
Pseudomonas diminuta	100	100	35	
Pseudomonas fluorescens	97 (3)	100	37	D-glucose 21 (9)%
Pseudomonas gladioli	100	97	32	
Pseudomonas-like group 2	97 (3)	47	33	Most strains were only positive with the Kovacs' method; reactions were frequently weak
Pseudomonas mallei	100	100	33	
Pseudomonas mendocina	88	25	34	
Pseudomonas pertucinogena	100	100	32	4 strains studied
Pseudomonas pickettii biovar 1	100	100	42	2 strains studied
Pseudomonas pickettii biovar 2	99 (1)	100	34	
Pseudomonas pseudomallei	100	100	34	
Pseudomonas putida	100	100	35	
Pseudomonas stutzeri	100	100	32	
"Pseudomonas thomasii"	100	100	35	
Pseudomonas vesicularis	100	100	34	
Saccharolytic Psychrobacter immobilis	43 (23)	98	37	
Roseomonas fauriae	100	100	22	7 strains studied
Roseomonas gilardii	60 (40)	100	31	D-glucose 20%; 5 strains studied
	43 (52)	52	30	D-glucose (43)%

(continued)

Glucose Oxidizers
(continued)

Table 2.6. MacConkey-positive/oxidase-positive (continued)

Organism	MacConkey %+	Oxidase %+	"Round-Table" Chart #	Comments
Shewanella putrefaciens biotype 1	100	100	39	D-glucose 17 (33)%
Sphingomonas paucimobilis/parapaucimobilis	10 (13)	75	28	Neither the type strain of S. paucimobilis nor two reference strains of S. parapaucimobilis grew on MacConkey agar; however, growth was observed with 31 of 134 phenotypically similar strains
Ic	100	100	37	
Vb-3	100	100	35	

Table 2.7. MacConkey-negative/oxidase-negative

Organism	MacConkey %+	Oxidase %+	"Round-Table" Chart #	Comments
Brucella canis	12 (29)	72	40, 44	
Francisella tularensis	3 (2)	0	41	When observed, growth on MacConkey agar was weak and occasionally delayed
O-1	6 (40)	77	28	
Pseudomonas mallei	88	25	34	8 strains studied
Roseomonas gilardii	43 (52)	52	30	D-glucose (43)%
Sphingomonas paucimobilis/parapaucimobilis	10 (13)	75	28	Neither the type strain of S. paucimobilis nor two reference strains of S. parapaucimobilis grew on MacConkey agar; however, growth was observed with 31 of 134 phenotypically similar strains

Glucose Oxidizers
(continued)

Table 2.8. MacConkey-negative/oxidase-positive

Organism	MacConkey %+	Oxidase %+	"Round-Table" Chart #	Comments
Acidovorax delafieldii	(100)	100	25	2 strains studied; although both strains grew on MacConkey, growth was weak and delayed
Acidovorax facilis	0	100	25	2 strains studied
Acidovorax temperans	50 (50)	100	25	2 strains studied; growth of one strain was weak and delayed on MacConkey
"*Agrobacterium* yellow group"	(50)	100	28	2 strains studied
Balneatrix alpica	0	100	13	1 strain studied
Brucella abortus	50 (20)	96	40, 44	D-glucose 50 (25)%
Brucella canis	12 (29)	72	40, 44	
Brucella melitensis	62	100	40, 44	
Brucella suis	23 (15)	95	40, 44	
EF-4b	65 (6)	100	10	
EO-2	64 (18)	100	22	
Flavobacterium meningosepticum	89 (3)	99	13	*F. meningosepticum* is a genetically heterogeneous species
Flavobacterium mizutaii	0	100	17	6 strains studied
Flavobacterium species (IIb)	54 (9)	96	14	This group may contain strains of *F. indologenes*, *F. gleum* or other unnamed species
Flavobacterium spiritivorum	(46)	100	17	
Flavobacterium yabuuchiae	(50)	100	17	2 strains studied
Francisella philomiragia	19 (13)	100	41	
Methylobacterium species	15	96	30	
Neisseria canis	(100)	100	8, 46	1 strain studied; growth on MacConkey was weak and delayed
Neisseria elongata subsp. *nitroreducens*	19 (35)	100	10, 46	D-glucose 23%
Neisseria gonorrhoeae	0	100	8, 46	
Neisseria lactamica	0	100	9, 46	

(continued)

Glucose Oxidizers *(continued)*

Table 2.8. MacConkey-negative/oxidase-positive *(continued)*

Organism	MacConkey %+	Oxidase %+	"Round-Table" Chart #	Comments
Neisseria meningitidis	0	100	8, 46	
Neisseria polysaccharea	0	100	8, 46	1 strain studied
O-1	6 (40)	77	28	
O-2	9	97	28	D-glucose 73 (11)%
Pseudomonas mallei	88	25	34	8 strains studied
Pseudomonas vesicularis	43 (23)	98	37	
Riemerella anatipestifer	0	100	12	
Roseomonas fauriae	60 (40)	100	31	D-glucose 20%
Roseomonas gilardii	43 (52)	52	30	D-glucose (43)%
Sphingomonas paucimobilis/parapaucimobilis	10 (13)	75	28	Neither the type strain of *S. paucimobilis* nor two reference strains of *S. parapaucimobilis* grew on MacConkey agar; however, growth was observed with 31 of 134 phenotypically similar strains
IIe	3	100	16	
IIh	0	100	16	
IIi	0	100	16	

Glucose Nonoxidizers

Table 2.9. MacConkey-positive/oxidase-negative

Organism	MacConkey %+	Oxidase %+	"Round-Table" Chart #	Comments
Nonhemolytic/asaccharolytic *Acinetobacter* species	73 (4)	<1	21	Species that may show these characteristics include *A. junii*, *A. johnsonii*, *A. lwoffii*, *A. radioresistens* [D-glucose 33%], and genomospecies 15 [D-glucose 50%]
ß-hemolytic/asaccharolytic *Acinetobacter* species	95	0	21	Species that may show these characteristics include genomospecies 15, 16, 17, *A. haemolyticus* [D-glucose 52%], *A. radioresistens* [D-glucose 33%], and genomospecies 6 [D-glucose 66%]
Bordetella parapertussis	100	0	21	
Brucella canis	12 (29)	72	40, 44	D-glucose 81 (14)%
NO-1	5 (15)	5	21	
Bordetella holmesii (NO-2)	77 (23)	0	21	
Roseomonas gilardii	43 (52)	52	30	D-glucose (43)%
Xanthomonas maltophilia	100	0	29	Using the Kovacs' method, 14 of 73 strains tested produced a weak reaction after 10 sec; D-glucose 85 (5)%, frequently weak

SECTION 2: KING'S KEY AND "ROUND-TABLE" CHARTS

Glucose Nonoxidizers
(continued)

Table 2.10. MacConkey-positive/oxidase-positive

Organism	MacConkey %+	Oxidase %+	"Round-Table" Chart #	Comments
Acidovorax delafieldii	(100)	100	25	2 strains studied; delayed (3–7 days) D-glucose oxidizer
Afipia felis	(44)	100	26	
"*Agrobacterium* yellow group"	(50)	100	28	2 strains studied; weak and delayed (3–7 days) D-glucose oxidizer
Alcaligenes eutrophus	+	+	23	1 strain studied
Alcaligenes faecalis	100	100	24	
Alcaligenes-like group 1	100	100	24	8 strains studied
Alcaligenes piechaudii	100	100	24	4 strains studied
Alcaligenes xylosoxidans subsp. *denitrificans*	100	100	24	
Alcaligenes xylosoxidans subsp. *xylosoxidans*	100	100	24	D-glucose 78%
Bordetella avium	+	+	23	1 strain studied
Bordetella bronchiseptica	100	100	23	
Brucella abortus	50 (20)	96	40, 44	D-glucose 50 (25)%, 4 strains tested
Brucella canis	12 (29)	72	40, 44	D-glucose 81 (14)%
Brucella melitensis	62	100	40, 44	D-glucose 33%, 3 strains tested
Campylobacter species	v	+	38	
Comamonas acidovorans	100	100	38	
Comamonas species	96 (4)	100	38	Strains within this group may include *C. testosteroni* or *C. terrigena*
Flavobacterium odoratum	91 (5)	99	15	
Gilardi rod group 1	93	100	20	
Methylobacterium species	15	96	30	D-glucose 40%
Moraxella atlantae	80 (20)	100	19	
Moraxella osloensis	70	100	18	

Glucose Nonoxidizers
(continued)

Moraxella phenylpyruvica	80 (6)	100	19	
Neisseria canis	(100)	100	8, 46	1 strain studied; delayed D-glucose oxidation (3–7 days) in CTA medium
Neisseria elongata subsp. elongata	13 (53)	100	10, 46	
Neisseria elongata subsp. glycolytica	50 (50)	100	10, 46	Weak acid production from D-glucose may be detected with the rapid sugar test
Neisseria elongata subsp. nitroreducens	19 (35)	100	10, 46	D-glucose 23%
Neisseria flavescens	67	100	9, 46	
Neisseria mucosa	57 (3)	100	9, 46	D-glucose 83 (10)%
Neisseria sicca	68	100	9, 46	D-glucose 78 (14)%
Neisseria subflava	47 (2)	100	9, 46	D-glucose 68 (6)%
Neisseria weaveri	27 (18)	100	10, 46	
Ochrobactrum anthropi	100	100	27	
Oligella ureolytica	62 (27)	100	20	
Oligella urethralis	96	100	20	
Pseudomonas alcaligenes	96	96	36	
Pseudomonas diminuta	97 (3)	100	37	D-glucose 21 (9)%
Pseudomonas pseudoalcaligenes	100	100	36	D-glucose 9%
Pseudomonas species CDC group 1	97 (3)	100	36	
Pseudomonas vesicularis	43 (23)	98	37	
Asaccharolytic Psychrobacter immobilis	40	100	19	5 strains studied
Roseomonas cervicalis	100	100	30	7 strains studied
Roseomonas fauriae	60 (40)	100	31	D-glucose 20%, 5 strains studied
Roseomonas genomospecies 4	100	100	30	3 strains studied
Roseomonas genomospecies 5	67 (33)	100	30	3 strains studied
Roseomonas genomospecies 6	100 w	100	31	1 strain studied

(continued)

Glucose Nonoxidizers (continued)

Table 2.10. MacConkey-positive/oxidase-positive (continued)

Organism	MacConkey %+	Oxidase %+	"Round-Table" Chart #	Comments
Roseomonas gilardii	43 (52)	52	30	D-glucose (43)%
Shewanella putrefaciens biotype 1	100	100	39	D-glucose 17 (33)%
Shewanella putrefaciens biotype 2	100	100	39	
Taylorella equigenitalis	(+)	+	15	1 strain studied; growth on MacConkey agar, was weak and late
IIg	100	100	16	
IVc-2	94 (6)	100	23	

Table 2.11. MacConkey-negative/oxidase-negative

Organism	MacConkey %+	Oxidase %+	"Round-Table" Chart #	Comments
Nonhemolytic/asaccharolytic *Acinetobacter* species	73 (4)	<1	21	Species that may show these characteristics include *A. junii, A. johnsonii, A. lwoffii, A. radioresistens* [D-glucose 33%], and genomospecies 15 [D-glucose 50%]
Bartonella bacilliformis	0	0	43	2 strains studied
Bartonella (formerly *Rochalimaea*) species	0	13 w	43	A weak positive oxidase reaction was observed using the Kovacs' modification with 2/15 strains studied; this reaction was observed with one *B. quintana* strain and the type strain of *B. vinsonii*
Brucella canis	12 (29)	72	40, 44	D-glucose 81 (14)%
Francisella tularensis	(2)	0	41	Acid production from D-glucose may be detected if a cysteine-enriched medium is used
NO-1	5 (15)	5	21	
Bordetella holmesii (NO-2)	77 (23)	0	21	
Roseomonas gilardii	43 (52)	52	30	D-glucose (43)%

Glucose Nonoxidizers
(continued)

Table 2.12. MacConkey-negative/oxidase-positive

Organism	MacConkey %+	Oxidase %+	"Round-Table" Chart #	Comments
Acidovorax delafieldii	(100)	100	25	2 strains studied; delayed (3–7 days) D-glucose oxidizer
Acidovorax facilis	0	100	25	2 strains studied; D-glucose 50 (50)%
Afipia broomeae	0	100	26	3 strains studied
Afipia clevelandensis	–	+	26	1 strain studied
Afipia felis	(44)	100	26	
Afipia genomospecies 1	0	100	26	2 strains studied
Afipia genomospecies 2	–	+	26	1 strain studied
Afipia genomospecies 3	–	+	26	1 strain studied
"*Agrobacterium* yellow group"	(50)	100	28	2 strains studied; weak and delayed D-glucose oxidizer
Bartonella (formerly *Rochalimaea*) species	0	13 w	43	A weak positive oxidase reaction was observed using the Kovacs' modification with 2/15 strains studied; this reaction was observed with one *B. quintana* strain and the type strain of *B. vinsonii*
Bordetella pertussis	0	94	42	
Brucella abortus	50 (20)	96	40, 44	D-glucose 50 (25)%
Brucella canis	12 (29)	72	40, 44	D-glucose 81 (14)%
Brucella melitensis	62	100	40, 44	D-glucose 33%, 3 strains studied
Campylobacter species	v	+		
Eikenella corrodens	0	100	18	
Kingella denitrificans	0	100	12	D-glucose 30 (62)%
Methylobacterium species	15	96	30	D-glucose 40%
Moraxella atlantae	80 (20)	100	19	
Moraxella bovis	0	100	19	7 strains studied
Moraxella (Branhamella) catarrhalis	5	100	18, 46	
Moraxella lacunata	2	100	18	

(continued)

Glucose Nonoxidizers (continued)

Table 2.12. MacConkey-negative/oxidase-positive (continued)

Organism	MacConkey %+	Oxidase %+	"Round-Table" Chart #	Comments
Moraxella lincolnii	-	+	19	1 strain studied
Moraxella nonliquefaciens	8 (2)	100	18	
Moraxella osloensis	70	100	18	
Moraxella phenylpyruvica	80 (6)	100	19	
Neisseria canis	(+)	+	8, 46	1 strain studied; D-glucose oxidation and growth on MacConkey agar were both delayed (3–7 days incubation)
Neisseria cinerea	0	100	8, 46	
Neisseria elongata subsp. *elongata*	13 (53)	100	10, 46	
Neisseria elongata subsp. *glycolytica*	50 (50)	100	10, 46	
Neisseria elongata subsp. *nitroreducens*	19 (35)	100	10, 46	D-glucose 23%
Neisseria flavescens	67	100	9, 46	
Neisseria mucosa	57 (3)	100	9, 46	D-glucose 83 (10)%
Neisseria sicca	68	100	9, 46	D-glucose 78 (14)%
Neisseria subflava	47 (2)	100	9, 46	D-glucose 68 (6)%
Neisseria weaveri	27 (18)	100	10, 46	
Oligella ureolytica	62 (27)	100	20	
O-2	9	97	28	D-glucose 73 (11)%
Pseudomonas vesicularis	43 (23)	98	37	
Asaccharolytic *Psychrobacter immobilis*	40	100	19	5 strains studied
Roseomonas fauriae	60 (40)	100	31	D-glucose 20%, 5 strains studied
Roseomonas gilardii	43 (52)	52	30	D-glucose (43)%
Roseomonas genomospecies 5	67 (33)	100	30	3 strains studied
Roseomonas genomospecies 6	+w	+	31	1 strain studied
Taylorella equigenitalis	(+)	+	15	1 strain studied; MacConkey growth was weak and late
Weeksella virosa	(10)	100	15	
Weeksella zoohelcum	2	100	15	

Acid Producers in OF Basal Medium Without Carbohydrate

Table 2.13. MacConkey-positive/oxidase-positive

Organism	MacConkey %+	Oxidase %+	"Round-Table" Chart #	Comments
OFBA-1	100	100	32	6 strains studied

Index to "Round-Table" Charts

Species	Page
Achromobacter group B	182
Achromobacter group E	182
Acidovorax delafieldii	178
Acidovorax facilis	178
Acidovorax temperans	178
Acinetobacter species: Saccharolytic strains	172
Acinetobacter species: Asaccharolytic strains	170
Actinobacillus actinomycetemcomitans	140
Actinobacillus equuli	136
Actinobacillus hominis	136
Actinobacillus lignieresii	136
Actinobacillus suis	136
Actinobacillus ureae	136
Afipia broomeae	180
Afipia clevelandensis	180
Afipia felis	180
Afipia genomospecies 1	180
Afipia genomospecies 2	180
Afipia genomospecies 3	180
Agrobacterium radiobacter biovar 1	182
"*Agrobacterium* yellow group"	184
Alcaligenes eutrophus	174
Alcaligenes faecalis	176
Alcaligenes-like Group 1	176
Alcaligenes piechaudii	176
Alcaligenes xylosoxidans subsp. *denitrificans*	176
Alcaligenes xylosoxidans subsp. *xylosoxidans*	176
Balneatrix alpica	154
Bartonella bacilliformis	214
Bartonella (formerly *Rochalimaea*) species	214
Bisgaard's taxon 16	132
Bordetella avium	174
Bordetella bronchiseptica	174
Bordetella holmesii (NO-2)	170
Bordetella parapertussis	170
Bordetella pertussis	212
Brucella abortus	208
Brucella canis	208
Brucella melitensis	208
Brucella suis	208
Capnocytophaga species (DF-1)	142
Capnocytophaga canimorsus	142
Capnocytophaga cynodegmi	142
Cardiobacterium hominis	138
Chromobacterium violaceum	130
Chryseomonas luteola	186
Comamonas acidovorans	204
Comamonas terrigena	204
Comamonas testosteroni	204
DF-3	138
DF-3-like	138
EF-4a	134
EF-4b	148
Eikenella corrodens	164
EO-2	172
EO-3	172
Flavimonas oryzihabitans	186
Flavobacterium breve and similar strains	156
Flavobacterium species (IIb), (*F. gleum* and *F. indologenes*)	156
Flavobacterium meningosepticum and similar strains	154
Flavobacterium mizutaii	162
Flavobacterium multivorum	162
Flavobacterium odoratum	158
Flavobacterium spiritivorum	162
Flavobacterium thalpophilum	162
Flavobacterium yabuuchiae	162
Francisella philomiragia	210
Francisella tularensis	210
Gardnerella vaginalis	138
Gilardi rod group 1	168
Haemophilus aphrophilus	140
Haemophilus ducreyi	218
"*Haemophilus felis*"	218
Haemophilus haemoglobinophilus	218
Haemophilus haemolyticus	218
Haemophilus influenzae	218
Haemophilus influenzae biogroup aegyptius	218
Haemophilus parahaemolyticus	218
Haemophilus parainfluenzae	218
Haemophilus paraphrophilus	218
Haemophilus segnis	218
Kingella denitrificans	152
Kingella kingae	152
Leptotrichia buccalis	142
Methylobacterium species	188
Moraxella atlantae	166
Moraxella bovis	166
Moraxella (*Branhamella*) *catarrhalis*	164
Moraxella lacunata	164
Moraxella lincolnii	166
Moraxella nonliquefaciens	164
Moraxella osloensis	164
Moraxella phenylpyruvica	166
Neisseria canis	144
Neisseria cinerea	144
Neisseria elongata subsp. *elongata*	148
Neisseria elonagata subsp. *glycolytica*	148
Neisseria elontata subsp. *nitroreducens*	148

Species	Page
Neisseria flavescens	146
Neisseria gonorrhoeae	144
Neisseria lactamica	146
Neisseria meningitidis	144
Neisseria mucosa	146
Neisseria polysaccharea	144
Neisseria sicca	146
Neisseria subflava	146
Neisseria weaveri	148
NO-1	170
Bordetella holmesii (NO-2)	170
Ochrobactrum anthropi	182
OFBA-1	192
Oligella ureolytica	168
Oligella urethralis	168
O-1	184
O-2	184
Pasteurella aerogenes	132
Pasteurella bettyae	132
Pasteurella canis	130
Pasteurella dagmatis	130
Pasteurella gallinarum	134
Pasteurella haemolytica	134
Pasteurella multocida	130
Pasteurella pneumotropica	132
Pasteurella stomatis	130
Pasteurella trehalosi	134
Pseudomonas aeruginosa	192
Pseudomonas alcaligenes	200
Pseudomonas cepacia	194
"*Pseudomonas denitrificans*"	198
Pseudomonas diminuta	202
Pseudomonas fluorescens	192
Pseudomonas gladioli	194
Pseudomonas-like group 2	194
Pseudomonas mallei	196
Pseudomonas mendocina	192
Pseudomonas pertucinogena	212
Pseudomonas pickettii	196
Pseudomonas pseudoalcaligenes	200
Pseudomonas pseudomallei	198
Pseudomonas putida	192
Pseudomonas species CDC Group 1	200
Pseudomonas stutzeri	198
"*Pseudomonas thomasii*"	196
Pseudomonas vesicularis	202
Psychrobacter immobilis: Asaccharolytic strains	166
Psychrobacter immobilis: Saccharolytic strains	172
Riemerella anatipestifer	152
Roseomonas cervicalis	188
Roseomonas fauriae	190
Roseomonas genomospecies 4	188
Roseomonas genomospecies 5	188
Roseomonas genomospecies 6	190
Roseomonas gilardii	188
Shewanella putrefaciens biotype 1	206
Shewanella putrefaciens biotype 2	206
Simonsiella species	150
Sphingomonas paucimobilis/parapaucimobilis	184
Streptobacillus moniliformis	140
Suttonella indologenes	152
Taylorella equigenitalis	158
Weeksella virosa	158
Weeksella zoohelcum	158
Xanthomonas maltophilia	186
Ic	202
IIe	160
IIg	160
IIh	160
IIi	160
IVc-2	174
Vb-3	198

CHART 1	Chromobacterium violaceum		Pasteurella multocida[a]		Pasteurella canis		Pasteurella stomatis		Pasteurella dagmatis	
NUMBER OF STRAINS	37		225		31		8		129	
TEST PERFORMED	SIGN	% +	SIGN	% +	SIGN	% +	SIGN	% +	SIGN	% +
Morphology	mrs		srs		srs		srs		srs	
Motility; flagella	[m;p,L]		[nm]		[nm]		[nm]		[nm]	
Gas from D-glucose	–[b]		[–]	0	–	0			v	25[c]
Action on blood	v	48 β	v	29 al	v	29 al	v	38 al	v	43 al
Fermentative or oxidative	[F]		[F]		[F]		[F]		[F]	
Carbohydrate base	F		F		F		F		F	
Acid from:										
D-Glucose	+	100	+	98 (2)	+	100	+	100	+	100
D-Xylose	–	0	v	81 (4)	–	0	[–]	0	[–]	0
D-Mannitol	[–]	0	+	97 (3)	–	0	–	0	[–]	0
Lactose	[–]	0	–	8 (<1)	[–]	0	–	0	–	3
Sucrose	v	20 (6)	+	96 (3)	[+]	100	[+]	100	+	100
Maltose	[–]	0 (3)	[–]	<1	[–]	0	[–]	0	[+]	100
Catalase	+	97	+	98	+	96	+	100	+	99
Oxidase	v	67	[+]	96	[+]	92	[+]	100	[+]	98
Growth on:										
MacConkey	[+]	100	[–]	1 (1)	[–]	0	[–]	0	[–]	1 (2)
SS	[+ or (+)]	71 (23)	–	0	–	0	–	0	–	0
Simmons citrate	v	68 (9)	–	0	–	0	–	0	–	0
Urea, Christensen's	v	5 (14)	[–]	0	–	0	[–]	0	[+]	95 (5)
Nitrate reduction	+	97	[+]	99	+	96	+	100	[+]	100
Gas from nitrate			–	0	–	4	–	0	–	0
Indole	v[d]	21	[+]	98	+	96	[+]	100	[+]	100

Test										
TSI slant, acid	−	8	+	99	+	96	+	100	+	100
TSI butt, acid	+	94	+	92	+	96	+	100	+	100
H$_2$S (TSI butt)	−	0	−	0	−	0	−	0	−	0
H$_2$S (Pb ac paper)	v	41	v	59	+	92	+	100	+	98 (1)
Gelatin hydrolysis[e]	v	86	−	0	−	0	−	0	v	19
Litmus milk	pep	97	−	6 A	v	14 IR	v	38 IR	−	4 IR
Growth at:										
25°C	+	100	v	82	v	62	v	38	v	87
35°C	+	100	+	100	+	100	+	100	+	97
42°C	v	85	v	29	−	0	v	25	v	29
Esculin hydrolysis	−	5[f]	−	0	−	0	−	0	−	0
Lysine decarboxylase	[−]	0	−	3	−	7	−	0	−	0
Arginine dihydrolase	[+]	100	−	3	−	0	−	0	−	0
Ornithine decarboxylase	[−]	0	[+]	93	[+]	100	[−]	0	−	0
Violet insoluble pigment	[+]	91	−	0	−	0	−	0	−	0
Nutrient broth, 0% NaCl	[+]	100	v	79 (1)	v	25	v	(25)	v	17 (4)
Nutrient broth, 6% NaCl	−	8	v	15	v	25	v	13	v	5

[a] Three subspecies have been described: *P. multocida* subsp. *multocida* (sorbitol-positive, dulcitol-negative); *P. multocida* subsp. *septica* (sorbitol-negative, dulcitol-negative); and *P. multocida* subsp. *gallicida* (sorbitol-positive, dulcitol-positive).
[b] Gas-producing strains have been described.
[c] Gas was observed directly in Durham tubes for 21 strains, and was observed by testing a thioglycolate culture with a hot needle for 21 strains.
[d] Nonpigmented colonies may be indole-positive.
[e] Incubation of 7–14 days.
[f] The two esculin-positive strains were not pigmented.

CHART 2

TEST PERFORMED	Bisgaard's taxon 16 SIGN	Bisgaard's taxon 16 % +	Pasteurella pneumotropica SIGN	Pasteurella pneumotropica % +	Pasteurella bettyae SIGN	Pasteurella bettyae % +	Pasteurella aerogenes SIGN	Pasteurella aerogenes % +
NUMBER OF STRAINS	30		107		88		16	
Morphology	srs		srs		tsrs		cc,srs	
Motility; flagella	[nm]		[nm]		[nm]		[nm]	
Gas from D-glucose	-	3	[-]	0	[+][a]	92	[+]	100
Action on blood	v	32 al	v	32 al	v	24 al	v	52 al
Fermentative or oxidative	[F]		[F]		[F]		[F]	
Carbohydrate base	F		F		F		F	
Acid from:								
D-Glucose	+	100	+	97 (3)	[+]	100	+	100
D-Xylose	[-]	0	[+ or (+)]	76 (19)	-	0	v	81
D-Mannitol	[-]	0	[-]	2 (1)	[-]	0	-	6
Lactose	-	0	v	14 (39)	-	0	v	19 (38)
Sucrose	+	100	+	97 (3)	[-]	0	+	94
Maltose	[+]	100	+	97 (3)	-	0	+	100
Catalase	+	100	+	100	-	1	+	100
Oxidase	[+]	100	[+]	99	v	61	[+]	100
Growth on:								
MacConkey	v	7 (10)	[v]	36 (17)	v	20 (24)	[+]	100
SS	-	0	-	0	-	0	-	6
Simmons citrate	-	0	-	0	-	0	-	0
Urea, Christensen's	[-]	0	[+][b]	95 (1)	[-]	0	[+]	100
Nitrate reduction	[+]	100	[+]	100	[+]	100	[+]	100
Gas from nitrate	-	0	-	0	-	2	-	0

Indole	[+]	100	[+]	90	[+]	100	[−]	0
TSI slant, acid	+	100	+	100	+	100	+	100
TSI butt, acid	+	100	+	97	+	100	−	0
H$_2$S (TSI butt)	−	0	−	0	−	0	−	0
H$_2$S (Pb ac paper)	+	100	+	100	+	94	+	100
Gelatin hydrolysis^c	−	0	−	0	−	0	−	0
Litmus milk	−	3 (3) A	−	9 A	−	5	v	56A
Growth at:								
25°C	v	80	v	74	v	79	+	94
35°C	+	100	+	98	+	96	+	100
42°C	v	23	−	8	v	13	+	94
Esculin hydrolysis	−	0	−	0	−	0	−	0
Lysine decarboxylase	−	0	v	33	−	0	−	0
Arginine dihydrolase	−	0	−	0	−	0	−	0
Ornithine decarboxylase	−	0	[+]	100	−	0	v	88
Nutrient broth, 0% NaCl	v	71	v	84	v	83	+	100
Nutrient broth, 6% NaCl	v	18	−	2	−	10	−	0

^{a}Volume of gas is frequently small.
^{b}May require a drop of serum on slant or a heavy inoculum.
^{c}Incubation of 7–14 days.

CHART 3

TEST PERFORMED	Pasteurella haemolytica		Pasteurella trehalosi		Pasteurella gallinarum		EF-4a[a]	
NUMBER OF STRAINS	28		11		10		97	
	SIGN	% +	SIGN	% +	SIGN	% +	SIGN	% +
Morphology	srs		srs		srs		cc, srs	
Motility; flagella	[nm]		[nm]		[nm]		[nm]	
Gas from D-glucose	-	0	-		-	0	[-]	0
Action on blood	v	75 β	v	27 β	v	30 al	v	34 ly
Fermentative or oxidative	[F]		[F]		[F]		[F]	
Carbohydrate base	F		F		F		F	
Acid from:								
D-Glucose	+	96 (4)	+	100	+	90 (10)	+	100
D-Xylose	[+]	96 (4)	-	0	v	33	-	0
D-Mannitol	[+]	96 (4)	[+ or (+)]	82 (18)	-	0	-	0
Lactose	v	7 (43)	-	0	-	0	[-]	0
Sucrose	+	96 (4)	+	100	+	100	-	0
Maltose	+	93 (7)	+ or (+)	82 (18)	+	100	[-]	0
Arabinose	v	23	[-]	0				
Galactose	[+]	92	[-]	0				
Trehalose	[-]	0	[+]	100				
Catalase	+	96	v	27	[+]	100	[+]	100
Oxidase	[+]	96	+	100	[+]	90	[+]	100
Growth on:								
MacConkey	v	64 (14)	[+]	100	v	20 (10)	[v]	42 (8)
SS	-	0	-	0	-	0	-	1
Simmons citrate	-	0	-	0	-	0	-	3 (1)

Urea, Christensen's	[-]	0	[-]	0	[-]	0	[-]	0
Nitrate reduction	+	100	+	100	[+]	100	[+]	97[b]
Gas from nitrate	-	0	-	0	-	0	[v]	62
Indole	-	0	[-]	0	[-]	0	[-]	0
TSI slant, acid	+	100	+	100	+	100	-	3
TSI butt, acid	+	100	+	100	+	100	v	73
H$_2$S (TSI butt)	-	0	-	0	-	0	-	0
H$_2$S (Pb ac paper)	+	100	+	100	+	90	+	95 (2)
Gelatin hydrolysis[c]	-	0	v	36	-	0	v	60
Litmus milk	v	19 A	-	9 A	-	0	-	2 (1) k
Growth at:								
25°C	v	77	v	55	v	88	v	89
35°C	+	100	+	100	+	100	+	100
42°C	v	14	v	27	v	67	v	70
Esculin hydrolysis	-	0	v	64	-	0	-	0
Lysine decarboxylase	-	0	-	0	-	0	-	0
Arginine dihydrolase	-	0	-	0	-	0	[v]	77 (2)
Ornithine decarboxylase	-	0	[-]	0	v	25	-	0
Nutrient broth, 0% NaCl	+	94	v	18	v	25	+	94 (2)
Nutrient broth, 6% NaCl	-	6	-	0	-	0	-	6 (2)

[a]Yellow to amber pigment, soluble or insoluble, produced by some strains.
[b]Thirteen strains (13%) completely reduced nitrate and nitrite without gas formation.
[c]Incubation of 7–14 days.

CHART 4

TEST PERFORMED	Actinobacillus lignieresii		Actinobacillus equuli		Actinobacillus suis		Actinobacillus hominis	Actinobacillus ureae	
NUMBER OF STRAINS	30		19		33		1	97	
	SIGN	% +	SIGN	% +	SIGN	% +	SIGN	SIGN	% +
Morphology	srs		srs		srs		srs	plr	
Motility; flagella	[nm]		[nm]		[nm]		[nm]	[nm]	
Gas from D-glucose			-	0	-	0	-		
Action on blood	v	17 al	v	11 gr	[v]	76 β	-	v	16 al
Fermentative or oxidative	[F]		[F]		[F]		[F]	[F]	
Carbohydrate base	F		F		F		F	F	
Acid from:									
D-Glucose	+	96 (4)	+	100	+	94 (6)	+	+	100
D-Xylose	+ or (+)	87 (13)	+	100	+	94 (6)	+	[-]	0
D-Mannitol	+	91 (9)	+	100	v	54	+	[+]	99 (1)
Lactose	v	17 (61)	+	95	+ or (+)	79 (18)	+	-	0
Sucrose	+	96 (4)	+	100	+	94 (6)	+	+	99 (1)
Maltose	+ or (+)	83 (17)	+	95	+	94 (6)	+	+	91 (5)
Trehalose	[-]	0	[+]	100	+	95	[+]		
Melibiose	[-]	0	[+ or (+)]	74 (26)	+ or (+)	82 (13)	[(+)]		
Raffinose	v	25 (46)	+	100	+	100	[+]		
Mannose	+	100	+	100	[+]	91 (9)	[-]		
Catalase	[v]	89	[v]	68 (5)	[v]	85	+	v	63
Oxidase	[+]	100	[+]ª	100	[+]	100	+	[+]	99
Growth on:									
MacConkey	v	67 (8)	v	84 (5)	[+ or (+)]	82 (12)	[+]	[-]	0
SS	-	0	-	0	-	3	-	-	0

Test								
Simmons citrate	–	0	–	0	–	0	–	0
Urea, Christensen's	[+]^b	100	[+]^b	100	[+]^b	97 (3)	+	[+]^b
Nitrate reduction	[+]	100	[+]	100	[+]	100	[+]	[+]
Gas from nitrate	–	0	–	0	–	0	–	–
Nitrite reduction					[+]^c		[+]^c	
Indole	–	0	–	0	[–]	0	–	[–]
TSI slant, acid	+	100	+	100	+	100	+	+
TSI butt, acid	+	100	+	100	+	100	+	+
H₂S (TSI butt)	–	0	–	0	–	0	–	–
H₂S (Pb ac paper)	+	100	+	100	+	100	+	+
Gelatin hydrolysis^d	–	0	v	47	–	6	–	–
Litmus milk	v	29 A	v	83 A	v	82 A	–	v
Growth at:								
25°C	v	82	v	89	+	94	+	v
35°C	+	100	+	100	+	97	+	v
42°C	v	14	v	31	v	31	–	–
Esculin hydrolysis	[–]	0	[–]	0	[+]	100	(+)	–
Lysine decarboxylase	–	0	–	0	–	0	–	–
Arginine dihydrolase	–	0	–	0	–	0	–	–
Ornithine decarboxylase	–	0	–	0	–	0	–	–
Nutrient broth, 0% NaCl	+	100	v	89	v	88	+	–
Nutrient broth, 6% NaCl	–	10	–	5	–	0	–	–

^a Described as oxidase-variable in *Bergey's Manual of Systematic Bacteriology*.
^b May require a drop of serum on slant or a heavy inoculum.
^c Nitrite reduction was observed in 0.01% nitrite but not in 0.1% nitrite.
^d Incubation of 7–14 days.

CHART 5	Gardnerella vaginalis		Cardiobacterium hominis		DF-3		DF-3-like	
NUMBER OF STRAINS	126		65		21		7	
TEST PERFORMED	SIGN	% +	SIGN	% +	SIGN	% +	SIGN	% +
Morphology	srs (Gram variable)		mrs		cc, srs		cc, srs	
Motility; flagella	nm		[nm]		nm		[nm]	
Gas from D-glucose	-	0	-	0	-	0		
Action on blood	[inc β]	93	v	12 gr	[-]	10 al	[β]	100
Fermentative or oxidative	F[a]		[F][a]		[F]		[F]	
Carbohydrate base	F[b]		F[b]		F[c]		F[b]	
Acid from:								
D-Glucose	[+ or (+)]	81 (13)	+ or (+)	80 (20)	[+ or (+)]	86 (14)	[+ or (+)]	43 (57)
D-Xylose	v	36 (9)	[-]	0	[+ or (+)]	86 (14)	[-]	0
D-Mannitol	[-]	0	[+ or (+)]	52 (43)	-	0	-	0
Lactose	-	0 (1)	[-]	0	[+ or (+)]	52 (43)	+ or (+)	57 (43)
Sucrose	v	11 (4)	+ or (+)	63 (35)	+ or (+)	62 (33)	[-]	0
Maltose	[+ or (+)]	87 (13)	+ or (+)	69 (31)	[+ or (+)]	81 (19)	+ or (+)	57 (43)
Catalase	[-]	2	[-]	1	[-]	0	v	29
Oxidase	[-]	6	[+]	100	[-]	0	[-]	0
Growth on:								
MacConkey	[-]	5	[-]	0	[-]	0	[-]	0
SS	-	0	-	0	-	0	-	0
Simmons citrate	-	0	-	0	-	0	-	0
Urea, Christensen's	-	0 (5)	[-]	0	-	0 (4)	-	0
Nitrate reduction	-	0	[-]	0	[-]	0	-	0

Gas from nitrate	–	0	–	0	0
Indole	–	0	[+]	100	71 (14)	[+]	100
TSI slant, acid	v	14	+	92 (1)	95	v	29
TSI butt, acid	v	14	v	81 (3)	100	v	29
H$_2$S (TSI butt)	–	0	–	0	0	–	0
H$_2$S (Pb ac paper)	v	31	+	90	90	v	86
Gelatin hydrolysis[d]	–	0	–	0	[–]	0	v	43
Litmus milk	v	32 (16) A	v	12 (5) Aw	v	76 A	v	71 pep
Growth at:						
25°C	–	5	v	35	v	50	v	14
35°C	+	100	+	100	v	85	v	57
42°C	v	14	v	20	v	20	v	14
Esculin hydrolysis	–	0	–	2	[+]	100	v	57 (14)
Lysine decarboxylase					–	0	–	0
Arginine dihydrolase					–	0	–	0
Ornithine decarboxylase					–	0	–	0
Nutrient broth, 0% NaCl	v	24	v	62	v	38	v	17
Nutrient broth, 6% NaCl	–	0	–	8	–	10	v	17

[a]TSI reaction not always indicative of fermentative activity.
[b]1–2 drops of rabbit serum per 3 ml of medium may be required.
[c]1–2 drops of rabbit serum per 3 ml of medium may be required; reactions may be obtained within 4 h by using the rapid sugar test.
[d]Incubation of 7–14 days.

CHART 6

TEST PERFORMED	Haemophilus aphrophilus		Actinobacillus actinomycetemcomitans		Streptobacillus moniliformis	
NUMBER OF STRAINS	203		120		48	
	SIGN	% +	SIGN	% +	SIGN	% +
Morphology	[rv or plr]		rv		[plr,ch]	
Motility; flagella	nm		[nm]		nm	
Gas from D-glucose	[+]a	98	v	28		
Action on blood	v	41 al	v	44 al	v	12 al
Fermentative or oxidative	[F]		[F]		F	
Carbohydrate base	F		Fb		Fc	
Number of strains	63	30	15	6	4	1	1
Acid from:						
D-Glucose	[+]	98 (2)	[+]	[+]	[+]	[+]	[-]	83 (16)	[+ or (+)]	79 (21)
D-Xylose	-	0	-	+	+	-	+	33 (9)	-	(4)
D-Mannitol	-	0	+	+	-	+	-	66 (16)	-	0
Lactose	[+]	96 (4)	[-]	[-]	[-]	[-]	[-]	0	-	0
Sucrose	+	99 (1)	[-]	[-]	[-]	[-]	[-]	0	-	0
Maltose	[+]	99 (1)	+	+	+	-	-	80 (15)	[+ or (+)]	81 (19)
Catalase	[-]	5	[+]					99	-	2
Oxidase	v	28	v					19	-	3
Growth on:						
MacConkey	v	19	[-]					4 (1)	[-]	0
SS	-	0	-					0	-	0
Simmons citrate	-	0	-					0	-	0
Urea, Christensen's	-	0	[-]					0	-	0
Nitrate reduction	[+]	100	[+]					100	[-]	0

Gas from nitrate	–	0	–	0
Indole	[–]	0	[–]	0
TSI slant, acid	+	100	+	29 (12)
TSI butt, acid	+	100	[+]	26 (7)
H$_2$S (TSI butt)	–	0	–	7
H$_2$S (Pb ac paper)	+	100	+	50
Gelatin hydrolysis[d]	–	0	–	0
Litmus milk	v	30 (7) A	–	8
Growth at:				
25°C	v	55	v	40
35°C	+	99	+	87
42°C	v	26	–	20
Esculin hydrolysis	[–]	0	[–]	47 (9)
Lysine decarboxylase	–	0		
Arginine dihydrolase	–	0		
Ornithine decarboxylase	–	0		
Nutrient broth, 0% NaCl	v	56	–	7
Nutrient broth, 6% NaCl	–	1	–	4

[a]Volume of gas is frequently small.
[b]One or two drops of rabbit serum may need to be added per 3 ml of medium; reactions may be obtained within 4 h using the rapid sugar test.
[c]Usually requires 20% serum or ascitic fluid.
[d]Incubation of 7–14 days.

SECTION 2: KING'S KEY AND "ROUND-TABLE" CHARTS

CHART 7	*Leptotrichia buccalis*		*Capnocytophaga* species (DF-1)[a]		*Capnocytophaga canimorsus*		*Capnocytophaga cynodegmi*	
NUMBER OF STRAINS	9		155		90		13	
TEST PERFORMED	SIGN	% +	SIGN	% +	SIGN	% +	SIGN	% +
Morphology	[ltrt]		[trt]		[m,ltrt]		[ltrt,fc]	
Motility; flagella	nm		v[b]		[gl][c]		[gl][c]	
Gas from D-glucose			[−]	0				
Action on blood	−	0	v	34 al or gr	v	10 gr	v	67 β
Fermentative or oxidative	[F]		[F][d]		[F][d]		[F]	
Carbohydrate base	F[e]		F		F[f]		F[g]	
Acid from:								
D-Glucose	[+ or (+)]	88 (12)	[+]	90 (10)	[v]	48 (32)	[+ or (+)]	69 (31)
D-Xylose	−	0	−	0	−	0	−	0
D-Mannitol	−	0	−	0	[−]	0	[−]	0
Lactose	v	44 (11)	v	75 (11)	[+ or (+)]	56 (39)	[+ or (+)]	69 (31)
Sucrose	[+ or (+)]	44 (56)	[+]	90 (9)	[−]	0	[v]	30 (46)
Maltose	[+ or (+)]	67 (33)	[+ or (+)]	86 (14)	[+ or (+)]	53 (39)	[+ or (+)]	69 (31)
Inulin					[−]	0	[v]	77 (11)
Raffinose					[−]	0	[v]	70 (10)
Catalase	[−]	0	[−]	7	[+]	100	[+]	100
Oxidase	[−]	0	[−]	7	[+]	100	[+]	100
Growth on:								
MacConkey	[−]	0	[−]	0	[−]	0	−	0
SS	−	0	−	0	−	0	−	0
Simmons citrate	−	0	−	0	−	0	−	0
Urea, Christensen's	−	0	−	0	−	0	−	0

Test						
Nitrate reduction	−	0	v	63	−	0
Gas from nitrate	−	0	−	0	−	0
Indole	−	0	−	0	[−]	0
TSI slant, acid	v	11 (33)	v	73	v	11
TSI butt, acid	v	11 (33)	v	55	v	12
H$_2$S (TSI butt)	−	0	−	0	−	0
H$_2$S (Pb ac paper)	v	11	v	50	v	23
Gelatin hydrolysis[h]	−	0	−	0	−	0
Litmus milk	v	44 A	v	49 (7) A	−	1 A w
Growth at:						
25°C	v	17	v	21	v	24
35°C	+	100	+	100	+	95
42°C	v	17	v	16	−	8
Esculin hydrolysis	v	89	[v]	81 (2)	v	24 (31)
Lysine decarboxylase	−	0	−	0	−	0
Arginine dihydrolase	−	0	−	8	[+]	95
Ornithine decarboxylase	−	0	−	0	−	0
Nutrient broth, 0% NaCl	−	0	v	21 (1)	−	4 (1)
Nutrient broth, 6% NaCl	−	0	−	1	−	1
ONPG					[+]	100

Test				
Nitrate reduction	−	0	−	0
Gas from nitrate	−	0	−	0
Indole	−	0	−	0
TSI slant, acid	−	8	−	8
TSI butt, acid	−	8	−	8
H$_2$S (TSI butt)	−	0	−	0
H$_2$S (Pb ac paper)	v	33	−	0
Gelatin hydrolysis[h]				
Litmus milk	−	9 IR		
Growth at:				
25°C	−	10		
35°C	v	70		
42°C	−	0		
Esculin hydrolysis	+ or (+)	77 (23)		
Lysine decarboxylase	−	0		
Arginine dihydrolase	[+]	100		
Ornithine decarboxylase	−	0		
Nutrient broth, 0% NaCl	−	8		
Nutrient broth, 6% NaCl	−	0		
ONPG	[+]	100		

[a]Yellow nonsoluble pigment produced by 61 of 155 strains.
[b]Gliding motility has been reported; we have observed delayed spreading in motility medium and an occasional single polar or lateral "flagellum."
[c]Gliding motility is observed by light microscopy; flagella are not detected.
[d]TSI reaction is not always indicative of fermentative activity.
[e]Carbohydrate reactions may require addition of 1–2 drops of normal rabbit serum per 3 ml of enteric fermentation broth.
[f]One or two drops of rabbit serum per 3 ml of medium may be required; reactions may be obtained within 4 h using a small volume rapid sugar technique.
[g]Acid production is enhanced by heavily inoculating small volumes of fermentation media.
[h]Incubation of 7–14 days.

CHART 8

TEST PERFORMED	Neisseria gonorrhoeae		Neisseria cinerea		Neisseria meningitidis		Neisseria polysaccharea	Neisseria canis
NUMBER OF STRAINS	197		58		375		1	1
	SIGN	% +	SIGN	% +	SIGN	% +	SIGN	SIGN
Morphology	[cd]		[cd]		[cd]		[cd]	[cd]
Motility; flagella	[nm]		[nm]		[nm]		[nm]	[nm]
Action on blood	-	0	-	0	-	2 ly	-	-
Fermentative or oxidative	O		n-o		O		O	O
Carbohydrate base	CTA		CTA		CTA		CTA	CTA
Acid from:								
D-Glucose	[+]a	90 (3)	[-]	0	[+]a	99	[+]	(+)
D-Mannitol	-	0	-	0	-	0	-	-
Lactose	-	0	-	0	[-]	0	[-]	-
Sucrose	-	0	-	0	[-]	0	(+w)	-
Maltose	[-]	0	-	0	[+]	99	[+]	-
Fructose	-	0	-	0	-	0	-	-
ONPG	-	0			[-]	0		
Catalase			+	100			+	[+]
Oxidase	[+]	100	[+]	100	[+]	100	[+]	[+]
Growth on:								
MacConkey	-	0	-	0	-	0		(+)
SS			-	0				-
Simmons citrate			-	0				
Urea, Christensen's			-	0				[-]
Nitrate reduction	[-]	0	[-]	0	-	0	-	[+]
Gas from nitrate			-	0				-

Nitrite reduction (0.1%)	–	0	+	100	v	29	–	
Nitrite reduction (0.01%)	v	78					+	–
Gas from nitrite			v[b]	22			–	
Indole			–	0			–	
TSI slant, acid			–	0			–	
TSI butt, acid			–	0				
H$_2$S (TSI butt)			–	0				
H$_2$S (Pb ac paper)			v	33			+	
Pigment on Loeffler	–	10 sl yel	v	57 lt yel	–	7 sl yel	sl yel	[yel]
Growth on nutrient agar at:								
25°C	–	0	v	74	–	0	(+)	
35°C	[–]	0	[+]	91	[–]	1	[+]	
Amylosucrase	–	0	[–]	0	–	0	[+]	
Growth on Thayer-Martin	[+]	96	[v]	20	[+]	100	[+]	

[a] Occasional negative reaction if caps on CTA tubes are not tightened before incubation.
[b] Only a small volume of gas is produced in 0.01% nitrite; many strains have not grown in 0.1% nitrite.

CHART 9

TEST PERFORMED	Neisseria lactamica 287		Neisseria subflava 153		Neisseria sicca 43		Neisseria mucosa 30		Neisseria flavescens 10	
	SIGN	% +	SIGN	% +	SIGN	% +	SIGN	% +	SIGN	% +
Morphology	[cd]		[cd]		[cd]		[cd]		[cd]	
Motility; flagella	[nm]		[nm]		[nm]		[nm]		[nm]	
Action on blood	-	0	v	56 ly	v	22 ly	v	25 ly	v	33 ly
Fermentative or oxidative	F		F		F		F		n-o	
Carbohydrate base	CTA		CTA		CTA		CTA		CTA	
Acid from:										
D-Glucose	[+]	96 (3)	v	68 (6)	[+ or (+)]	78 (14)	[+ or (+)]	83 (10)	[-]	0
D-Mannitol	-	0	-	0	-	0	-	0	-	0
Lactose	[+]	95 (5)	[-]	0	[-]	0	[-]	0	-	0
Sucrose	-	0	[v]	56 (5)	[+]	92 (8)	[+]	90 (10)	-	0
Maltose	[+]	98 (2)	[+]	99 (1)	[+]	95 (5)	+	90 (10)	-	0
Fructose	-	0	[v]	77 (5)	[+]	100	[+]	90 (10)	-	0
ONPG	[+]	100	-	0	-	0				
Catalase	+	100	v	80	v	80	+	100	+	100
Oxidase	[+]	100	[+]	100	[+]	100	[+]	100	[+]	100
Growth on:										
MacConkey	-	0	v	47 (2)	v	68	v	57 (3)	v	67
SS	-	0	-	0	-	0	-	0	-	0
Simmons citrate	-	0	-	0	-	0	-	0	-	0
Urea, Christensen's	-	0	-	0	-	0	-	0	-	0
Nitrate reduction	[-]	0	[-]	0	[-]	0	[+]	100	[-]	0
Gas from nitrate							[+]	100		

Nitrite reduction	+	100	v	80	+	100	[+]	100	+	100
TSI slant, acid	v	30	v	68	+	100	+	100	-	0
TSI butt, acid	v	19	v	56 (4)	+ or (+)	82 (9)	v	88	-	0
H$_2$S (TSI butt)	-	0	-	0	-	0	-	0	-	0
H$_2$S (Pb ac paper)	v	88	+	100	+	100	+	100	+	100
Pigment on Loeffler	v	38 sl yel	[yel]	99	-	0	v	50 sl yel	[yel]	90
Phenylalanine deaminase	-	0	v	32						
Growth on nutrient agar at:										
25°C	v	22	+	95	+	100	+	96	v	86
35°C	v	74	[+]	95	[+]	90 (10)	[+]	100	[+]	100
Amylosucrase	-	0	v	73	+	100	+	100	[+]	100
Growth on Thayer-Martin	[+]	100	[-]	8	[-]	0	[v]	15 w	[-]	0

CHART 10	Neisseria elongata subsp. elongata		Neisseria elongata subsp. glycolytica		Neisseria elongata subsp. nitroreducens		Neisseria weaveri		EF-4b	
NUMBER OF STRAINS	15		2		26		132		34	
TEST PERFORMED	SIGN	% +	SIGN	% +	SIGN	% +	SIGN	% +	SIGN	% +
Morphology	[cc, srs]		[cc, srs]		[cc, srs]		[mrs]		cc, srs	
Motility; flagella	[nm]		[nm]		[nm]		[nm]		[nm]	
Action on blood	-	0	-	0	-	0	v	39 ly	v	41 ly
Fermentative or oxidative	n-o		[O]		n-o or O		n-o		[O]	
Carbohydrate base	OF		OF		OF		OF		OF	
Acid from:										
D-Glucose	[-]	0	[+ or (+)]	0	v	23	[-]	0	+ or (+)	70 (26)
D-Xylose	-	0	-	0	-	0	[-]	0	-	0
D-Mannitol	-	0	-	0	-	0	-	0	-	0
Lactose	-	0	-	0	-	0	-	0	[-]	0
Sucrose	-	0	-	0	-	0	-	0	-	0
Maltose	-	0	-	0	-	0	-	0	[-]	0
Catalase	[-]	0	[+]	100	[-]	0	[+]	100	[+]	100
Oxidase	[+]	100	[+]	100	[+]	100	[+]	100	+	100
Growth on:										
MacConkey	v	13 (53)	[+ or (+)]	50 (50)	v	19 (35)	v	27 (18)	[v]	65 (6)
SS	-	0	-	0	-	0	-	0	-	0
Cetrimide	-	0	-	0	-	0	-	0	-	0
Simmons citrate	-	0	-	0	-	0	-	0	v	14 (6)
Urea, Christensen's	-	0	-	0	-	0	-	0	[-]	0
Nitrate reduction	[-]	0	[-]	0	[+]	100	[-]	0	[+][b]	97
Gas from nitrate	-	0	-	0	-	0	-	0	[-]	0
Nitrite reduction	[+]	92	v	50	[+]	100	[v][c]	72	[-]	0
Indole	-	0	-	0	-	0	-	0	[-]	0
TSI slant, acid	-	0	-	0	-	0	-	0	-	0

TSI butt, acid	−	0	−	0	−	0	−	6
H$_2$S (TSI butt)	−	0	−	0	−	0	−	0
H$_2$S (Pb ac paper)	v	67	+	100	+	86	v	88
Gelatin hydrolysis[d]	−	0	−	0	−	0	−	9
Litmus milk	−	0	−	0	−	6 k	−	6k
Pigment	v	13 yel sol	v	50 yel sol	v	24 yel-ta sol		
Growth at:								
25°C	v	67	+	100	+	94	v	88
35°C	+	100	+	100	+	100	+	100
42°C	v	27	−	0	v	41	v	69
Esculin hydrolysis	−	0	−	0	−	0	−	0
Nutrient broth, 0% NaCl	+	100	+	100	v	85	+ or (+)	89 (7)
Nutrient broth, 6% NaCl	−	0	−	0	v	18	−	0
Phenylalanine deaminase			−	0	v[e]	71		
Arginine dihydrolase			−	0	−	0	[−]	0
Alkalinization of:								
Acetamide	−	0	−	0	−	2		
Serine	−	0	−	0	−	0		
Tartrate	−	0	−	0	−	0		
Sodium acetate	v	36 (9)	+	100	−	4		
Penicillin sensitivity					+	97		

[a]Acid production may be detected in rapid sugar test base.
[b]Sixteen strains reduced nitrate and nitrite without gas formation.
[c]Several strains are negative in 0.1% nitrite but positive in 0.01% nitrite at 48 h of incubation.
[d]Incubation of 7–14 days.
[e]Reaction is enhanced when using strains grown on trypticase soy agar with 5% sheep blood.

CHART 11

TEST PERFORMED	Simonsiella crassa Reference Strain		Simonsiella-Like Strains	
NUMBER OF STRAINS	1		3	
	SIGN		SIGN	% +
Morphology	[dskfil][a]		[dskfil]	
Motility; flagella	nm[b]		nm[b]	
Action on blood	lys		v	67 β
Fermentative or oxidative	F[c]		F[c]	
Carbohydrate base	F[c]		F[c]	
Acid from:				
D-Glucose	[+w]		[+]	100
D-Xylose	[−]		[−]	0
D-Mannitol	−		−	0
Lactose	−		−	0
Sucrose	+w		−	0
Maltose	−		+	100
Catalase	−		+	100
Oxidase	+		+	100
Growth on:				
MacConkey	−		−	0
SS	−		−	0
Simmons citrate	−		−	0
Urea, Christensen's	−		−	0
Nitrate reduction	+		v	67
Gas from nitrate	+		v	67
Indole	−		−	0
TSI slant, acid	−		−	0
TSI butt, acid	−		−	0
H₂S (TSI butt)	−		−	0
H₂S (Pb ac paper)	−		+	100
Gelatin hydrolysis[d]	−		−	0

	a	
Litmus milk	–	–
Growth at:		
25°C	–	v
35°C	+	+
42°C	–	–
Esculin hydrolysis	–	–
Nutrient broth, 0% NaCl	+	+
Nutrient broth, 6% NaCl	+	–

	0
	67
	100
	0
	0
	100
	0

[a]Multicellular filaments made up of disk-like cells in a "roll of coins" arrangement.
[b]Gliding motility of multicellular filaments is observed; individual cells do not express flagella.
[c]Acid production from carbohydrates was not observed in OF or TSI agar media.
[d]Incubation of 7–14 days.

SECTION 2: KING'S KEY AND "ROUND-TABLE" CHARTS

CHART 12	Riemerella anatipestifer		Kingella kingae		Kingella denitrificans		Suttonella indologenes	
NUMBER OF STRAINS	5		137		60		6	
TEST PERFORMED	SIGN	% +	SIGN	% +	SIGN	% +	SIGN	% +
Morphology	srs		cc, pr, chs		cc, pr, chs		vbr	
Motility; flagella	[nm]		[nm]		[nm]		[nm]	
Action on blood	v	60 LG	v	81 β	-	0	v	17 al
Fermentative or oxidative	[O]		F[a]		[F][a]		F	
Carbohydrate base	OF		F[b]		F[b]		F	
Acid from:								
D-Glucose	[+ or (+)]	60 (40)	[+ or (+)]	41 (54)	[+ or (+)]	30 (62)	[+]	100
D-Xylose	-	0	-	0	-	0	-	0
D-Mannitol	-	0	-	0	-	0	-	0
Lactose	[-]	0	-	0	-	0	-	0
Sucrose	-	0	[-]	0	-	0	[+]	100
Maltose	[+ or (+)]	40 (60)	[+ or (+)]	71 (29)	[-]	0	[+ or (+)]	17 (83)
Catalase	+	100	[-]	0	[-]	10	v	17
Oxidase	+	100	[+]	100	[+]	100	[+]	100
Growth on:								
MacConkey	-	0	-	9 (1)	[-]	0	v	17
SS	-	0	-	0	-	0	-	0
Simmons citrate	-	0	-	0	-	0	-	0
Urea, Christensen's	v	40 (20)	-	0	-	0	-	0
Nitrate reduction	[-]	0	-	3 (1)	[+]	93	-	0
Gas from nitrate	-	0	-	0	[v]	88	-	0
Nitrite reduction			v[c]	50				
Indole	[-]	0	-	0	-	0	[+]	100
TSI slant, acid	-	0	v	21	-	2	+	100
TSI butt, acid	-	0	v	11	-	0	v	33 (50)
H$_2$S (TSI butt)	-	0	-	0	-	0	-	0
H$_2$S (Pb ac paper)	+	100	v	60	v	73	+	100

Gelatin hydrolysis[d]	+	100	-	1	-	0	-	0
Litmus milk	v	60 pep	v	82 pep	-	6 IR	-	0
Pigment:								
Soluble	v	80 ta-yel	-	2 br-yel	-	2 yel	-	0
Insoluble						2 yel		0
Growth at:								
25°C	+	100	v	34	v	40	v	50
35°C	+	100	+	98	+	100	+	100
42°C	-	0	-	10	v	48	-	0
Esculin hydrolysis	-	0	-	0	-	0	-	0
Lysine decarboxylase			-	0			-	0
Arginine dihydrolase			-	0			-	0
Ornithine decarboxylase			-	0			-	0
Nutrient broth, 0% NaCl			v	60	v	78 (4)	-	0
Nutrient broth, 6% NaCl			-	9	-	0	v	17

[a]TSI reaction is not always indicative of fermentative activity.
[b]1–2 drops of rabbit serum per 3 ml of medium is required.
[c]Some strains reduced 0.01% nitrite but not 0.1% nitrite; some of these strains failed to grow in the higher nitrite concentration.
[d]Incubation of 7–14 days.

CHART 13

	Flavobacterium meningosepticum TYPE STRAIN	Flavobacterium meningosepticum-like strains 148		Balneatrix alpica TYPE STRAIN
NUMBER OF STRAINS				
TEST PERFORMED	SIGN	SIGN	% +	SIGN
Morphology	lrs,II	lrs,II		srs, mrs
Motility; flagella	[nm]	[nm][a]		[m;p]
Action on blood	LG	v	75 LG	-
Fermentative or oxidative	[O]	[O]		[O]
Carbohydrate base	OF	OF		OF
Acid from:				
D-Glucose	[+]	[+]	95 (4)	[+]
D-Xylose	-	-	2 (1)	-
D-Mannitol	[+]	[+]	91 (8)	[(+)]
Lactose	+	v	42 (15)	-
Sucrose	-	-	0	-
Maltose	+	+	93 (7)	+
Starch		-	0	
Trehalose	+	+	100	-
ONPG	+	+	100	
Catalase	+	+	100	+[b]
Oxidase	[+]	[+]	99	+
Growth on:				
MacConkey	(+)	[+ or (+)]	89 (3)	-
SS	-	-	1	-
Simmons citrate	-	v	9 (3)	+
Urea, Christensen's	-	-	3 (5)	-
Nitrate reduction	[-]	[-]	0	[+][c]
Gas from nitrate				-
Nitrite reduction				-
Indole	[+]	[+]	100	[+]
TSI slant, acid	-	-	0	-

Test			
TSI butt, acid	-	-	-
H$_2$S (TSI butt)	-	-	-
H$_2$S (Pb ac paper)	+	+	98	+
Gelatin hydrolysis[d]	+	+	91	-
Litmus milk	[pep]	[pep]	98	k
Pigment		v[e]	(37) ta-br sol	ta-yel sol
Growth at:				
25°C	+	+	100	+
35°C	+	+	100	+
42°C	+	v	45	+
Esculin hydrolysis	[+]	[+]	99	[-]
Lysine decarboxylase				
Arginine dihydrolase				
Ornithine decarboxylase				
Nutrient broth, 0% NaCl	+	+	100	+
Nutrient broth, 6% NaCl	-	-	7	-
Assimilation of:				
Acetamide				-
Serine				+
Tartrate				-
Sodium acetate				-
Phenylalanine deaminase				+

[a]Polar and lateral flagella have been demonstrated on some strains.
[b]Weak reaction when tested on heart infusion agar-grown cells at 24 h of incubation.
[c]Reaction observed in infusion base, no growth occurs in routine nitrate broth.
[d]Incubation of 7–14 days.
[e]Some strains exhibit slight yellow growth pigment.

CHART 14

Flavobacterium species (IIb), including *F. indologenes* and *F. gleum*

CHART 14	*Flavobacterium breve* TYPE STRAIN	*Flavobacterium breve*-like		*Flavobacterium indologenes* TYPE STRAIN	*Flavobacterium gleum* TYPE STRAIN	*Flavobacterium* species (IIb)	
NUMBER OF STRAINS		6				155	
TEST PERFORMED	SIGN	SIGN	% +	SIGN	SIGN	SIGN	% +
Morphology	s-lr	s-lr, fc		mrs, II	mrs, II	mrs, II	
Motility; flagella	[nm]	[nm]		[nm]	[nm]	[nm]	
Action on blood	ly	v	40 ly	-	-	v	26 LG
Fermentative or oxidative	[O]	[O]		[O]	[O]	[O]	
Carbohydrate base	OF	OF		OF	OF	OF	
Acid from:							
D-Glucose	[+]	[+ or (+)]	83 (17)	(+)	(+)	[+]	92 (6)
D-Xylose	-	-	0	-	(+)	v	30 (1)
D-Mannitol	[-]	[-]	0	[-]ª	[-]	[-]	10
Lactose	-	-	0	-	-	-	0
Sucrose	-	-	0	-	-	v	13 (1)
Maltose	+	+ or (+)	83 (17)	(+)	(+)	+	92 (6)
Starch	+	v	75	(+)	-	+	100
Trehalose	[-]	[-]	0	[(+)]	[(+)]	[+]	100
ONPG	[-]	[-]	0	-	-	v	57
Catalase	+	+	100	+	+	+	99
Oxidase	[+]	[+]	100	[+]	[+]	[+]	96
Growth on:							
MacConkey	[+]	[+]	100	+ª	+	v	54 (9)
SS	-	-	0	-	-	-	0
Simmons citrate	-	-	0	+	+ᵇ	-	2 (1)
Urea, Christensen's	-	-	0	-	(+)	v	14 (28)
Nitrate reduction	-	-	0	-ª	+	v	22
Gas from nitrate	-	-		-ª	-	-	0
Nitrite reduction				-	+	v	20
Indole	[+]	[+]	100	[+]	[+]	[+]	98

TSI slant, acid	–	0	–	–	–	1
TSI butt, acid	–	0	–	–	–	10
H$_2$S (TSI butt)	–	0	–	–	–	1
H$_2$S (Pb ac paper)	+	100	+	+	+	99
Gelatin hydrolysis[c]	+	100	+	+	v	78
Litmus milk	IR	50 IR	[pep]	[pep]	[pep]	93
Pigment:						
Insoluble	yel	66 sl yel	[yel]	[yel]	[yel]	99
Soluble	(br)	(50 br)		[br]	v	11 yel-br
Growth at:						
25°C	+	100	+	+	+	100
35°C	+	100	+	+	+	100
42°C	–	0	–	+	v	42
Esculin hydrolysis	[–]	0	+	+	v	70
Lysine decarboxylase	–	0				
Arginine dihydrolase	–	0				
Ornithine decarboxylase	–	0				
Nutrient broth, 0% NaCl	+	100	+	+	+	100
Nutrient broth, 6% NaCl	–	0	–	–	–	0

[a]In the original description of *F. indologenes* [reference 3 of the *Flavobacterium* species (IIb) section], 30% of strains (4/13) oxidized D-mannitol, 46% of strains (6/13) grew on MacConkey agar, and 38% of strains (5/13) reduced nitrate to gas.
[b]The original description of *F. gleum* lists the type strain as citrate negative.
[c]Incubation of 7–14 days.

SECTION 2: KING'S KEY AND "ROUND-TABLE" CHARTS

CHART 15

TEST PERFORMED	Flavobacterium odoratum[a]		Weeksella zoohelcum		Weeksella virosa		Taylorella equigenitalis REFERENCE STRAIN
NUMBER OF STRAINS	74		41		87		
	SIGN	% +	SIGN	% +	SIGN	% +	SIGN
Morphology	plr		s-lr, II		srs, II		srs
Motility; flagella	[nm]		[nm]		[nm]		nm
Action on blood	v	77 LG	v	36 LG	v	76 LG	-
Fermentative or oxidative	[n-o]		[n-o]		[n-o]		[n-o]
Carbohydrate base	OF		OF		OF		OF
Acid from:							
D-Glucose	[-]	0	[-]	0	[-]	0	[-]
D-Xylose	[-]	0	[-]	0	[-]	0	-
D-Mannitol	-	0	-	0	-	0	-
Lactose	-	0	-	0	-	0	-
Sucrose	-	0	-	0	-	0	-
Maltose	-	0	-	0	-	0	-
Catalase	+	100	+	100	+	98	+
Oxidase	[+]	99	[+]	100	[+]	100	+
Growth on:							
MacConkey	+	91 (5)	[-]	2	[-]	0 (10)	(+)[b]
SS	v	30 (11)	-	0	-	0	-
Simmons citrate	-	0	-	0	[-]	0	-
Urea, Christensen's	[+]	100	[+]	100	[-][c]	0	[-]
Nitrate reduction	[-]	0	[-]	0	-	0	-
Gas from nitrate			-	0	-	0	-
Nitrite reduction	v	83	[+]	98	[+]	100	[-]
Indole	[-]	0	-	0	-	0	-
TSI slant, acid	-	0	-	0	-	0	-
TSI butt, acid	-	0	-	0	-	0	-
H$_2$S (TSI butt)	[-]	0	-	0	-	0	-
H$_2$S (Pb ac paper)	v	16	v	59	+	95	-

Gelatin hydrolysis[a]	[+]	96	[+]	98	[+]	-
Litmus milk	pep	93	v	18 pep	v	-
Pigment:						
Insoluble	v	85 yel	-	0	-	-
Soluble	v	28 ta-br	ta-yel	100	br-ta	-
Growth at:						
25°C	+	100	v	30	v	+
35°C	+	100	+	95	+	+
42°C	v	31	-	10	v	-
Esculin hydrolysis	-	0	-	0	-	-
Lysine decarboxylase	-	0	-	0	-	-
Arginine dihydrolase	-	0 (9)	+	100	-	-
Ornithine decarboxylase	-	0	-	0	-	[-]
Nutrient broth, 0% NaCl	+	100	v	15	+	-
Nutrient broth, 6% NaCl	v	20 (5)	-	0	-	+

[a]The type strain reduces nitrite with gas formation and does not grow on MacConkey agar.
[b]Very light growth observed at 7 days of incubation.
[c]In 3–7 days, 27 of 86 strains produced pink reactions that might be interpreted as weakly positive.
[d]Incubation of 7–14 days.

CHART 16	IIe		IIg		IIh		IIi	
NUMBER OF STRAINS	30		12		21		23	
TEST PERFORMED	SIGN	% +	SIGN	% +	SIGN	% +	SIGN	% +
Morphology	srs, II		cc, rv		srs, II		srs, II	
Motility; flagella	[nm]		[nm]		[nm]		[nm]	
Action on blood	v	7 br	v	42 ly	v	32 LG	v	39 LG
Fermentative or oxidative	[O]		n-o		[O]		[O]	
Carbohydrate base	OF		OF		OF		OF	
Acid from:								
D-Glucose	[+ or (+)]	83 (17)	[-]	0	[+ or (+)]	85 (15)	[+]	91 (9)
D-Xylose	-	0	-	0	-	5	+ or (+)	87 (13)
D-Mannitol	[-]	0	-	0	[-]	0	[-]	0
Lactose	[-]	0	-	0	[-]	0	+	91 (9)
Sucrose	-	0	-	0	-	0	[+]	91 (9)
Maltose	[+]	97 (3)	+	92	[+]	95	+	91 (9)
Catalase	+	100	+	100	+	100	[+]	100
Oxidase	+	100	+	100	[+]	100	[+]	100
Growth on:								
MacConkey	[-]	3	[+]	100	[-]	0	[-]	0
SS	-	0	-	0	-	0	-	0
Simmons citrate	-	0	-	8	-	0	-	0
Urea, Christensen's	-	0	-	0	-	0	v	14 (18)
Nitrate reduction	[-]	0	[-]	0	[-]	0	-	0
Indole	[+]	100	[+]	100	[+]	100	[+]	100
TSI slant, acid	-	0	-	0	-	0	-	0
TSI butt, acid	-	0	-	0	-	5	[-]	0
H$_2$S (TSI butt)	-	0	-	0	-	0	-	0
H$_2$S (Pb ac paper)	v	87	v	50	+	100	v	70
Gelatin hydrolysis[a]	[-]	3	[-]	0	[-]	7	[-]	0
Litmus milk	v	14 IR	v	18 k	-	7 IR	-	4 A

Pigment:					
Insoluble	–	7 w-yel	v	17 ta	–	v	22 yel
Soluble	v	62 br-ta-yel	v	17 br-yel	br	v	48 yel-ta
Growth at:						
25°C	+	90	+	100	+	+	100
35°C	+	100	+	100	+	+	100
42°C	–	0	+	90	–	v	36
Esculin hydrolysis	[–]	0	[–]	0	[+]	+	96
Lysine decarboxylase	–	0				–	0
Arginine dihydrolase	–	0				–	0
Ornithine decarboxylase	–	0				–	0
Nutrient broth, 0% NaCl	+	97	+	100	v	86	+	100
Nutrient broth, 6% NaCl	–	3	–	0	–	5	–	9

[a]Incubation of 7–14 days.

CHART 17

TEST PERFORMED	Flavobacterium mizutaii		Flavobacterium multivorum		Flavobacterium thalpophilum		Flavobacterium spiritivorum[a]		Flavobacterium yabuuchiae[a]	
NUMBER OF STRAINS	6		22		10		13		2	
	SIGN	% +	SIGN	% +	SIGN	% +	SIGN	% +	SIGN	% +
Morphology	srs, II		srs		srs, II		srs		srs	
Motility; flagella	[nm]		[nm]		[nm]		[nm]		[nm]	
Action on blood	v	17 ly	v	14 gr	v	10 lav	v	15 LG	–	0
Fermentative or oxidative	[O]		[O]		[O]		[O]		[O]	
Carbohydrate base	OF		OF		OF		OF		OF	
Acid from:										
D-Glucose	[+ or (+)]	67 (33)	[+]	100	+	100	[+]	100	+	100
D-Xylose	[(+)]	0 (100)	+	100	+	100	+	92 (8)	+	100
D-Mannitol	[–]	0	[–]	0	[–]	0	[+]	100	[+]	100
Lactose	+	100	+	100	+	100	+	92 (8)	+	100
Sucrose	+ or (+)	50 (50)	+	100	+	100	+	100	+	100
Maltose	+ or (+)	50 (50)	[+]	100	+	100	+	92 (8)	+	100
Catalase	+	100	+	100	+	100	+	100	+	100
Oxidase	+	100	[+]	100	+	100	[+]	100	+	100
Growth on:										
MacConkey	[–]	0	[+]	100	[+]	100	[v]	0 (46)	v	0 (50)
SS	–	0	–	0	–	0	–	0	–	0
Simmons citrate	–	0	–	0	–	0	–	0	–	0
Urea, Christensen's	–	0	[+]	95	+	90 (10)	[+ or (+)]	62 (38)	v	50
Nitrate reduction	–	0	–	0	[+]	100	–	0	–	0
Gas from nitrate	–	0	–	0	–	0	–	0	–	0
Nitrite reduction	+[b]	100								
Indole	[–][c]	0	[–]	0	[–]	0	[–]	0	[–]	0
TSI slant, acid	v	17	v	55 (5)	[+]	100	–	0	–	0
TSI butt, acid	v	17	v	5 (76)	v	10 (70)	–	0	–	0
H$_2$S (TSI butt)	–	0	–	0	–	0	–	0	–	0
H$_2$S (Pb ac paper)	+	100	+	86 (5)	+	100	v	56	–	0

Gelatin hydrolysis[d]	-[e]	0	-	0	v	15	-	0
Litmus milk	v	50 IR	v	0 (10) Aw	v	0	-	0
Pigment:								
Insoluble	v[f]	33 yel	v	57 sl yel	v	54 pale yel	v	50 sl yel
Soluble	v	33 ta	-	5 ta			v	50 br
Growth at:								
25°C	+	100	+	100	+	100	+	100
35°C	+	100	+	100	+	100	+	100
42°C	-	0	-	0	[+]	9	-	0
Esculin hydrolysis	+	100	[+]	100	+	100	+	100
Lysine decarboxylase	-	0	[-]	0	-	0	-	0
Arginine dihydrolase	v	25	[-]	0	v	25	v	50
Ornithine decarboxylase	-	0	-	0	-	0	-	0
Nutrient broth, 0% NaCl	+	100	+	100	+	100	+	100
Nutrient broth, 6% NaCl	-	0	v	25	-	10	-	0

[a] *F. spiritivorum* can be differentiated from *F. yabuuchiae* with the acid phosphatase test (Int. J. Syst. Bacteriol. 38: 348–353, 1988).
[b] With the type strain, a partial reaction is observed at 2 days in 0.1% nitrite, and a complete reaction is observed in 0.01% nitrite.
[c] Very weak pink color develops in the xylene layer.
[d] Incubation of 7–14 days.
[e] The rapid gelatin test was positive in four of four strains tested.
[f] Yellow growth pigment was observed at room temperature incubation.

CHART 18	Eikenella corrodens[a]		Moraxella (Branhamella) catarrhalis		Moraxella lacunata		Moraxella nonliquefaciens		Moraxella osloensis	
NUMBER OF STRAINS	506		74		66		243		163	
TEST PERFORMED	SIGN	% +	SIGN	% +	SIGN	% +	SIGN	% +	SIGN	% +
Morphology	mrs		[cd]		cc, mr		cc, sbr		cc, sbr	
Motility; flagella	[nm]		[nm]		[nm]		[nm]		[nm]	
Action on blood	v	17 ly	v	32 ly	v	24 al	v	10 ly	v	7 al
Fermentative or oxidative	[n-o]		n-o		n-o		n-o		n-o	
Carbohydrate base	OF		CTA		OF[b]		OF[b]		OF	
Acid from:										
D-Glucose	[-]	0	[-]	0	[-]	0	[-]	0	[-]	0
D-Xylose	[-]	0	-	0	-	0	-	0	-	0
D-Mannitol	-	0	-	0	-	0	-	0	-	0
Lactose	-	0	-	0	-	0	-	0	-	0
Sucrose	-	0	-	0	-	0	-	0	-	0
Maltose	-	0	[-]	0	-	0	-	0	-	0
Catalase	[-]	8	+	100	+	100	+	95	+	95
Oxidase	[+]	100	[+]	100	[+]	100	[+]	100	[+]	100
Growth on:										
MacConkey	[-]	0	-	5	-	2	[-]	8 (2)	v	70
SS	-	0	-	0	-	0	-	0	-	0
Simmons citrate	-	0	-	0	-	0	-	0	-	0
Urea, Christensen's	[-]	0	-	0	[-]	0	[-]	0	[-]	0
Phenylalanine deaminase			v	68	v	17			v	14
Nitrate reduction	[+]	99	[+]	92	[+]	98	[+]	95	v	24
Gas from nitrate	[-]	0	-	0	-	0	-	0	-	0
Nitrite reduction			+	91	-	0	-	0	-	0
Indole	-	0	-	0	-	0	-	0	-	0
TSI slant, acid	-	0	-	0	-	0	-	0	-	0
TSI butt, acid	-	0	-	0	-	0	-	0	-	0
H₂S (TSI butt)	-	0	-	0	-	0	-	0	-	0

H$_2$S (Pb ac paper)	v	73	v	73	v	34	v	83	v	74
Gelatin hydrolysis[c]	-	0	-	0	v	42	-	0	-	0
Litmus milk	-	1 k	-	6 k	v	48 pep	-	2 (1) k	-	9 k
Loeffler slant, digestion					[+]	100	[-]	0		
Pigment:										
Soluble					-	2 br	-	3 yel		6 yel-br
Insoluble	v	(70) sl yel	v	25 sl yel						
Growth at:										
25°C	v	15	v	85	v	33	+	93	+	96
35°C	+	99	+	97	v	73	v	88	+	98
42°C	v	52	v	23	-	0	v	15	v	51
Esculin hydrolysis	-	0	-	0	-	0	-	0	-	0
Lysine decarboxylase	v	82								
Arginine dihydrolase	-	0								
Ornithine decarboxylase	[+]	98								
Nutrient broth, 0% NaCl	v	71	v	47	[-]	5	v	22	[+]	98
Nutrient broth, 6% NaCl	-	1			-	2	-	0	v	12
Utilization of:										
Sodium acetate					[-]	0	[-]	0	[+]	100
Acetamide					-	0				
Serine					-	0				
Tartrate					-	0				
Penicillin sensitivity[d]					[+]	95	+	99	+	92

[a] Candle jar (CO$_2$) required or enhances growth.
[b] Reactions neutral (does not grow or grows poorly) in King's OF medium.
[c] Incubation of 7–14 days.
[d] Based on results obtained by streaking a blood agar plate with growth from an 18–36 h culture and then placing a 10-unit penicillin disc on the streaked area; a positive reaction is indicated by the appearance of a zone of inhibition.

CHART 19

TEST PERFORMED	Moraxella bovis		Moraxella atlantae		Moraxella phenylpyruvica		Moraxella lincolnii		Asaccharolytic Psychrobacter immobilis	
NUMBER OF STRAINS	7		73		50		1		5	
	SIGN	% +	SIGN	% +	SIGN	% +	SIGN	% +	SIGN	% +
Morphology	cc, sbr		sbr, cc		[cc, br]		[cc, ch]		cc	
Motility; flagella	[nm]		[nm]		[nm]		[nm]		[nm]	
Action on blood	[B]	100	–	3 al	v	16 ly	–		–	0
Fermentative or oxidative	[n-o]		[n-o]		[n-o]		[n-o]		[n-o]	
Carbohydrate base	OF		OF		OF		OF		OF	
Acid from:										
D-Glucose	[–]	0	[–]a	0	[–]	0	[–]		[–]	0
D-Xylose	–	0	–	0	–	0	–		–	0
D-Mannitol	–	0	–	0	–	0	–		–	0
Lactose	–	0	–	0	–	0	–		–	0
Sucrose	–	0	–	0	–	0	–		–	0
Maltose	–	0	–	0	–	0	–		–	0
Catalase	v	14	+	95	+	90	+		+	100
Oxidase	[+]	100	[+]	100	[+]	100	[+]		[+]	100
Growth on:										
MacConkey	[–]	0	[+ or (+)]	80 (20)	v	80 (6)	–		v	40
SS	–	0	–	0	–	0	–		–	0
Simmons citrate	–	0	–	0	–	0	–		v	20
Urea, Christensen's	[–]	0	[–]	0	[+]	100	–		v	(20)
Phenylalanine deaminase	–	0	–	0	[+]	97	–		–	
Nitrate reduction	v	14	[–]	0	v	68	[–]		v	40 (20)
Gas from nitrate	–		–	0	–	0	–		–	0
Nitrite reduction			–	3	–	0	–b		–	
Indole	–	0	–	0	–	0	–		[–]	0
TSI slant, acid	–	0	–	0	–	0	–c		–	0
TSI butt, acid	–	0	–	0	–	0	–c		–	0
H$_2$S (TSI butt)	–	0	–	0	–	0	–		–	0

H₂S (Pb ac paper)	+	100	v	61	v	47		v	20 (20)
Gelatin hydrolysis[d]	[+]	100	-	0	-	0	-	-	0
Litmus milk	pep	100	-	3 IR	v	22 (20) k		-	0
Pigment			-	0	-	8 yel-br sol	-	[-]	0
Growth at:									
25°C	+	100	v	51	v	85	+	[+][e]	100
35°C	+	100	+	99	+	100	+	[v]	40
42°C	-	0	v	46	v	29	-	v	20
Esculin hydrolysis			-	0	-	0	-	-	0
Nutrient broth, 0% NaCl			-	0	v	53	[-]	+	100
Nutrient broth, 6% NaCl			-	0	v	19	-	v	60
Alkalinization of:									
Sodium acetate			v		v	43	[-]		
Penicillin sensitivity[f]	+	100	+	100	v	73			
Psychrobacter transformation								[+]	100

[a] Usually does not grow in OF medium.
[b] Nitrite-positive strains have been described.
[c] No growth was detected.
[d] Incubation of 7–14 days.
[e] Usually prefers 25° C.
[f] Based on results obtained by streaking a blood agar plate with growth from an 18- to 36-h culture and then placing a 10-U penicillin disk on the streaked area; a positive reaction is indicated by the appearance of a zone of inhibition.

SECTION 2: KING'S KEY AND "ROUND-TABLE" CHARTS

CHART 20

TEST PERFORMED	Oligella urethralis		Oligella ureolytica		Gilardi rod group 1	
NUMBER OF STRAINS	22		37		15	
	SIGN	% +	SIGN	% +	SIGN	% +
Morphology	[cc]		srs		[sbr]	
Motility; flagella	[nm]		[m;pe]a		nm	
Action on blood	v	18 ly	v	71 ly	v	20 gr & br
Fermentative or oxidative	n-o		[n-o]		n-o	
Carbohydrate base	OF		OF		OF	
Acid from:						
D-Glucose	[-]	0	[-]	0	[-]	0
D-Xylose	[-]	0	[-]	0	-	0
D-Mannitol	-	0	-	0	-	0
Lactose	-	0	-	0	-	0
Sucrose	-	0	-	0	-	0
Maltose	-	0	-	0	-	0
Catalase	+	100	+	100	+	100
Oxidase	[+]	100	[+]	100	+	100
Growth on:						
MacConkey	[+]	96	[v]	62 (27)	[+]	93
SS	-	9	[-]	5	v	80
Simmons citrate	v	46	v	14 (16)	-	0
Urea, Christensen's	[-]	0	[+]	97	[-]	0
Phenylalanine deaminase	[+]	100	[+]	100	[+]	100
Nitrate reduction	[-]	0	[+]	100	[-]	0
Gas from nitrate			[v]	60		
Nitrite reduction	[+]	100	-	0	[-]	0
Indole	-	0	-	0	-	0
TSI slant, acid	-	0	-	0	-	0
TSI butt, acid	-	0	-	0	-	0

H₂S (TSI butt)	−	−	0	0
H₂S (Pb ac paper)	−	9	38	87
Gelatin hydrolysis[b]	−	0	0	0
Litmus milk	v	36 k	65 k	20 k
Pigment:				
Soluble	−	4 amb	3 yel	60 amb
Insoluble				67 amb
Growth at:				
25°C	v	50	67	100
35°C	+	100	88	100
42°C	v	59	18	80
Esculin hydrolysis	−	0	0	0
Nutrient broth, 0% NaCl	+	96	19 (3)	100
Nutrient broth, 6% NaCl	v	59	15 (5)	7 (13)
Penicillin sensitivity[c]	+	100		

[a] Motility may be delayed or difficult to demonstrate.
[b] Incubation of 7–14 days.
[c] Based on results obtained by streaking a blood agar plate with growth from an 18- to 36-h culture and then placing a 10-U penicillin disk on the streaked area; a positive reaction is indicated by the appearance of a zone of inhibition.

CHART 21

TEST PERFORMED	Nonhemolytic/ Asaccharolytic Acinetobacter species 249		β-hemolytic/ Asaccharolytic Acinetobacter species 21		NO-1 22		Bordetella holmesii (NO-2) 13		Bordetella parapertussis 12	
NUMBER OF STRAINS	SIGN	%+	SIGN	%+	SIGN	%+	SIGN	%+	SIGN	%+
Morphology	cc		cc		rs		cc, srs, occas. lr		srs	
Motility; flagella	[nm]		[nm]		[nm]		[nm]		[nm]	
Action on blood	v	22 ly	[β]	100	v	30 ly	v	38 gr	[β]	100
Fermentative or oxidative	[n-o]		[n-o]		[n-o]		[n-o]		[n-o]	
Carbohydrate base	OF		OF		OF		OF		OF	
Acid from:										
d-Glucose	[-]	0	[-]	0	[-]	0	[-]	0	[-]	0
d-Xylose	-	0	-	5	-	0	-	0	-	0
d-Mannitol	-	0	-	0	-	0	-	0	-	0
Lactose	-	0	-	0	-	0	-	0	-	0
Sucrose	-	0	-	0	-	0	-	0	-	0
Maltose	-	0	-	0	-	0	-	0	-	0
Catalase	+	99	+	100	+	100	v	38	+	100
Oxidase	[-]	0.8 w	[-]	0	[-]	5	[-]	0	[-]	0
Growth on:										
MacConkey	v	73 (4)	[+]	95	[v]	5 (15)	[+ or (+)]a	77 (23)	[+]	100
SS	-	7 (2)	v	37	-	0	-	0	-	0
Simmons citrate	v	27 (3)	v	63 (5)	-	0	-	0	v	67 (8)
Urea, Christensen's	v	5 (14)	v	21 (10)	-	5	-	0	[+]	100
Nitrate reduction	-	5	-	10	[+]	100	[-]	0	[-]	0
Gas from nitrate	-	0	-	0	-	0	-	0	-	0
Nitrite reduction	-	0	-	0	-	0	-	0	-	0
Indole	-	0	-	0	-	0	-	0	-	0
TSI slant, acid	-	0	-	0	-	0	-	0	-	0
TSI butt, acid	-	0	-	0	-	0	-	0	-	0
H$_2$S (TSI butt)	-	0.4	-	0	-	0	-	0	[-]	0

H$_2$S (Pb ac paper)	v	v	v	v	v	v	v	v
Gelatin hydrolysis[b]	-	3	-	-	-	0	0	17
Litmus milk	v	19 k	v	v	(20 k)	v	31 k	63 (13) k
Pigment	v	10 br sol	v	50 yel-tan sol	-	0	100	100
Growth at:								
25°C	+	99	+	95	v	20	v	60
35°C	+	98	+	100	+	100	+	100
42°C	v	48	v	63	v	15	v	18
Esculin hydrolysis	-	0	-	0	-	0	-	
Lysine decarboxylase	-	0	-	0	-	0	-	
Arginine dihydrolase	v	14	v	20	-	0	-	
Ornithine decarboxylase	-	0	-	0	-	0	-	
Nutrient broth, 0% NaCl	v	86	+	95	v	10 (5)	v	92
Nutrient broth, 6% NaCl	v	18	v	26	-	0	-	0
Acinetobacter transformation	[+]	100	[+]	100	[-]	0		

[a]Light growth.
[b]Incubation of 7–14 days.

SECTION 2: KING'S KEY AND "ROUND-TABLE" CHARTS

CHART 22

TEST PERFORMED	Nonhemolytic/Saccharolytic Acinetobacter species		β-hemolytic/Saccharolytic Acinetobacter species		EO-2[a]		EO-3		Saccharolytic Psychrobacter immobilis	
NUMBER OF STRAINS	65		12		11		7		7	
	SIGN	% +	SIGN	% +	SIGN	% +	SIGN	% +	SIGN	% +
Morphology	cc		cc		cc		cc		cc	
Motility; flagella	[nm]		[nm]		[nm]		[nm]		[nm]	
Action on blood	v	35 ly	[β]	100	-	9 ly	-	0	v	29 ly
Fermentative or oxidative	[O]		[O]		[O]		[O]		[O]	
Carbohydrate base	OF		OF		OF		OF		OF	
Acid from:										
D-Glucose	[+]	97 (3)	[+ or (+)]	83 (17)	[+]	91 (9)	[+]	100	[+ or (+)]	57 (43)
D-Xylose	-	0	v	88	[+]	91 (9)	+	100	[+ or (+)]	57 (43)
D-Mannitol	-	0	-	0	-	0	[+ or (+)]	57 (43)	-	0
Lactose	v	70 (14)	v	63 (13)	+ or (+)	45 (55)	+ or (+)	71 (29)	[+ or (+)]	57 (43)
Sucrose	-	0	-	0	[-]	0	-	0	-	0
Maltose	v	18 (35)	v	(38)	-	(9)	v	14 (14)	-	0
Catalase	+	100	+	100	v	82	+	100	+	100
Oxidase	[-]	2	[-]	0	[+]	100	[+]	100	[+]	100
Growth on:										
MacConkey	[+]	95	[+]	100	v	64 (18)	+	100	+	100
SS	-	5 (2)	v	38	-	0	-	0	-	0
Simmons citrate	v	62 (3)	v	71	+ or (+)	64 (36)	+ or (+)	29 (71)	-	0
Urea, Christensen's	v	25 (25)	v	(13)	+ or (+)	36 (55)	+ or (+)	14 (86)	v	(43)
Nitrate reduction	-	9	-	0	[+]	100	[-]	0	v	86
Gas from nitrate	-	0	-	0	v	18	-	0	-	0
Nitrite reduction	-	0	-	0						
Indole	-	0	-	0	[-]	0	-	0	-	0
TSI slant, acid	v	11	-	0	-	0	-	0	-	0
TSI butt, acid	-	0	-	0	-	0	-	0	-	0
H$_2$S (TSI butt)	-	0	-	0	-	0	-	0	-	0

Test	1	2	3	4	5	6	7
H₂S (Pb ac paper)	v	v	v	v	+	100	v 43 (14)
Gelatin hydrolysis[b]	–	33 0	v	–	64 (9) 0	– 0	– 0
Litmus milk	v 35 A	v	v 25 A, cl	v	v 82 k	v 43 szf	v 14 k
Pigment:							
Soluble	v 24 br-tan-yel-amb	–	v 20 br	v	v 55 yel		v 43 tan
Insoluble		5 (5)	[+] 100			[yel] 100	
Growth at:							
25°C	+ 100	+ 100	+ 100	v	+ 73	+ 100	[+] 100
35°C	+ 100	+ 100	+ 100	[+] 100	+ 100	+ 100	[v] 57
42°C	v 73	v	v 63	v 36	v 36	v 14	– 0
Esculin hydrolysis	– 0	–	– 0	– 0	– 0	– 0	– 0
Lysine decarboxylase	– 5 (5)	–	– 0				–
Arginine dihydrolase	v 16 (5)	v	– 0				v 29
Ornithine decarboxylase	– 5	–	– 0				– 0
Nutrient broth, 0% NaCl	+ 100	+ 88	v 88	v 64	v 64	v 71	+ or (+) 43 (57)
Nutrient broth, 6% NaCl	v 15	– 0	– 0	v 36	v 36	v 43	+ 100
Acinetobacter transformation	[+] 100	[+] 100	[+] 100	[–] 0	[–] 0	[–] 0	[+] 100
Psychrobacter transformation	100	100	100				100

[a]Colonies frequently mucoid.
[b]Incubation of 7–14 days.
[c]Usually prefers 25° C.

SECTION 2: KING'S KEY AND "ROUND-TABLE" CHARTS

CHART 23

TEST PERFORMED	Bordetella bronchiseptica SIGN	Bordetella bronchiseptica % +	Bordetella avium TYPE STRAIN SIGN	IVc-2 SIGN	IVc-2 % +	Alcaligenes eutrophus REFERENCE STRAIN SIGN
NUMBER OF STRAINS	85			36		
Morphology	mrs		mrs	srs		srs
Motility; flagella	[m;pe]		[m;pe]	[m;pe]		[m;pe]
Action on blood	v	46 ly	β	v	53 ly	-
Fermentative or oxidative	[n-o]		[n-o]	[n-o]		[n-o]
Carbohydrate base	OF		OF	OF		OF
Acid from:						
D-Glucose	[-]	0	-	[-]	0	[-][a]
D-Xylose	[-]	7	-	[-]	0	-
D-Mannitol	-	1	-	-	0	-
Lactose	-	0	-	-	0	-
Sucrose	-	0	-	-	0	-
Maltose	-	0	-	-	0	-
Catalase	+	100	+	+	100	+
Oxidase	[+]	100	+	[+]	100	[+]
Growth on:						
MacConkey	[+]	100	[+]	[+]	94 (6)	[+]
SS	[+]	99	[+]	[-]	3 (6)	[-]
Cetrimide	-	0	-	-	0 (3)	-
Simmons citrate	+	98 (1)	+w	+	100	+
Urea, Christensen's	[+]	99	-	[+]	100	(+)
Nitrate reduction	[+]	92	[-]	[v]	11	[+]
Gas from nitrate	[-]	0	-	[-]	0	-
Nitrite reduction	-		[-]	-		-
Indole	-	0	-	-	0	-
TSI slant, acid	-	0	-	-	0	-
TSI butt, acid	-	0	-	-	0	-
H$_2$S (TSI butt)	[-]	0	-	[-]	0	-

H$_2$S (Pb ac paper)	v	74	-	v	51	[-]
Gelatin hydrolysis[b]	-	0	-	-	0	k
Litmus milk	k	98		k	97	-
Pigment	-	0	w amb sol	v	22 yel-ta sol	
Growth at:						
25°C	+	99	+	+	94	+
35°C	+	100	+	+	100	+
42°C	v	78	+	v	86	-
Esculin hydrolysis	-	0	-	-	0	-
Lysine decarboxylase			-			-
Arginine dihydrolase			-			-
Ornithine decarboxylase						
Nutrient broth, 0% NaCl	+	100	+	+	100	+
Nutrient broth, 6% NaCl	v	82 (5)	-	v	11	-
Alkalinization of tartrate	[-]	0		[+]	100	

[a]Weak acid may be observed at 7–14 days of incubation.
[b]Incubation of 7–14 days.

CHART 24

TEST PERFORMED	Alcaligenes faecalis		Alcaligenes piechaudii		Alcaligenes-like group 1		Alcaligenes xylosoxidans subsp. denitrificans		Alcaligenes xylosoxidans subsp. xylosoxidans	
NUMBER OF STRAINS	49		4		8		4		135	
	SIGN	% +	SIGN	% +	SIGN	% +	SIGN	% +	SIGN	% +
Morphology	mrs		rs		mrs		mrs		mrs	
Motility; flagella	[m;pe]		[m;pe]		[m;pe]		[m;pe]		[m;pe]	
Action on blood	v	48 br & gr	v	75 ly	v	25 ly	v	25 β	v	42 ly
Fermentative or oxidative	[n-o]		[n-o]		[n-o]		[n-o]		[O]	
Carbohydrate base	OF		OF		OF		OF		OF	
Acid from:										
D-Glucose	[-]	0	[-]	0	[-]	0	[-]	0	[v]	78
D-Xylose	[-]	0	-	0	[-]	0	[-]	0	[+]	99
D-Mannitol	-	0	-	0	-	0	-	0	-	0
Lactose	-	0	-	0	-	0	-	0	-	0
Sucrose	-	0	-	0	-	0	-	0	-	0
Maltose	-	0	-	0	-	0	-	0	-	0
Catalase	+	98	+	100	+	100	+	100	+	98
Oxidase	[+]	100	+	100	[+]	100	[+]	100	+	100
Growth on:										
MacConkey	[+]	100	[+]	100	[+]	100	[+]	100	[+]	100
SS	[+]	100	+	100	[v]	13	[+]	100	[+]	98
Cetrimide	v	59	+	100	-	0	v	25 (25)	+	95 (1)
Simmons citrate	+	100	+	100	+	100	+	100	+	95
Urea, Christensen's	[-]	2	-	0	v	75	-	0	[-]	0
Nitrate reduction	[-]	0	[+]	100	[+]	100	[+]	100	[+]	100
Gas from nitrate	-	0	[-]	0	[+]	100	[+]	100	[v]	60
Nitrite reduction	[+]	100	[-]	0	-	0	-	0	-	0
Indole	-	0	-	0	-	0	-	0	-	0
TSI slant, acid	-	0	-	0	-	0	-	0	-	0
TSI butt, acid	-	0	-	0	-	0	-	0	-	0
H₂S (TSI butt)	-	0	−a	0	[-]	0	[-]	0	[-]	0

H₂S (Pb ac paper)[a]	−	8	+	100	v	13	v	25	−	0
Gelatin hydrolysis[b]	v	22	−	0	−	0	−	0	−	0
Litmus milk	k	96	k	100	v	88 k	v	75 k	k	95
Pigment	v	22 yel sol	−	0	−	0	v	25 yel sol	−	5 br sol
Growth at:										
25°C	+	100	+	100	+	100	+	100	+	98
35°C	+	100	+	100	+	100	+	100	+	100
42°C	v	18	v	75	−	58	v	25	v	84
Esculin hydrolysis	−	0	−	0	−	0	−	0	−	0
Lysine decarboxylase	−	0			−	0			−	0
Arginine dihydrolase	−	0			v	(33)			v	13
Ornithine decarboxylase	−	0			−	0			−	0
Nutrient broth, 0% NaCl	+	100	+	100	+	100	+	100	+	100
Nutrient broth, 6% NaCl	+	98 (2)	+[c]	100	v	13	v	25	v	69
Alkalinization of:										
Acetamide			−	0	−	0	v	33 (33)		
Serine			v	50 (25)	v	(75)	+ or (+)	33 (67)		
Tartrate			+	100	[−]	0	[+ or (+)]	67 (33)		

[a] The line of the stab became dark in 3–7 days with three of the four strains.
[b] Incubation of 7–14 days.
[c] Light growth at 48 h, heavier at 7 days.

CHART 25

TEST PERFORMED	Acidovorax delafieldii		Acidovorax facilis		Acidovorax temperans	
NUMBER OF STRAINS	2		2		2	
	SIGN	% +	SIGN	% +	SIGN	% +
Morphology	mrs		mrs		mrs	
Motility; flagella	[m;p,1-2]		[m;p,1-2]		[m;p,1-2]	
Action on blood	-	0	-	0	lav	100
Fermentative or oxidative	[O]		[O]		[O]	
Carbohydrate base	OF		OF		OF	
Acid from:						
D-Glucose	[(+)]	(100)	[+ or (+)]	50 (50)	[+]	100
D-Xylose	[(+)]	(100)	[(+)]	(100)	[-]	0
D-Mannitol	[+ or (+)]	50 (50)	[(+)]	(100)	[v]	50
Lactose	-	0	-	0	-	0
Sucrose	-	0	-	0	-	0
Maltose	-	0	-	0	-	0
Catalase	-ᵃ	0	+	100	+ᵃ	100
Oxidase	[+]	100	[+]	100	[+]	100
Growth on:						
MacConkey	(+)	(100)	[-]	0	[+ or (+)]	50 (50)
SS	-	0	-	0	-	0
Cetrimide	-	0	-	0	-	0
Simmons citrate	[+]	100	[-]	0	[-]	0
Urea, Christensen's	+	100	(+)	(100)	v	50
Nitrate reduction	+	100	+	100	+	100
Gas from nitrate	-	0	-	0	[+]ᵇ	100
Indole	-	0	-	0	-	0
TSI slant, acid	-	0	-	0	-	0
TSI butt, acid	-	0	-	0	-	0
H₂S (TSI butt)	-	0	-	0	-	0
H₂S (Pb ac paper)	+	100	+	100	+	100

	[−]	[0]	[+]	100	[−]	0
Gelatin hydrolysis[c]						
Litmus milk	k	100	k	100	k	100
Pigment	yel sol	50 (50)	−	0	yel-sol	100
Growth at:						
25°C	+	100	+	100	+	100
35°C	+	100	+	100	+	100
42°C	v	50	−	0	+	100
Esculin hydrolysis	−	0	−	0	−	0
Lysine decarboxylase	−	0	−	0	−	0
Arginine dihydrolase	(+)[d]	(100)	+[d]	100	−	0
Ornithine decarboxylase	−	0	−	0	−	0
Nutrient broth, 0% NaCl	+	100	+	100	+	100
Nutrient broth, 6% NaCl	−	0	−	0	−	0

[a]The original description of *A. delafieldii* and *A. temperans* lists these species as catalase-variable (Willems, et al., Int. J. Syst. Bacteriol. *40*: 384–398, 1990).
[b]A small amount.
[c]Incubation of 7–14 days.
[d]The original description of *A. delafieldii* and *A. facilis* lists these species as arginine dihydrolase-negative (Willems, et al., Int. J. Syst. Bacteriol. *40*: 384–398, 1990).

CHART 26

TEST PERFORMED[a]	Afipia felis		Afipia broomeae		Afipia clevelandensis	Afipia genomospecies 1		Afipia genomospecies 2	Afipia genomospecies 3
NUMBER OF STRAINS	16		3		TYPE STRAIN	2		TYPE STRAIN	TYPE STRAIN
	SIGN	% +	SIGN	% +	SIGN	SIGN	% +	SIGN	SIGN
Morphology	srs,plr		srs		srs	srs		srs	srs
Motility; flagella	[m;p,1-2]		[m;p,1-2]		[m;p,1-2]	[m;p,1-2]		[m;p,1-2]	[m;p,1-2]
Action on blood	-	0	-	0	-	-	0	-	-
Fermentative or oxidative	[n-o]		[n-o]		[n-o]	[n-o]		[n-o]	[n-o]
Carbohydrate base	OF		OF		OF	OF		OF	OF
Acid from:									
D-Glucose	[-]	0	[-]	0	[-]	[-]	0	[-]	[-]
D-Xylose	[(+)]	(93)	[(+)]	(100)	[-]	[(+)]	(100)	[(+)]	[(+)]
D-Mannitol	-	0	-	0	-	[(+)]	(100)	-	-
Lactose	-	0	-	0	-	-	0	-	-
Sucrose	-	0	-	0	-	-	0	-	-
Maltose	-	0	-	0	-	-	0	-	-
Catalase	v	37	[+]	100	[-]	v	50	+	-
Oxidase	+	100	+	100	+	+	100	+	+
Growth on:									
MacConkey	v	(44)	-	0	-	-	0	-	-
SS	-	0	-	0	-	-	0	-	-
Cetrimide	-	0	-	0	-	-	0	-	-
Simmons citrate	-	0	-	0	-	+	100	-	-
Urea, Christensen's[b]	[+ or (+)]	88 (12)	[+ or (+)]	33 (67)	[(+)]	[+]	100	[+]	[(+)]
Nitrate reduction	[+]	100	[-]	0	[-]	[-]	0	[-]	[-]
Gas from nitrate	-	0	-	0	-	-	0	-	-
Nitrite reduction	-	0	-	0	-	-	0	-	-
Indole	-	0	-	0	-	-	0	-	-
TSI slant, acid	-	0	-	0	-	-	0	-	-
TSI butt, acid	-	0	-	0	-	-	0	-	-
H₂S (TSI butt)	-	0	-	0	-	-	0	-	-

Test						
H$_2$S (Pb ac paper)	-	-	-	+ or (+)	-	(+)
Gelatin hydrolysis[c]	-	0	0	-	-	-
Litmus milk	k or (k)	50 (44)	(k)	(k)	(k)	(k)
Pigment	v	(13 sl ta sol)	-	-	-	-
Growth at:						
25°C	+	100	+	+	+	+
30°C	+	100	+	+	+	+
35°C	+	100	v	+	+	+
42°C	-	0	-	-	-	-
Esculin hydrolysis	-	0	-	-		
Lysine decarboxylase	-	0	-			
Arginine dihydrolase	-	0	-			
Ornithine decarboxylase	-	0	-			
Phenylalanine deaminase	+	100	-	+	-	+
Nutrient broth, 0% NaCl	+	100	+	+	+	+
Nutrient broth, 6% NaCl	-	0	-	-	-	-
Alkalinization of:						
Acetamide	-	(7)	-	-	-	-
Serine	-	0	-	-	-	-
Tartrate	v	7 (13)	-	(+)	-	-
Sodium acetate	+ or (+)	82 (18)	v	-	-	(+)

[a]Incubations were at 30°C, unless indicated otherwise.
[b]The slant was inoculated heavily.
[c]Incubation of 7–14 days.

CHART 27	"Achromobacter" group B		"Achromobacter" group E		Ochrobactrum anthropi		Agrobacterium radiobacter biovar 1	
NUMBER OF STRAINS	3		2		14		66	
TEST PERFORMED	SIGN	% +	SIGN	% +	SIGN	% +	SIGN	% +
Morphology	mrs		mrs		mrs		spr	
Motility; flagella	[m;pL]	100	[m;pL]		[m;pe]a		[m;pe]a	
Action on blood	LG	100	LG	100	v	43 ly	v	15 ly
Fermentative or oxidative	[O]		[O]		[O]		[O]	
Carbohydrate base	OF		OF		OF		OF	
Acid from:								
D-Glucose	[+]	100	[+]	100	[+]	93 (7)	[+]	94 (6)
D-Xylose	+	100	+	100	+	100	[+]	97 (3)
D-Mannitol	v	(67)	[−]	0	v	43 (14)	+	94 (6)
Lactose	v	(33)	(+)b	(100)	[−]	0	[+ or (+)]	79 (21)
Sucrose	(+)	(100)	(+)	(100)	v	50	+	95 (5)
Maltose	+ or (+)	33 (67)	+	100	v	64	[+]	97 (3)
Rhamnose	+	100	+	100				
Sorbitol	[+ or (+)]	67 (33)	[−]	0				
Catalase	+	100	+	100	+	100	+	98
Oxidase	[+]	100	+	100	[+]	100	[+]	100
Growth on:								
MacConkey	[+]	100	[+]	100	[+]	100	[+]	100
SS	[+]	100	[+]	100	[+]	100	v	20 (5)
Cetrimide	−	0	+ or (+)	50 (50)	−	0	−	0
Simmons citrate	+	100	+	100	v	64	+	97 (3)
Urea, Christensen's	+	100	+	100	[+]	100	[+ or (+)]	88 (9)
Nitrate reduction	[+]	100	[+]c	100	v	86	v	83
Gas from nitrate	[+]	100	[+]	100	v	43	[−]	5
Nitrite reduction	+	100	+d	100			v	38
Indole	−	0	−	0	[−]	0	[−]	0
TSI slant, acid	−	0	−	0	−	0	−	0

TSI butt, acid	−	−	−	−	−
H$_2$S (TSI butt)	−	0	(50)	v	v	11 (3)
H$_2$S (Pb ac paper)	(+)	(100)	100	+	+	100
Gelatin hydrolysis[e]	−	0	0	−	−	2
Litmus milk	k	100	100	k	93	78 szf
Pigment	v	67 yel ta sol	(50 ta sol)	v	21 br yel sol	21 br yel sol
Growth at:						
25°C	+	100	100	+	100	100
35°C	+	100	100	+	100	100
42°C	+	100	100	v	64	32
Esculin hydrolysis	[+]	100	100	v	29 (7)	100
Lysine decarboxylase	−	0	0	−	0	0
Arginine dihydrolase	+	100	100	v	71	8
Ornithine decarboxylase	−	0	0	−	0	0
Nutrient broth, 0% NaCl	+	100	100	+	100	100
Nutrient broth, 6% NaCl	+	100	100	v	60	18
Alkalinization of:						
Acetamide	−	0	0			
Serine	v	(67)	(50)			
Tartrate	−	0	0			
3-Ketolactonate	[−]	0	0	[−]	0	100

[a]Frequently, individual cells have only a single flagellum, either polar, subpolar, or lateral.
[b]Acid production from lactose was detected only after 15 to 21 days of incubation.
[c]When tested at 48 h of incubation, nitrate reduction is only observed in heart infusion broth base (see *Media and Methods*). Residual nitrate (as evidenced by small amounts of red precipitate after the addition of zinc dust) may be observed in tubes that display the reduction of nitrate to gas.
[d]When tested at 48 h of incubation, nitrite reduction may only be observed in media containing ≤0.01% nitrite.
[e]Incubation of 7–14 days.

SECTION 2: KING'S KEY AND "ROUND-TABLE" CHARTS

CHART 28	"Agrobacterium yellow group"		Sphingomonas paucimobilis TYPE STRAIN		Sphingomonas parapaucimobilis		O-1		O-2	
NUMBER OF STRAINS	2				2		62		66	
TEST PERFORMED	SIGN	% +	SIGN		SIGN	% +	SIGN	% +	SIGN	% +
Morphology	lrs		mrs		mrs		rs		srs, II	
Motility; flagella	[m;p,1-2]a		[m;p,1-2]a		[m;p,1-2]a	50 ly	[m;p,1-2]a	8 LG	[v,p,1-2 or p,L]b	20
Action on blood	-	0	ly		v		v		v	14 β
Fermentative or oxidative	[O]		[O]		[O]		[O]		[O or n-o]	
Carbohydrate base	OF		OF		OF		OF		OF	
Acid from:										
D-Glucose	(+)	(100)	+ w		[(+)]	(100)	[+ or (+)]	69 (31)	v	73 (11)
D-Xylose	+ or (+)	50 (50)	+		+	100	-	-	-	2
D-Mannitol	[-]	0	[-]		[-]	0	[-]	0	-	2
Lactose	[(+)]	(100)	[+]		[+ or (+)]	50 (50)	-	0	-	2
Sucrose	(+)	(100)	+		+	100	[-]	0	[+ or (+)]	64 (36)
Maltose	[+]	100	[+]		[+ or (+)]	50 (50)	[-]	0	[+ or (+)]	71 (27)
Glycerol			[-]		[+]	100				
Rhamnose			[-]		[+]	100				
Catalase	+	100	+		+	100	+	98	+	91
Oxidase	+	100	+ w		v	50 w	v	77	+	97
Growth on:										
MacConkey	v	(50)	-		-	0	v	6 (40)	-	9
SS	-	0	-		-	0	-	0	-	0
Cetrimide	-	0	-		-	0	-	0	-	0
Simmons citrate	-	0	-		-	0	-	0	-	0
Urea, Christensen's	(+)	(100)	[-]		-	0	-	2	v	12
Nitrate reduction	-	0	-		-	0	-	0	v	15
Gas from nitrate	-	0	-		-	0	-	0	-	0
Nitrite reduction	-	0	-		-	0	-	0	-	0
Indole	-	0	-		-	0	-	0	-	0
TSI slant, acid	-	0	-		+ w	100 w	-	0	v	18

TSI butt, acid	−	−	−	−	−	20
H$_2$S (TSI butt)	0	−	−	0	−	0
H$_2$S (Pb ac paper)	(+)	[−]	[+]	100	+	91
Gelatin hydrolysis[c]	−	−	−	0	v	38
Litmus milk	v	k/IR	(IR)	100	−	48 pep
Pigment:						
Insoluble	[yel]	[yel]	[yel]	100	[yel-or]	100
Soluble		(wk tan)	tan-br	100	v	18 ta-br
Growth at:						
25°C	+	+	+	100	+	89
35°C	+	+	+	100	+	100
42°C	−	+ w	−	0	v	31
Esculin hydrolysis	50 (50)	+	+	100	v	64
Lysine decarboxylase	−		−		−	0
Arginine dihydrolase	−		−		v	22
Ornithine decarboxylase	−		−		−	6
Nutrient broth, 0% NaCl	+ or (+)	+	+	100	+	92
Nutrient broth, 6% NaCl	−	−	−	50	−	22
3-Ketolactonate	[+]	[−]	[−]	0	v	

[a]Usually appears nonmotile in motility medium, motility may be demonstrated by the wet mount method.
[b]Motility was detected by wet preparation only with two strains, by flagella stain only with seven strains, and in OF medium only in one strain.
[c]Incubation of 7–14 days.

CHART 29

TEST PERFORMED	Xanthomonas maltophilia		Chryseomonas luteola		Flavimonas oryzihabitans	
NUMBER OF STRAINS	228		34		36	
	SIGN	% +	SIGN	% +	SIGN	% +
Morphology	mrs		mrs		srs	
Motility; flagella	[m;p,>2]		[m;p,>2]		[m;p,1-2]	
Action on blood	v	81 LG	v	41 ly	v	43 ly
Fermentative or oxidative	[O]		[O]		[O]	
Carbohydrate base	OF		OF		OF	
Acid from:						
D-Glucose	[+ or (+)]	85 (5)	[+]	100	[+]	100
D-Xylose	v	35 (1)	+	100	+	100
D-Mannitol	[-]	0	[+ or (+)]	76 (18)	[+]	100
Lactose	v	60 (1)	v	3 (24)	v	14 (22)
Sucrose	v	63 (1)	v	12	v	25
Maltose	[+]a	100	+	100	+	97 (3)
Catalase	[+]	100	+	100	+	94
Oxidase	[-]b	0	[-]c	0	[-]c	0
Growth on:						
MacConkey	[+]	100	[+]	100	[+]	100
SS	v	22 (21)	v	68	v	22
Cetrimide	-	2 (7)	-	0	v	25 (28)
Simmons citrate	v	34 (12)	+	100	+	97
Urea, Christensen's	v	3 (12)	v	26 (38)	v	77
Nitrate reduction	v	39	v	62	[-]	6
Gas from nitrate	[-]	0	-	0	-	0
Indole	-	0	-	0	-	0
TSI slant, acid	-	0	-	0	-	0
TSI butt, acid	-	0	-	0	-	0
H_2S (TSI butt)	[-]	0	-	0	-	0
H_2S (Pb ac paper)	+	95	v	12	+	97

Gelatin hydrolysis[d]	[+]	93	v	17
Litmus milk	pep	96	v	57 k
Pigment:				
Soluble	br-ta	98		
Insoluble			[yel]	100
Growth at:				
25°C	+	99	+	100
35°C	+	99	+	100
42°C	v	48	+	33
Esculin hydrolysis	v	39	[+]	0
Lysine decarboxylase	[+]	93	[−]	7
Arginine dihydrolase	−	0	[+]	14
Ornithine decarboxylase	−	0	−	3
Nutrient broth, 0% NaCl	+	100	+	100
Nutrient broth, 6% NaCl	v	22	v	62

[a] Acid reaction in maltose usually stronger than acid reactions in the other carbohydrates listed.
[b] Fourteen of 73 strains that were negative when the reagent was applied directly to growth on a blood agar plate were positive when tested by Kovacs' method; 10 of 14 were recorded as only weakly positive.
[c] A few strains were weakly positive by the Kovacs' method.
[d] Incubation of 7–14 days.

CHART 30

TEST PERFORMED	Methylobacterium species		Roseomonas gilardii		Roseomonas cervicalis		Roseomonas genomospecies 3		Roseomonas genomospecies 5	
NUMBER OF STRAINS	90		21		7		3		3	
	SIGN	% +	SIGN	% +	SIGN	% +	SIGN	% +	SIGN	% +
Morphology	[vps-mr]		cc		cc		cc		cc	
Motility; flagella	[m;p,1]	8 ly	v^a	33	[m;p,1-2]	14 β	v^b	67	[nm]	67
Action on blood	-		-	5 ly	v		v	33 lav	-	
Fermentative or oxidative	[n-o or O]		[n-o or O]		[n-o]		[n-o]		[n-o]	
Carbohydrate base	OF		OF		OF		OF		OF	
Acid from:										
D-Glucose	v	40	v	(43)	[-]	0	[-]	0	[-]	0
D-Xylose	[+]	94	v	19 (57)	v	43	[+]	100	v	67
D-Mannitol	-	2	[v]	14 (38)	[-]	0	[-]	0	[-]	0
Lactose	[-]	0	-	0	-	0	-	0	-	0
Sucrose	-	0	-	0	-	0	-	0	-	0
Maltose	-	2	-	0	-	0	-	0	-	0
Catalase	+	100	+	100	+	100	+	100	+	100
Oxidase	[+]	96	v	52	[+]	100	[+]	100	+	100
Growth on:										
MacConkey	v	15	+ or (+)	43 (52)	+	100	+	100	+ or (+)	67 (33)
SS	-	0	-	0	-	0	-	0	-	0
Cetrimide	-	0	-	0	-	0	-	0	-	0
Simmons citrate	-	2 (3)	[+]	100	+ or (+)	86 (14)	[-]	0	[v]	33
Urea, Christensen's	v	29 (26)	+ or (+)	71 (29)	+ or (+)	86 (14)	+ or (+)	67 (33)	+	100
Nitrate reduction	v	25	-	5	-	0	[+]	100	-	0
Gas from nitrate	-	0	-	0	-	0	-	0	-	0
Indole	[-]	0	-	0	-	0	-	0	-	0
TSI slant, acid	-	0	-	0	-	0	-	0	-	0
TSI butt, acid	-	0	-	0	-	0	-	0	-	0
H$_2$S (TSI butt)	-	0	-	0	-	0	-	0	-	0

H₂S (Pb ac paper)	v	+	+	+	+	+	+	100
Gelatin hydrolysis[c]	-	-	-	-	-	-	-	0
Litmus milk	v	k or (k)	76 (19)	v	86 k	k	v	67 k
Pigment: Insoluble	[pk-co]	[pk]	100	[pk]	100	[pk]	[pk]	100
Growth at:								
25°C	+	+	90	+	100	+	+	100
35°C	+	+	95	+	100	+	+	100
42°C	v	v	67	+	100	+	v	67
Esculin hydrolysis	-	[-]	0	[-]	0	[-]	[-]	0
Nutrient broth, 0% NaCl	+	+	100	+	100	+	+	100
Nutrient broth, 6% NaCl	-	v	24	-	0	v	-	0

[a]Motility was more easily demonstrated in OF medium than in motility medium. Motile strains demonstrated either 1–2 polar flagella, or detached flagella.
[b]Motile strains demonstrated 1–2 polar flagella.
[c]Incubation of 7–14 days.

CHART 31	Roseomonas fauriae		Roseomonas genomospecies 6
NUMBER OF STRAINS	5		REFERENCE STRAIN
TEST PERFORMED	SIGN	% +	SIGN
Morphology	cc, srs		cc, srs
Motility; flagella	[m;p,1-2]		[m;p,1-2]
Action on blood	v	25 ly	-
Fermentative or oxidative	[n-o or O]		[n-o]
Carbohydrate base	OF		OF
Acid from:			
D-Glucose	v	20	[-]
D-Xylose	[+ or (+)]	80 (20)	-
D-Mannitol	-	0	[-]
Lactose	-	0	-
Sucrose	-	0	-
Maltose	-	0	-
Catalase	+	100	+
Oxidase	[+]	100	+
Growth on:			
MacConkey	+ or (+)	60 (40)	(+ w)
SS	v	20	-
Cetrimide	-	0	-
Simmons citrate	v	60 (20)	(+)
Urea, Christensen's	+	100	+
Nitrate reduction	[+]	100	+
Gas from nitrate	v	20	-
Indole	-	0	-
TSI slant, acid	-	0	-
TSI butt, acid	-	0	-
H$_2$S (TSI butt)	-	0	-
H$_2$S (Pb ac paper)	+	100	+

Gelatin hydrolysis[a]	−	−
Litmus milk	k	k
Pigment: Insoluble	[pk]	[pk]
Growth at:		
25°C	+	+
35°C	+	+
42°C	+	+
Esculin hydrolysis	[+]	[+]
Nutrient broth, 0% NaCl	+	+
Nutrient broth, 6% NaCl	v	−

[a]Incubation of 7–14 days.

CHART 32	Pseudomonas aeruginosa		Pseudomonas fluorescens		Pseudomonas putida		OFBA-1		Pseudomonas mendocina	
NUMBER OF STRAINS	201		155		16		6		4	
TEST PERFORMED	SIGN	% +	SIGN	% +	SIGN	% +	SIGN	% +	SIGN	% +
Morphology	mrs		mrs		mrs		mrs		mrs	
Motility; flagella	[m;p,1-2]		[m;p, >2]		[m;p >2]		[m;p,1-2]		[m;p,1-2]	
Action on blood	v	42 LG	v	50 ly	v	38 gr	β	100	ly	100
Fermentative or oxidative	[O]		[O]		[O]		[unknown]		[O]	
Carbohydrate base	OF		OF		OF		OF		OF	
Acid from:										
D-Glucose	[+]	97 (1)	[+]	100	[+]	100	[+]	100	[+]	100
D-Xylose	+	90 (1)	+	100	+	100	+	100	+ or (+)	75 (25)
D-Mannitol	v	70 (3)	v	53 (2)	v	25	+ or (+)	67 (34)	-	0
Lactose	-	0 (<1)	v	24 (3)	v	25 (13)	+ or (+)	67 (34)	[-]	0
Sucrose	[-]	0	[v]	48	[-]	0	+ or (+)	67 (34)	-	0
Maltose	-	0 (<1)	[-]	2	[v]	31	+ or (+)	67 (34)	[-]	0
Blank							[+ or (+)]	67 (34)		
Catalase	+	100	+	99 (1)	+	100	+	100	+	100
Oxidase	[+]	99	[+]	97	[+]	100	[+]	100	[+]	100
Growth on:										
MacConkey	[+]	100	[+]	100	[+]	100	[+]	100	[+]	100
SS	+	96	[+ or (+)]	86 (7)	+	100	+	100	[+]	100
Cetrimide	[+]	94 (2)	v	89	v	81 (6)	+	100	+ or (+)	75 (25)
Simmons citrate	+	95 (1)	+	93 (3)	+	94 (6)	v	33 (33)	+	100
Urea, Christensen's	v	48 (9)	v	21 (31)	v	13 (44)	v	17 (33)	v	50
Nitrate reduction	+	98	v	19	[-]	0	+	100	+	100
Gas from nitrate	[+]	93	[-]	3	-	0	[+]	100	[+]	100
Indole	-	0	[-]	0	-	0	-	0	-	0
TSI slant, acid	-	0	-	0	-	0	+	100	-	0
TSI butt, acid	-	0	-	0	-	0	+ or (+)	33 (67)	-	0
H₂S (TSI butt)	-	0	-	0	-	0	-	0	-	0

Test	1	2	3	4	5
H$_2$S (Pb ac paper)	[-]	v	-	v	[+]
Gelatin hydrolysis[a]	[v] 82	[+] 100	[-] 0	[+] 50	[-] 0
Litmus milk	[v] 89 pep	[pep] 95	v 62 k	v 50 pep	k or (k) 25 (75)
Pigment:					
Pyoverdin	[v] 65	[+] 96	[+] 93	[-] 0	
Pyocyanin	[v] 46	[-] 0	[-] 0	[-] 0	
Pyorubrin	v 25	- 0	- 0	[-] 0	
Other	v 23 pyomelanin	- 4 br sol		[-] 0	sl yel ins 100
Growth at:					
5°C		+ 95 (1)			
25°C	+ 100	+ 100	+ 100	+ 100	+ 100
35°C	+ 100	+ 90	+ 100	+ 100	+ 100
42°C	[+] 100	[-] 0	[-] 0	+ 100	+ 100
Esculin hydrolysis	- 0	- 0	- 0	- 0	- 0
Lysine decarboxylase	[-] 0	[-] 0	[-] 0	- 0	[-] 0
Arginine dihydrolase	[+] 100	[+] 97	[+] 100	+ 100	[+] 100
Ornithine decarboxylase	- 0	- 0	- 0	- 0	- 0
Nutrient broth, 0% NaCl	+ 100	+ 99	+ 100	+ 100	+ 100
Nutrient broth, 6% NaCl	v 65	v 43	+ 100	v 75	+ 100
Acetamide alkalinization	[+] 100	v 6 (12)	[-] 0	- 0	[-] 0
2-Ketogluconate	+ 96				[-] 0

[a] Incubation of 7–14 days.

CHART 33	Pseudomonas cepacia		Pseudomonas gladioli		Pseudomonas-like group 2	
NUMBER OF STRAINS	159		58		11	
TEST PERFORMED	SIGN	% +	SIGN	% +	SIGN	% +
Morphology	mrs		mrs		mrs	
Motility; flagella	[m;p, >2]	47 ly	[m;p, >2][a]	27 gr	[m;p, >2]	20 ly
Action on blood	v		v		v	
Fermentative or oxidative	[O]		[O]		[O]	
Carbohydrate base	OF		OF		OF	
Acid from:						
D-Glucose	[+]	100	[+]	98 (2)	[+]	100
D-Xylose	+	100	+	98 (2)	+	100
D-Mannitol	[+]	100	[+]	91 (9)	[+]	100
Lactose	[+]	99 (1)	v	9 (28)	[+]	100
Sucrose	v	86 (1)	-	0	-	0
Maltose	[+]	99 (1)	[-]	0	[-]	0
Dulcitol			[+ or (+)]	78 (22)	[-]	0
Inositol			[+]	96 (4)	[-]	0
Erythritol			[-]	0	[v]	73
Catalase	+	99	+	98	+	100
Oxidase	v	86	v[b]	47	[+]	100
Growth on:						
MacConkey	[+]	100	[+]	97 (3)	+	100
SS	v	6 (18)	-	3 (1)	-	0
Cetrimide	v	44 (22)	-	3 (1)	-	0
Simmons citrate	+	94 (5)	+	93 (5)	+	100
Urea, Christensen's	v	60 (18)	v	30 (28)	+	91 (9)
Nitrate reduction	v	57	v	43	v	18
Gas from nitrate	[-]	0	-	0	-	0
Indole	-	0	-	0	-	0
TSI slant, acid	-	1	-	0	-	0

Test						
TSI butt, acid	-	0	-	0	-	0
H$_2$S (TSI butt)	-	0	-	0	-	0
H$_2$S (Pb ac paper)	[-]	0	v	31 (5)	v	89
Gelatin hydrolysis[c]	v	20	v	12	-	0
Litmus milk	pep	94	v	71 pep	v	73 k
Pigment	v	60 yel (TSI)[d]	v	41 yel sol	-	0
Growth at:						
25°C	+	98	+	97	v	80
35°C	+	100	+	100	+	100
42°C	v	83	-	9	v	20
Esculin hydrolysis	v	63 (6)	-	0	-	0
Lysine decarboxylase	[v]	80	[-]	0	-	0
Arginine dihydrolase	[-]	0	[-]	2	v	30
Ornithine decarboxylase	v	48	-	0	-	0
Nutrient broth, 0% NaCl	+	100	+	100	+	100
Nutrient broth, 6% NaCl	-	7	-	10	-	0

[a]Frequently difficult to demonstrate more than 1–2 polar flagella.
[b]Most strains studied were only positive with the Kovacs' method. Reactions were frequently weak.
[c]Incubation of 7–14 days.
[d]60% of the *P. cepacia* strains produced a yellow pigment which occurs only on iron-containing media (e.g., TSI). This usually is not evident before 42–48 h. Sometimes a chartreuse pigment occurs on ordinary media.

CHART 34	Pseudomonas pickettii biovar 1		Pseudomonas pickettii biovar 2		"Pseudomonas thomasii"		Pseudomonas mallei	
NUMBER OF STRAINS	70		54		31		8	
TEST PERFORMED	SIGN	% +	SIGN	% +	SIGN	% +	SIGN	% +
Morphology	mrs		mrs		mrs		cc	
Motility; flagella	[m;p,1-2]a		[m;p,1-2]a		[m;p,1-2]		[nm]	
Action on blood	v	56 ly	v	21 ly	v	24 gr	ly	100
Fermentative or oxidative	[O]		[O]		[O]		[O]	
Carbohydrate base	OF		OF		OF		OF	
Acid from:								
D-Glucose	[+]	100	[+]	100	[+]	100	[+]	100
D-Xylose	[+]	100	[+]	100	+	100	v	12 (50)b
D-Mannitol	-	0	-	0	[+]	100	v	62 (14)b
Lactose	[+]	100	[-]	0	[+]	100	v	12 (62)b
Sucrose	-	0	-	0	[-]	0	[-]	0
Maltose	[+]	100	[-]	0	[+]	100	v	0 (75)b
Catalase	+	100	+	100	v	87	+	100
Oxidase	[+]	100	[+]	100	[+]	100	v	25
Growth on:								
MacConkey	[+]	99 (1)	[+]	100	[+]	100	v	88
SS	-	0	-	0	-	0	-	0
Cetrimide	-	1	-	0	-	0 (3)	-	0
Simmons citrate	+	99	+	100	+	100	[-]	0
Urea, Christensen's	[+]	100	[+]	100	+ or (+)	81 (19)	v	12
Nitrate reduction	+	100	+	100	v	13	[+]	100
Gas from nitrate	[v]c	86	[+]c	100	[-]	0	[-]	0
Indole	-	0	-	0	-	0	-	0
TSI slant, acid	-	0	-	0	-	0	-	0
TSI butt, acid	-	0	-	0	-	0	-	0
H2S (TSI butt)	-	0	-	0	-	0	-	0
H2S (Pb ac paper)	v	67	v	38	v	23	+	100

Gelatin hydrolysis[d]	v	12	v	33	v	30	-	0
Litmus milk	v	83 k	k	90	v	52 k	-	0
Pigment	v	36 yel-ta sol			v	33 yel-ta sol	-	0
Growth at:								
25°C	+	100	+	100	+	97	+	100
35°C	+	100	+	100	+	100	+	100
42°C	v	83	+	94	v	84	-	0
Esculin hydrolysis	-	0	-	0	-	0	-	0
Lysine decarboxylase	[-]	0	[-]	0	[-]	0	[-]	0
Arginine dihydrolase	[-]	6	[-]	0	[-]	3	[+]	100
Ornithine decarboxylase	-	0	-	0	-	0	-	0
Nutrient broth, 0% NaCl	+	100	+	100	+	100	+	100
Nutrient broth, 6% NaCl	-	4	-	3	-	0	-	0

[a]Weakly motile, may be difficult to demonstrate.
[b]Some delayed carbohydrate reactions required 7–14 days of incubation.
[c]May be negative if not tested in semi-aerobic nitrate medium.
[d]Incubation of 7–14 days.

CHART 35

TEST PERFORMED	Pseudomonas pseudomallei 70 SIGN	% +	Pseudomonas stutzeri 28 SIGN	% +	"Pseudomonas denitrificans" REFERENCE STRAIN SIGN	28 SIMILAR STRAINS SIGN	% +	Vb-3 65 SIGN	% +
Morphology	mrs		mrs		mrs	mrs		mrs	
Motility; flagella	[m;p, >2]		[m;p,1-2]		[m;p,1-2]	[m;p,1-2]		[m;p,1-2]	
Action on blood	v	40 gr	v	26 ly	ly	v	31 ly	v	32 gr
Fermentative or oxidative	[O]		[O]		O	O		[O]	
Carbohydrate base	OF		OF		OF	OF		OF	
Acid from:									
D-Glucose	[+]	100	[+]	96 (4)	[+]	[+]	100	[+]	100
D-Xylose	+ or (+)	86 (14)	+	93 (7)	(+)	+ or (+)	43 (57)	+	97 (3)
D-Mannitol	+	94 (6)	+ or (+)	89 (4)	-	-	0	v	65 (9)
Lactose	[+]	99 (1)	[-]	0	+	+	96 (4)	[-]	0
Sucrose	v	66 (4)	-	0	-	-	0	-	0
Maltose	[+]	99 (1)	[+]	100	+	+	100	[+]	95 (5)
Catalase	+	100	+	100	+	+	100	+	98
Oxidase	[+]	100	[+]	100	+	+	100	[+]	100
Growth on:									
MacConkey	[+]	100	[+]	100	[+]	[+]	100	[+]	100
SS	v	8 (31)	v	54	[+]	[+ or (+)]	89 (4)	+	100
Cetrimide	-	7 (3)	-	4	+	+	100	v	18 (2)
Simmons citrate	v	77 (4)	+ or (+)	82 (14)	+	+	96	+	92 (6)
Urea, Christensen's	v	13 (8)	v	33 (22)	-	v	25 (18)	v	12 (45)
Nitrate reduction	[+]	100	+	100	[+]	[+]	100	+	100
Gas from nitrate	[+]a	100	[+]	100	[+]	[v]	71	[+]	100
Indole	-	0	-	0	-	-	0	-	0
TSI slant, acid	v	72	-	0	-	-	0	-	0
TSI butt, acid	-	0	-	0	-	-	0	-	0
H$_2$S (TSI butt)	-	0	-	0	-	-	7	-	0
H$_2$S (Pb ac paper)	v	26	v	36	+	+	100	v	5 (14)

Gelatin hydrolysis[b]	v	79	[−]	0	−	−	−	4
Litmus milk	pep	96	v	57 k	k	v	82 k	69 k
Pigment:								
Soluble					tan	v	44 yel-ta	
Insoluble	v	51 white	v	86 sl yel		v	44 yel-ta	95
Growth at:								
25°C	+	100	+	100	+	+	93	98
35°C	+	100	+	100	+	+	100	100
42°C	+	100	v	69	+	v	39	0
Esculin hydrolysis	v	59	−	0	−	−	0	0
Lysine decarboxylase	[−]	0	[−]	0	−	−	0	[−]
Arginine dihydrolase	[+]	100	[−]	0	[+]	[+]	100	[+]
Ornithine decarboxylase	−	0	−	0	−	−	0	−
Nutrient broth, 0% NaCl	+	100	+	96	+	+	100	+
Nutrient broth, 6% NaCl	v	12	+ or (+)	80 (16)	+	v	14	85 (3)
Acetamide alkalinization			−	0	−	−	0	−
2-Ketogluconate			−	0				10

[a]The volume of gas may be small; gas may not be detected unless incubated at 25°C.
[b]Incubation of 7–14 days.

CHART 36

TEST PERFORMED	Pseudomonas alcaligenes SIGN	Pseudomonas alcaligenes % +	Pseudomonas pseudoalcaligenes REFERENCE STRAIN SIGN	Pseudomonas pseudoalcaligenes PHENOTYPICALLY SIMILAR STRAINS SIGN	Pseudomonas pseudoalcaligenes PHENOTYPICALLY SIMILAR STRAINS % +	Pseudomonas species CDC Group 1 SIGN	Pseudomonas species CDC Group 1 % +
NUMBER OF STRAINS	26			34		31	
Morphology	mrs		mrs	mrs		mrc	
Motility; flagella	[m;p,1-2]		[m,1 polar]	[m;p,1-2]		[m;p,1-2]	
Action on blood	v	31 ly	sl LG	v	65 ly	v	33 ly or gr
Fermentative or oxidative	[n-o]		[n-o]	[n-o]		[n-o]	
Carbohydrate base	OF		OF	OF		OF	
Acid from:							
D-Glucose	[−]	0	[−]	[−]	9	[−]	0
D-Xylose	[−]	0	+w	v	18 (12)	[−]	0
D-Mannitol	−	0	−	−	0	−	0
Lactose	−	0	−	−	0	−	0
Sucrose	−	0	−	−	0	−	0
Maltose	[−]	0	[−]	[−]	0	[−]	0
Fructose	[−]	0	[+]	[+ or (+)]	79 (21)	v	81
Catalase	+	92	+w	+	97	[+]	100
Oxidase	[+]	96	[+]	[+]	100	[+]	100
Growth on:							
MacConkey	[+]	96	[+]	[+]	100	[+]	97 (3)
SS	v	38 (8)	−	v	90 (3)	v	30 (6)
Cetrimide	v	15 (4)	+	v	56 (18)	v	13 (6)
Simmons citrate	v	57 (8)	−	v	26 (9)	v	42 (6)
Urea, Christensen's	[−]	0	[−]	[−]	3 (6)	[−]	3 (7)
Nitrate reduction	v	54	+	+	100	+	100
Gas from nitrate	[−]	0	[−]	[−]	0	[+]	100
Indole	−	0	−	−	0	−	0
TSI slant, acid	−	0	−	−	0	−	0
TSI butt, acid	−	0	−	−	0	−	0
H₂S (TSI butt)	−	0	[−]	[−]	0	−	0

H$_2$S (Pb ac paper)	v	65	+	97	v	68
Gelatin hydrolysis[a]	-	0	-	0	-	4
Litmus milk	v	46 k	k	38 k	v	39 k
Pigment	v	32 ta sol	-	18 yel-br sol	v	60 yel-br sol
Growth at:						
25°C	+	100	+	100	+	97
35°C	+	100	+	100	+	97
42°C	-	0	[+ w]	94	v	48
Esculin hydrolysis	-	0	-	0	-	0
Lysine decarboxylase	-	0	-	0	-	0
Arginine dihydrolase	v	12	+	78	v	33
Ornithine decarboxylase	-	0	-	0	-	0
Nutrient broth, 0% NaCl	+	95	+	100	+	93
Nutrient broth, 6% NaCl	v	41	+	62 (6)	v	14

[a]Incubation of 7–14 days.

CHART 37	Pseudomonas diminuta		Pseudomonas vesicularis		Ic	
NUMBER OF STRAINS	68		94		34	
TEST PERFORMED	SIGN	% +	SIGN	% +	SIGN	% +
Morphology	rs		mrs		t,s-lr	
Motility; flagella	[m;p,1-2][b]		[m;p,1-2][b]		[m;p,1-2][c]	
Action on blood	v	69 LG or gr	v	7 LG	v	42 ly
Fermentative or oxidative	[n-o or O]		[O]		[O]	
Carbohydrate base	OF		OF		OF	
Acid from:						
D-Glucose	[v]	21 (9)	[+ or (+)]	87 (12)	[+]	97
D-Xylose	[-]	0	[v]	27 (9)	-	0
D-Mannitol	-	0	-	0	-	0
Lactose	-	0	-	0	-	0
Sucrose	-	0	-	0	-	0
Maltose	[-]	0	[+]	94 (6)	[+]	100
Catalase	+	98	v	83	+	97
Oxidase	[+]	100	[+]	98	[+]	100
Growth on:						
MacConkey	[+]	97 (3)	[v]	43 (23)	[+]	100
SS	-	1 (1)	-	1	+	100
Cetrimide	-	0 (1)	-	0	[+]	94
Simmons citrate	-	1	-	1 (1)	v	41 (15)
Urea, Christensen's	v	0 (13)	-	2 (5)	v	18 (15)
Nitrate reduction	[-]	3	[-]	5	[+]	100
Gas from nitrate	[-]	0	[-]	0	-	0
Indole	[-]	0	-	0	-	0
TSI slant, acid	-	0	-	0	-	0
TSI butt, acid	-	0	-	0	-	0
H$_2$S (TSI butt)	-	0	-	0[a]	-	3
H$_2$S (Pb ac paper)	v	34	v	49 (6)	[+]	100

Gelatin hydrolysis[d]	v	v	v	-	0
Litmus milk	v	35 k	v	17 IR	24 (3) k
Pigment:					
Soluble	[br-ta]	96	v	60 br-ta	32 tan-br
Insoluble			v	52 yel-or	9 w pk
Growth at:					
25°C	+	100	+	97	100
35°C	+	100	+	97	100
42°C	v	38	v	19	97
Esculin hydrolysis	-	5 (1)	[+ or (+)]	88 (6)	0
Lysine decarboxylase	-	0	-	0	0
Arginine dihydrolase	-	0	-	7	100
Ornithine decarboxylase	-	0	-	0	0
Nutrient broth, 0% NaCl	+	100	+	95	100
Nutrient broth, 6% NaCl	v	21 (2)	v	23 (2)	91
Alkalinization of acetamide					12 (6)

[a]Black color was observed around the stab in the TSI butt at 7 days incubation for one strain.
[b]Flagella have a short wavelength with low amplitude.
[c]Medium wavelength.
[d]Incubation of 7–14 days.

CHART 38	Comamonas acidovorans		Comamonas testosteroni		Comamonas terrigena		Comamonas species	
NUMBER OF STRAINS	69		TYPE STRAIN		TYPE STRAIN		28	
TEST PERFORMED	SIGN	% +	SIGN		SIGN		SIGN	% +
Morphology	mrs		[pr, r]		[lr, fc]		[pr, r]	
Motility; flagella	[m;p, >2]		[m;p, >2]		[m;p, >2]		[m;p, >2]	
Action on blood	v	35 ly	-		-		v	32 ly
Fermentative or oxidative	[n-o]		[n-o]		[n-o]		[n-o]	
Carbohydrate base	OF		OF		OF		OF	
Acid from:								
D-Glucose	[-]	0	[-]		[-]		[-]	0
D-Xylose	[-]	0	[-]		[-]		[-]	0
D-Mannitol	[+]	100	[-]		[-]		[-]	0
Lactose	-	0	-		-		-	0
Sucrose	-	0	-		-		-	0
Maltose	-	0	-		-		-	0
Fructose	+	100	[-]		[-]		[-]	6[a]
Catalase	+	100	+		+		+	96
Oxidase	[+]	100	[+]		[+]		[+]	100
Growth on:								
MacConkey	[+]	100	[+]		[(+)]		[+]	96 (4)
SS	v	63 (20)	-		-		v	21 (11)
Cetrimide	-	4 (3)	-		-		-	(4)
Simmons citrate	+	94 (4)	[+]		[-]		[v]	47 (7)
Urea, Christensen's	[-]	(4)	[-]		[+]		[v]	7 (14)
Nitrate reduction	+	99	+		[+]		[+]	96
Gas from nitrate	[-]	0	[-]		[-]		[-]	0
Indole	-	0	-		-		-	0
TSI slant, acid	-	0	-		-		-	0
TSI butt, acid	-	0	-		-		-	0

Test				
H₂S (TSI butt)	−	0	−	−	0
H₂S (Pb ac paper)	v	57	−	+w	v	61
Gelatin hydrolysis[b]	v	11	−	−	−	0
Litmus milk	k	97	k	k	v	86 k
Pigment:						
Fluorescent	v	26	−			
Soluble	v	44 yel-ta	−	tan	v	27 yel-br
Growth at:						
25°C	+	100	+	+	+	100
35°C	+	100	+	+w	+	100
42°C	v	29 (10)	+w	−	v	68
Esculin hydrolysis	−	0	−	−	−	0
Lysine decarboxylase	−	0	−	−	−	0
Arginine dihydrolase	−	0	−	−	−	0
Ornithine decarboxylase	−	0	−	−	−	0
Nutrient broth, 0% NaCl	+	100	+	+	+	100
Nutrient broth, 6% NaCl	−	6	−	−	v	14
Alkalinization of:						
Acetamide	v	80 (8)	[−]	[−]	[−]	6
Serine	−	4	[−]	[−]	[v]	17
Tartrate	v	88	[−]	[−]	[−]	0
Sodium acetate			[+]	[−]		

[a]Fructose reaction weak.
[b]Incubation of 7–14 days.

SECTION 2: KING'S KEY AND "ROUND-TABLE" CHARTS

CHART 39		*Shewanella putrefaciens* biotype 1		*Shewanella putrefaciens* biotype 2	
NUMBER OF STRAINS		24		26	
TEST PERFORMED		SIGN	% +	SIGN	% +
Morphology		mrs		mrs	
Motility; flagella		[m;p,1-2]		[m;p,1-2]	
Action on blood		v	75 LG	LG	92
Fermentative or oxidative		[O or n-o]		[n-o]	
Carbohydrate base		OF		OF	
Acid from:					
D-Glucose		[v]	17 (33)	[-]	0
D-Xylose		[-]	0	[-]	0
D-Mannitol		-	0	-	0
Lactose		-	0	-	0
Sucrose		[+]	96 (4)	-	0
Maltose		+	92 (8)	-	0
Catalase		+	100	+	100
Oxidase		[+]	100	[+]	100
Growth on:					
MacConkey		[+]	100	[+]	100
SS		-	0 (8)	+	92 (4)
Cetrimide		-	4	-	8
Simmons citrate		-	4 (4)	-	8
Urea, Christensen's		v	4 (8)	v	42
Nitrate reduction		[+]	100	[+]	100
Gas from nitrate		[-]	0	[-]	0
Indole		-	0	-	0
TSI slant, acid		-	0	-	0
TSI butt, acid		-	0	-	0
H₂S (TSI butt)		[+]	96	[+]	100
H₂S (Pb ac paper)		+	100	+	100

Gelatin hydrolysis[a]	v	65	+	100
Litmus milk	v	61 pep	v	50 IR & pep
Pigment	v	71 br-ta sol	br-yel sol	100
Growth at:				
25°C	+	100	+	100
35°C	+	96	+	100
42°C	v	38	v	23
Esculin hydrolysis	−	0	−	0
Lysine decarboxylase	−	0	−	0
Arginine dihydrolase	−	0	−	0
Ornithine decarboxylase	[+]	100	[+]	100
Nutrient broth, 0% NaCl	[+]	100	[−]	0
Nutrient broth, 6% NaCl	v	43	[+]	100

[a]Incubation of 7–14 days.

CHART 40

TEST PERFORMED	Brucella abortus		Brucella melitensis		Brucella suis		Brucella canis[a]	
NUMBER OF STRAINS	81		48		190		28	
	SIGN	% +	SIGN	% +	SIGN	% +	SIGN	% +
Morphology	[rv]		[rv]		[rv]		[rv]	
Motility; flagella	[nm]		[nm]		[nm]		[nm]	
Action on blood	v	25 ly	-	0	v	17 al	v	20 sl ly
Fermentative or oxidative	[O]		[O]		[O]		[O]	
Carbohydrate base	OF		OF		OF		OF	
Acid from:								
D-Glucose	[v]	50 (25)	[v]	33	[+]	100	[+ or (+)]	81 (14)
D-Xylose	[+]	100	[+]	100	[+]	100	[+ or (+)]	81 (19)
D-Mannitol	-	0	-	0	-	0	-	0
Lactose	-	0	-	0	-	0	-	0
Sucrose	[-]	0	[-]	0	[-]	0	[-]	0
Maltose	-	0	-	0	-	0	-	0
Catalase	+	100	+	100	+	100	+	100
Oxidase	[+]	96	[+]	100	[+]	95	[v]	72
Growth on:								
MacConkey	v	50 (20)	v	62	v	23 (15)	v	12 (29)
SS	-	0	-	0	-	0	-	0
Simmons citrate	[-]	0	[-]	0	[-]	0	[-]	0
Urea, Christensen's	[+]	98	[+]	100	[+]	100	[+]	100
Nitrate reduction	[+]	100	[+]	100	[+]	100	[+]	100
Gas from nitrate	-	0	-	0	[v]	52	[v]	77
Indole	[-]	0	[-]	0	[-]	0	[-]	0
TSI slant, acid	-	0	-	0	-	0	-	0
TSI butt, acid	-	0	-	0	-	0	-	0
H$_2$S (TSI butt)	-	0	-	0	-	0	-	0
H$_2$S (Pb ac paper)[b]	+ or (+)	89 (4)	v	39	+	95	v	82
Gelatin hydrolysis[c]	-	0	-	0	-	0	-	0

Growth at:					
25°C	v	v	v	v	v
35°C	+	+	+	+	+
42°C	v	v	v	v	v
Esculin hydrolysis	–	0	0	0	0
Nutrient broth, 0% NaCl	+	+	+	+	+
Nutrient broth, 6% NaCl	–	0	0	0	0
Requires CO_2 atmosphere	v	–[d]	0	0	0

[a] A heavy suspension in 0.5% phenolized saline usually forms a "gel" within 30 min.
[b] Generally *Brucella* strains are tested with heart infusion agar.
[c] Incubation of 7–14 days.
[d] Growth of four of 20 strains tested was enhanced in a candle jar.

SECTION 2: KING'S KEY AND "ROUND-TABLE" CHARTS

CHART 41

TEST PERFORMED	Francisella philomiragia		Francisella tularensis biogroup tularensis		Francisella tularensis biogroup palaearctica		Francisella tularensis biogroup novicida	
NUMBER OF STRAINS	16		69		43		3	
	SIGN	% +	SIGN	% +	SIGN	% +	SIGN	% +
Morphology	[rv]		[rv]		[rv]		[rv]	
Motility; flagella	[nm]		[nm]		[nm]		[nm]	
Fermentative or oxidative	[O]							
Carbohydrate base	OF[a]		CA		CA		CA	
Acid from:								
D-Glucose	[+ or (+)]	63 (37)	+	92 (8)	+ or (+)	79 (21)	(+)	(100)
D-Xylose	[-]	0						
D-Mannitol	-	0						
Lactose	-	0						
Sucrose	[+ or (+)]	63 (37)	[-]	0	[-]	0	[+]	100
Maltose	[+ or (+)]	63 (37)	[+ or (+)]	71 (29)	[-]	0	v	66
Glycerol								
Catalase	+	94	[-]	0	[-]	0	[-]	0
Oxidase	[+]	100	[-]	0	[-]	0	[-]	0
Growth on:								
MacConkey	v	19 (13)	-	6	-	0	v	(66)
Simmons citrate	-	0						
Urea, Christensen's	-	0	[-]	0	[-]	0	[-]	0
Nitrate reduction	-	0	[-]	0	[-]	0	[-]	0
Indole	-	0	-	0	-	0	-	0
TSI slant, acid	-	0	-	0	-	0	-	0
TSI butt, acid	-	0	-	0	-	0	-	0
H$_2$S (TSI butt or slant)	[+ or (+)]	56 (44)	-	0	-	0	[-]	0
H$_2$S (Pb ac paper)	+	100						
Gelatin hydrolysis[b]	v	75	-	0	-	0	-	0
Litmus milk	-	0						

Growth at:								
25°C	+	100						
35°C	+	100						
42°C	v	19						
Esculin hydrolysis	-	0						
Lysine decarboxylase	-	0						
Arginine dihydrolase	-	0						
Ornithine decarboxylase	-	0						
Nutrient broth, 0% NaCl	[-]	0	[-]	10	-	0	-	0
Nutrient broth, 6% NaCl	[+ or (+)]	88 (6)	[-]	0	[-]	0	[v]	66

[a] Reactions performed in Difco OF medium.
[b] Incubation of 7–14 days.

CHART 42		Bordetella pertussis[a]		Pseudomonas pertucinogena[b]	
NUMBER OF STRAINS		51		2	
TEST PERFORMED		SIGN	% +	SIGN	% +
Morphology		srs		mrs	
Motility; flagella		[nm]		[m;p,1-2]	
Action on blood		[v]	81β	-	0
Fermentative or oxidative				[O]	
Carbohydrate base				OF	
Acid from:					
	D-Glucose			+ or (+)	50 (50)
	D-Xylose			[(+)]	(100)
	D-Mannitol			-	0
	Lactose			-	0
	Sucrose			-	0
	Maltose			-	0
Catalase		+	100	+	100
Oxidase		[+]	94	+	100
Growth on:					
	MacConkey			[+]	100
	SS			-	0
	Cetrimide			-	0
	Simmons citrate			-	0
Urea, Christensen's		[-][c]	0	-	0
Nitrate reduction				[-]	0
Gas from nitrate				-	0
Indole				-	0
TSI slant, acid				-	0
TSI butt, acid				-	0
H_2S (TSI butt)				-	0
H_2S (Pb ac paper)				+	100

Gelatin hydrolysis[d]		–	0
Litmus milk		–	0
Pigment	0	–	0
Growth at:			
25°C		+	100
35°C		+	100
42°C		+	100
Esculin hydrolysis		–	0
Lysine decarboxylase		–	0
Arginine dihydrolase		–	0
Ornithine decarboxylase		–	0
Nutrient broth, 0% NaCl		+	100
Nutrient broth, 6% NaCl		+	100

[a]Generally does not grow on the usual laboratory media; Bordet-Gengou and blood-charcoal agars are the media of choice.
[b]Colonial morphology mimics that of rough strains of Bordetella pertussis, but motility and flagellar morphology distinguish them from B. pertussis.
[c]Heavy inoculum.
[d]Incubation of 7–14 days.

CHART 43

Bartonella (formerly *Rochalimaea*) species

TEST PERFORMED	*Bartonella bacilliformis*[a]		*Bartonella henselae*		*Bartonella quintana*		*Bartonella vinsonii*	*Bartonella elizabethae*
NUMBER OF STRAINS	2		14		2		1	1
	SIGN	% +	SIGN	% +	SIGN	% +	SIGN	SIGN
Morphology	srs		srs		srs		srs	srs
Motility; flagella	[m;p,1–10]		[-]		[-]		[-]	[-]
Action on blood	-		v	(17β)	-	0	-	inc β
Fermentative or oxidative	[n-o]		[n-o]		[n-o]		[n-o]	[n-o]
Carbohydrate base	RFT		RST		RST		RST	RST
Acid from:								
D-Glucose	-	0	-	0	-	0	-	-
D-Xylose	-	0	-	0	-	0	-	-
D-Mannitol	-	0	-	0	-	0	-	-
Lactose	-	0	-	0	-	0	-	-
Sucrose	-	0	-	0	-	0	-	-
Maltose	-	0	-	0	-	0	-	-
Catalase	[+][b]	100						
Oxidase	-	0	-	0	+ w[c]	0	+ w[c]	-
Growth on:								
MacConkey	-	0						
SS	-	0						
Cetrimide	-	0						
Simmons citrate	-	0						
Growth on HIA with:								
X factor	-	0	v	64	v	50	(+)	+
V factor	[-]	0	[-]	0	[-]	0	[-]	[-]
Rabbit blood	[+]	100	[+]	100	[+]	100	[+]	[+]
Indole	-[d]	0	-	0	-	0	-	-
Growth at:[e]								
25°C	+	100						

35°C	+	100	(+)	(100)	(+)	(100)	(+)	(+)
42°C	-	0	-	0	-	0	-	-
Ornithine decarboxylase	-	0	-	0	-	0	-	-
Urea, Christensen's[f]	-	0	-	0	-	0	-	-

[a]Did not grow on media used for esculin hydrolysis, nitrate reduction, H$_2$S production, and gelatin hydrolysis.
[b]Performed on cells grown on rabbit blood agar.
[c]Kovacs' modification weakly positive, routine method negative.
[d]Spot indole test.
[e]Rabbit blood agar.
[f]Heavy inoculum.

SECTION 2: KING'S KEY AND "ROUND-TABLE" CHARTS

Species and biotypes of *Brucella* associated with human infections[a]

CHART 44 Species	Biotype	CO₂ Required	H₂S Production[b]	Urease[c] <5 min	Urease[c] >5 min	Growth on Dye Media Thionin 1:25,000	Thionin 1:50,000	Thionin 1:100,000
B. melitensis	1	-	-	v	+	-	+	+
	2	-	-	v	+	-	+	+
	3	-	-	v	+	-	+	+
B. abortus	1	v	+	-	+	-	-	-
	2	+	+	-	+[f]	-	-	-
	3	v	+	-	+	+	+	+
	4	v	+	-	+	-	-	-
	5	-	-	-	+	-	+	+
	6	-	v	-	+	-	+	+
	7[g]	v	+	-	+	-	+	+
B. suis	1	-	+	+[h]		+	+	+
	2	-	-	+[h]		-	+	+
	3	-	-	+[h]		+	+	+
	4	-	-	+[h]		+	+	+
	5	-	-	+[h]		+	+	
B. canis		-	-	+[h]		+	+	+

[a]Reactions based on those obtained in the Special Bacteriology Reference Laboratory, CDC, those given in Laboratory Techniques in "Brucellosis," 2nd ed., Geneva: World Health Organization, 1975, and the Manual of Clinical Microbiology, 5th edition (Ref. 6 in the *Brucella abortus* description). {+} = most strains positive; {−} = most strains negative.
[b]Heart infusion agar with lead acetate paper.
[c]Warm urea slant, inoculated heavily.
[d]Agglutination and phage procedures given in "Laboratory Techniques in Brucellosis," 2nd ed., World Health Organization, Geneva, 1975.
[e]Tbilisi, a *Brucella* phage originally isolated in the former USSR, has been designated as the reference phage.
[f]Rare strain urease-negative.
[g]Formerly *B. abortus* biovar 9; biovars 7 and 8 were deleted by the International Committee on Systematic Bacteriology, Subcommittee on Taxonomy of *Brucella* (Ref. 6 in the *Brucella abortus* description).
[h]Uusally immediate or instant reaction.
[i]Stated as "most strains negative" in A. Balows, et al. (Eds.), *"Manual of Clinical Microbiology."* 5th ed., American Society for Microbiology, Washington, 1991.
[j]Some strains studied in Special Bacteriology Reference Laboratory, CDC, grew at this concentration.
[k]*B. canis* forms a stringy mass or "gel" of increased viscosity when suspended in phenolized saline and agglutinates in specific antiserum.

Growth on Dye Media		Agglutination in Mono-specific Serum[d]		Acriflavin	Lysis by Phage Tb[d, e]	
Basic Fuchsin						
1:50,000	1:10,0000	B. abortus	B. melitensis		RTD	RTD x 10^4
+	+	−	+	−	−	−
+	+	+	−	−	−	−
+	+	+	+	−	−	−
+	+	+	−	−	+	+
−	−	+	−	−	+	+
+	+	+	−	−	+	+
{+}	{+}	−	+	−	+	+
+	+	−	+	−	+	+
+	+	+	−	−	+	+
+	+	−	+	−	+	+
{−}	{−}	+	−	−	−	+
−	−	+	−	−	−	+
+	+	+	−	−	−	+
+[i]	+[i]	+	+	−	−	+
−	−	−	+	−	−	+
−[j]	v	−	−	+[k]	−	−

SECTION 2: KING'S KEY AND "ROUND-TABLE" CHARTS

Chart 45. Species and biotypes of X- or V-factor-dependent *Haemophilus* isolated from humans and two animal sources

SPECIES AND BIOTYPES	Number of Strains	Factor requirement X	Factor requirement V	Porphyrin test	Hemolysis	D-Glucose, acid	Lactose, acid	Urease	Indole	Ornithine decarboxylase	Catalase[a]	Oxidase
"*H. felis*"	1	–	+	+	–	+	+	–	–	–	+	–
H. influenzae	154										+	v
biotype I	34	+	+	–	–	+	–	+	+	+		
biotype II	62	+	+	–	–	+	–	+	+	–		
biotype III[b]	36	+	+	–	–	+	–	+	–	–		
biotype IV	14	+	+	–	–	+	–	+	+	+		
biotype V	8	+	+	–	–	+	–	–	+	+		
biotype VI	11	+	+	–	–	+	–	–	–	+		
biotype VII	5	+	+	–	–	+	–	–	+	–		
biotype VIII	6	+	+	–	–	+	–	–	–	–		
H. haemolyticus	18	+	+	–	+	+	–	+	v	–	+	+
H. parahaemolyticus	54	–	+	+	v	+	–	+	–	–	+	v
H. parainfluenzae:	41			+	v						v	v
biotype I	9	–	+	+	–	+	–	–	–	+		
biotype II	14	–	+	+	–	+	–	+	–	+		
biotype III	18	–	+	+	–	+	–	+	–	–		
biotype IV	10	–	+	+	–	+	–	+	+	+		
biotype VI[a]	3	–	+	+	–	+	–	+	+	+		
biotype VII	3	–	+	+	–	+	–	+	+	–		
biotype VIII	2	–	+	+	–	+	–	+	+	–		
H. paraphrophilus	16	–	+	+	–	+	+	–	–	–	–	v[c]
H. segnis[d]	1	–	+	+	–	w	–	–	–	–	v	–
H. ducreyi[e]	31	+	–	–	–	–	–	–	–	–	–	+
H. haemoglobinophilus (*H. canis*)	2	+	–	–	–	+	–	–	+	–		

[a] Catalase reactions and biochemical reactions for *H. parainfluenzae* biotype VI are from Kilian's description ("Manual of Clinical Microbiology, 5th ed., pp. 463-470).

[b] *H. influenzae* biogroup aegyptius (*H. aegyptius*) has the characteristics of *H. influenzae* biotype III but according to Mazloum, et al. (Acta. Pathol. Microbiol. Immunol. Scand. Sect. B *90*: 109–112, 1982), *H. aegyptius* can be differentiated from strains of *H. influenzae* by the following characteristics: no growth on tryptic soy agar in the presence of X and V factors, inhibition by trooleandomycin, distinct bacillary morphology (vs. coccobacillary morphology), and no acid production from D-xylose (some biotype III strains of *H. influenzae* also are negative). However, none of these characters will absolutely distinguish the two species. Distinct outer membrane protein profiles of these organisms have been reported as useful to differentiate them (J. Clin. Microbiol. 22: 708–713, 1985).

[c] Described as oxidase-positive in "*Bergey's Manual of Systematic Bacteriology*."

[d] Based on Kilian's description (J. Gen. Microbiol. 93: 9–62, 1976). Oberhofer's *H. parainfluenzae* biotype IV has these same characteristics (J. Clin. Microbiol. *10*: 168–174, 1979).

[e] Alkaline phosphatase: *H. ducreyi* positive; *H. haemoglobinophilus*-negative.

Differentiation of *Moraxella* (*Branhamella*) *catarrhalis* and *Neisseria* species with coccoid morphology

CHART 46

TEST PERFORMED	*Moraxella catarrhalis*		*Neisseria canis*		*Neisseria cinerea*	
NUMBER OF STRAINS	74		1		58	
	SIGN	% +	SIGN	% +	SIGN	% +
Oxidase	+	100	+		+	100
Acid from carbohydrates in CTA base:						
D-Glucose	-	0	(+)		-	0
Lactose	-	0	-		-	0
Sucrose	-	0	-		-	0
Maltose	-	0	-		-	0
Fructose	-	0			-	0
Growth on:						
Nutrient agar, 0% NaCl, 35°C	+	97	+		+	91
Nutrient agar, 0% NaCl, 25°C	v	86			v	74
Thayer-Martin medium, modified (>10 colonies)	v	17			v	20
Nitrate reduction	+	92	+		-	0
Gas from nitrate	-	0	-		-	0
Nitrite reduction[a]	v	17			v	35
Pigmentation on Loeffler slant	v	25 sl yel			v	lt yel
Amylosucrase	-	0			-	0

CHART 46 (continued)

TEST PERFORMED	Neisseria flavescens		Neisseria gonorrhoeae		Neisseria lactamica		Neisseria meningitidis	
NUMBER OF STRAINS	10		197		287		375	
	SIGN	% +	SIGN	% +	SIGN	% +	SIGN	% +
Oxidase	+	100	+	100	+	100	+	100
Acid from carbohydrates in CTA base:								
D-Glucose	-	0	+	90 (3)	+	96 (3)	+	99
Lactose	-	0	-	0	+[b]	95 (3)	-	0
Sucrose	-	0	-	0	-	0	-	0
Maltose	-	0	-	0	+	98 (2)	+	99
Fructose	-	0	-	0	-	0	-	0
Growth on:								
Nutrient agar, 0% NaCl, 35°C	+	100	-	0	v	74	-	1
Nutrient agar, 0% NaCl, 25°C	v	86[c]	-	0	v	22	-	0
Thayer-Martin medium, modified (>10 colonies)	-	0[a]	+	96	+	100	+	100
Nitrate reduction	-	0	-	0	-	0	-	0
Gas from nitrate	-	0	-	0	-	0	-	0
Nitrite reduction	v	50[c]	-	0	v	67[c]	-	5[c]
Pigmentation on Loeffler slant	yel	90	10 sl yel		38 sl yel		7 sl yel	
Amylosucrase	+	100	-	0	-	0[b]	-	0

[a]Results using 0.1% nitrite. *N. gonorrhoeae* and some strains of other *Neisseria* species that are negative in 0.1% nitrite can reduce 0.01% nitrite.
[b]ONPG +, 100%.
[c]Less than 10 strains tested.

CHART 46 (*continued*)

TEST PERFORMED	Neisseria mucosa		Neisseria polysaccharea		Neisseria sicca		Neisseria subflava	
Number of strains	30		1		43		153	
	SIGN	% +	SIGN	% +	SIGN	% +	SIGN	% +
Oxidase	+	100	+		+	100	+	100
Acid from carbohydrates in CTA base:								
D-Glucose	+ or (+)	83 (10)	+		+ or (+)	78 (14)	v	68 (6)
Lactose	–	0	–		–	0	–	0
Sucrose	+	90 (10)	+ w		+	92 (8)	v	56 (5)
Maltose	+	90 (10)	+		+	95 (5)	+	99 (1)
Fructose	+	90 (10)	–		+	100	v	77 (5)
Growth on:								
Nutrient agar, 0% NaCl, 35°C	+	100	+		+	90 (10)	+	95
Nutrient agar, 0% NaCl, 25°C	+	96	(+)		+	100	+	95
Thayer-Martin medium, modified (>10 colonies)	–	0	+		–	0[b]	–	8
Nitrate reduction	+	100	–		–	0	–	0
Gas from nitrate	+	100	–		–	0	–	0
Nitrite reduction[a]	+	100	+		+	100[c]	v	80[c]
Pigmentation on Loeffler slant	v	sl yel	sl yel		–	0	yel	99
Amylosucrase	+	100[c]			+	100	v	73

[a]Results using 0.1% nitrite. *N. gonorrhoeae* and some strains of other *Neisseria* species that are negative in 0.1% nitrite can reduce 0.01% nitrite.
[b]ONPG +, 100%.
[c]Less than 10 strains tested.

Description of Species

"Achromobacter" group B

Gram-negative rod; motile with polar and lateral flagella; oxidizes glucose, xylose, sucrose, and maltose (these reactions may be delayed); positive reactions for oxidase, catalase, urease, esculin, and citrate; grows on MacConkey and SS agars; reduces nitrate with gas formation; indole-negative.

Source (3 strains)

Blood.

Reference strain

G6751 ← NCTC 12246, from blood.

Literature

1. Holmes B., and C.A. Dawson. 1983. Numerical taxonomic studies on *Achromobacter* isolates from clinical material. *In* H. Leclerc (Ed.), Gram-negative bacteria of medical and public health importance: taxonomy-identification–applications, Proceedings of Symposium, Lille, May 25–27. Les Editions INSERM *114*: 331–341.

2. Holmes B., M. Costas, A.C. Wood, and K. Kersters. 1990. Numerical analysis of electrophoretic protein patterns of *"Achromobacter"* group B, E, and F strains from human blood. J. Appl. Bacteriol. *68*: 495–504.

3. Cieslak T.J., M.L. Robb, C.J. Drabick, and G.W. Fischer. 1992. Catheter-associated sepsis caused by *Ochrobactrum anthropi*: report of a case and review of related nonfermentative bacteria. Clin. Infect. Dis. *14*: 902–907.

Comments

The term *"Achromobacter"* currently represents a recognized genus with no recognized species, hence the use of quotes when describing these organisms. *Achromobacter xylosoxidans*, formerly the only recognized *Achromobacter* species, has been renamed *Alcaligenes xylosoxidans* subsp. *xylosoxidans*. The *"Achromobacter"* group originally included six biovars, A through F. Biovars A, C, and D recently have been redefined as biovars of the new species *Ochrobactrum anthropi*. The remaining *"Achromobacter"* biovars (B, E, and F) usually can be differentiated from *O. anthropi* by their hydrolysis of esculin and flagellar morphology. Esculin-positive *O. anthropi* strains can be differentiated from *"Achromobacter"* by cellular fatty acid analysis. Mannitol (when positive) and dulcitol oxidation (when negative) are useful tests in differentiating *"Achromobacter"* group B from the other *"Achromobacter"* groups. Holmes and Dawson (1) indicate at least 85% of 15 strains of *"Achromobacter"* group B were mannitol-positive. One of three strains tested at CDC was mannitol-negative. Information on the cellular fatty acid composition of this group is presented in Table 4.2 and Figure 4.7 of the cellular fatty acid section.

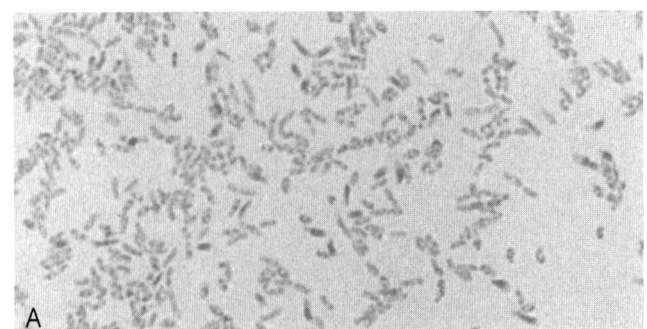

Achromobacter" group B strain G6751.
Gram stain BAP 35° C 24 h x 1700.

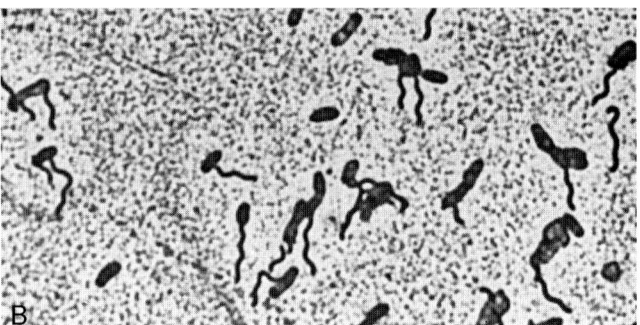

"Achromobacter" group B strain G6751. Flagella stain
(Ryu method) TGY 25° C 24 h x 1700.

"Achromobacter" group B

TEST PERFORMED	SIGN	% +	#+/T
	NUMBER OF STRAINS 3		
Morphology	mrs		
Motility; flagella	[m;p,L]		3/3
Action on blood	LG	100	3/3
Fermentative or oxidative	[O]		
Carbohydrate base	OF		
Acid from:			
D-Glucose	[+]	100	3/3
D-Xylose	+	100	3/3
D-Mannitol	v	(67)	(2)/3
Lactose	v	(33)	(1w)/3
Sucrose	(+)	(100)	(3)/3
Maltose	+ or (+)	33 (67)	1, (2)/3
L-Rhamnose	+	100	3/3
D-Sorbitol	[+ or (+)]	67 (33)	2 (1)/3
Cellobiose	+ or (+)	33 (67)	1, (1), (1w)/3
Catalase	+	100	3/3
Oxidase	[+]	100	3/3
Growth on:			
MacConkey	[+]	100	3/3
SS	[+]	100	3/3
Cetrimide	-	0	0/3
Simmons citrate	+	100	3/3
Urea, Christensen's	+	100	3/3
Nitrate reduction	[+]	100	3/3
Gas from nitrate	[+]	100	3/3
Nitrite reduction[a]	+	100	3/3
Indole	-	0	0/3
TSI slant, acid	-	0	0/3
TSI butt, acid	-	0	0/3
H_2S (TSI butt)	-	0	0/3
H_2S (Pb ac paper)	(+)	(100)	(3)/3
Gelatin hydrolysis[b]	-	0	0/3
Litmus milk	k	100k	3k/3
Pigment	v	67 yel ta sol	2/3
Growth at:			
25°C	+	100	3/3
35°C	+	100	3/3
42°C	+	100	3/3
Esculin hydrolysis	[+]	100	3/3
Lysine decarboxylase	-	0	0/3
Arginine dihydrolase	+	100	3/3
Ornithine decarboxylase	-	0	0/3
Nutrient broth, 0% NaCl	+	100	3/3
Nutrient broth, 6% NaCl	+	100	3/3
Alkalinization of:			
Acetamide	-	0	0/3
Serine	v	(67)	(2)/3
Tartrate	-	0	0/3
3-Ketolactonate	[-]	0	0/1

[a]When tested at 48 h of incubation, nitrite reduction may only be observed in media containing ≤0.01% nitrite.
[b]Incubation of 7–14 days.

"Achromobacter" group E

Gram-negative rod; motile with polar and lateral flagella; oxidizes glucose, xylose, sucrose, maltose, and lactose (the lactose reaction is weak and occurs at 15–21 days of incubation), but not mannitol or sorbitol; positive reactions for oxidase, catalase, urease, esculin, and citrate; reduces nitrate with gas formation; indole-negative.

Source (2 strains)
Blood.

Reference strain
G7626 ← NCTC 12247, from blood.

Literature

1. Holmes, B., and C.A. Dawson. 1983. Numerical taxonomic studies on *Achromobacter* isolates from clinical material. *In* H. Leclerc (Ed.), Gram-negative bacteria of medical and public health importance: taxonomy-identification–applications. Proceedings of Symposium, Lille, May 25–27. Les Editions INSERM *114:* 331–341.

2. Holmes, B., M. Costas, A.C. Wood, and K. Kersters. 1990. Numerical analysis of electrophoretic protein patterns of *"Achromobacter"* group B, E, and F strains from human blood. J. Appl. Bacteriol. *68:* 495–504.

3. Cieslak, T.J., M.L. Robb, C.J. Drabick, and G.W. Fischer. 1992. Catheter-associated sepsis caused by *Ochrobactrum anthropi*: report of a case and review of related nonfermentative bacteria. Clin. Infect. Dis. *14:* 902–907.

Comments

The term *"Achromobacter"* currently represents a recognized genus with no recognized species, hence the use of quotes when describing these organisms. *Achromobacter xylosoxidans*, formerly the only recognized *Achromobacter* species, has been renamed *Alcaligenes xylosoxidans* subsp. *xylosoxidans*. The *"Achromobacter"* group originally included six biovars, A through F. Biovars A, C, and D recently have been redefined as biovars of the new species *Ochrobactrum anthropi*. The remaining *"Achromobacter"* biovars (B, E, and F) usually can be differentiated from *O. anthropi* by their flagellar morphology and hydrolysis of esculin. Esculin-positive *O. anthropi* strains can be differentiated from *"Achromobacter"* by cellular fatty acid analysis. *"Achromobacter"* group E can be differentiated from *"Achromobacter"* group B by its failure to oxidize mannitol and ethanol. According to Holmes and Dawson (1), *"Achromobacter"* group E can be differentiated from *"Achromobacter"* group F by its positive reaction for rhamnose oxidation, growth on MacConkey agar, and failure to oxidize mannitol. Information on the cellular fatty acid composition of this group is presented in Table 4.2 and Figure 4.7 of the cellular fatty acid section.

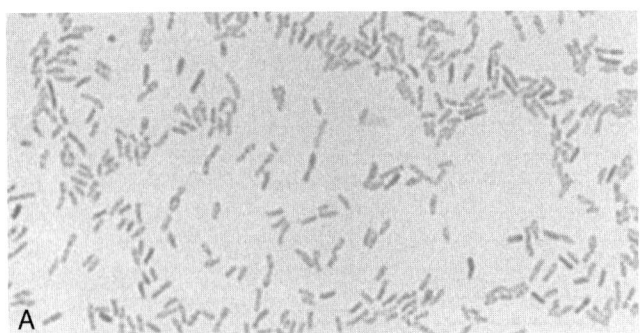

Achromobacter" group E strain G7626.
Gram stain BAP 35° C 24 h x 1700.

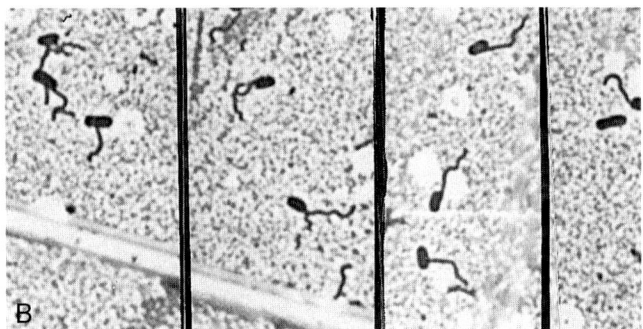

"*Achromobacter"* group E strain G7626. Flagella stain
(Ryu method) TGY 25° C 24 h x 1700.

"Achromobacter" group E

TEST PERFORMED	SIGN	% +	#+/T
		NUMBER OF STRAINS 2	
Morphology	mrs		
Motility; flagella	[m:p,L]		2/2
Action on blood	LG	100	2/2
Fermentative or oxidative	[O]		
Carbohydrate base	OF		
Acid from:			
D-Glucose	[+]	100	2/2
D-Xylose	+	100	2/2
D-Mannitol	[-]	0	0/2
Lactose	(+)[a]	(100)	(2)/2
Sucrose	(+)	(100)	(1), (1w)/2
Maltose	+	100	1, 1w/2
D-Sorbitol	[-]	0	0/2
L-Rhamnose	+	100	2/2
Catalase	+	100	2/2
Oxidase	+	100	2/2
Growth on:			
MacConkey	[+]	100	2/2
SS	[+]	100	2/2
Cetrimide	+ or (+)	50 (50)	1, (1w)/2
Simmons citrate	+	100	2/2
Urea, Christensen's	+	100	2/2
Nitrate reduction[b]	[+]	100	2w/2
Gas from nitrate	[+]	100	1w, 1/2
Nitrite reduction[c]	+	100	2/2
Indole	-	0	0/2
TSI slant, acid	-	0	0/2
TSI butt, acid	-	0	0/2
H_2S (TSI butt)	v	(50)	(1w)/2
H_2S (Pb ac paper)	+	100	2/2
Gelatin hydrolysis[d]	-	0	0/2
Litmus milk	k	100	2/2
Pigment	v	(50 ta sol)	(1 ta, w, sol)/2
Growth at:			
25°C	+	100	2/2
35°C	+	100	2/2
42°C	+	100	2/2
Esculin hydrolysis	[+]	100	2/2
Lysine decarboxylase	-	0	0/2
Arginine dihydrolase	+	100	2/2
Ornithine decarboxylase	-	0	0/2
Nutrient broth, 0% NaCl	+	100	2/2
Nutrient broth, 6% NaCl	+	100	2/2
ONPG	+	100	1/1
Alkalinization of:			
Acetamide	-	0	0/2
Serine	v	(50)	(1w)/2
Tartrate	-	0	0/2
3-Ketolactonate	[-]	0	0/1

[a] Acid production from lactose was detected only after 15 to 21 days of incubation.
[b] When tested at 48 h of incubation, nitrate reduction is only observed in heart infusion broth base (see *Media and Methods*). Residual nitrate (as evidenced by small amounts of red precipitate after the addition of zinc dust) may be observed in tubes that display the reduction of nitrate to gas.
[c] When tested at 48 h of incubation, nitrite reduction is observed only in media containing ≤0.01% nitrite.
[d] Incubation of 7–14 days.

Acidovorax delafieldii

Gram-negative; motile (single polar flagellum); oxidizes glucose, xylose, and mannitol (usually weak and late); grows weak and late on MacConkey agar; oxidase- and nitrate-positive.

Sources (2 strains)

Soil and central venous catheter.

Reference strains

G8094 ← CCUG 1779 = ATCC 17505, type strain, from soil.
G8095 ← CCUG 23830B, from central venous catheter.

Literature

1. Willems, A., E. Falsen, B. Pot, E. Jantzen, B. Hoste, P. VanDamme, M. Gillis, K. Kersters, and J. De Ley. 1990. *Acidovorax*, a new genus for *Pseudomonas facilis*, *Pseudomonas delafieldii*, E. Falsen (EF) group 13, EF group 16, and several clinical isolates, with the species *Acidovorax facilis* comb. nov., *Acidovorax delafieldii* comb. nov., and *Acidovorax temperans* sp. nov. Int. J. Syst. Bacteriol. 40: 384–398.

2. Hollis, D.G., R.E. Weaver, C.W. Moss, M.I. Daneshvar, and P.L. Wallace. 1992. Chemical and cultural characterization of CDC group WO-1, a weakly oxidative gram-negative group of organisms isolated from clinical sources. J. Clin. Microbiol. 30: 291–295.

Comments

CDC group WO-1 strains that oxidize xylose and are citrate-positive are probably *A. delafieldii*. The reactions presented for the reference strains were obtained using methods presented in this manual. Using API galleries, Willems et al. (1) found assimilation of glucose, xylose, and citrate as sole carbon sources to be negative, negative, and variable, respectively. Information on the cellular fatty acid composition of this group is presented in Table 4.34 and Figure 4.39 of the cellular fatty acid section.

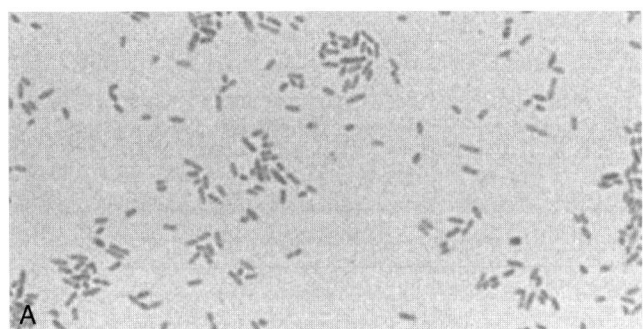

Acidovorax delafieldii strain G8094.
Gram stain BAP 35° C 24 h x 1700.

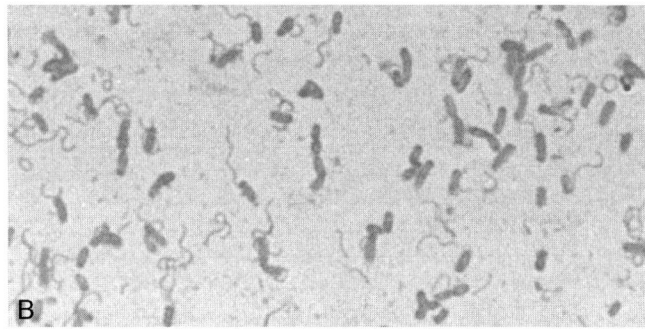

Acidovorax delafieldii strain G8094. Flagella stain (Ryu method) TGY 25° C 24 h x 1700.

Acidovorax delafieldii

TEST PERFORMED	SIGN	%+	#+/T
	2 REFERENCE STRAINS		
Morphology	mrs		
Motility; flagella	[m;p,1-2]		
Action on blood	-	0	0/2
Fermentative or oxidative	[O]		
Carbohydrate base	OF		
Acid from:			
D-Glucose	[(+)]	(100)	(2)/2
D-Xylose	[(+)]	(100)	(2)/2
D-Mannitol	[+ or (+)]	50 (50)	1 (1)/2
Lactose	-	0	0/2
Sucrose	-	0	0/2
Maltose	-	0	0/2
Catalase	-[a]	0	0/2
Oxidase	[+]	100	2/2
Growth on:			
MacConkey	(+)	(100)	(2w)/2
SS	-	0	0/2
Cetrimide	-	0	0/2
Simmons citrate	[+]	100	2/2
Urea, Christensen's	+	100	2/2
Nitrate reduction	+	100	2/2
Gas from nitrate	-	0	0/2
Indole	-	0	0/2
TSI slant, acid	-	0	0/2
TSI butt, acid	-	0	0/2
H_2S (TSI butt)	-	0	0/2
H_2S (Pb ac paper)	+	100	2/2
Gelatin hydrolysis[b]	-	0	0/2
Litmus milk	k	100	2/2
Pigment	yel sol	50 (50)	1 (1)/2
Growth at:			
25°C	+	100	2/2
35°C	+	100	2/2
42°C	v	50	1w/2
Esculin hydrolysis	-	0	0/2
Lysine decarboxylase	-	0	0/2
Arginine dihydrolase	(+)[c]	(100)	(2)/2
Ornithine decarboxylase	-	0	0/2
Nutrient broth, 0% NaCl	+	100	2/2
Nutrient broth, 6% NaCl	-	0	0/2

[a]Reported to be variable by Willems et al. (1).
[b]Incubation of 7–14 days.
[c]Reported to be negative by Willems et al. (1).

Acidovorax facilis

Gram-negative; motile (single polar flagellum); oxidizes glucose, xylose, and mannitol (usually weak and late); does not grow on MacConkey agar; oxidase- and nitrate-positive; citrate- and nitrite-negative; gelatinase-positive.

Sources (2 strains)

Soil.

Reference strains

G8096 ← CCUG 2113 = ATCC 11228, type strain, from soil.
G8097 ← CCUG 15919 = ATCC 17695, from soil.

Literature

1. Willems, A., E. Falsen, B. Pot, E. Jantzen, B. Hoste, P. VanDamme, M. Gillis, K. Kersters, and J. De Ley. 1990. *Acidovorax*, a new genus for *Pseudomonas facilis, Pseudomonas delafieldii*, E. Falsen (EF) group 13, EF group 16, and several clinical isolates, with the species *Acidovorax facilis* comb. nov., *Acidovorax delafieldii* comb. nov., and *Acidovorax temperans* sp. nov. Int. J. Syst. Bacteriol. 40: 384–398.

Comment

Using carbon source assimilation, Willems et al. (1) found both reference strains to be xylose-negative.

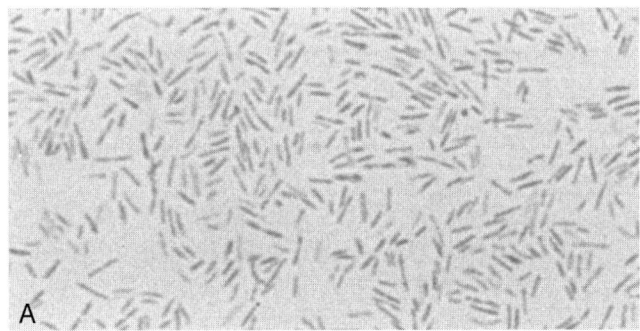

Acidovorax facilis strain G8096.
Gram stain BAP 35° C 24 h x 1700.

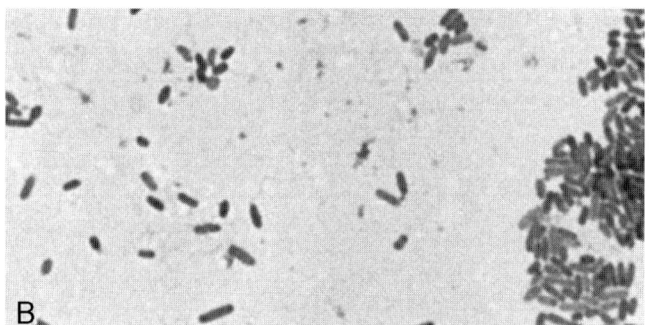

Acidovorax facilis strain G8096. Flagella stain
(Ryu method) TGY 25° C 24 h x 1700.

Acidovorax facilis

TEST PERFORMED	SIGN	%+	#+/T
Morphology	mrs		
Motility; flagella	[m;p,1-2]		
Action on blood	-	0	0/2
Fermentative or oxidative	[O]		
Carbohydrate base	OF		
Acid from:			
D-Glucose	[+ or (+)]	50 (50)	1 (1)/2
D-Xylose	[(+)]	(100)	(2)/2
D-Mannitol	[(+)]	(100)	(2)/2
Lactose	-	0	0/2
Sucrose	-	0	0/2
Maltose	-	0	0/2
Catalase	+	100	2/2
Oxidase	[+]	100	2/2
Growth on:			
MacConkey	[-]	0	0/2
SS	-	0	0/2
Cetrimide	-	0	0/2
Simmons citrate	[-]	0	0/2
Urea, Christensen's	(+)	(100)	(2)/2
Nitrate reduction	+	100	2/2
Gas from nitrate	-	0	0/2
Indole	-	0	0/2
TSI slant, acid	-	0	0/2
TSI butt, acid	-	0	0/2
H_2S (TSI butt)	-	0	0/2
H_2S (Pb ac paper)	+	100	1, 1w/2
Gelatin hydrolysis[a]	[+]	100	2/2
Litmus milk	k	100	2/2
Pigment	-	0	0/2
Growth at:			
25°C	+	100	2/2
35°C	+	100	2/2
42°C	-	0	0/2
Esculin hydrolysis	-	0	0/2
Lysine decarboxylase	-	0	0/2
Arginine dihydrolase	+[b]	100	2/2
Ornithine decarboxylase	-	0	0/2
Nutrient broth, 0% NaCl	+	100	2/2
Nutrient broth, 6% NaCl	-	0	0/2

[a]Incubation of 7–14 days.
[b]Reported to be negative by Willems et al. (1).

Acidovorax temperans

Gram-negative; motile (single polar flagellum); oxidizes glucose, but not xylose; grows on MacConkey agar (growth may be weak and late); oxidase- and catalase-positive; citrate- and gelatin-negative; nitrate-positive.

Sources (2 strains)

Urine and wound.

Reference strains

G8098 ← CCUG 11779B = ATCC 49665, type strain, from urine.
G8099 ← CCUG 22215, from wound.

Literature

1. Willems, A., E. Falsen, B. Pot, E. Jantzen, B. Hoste, P. VanDamme, M. Gillis, K. Kersters, and J. De Ley. 1990. *Acidovorax*, a new genus for *Pseudomonas facilis, Pseudomonas delafieldii*, E. Falsen (EF) group 13, EF group 16, and several clinical isolates, with the species *Acidovorax facilis* comb. nov., *Acidovorax delafieldii* comb. nov., and *Acidovorax temperans* sp. nov. Int. J. Syst. Bacteriol. 40: 384–398.

2. Hollis, D.G., R.E. Weaver, C.W. Moss, M.I. Daneshvar, and P.L. Wallace. 1992. Chemical and cultural characterization of CDC group WO-1, a weakly oxidative Gram-negative group of organisms isolated from clinical sources. J. Clin. Microbiol. 30: 291–295.

Comment

CDC group WO-1 strains that are xylose-negative and citrate-negative are probably *A. temperans*. The cellular fatty acid profiles of the two reference strains are similar to the profiles presented in Figure 4.39 and Table 4.34 of the cellular fatty acid section.

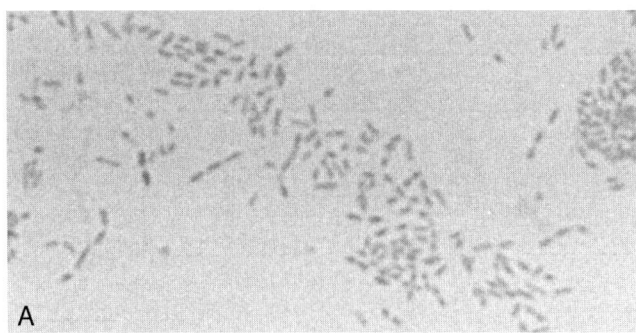

Acidovorax temperans strain G8099.
Gram stain BAP 35° C 24 h x 1700.

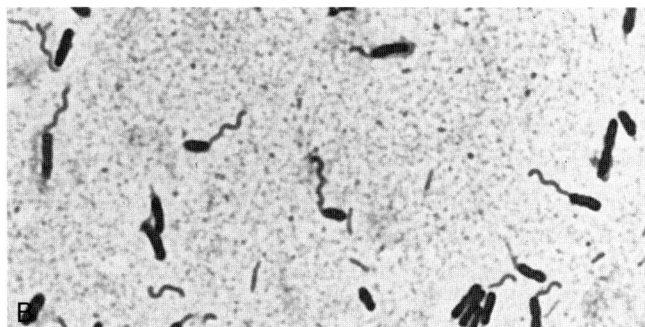

Acidovorax temperans strain G8099. Flagella stain
(Ryu method) TGY 25° C 24 h x 1700.

Acidovorax temperans

TEST PERFORMED	SIGN	% +	#+/T
Morphology	mrs		
Motility; flagella	[m;p,1-2]		
Action on blood	lav	100	2/2
Fermentative or oxidative	[O]		
Carbohydrate base	OF		
Acid from:			
D-Glucose	[+]	100	1, 1w/2
D-Xylose	[-]	0	0/2
D-Mannitol	[v]	50	1w/2
Lactose	-	0	0/2
Sucrose	-	0	0/2
Maltose	-	0	0/2
Catalase	+[a]	100	1, 1w/2
Oxidase	[+]	100	2/2
Growth on:			
MacConkey	[+ or (+)]	50 (50)	1 (1w)/2
SS	-	0	0/2
Cetrimide	-	0	0/2
Simmons citrate	[-]	0	0/2
Urea, Christensen's	v	50	1w/2
Nitrate reduction	+	100	2/2
Gas from nitrate	[+][b]	100	2w/2
Indole	-	0	0/2
TSI slant, acid	-	0	0/2
TSI butt, acid	-	0	0/2
H$_2$S (TSI butt)	-	0	0/2
H$_2$S (Pb ac paper)	+	100	2/2
Gelatin hydrolysis[c]	[-]	0	0/2
Litmus milk	k	100	2/2
Pigment	yel-sol	100	2/2
Growth at:			
25°C	+	100	2/2
35°C	+	100	2/2
42°C	+	100	2/2
Esculin hydrolysis	-	0	0/2
Lysine decarboxylase	-	0	0/2
Arginine dihydrolase	-	0	0/2
Ornithine decarboxylase	-	0	0/2
Nutrient broth, 0% NaCl	+	100	2/2
Nutrient broth, 6% NaCl	-	0	0/2

[a]Reported as variable by Willems et al. (1).
[b]A small amount.
[c]Incubation of 7–14 days.

Acinetobacter

Gram-negative, thick coccoid rod occurring in pairs or short chains, with some filaments and pleomorphism, nonflagellated (some strains show a twitching motility under special conditions); capsules and fimbriae variable, oxidase-negative, catalase-positive, usually grows on MacConkey agar, may or may not produce acid from glucose, xylose, lactose, and maltose; positive in the *Acinetobacter* transformation assay of Juni (see *Media and Methods*). Although *Acinetobacter* are considered nonfastidious, several fastidious isolates received in our laboratory have been identified as *Acinetobacter* with the transformation assay. These fastidious *Acinetobacter* strains usually produce punctate translucent colonies on blood agar at 18–24 h, grow lightly or late on MacConkey agar (if at all), produce a weakly alkaline reaction or no change on King's OF medium, and usually do not grow in nutrient broth.

Sources (197 nonhemolytic/asaccharolytic strains)

Blood (25%), wound (12%), sputum (10%), cerebrospinal fluid (10%), urine (7%), throat (4%), exudate (4%), tissue (3%), skin (2%), eye (1%), stool (1%), other (24%).

Sources (53 nonhemolytic/saccharolytic strains)

Sputum (19%), wound (17%), blood (11%), urine (11%), exudate (4%), skin (2%), foot exudate (2%), cerebrospinal fluid (2%), brain (2%), throat (2%), other (29%).

Sources (15 β-hemolytic/asaccharolytic strains)

Sputum (33%), urine (13%), wound (13%), throat (7%), exudate (7%), other (27%).

Sources (11 β-hemolytic/saccharolytic strains)

Sputum (36%), blood (27%), urine (18%), ear (9%), other (9%).

Reference strains

See next table.

Literature

1. Bouvet, P.J.M., and P.A.D. Grimont. 1986. Taxonomy of the genus *Acinetobacter* with the recognition of *Acinetobacter baumannii* sp. nov., *Acinetobacter haemolyticus* sp. nov., *Acinetobacter johnsonii* sp. nov., and *Acinetobacter junii* sp. nov. and emended descriptions of *Acinetobacter calcoaceticus* and *Acinetobacter lwoffii*. Int. J. Syst. Bacteriol. *36:* 228–240.

2. Tjernberg, I., and J. Ursing. 1989. Clinical strains of *Acinetobacter* classified by DNA-DNA hybridization. APMIS *97:* 595–605.

3. Bouvet, P.J.M., and S. Jeanjean. 1989. Delineation of new proteolytic genomic species in the genus *Acinetobacter*. Res. Microbiol. *140:* 291–299.

4. Nishimura, Y., I. Takeshi, and H. Hiroshi. 1988. *Acinetobacter radioresistens* sp. nov. isolated from cotton and soil. Int. J. Syst. Bacteriol. *38:* 209–211.

5. Gerner-Smidt, P., I. Tjernberg, and J. Ursing. 1991. Reliability of phenotypic tests for identification of *Acinetobacter* species. J. Clin. Microbiol. *29:* 277–282.

6. Dijkshoorn, L., and J. van der Toorn. 1992. *Acinetobacter* species: which do we mean? Clin. Infect. Dis. *15:* 748–749.

Comments

One of the more frequently encountered nonfermenters in the clinical microbiology laboratory, the genus *Acinetobacter* represents a complex group of genomospecies that cannot be reliably identified to the species level with biochemical tests. Traditionally, organisms within this genus have been divided into glucose oxidizers and nonoxidizers. Synonyms for the glucose oxidizer group include: *Herellea vaginicola*, *A. anitratis*, *A. calcoaceticus* var anitratus, and *Bacterium anitratum*. Synonyms for the nonoxidizer group include *Mima polymorpha*, *A. lwoffii*, and *A. calcoaceticus* var lwoffii. A fish-like odor may be detected in some strains. In some strains, rod forms may be rare. Recent genetic studies have identified at least 16 separate *Acinetobacter* genomospecies, seven of which have been named. These genomospecies can be grouped roughly according to their ability to produce a complete β-like hemolysis of erythrocytes and their ability to oxidize glucose. The next table lists the described genomospecies with characteristics provided by the authors in the original descriptions. Information on the cellular fatty acid composition of this group is presented in Table 4.3 and Figure 4.8 of the cellular fatty acid section.

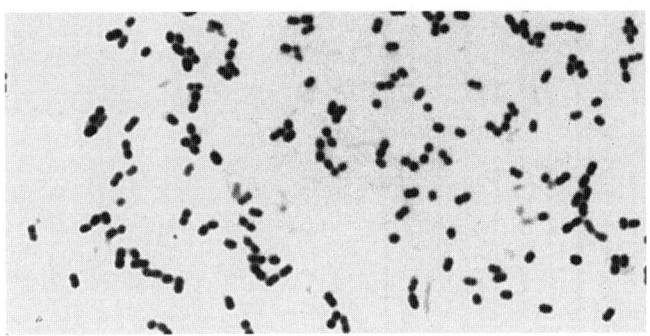

Acinetobacter species strain KC1125. Gram stain BAP 35° C 24 h x 1700.

Some common characteristics of described *Acinetobacter* genomospecies.

Genomospecies number	Proposed name	(Literature No.)	Reference strain[a]	Glucose oxidation[b]	Hemolysis	Gelatin	Growth temp (°C)		
							37	41	44
Genomospecies 1	*A. calcoaceticus*	(1)	ATCC 23055[T]	+	−	−	+	−	−
Genomospecies 2	*A. baumannii*	(1)	ATCC 19606[T]	95	−	−	+	+	+
Genomospecies 3		(1)	ATCC 17922	+	−	−	+	+	−
Genomospecies 13		(2)		+	−	−	+	+	50
Genomospecies 10		(1)	ATCC 17924	+	−	−	+	−	−
Genomospecies 14 and Genomospecies 13		(2) (3)		+	+	75	25	−	−
Genomospecies 15		(2)		50	−	−	+	50	−
Genomospecies 14		(3)		+	+	+	+	−	−
Genomospecies 4	*A. haemolyticus*	(1)	ATCC 17906[T]	52	+	96	+	−	−
Genomospecies 6		(1)	ATCC 17979	66	+	+	+	−	−
Genomospecies 12	*A. radioresistens*[c]	(1)	ATCC 43998[T]	33	−	−	+	−	−
Genomospecies 5	*A. junii*	(1)	ATCC 17908[T]	−	−	−	+	90	−
Genomospecies 7	*A. johnsonii*	(1)	ATCC 17909[T]	−	−	−	−	−	−
Genomospecies 8	*A. lwoffii*	(1)	ATCC 15309[T]	−	−	−	+	−	−
Genomospecies 9		(1)	ATCC 9957	−	−	−	+	−	−
Genomospecies 11		(1)	ATCC 11171	−	+	+	+	−	−
Genomospecies 15		(3)		−	+	+	+	−	−
Genomospecies 16		(3)		−	−	−	+	−	−
Genomospecies 17		(3)		−	+	+	+	−	−

[a]T Type strain
[b]Numbers represent percentage of positives as reported by the authors; + and − represent 100% positive and 100% negative, respectively.
[c]This name was proposed by Nishimura et al. (4).

Nonhemolytic/Asaccharolytic *Acinetobacter*

TEST PERFORMED	SIGN	% +	#+/T
	NUMBER OF STRAINS 249		
Morphology	cc		
Motility; flagella	[nm]		
Action on blood	v	22 ly	38/249; 3 α, 4 br, 8 gr, 2 lav
Fermentative or oxidative	[n-o]		
Carbohydrate base	OF		
Acid from:			
D-Glucose	[-]	0	0/249
D-Xylose	-	0	0/241
D-Mannitol	-	0	0/241
Lactose	-	0	0/240
Sucrose	-	0	0/240
Maltose	-	0	0/239
Catalase	+	99	238/239
Oxidase	[-]	0.8w	2w/241
Growth on:			
MacConkey	v	73 (4)	176 (10)/242
SS	-	7 (2)	16(4)/240
Cetrimide	-	(0.5)	1/221
Simmons citrate	v	27 (3)	65 (8)/241
Urea, Christensen's	v	5 (14)	13 (34)/192
Nitrate reduction	-	5	13/245
Gas from nitrate	-	0	0/245
Nitrite reduction	-	0	0/10
Indole	-	0	0/236
TSI slant, acid	-	0	0/245
TSI butt, acid	-	0	0/244
H_2S (TSI butt)	-	0.4	1/234
H_2S (Pb ac paper)	v	52	123/237
Gelatin hydrolysis[a]	-	3	5/185
Litmus milk	v	19 k	46/234; 7 IR, 3 dirty, 2 A, 1 szf
Pigment	v	10 br-sol	10/99; 11 tan-yel-sol, 3 yel
Growth at:			
25°C	+	99	239/242
35°C	+	98	238/242
42°C	v	48	114/240
Esculin hydrolysis	-	0	0/235
Lysine decarboxylase	-	0	0/27
Arginine dihydrolase	v	14	2, 2w/28
Ornithine decarboxylase	-	0	0/27
Nutrient broth, 0% NaCl	v	86	200, (2)/234
Nutrient broth, 6% NaCl	v	18	43/233
Acinetobacter transformation	[+]	100	249/249

[a]Incubation of 7–14 days.

Nonhemolytic/Saccharolytic *Acinetobacter*

	NUMBER OF STRAINS 65		
TEST PERFORMED	SIGN	%+	#+/T
Morphology	cc		
Motility; flagella	[nm]		
Action on blood	v	35 ly	23/65; 2 α, 2 lav, 2 gr, 1 br
Fermentative or oxidative	[O]		
Carbohydrate base	OF		
Acid from:			
D-Glucose	[+]	97 (3)	63 (2)/65
D-Xylose	-	0	0/60
D-Mannitol	-	0	0/60
Lactose	v	70 (14)	41, 1w, (7), (1w)/60
Sucrose	-	0	0/60
Maltose	v	18 (35)	11 (21)/60
Catalase	+	100	59/59
Oxidase	[-]	2	1w/63
Growth on:			
MacConkey	[+]	95	57/60
SS	-	5 (2)	3 (1)/54
Cetrimide	-	2	1/52
Simmons citrate	v	62 (3)	37 (2)/60
Urea, Christensen's	v	25 (25)	15 (15)/59
Nitrate reduction	-	9	6/64
Gas from nitrate	-	0	0/64
Nitrite reduction	-	0	0/4
Indole	-	0	0/59
TSI slant, acid	v	11	7/64
TSI butt, acid	-	0	0/64
H_2S (TSI butt)	-	0	0/58
H_2S (Pb ac paper)	v	33	20/60
Gelatin hydrolysis[a]	-	0	0/45
Litmus milk	v	35 A	19/55; 10 IR, 7 Acl, 4 k, 1 Aw
Pigment	v	24 br-tan-yel-amb-sol	6/28; 3 yel
Growth at:			
25°C	+	100	58/58
35°C	+	100	60/60
42°C	v	73	43/59
Esculin hydrolysis	-	0	0/58
Lysine decarboxylase	-	5 (5)	(1), 1w/19
Arginine dihydrolase	v	16 (5)	2, 1w, (1)/19
Ornithine decarboxylase	-	5	1w/19
Nutrient broth, 0% NaCl	+	100	59/59
Nutrient broth, 6% NaCl	v	15	9/59
Acinetobacter transformation	[+]	100	65/65

[a]Incubation of 7–14 days.

β-hemolytic/Asaccharolytic *Acinetobacter*

TEST PERFORMED	SIGN	% +	# +/T
	NUMBER OF STRAINS 21		
Morphology	cc		
Motility; flagella	[nm]		
Action on blood	[β]	100	21/21
Fermentative or oxidative	[n-o]		
Carbohydrate base	OF		
Acid from:			
D-Glucose	[−]	0	0/21
D-Xylose	−	5	1/20
D-Mannitol	−	0	0/19
Lactose	−	0	0/19
Sucrose	−	0	0/19
Maltose	−	0	0/19
Catalase	+	100	19/19
Oxidase	[−]	0	0/20
Growth on:			
MacConkey	[+]	95	18/19
SS	v	37	7/19
Cetrimide	−	0	0/19
Simmons citrate	v	63 (5)	12 (1)/19
Urea, Christensen's	v	21 (10)	4 (2)/19
Nitrate reduction	−	10	2/20
Gas from nitrate	−	0	0/20
Indole	−	0	0/19
TSI slant, acid	−	0	0/21
TSI butt, acid	−	0	0/21
H_2S (TSI butt)	−	0	0/19
H_2S (Pb ac paper)	v	33	6/18
Gelatin hydrolysis[a]	−	6	1/17
Litmus milk	v	35 k	6k/17
Pigment	v	50 yel-tan-sol	2/4
Growth at:			
25°C	+	95	18/19
35°C	+	100	19/19
42°C	v	63	12/19
Esculin hydrolysis	−	0	0/17
Lysine decarboxylase	−	0	0/5
Arginine dihydrolase	v	20	1/5
Ornithine decarboxylase	−	0	0/5
Nutrient broth, 0% NaCl	+	95	18/19
Nutrient broth, 6% NaCl	v	26	5/19
Acinetobacter transformation	[+]	100	21/21

[a]Incubation of 7–14 days.

β-hemolytic/Saccharolytic *Acinetobacter*

TEST PERFORMED	SIGN	% +	#+/T
	NUMBER OF STRAINS 12		
Morphology	cc		
Motility; flagella	[nm]		
Action on blood	[β]	100	12/12
Fermentative or oxidative	[O]		
Carbohydrate base	OF		
Acid from:			
D-Glucose	[+ or (+)]	83 (17)	10 (2)/12
D-Xylose	v	88	7/8
D-Mannitol	−	0	0/8
Lactose	v	63 (13)	5 (1)/8
Sucrose	−	0	0/8
Maltose	v	(38)	(3)/5
Catalase	+	100	7/7
Oxidase	[−]	0	0/10
Growth on:			
MacConkey	[+]	100	9/9
SS	v	38	3/8
Cetrimide	−	0	0/8
Simmons citrate	v	71	5/7
Urea, Christensen's	v	(13)	(1)/8
Nitrate reduction	−	0	0/9
Gas from nitrate	−	0	0/9
Indole	−	0	0/8
TSI slant, acid	−	0	0/11
TSI butt, acid	−	0	0/11
H_2S (TSI butt)	−	0	0/7
H_2S (Pb ac paper)	v	75	6/8
Gelatin hydrolysis[a]	v	50	3/6
Litmus milk	v	25 Acl	2/8; 1 A, 1 IR, 1 kw
Pigment	v	20 br-sol	1/4
Growth at:			
25°C	+	100	8/8
35°C	+	100	8/8
42°C	v	63	5/8
Esculin hydrolysis	−	0	0/7
Lysine decarboxylase	−	0	0/1
Arginine dihydrolase	−	0	0/2
Ornithine decarboxylase	−	0	0/2
Nutrient broth, 0% NaCl	v	88	7/8
Nutrient broth, 6% NaCl	−	0	0/8
Acinetobacter transformation	[+]	100	12/12

[a]Incubation of 7–14 days.

Actinobacillus actinomycetemcomitans

Gram-negative very small rod, nonmotile; fastidious; requires or is enhanced by candle-jar atmosphere; ferments glucose, sometimes with slight gas, but not lactose and sucrose; (CDC biotypes are based on fermentation of xylose, mannitol, and maltose); catalase- and nitrate-positive; MacConkey-negative, urease-negative, indole-negative.

Sources (120 strains)

Blood (56%), neck (7%), lung tissue (5%), mandibular and submandibular abscess (4%), abscess-site unknown (4%), wound, rib, and facial sinus (2% each), other (15%), unknown (3%).

Reference strain

KC517 ← NCTC 9710 = ATCC 3384, type strain, from abscess.

Literature

1. Topley, W.W.C., and G.S. Wilson. 1929. Principles of Bacteriology and Immunity, 1st ed., Edward Arnold, London, p. 256.

2. King, E.O., and H.W. Tatum. 1962. *Actinobacillus actinomycetemcomitans* and *Haemophilus aphrophilus*. J. Infect. Dis. *111:* 85–94.

3. Page, M.I., and E.O. King. 1966. Infections due to *Actinobacillus actinomycetemcomitans* and *Haemophilus aphrophilus*. N. Engl. J. Med. *275:* 181–188.

4. Potts, T.V., J.J. Zambon, and R.J. Genco. 1985. Reassignment of *Actinobacillus actinomycetemcomitans* to the genus *Haemophilus* as *Haemophilus actinomycetemcomitans* comb. nov. Int. J. Syst. Bacteriol. *35:* 337–341.

5. Frederiksen, W. 1987. International Committee on Systematic Bacteriology Subcommittee on *Pasteurellaceae* and related organisms. Minutes of the meetings, 6 and 10 September 1986, Manchester, England. Int. J. Syst. Bacteriol. *37:* 474.

Comments

Growth in broth is granular, with colonies adhering to sides of tube; star-like colonies may be observed on prolonged incubation. Often found in association with *Actinomyces*, especially *A. israelii*. It has been proposed to transfer these organisms to the genus *Haemophilus*; however, this proposal was not favored by the International Committee on Systematic Bacteriology Subcommittee on *Pasteurellaceae* and Related Organisms (4, 5). Information on the cellular fatty acid composition of this species is presented in Table 4.4A and Figure 4.9 of the cellular fatty acid section.

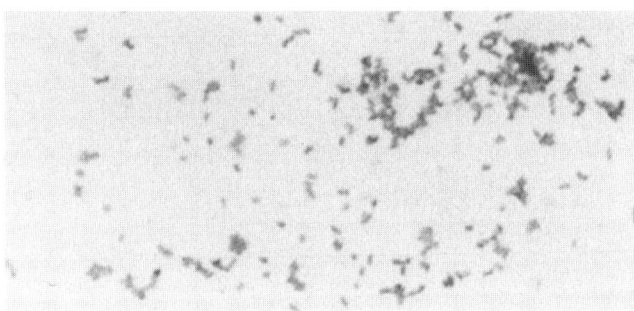

Actinobacillus actinomycetemcomitans strain KC517.
Gram stain BAP 35° C 24 h x 1700.

Actinobacillus actinomycetemcomitans

TEST PERFORMED				SIGN				% +	#+/T
NUMBER OF STRAINS 120									
Morphology				rv					
Motility				[nm]					
Action on blood				v				44 al	32/72; 1 sl lav
Gas from glucose				v				28	33/120
Fermentative or oxidative				[F]					
Carbohydrate base				F[a]					
Number of strains	63	30	15	6	4	1	1		
Acid from:									
D-Glucose	[+]	[+]	[+]	[+]	[+]	[+]	[-]	83 (16)	100 (19)/120
D-Xylose	-	+	+	-	+	-	+	33 (9)	39 (11)/120
D-Mannitol	+	+	-	-	+	+	-	66 (16)	79 (19)/120
Lactose	[-]	[-]	[-]	[-]	[-]	[-]	[-]	0	0/120
Sucrose	[-]	[-]	[-]	[-]	[-]	[-]	[-]	0	0/120
Maltose	+	+	+	+	-	-	-	80 (15)	96 (18)/120
Catalase				[+]				99	110, 9w/120
Oxidase				v				19	6, 15w/111
Growth on:									
MacConkey				[-]				4 (1)	5w (1)/120
SS				-				0	0/120
Simmons citrate				-				0	0/120
Urea, Christensen's				[-]				0	0/120
Nitrate reduction				[+]				100	118, 1w/119
Gas from nitrate				-				0	0/120
Indole				[-]				0	0/120
TSI slant, acid				+				100	101, 19w/120
TSI butt, acid				[+]				100	101, 19w/120
H_2S (TSI butt)				-				0	0/119
H_2S (Pb ac paper)				+				91	88, 20w/119
MR				-				8	4, 1w/60
VP				-				0	0/55
Gelatin hydrolysis[b]				-				0	0/45
Litmus milk				-				4 IR	1/118; 1 A
Growth at:									
25°C				v				24	24/98
35°C				+				100	112/112
42°C				-				4	4/98
Esculin hydrolysis				[-]				0	0/120
Nutrient broth, 0% NaCl				-				7	4, 3w/98
Nutrient broth, 6% NaCl				-				0	0/98

[a]Rabbit serum (1–2 drops) may need to be added per 3 mL of medium; reactions may be obtained within 4 h using the Rapid Sugar Test.
[b]Incubation of 7–14 days.

Actinobacillus equuli

Gram-negative coccobacillary to rod form, nonmotile; ferments glucose, xylose, mannitol, lactose, sucrose, maltose, raffinose, melibiose, and trehalose; oxidase- and urease-positive; indole- and esculin-negative; usually produces catalase; nitrate-positive.

Sources (27 strains)

Horse (27%), human (animal bite or wound) (19%), pig (12%), other animals (11%), unknown (31%).

Reference strain

KC601 ← NCTC 3365, from foal, pyemic nephritis.

Literature

1. Valleé, A., P. Thibault, and L. Second. 1963. Contribution à l'étude d'*A. lignieresii* et d'*A. equuli*. Ann. Inst. Pasteur, Paris *104:* 108–114.
2. Sneath, P.H.A., and V.B.D. Skerman. 1966. A list of type and reference strains of bacteria. Int. J. Syst. Bacteriol. *16:* 1–113.
3. Peel, M.M., K.A. Hornidge, M. Luppino, A.M. Stacpoole, and R.E. Weaver. 1991. *Actinobacillus* spp. and related bacteria in infected wounds of humans bitten by horses and sheep. J. Clin. Microbiol. *29:* 2535–2538.
4. Phillips, J.E. 1984. *Actinobacillus. In* N.R. Krieg and J.G. Holt (Eds.), Bergey's Manual of Systematic Bacteriology, Vol. 1, Williams & Wilkins, Baltimore, pp. 570–575.

Comments

Pathogenic for horses and pigs. Colonies may be sticky. *Bergey's Manual of Systematic Bacteriology* describes these organisms as oxidase-variable (4). An *Actinobacillus equuli*-like bacterium has been described with a biochemical profile consistent with *A. equuli*, except that a small volume of gas from glucose was observed (3). Information on the cellular fatty acid composition of this species is presented in Table 4.4A and Figure 4.9 of the cellular fatty acid section.

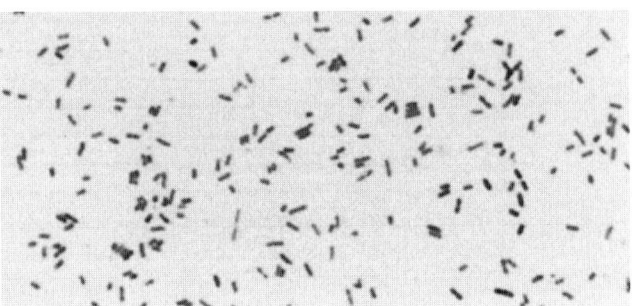

Actinobacillus equuli strain KC601. Gram stain BAP 35° C 24 h x 1700.

Actinobacillus equuli

TEST PERFORMED	SIGN	% +	#+/T
Morphology	srs		
Motility	[nm]		
Gas form glucose	−	0	0/19
Action on blood	v	11 gr	2/19; 2 al, 1 ly
Fermentative or oxidative	[F]		
Carbohydrate base	F		
Acid from:			
D-Glucose	+	100	19/19
D-Xylose	+	100	19/19
D-Mannitol	+	100	19/19
Lactose	+	95	18/19
Sucrose	+	100	19/19
Maltose	+	95	18/19
Trehalose	[+]	100	19/19
Melibiose	[+ or (+)]	74 (26)	15 (4)/19
Raffinose	+	100	19/19
D-Mannose	+	100	19/19
Catalase	[v]	68 (5)	10, 3w (1)/ 19
Oxidase	[+][a]	100	17, 2w/19
Growth on:			
MacConkey	v	84 (5)	16 (1)/19
SS	−	0	0/19
Simmons citrate	−	0	0/19
Urea, Christensen's	[+][b]	100	19/19
Nitrate reduction	[+]	100	19/19
Gas from nitrate	−	0	0/19
Indole	−	0	0/19
TSI slant, acid	+	100	19/19
TSI butt, acid	+	100	19/19
H_2S (TSI butt)	−	0	0/19
H_2S (Pb ac paper)	+	100	19/19
MR	v	87	12, 2w/16
VP	v	12	2w/17
Gelatin hydrolysis[c]	v	47	8/17
Litmus milk	v	83 A	15/18; 2 IR
Growth at:			
25°C	v	89	17/19
35°C	+	100	19/19
42°C	v	31	6/19
Esculin hydrolysis	[−]	0	0/19
Lysine decarboxylase	−	0	0/9
Arginine dihydrolase	−	0	0/9
Ornithine decarboxylase	−	0	0/9
Nutrient broth, 0% NaCl	v	89	17/19
Nutrient broth, 6% NaCl	−	5	1/19

[a]Described as oxidase-variable in *Bergey's Manual of Systematic Bacteriology* (4).
[b]May require a drop of serum on slant or a heavy inoculum.
[c]Incubation of 7–14 days.

Actinobacillus hominis

Gram-negative coccobacillary to rod form, nonmotile; ferments glucose, xylose, mannitol, lactose, sucrose, maltose, trehalose, raffinose, melibiose (late), but usually not D-mannose; oxidase-, urease-, esculin-, and nitrate-positive; indole-negative; catalase-positive.

Source (1 strain)

Sputum.

Reference strain

KC1702 ← NCTC 11529 = ATCC 49457, type strain.

Literature

1. Friis-Møller, A. 1981. A new *Actinobacillus* species from the human respiratory tract: *Actinobacillus hominis* nov. sp. In M. Kilian, W. Frederiksen, and E.L. Biberstein (Eds.), *Haemophilus, Pasteurella,* and *Actinobacillus,* Academic Press, New York, pp. 151–157.

2. Wust, J., J. Gubler, W. Mannheim, and A. von Graevenitz. 1991. *Actinobacillus hominis* as a causative agent of septicemia in hepatic failure. Eur. J. Clin. Microbiol. Infect. Dis. *10:* 693–694.

3. Eckert, F., A. Stenzel, R. Mutters, W. Frederiksen, and W. Mannheim. 1991. Some unusual members of the family *Pasteurellaceae* isolated from human sources—phenotypic features and genomic relationships. Zbl. Bakt. *275:* 143–155.

Comments

D-Mannose-positive *A. hominis* strains have been described (3). These strains are difficult to differentiate from *A. equuli* and *A. suis* without DNA hybridization studies. Information on the cellular fatty acid composition of this species is presented in Table 4.4A and Figure 4.9 of the cellular fatty acid section.

Actinobacillus hominis

TEST PERFORMED	REFERENCE STRAIN SIGN
Morphology	srs
Motility	[nm]
Action on blood	−
Fermentative or oxidative	[F]
Carbohydrate base	F
Acid from:	
D-Glucose	+
D-Xylose	+
D-Mannitol	+
Lactose	+
Sucrose	+
Maltose	+
Trehalose	[+]
Melibiose	[(+)]
Raffinose	[+]
D-Mannose	[−]
Catalase	+
Oxidase	+
Growth on:	
MacConkey	[+]
SS	−
Simmons citrate	−
Urea, Christensen's	+
Nitrate reduction	[+]
Gas from nitrate	−
Nitrite reduction	[+][a]
Indole	−
TSI slant, acid	+
TSI butt, acid	+
H_2S (TSI butt)	−
H_2S (Pb ac paper)	+
Gelatin hydrolysis[b]	−
Litmus milk	−
Pigment	−
Growth at:	
25°C	+
35°C	+
42°C	−
Esculin hydrolysis	(+)
Lysine decarboxylase	−
Arginine dihydrolase	−
Ornithine decarboxylase	−
Nutrient broth, 0% NaCl	+
Nutrient broth, 6% NaCl	−

[a]Nitrite reduction was observed in 0.01% nitrite but not in 0.1% nitrite.
[b]Incubation of 7–14 days.

Actinobacillus lignieresii

Gram-negative coccobacillary to rod form, nonmotile; ferments glucose, xylose, mannitol, lactose (late), sucrose, and maltose, not trehalose or melibiose; oxidase- and urease-positive; indole- and esculin-negative; usually produces catalase; nitrate-positive.

Sources (30 strains)

Bovine (17%), human wound (9%), horse bite (9%), human finger (4%), other animals (dog, sheep, rat) (13%), unknown (48%).

Reference strain

KC208 ← NCTC 4975, from bovine lesions.

Literature

1. Valleé, A., P. Thibault, and L. Second. 1963. Contribution à l'étude d'*A. lignieresii* et d'*A. equuli.* Ann. Inst. Pasteur, Paris *104:* 108–114.

2. Sneath, P.H.A., and V.B.D. Skerman. 1966. A list of type and reference strains of bacteria. Int. J. Syst. Bacteriol. *16:* 1–113.

3. Samitz, E.M., and E.L. Biberstein. 1991. *Actinobacillus suis*-like organisms and evidence of hemolytic strains of *Actinobacillus lignieresii* in horses. Am. J. Vet. Res. *52:* 1245–1251.

Comments

Pathogenic for sheep and cattle. Rare hemolytic strains have been described (3). Information on the cellular fatty acid composition of this species is presented in Table 4.4A and Figure 4.9 of the cellular fatty acid section.

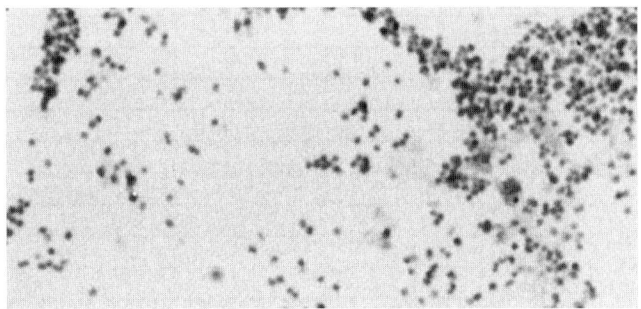

Actinobacillus lignieresii strain KC208. Gram stain BAP 35° C 24 h x 1700.

Actinobacillus lignieresii

TEST PERFORMED	SIGN	%+	#+/T
Morphology	srs		
Motility	[nm]		
Action on blood	v	17 al	4/23; 1 β
Fermentative or oxidative	[F]		
Carbohydrate base	F		
Acid from:			
D-Glucose	+	96 (4)	22 (1)/23
D-Xylose	+ or (+)	87 (13)	20 (3)/23
D-Mannitol	+	91 (9)	19, 1w (2)/22
Lactose	v	17 (61)	3, 1w (14)/23
Sucrose	+	96 (4)	23 (1)/24
Maltose	+ or (+)	83 (17)	19 (4)/23
Trehalose	[-]	0	0/24
Melibiose	[-]	0	0/24
Raffinose	v	25 (46)	6 (11)/24
D-Mannose	+	100	23, 1w/24
Catalase	[v]	89	16, 8w/27
Oxidase	[+]	100	19, 3w/22
Growth on:			
MacConkey	v	67 (8)	16 (2)/24
SS	-	0	0/22
Simmons citrate	-	0	0/23
Urea, Christensen's	[+][a]	100	22/22
Nitrate reduction	[+]	100	23/23
Gas from nitrate	-	0	0/23
Indole	-	0	0/23
TSI slant, acid	+	100	23/23
TSI butt, acid	+	100	23/23
H_2S (TSI butt)	-	0	0/23
H_2S (Pb ac paper)	+	100	23/23
MR	+	100	9, 6w/15
VP	-	0	0/21
Gelatin hydrolysis[b]	-	0	0/23
Litmus milk	v	29 A	7/24; 3 IR, 2 w
Growth at:			
25°C	v	82	18/22
35°C	+	100	22/22
42°C	v	14	3/22
Esculin hydrolysis	[-]	0	0/23
Lysine decarboxylase	-	0	0/7
Arginine dihydrolase	-	0	0/7
Ornithine decarboxylase	-	0	0/7
Nutrient broth, 0% NaCl	+	100	9/9
Nutrient broth, 6% NaCl	-	10	1/9

[a]May require a drop of serum on slant or a heavy inoculum.
[b]Incubation of 7–14 days.

Actinobacillus suis

Gram-negative coccobacillary to rod form, nonmotile; ferments glucose, xylose, lactose, sucrose, maltose, D-mannose, and sometimes mannitol; oxidase-, urease-, and esculin-positive; indole-negative, usually produces a clear zone of hemolysis; usually grows on MacConkey agar; reduces nitrate, usually produces catalase.

Sources (33 strains)

Horse (22%), human wound (15%), sputum and upper respiratory (15%), cattle (10%), pig (7%), animal bite (donkey, hamster, zebra) (7%), human blood (5%), finger lesion (2%), other animals (12%), unknown (5%).

Reference strain

KC560 (P.W. Wetmore 1627) from hock of piglet.

Literature

1. van Dorssen, C.A., and F.H.J. Jaartsveld. 1962. *Actinobacillus suis* (novo species) een bij het varken voorko mende bacterie. Tijdschr. Diergeneesk. *87:* 450–458.

2. Zimmerman, T. 1964. Untersuchungen uber die Actinobazillose des schweines 1. Mitteilung: Isolierun und charakterisierung der erreger. Dtsch. Tieraerztl. Wochenschr. *71:* 457–461.

3. Kilian, M., and W. Frederiksen. 1981. Identification tables for the *Haemophilus-Pasteurella-Actinobacillus* group. *In* M. Kilian, W. Frederiksen, and E.L. Biberstein (Eds.), *Haemophilus, Pasteurella,* and *Actinobacillus.* Academic Press, New York, pp. 286–287.

4. Bisgaard, M., K. Piechulla, Y.-T. Ying, W. Frederiksen, and W. Mannheim. 1984. Prevalence of organisms described as *Actinobacillus suis* or haemolytic *Actionbacillus equuli* in the oral cavity of horses. Comparative investigations of strains obtained and porcine strains of *A. suis sensu stricto*. Acta. Pathol. Microbiol. Immunol. Scand. Sect. B, *92:* 291–298.

Comments

Van Dorssen and Jaartsveld (1), Zimmerman (2), and Kilian and Frederiksen (3) have variously described *Actinobacillus suis*. *Bergey's Manual of Systematic Bacteriology* describes *A. suis* as mannitol-negative. Strains we have encountered and considered to be *A. suis* may represent a heterogenous group and should be studied further. Information on the cellular fatty acid composition of this species is presented in Table 4.4A and Figure 4.9 of the cellular fatty acid section.

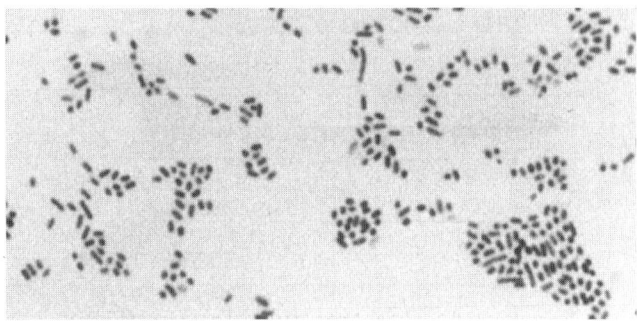

Actinobacillus suis strain KC560. Gram stain BAP 35° C 24 h x 1700.

Actinobacillus suis

TEST PERFORMED	SIGN	% +	#+/T
Morphology	srs		
Motility	[nm]		
Gas from glucose	−	0	0/33
Action on blood	[v]	76 ß	25/33; 2 al, 1 ly, 1 LG
Fermentative or oxidative	[F]		
Carbohydrate base	F		
Acid from:			
D-Glucose	+	94 (6)	30, 1w (2)/33
D-Xylose	+	94 (6)	30, 1w (2)/33
D-Mannitol	v	54	18/33
Lactose	+ or (+)	79 (18)	25, 1w (6)/33
Sucrose	+	94 (6)	30, 1w (2)/33
Maltose	+	94 (6)	30, 1w (2)/33
Trehalose	+	95	21/22
Melibiose	+ or (+)	82 (13)	18 (3)/22
Raffinose	+	100	22/22
D-Mannose	[+]	91 (9)	20 (2)/22
Catalase	[v]	85	14, 14w/33
Oxidase	[+]	100	30, 3w/33
Growth on:			
MacConkey	[+ or (+)]	82 (12)	20, 7w (4)/33
SS	−	3	1/33
Simmons citrate	−	0	0/33
Urea, Christensen's	[+][a]	97 (3)	31, 1w (1)/33
Nitrate reduction	[+]	100	33/33
Gas from nitrate	−	0	0/33
Indole	[−]	0	0/33
TSI slant, acid	+	100	32, 1w/33
TSI butt, acid	+	100	32, 1w/33
H$_2$S (TSI butt)	−	0	0/33
H$_2$S (Pb ac paper)	+	100	29, 4w/33
MR	v	24	7w/33
VP	v	50	6, 8w/28
Gelatin hydrolysis[b]	−	6	2/31
Litmus milk	v	82 A	14, 13w/33; 1 kw, 1 IR
Growth at:			
25°C	+	94	31/33
35°C	+	97	32/33
42°C	v	31	10/32
Esculin hydrolysis	[+]	100	33/33
Lysine decarboxylase	−	0	0/7
Arginine dihydrolase	−	0	0/7
Ornithine decarboxylase	−	0	0/7
Nutrient broth, 0% NaCl	v	88	22/25
Nutrient broth, 6% NaCl	−	0	0/25

[a]May require a drop of serum on slant or a heavy inoculum.
[b]Incubation of 7–14 days.

Actinobacillus ureae

Gram-negative pleomorphic rod; nonmotile; ferments glucose, mannitol, sucrose, and maltose, not xylose; no growth on MacConkey agar; oxidase- and urease-positive (oxidase often weak); indole-negative; frequently catalase-negative; reduces nitrate.

Sources (97 strains)

Sputum (67%), bronchi (6%), throat (4%), trachea (4%), nasal (3%), sinus (2%), blood (2%), exudate (2%), cerebrospinal fluid, eye, and otitis media (1% each), unknown (7%).

Reference strain

KC518 S. D. Henriksen, 218/60.

Literature

1. Jones, D.M. 1962. A *Pasteurella*-like organism from the human respiratory tract. J. Pathol. Bacteriol. *83:* 143–151.

2. Jones, D.M., and P.M. O'Connor. 1962. *Pasteurella haemolytica* var ureae from human sputum. J. Clin. Pathol. *15:* 247–248.

3. Mutters, R., S. Pohl, and W. Mannheim. 1986. Transfer of *Pasteurella ureae* Jones 1962 to the genus *Actinobacillus* Brumpt 1910: *Actinobacillus ureae* comb. nov. Int. J. Syst. Bacteriol. *36:* 343–344.

Comments

Formerly *Pasteurella ureae*. Urea should be inoculated heavily or serum added to slant surface. Associated mainly with respiratory sources; all strains received by CDC have been from human sources. Information on the cellular fatty acid composition of this species is presented in Table 4.4A and Figure 4.9 of the cellular fatty acid section.

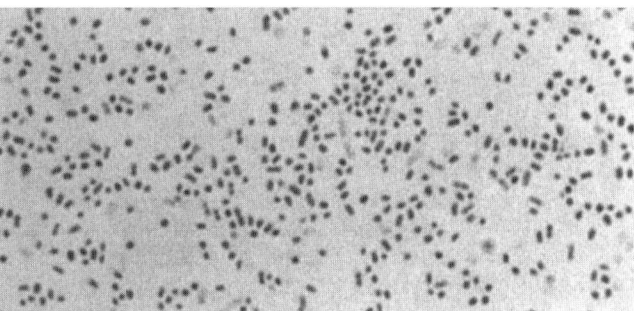

Actinobacillus ureae strain KC518. Gram stain BAP 35° C 24 h x 1700.

Actinobacillus ureae

TEST PERFORMED	SIGN	% +	#+/T
NUMBER OF STRAINS 97			
Morphology	plr		
Motility; flagella	[nm]		
Action on blood	v	16 al	
Fermentative or oxidative	[F]		
Carbohydrate base	F		
Acid from:			
D-Glucose	+	100	87, 10w/97
D-Xylose	[−]	0	0/97
D-Mannitol	[+]	99 (1)	84, 12w (1)/97
Lactose	−	0	0/97
Sucrose	+	99 (1)	83, 13w (1)/97
Maltose	+	91 (5)	75, 12w (5)/96
Catalase	v	63	34, 27w/97
Oxidase	[+]	99	83, 10w/94
Growth on:			
MacConkey	[−]	0	0/97
SS	−	0	0/97
Simmons citrate	−	0	0/97
Urea, Christensen's	[+][a]	100	97/97
Nitrate reduction	[+]	99	93, 1w/95
Gas from nitrate	−	0	0/95
Indole	[−]	0	0/93
TSI slant, acid	+	100	95/95
TSI butt, acid	+	99	94/95
H_2S (TSI butt)	−	0	0/97
H_2S (Pb ac paper)	+	91 (1)	68, 20w (1)/97
Gelatin hydrolysis[b]	−	0	0/76
Litmus milk	v	9 (1) k	9 (1)/96; 1 IR
Growth at:			
25°C	v	88	77/87
35°C	v	82	69/84
42°C	−	5	4/87
Esculin hydrolysis	−	0	0/94
Lysine decarboxylase	−	0	0/14
Arginine dihydrolase	−	0	0/14
Ornithine decarboxylase	−	0	0/14
Nutrient broth, 0% NaCl	−	0	0/59
Nutrient broth, 6% NaCl	−	0	0/59

[a]May require a drop of serum on slant or a heavy inoculum.
[b]Incubation of 7–14 days.

Afipia broomeae

Fastidious nonfermentative Gram-negative rod; motile (one to two polar flagella); oxidase-positive; no growth occurs on MacConkey agar; weakly catalase-positive; urease-positive; nitrate-negative; oxidizes xylose weakly and delayed (acid observed at 3 to 14 days of incubation); optimal growth is obtained at 30° C; grows well on buffered charcoal-yeast extract agar and in nutrient broth.

Sources (3 strains)

Sputum, bone marrow, abscess.

Reference strain

F186 = ATCC 49717, type strain, from sputum.

Literature

1. Brenner, D.J., D.G. Hollis, C.W. Moss, et al. 1991. Proposal of *Afipia* gen. nov. with *Afipia felis* sp. nov. (formerly the cat scratch disease bacillus), *Afipia clevelandensis* sp. nov. (formerly the Cleveland Clinic Foundation strain), *Afipia broomeae* sp. nov. and three unnamed genospecies. J. Clin. Microbiol. 29: 2450–2460.

Comments

Afipia broomeae cannot be differentiated from *Afipia* unnamed genomospecies 2 without DNA hybridization studies. Information on the cellular fatty acid composition of this species is presented in Table 4.6 and Figure 4.11 of the cellular fatty acid section.

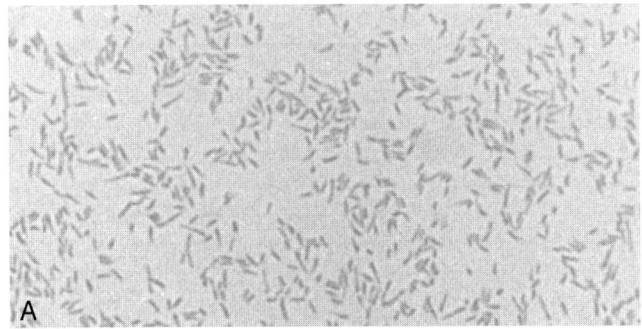 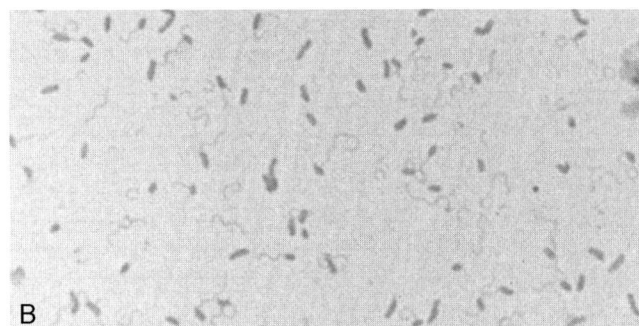

Afipia broomeae strain F186. Gram stain BCYE 30° C 48 h x 1700.

Afipia broomeae strain F186. Flagella stain (Ryu method) TGY 30° C 48 h x 1700.

Afipia broomeae

TEST PERFORMED[a]	SIGN	% +	#+/T
		NUMBER OF STRAINS 3	
Morphology	srs		
Motility; flagella	[m;p,1-2]		3/3
Action on blood	−	0	0/3
Fermentative or oxidative	[n-o]		
Carbohydrate base	OF		
Acid from:			
D-Glucose	[−]	0	0/3
D-Xylose	[(+)]	(100)	(3w)/3
D-Mannitol	−	0	0/3
Lactose	−	0	0/3
Sucrose	−	0	0/3
Maltose	−	0	0/3
Catalase	[+]	100	3/3
Oxidase	+	100	3/3
Growth on:			
MacConkey	−	0	0/3
SS	−	0	0/3
Cetrimide	−	0	0/3
Simmons citrate	−	0	0/3
Urea, Christensen's[b]	[+ or (+)]	33(67)	1,(2)/3
Nitrate reduction	[−]	0	0/3
Gas from nitrate	−	0	0/3
Nitrite reduction	−	0	0/2
Indole	−	0	0/3
TSI slant, acid	−	0	0/3
TSI butt, acid	−	0	0/3
H$_2$S (TSI butt)	−	0	0/3
H$_2$S (Pb ac paper)	−	0	0/3
Gelatin hydrolysis[c]	−	0	0/3
Litmus milk	(k)	(100)	(3)/3
Pigment	−	0	0/3
Growth at:			
25°C	+	100	3/3
30°C	+	100	3/3
35°C	v	33	1/3
42°C	−	0	0/3
Esculin hydrolysis	−	0	0/3
Lysine decarboxylase	−	0	0/1
Arginine dihydrolase	−	0	0/1
Ornithine decarboxylase	−	0	0/1
Phenylalanine deaminase	−	0	0/2
Nutrient broth, 0% NaCl	+	100	3/3
Nutrient broth, 6% NaCl	−	0	0/3
Alkalinization of:			
Acetamide	−	0	0/3
Serine	−	0	0/3
Tartrate	−	0	0/3
Sodium acetate	v	(50)	(1)/2

[a]Incubations were at 30° C, unless indicated otherwise.
[b]The slant was inoculated heavily.
[c]Incubation of 7–14 days.

Afipia clevelandensis

Fastidious nonfermentative Gram-negative rod; motile (1-2 polar flagella); oxidase-positive; no growth occurs on MacConkey agar; urease-positive; nitrate-negative; asaccharolytic; catalase-negative; optimal growth is obtained at 25–30° C; grows well on buffered charcoal-yeast extract agar and in nutrient broth; the mol% G+C of the DNA is 64.

Source (1 strain)

Tibial biopsy.

Reference strain

G1849 = ATCC 49720, type strain, from tibial biopsy.

Literature

1. Brenner, D.J., D.G. Hollis, C.W. Moss, et al. 1991. Proposal of *Afipia* gen. nov. with *Afipia felis* sp. nov. (formerly the cat scratch disease bacillus), *Afipia clevelandensis* sp. nov. (formerly the Cleveland Clinic Foundation strain), *Afipia broomeae* sp. nov. and three unnamed genospecies. J. Clin. Microbiol. 29: 2450–2460.

2. Hall, G.S., K. Pratt-Rippin, and J.A. Washington. 1991. Isolation of agent associated with cat scratch disease bacillus from pretibial biopsy. Diagn. Microbiol. Infect. Dis. 14: 511–513.

Comments

A. clevelandensis can be distinguished from other *Afipia* species by its failure to produce acid from xylose. Information on the cellular fatty acid composition of this species is presented in Table 4.6 and Figure 4.11 of the cellular fatty acid section.

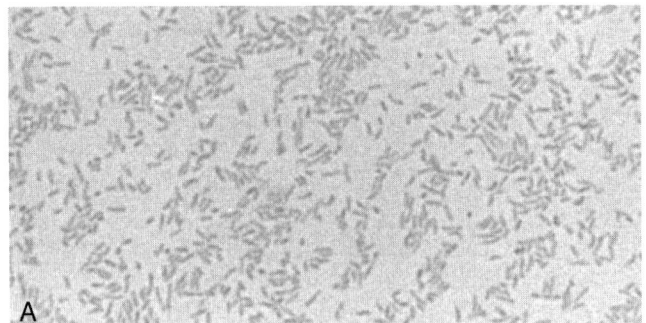

Afipia clevelandensis strain G1849. Gram stain BCYE 30° C 48 h x 1700.

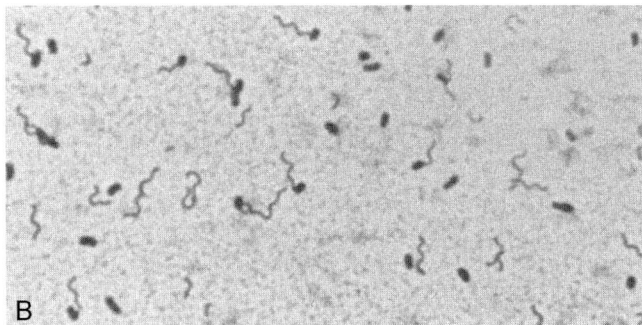

Afipia clevelandensis strain G1849. Flagella stain (Ryu method) TGY 30° C 48 h x 1700.

Afipia clevelandensis

TEST PERFORMED[a]	TYPE STRAIN SIGN
Morphology	srs
Motility; flagella	[m;p,1-2]
Action on blood	–
Fermentative or oxidative	[n-o]
Carbohydrate base	OF
Acid from:	
D-Glucose	[–]
D-Xylose	[–]
D-Mannitol	–
Lactose	–
Sucrose	–
Maltose	–
Catalase	[–]
Oxidase	+
Growth on:	
MacConkey	–
SS	–
Cetrimide	–
Simmons citrate	–
Urea, Christensen's[b]	[(+)]
Nitrate reduction	[–]
Gas from nitrate	–
Nitrite reduction	–
Indole	–
TSI slant, acid	–
TSI butt, acid	–
H_2S (TSI butt)	–
H_2S (Pb ac paper)	–
Gelatin hydrolysis[c]	–
Litmus milk	(k)
Pigment	–
Growth at:	
25°C	+
30°C	+
35°C	+
42°C	–
Esculin hydrolysis	–
Lysine decarboxylase	
Arginine dihydrolase	
Ornithine decarboxylase	
Phenylalanine deaminase	–
Nutrient broth, 0% NaCl	+
Nutrient broth, 6% NaCl	–
Alkalinization of:	
Acetamide	–
Serine	–
Tartrate	
Sodium acetate	(+)

[a]Incubations were performed at 30° C, unless otherwise indicated.
[b]The slant was inoculated heavily.
[c]Incubation of 7–14 days.

Afipia felis

Fastidious nonfermentative Gram-negative rod; motile by means of a single polar flagellum; oxidase-positive; growth on MacConkey agar is variable and delayed; urease-positive; reduces nitrate to nitrite; oxidizes xylose weakly and delayed (acid observed at 3 to 14 days of incubation); optimal growth is obtained at 25° or 30° C; grows well on buffered charcoal-yeast extract agar and in nutrient broth.

Sources (16 strains)

Lymph node (75%), splenic lesion (6%), unknown (19%).

Reference strain

G1492 = ATCC 53690, type strain, from lymph node.

Literature

1. Brenner, D.J., D.G. Hollis, C.W. Moss, et al. 1991. Proposal of *Afipia* gen. nov. with *Afipia felis* sp. nov. (formerly the cat scratch disease bacillus), *Afipia clevelandensis* sp. nov. (formerly the Cleveland Clinic Foundation strain), *Afipia broomeae* sp. nov., and three unnamed genospecies. J. Clin Microbiol. *29:* 2450–2460.

2. English, C.K., D.J. Wear, A.M. Margileth, C.R. Lissner, and G.P. Walsh. 1988. Cat-scratch disease. Isolation and culture of the bacterial agent. J. Am. Med. Assoc. *259:* 1347–1352.

Comments

Afipia felis has been isolated from lymph node specimens of individuals with cat scratch disease. The role of *A. felis* in the pathogenesis of cat scratch disease is not well understood. The original description of *A. felis* (1) lists the organism as catalase-negative; however, some additional strains studied since the original description have demonstrated weak levels of catalase activity. *A. felis* can be differentiated from other *Afipia* species by its ability to reduce nitrate. Information on the cellular fatty acid composition of this species is presented in Table 4.6 and Figure 4.11 of the cellular fatty acid section.

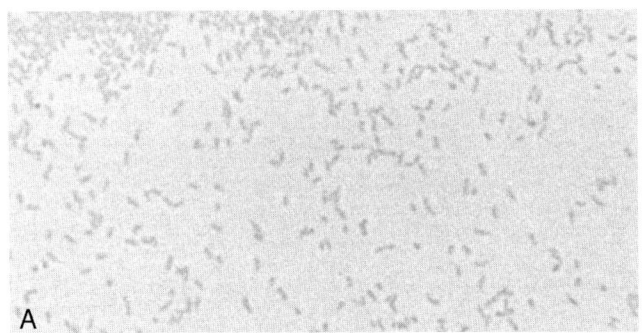

Afipia felis strain G1492. Gram stain BCYE 30° C 48 h x 1700.

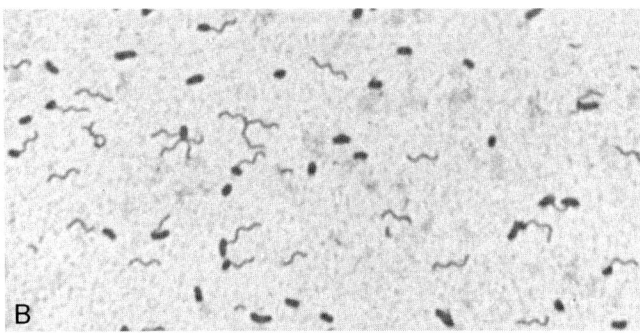

Afipia felis strain G1492. Flagella stain (Ryu method) TGY 30° C 48 h x 1700.

Afipia felis

TEST PERFORMED[a]	SIGN	% +	#+/T
	NUMBER OF STRAINS 16		
Morphology	srs, plr		
Motility; flagella	[m;p,1-2]		16/16
Action on blood	−	0	0/16
Fermentative or oxidative	[n-o]		
Carbohydrate base	OF		
Acid from:			
D-Glucose	[−]	0	0/16
D-Xylose	[(+)]	(93)	(15w)/16
D-Mannitol	−	0	0/16
Lactose	−	0	0/16
Sucrose	−	0	0/16
Maltose	−	0	0/16
Catalase	v	37	6w/16
Oxidase	+	100	16/16
Growth on:			
MacConkey	v	(44)	(7w)/16
SS	−	0	0/16
Cetrimide	−	0	0/16
Simmons citrate	−	0	0/16
Urea, Christensen's[b]	[+ or (+)]	88 (12)	14, (2)/16
Nitrate reduction	[+]	100	16/16
Gas from nitrate	−	0	0/16
Nitrite reduction	−	0	0/1
Indole	−	0	0/16
TSI slant, acid	−	0	0/16
TSI butt, acid	−	0	0/16
H_2S (TSI butt)	−	0	0/16
H_2S (Pb ac paper)	−	0	0/16
Gelatin hydrolysis[c]	−	0	0/16
Litmus milk	k or (k)	50 (44)	8, (7)/16
Pigment	v	(13 sl ta sol)	(2)/16
Growth at:			
25°C	+	100	16/16
30°C	+	100	16/16
35°C	+	100	16/16
42°C	−	0	0/16
Esculin hydrolysis	−	0	0/16
Lysine decarboxylase	−	0	0/13
Arginine dihydrolase	−	0	0/13
Ornithine decarboxylase	−	0	0/13
Phenylalanine deaminase	+	100	6, 6w/12
Nutrient broth, 0% NaCl	+	100	16/16
Nutrient broth, 6% NaCl	−	0	0/16
Alkalinization of:			
Acetamide	−	(7)	(1)/15
Serine	−	0	0/15
Tartrate	v	7 (13)	1, (2)/15
Sodium acetate	+ or (+)	82 (18)	9, (2)/11

[a]Incubations were at 30° C, unless indicated otherwise.
[b]The slant was inoculated heavily.
[c]Incubation of 7–14 days.

Afipia genomospecies 1

Fastidious nonfermentative Gram-negative rod; motile by means of a single polar flagellum; no growth occurs on MacConkey agar; oxidase-positive; urease-positive; oxidizes xylose and mannitol weakly; citrate-positive; optimal growth is obtained at 30° C; grows well on buffered charcoal-yeast extract agar and in nutrient broth; the mol% G+C of the DNA is 69.

Sources (2 strains)
Pulmonary fluid, pleural fluid.

Reference strain
F872 = ATCC 49721, type strain, from pleural fluid.

Literature
1. Brenner, D.J., D.G. Hollis, C.W. Moss, et al. 1991. Proposal of *Afipia* gen. nov. with *Afipia felis* sp. nov. (formerly the cat scratch disease bacillus), *Afipia clevelandensis* sp. nov. (formerly the Cleveland Clinic Foundation strain), *Afipia broomeae* sp. nov. and three unnamed genospecies. J. Clin. Microbiol. 29: 2450–2460.

Comments
The three *Afipia* genomospecies originally were called genospecies. Since the term genospecies has been used previously in another context, these organisms are now called genomospecies. *Afipia* genomospecies 1 can be differentiated from other *Afipia* species by its oxidation of mannitol and utilization of citrate. Information on the cellular fatty acid composition of this group is presented in Table 4.6 and Figure 4.11 of the cellular fatty acid section.

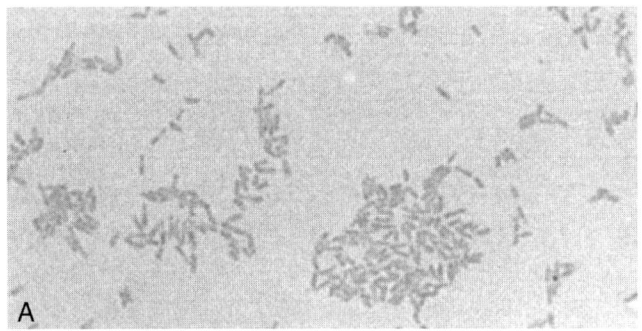

Afipia genomospecies 1 strain F872. Gram stain BCYE 30° C 48 h x 1700.

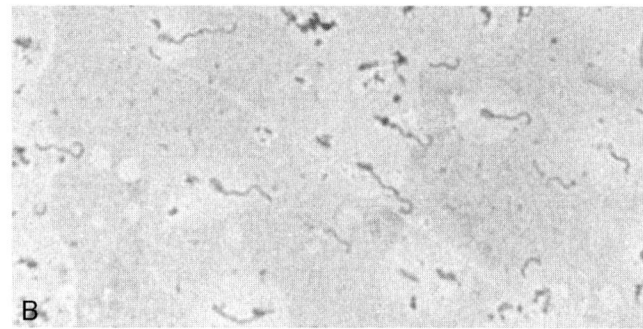

Afipia genomospecies 1 strain F872. Flagella stain (Ryu method) TGY 30° C 48 h x 1700.

Afipia genomospecies 1

TEST PERFORMED[a]	SIGN	% +	#+/T
	NUMBER OF STRAINS 2		
Morphology	srs		
Motility; flagella	[m;p,1-2]		2/2
Action on blood	-	0	0/2
Fermentative or oxidative	[n-o]		
Carbohydrate base	OF		
Acid from:			
D-Glucose	[-]	0	0/2
D-Xylose	[(+)]	(100)	(2)/2
D-Mannitol	[(+)]	(100)	(2)/2
Lactose	-	0	0/2
Sucrose	-	0	0/2
Maltose	-	0	0/2
Catalase	v	50	1w/2
Oxidase	+	100	1/1
Growth on:			
MacConkey	-	0	0/2
SS	-	0	0/2
Cetrimide	-	0	0/2
Simmons citrate	+	100	2/2
Urea, Christensen's[b]	[+]	100	2/2
Nitrate reduction	[-]	0	0/2
Gas from nitrate	-	0	0/2
Indole	-	0	0/2
TSI slant, acid	-	0	0/2
TSI butt, acid	-	0	0/2
H_2S (TSI butt)	-	0	0/2
H_2S (Pb ac paper)	+ or (+)	50 (50)	1, (1)/2
Gelatin hydrolysis[c]	-	0	0/2
Litmus milk	(k)	(100)	(2)/2
Pigment	-	0	0/2
Growth at:			
25°C	+	100	2/2
30°C	+	100	2/2
35°C	+	100	2/2
42°C	-	0	0/2
Esculin hydrolysis	-	0	0/2
Phenylalanine deaminase	+	100	1w/1
Nutrient broth, 0% NaCl	+	100	2/2
Nutrient broth, 6% NaCl	-	0	0/2
Alkalinization of:			
Acetamide	-	0	0/1
Serine	-	0	0/1
Tartrate	(+)	(100)	(1)/1
Sodium acetate	-	0	0/1

[a]Incubations were at 30° C, unless indicated otherwise.
[b]The slant was inoculated heavily.
[c]Incubation of 7–14 days.

Afipia genomospecies 2

Fastidious nonfermentative Gram-negative rod; motile by a single polar flagellum; oxidase-positive; no growth occurs on MacConkey agar; urease-positive; oxidizes xylose weakly; optimal growth is obtained at 30° C; grows well on buffered charcoal-yeast extract agar and nutrient broth.

Source (1 strain)

Bronchial wash.

Reference strain

G4438 = ATCC 49722, type strain, from bronchial wash.

Literature

1. Brenner, D.J., D.G. Hollis, C.W. Moss, et al. 1991. Proposal of *Afipia* gen. nov. with *Afipia felis* sp. nov. (formerly the cat scratch disease bacillus), *Afipia clevelandensis* sp. nov. (formerly the Cleveland Clinic Foundation strain), *Afipia broomeae* sp. nov. and three unnamed genospecies. J. Clin. Microbiol. *29:* 2450–2460.

Comments

The three *Afipia* genomospecies originally were called genospecies. Since the term genospecies has been used previously in another context, these organisms are now called genomospecies. *Afipia* genomospecies 2 cannot be differentiated from *Afipia broomeae* without DNA hybridization studies. Information on the cellular fatty acid composition of this group is presented in Table 4.6 and Figure 4.11 of the cellular fatty acid section.

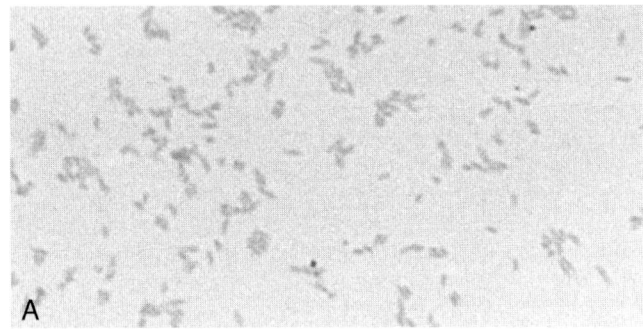

Afipia genomospecies 2 strain G4438. Gram stain BCYE 30° C 48 h x 1700.

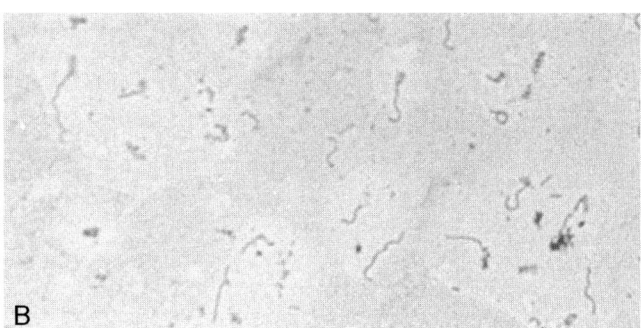

Afipia genomospecies 2 strain G4438. Flagella stain (Ryu method) TGY 30° C 48 h x 1700.

Afipia genomospecies 2

TEST PERFORMED[a]	REFERENCE STRAIN SIGN
Morphology	srs
Motility; flagella	[m;p,1-2]
Action on blood	−
Fermentative or oxidative	[n-o]
Carbohydrate base	OF
Acid from:	
D-Glucose	[−]
D-Xylose	[(+)]
D-Mannitol	−
Lactose	−
Sucrose	−
Maltose	−
Catalase	+
Oxidase	+
Growth on:	
MacConkey	−
SS	−
Cetrimide	−
Simmons citrate	−
Urea, Christensen's[b]	[+]
Nitrate reduction	[−]
Gas from nitrate	−
Indole	−
TSI slant, acid	−
TSI butt, acid	−
H_2S (TSI butt)	−
H_2S (Pb ac paper)	−
Gelatin hydrolysis[c]	−
Litmus milk	(k)
Pigment	−
Growth at:	
25°C	+
30°C	+
35°C	+
42°C	−
Esculin hydrolysis	−
Phenylalanine deaminase	−
Nutrient broth, 0% NaCl	+
Nutrient broth, 6% NaCl	−
Alkalinization of:	
Acetamide	−
Serine	−
Tartrate	−
Sodium acetate	(+)

[a]Incubations were at 30° C, unless indicated otherwise.
[b]The slant was inoculated heavily.
[c]Incubation of 7–14 days.

Afipia genomospecies 3

Fastidious nonfermentative Gram-negative rod; motile by means of a single polar flagellum; oxidase-positive; no growth occurs on MacConkey agar; urease-positive; produces small amounts of H_2S (detectable by the lead acetate paper method); oxidizes xylose weakly; optimal growth is obtained at 25° to 30° C; grows well on buffered charcoal-yeast extract agar and in nutrient broth.

Source (1 strain)

Water.

Reference strain

G5357 = ATCC 49723, type strain, from water.

Literature

1. Brenner, D.J., D.G. Hollis, C.W. Moss, et al. 1991. Proposal of *Afipia* gen. nov. with *Afipia felis* sp. nov. (formerly the cat scratch disease bacillus), *Afipia clevelandensis* sp. nov. (formerly the Cleveland Clinic Foundation strain), *Afipia broomeae* sp. nov. and three unnamed genospecies. J. Clin. Microbiol. 29: 2450–2460.

Comments

The three *Afipia* genomospecies originally were called genospecies. Since the term genospecies has been used previously in another context, these organisms are now called genomospecies. Information on the cellular fatty acid composition of this group is presented in Table 4.6 and Figure 4.11 of the cellular fatty acid section.

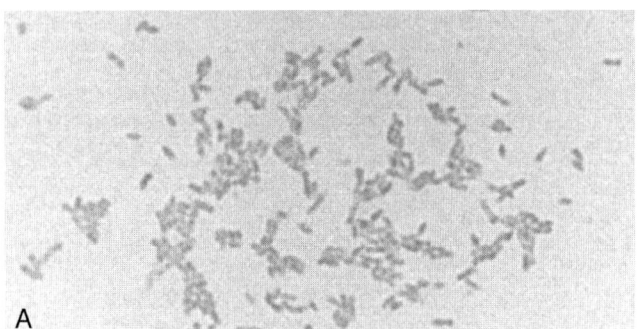

Afipia genomospecies 3 strain G5357. Gram stain BCYE 30° C 48 h x 1700.

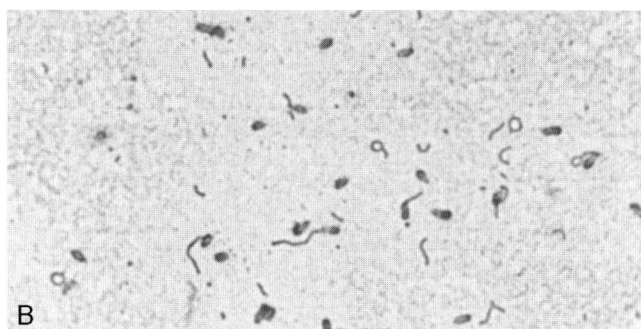

Afipia genomospecies 3 strain G5357. Flagella stain (Ryu method) TGY 30° C 48 h x 1700.

Afipia genomospecies 3

TEST PERFORMED[a]	REFERENCE STRAIN SIGN
Morphology	srs
Motility; flagella	[m;p,1-2]
Action on blood	–
Fermentative or oxidative	[n-o]
Carbohydrate base	OF
Acid from:	
D-Glucose	[–]
D-Xylose	[(+)]
D-Mannitol	–
Lactose	–
Sucrose	–
Maltose	–
Catalase	–
Oxidase	+
Growth on:	
MacConkey	–
SS	–
Cetrimide	–
Simmons citrate	–
Urea, Christensen's[b]	[(+)]
Nitrate reduction	[–]
Gas from nitrate	–
Nitrite reduction	–
Indole	–
TSI slant, acid	–
TSI butt, acid	–
H_2S (TSI butt)	–
H_2S (Pb ac paper)	(+)
Gelatin hydrolysis[c]	–
Litmus milk	(k)
Pigment	–
Growth at:	
25°C	+
30°C	+
35°C	+
42°C	–
Esculin hydrolysis	–
Phenylalanine deaminase	+
Nutrient broth, 0% NaCl	+
Nutrient broth, 6% NaCl	–
Alkalinization of:	
Acetamide	–
Serine	–
Tartrate	–
Sodium acetate	(+)

[a]Incubations were at 30° C, unless indicated otherwise.
[b]The slant was inoculated heavily.
[c]Incubation of 7–14 days.

Agrobacterium radiobacter biovar 1

Gram-negative rod, motile (peritrichous flagella; if only one flagellum is observed, it is usually polarly, subpolarly, or laterally located); oxidizes glucose, xylose, mannitol, lactose, sucrose, and maltose; grows on MacConkey agar; oxidase- and urease-positive, 3-ketolactonate- and nitrate-positive (most strains also reduce nitrite, usually without gas production); does not produce indole; lysine and ornithine decarboxylase-negative; arginine dihydrolase usually negative.

Sources (66 strains)

Sputum (21%), blood (17%), wound (11%), genitourinary (9%), eye (6%), bronchial wash (3%), other (32%), unknown (1%).

Reference strain

KC1283 ← ATCC 19358, type strain.

Literature

1. Riley, P.S. and R.E. Weaver. 1977. Comparison of thirty-seven strains of Vd-3 bacteria with *Agrobacterium radiobacter*: morphological and physiological observations. J. Clin. Microbiol. *5:* 172–177.

2. Kersters, K., and J. De Ley. 1984. Genus *Agrobacterium* Conn 1942, 359.[AL] *In* N.R. Krieg and J.G. Holt (Eds.), Bergey's Manual of Systematic Bacteriology, Vol. 1, Williams & Wilkins, Baltimore, pp. 244–254.

3. Sawada, H., H. Ieki, H. Oyaizu, and S. Matsumoto. 1993. Proposal for rejection of *Agrobacterium tumefaciens* and revised descriptions for the genus *Agrobacterium* and for *Agrobacterium radiobacter* and *Agrobacterium rhizogenes*. Int. J. Syst. Bacteriol. *43:* 694–702.

Comments

Formerly Vd-3. Distinctive clearing zone is produced by most strains at the surface of litmus milk in 24–48 h; a strongly positive H_2S test results with lead acetate paper placed over a TSI agar slant. *A. radiobacter* biovar 2 differs from *A. radiobacter* biovar 1 by its failure to grow at 35° C and negative 3-ketolactonate reaction. We have received multiple strains that resemble *A. radiobacter* biovar 2 except that they grow lightly at 35° C. Sawada et al. (3) have proposed that *A. radiobacter* biovar 1 be called *A. radiobacter* and that *A. radiobacter* biovar 2 be called *A. rhizogenes*. Information on the cellular fatty acid composition of this group is presented in Table 4.7 and Figure 4.12 of the cellular fatty acid section.

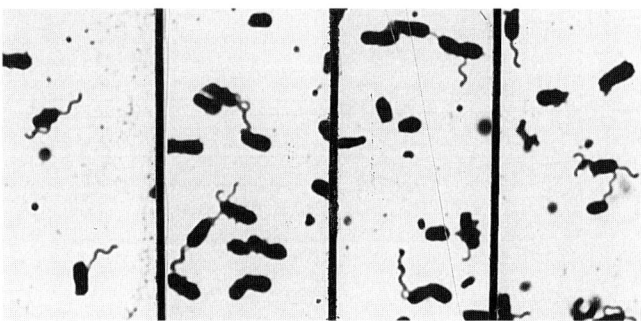

Agrobacterium radiobacter biovar 1. Flagella stain FB 25° C 24 h x 1700.

Agrobacterium radiobacter biovar 1

TEST PERFORMED	SIGN	% +	#+/T
	NUMBER OF STRAINS 66		
Morphology	spr		
Motility; flagella	[m;pe]a		
Action on blood	v	15 ly	10/66; 7 gr, 6 LG, 5 lav, 1 ß
Fermentative or oxidative	[O]		
Carbohydrate base	OF		
Acid from:			
D-Glucose	[+]	94 (6)	56, 6w (4)/66
D-Xylose	[+]	97 (3)	64 (2)/66
D-Mannitol	+	94 (6)	50, 12w(4)/66
Lactose	[+ or (+)]	79 (21)	37, 15w (2) (12w)/66
Sucrose	+	95 (5)	61, 2w (2)/66
Maltose	[+]	97 (3)	62, 2w (2)/66
Catalase	+	98	58, 7w/66
Oxidase	[+]	100	66/66
Growth on:			
MacConkey	[+]	100	66/66
SS	v	20 (5)	3, 10w (2) (1w)/66
Cetrimide	-	0	0/66
Simmons citrate	+	97 (3)	64 (2)/66
Urea, Christensen's	[+ or (+)]	88 (9)	58 (6)/66
Nitrate reduction	v	83	55/66
Gas from nitrate	[-]	5	3/66
Nitrite reduction	v	38	11/29
Indole	[-]	0	0/66
TSI slant, acid	-	0	0/66
TSI butt, acid	-	0	0/66
H_2S (TSI butt)	v	11 (3)	4w, (3) (2w)/66
H_2S (Pb ac paper)	+	100	66/66
Gelatin hydrolysis[b]	-	2	1/57
Litmus milk	v	78 szf	52/66; 2 pep, 3 dirty, 3 k
Pigment	v	21 br yel sol	14/66
Growth at:			
25°C	+	100	56, 10w/66
35°C	+	100	64, 2w/66
42°C	v	32	6, 15w/66
Esculin hydrolysis	+	100	66/66
Lysine decarboxylase	[-]	0	0/48
Arginine dihydrolase	[-]	8	2, 2w/48
Ornithine decarboxylase	-	0	0/48
Nutrient broth, 0% NaCl	+	100	59, 7w/66
Nutrient broth, 6% NaCl	v	18	10, 2w/66
3-Ketolactonate	[+]	100	66/66

[a]Frequently individual cells have only a single flagellum, either polar, subpolar, or lateral.
[b]Incubation of 7–14 days.

"*Agrobacterium* yellow group"

Gram-negative, slender to medium width, medium sometimes a long straight rod; motile with single polar flagellum; motility difficult to demonstrate in motility medium; produces a yellow insoluble growth pigment; aerobic; variable growth on MacConkey agar; oxidizes glucose, xylose, lactose, sucrose, and maltose, but not mannitol; carbohydrate reactions may be weak and/or late; oxidase- and catalase-positive; urease-positive (weak and late); nitrate-negative; 3-ketolactonate-positive.

Sources (2 strains)

Blood (1), unknown (1).

Reference strain

CDC G618 ← B. Holmes (T225/82), NCTC, England.

Literature

1. Holmes, B., and P. Roberts. 1981. The classification, identification, and nomenclature of agrobacteria. J. Appl. Bacteriol. *50:* 443–467.

2. Swann, R.A., S.J. Foulkes, B. Holmes, J.B. Young, R.G. Mitchell, and S.T. Reeders. 1985. "*Agrobacterium* yellow group" and *Pseudomonas paucimobilis* causing peritonitis in patients receiving continuous ambulatory peritoneal dialysis. J. Clin. Pathol. *38:* 1293–1299.

Comments

This organism closely resembles *Sphingomonas paucimobilis*. The 3-ketolactonate test is useful in differentiation. Lateral or subpolar flagella may be observed on rare cells.

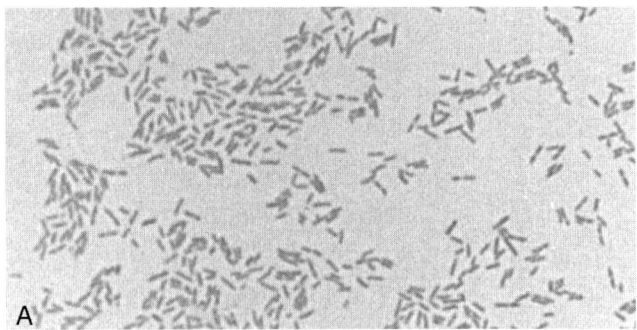

"*Agrobacterium* yellow group" strain G618. Gram stain BAP 35° C 24 h x 1700.

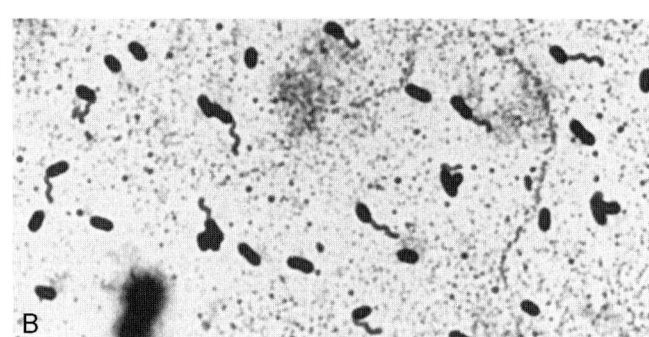

"*Agrobacterium* yellow group" strain G618. Flagella stain (Ryu method) TGY 25° C 24 h x 1700.

"*Agrobacterium* yellow group"

TEST PERFORMED	SIGN	% +	#+/T
		NUMBER OF STRAINS 2	
Morphology	lrs		
Motility; flagella	[m;p,1-2]a		2/2
Action on blood	–	0	0/2
Fermentative or oxidative	[O]		
Carbohydrate base	OF		
Acid from:			
D-Glucose	(+)	(100)	(2w)/2
D-Xylose	+ or (+)	50 (50)	1, (1)/2
D-Mannitol	[–]	0	0/2
Lactose	[(+)]	(100)	(2)/2
Sucrose	(+)	(100)	(1), (1w)/2
Maltose	[+]	100	2/2
Catalase	+	100	2/2
Oxidase	+	100	2/2
Growth on:			
MacConkey	v	(50)	(1w)/2
SS	–	0	0/2
Cetrimide	–	0	0/2
Simmons citrate	–	0	0/2
Urea, Christensen's	(+)	(100)	(2w)/2
Nitrate reduction	–	0	0/2
Gas from nitrate	–	0	0/2
Nitrite reduction	–	0	0/1
Indole	–	0	0/2
TSI slant, acid	–	0	0/2
TSI butt, acid	–	0	0/2
H_2S (TSI butt)	–	0	0/2
H_2S (Pb ac paper)	(+)	(100)	(1), (1w)/2
Gelatin hydrolysisb	–	0	0/2
Litmus milk	v	(50 IR)	(1 wIR)/2
Pigment:			
Insoluble	[yel]	100	2/2
Growth at:			
25°C	+	100	2/2
35°C	+	100	2/2
42°C	–	0	0/2
Esculin hydrolysis	[+ or (+)]	50 (50)	1, (1)/2
Lysine decarboxylase	–	0	0/1
Arginine dihydrolase	–	0	0/1
Ornithine decarboxylase	–	0	0/1
Nutrient broth, 0% NaCl	+ or (+)	50 (50)	1, (1)/2
Nutrient broth, 6% NaCl	–	0	0/2
3-Ketolactonate	[+]	100	2/2

aMotility may be difficult to demonstrate.
bIncubation of 7–14 days.

Alcaligenes eutrophus

Gram-negative short straight rod; motile with peritrichous flagella; motility is difficult to demonstrate in motility medium but can be observed in wet mounts of overnight broth cultures; aerobic; grows on MacConkey agar; does not usually oxidize carbohydrates, although weak glucose oxidization can be observed with prolonged incubation (14–21 days); oxidase-, catalase- and urease-positive; reduces nitrate without gas formation.

Source (1 strain)

Soil.

Reference strain

Alcaligenes eutrophus strain KC1340 ← ATCC 17697, type strain, from soil.

Literature

1. Rossau, R., K. Kersters, E. Falsen, E. Jantzen, P. Segers, A. Union, L. Nehls, and J. De Ley. 1987. *Oligella*, a new genus including *Oligella urethralis* comb. nov. (formerly *Moraxella urethralis*) and *Oligella ureolytica* sp. nov. (formerly CDC group IVe); relationship to *Taylorella equigenitalis* and related taxa. Int. J. Syst. Bacteriol. *37*: 198–210.

Comments

Cells with polar and/or subpolar flagella also may be observed. This organism closely resembles CDC group IVc-2 and *Bordetella bronchiseptica*. Growth on SS agar and nitrate reduction are useful differentiating characteristics.

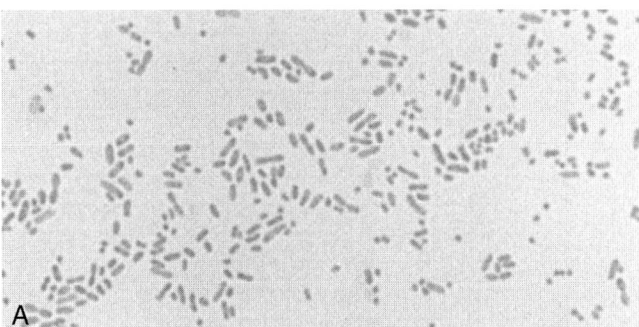

Alcaligenes eutrophus strain KC1340. Gram stain BAP 35° C 24 h x 1700.

Alcaligenes eutrophus

TEST PERFORMED	SIGN
REFERENCE STRAIN	
Morphology	srs
Motility; flagella	[m;pe]
Action on blood	-
Fermentative or oxidative	[n-o]
Carbohydrate base	OF
Acid from:	
D-Glucose[a]	[-]
D-Xylose	-
D-Mannitol	-
Lactose	-
Sucrose	-
Maltose	-
Catalase	+
Oxidase	[+]
Growth on:	
MacConkey	[+]
SS	[-]
Cetrimide	-
Simmons citrate	+
Urea, Christensen's	(+)
Nitrate reduction	[+]
Gas from nitrate	-
Indole	-
TSI slant, acid	-
TSI butt, acid	-
H_2S (TSI butt)	-
H_2S (Pb ac paper)	[-]
Gelatin hydrolysis[b]	-
Litmus milk	k
Pigment	-
Growth at:	
25°C	+
35°C	+
42°C	+
Esculin hydrolysis	-
Lysine decarboxylase	-
Arginine dihydrolase	-
Ornithine decarboxylase	-
Nutrient broth, 0% NaCl	+
Nutrient broth, 6% NaCl	-

[a]Weak acid may be observed at 7–14 days' incubation.
[b]Incubation of 7–14 days.

Alcaligenes faecalis

Gram-negative coccoid to medium-length rod; motile (peritrichous flagella); does not use carbohydrates; grows on MacConkey and SS agars; oxidase-positive, urease-negative; does not reduce nitrate; reduces nitrite to gas.

Sources (49 strains)

Urine (43%), ear (16%), wound (8%), toe (4%), leg ulcer (4%), other (21%).

Reference strain

KC494 ← ATCC 8750, type strain.

Literature

1. Málek, I., M. Radochová, and O. Lysenko. 1963. Taxonomy of the species *Pseudomonas odorans*. J. Gen. Microbiol. 33: 349–355.
2. Kersters, K., K.-H. Hinz, A. Hertle, P. Segers, A. Lievens, O. Siegmann, and J. De Ley. 1984. *Bordetella avium* sp. nov., isolated from the respiratory tracts of turkeys and other birds. Int. J. Syst. Bacteriol. 34: 56–70.
3. Kersters, K., and J. De Ley. 1984. Genus *Alcaligenes* Castellani and Chalmers 1919, 936. *In* N.R. Krieg and J.G. Holt (Eds.), Bergey's Manual of Systematic Bacteriology, Vol. 1. Williams & Wilkins, Baltimore, pp. 361–373.

Comments

Formerly CDC group VI. "*A. odorans*" is a synonym for *A. faecalis*. *A. faecalis* is, however, the accepted name because it was validly published prior to the description of "*A. odorans*." Produces a distinctive fruity odor. Growth is dull, spreading. Biochemical profile of *A. faecalis* is similar to *Bordetella avium*. Nitrite reduction is a useful discriminating characteristic. Information on the cellular fatty acid composition of this species is presented in Table 4.8 and Figure 4.13 of the cellular fatty acid section.

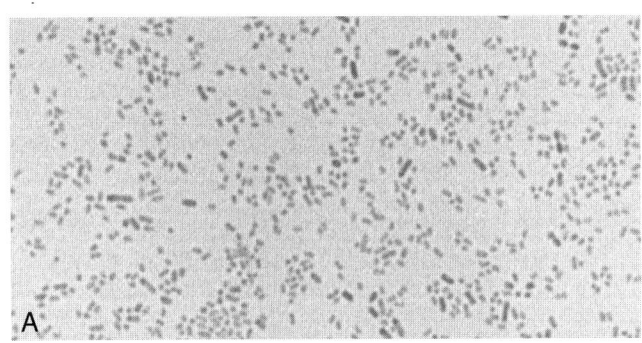

Alcaligenes faecalis strain KC494. Gram stain BAP 35° C 24 h x 1700.

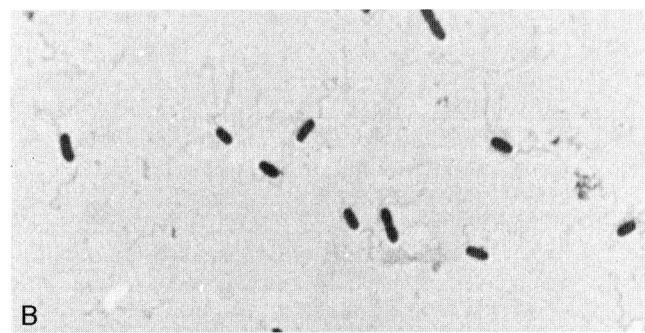

Alcaligenes faecalis strain KC494. Flagella stain (Ryu method) TGY 25° C 24 h x 1700.

Alcaligenes faecalis

TEST PERFORMED	SIGN	% +	#+/T
	NUMBER OF STRAINS 49		
Morphology	mrs		
Motility; flagella	[m;pe]		
Action on blood	v	48 br & gr	23/48; 8 gr, 3 ly, 3 br
Fermentative or oxidative	[n-o]		
Carbohydrate base	OF		
Acid from:			
D-Glucose	[−]	0	0/49
D-Xylose	[−]	0	0/49
D-Mannitol	−	0	0/49
Lactose	−	0	0/49
Sucrose	−	0	0/49
Maltose	−	0	0/49
Catalase	+	98	48/49
Oxidase	[+]	100	48, 1w/49
Growth on:			
MacConkey	[+]	100	49/49
SS	[+]	100	47, 2w/49
Cetrimide	v	59	25, 4w/49
Simmons citrate	+	100	49/49
Urea, Christensen's	[−]	2	1w/49; 13 pk
Nitrate reduction	[−]	0	0/49
Gas from nitrate	−	0	0/49
Nitrite reduction	[+]	100	49/49
Indole	−	0	0/49
TSI slant, acid	−	0	0/49
TSI butt, acid	−	0	0/49
H_2S (TSI butt)	−	0	0/49
H_2S (Pb ac paper)	−	8	4w/49
Gelatin hydrolysis[a]	v	22	9/41
Litmus milk	k	96	47/49
Pigment	v	22 yel sol	11/49
Growth at:			
25°C	+	100	49/49
35°C	+	100	49/49
42°C	v	18	6, 3w/49
Esculin hydrolysis	−	0	0/49
Lysine decarboxylase	−	0	0/9
Arginine dihydrolase	−	0	0/9
Ornithine decarboxylase	−	0	0/9
Nutrient broth, 0% NaCl	+	100	49/49
Nutrient broth, 6% NaCl	+	98 (2)	45, 3w (1)/49

[a]Incubation of 7–14 days.

CDC *Alcaligenes*-like Group 1

Gram-negative rod, motile (peritrichous flagella); does not utilize carbohydrates; grows on MacConkey agar but usually not on SS agar; oxidase-positive; urease-variable; reduces both nitrate and nitrite to gas; does not produce H_2S in TSI agar.

Sources (8 strains)

Blood (3 strains), brain abscess (1 strain), urine (1 strain), bronchial wash (1 strain), knee joint (1 strain), water (1 strain).

Reference strain

F6382, from blood.

Literature

None.

Comments

Similar to *Alcaligenes xylosoxidans* subsp. *denitrificans*. Differential tests include weak to no growth on SS and/or cetrimide agar, alkalinization of tartrate, and cellular fatty acid analysis. The cellular fatty acid profile for these organisms is listed on Figure 4.15 and Table 4.10 in the cellular fatty acid section.

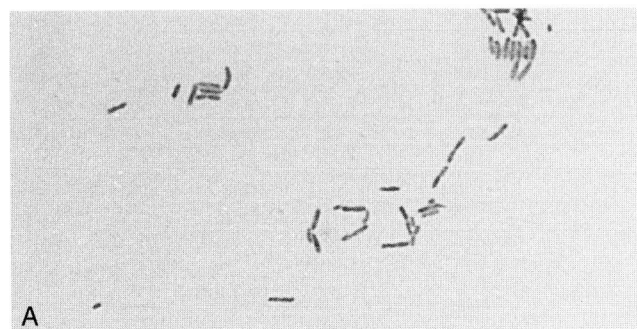

Alcaligenes-like group 1 strain F6382. Gram stain BAP 35° C 24 h x 1700.

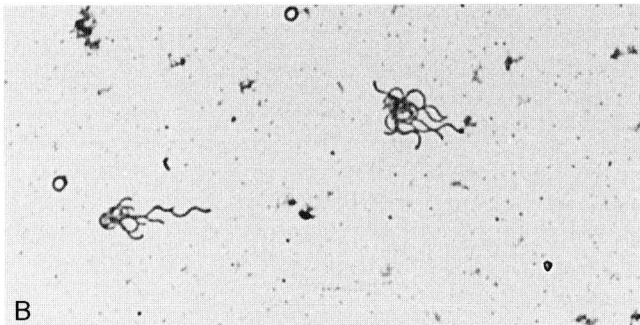

Alcaligenes-like group 1 strain F6382. Flagella stain (Ryu method) TGY 25° C 24 h x 1700.

CDC *Alcaligenes*-like Group 1

TEST PERFORMED	SIGN	% +	#+/T
	NUMBER OF STRAINS 8		
Morphology	mrs		
Motility; flagella	[m;pe]		
Action on blood	v	25 ly	2/8; 1 gr, 1 lav
Fermentative or oxidative	[n-o]		
Carbohydrate base	OF		
Acid from:			
D-Glucose	[-]	0	0/8
D-Xylose	[-]	0	0/8
D-Mannitol	-	0	0/8
Lactose	-	0	0/8
Sucrose	-	0	0/8
Maltose	-	0	0/8
Catalase	+	100	8/8
Oxidase	[+]	100	8/8
Growth on:			
MacConkey	[+]	100	8/8
SS	[v]	13	1w/8
Cetrimide	-	0	0/8
Simmons citrate	+	100	8/8
Urea, Christensen's	v	75	5, 1w/8
Nitrate reduction	[+]	100	8/8
Gas from nitrate	[+]	100	8/8
Indole	-	0	0/8
TSI slant, acid	-	0	0/8
TSI butt, acid	-	0	0/8
H$_2$S (TSI butt)	[-]	0	0/8
H$_2$S (Pb ac paper)	v	13	1w/8
Gelatin hydrolysis[a]	-	0	0/8
Litmus milk	v	88k	7/8
Pigment	-	0	0/8
Growth at:			
25°C	+	100	8/8
35°C	+	100	8/8
42°C	v	50	4/8
Esculin hydrolysis	-	0	0/8
Lysine decarboxylase	-	0	0/3
Arginine dihydrolase	v	(33)	(1w)/3
Ornithine decarboxylase	-	0	0/3
Alkalinization of:			
Acetamide	-	0	0/4
Serine	v	(75)	(3)/4
Tartrate	[-]	0	0/4
Nutrient broth, 0% NaCl	+	100	8/8
Nutrient broth, 6% NaCl	v	13	1w/8

[a]Incubation of 7–14 days.

Alcaligenes piechaudii

Gram-negative rod, motile (peritrichous flagella); does not use carbohydrates; grows on MacConkey agar; oxidase- and catalase-positive; urease-negative; reduces nitrate without gas formation; does not reduce nitrite.

Sources (1 strain each)

Pharyngeal swab, nose, nose wound, unknown.

Reference strains

KC1949 = ATCC 43552, type strain, from pharyngeal swab.
KC1964 = LMG 6100, from nose wound.
KC1965 = LMG 6102, from nose.
KC1966 = LMG 6103.

Literature

1. Kiredjian, M., B. Holmes, K. Kersters, I. Guilvout, and J. De Ley. 1986. *Alcaligenes piechaudii*, a new species from human clinical specimens and the environment. Int. J. Syst. Bacteriol. *36:* 282–287.

2. Peel, M.M., A.J. Hibberd, B.M. King, and H.G. Williamson. 1988. *Alcaligenes piechaudii* from chronic ear discharge. J. Clin. Microbiol. *26:* 1580–1581.

Comments

A. piechaudii is biochemically similar to *A. faecalis* and *Bordetella avium*. Nitrate and nitrite reduction are useful differential characteristics. Information on the cellular fatty acid composition of this species is presented in Table 4.9 and Figure 4.14 of the cellular fatty acid section.

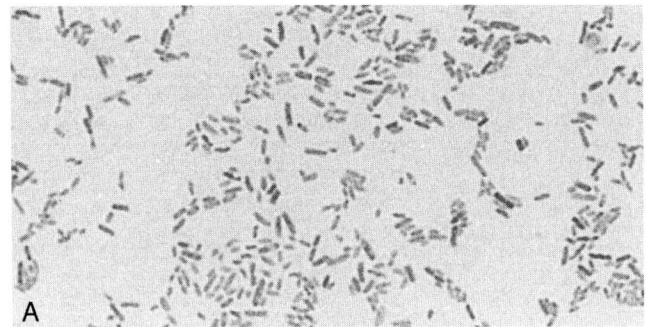

Alcaligenes piechaudii strain KC1949. Gram stain BAP 35° C 24 h x 1700.

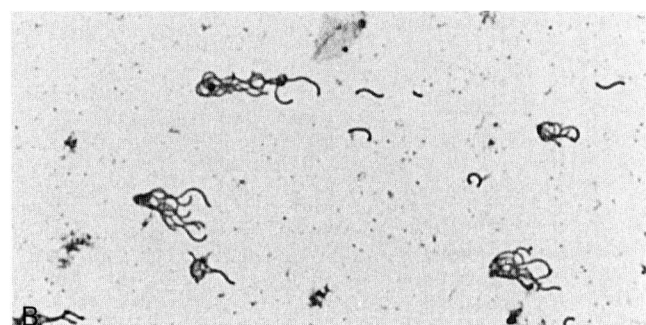

Alcaligenes piechaudii strain KC1949. Flagella stain (Ryu method) TGY 25° C 24 h x 1700.

Alcaligenes piechaudii

TEST PERFORMED	NUMBER OF STRAINS 4		
	SIGN	%+	#+/T
Morphology	rs		
Motility; flagella	[m;pe]		
Action on blood	v	75 ly	3/4
Fermentative or oxidative	[n-o]		
Carbohydrate base	OF		
Acid from:			
D-Glucose	[-]	0	0/4
D-Xylose	-	0	0/4
D-Mannitol	-	0	0/4
Lactose	-	0	0/4
Sucrose	-	0	0/4
Maltose	-	0	0/4
Catalase	+	100	4/4
Oxidase	+	100	4/4
Growth on:			
MacConkey	[+]	100	4/4
SS	+	100	4/4
Cetrimide	+	100	4/4
Simmons citrate	+	100	4/4
Urea, Christensen's	-	0	0/4
Nitrate reduction	[+]	100	4/4
Gas from nitrate	[-]	0	0/4
Nitrite reduction	[-]	0	0/4
Indole	-	0	0/4
TSI slant, acid	-	0	0/4
TSI butt, acid	-	0	0/4
H_2S (TSI butt)	-[a]	0	0/4
H_2S (Pb ac paper)	+	100	4/4
Gelatin hydrolysis[b]	-	0	0/4
Litmus milk	k	100	4/4
Pigment	-	0	0/4
Growth at:			
25°C	+	100	4/4
35°C	+	100	4/4
42°C	v	75	3/4
Esculin hydrolysis	-	0	0/4
Nutrient broth, 0% NaCl	+	100	4/4
Nutrient broth, 6% NaCl	+[c]	100	4/4
Alkalinization of:			
Acetamide	-	0	0/3
Serine	v	50 (25)	2 (1)/4
Tartrate	+	100	4/4

[a]The line of the stab became dark in 3–7 days with 3 of the 4 strains.
[b]Incubation of 7–14 days.
[c]Light growth at 48 h, heavier at 7 days.

Alcaligenes xylosoxidans subsp. *denitrificans*

Gram-negative rod, motile (peritrichous flagella); does not utilize carbohydrates; grows on MacConkey and SS agar; oxidase-positive, urease-negative; reduces both nitrate and nitrite to gas; does not produce H_2S in TSI agar.

Sources (3 strains)

Wound (2 strains), soil (1 strain).

Reference strain

KC367 ← E. Leifson, 412 = ATCC 15173, type strain, from soil.

Literature

1. Leifson, E., and R. Hugh. 1954. *Alcaligenes denitrificans* n.sp. J. Gen. Microbiol. *11:* 512–513.
2. Kiredjian, M., B. Holmes, K. Kersters, I. Guilvout, and J. De Ley. 1986. *Alcaligenes piechaudii*, a new species from human clinical specimens and the environment. Int. J. Syst. Bacteriol. *36:* 282–287.

Comments

Former names are *Alcaligenes denitrificans* and *A. denitrificans* subsp. *denitrificans*. The present name was assigned after it was demonstrated that *A. denitrificans* has a high degree of genetic relatedness with *Alcaligenes xylosoxidans* subsp. *xylosoxidans* (formerly *Achromobacter xylosoxidans*) (2). The species name, *denitrificans*, was corrected to *xylosoxidans* because the name *xylosoxidans* has priority (it was proposed earlier than *denitrificans*) (2). The cellular fatty acid profile of these strains presented in Figure 4.13 and Table 4.8 of the CFA section. This group is similar to CDC *Alcaligenes*-like group 1, but can be differentiated by its ability to grow well on SS agar and its cellular fatty acid profile.

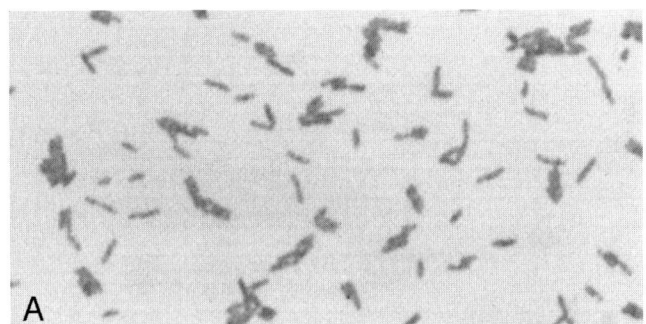

Alcaligenes xylosoxidans subsp. *denitrificans* strain KC367. Gram stain BAP 35° C 24 h x 1700.

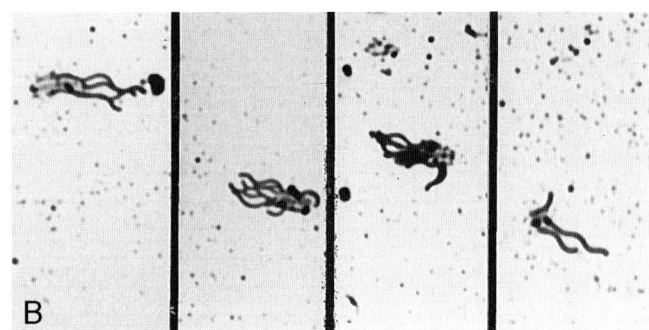

Alcaligenes xylosoxidans subsp. *denitrificans* strain KC367. Flagella stain FB 25° C 24 h x 1700.

Alcaligenes xylosoxidans subsp. *denitrificans*

TEST PERFORMED	TYPE STRAIN SIGN	3 PHENOTYPICALLY SIMILAR STRAINS		
		SIGN	% +	#+/T
Morphology	mrs	mrs		
Motility; flagella	[m;pe]	[m;pe]		
Action on blood	ly	v	33 ß	1/3
Fermentative or oxidative	[n-o]	[n-o]		
Carbohydrate base	OF	OF		
Acid from:				
D-Glucose	[-]	[-]	0	0/3
D-Xylose	[-]	[-]	0	0/3
D-Mannitol	-	-	0	0/3
Lactose	-	-	0	0/3
Sucrose	-	-	0	0/3
Maltose	-	-	0	0/3
Catalase	+	+	100	3/3
Oxidase	[+]	[+]	100	3/3
Growth on:				
MacConkey	[+]	[+]	100	3/3
SS	[+]	[+]	100	3/3
Cetrimide	+	v	(33)	(1)/3
Simmons citrate	+	+	100	3/3
Urea, Christensen's	-	-	0	0/3
Nitrate reduction	[+]	[+]	100	3/3
Gas from nitrate	[+]	[+]	100	3/3
Indole	-	-	0	0/3
TSI slant, acid	-	-	0	0/3
TSI butt, acid	-	-	0	0/3
H_2S (TSI butt)	[-]	[-]	0	0/3
H_2S (Pb ac paper)	-	v	33	1w/3
Gelatin hydrolysis[a]	-	-	0	0/3
Litmus milk	k	v	67 k	2/3; 1 IR
Pigment	-	v	33 yel-sol	1w/3
Growth at:				
25°C	+	+	100	3/3
35°C	+	+	100	3/3
42°C	+w	-	0	0/3
Esculin hydrolysis	-	-	0	0/3
Alkalinization of:				
Acetamide	-	+ or (+)	50 (50)	1 (1)/2
Serine	(+)	+ or (+)	50 (50)	1 (1)/2
Tartrate	[+]	[+ or (+)]	67 (33)	2 (1)/3
Nutrient broth, 0% NaCl	+	+	100	3/3
Nutrient broth, 6% NaCl	-	v	33	1/3

[a] Incubation of 7–14 days.

Alcaligenes xylosoxidans subsp. *xylosoxidans*

Gram-negative short to medium straight rod, motile (peritrichous flagella); oxidase-positive; oxidizes glucose (weakly) and xylose; grows on MacConkey and SS agars; citrate-positive, urease-negative; nitrate-positive, sometimes with gas; H_2S-negative (TSI).

Sources (135 strains)

Blood (14%), cerebrospinal fluid (11%), urine (10%), sputum (6%), wound (5%), ear and bubo (3% each), saline, water, skin, throat, mouth, and stool (2% each), other (15%), unknown (21%).

Reference strain

KC1064 ← E. Yabuuchi, KM453 = ATCC 27061, type strain, from ear discharge.

Literature

1. Yabuuchi, E., and A. Ohyama. 1971. *Achromobacter xylosoxidans* sp. n. from human ear discharge. Jpn. J. Microbiol. *15:* 477–481.

2. Kersters, K., and J. De Ley. 1984. Genus *Alcaligenes* Castellani and Chalmers 1919, 936[AL]. *In* N.R. Krieg and J.G. Holt (Eds.), Bergey's Manual of Systemic Bacteriology, Vol. 1, Williams & Wilkins, Baltimore, pp. 361–373.

3. Kiredjian, M., B. Holmes, K. Kersters, I. Guilvout, and J. De Ley. 1986. *Alcaligenes piechaudii*, a new species from human clinical specimens and the environment. Int. J. Syst. Bacteriol. *36:* 282–287.

Comments

Formerly named *Achromobacter xylosoxidans*, this group has been placed in the genus *Alcaligenes* based on a high degree of genetic relatedness with *Alcaligenes xylosoxidans* subsp. *denitrificans*. This group was originally renamed *Alcaligenes denitrificans* subsp. *xylosoxydans* (2), and then corrected to its current form, which recognizes the priority of the epithet *xylosoxidans* (3). Information on the cellular fatty acid composition of this group is presented in Table 4.8 and Figure 4.13 of the cellular fatty acid section.

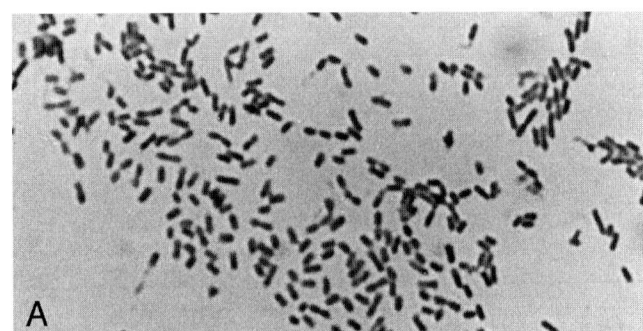

Alcaligenes xylosoxidans subsp. *xylosoxidans* strain KC1064. Gram stain BAP 35° C 24 h x 1700.

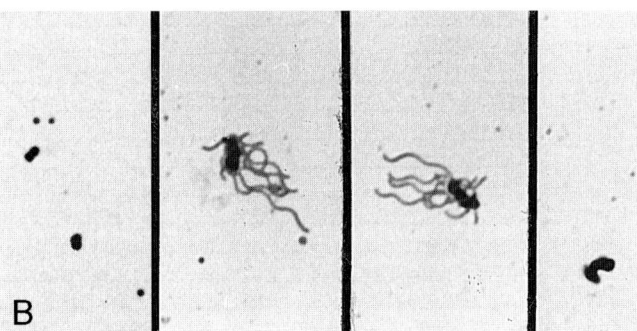

Alcaligenes xylosoxidans subsp. *xylosoxidans* strain KC1064. Flagella stain FB 25° C 24 h x 1700.

Alcaligenes xylosoxidans subsp. **xylosoxidans**

TEST PERFORMED	SIGN	% +	#+/T
		NUMBER OF STRAINS 135	
Morphology	mrs		
Motility; flagella	[m;pe]		
Action on blood	v	42 ly	51/122; 24 LG, 21 gr, 5 lav, 2 al
Fermentative or oxidative	[O]		
Carbohydrate base	OF		
Acid from:			
D-Glucose	[v]	78	2, 97w/127
D-Xylose	[+]	99	123, 1w/125
D-Mannitol	-	0	0/129
Lactose	-	0	0/129
Sucrose	-	0	0/130
Maltose	-	0	0/128
Catalase	+	98	126/128
Oxidase	+	100	133/133
Growth on:			
MacConkey	[+]	100	133/133
SS	[+]	98	130/133
Cetrimide	+	95(1)	119, 4w(1)/129
Simmons citrate	+	95	128/130
Urea, Christensen's	[-]	0	0/128; 2 pk
Nitrate reduction	[+]	100	133/133
Gas from nitrate	[v]	60	80/133
Indole	-	0	0/134
TSI slant, acid	-	0	0/132
TSI butt, acid	-	0	0/132
H_2S (TSI butt)	[-]	0	0/133
H_2S (Pb ac paper)	-	0	0/133
Gelatin hydrolysis[a]	-	0	0/112
Litmus milk	k	95	122/128
Pigment	-	5 br sol	6/112
Growth at:			
25°C	+	98	120/122
35°C	+	100	122/122
42°C	v	84	102/122
Esculin hydrolysis	-	0	0/118
Lysine decarboxylase	-	0	0/16
Arginine dihydrolase	v	13	2/16
Ornithine decarboxylase	-	0	0/16
Nutrient broth, 0% NaCl	+	100	118/118
Nutrient broth, 6% NaCl	v	69	82/118

[a]Incubation of 7–14 days.

Balneatrix alpica

Gram-negative short to medium length rod, motile (single polar flagellum); produces acid (may be weak or late) from glucose, mannitol and maltose, but not from xylose, lactose, or sucrose; oxidase-positive; does not grow on MacConkey or SS agars; utilizes citrate; indole-positive; urease-negative.

Source

Cerebrospinal fluid.

Reference strain

KC1954 ← CIP 103589, type strain, from cerebrospinal fluid.

Literature

1. Dauga, C., M. Gillis, et al. 1993. *Balneatrix alpica* gen. nov., sp. nov., a bacterium associated with pneumonia and meningitis in a spa therapy centre. Res. Microbiol. *144:* 35–46.
2. Casalta, J.P, Y. Peloux, et al. 1989. Pneumonia and meningitis caused by a new nonfermentative unknown gram-negative bacterium. J. Clin. Microbiol. *27:* 1446–1448.

Comments

Biochemically similar to *Flavobacterium meningosepticum,* with motility and failure to hydrolyze ONPG useful differential characteristics. Motility may be difficult to demonstrate. Has been associated with pneumonia and meningitis. Information on the cellular fatty acid composition of this species is presented in Table 4.13 and Figure 4.18 of the cellular fatty acid section.

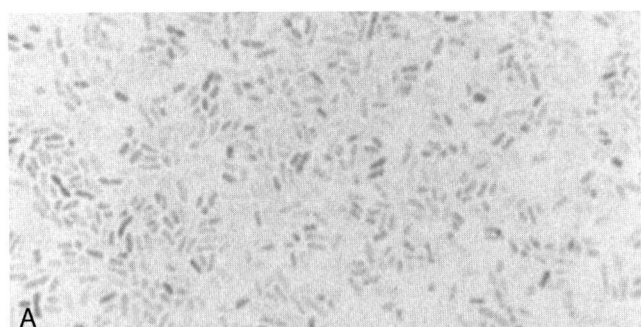

Balneatrix alpica strain G8608. Gram stain BAP 35° C 24 h x 1700.

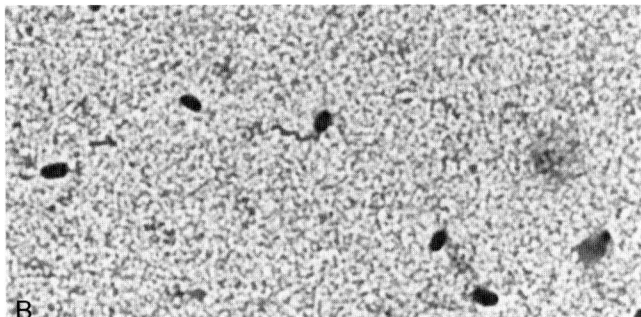

Balneatrix alpica strain G8608. Flagella stain (Ryu method) TGY 25° C 24 h x 1700.

Balneatrix alpica

TEST PERFORMED	TYPE STRAIN SIGN
Morphology	srs, mrs
Motility; flagella	[m;p]
Action on blood	−
Fermentative or oxidative	[O]
Carbohydrate base	OF
Acid from:	
D-Glucose	[+]
D-Xylose	−
D-Mannitol	[(+)]
Lactose	−
Sucrose	−
Maltose	+
Catalase[a]	+
Oxidase	+
Growth on:	
MacConkey	−
SS	−
Cetrimide	−
Simmons citrate	+
Urea, Christensen's	−
Nitrate reduction[b]	[+]
Gas from nitrate	−
Nitrite reduction	−
Indole	[+]
TSI slant, acid	−
TSI butt, acid	−
H_2S (TSI butt)	−
H_2S (Pb ac paper)	+
Gelatin hydrolysis[c]	−
Litmus milk	k
Pigment	ta-yel sol
Growth at:	
25°C	+
35°C	+
42°C	+
Esculin hydrolysis	[−]
Nutrient broth, 0% NaCl	+
Nutrient broth, 6% NaCl	−
Alkalinization of:	
Acetamide	−
Serine	+
Tartrate	−
Sodium acetate	−
Phenylalanine deaminase	+
ONPG	−

[a]Weak reaction when tested on heart infusion agar-grown cells at 24 h of incubation.
[b]Reaction observed in infusion base, no growth occurs in routine nitrate broth.
[c]Incubation of 7–14 days.

Bartonella bacilliformis

Gram-negative, fastidious, motile (1–10 polar flagella; subpolar and/or lateral flagella may be observed on rare cells), grows slowly (3–5 days' incubation) on heart infusion agar with 5% rabbit blood; grows better at 25° C than at 35° C; aerobic; no growth on MacConkey or nutrient agar; does not use carbohydrates; oxidase-negative; catalase-positive; citrate-, indole-, and urease-negative.

Sources (2 strains)

Unknown.

Reference strain

KC583 ← Herrer 020/F12,63 ← ATCC 35685, type strain.

Literature

1. Brenner, D.J., S.P. O'Connor, D.G. Hollis, R.E. Weaver, and A.G. Steigerwalt. 1991. Molecular characterization and proposal of a neotype strain for *Bartonella bacilliformis*. J. Clin. Microbiol. *29:* 1229–1302.

2. O'Connor, S.P., M. Dorsch, A.G. Steigerwalt, D.J. Brenner, and E. Stackebrandt. 1991. 16S rRNA sequences of *Bartonella bacilliformis* and cat scratch disease bacillus reveal phylogenetic relationships with the alpha-2 subgroup of the class *Proteobacteria*. J. Clin. Microbiol. *29:* 2144–2150.

3. D. Weinman. 1981. Bartonellosis and anemia associated with *Bartonella*-like structures, *In* A. Balows and W. Hausler, Jr. (Eds.), Diagnostic Procedures for: Bacterial, Mycotic, and Parasitic Infections, 6th ed., American Public Health Association, Washington, pp. 235–248.

4. Relman, D.A., P.W. Lepp, K.N. Sadler, and T.M. Schmidt. 1992. Phylogenetic relationships among the agent of bacillary angiomatosis, *Bartonella bacilliformis,* and other alpha-proteobacteria. Mol. Microbiol. *6:* 1801–1807.

5. Gray, G.C., et al. 1990. An epidemic of oroya fever in the Peruvian Andes. Am. J. Trop. Med. Hyg. *42:* 215–221.

Comments

B. bacilliformis is the causative agent of bartonellosis, a febrile disease limited to remote mountaneous regions of South America. Optimal growth on rabbit blood agar is obtained by incubation at room temperature in a jar with a small beaker of distilled water. Closely related to *Bartonella quintana* (1, 2). Information on the cellular fatty acid composition of this species is presented in Table 4.14 and Figure 4.19 of the cellular fatty acid section.

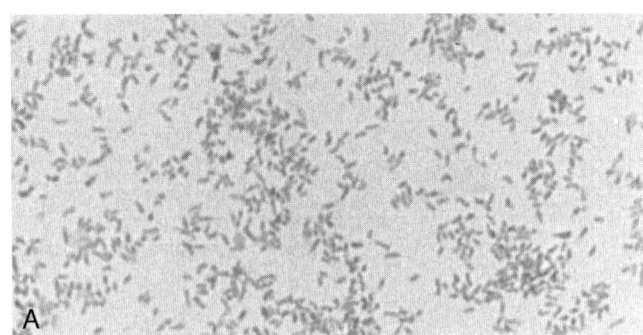

Bartonella bacilliformis strain KC583. Gram stain BAP 25° C 5 days x 1700.

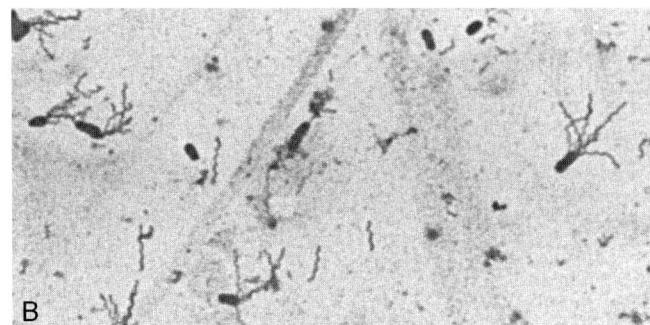

Bartonella bacilliformis strain KC583. Flagella stain (Ryu method) BAP 25° C 5 days x 1700.

Bartonella bacilliformis[a]

TEST PERFORMED	SIGN	% +	#+/T
	NUMBER OF STRAINS 2		
Morphology	srs		
Motility; flagella	[m;p,1-10]		
Action on blood	−		
Fermentative or oxidative	[n-o]		
Carbohydrate base	RFT		
Acid from:			
D-Glucose	−	0	0/2
D-Xylose	−	0	0/2
D-Mannitol	−	0	0/2
Lactose	−	0	0/2
Sucrose	−	0	0/2
Maltose	−	0	0/2
Catalase[b]	[+]	100	2/2
Oxidase	−	0	0/2
Growth on:			
MacConkey	−	0	0/2
SS	−	0	0/2
Cetrimide	−	0	0/2
Simmons citrate	−	0	0/2
Urea, Christensen's[c]	−	0	0/2
Indole[d]	−	0	0/1
Growth on HIA with:			
X factor	−	0	0/2
V factor	[−]	0	0/2
Rabbit blood	[+]	100	2/2
Ornithine decarboxylase	−	0	0/2
Growth at:[e]			
25°C	(+)	(100)	(2)/2
35°C	(+)	(100)	(2w)/2
42°C	−	0	0/2
Nutrient broth, 0% NaCl	−	0	0/2
Nutrient broth, 6% NaCl	−	0	0/2

[a]Did not grow on media used for esculin hydrolysis, nitrate reduction, H_2S production, and gelatin hydrolysis.
[b]Performed on cells grown on rabbit blood agar.
[c]Heavy inoculum.
[d]Spot indole test.
[e]Rabbit blood agar.

Bartonella (formerly *Rochalimaea*) species

Gram-negative; fastidious; nonmotile; grows slowly (3–5 days' incubation); grows best at 35–37° C; growth and biochemical reactivity may be stimulated by hemin; asaccharolytic; catalase-, urease-, and alkaline phosphatase-negative.

Sources

B. henselae (14 strains)
    Blood (11 strains), lymph node (2 strains), bone marrow (1 strain).

B. quintana (2 strains)
    Blood (1 strain), unknown (1 strain).

B. vinsonii (1 strain)
    Canadian vole.

B. elizabethae (1 strain)
    Blood.

Reference strains

G5436 ← R. Regnery, Houston-1 = ATCC 49882, *B. henselae* type strain, from blood.
G6129 ← R. Regnery, 021047 = ATCC VR 358, *B. quintana* type strain, from blood.
G6130 ← R. Regnery, 021046 = ATCC VR 152, *B. vinsonii* type strain, from a vole.
F9251 = ATCC 49927, *B. elizabethae* type strain, from blood.

Literature

1. Schmincke, A. 1917. Histopathologischer Befund in Roseolen der Haut bei wolhynischem. Fieber. Muench. Med. Wochschr. *64:* 91.

2. Varela, G., J.W. Vinson, and C. Molina-Pasquel. 1969. Trench fever. II. Propagation of *Rickettsia quintana* on a cell-free medium from the blood of two patients. Am. J. Trop. Med. Hyg. *18:* 708–712.

3. Weiss, E., and G.A. Dasch. 1982. Differential characteristics of strains of *Rochalimaea: Rochalimaea vinsonii* sp. nov., the Canadian vole agent. Int. J. Syst. Bacteriol. *32:* 305–314.

4. Regnery, R.L., B.E. Anderson, J.E. Clarridge III, M. C. Rodreguez-Barradas, D.C. Jones, and J.H. Carr. 1992. Characterization of a novel *Rochalimaea* species, *R. henselae* sp. nov., isolated from blood of a febrile, human immunodeficiency virus-positive patient. J. Clin. Microbiol. *30:* 265–274.

5. Welch, D.F., D.A. Pickett, L.N. Slater, A.G. Steigerwalt, and D.J. Brenner. 1992. *Rochalimaea henselae* sp. nov., a cause of septicemia, bacillary angiomatosis, and parenchymal bacillary peliosis. J. Clin. Microbiol. *30:* 275–280.

6. Regnery, R.L., J.G. Olsen, B.A. Perkins, and W. Bibb. 1992. Serological response to "*Rochalimaea henselae*" antigen in suspected cat-scratch disease. Lancet *339:* 1443–1445.

7. Daly, J.S., M.G. Worthington, D.J. Brenner, et al. 1993. *Rochalimaea elizabethae* sp. nov. isolated from a patient with endocarditis. J. Clin. Microbiol. *31:* 872–881.

8. Brenner, D.J., S.P. O'Connor, H.H. Winkler, and A.G. Steigerwalt. 1993. Proposals to unify the genera *Bartonella* and *Rochalimaea*, with the descriptions of *Bartonella quintana* comb. nov., *Bartonella vinsonii* comb. nov., *Bartonella henselae* comb. nov., and *Bartonella elizabethae* comb. nov., and to remove the family *Bartonellaceae* from the order *Rickettsiales*. Int. J. Syst. Bacteriol. *43:* 777–786.

9. Drancourt, M., and D. Raoult. 1993. Proposed tests for the routine differentiation of *Rochalimaea* species. Eur. J. Clin. Microbiol. Infect. Dis. *12:* 710–713.

Comments

Until recently, these organisms were classified as rickettsiae; however, phylogenetic analysis of 16S rRNA sequences shows them to be more closely related to *Bartonella bacilliformis* than to any of the known *Rickettsia* species (8). *B. henselae* is an etiologic agent associated with severe disseminated disease in immunosuppressed patients, including bacillary angiomatosis and parenchymal bacillary peliosis. This organism has also been proposed as the etiologic agent of cat scratch disease (6). The only known strain of *B. elizabethae* was isolated from blood cultures of a patient with endocarditis (7). *B. quintana* is the etiologic agent of trench fever (2). *B. vinsonii* has been isolated from a Canadian vole, but not from any human source. Isolation of these organisms on a solid medium usually requires extended (5–7 days) incubation. Identification to the genus level can be obtained by biochemical, preformed enzyme, and cellular fatty acid analysis (4, 7). At CDC, species-level identification is obtained by either DNA-DNA hybridization or restriction analysis of PCR-amplified gene fragments (4, 7). A biochemical approach for the differentiation of *B. henselae*, *B. quintana*, and *B. vinsonii* has been described (9), but we have not applied this approach to *B. elizabethae*. Information on the cellular fatty acid composition of this group is presented in Table 4.15 and Figure 4.20 of the cellular fatty acid section.

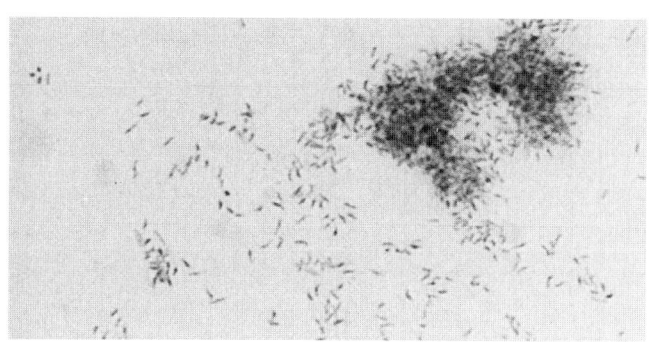

Bartonella henselae strain G5436. Gram stain BAP 35° C 3 days x 1700.

Bartonella (formerly *Rochalimaea*) species

	Bartonella (formerly *Rochalimaea*) species						
	B. henselae 14 strains		*B. quintana* 2 strains		*B. vinsonii* 1 strain		*B. elizabethae* 1 strain
TEST PERFORMED	SIGN	#+/T	SIGN	#+/T	SIGN		SIGN
Morphology	srs		srs		srs		srs
Motility; flagella	[nm]		[nm]		[nm]		[nm]
Action on blood	v	(1β)/6	–	0/1	–		inc β
Fermentative or oxidative	[n-o]		[n-o]		[n-o]		[n-o]
Carbohydrate base	RST		RST		RST		RST
D-Glucose	–	0/11	–	0/1	–		–
D-Xylose	–	0/8	–	0/1	–		–
D-Mannitol	–	0/7	–	0/1	–		–
Lactose	–	0/8	–	0/1	–		–
Sucrose	–	0/6	–	0/1	–		–
Maltose	–	0/6	–	0/1	–		–
Oxidase	–	0/12	+w[a]	1w/1	+w[a]		–
Growth on HIA with:							
X factor	v	9/14	v	1/2	(+)		+
V factor	[–]	0/14	[–]	0/2	[–]		[–]
Rabbit blood	[+]	3/3	[+]	1/1	[+]		[+]
Indole (spot test)	–	0/7	–	0/2	–		–
Ornithine decarboxylase	–	0/8	–	0/2	–		–
Urea, Christensen's[b]	–	0/10					
Growth on rabbit blood at:[c]							
35°C	(+)	(14)/14	(+)	(2)/2	(+)		(+)
Hydrolysis of:[d]							
p-Nitrophenyl-β,D-disaccharide	–	0/13	–	0/1	–		–
p-Nitrophenyl-α,L-arabinoside	–	0/13	–	0/1	–		–
ONPG	–	0/13	–	0/1	–		–
p-Nitrophenyl-α,D-glucoside	–	0/13	–	0/1	–		–
p-Nitrophenyl-β,D-glucoside	–	0/13	–	0/1	–		–
p-Nitrophenyl-α,D-galactoside	–	0/13	–	0/1	–		–
p-Nitrophenyl-α,L-fucoside	–	0/13	–	0/1	–		–
p-Nitrophenyl-N-acetyl-β,D-glucosaminide	–	0/13	–	0/1	–		–
p-nitrophenyl-phosphate	–	0/13	–	0/1	–		–
Leucyl-glycine-β-naphthylamide	+	13/13	+	1/1	+		+
Glycine-β-naphthylamide	+	13/13	+	1/1	+		+
Proline-β-naphthylamide	+	6, 6w/13	+	1/1	+		+
Phenylalanine-β-naphthylamide	+	13/13	+	1/1	+		+
Arginine-β-naphthylamide	+	13/13	+	1/1	+		+
Serine-β-naphthylamide	+	13/13	+	1/1	+		+
Pyrrolidonyl-β-naphthylamide	–	0/13	–	0/1	–		–

[a] Kovacs' modification weakly positive, routine method negative.
[b] Heavy inoculum.
[c] The preferred growth temperature for these organisms is 35°C versus room temperature which is preferred by *Bartonella bacilliformis*.
[d] Hydrolysis results were obtained with the RapID ANAII system (Innovative Diagnostic Systems, Atlanta, GA) using growth from 3- to 4-day rabbit blood agar cultures.

Bisgaard's taxon 16

Gram-negative rod; nonmotile; ferments glucose, sucrose and maltose (sometimes produces slight gas), xylose- and mannitol-negative; no growth on MacConkey agar; oxidase- and indole-positive (oxidase often weak); urease-negative; reduces nitrate; ornithine decarboxylase-negative.

Sources (30 strains)

Dog bite (26%); hand wound associated with dog bite (24%); wound (hand, forearm) (18%); animal bite (6%); other, including cervical swab, cellulitis swab, leg ulcer, tongue, dog vagina (26%).

Reference strain

A208.

Literature

1. Bisgaard, M., and R. Mutters. 1986. Characterization of some previously unclassified *"Pasteurella"* spp. obtained from the oral cavity of dogs and cats and description of a new species called "taxon 16." Acta. Pathol. Microbiol. Immunol. Scand. Sect. B *94:* 177–184.

2. Biberstein, E.L., S.S. Jang, P.H. Kass, and D.C. Hirsh. 1991. Distribution of indole-producing urease-negative pasteurellas in animals. J. Vet. Diag. Invest. *3:* 319–323.

3. Escande, F., and C. Lion. 1993. Epidemiology of human infections by *Pasteurella* and related groups in France. Zentrahbl. Bakeriol. *279:* 131–139.

Comments

Similar to *Pasteurella dagmatis,* except for negative urease reaction. Associated with animal bite wounds.

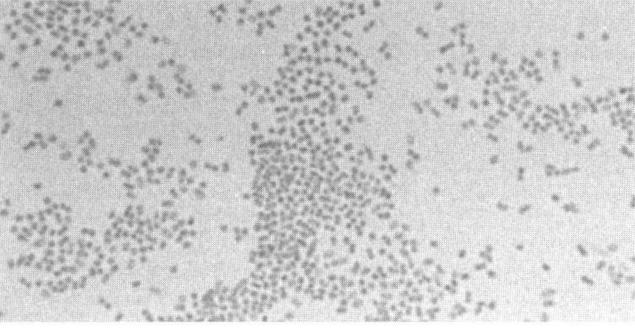

Bisgaard's taxon 16 strain A208. Gram stain BAP 35° C 24 h x 1700.

Bisgaard's taxon 16

TEST PERFORMED	SIGN	% +	#+/T
NUMBER OF STRAINS 30			
Morphology	srs		
Motility; flagella	[nm]		
Gas from glucose	-	3	1/30
Action on blood	v	32 al	8/25; 1 ly, 1 gr
Fermentative or oxidative	[F]		
Carbohydrate base	F		
Acid from:			
D-Glucose	+	100	28, 3w/30
D-Xylose	[-]	0	0/30
D-Mannitol	[-]	0	0/30
Lactose	-	0	0/30
Sucrose	+	100	27, 3w/30
Maltose	[+]	100	27, 3w/30
Catalase	+	100	28, 2w/30
Oxidase	[+]	100	20, 10w/30
Growth on:			
MacConkey	v	7 (10)	2w, (1), (2w)/30
SS	-	0	0/30
Simmons citrate	-	0	0/30
Urea, Christensen's	[-]	0	0/30
Nitrate reduction	[+]	100	30/30
Gas from nitrate	-	0	0/30
Indole	[+]	100	29, 1w/30
TSI slant, acid	+	100	30/30
TSI butt, acid	+	100	30/30
H_2S (TSI butt)	-	0	0/30
H_2S (Pb ac paper)	+	100	26, 4w/30
Gelatin hydrolysis[a]	-	0	0/25
Litmus milk	-	3 (3) a	1 (1)/30; 1 IR
Growth at:			
25°C	v	80	23, 1w/30
35°C	+	100	30/30
42°C	v	23	6, 1w/30
Esculin hydrolysis	-	0	0/30
Lysine decarboxylase	-	0	0/22
Arginine dihydrolase	-	0	0/22
Ornithine decarboxylase	-	0	0/22
Nutrient broth, 0% NaCl	v	71	17, 3w/28
Nutrient broth 6% NaCl	v	18	4, 1w/28

[a]Incubation of 7–14 days.

Bordetella avium

Gram-negative rod, motile (peritrichous flagella); grows well on MacConkey and SS agars; oxidase- and catalase-positive; urease-, nitrate-, and nitrite-negative.

Source (type strain)

Turkey respiratory tract.

Reference strain

KC1777 ← ATCC 35086, type strain, from turkey air sac exudate.

Literature

1. Kersters, K., K.-H. Hinz, A. Hertle, P. Segers, A. Lievens, O. Siegmann, and J. De Ley. 1984. *Bordetella avium* sp. nov., isolated from the respiratory tracts of turkeys and other birds. Int. J. Syst. Bacteriol. *34:* 56–70.
2. Hinz, K.-H., G. Glünder, and H. Lüders. 1978. Acute respiratory disease in turkey poults caused by *Bordetella bronchiseptica*-like bacteria. Vet. Rec. *103:* 262–263.

Comments

The biochemical profile of *B. avium* is similar to *Alcaligenes faecalis*. Nitrite reduction and cellular fatty acid chromatography are useful discriminating tests. Information on the cellular fatty acid composition of this species is presented in Table 4.16 and Figure 4.21 of the cellular fatty acid section.

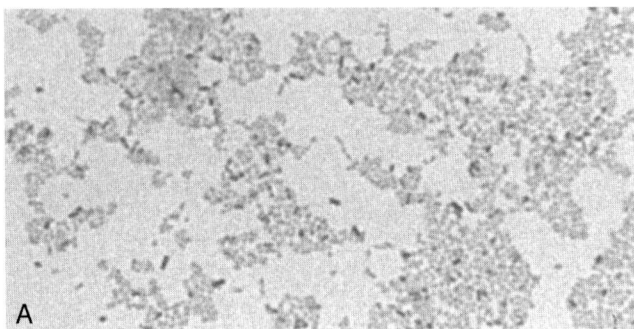

Bordetella avium strain KC1777. Gram stain BAP 35° C 24 h x 1700.

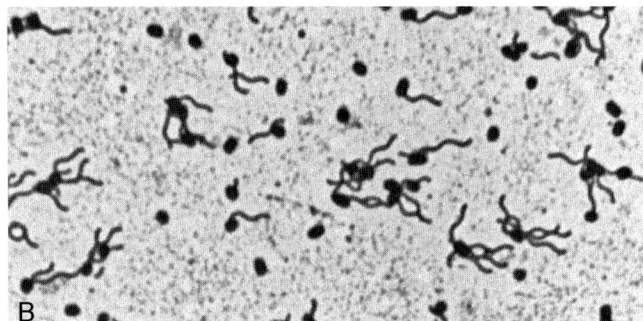

Bordetella avium strain KC1777. Flagella stain (Ryu method) TGY 25° C 24 h x 1700.

Bordetella avium

TEST PERFORMED	TYPE STRAIN SIGN
Morphology	mrs
Motility; flagella	[m;pe]
Action on blood	ß
Fermentative or oxidative	[n-o]
Carbohydrate base	OF
Acid from:	
D-Glucose	−
D-Xylose	−
D-Mannitol	−
Lactose	−
Sucrose	−
Maltose	−
Catalase	+
Oxidase	+
Growth on:	
MacConkey	[+]
SS	[+]
Cetrimide	−
Simmons citrate	+w
Urea, Christensen's	−
Nitrate reduction	[−]
Gas from nitrate	−
Nitrite reduction	[−]
Indole	−
TSI slant, acid	−
TSI butt, acid	−
H$_2$S (TSI butt)	−
H$_2$S (Pb ac paper)	−
Gelatin hydrolysis[a]	−
Litmus milk	−
Pigment	w amb sol
Growth at:	
25°C	+
35°C	+
42°C	+
Esculin hydrolysis	−
Lysine decarboxylase	−
Arginine dihydrolase	−
Ornithine decarboxylase	−
Nutrient broth, 0% NaCl	+
Nutrient broth, 6% NaCl	−

[a]Incubation of 7–14 days.

Bordetella bronchiseptica

Gram-negative rod, motile (peritrichous flagella); does not use carbohydrates; grows well on MacConkey and SS agars; oxidase-positive, usually urease- and nitrate-positive without gas formation; does not produce H_2S on TSI agar.

Sources (85 strains)

Guinea pig respiratory tract (14%), sputum (12%), human respiratory tract (12%), rabbit respiratory tract (7%), throat (6%), dog respiratory tract (5%), horse respiratory tract (4%), cerebrospinal fluid (2%), cat respiratory tract (2%), rat respiratory tract (2%), other (18%), unknown (16%).

Reference strain

KC1390 ← R. Hugh, 549 = ATCC 10580, type strain, from dog lung.

Literature

1. Moreno-López, M. 1952. El género *Bordetella*. Microbiol. Esp. *5:* 117–181.
2. Sneath, P.H.A., and V.B.D. Skerman. 1966. A list of type and reference strains of bacteria. Int. J. Syst. Bacteriol. *16:* 1–113.
3. Pittman, M. 1984. Genus *Bordetella* Moreno-López 1952,178.[AL] *In* N.R. Krieg and J. G. Holt (Eds.), Bergey's Manual of Systematic Bacteriology, Vol. 1, Williams & Wilkins, Baltimore, pp. 388–393.

Comments

Respiratory pathogen of dogs, swine, and laboratory animals including rabbits and guinea pigs. It may be transmitted from animals to humans. Information on the cellular fatty acid composition of this species is presented in Table 4.8 and Figure 4.13 of the cellular fatty acid section.

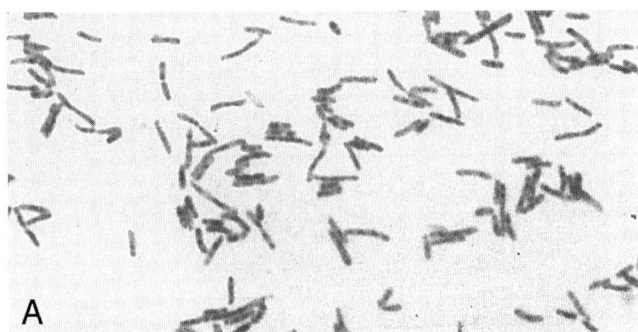 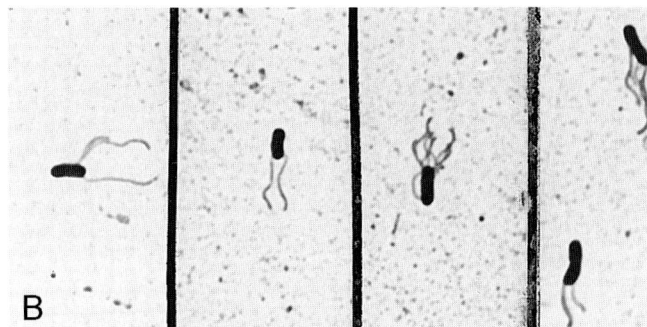

Bordetella bronchiseptica strain KC1390. Gram stain BAP 35° C 24 h x 1700.

Bordetella bronchiseptica strain KC1390. Flagella stain BAP 35° C 24 h x 1700.

Bordetella bronchiseptica

TEST PERFORMED	SIGN	% +	#+/T
NUMBER OF STRAINS 85			
Morphology	mrs		
Motility; flagella	[m;pe]		
Action on blood	v	46 ly	33/72; 10 ß, 8 gr, 7 LG, 2 br
Fermentative or oxidative	[n-o]		
Carbohydrate base	OF		
Acid from:			
D-Glucose	[-]	0	0/79
D-Xylose	[-]	7	5w[a]/70
D-Mannitol	-	1	1w/71
Lactose	-	0	0/71
Sucrose	-	0	0/71
Maltose	-	0	0/71
Catalase	+	100	70/70
Oxidase	[+]	100	84, 1w/85
Growth on:			
MacConkey	[+]	100	71/71
SS	[+]	99	78/79
Cetrimide	-	0	0/70
Simmons citrate	+	98 (1)	83 (1)/85
Urea, Christensen's	[+]	99	84/85
Nitrate reduction	[+]	92	78/85
Gas from nitrate	[-]	0	0/85
Indole	-	0	0/70
TSI slant, acid	-	0	0/79
TSI butt, acid	-	0	0/79
H$_2$S (TSI butt)	[-]	0	0/79
H$_2$S (Pb ac paper)	v	74	8, 49w/77
Gelatin hydrolysis[b]	-	0	0/55
Litmus milk	k	98	83/85
Pigment	-	0	0/50
Growth at:			
25°C	+	99	69/70
35°C	+	100	70/70
42°C	v	78	54/69
Esculin hydrolysis	-	0	0/62
Nutrient broth, 0% NaCl	+	100	61/61
Nutrient broth, 6% NaCl	v	82 (5)	53 (3)/65
Alkalinization of tartrate	[-]	0	0/15

[a]Reaction is neutral compared with the alkaline of the control.
[b]Incubation of 7–14 days.

Bordetella parapertussis

Gram-negative small rod, nonmotile; does not use carbohydrates; MacConkey-positive, oxidase-negative, urease-positive, nitrate-negative; grows slowly on blood agar, producing miniscule colonies, beta hemolysis, and a brown, water-soluble pigment (attributed to tyrosine); does not produce H_2S on TSI agar; does not hydrolyze gelatin.

Sources (8 strains)

Nasopharynx (4), cough plate (1), throat (1), unknown (2).

Reference strain

KC431 ← E. Updyke, 13; C8872, from nasopharynx.

Literature

1. Moreno-López, M. 1952. El género *Bordetella*. Microbiol. Esp. *5*: 117–181.
2. Eldering, G., and P. Kendrick. 1938. *Bacillus parapertussis*: a species resembling both *Bacillus pertussis* and *Bacillus bronchisepticus* but identical with neither. J. Bacteriol. *35*: 561–572.

Comments

Causes a pertussis-like syndrome; sometimes isolated when not expected; may be confused with fastidious *Acinetobacter*. Produces a brown soluble pigment that is enhanced on tyrosine-containing media. Fluorescent antibody test and serology are useful for confirmation. Information on the cellular fatty acid composition of this species is presented in Table 4.17 and Figure 4.22 of the cellular fatty acid section.

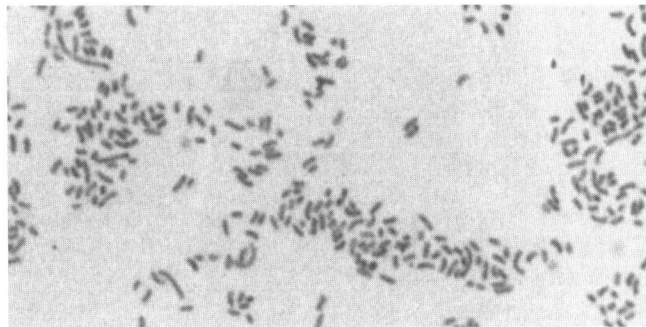

Bordetella parapertussis strain KC431. Gram stain BAP 35° C 24 h x 1700.

Bordetella parapertussis

TEST PERFORMED	SIGN	%+	#+/T
Morphology	srs		
Motility; flagella	[nm]		
Action on blood	[β]	100	12/12
Fermentative or oxidative	[n-o]		
Carbohydrate base	OF		
Acid from:			
D-Glucose	[-]	0	0/9
D-Xylose	-	0	0/9
D-Mannitol	-	0	0/9
Lactose	-	0	0/9
Sucrose	-	0	0/9
Maltose	[-]	0	0/9
Catalase	+	100	11/11
Oxidase	[-]	0	0/11
Growth on:			
MacConkey	[+]	100	11, 1w/12
SS	-	0	0/12
Cetrimide	-	0	0/12
Simmons citrate	v	67 (8)	8 (1)/12
Urea, Christensen's	[+]	100	12/12
Nitrate reduction	[-]	0	0/12
Gas from nitrate	-	0	0/12
Indole	-	0	0/12
TSI slant, acid	-	0	0/12
TSI butt, acid	-	0	0/12
H_2S (TSI butt)	[-]	0	0/12
H_2S (Pb ac paper)	v	17	2w/12
Gelatin hydrolysis[a]	[-]	0	0/6
Litmus milk	v	63 (13) k	5 (1)/8
Pigment	[br sol]	100	12/12
Growth at:			
25°C	v	60	6/10
35°C	+	100	12/12
42°C	v	18	1, 1w/11
Nutrient broth, 0% NaCl	+	92	11/12
Nutrient broth, 6% NaCl	-	0	0/12

[a]Incubation of 7–14 days.

Bordetella pertussis[a]

Gram-negative small rod, nonmotile; does not grow on usual laboratory media, although it may adapt by repeated transfer; grows well on charcoal-blood agar or Bordet-Gengou agar (addition of 5–40 µg cephalexin enhances isolation); usually produces β-like action on blood, but not as distinct as that produced by *B. parapertussis;* urease-negative with heavy inoculum; produces no pigment; oxidase-positive.

Sources (38 strains)

Nasopharynx (32%), cough plate (18%), throat (18%), nasal (5%), lung (3%), unknown (24%).

Reference strain

KC437 ← M. Pittman, 18323 = ATCC 9797, type strain

Literature

1. Moreno-López, M. 1952. El género *Bordetella*. Microbiol. Esp. *5:* 117–181.

2. Eldering, G., C. Hornbeck, and J. Baker. 1957. Serological study of *Bordetella pertussis* and related species. J. Bacteriol. *74:* 133–136.

3. Lautrop, H., with annex by B.W. Lacy. 1960. Laboratory diagnosis of whooping-cough or *Bordetella* infections. Bull. WHO *23:* 15–35.

4. Gilchrist, M.J.R. 1991. *Bordetella. In* A. Balows, et al. (Eds.), Manual of Clinical Microbiology, 5th ed., American Society for Microbiology, Washington, pp. 471–477.

Comments

Agglutination and fluorescent antibody testing are particularly useful for identification. Information on the cellular fatty acid composition of this species is presented in Table 4.18 and Figure 4.23 of the cellular fatty acid section.

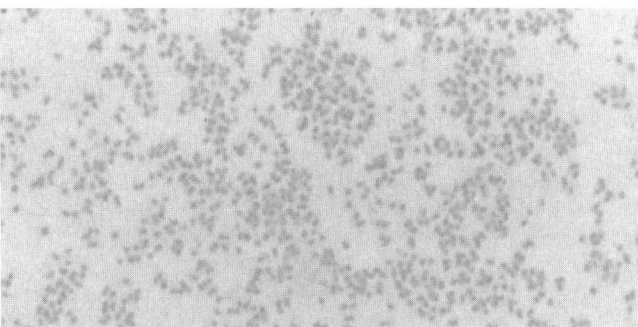

Bordetella pertussis strain KC437. Gram stain BG 35° C 48 h x 1700.

[a]Final identification of potential *B. pertussis* isolates sent to CDC is performed by the Childhood and Respiratory Diseases Branch, Division of Bacterial and Mycotic Diseases, National Centers for Infectious Diseases.

Bordetella pertussis[a]

TEST PERFORMED	SIGN	% +	#+/T
NUMBER OF STRAINS 51			
Morphology	srs		
Motility	[nm]		
Action on blood	[v]	81β	35/43; 6 ly
Fermentative or oxidative			
Carbohydrate base			
Acid from:			
D-Glucose			
D-Xylose			
D-Mannitol			
Lactose			
Sucrose			
Maltose			
Catalase	+	100	2/2
Oxidase	[+]	94	8, 8w/17
Growth on:			
MacConkey			
SS			
Cetrimide			
Simmons citrate			
Urea, Christensen's	[−][b]	0	0/37
Nitrate reduction			
Gas from nitrate			
Nitrite reduction			
Indole			
TSI slant, acid			
TSI butt, acid			
H$_2$S (TSI butt)			
H$_2$S (Pb ac paper)			
Gelatin hydrolysis			
Litmus milk			
Pigment	−	0	0/51
Growth at:			
25°C			
35°C			
42°C			
Esculin hydrolysis			
Lysine decarboxylase			
Arginine dihydrolase			
Ornithine decarboxylase			
Nutrient broth, 0% NaCl			
Nutrient broth, 6% NaCl			

[a]Generally does not grow on the usual laboratory media; Bordet-Gengou and blood-charcoal agars are the media of choice.
[b]Heavy inoculum.

SECTION 3: DESCRIPTION OF SPECIES

Brucella abortus

Gram-negative small coccoid rod, nonmotile; capnophilic to aerobic; oxidizes xylose and usually glucose; growth on MacConkey agar variable; catalase-positive, urease-positive (but not immediately); oxidase-positive, reduces nitrate, citrate-negative, indole-negative, usually resistant to basic fuchsin dye and sensitive to thionin dye (1:50,000 and 1:100,000 dilutions of the dyes); *Brucella* phage Tbilisi lyses smooth strains at routine test dilution (RTD) and 10,000 × RTD.

Sources (81 strains)

Blood (63%), bone marrow (5%), bovine tissue (2%), abscess (2%), other (4%), unknown (24%).

Reference strain

KC797 ← L. Jones, R19.

Literature

1. Feusier, M.L., and K.F. Meyer. 1920. Principles in serologic grouping of *B. abortus* and *B. melitensis*. Correlation between absorption and agglutination tests. J. Infect. Dis. 27: 173–184.

2. Stableforth, A.W., and L.M. Jones. 1963. Report of the subcommittee on taxonomy of the genus *Brucella*. Speciation in the genus *Brucella*. Int. Bull. Bacteriol. Nomencl. Taxon. 13: 145–148.

3. Corbel, M.J., and W.J. Brinley-Morgan. 1984. Genus Brucella. In N.R. Krieg and J.G. Holt (Eds.), Bergey's Manual of Systematic Bacteriology, Vol. 1, Williams & Wilkins, Baltimore, pp. 377–388.

4. Jones, L.M., and W. Wundt. 1971. International Committee on Nomenclature of Bacteria Sub-committee on the Taxonomy of *Brucella*. Minutes of Meeting, 7 August 1970. Int. J. Syst. Bacteriol. 21: 126–128.

5. Alton, G.G., L.M. Jones, and D.E. Pietz. 1975. Laboratory techniques in brucellosis, 2nd ed. WHO Monogr. Ser. 55.

6. Moyer, N.P., L.A. Holcomb, and W.J. Hausler. 1991. Brucella. In A. Balows, et al. (Eds.), Manual of Clinical Microbiology, 5th ed., American Society for Microbiology, Washington, pp. 457–462.

Comments

Pathogenic for cattle and other species, including humans. Growth on agar plate after 18- to 24- h incubation is scant and usually only occurs in areas where inoculum is heaviest; discrete colonies usually develop after further incubation. An occasional strain may give a "fuzzy" appearance in motility medium, suggesting motility; however, no flagella have been demonstrated. Seven biotypes are recognized (6). Information on the cellular fatty acid composition of this species is presented in Table 4.19 and Figure 4.24 of the cellular fatty acid section.

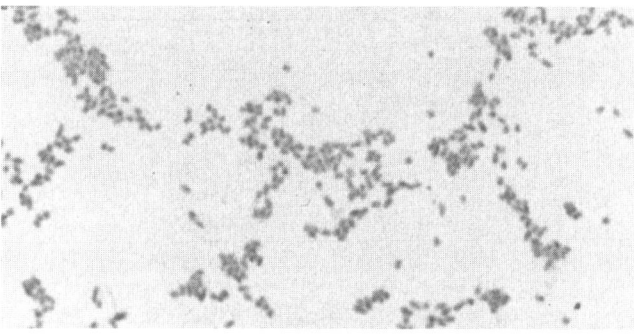

Brucella abortus strain KC797. Gram stain BAP 35° C 24 h x 1700.

Brucella abortus

TEST PERFORMED	SIGN	% +	#+/T
NUMBER OF STRAINS 81			
Morphology	[rv]		
Motility	[nm]		
Action on blood	v	25 ly	1/4
Fermentative or oxidative	[O]		
Carbohydrate base	OF		
Acid from:			
D-Glucose	[v]	50 (25)	2w (1)/4
D-Xylose	[+]	100	4/4
D-Mannitol	−	0	0/4
Lactose	−	0	0/4
Sucrose	[−]	0	0/4
Maltose	−	0	0/4
Catalase	+	100	14, 2w/16
Oxidase	[+]	96	22/23
Growth on:			
MacConkey	v	50 (20)	2, 3w (2)/10
SS	−	0	0/10
Simmons citrate	[−]	0	0/10
Urea, Christensen's	[+]	98	79/81
Nitrate reduction	[+]	100	13/13
Gas from nitrate	−	0	0/13
Indole	[−]	0	0/13
TSI slant, acid	−	0	0/5
TSI butt, acid	−	0	0/5
H$_2$S (TSI butt)	−	0	0/6
H$_2$S (Pb ac paper)[a]	+ or (+)	89 (4)	50, 20w (3)/79
Gelatin hydrolysis[b]	−	0	0/3
Growth at:			
25°C	v	43	1, 2w/7
35°C	+	100	81/81
42°C	v	14	1w/7
Esculin hydrolysis	−	0	0/6
Nutrient broth, 0% NaCl	+	100	7/7
Nutrient broth, 6% NaCl	−	0	0/7
Requires CO$_2$ atmosphere	v	67	41/61

[a] Generally *Brucella* strains are tested with heart infusion agar.
[b] Incubation of 7–14 days.

Brucella canis

Gram-negative small coccoid rod, nonmotile; biochemically similar to *Brucella suis*, but on primary isolation cultures are in the rough phase and show strong agglutination in acriflavine; oxidase test is variable; gives positive string test and usually forms "gel" in phenolized saline; agglutinates in specific antiserum.

Sources (28 strains)

Human blood (43%), canine blood (11%), canine cervix (7%), other canine sources (39%).

Reference strain

KC794 ← L.E. Carmichael RM-666 = ATCC 23365, type strain, allantoic fluid of aborted Beagle puppy.

Literature

1. Carmichael, L.E., and D.W. Bruner. 1968. Characteristics of a newly-recognized species of *Brucella* responsible for infectious canine abortion. N.S. Cornell Vet. *58:* 578–592.
2. Diaz, R., L.M. Jones, and J.B. Wilson. 1968. Antigenic relationship of the gram-negative organism causing canine abortion to smooth and rough brucellae. J. Bacteriol. *95:* 618–624.
3. Alton, G.G., L.M. Jones, and D.E. Pietz. 1975. Laboratory techniques in brucellosis, 2nd ed. WHO Monogr. Ser. 55.
4. Dees, S.B., D.G. Hollis, R.E. Weaver, and C.W. Moss. 1981. Cellular fatty acids of *Brucella canis* and *Brucella suis*. J. Clin. Microbiol. *14:* 111–112.
5. Moyer, N.P., L.A. Holcomb, and W.J. Hausler. 1991. *Brucella*. *In* A. Balows, et al. (Eds.), Manual of Clinical Microbiology, 5th ed., American Society for Microbiology, Washington, pp. 457-462.

Comments

Pathogenic for dogs and humans. Growth on agar plates after 18–24 h is scant and usually occurs only in areas where inoculum is heaviest; discrete colonies usually develop after further incubation. Cellular fatty acid profile is useful in distinguishing *B. canis* from other antigenically rough *Brucella* strains. Information on the cellular fatty acid composition of this species is presented in Table 4.20 and Figure 4.25 of the cellular fatty acid section.

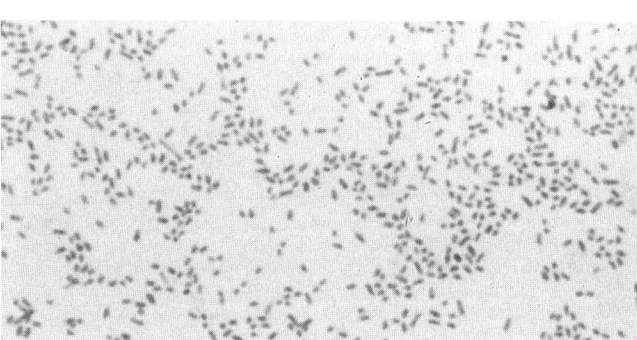

Brucella canis strain D2773. Gram stain BAP 35° C 24 h x 1700.

Brucella canis[a]

TEST PERFORMED	SIGN	%+	#+/T
	NUMBER OF STRAINS 28		
Morphology	[rv]		
Motility	[nm]		
Requires CO_2 atmosphere	−	0	0/13
Action on blood	v	20 sl ly	3/15; 1 gr
Fermentative or oxidative	[O]		
Carbohydrate base	OF		
Acid from:			
D-Glucose	[+ or (+)]	81 (14)	13, 4w (3)/21
D-Xylose	[+ or (+)]	81 (19)	15, 2w (4)/21
D-Mannitol	−	0	0/21
Lactose	−	0	0/21
Sucrose	[−]	0	0/21
Maltose	−	0	0/21
Catalase	+	100	15, 2w/17
Oxidase	[v]	72	4, 9w/18
Growth on:			
MacConkey	v	12 (29)	2w (5)/17
SS	−	0	0/17
Simmons citrate	[−]	0	0/17
Urea, Christensen's	[+]	100	28/28
Nitrate reduction	[+]	100	22/22
Gas from nitrate	[v]	77	17/22
Indole	[−]	0	0/18
TSI slant, acid	−	0	0/20
TSI butt, acid	−	0	0/20
H_2S (TSI butt)	−	0	0/19
H_2S (Pb ac paper)[b]	v	82	9, 9w/22
Gelatin hydrolysis[c]	−	0	0/7
Growth at:			
25°C	v	42	5w/12
35°C	+	100	28/28
42°C	v	33	4w/12
Esculin hydrolysis	−	0	0/15
Nutrient broth, 0% NaCl	+	100	13, 1w/14
Nutrient broth, 6% NaCl	−	0	0/14

[a]A heavy suspension in 0.5% phenolized saline usually forms a "gel" within 30 min.
[b]Generally *Brucella* strains are tested with heart infusion agar.
[c]Incubation of 7–14 days.

Brucella melitensis

Gram-negative small coccoid rod, nonmotile; oxidizes xylose and usually glucose; growth on MacConkey agar variable; urease-positive (immediately or delayed), oxidase-positive; reduces nitrate; indole- and citrate-negative; resistant to basic fuchsin and thionin dyes (1:50,000 and 1:100,000 dilutions); *Brucella* phage strain Tbilisi does not lyse *B. melitensis* at routine test dilution (RTD) and 10,000 × RTD.

Sources (48 strains)

Blood (69%), bone marrow (6%), liver (4%), spleen (2%), unknown (19%).

Reference strain

KC341 ← J. Schubert, M-1 ← H. Bauer.

Literature

1. Feusier, M.L., and K.F. Meyer. 1920. Principles in serologic grouping of *B. abortus* and *B. melitensis*. Correlation between absorption and agglutination tests. J. Infect. Dis. 27: 173–184.

2. Jones, L.M., and W. Wundt. 1971. International Committee on Nomenclature of Bacteria Sub-committee on the Taxonomy of *Brucella*. Minutes of Meeting, 7 August 1970. Int. J. Syst. Bacteriol. *21:* 126–128.

3. Alton, G.G., L.M. Jones, and D.E. Pietz. 1975. Laboratory techniques in brucellosis, 2nd ed. WHO Monogr. Ser. 55.

4. Staszkiewicz, J., C.M. Lewis, J. Coleville, M. Zervos, and J. Band. 1991. Outbreak of *Brucella melitensis* among microbiology laboratory workers in a community hospital. J. Clin. Microbiol. *29:* 287–290.

5. Moyer, N.P., L.A. Holcomb, and W.J. Hausler. 1991. *Brucella. In* A. Balows, et al. (Eds.), Manual of Clinical Microbiology, 5th ed., American Society for Microbiology, Washington, pp. 457–462.

Comments

Pathogenic for goats, sheep, and other species, including humans. Associated with laboratory-acquired infections (4). Growth on agar plate after 18–24 h is scant and usually occurs only in areas where inoculum is heaviest; discrete colonies usually develop after further incubation. Three biotypes are recognized (5). Information on the cellular fatty acid composition of this species is presented in Table 4.19 and Figure 4.24 of the cellular fatty acid section.

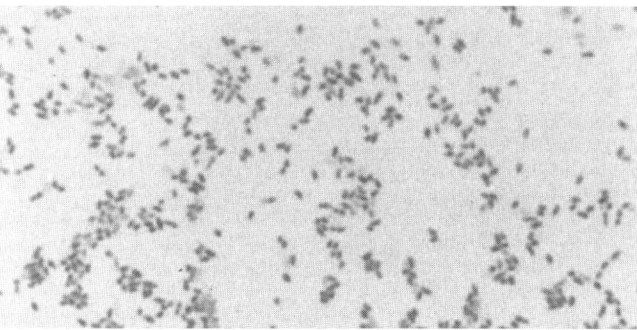

Brucella melitensis strain KC341. Gram stain BAP 35° C 24 h x 1700.

Brucella melitensis

	NUMBER OF STRAINS 48		
TEST PERFORMED	SIGN	%+	#+/T
Morphology	[rv]		
Motility	[nm]		
Action on blood	−	0	0/4
Fermentative or oxidative	[O]		
Carbohydrate base	OF		
Acid from:			
D-Glucose	[v]	33	1/3
D-Xylose	[+]	100	3/3
D-Mannitol	−	0	0/3
Lactose	−	0	0/3
Sucrose	[−]	0	0/3
Maltose	−	0	0/3
Catalase	+	100	7/7
Oxidase	[+]	100	13, 2w/15
Growth on:			
MacConkey	v	62	1w (4)/8
SS	−	0	0/8
Simmons citrate	[−]	0	0/8
Urea, Christensen's	[+]	100	48/48
Nitrate reduction	[+]	100	8/8
Gas from nitrate	−	0	0/8
Indole	[−]	0	0/8
TSI slant, acid	−	0	0/3
TSI butt, acid	−	0	0/3
H_2S (TSI butt)	−	0	0/3
H_2S (Pb ac paper)[a]	v	39	6, 11w (1)/44
Gelatin hydrolysis[b]	−	0	0/2
Growth at:			
25°C	v	17	1/6
35°C	+	100	48/48
42°C	v	50	3w/6
Esculin hydrolysis	−	0	0/2
Nutrient broth, 0% NaCl	+	100	7, 1w/8
Nutrient broth, 6% NaCl	−	0	0/8
Requires CO_2 atmosphere	−[c]	0	0/20

[a] Generally *Brucella* strains are tested with heart infusion agar.
[b] Incubation of 7–14 days.
[c] Growth of 4 of 20 strains tested was enhanced in a candle jar.

Brucella suis

Gram-negative small rod, nonmotile; oxidizes xylose and glucose; growth on MacConkey agar variable; urease-positive immediately with heavy inoculum, oxidase-positive; reduces nitrate, frequently with slight gas production, does not utilize citrate; does not produce indole; usually sensitive to basic fuchsin and resistant to thionin dyes (1:50,000 and 1:100,000 dilutions); *Brucella* phage strain Tbilisi lyses smooth *B. suis* strains at 10,000 × routine test dilution (RTD), but not at RTD.

Sources (190 strains)

Blood (74%), bone marrow (5%), lung tissue (2%), spleen (1%), other (6%), unknown (12%).

Reference strain

KC322 ← J. Schubert, S-13 ← G. Eldridge.

Literature

1. Huddleston, I.F. 1929. The differentiation of the species of the genus *Brucella*. Bull. Mich. Agric. Exp. Sta. *100*: 1–16.
2. Alton, G.G., L.M. Jones, and D.E. Pietz. 1975. Laboratory techniques in brucellosis, 2nd ed. WHO Monogr. Ser. 55.
3. Moyer, N.P., L.A. Holcomb, and W.J. Hausler. 1991. *Brucella*. *In* A. Balows, et al. (Eds.), Manual of Clinical Microbiology, 5th ed., American Society for Microbiology, Washington, pp. 457–462.

Comments

Pathogenic for pig, hare, reindeer, and other species, including humans. Growth on agar plate after 18- to 24- h incubation is scant and usually occurs only in areas where inoculum is heaviest; discrete colonies usually develop after further incubation. Five biotypes are recognized (3). Information on the cellular fatty acid composition of this species is presented in Table 4.19 and Figure 4.24 of the cellular fatty acid section.

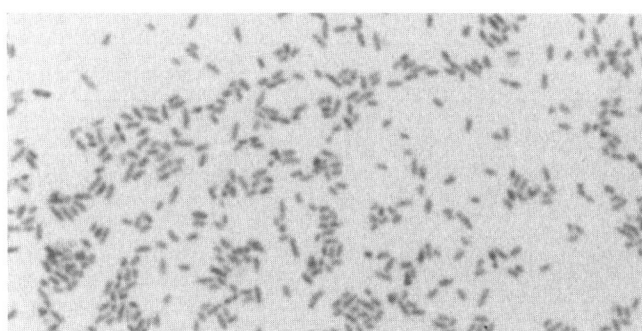

Brucella suis strain KC322. Gram stain BAP 35° C 24 h x 1700.

Brucella suis

TEST PERFORMED	SIGN	% +	#+/T
NUMBER OF STRAINS 190			
Morphology	[rv]		
Motility	[nm]		
Requires CO_2 atmosphere	–	0	0/99
Action on blood	v	17 al	2/12
Fermentative or oxidative	[O]		
Carbohydrate base	OF		
Acid from:			
D-Glucose	[+]	100	8, 4w/12
D-Xylose	[+]	100	11, 1w/12
D-Mannitol	–	0	0/12
Lactose	–	0	0/12
Sucrose	[–]	0	0/12
Maltose	–	0	0/12
Catalase	+	100	37, 2w/39
Oxidase	[+]	95	40, 13w/56
Growth on:			
MacConkey	v	23 (15)	1, 2w (2)/13
SS	–	0	0/13
Simmons citrate	[–]	0	0/13
Urea, Christensen's	[+]	100	190/190
Nitrate reduction	[+]	100	26, 1w/27
Gas from nitrate	[v]	52	14/27
Indole	[–]	0	0/16
TSI slant, acid	–	0	0/7
TSI butt, acid	–	0	0/7
H_2S (TSI butt)	–	0	0/7
H_2S (Pb ac paper)[a]	+	95	142, 25w/176
Gelatin hydrolysis[b]	–	0	0/2
Growth at:			
25°C	v	70	2, 5w/10
35°C	+	100	190/190
42°C	v	50	1, 4w/10
Esculin hydrolysis	–	0	0/2
Nutrient broth, 0% NaCl	+	100	12, 1w/13
Nutrient broth, 6% NaCl	–	0	0/13

[a]Generally *Brucella* strains are tested with heart infusion agar.
[b]Incubation of 7–14 days.

Capnocytophaga species (DF-1)

Gram-negative thin rod with tapered ends; gliding motility has been observed; some strains appear motile late in motility medium, with only an occasional, very delicate, single polar or lateral "flagellum" detectable (in our laboratory); facultatively anaerobic; fastidious; requires or is enhanced by candle-jar atmosphere; ferments glucose (without gas), sucrose, maltose, and usually lactose; frequently produces yellow-to-orange growth pigment; catalase-negative, oxidase-negative; no growth on MacConkey agar; esculin hydrolysis is variable; arginine dihydrolase is usually negative.

Sources (172 strains)

Blood (23%), sputum (18%), throat (16%), vaginal and cervical (6%), bronchial (4%), cerebrospinal fluid (4%), tracheal (2%), pleural fluid (2%), thoracentesis (2%), eye (2%), amniotic fluid (2%), other (14%), unknown (5%).

Reference strains

B2906, from blood.
B4267, from blood.
B7310, from throat.

Literature

1. Holt, S.C., E.R. Leadbetter, and S.S. Socransky. 1979. *Capnocytophaga*: new genus of gram-negative gliding bacteria. I. General characteristics, taxonomic considerations and significance. Arch. Microbiol. 122: 9-39.

2. Williams, B.L., D. Hollis, and L.V. Holdeman. 1979. Synonymy of strains of Center for Disease Control Group DF-1 with species of *Capnocytophaga*. J. Clin. Microbiol. 10: 550–562.

3. Holt, S.C., and S.A. Kinder. 1989. *Capnocytophaga*. In J.T. Staley, M.P. Bryant, N. Pfennig, and J.G. Holt (Eds.), Bergey's Manual of Systematic Bacteriology, Vol. 3, Williams & Wilkins, Baltimore, pp. 2050–2058.

4. Yamamoto, T., S. Kajiura, Y. Hirai, and T. Watanabe. 1994. *Capnocytophaga haemolytica* sp. nov. and *Capnocytophaga granulosa* sp. nov., from human dental plaque. Int. J. Syst. Bacteriol. 44: 324–329.

Comments

The organisms in this group previously have been designated CDC Group DF-1. The reactions listed for this group represent a combination of three *Capnocytophaga* species: *C. ochracea*, *C. gingivalis*, and *C. sputigena* (1). It is difficult to differentiate these three species by phenotypic characteristics. Starch and gelatin hydrolysis, and the ONPG reaction, however, may be useful in speciation of these organisms (3). *C. ochracea* is ONPG- and usually starch hydrolysis-positive but gelatin-negative. *C. sputigena* is ONPG and sometimes gelatin-positive but starch-negative. *C. gingivalis* is negative for all three reactions. Distinctive fringe-like or pseudo-pitting colonies frequently grow into the agar. Usually (17 of 28) grow on Thayer-Martin selective medium. Information on the cellular fatty acid composition of this group is presented in Table 4.26 and Figure 4.31 of the cellular fatty acid section. Recently two additional *Capnocytophaga* species, *C. haemolytica* and *C. granulosa*, isolated from human dental plaque have been described (4). Distinguishing characteristics described for *C. haemolytica* are the presence of hemolytic activity and lack of aminopeptidase activity. Distinguishing characteristics of *C. granulosa* include the presence of intracellular granular inclusions and aerobic growth.

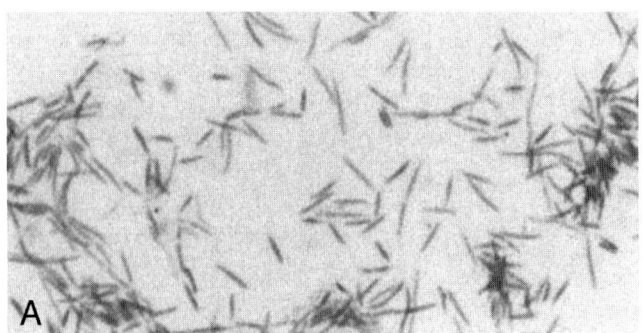

Capnocytophaga species (DF-1) strain B7310. Gram stain BAP 35° C 24 h x 1700.

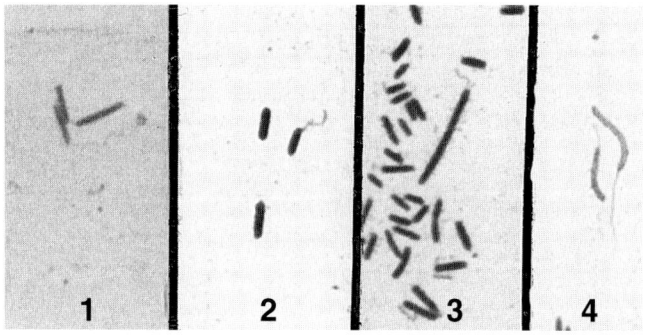

Capnocytophaga species strain B2906 (plate 1). Flagella stain FB 35° C 24 h x 1700. Strain B4267 (plates 2-4). Flagella stain FB 25° C 24 h x 1700.

Capnocytophaga species[a] (DF-1)

TEST PERFORMED	SIGN	% +	#+/T
NUMBER OF STRAINS 155			
Morphology	[trt]		
Motility; flagella	v[b]		
Gas from glucose	[-]	0	0/155
Action on blood	v	34 al or gr	40/116; 1 ly
Fermentative or oxidative	[F][c]		
Carbohydrate base	F		
Acid from:			
D-Glucose	[+]	90 (10)	134, 5w (16)/155
D-Xylose	-	0	0/155
D-Mannitol	-	0	0/155
Lactose	v	75 (11)	113, 4w (17)/155
Sucrose	[+]	90 (9)	134, 6w (14)/155
Maltose	[+ or (+)]	86 (14)	128, 6w (21)/155
Catalase	[-]	7	3, 9w/155
Oxidase	[-]	7	10, 1w/155
Growth on:			
MacConkey	[-]	0	0/155
SS	-	0	0/155
Simmons citrate	-	0	0/155
Urea, Christensen's	-	0	0/120
Nitrate reduction	v	63	98/155
Gas from nitrate	-	0	0/98
Indole	-	0	0/155
TSI slant, acid	v	73	112/154
TSI butt, acid	v	55	85/154
H_2S (TSI butt)	-	0	0/155
H_2S (Pb ac paper)	v	50	39, 38w/155
Gelatin hydrolysis[d]	-	0	0/87
Litmus milk	v	49 (7) A	44, 30w (10)/152; 13 IR
Growth at:			
25°C	v	21	5, 20w/119
35°C	+	100	34, 85w/119
42°C	v	16	8, 11w/119
Esculin hydrolysis	[v]	81 (2)	125 (3)/155
Lysine decarboxylase	-	0	0/11
Arginine dihydrolase	-	8	2w/26
Ornithine decarboxylase	-	0	0/11
Nutrient broth, 0% NaCl	v	21 (1)	3, 28w (2)/145
Nutrient broth, 6% NaCl	-	1	2w/145

[a]Yellow nonsoluble pigment produced by 61 of 155 strains.
[b]Gliding motility has been reported; we have observed delayed spreading in motility medium and an occasional single polar or lateral "flagellum."
[c]TSI reaction is not always indicative of fermentative activity.
[d]Incubation of 7–14 days.

Capnocytophaga canimorsus

Gram-negative medium to long thin rod, frequently slightly tapered, and with occasional spindle forms; exhibits gliding motility without flagella; fastidious; requires or is enhanced by candle-jar atmosphere; ferments glucose, lactose, and maltose (serum may enhance reaction); does not ferment raffinose and inulin; no growth on MacConkey agar; oxidase-positive (may be weak); catalase-positive; urease-negative; nitrate- and indole-negative; arginine dihydrolase-positive; ONPG-positive.

Sources (90 strains)

Blood (82%), cerebrospinal fluid (7%), other (6%), unknown (5%).

Reference strain

7120 ← ATCC 35979, type strain, from human blood.

Literature

1. Butler, T., R.E. Weaver, T.K.V. Ramani, C.T. Uyeda, R.A. Bóbo, J.S. Ryu, and R.B. Kohler. 1977. Unidentified gram-negative rod infection, a new disease of man. Ann. Intern. Med. *86:* 1–5.

2. Bailie, W.E., E.C. Stowe, and A.M. Schmitt. 1978. Aerobic bacterial flora of oral and nasal fluids of canines with reference to bacteria associated with bites. J. Clin. Microbiol. *7:* 223–231.

3. Brenner, D.J., D.G. Hollis, G.R. Fanning, and R.E. Weaver. 1989. *Capnocytophaga canimorsus* sp. nov. (Formerly CDC Group DF-2), a cause of septicemia following dog bite, and *C. cynodegmi* sp. nov., a cause of localized wound infection following dog bite. J. Clin. Microbiol. *27:* 231–235.

Comments

These organisms formerly were designated CDC Group DF-2. In some cases, these organisms have been seen in peripheral blood smears. They have been associated with dog bites and have been isolated from canine oral cavities; several patients have had histories of splenectomy. Colonies on heart infusion agar with 5% rabbit or sheep blood at 18–24 h are usually <0.5 mm in diameter, but after 48 h of incubation, colonies are 1–3.5 mm in diameter. The surface of young colonies is raised or slightly uneven, usually with a flat, spreading edge. Grows poorly on trypticase soy agar. Information on the cellular fatty acid composition of this species is presented in Table 4.26 and Figure 4.31 of the cellular fatty acid section.

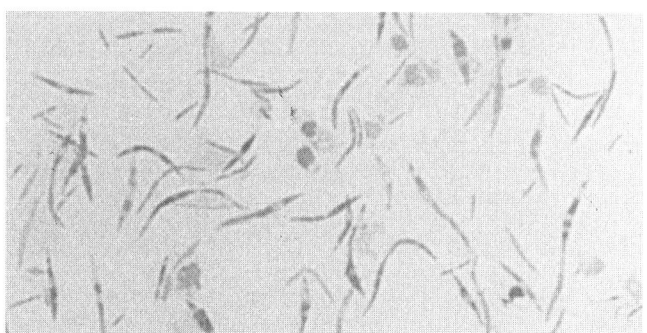

Capnocytophaga canimorsus strain 7120. Gram stain BAP 35° C 24 h x 1700.

Capnocytophaga canimorsus

TEST PERFORMED	SIGN	% +	#+/T
	NUMBER OF STRAINS 90		
Morphology	[m, ltrt]		
Motility; flagella	[gl]		
Action on blood	v	10 gr	8/83; 5 lav, 3 al, 3 ly, 1 ß
Fermentative or oxidative	[F][a]		
Carbohydrate base	F[b]		
Acid from:			
D-Glucose	[v]	48 (32)	22, 21w, (25), (3w)/ 89
D-Xylose	-	0	0/89
D-Mannitol	[-]	0	0/89
Lactose	[+ or (+)]	56 (39)	27, 20w, (30), (4w)/ 89
Sucrose	[-]	0	0/89
Maltose	[+ or (+)]	53 (39)	27, 20w, (30), (4w)/ 89
Inulin	[-]	0	0/29
Raffinose	[-]	0	0/39
Catalase	[+]	100	89/89
Oxidase	[+]	100	67, 19w/86
Growth on:			
MacConkey	[-]	0	0/79
SS	-	0	0/79
Simmons citrate	-	0	0/76
Urea, Christensen's	-	0	0/88
Nitrate reduction	-	0	0/89
Gas from nitrate	-	0	0/89
Nitrite reduction	v	40	2/5
Indole	[-]	0	0/72
TSI slant, acid	v	11	9/84
TSI butt, acid	v	12	8/69
H_2S (TSI butt)	-	0	0/64
H_2S (Pb ac paper)	v	23	14/60
Gelatin hydrolysis[c]	-	0	0/54
Litmus milk	-	1 Aw	1 Aw/78
Growth at:			
25°C	v	24	9/38
35°C	+	95	76/80
42°C	-	8	3/36
Esculin hydrolysis	v	24 (31)	21, (25), (2w)/86
Lysine decarboxylase	-	0	0/54
Arginine dihydrolase	[+]	95	54, 4w/61
Ornithine decarboxylase	-	0	0/54
Nutrient broth, 0% NaCl	-	4(1)	3, (1)/79
Nutrient broth, 6% NaCl	-	1	1/79
ONPG	[+]	100	50/50

[a]TSI reaction is not always indicative of fermentative activity.
[b]Rabbit serum (1–2 drops/3 ml of medium) may be required; reactions may be obtained within 4 h by heavily inoculating small volumes of fermentation medium.
[c]Incubation of 7–14 days.

Capnocytophaga cynodegmi

Gram-negative, thin, medium to long rod (slightly curved filaments may be observed); motile (gliding motility as observed by bright field microscopy, no flagella detected); requires or is enhanced by candle-jar atmosphere; no growth on MacConkey agar; produces acid from glucose, lactose, and usually sucrose and maltose, inulin, and raffinose (acid production may be weak and/or delayed); no acid production from xylose and mannitol; oxidase-, catalase-, esculin-, and ONPG-positive; urease- and nitrate-negative.

Sources (13 strains)

Wound (4) (most associated with dog bite), dog oral cavity (3), eye (2), blood (1), other (2), unknown (1).

Reference strain

E6447 ← ATCC 49044, type strain, from dog mouth.

Literature

1. Brenner, D.J., D.G. Hollis, G.R. Fanning, and R.E. Weaver. 1989. *Capnocytophaga canimorsus* sp. nov. (Formerly CDC Group DF-2), a cause of septicemia following dog bite, and *C. cynodegmi* sp. nov., a cause of localized wound infection following dog bite. J. Clin. Microbiol. *27:* 231–235.

Comments

Formerly named CDC Group DF-2-like, *C. cynodegmi* usually is associated with localized infection following dog bites or cat scratches. β-like hemolysis often is observed, and is best demonstrated by stabbing the inoculum into the agar. Growth is often adherent to the agar. Information on the cellular fatty acid composition of this species is presented in Table 4.26 and Figure 4.31 of the cellular fatty acid section.

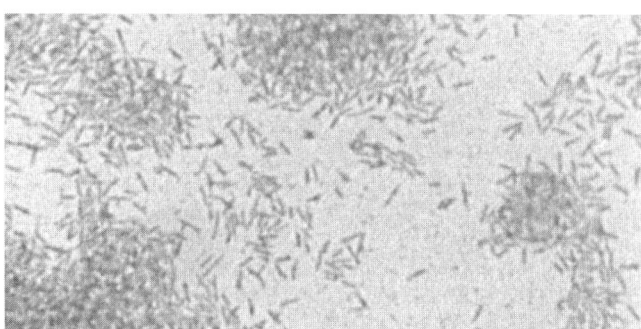

Capnocytophaga cynodegmi strain E6447. Gram stain BAP 35° C 24 h x 1700.

Capnocytophaga cynodegmi

TEST PERFORMED	SIGN	%+	#+/T
Morphology	[ltrt,fc]		
Motility; flagella[a]	[gl]		
Action on blood	v	67 ß	8 ß/12
Fermentative or oxidative	[F]		
Carbohydrate base	F[b]		
Acid from:			
D-Glucose	[+ or (+)]	69 (31)	5, 4w (4w)/13
D-Xylose	-	0	0/13
D-Mannitol	[-]	0	0/13
Lactose	[+ or (+)]	69 (31)	5 (1), 4w (4w)/13
Sucrose	[v]	30 (46)	2 (2), 3w (3w)/13
Maltose	[+ or (+)]	69 (31)	4 (2), 5w (2w)/13
Inulin	[v]	77 (11)	3, 4w (1w)/9
Raffinose	[v]	70 (10)	3, 4w (1w)/10
Catalase	[+]	100	11, 1w/12
Oxidase	[+]	100	13/13
Growth on:			
MacConkey	[-]	0	0/12
SS	-	0	0/12
Simmons citrate	-	0	0/12
Urea, Christensen's	-	0	0/8
Nitrate reduction	-	0	0/12
Gas from nitrate	-	0	0/12
Nitrite reduction	v	67	4/6
Indole	[-]	0	0/12
TSI slant, acid	-	8	1w/13
TSI butt, acid	-	8	1w/13
H$_2$S (TSI butt)	-	0	0/9
H$_2$S (Pb ac paper)	v	33	2, 1w/9
Gelatin hydrolysis[c]	-	0	0/9
Litmus milk	-	9 IR	1 IR/11
Pigment	-	0	0/8
Growth at:			
25°C	-	10	1/10
35°C	v	70	7/10
42°C	-	0	0/10
Esculin hydrolysis	+ or (+)	77 (23)	10 (3)/13
Lysine decarboxylase	-	0	0/8
Arginine dihydrolase	[+]	100	8/8
Ornithine decarboxylase	-	0	0/8
Nutrient broth, 0% NaCl	-	8	1/12
Nutrient broth, 6% NaCl	-	0	0/12
ONPG	[+]	100	5/5

NUMBER OF STRAINS 13

[a]Gliding motility is observed by light microscopy. Flagella are not detected.
[b]Acid production is enhanced by heavily inoculating small volumes of fermentation medium.
[c]Incubation of 7–14 days.

Cardiobacterium hominis

Gram-negative, pleomorphic rod that may form rosettes and may retain crystal violet stain, nonmotile; fastidious, and growth may be enhanced by candle-jar atmosphere; ferments glucose, mannitol, sucrose, and maltose, not xylose nor lactose; does not grow on MacConkey agar; oxidase-positive; catalase-, urease-, and nitrate-negative; indole-positive (may be weak).

Sources (32 strains)

Blood (88%), other (12%).

Reference strain

KC1408 ← ATCC 29308, from blood.

Literature

1. Slotnick, I.J., and M. Dougherty. 1964. Further characterization of an unclassified group of bacteria causing endocarditis in man: *Cardiobacterium hominis* gen. et sp. n. Antonie van Leeuwenhoek J. Microbiol. Serol. *30:* 261–272.

2. Midgley, J., S.P. LaPage, B.A.G. Jenkins, G.I. Barrow, M.E. Roberts, and A.G. Buck. 1970. *Cardiobacterium hominis* endocarditis. J. Med. Microbiol. *3:* 91–98.

3. Snell, J.J.S., and S.P. Lapage. 1976. Transfer of some saccharolytic *Moraxella* species to *Kingella* Henriksen and Bøvre 1976, with descriptions of *Kingella indologenes* sp. nov. and *Kingella denitrificans* sp. nov. Int. J. Syst. Bacteriol. *26:* 451–458.

4. Weaver, R.E. 1984. Genus *Cardiobacterium* Slotnick and Dougherty 1964, 271.[AL] *In* N.R. Krieg and J.G. Holt (Eds.), Bergey's Manual of Systematic Bacteriology, Vol. 1, Williams & Wilkins, Baltimore, pp. 583–585.

5. Dewhirst, F.E., B.J. Paster, S. LaFontaine, and J.I. Hood. 1990. Transfer of *Kingella indologenes* (Snell and Lapage 1976) to the genus *Suttonella* gen. nov. as *Suttonella indologenes* comb. nov.; transfer of *Bacteroides nodosus* (Beveridge 1941) to the genus *Dichelobacter* gen. nov. as *Dichelobacter nodosus* comb. nov.; and assignment of the genera *Cardiobacterium*, *Dichelobacter*, and *Suttonella* to *Cardiobacteriaceae* fam. nov. in the gamma division of *Proteobacteria* on the basis of 16S rRNA sequence comparisons. Int. J. Syst. Bacteriol. *40:* 426–433.

Comments

Colonies frequently pit the agar substrate. Addition of yeast extract to the medium is reported to render cells uniform in appearance. The rare mannitol-negative strain would be difficult to differentiate from *Suttonella indologenes*. Snell and Lapage reported (3) that casein digestion, phosphate production, and Tween 20 and Tween 40 hydrolysis are negative for *C. hominis*, but positive for *Suttonella indologenes*. Found in human nose and throat; a cause of endocarditis in humans. Information on the cellular fatty acid composition of this species is presented in Table 4.27 and Figure 4.32 of the cellular fatty acid section.

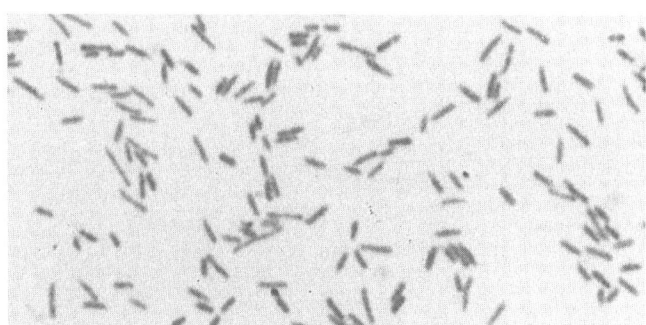

Cardiobacterium hominis strain KC1408. Gram stain BAP 35° C 24 h x 1700.

Cardiobacterium hominis

TEST PERFORMED	SIGN	%+	#+/T
NUMBER OF STRAINS 65			
Morphology	mrs		
Motility	[nm]		
Gas from glucose	−	0	0/65
Action on blood	v	12 gr	8/65
Fermentative or oxidative	[F]a		
Carbohydrate base	Fb		
Acid from:			
D-Glucose	+ or (+)	80 (20)	43, 9w (13)/65
D-Xylose	[−]	0	0/65
D-Mannitol	[+ or (+)]	52 (43)	29, 5w (28)/65
Lactose	[−]	0	0/65
Sucrose	+ or (+)	63 (35)	35, 6w (23)/65
Maltose	+ or (+)	69 (31)	37, 8w (20)/65
Catalase	[−]	1	1w/65
Oxidase	[+]	100	65/65
Growth on:			
MacConkey	[−]	0	0/65
SS	−	0	0/65
Simmons citrate	−	0	0/65
Urea, Christensen's	[−]	0	0/58
Nitrate reduction	[−]	0	0/65
Indole	[+]	100	64/64
TSI slant, acid	+	92 (1)	58 (1)/63
TSI butt, acid	v	81 (3)	51 (2)/61
H_2S (TSI butt)	−	0	0/61
H_2S (Pb ac paper)	+	90	55/61
MR	v	50	5/10
VP	−	0	0/6
Gelatin hydrolysisc	−	0	0/38
Litmus milk	v	12 (5) Aw	7w (3)/58; 6 IR
Growth at:			
25°C	v	35	15/43
35°C	+	100	55/55
42°C	v	20	8/40
Esculin hydrolysis	−	2	1/52
Nutrient broth, 0% NaCl	v	62	39/63
Nutrient broth, 6% NaCl	−	8	5/62

aTSI reaction is not always indicative of fermentative activity.
bRabbit serum (1–2 drops 3 ml of medium) may be required; reactions may be obtained within 4 h using the rapid sugar test.
cIncubation of 7–14 days.

Chromobacterium violaceum

Gram-negative, usually straight rod, motile (both 1 polar and 1–4 lateral flagella); ferments glucose, usually without gas, and sometimes sucrose, but not mannitol, or maltose; grows on MacConkey and SS agars; usually produces ethanol-soluble violet pigment; growth best at 30–35° C, no growth at 4° C; nonpigmented colonies frequently indole-positive; does not require NaCl for growth; lysine and ornithine decarboxylase-negative, arginine dihydrolase-positive.

Sources (38 strains)

Water (18%), blood (13%), wound (11%), skin lesion (10%), pine shavings (8%), urine (6%), ear (5%), autopsy material (5%), other (16%), unknown (8%).

Reference strains

KC531 ← ATCC 12472, type strain, from fresh water.
F4417, from leg abscess.

Literature

1. Wheat, R.P., A. Zuckerman, and L.A. Rantz. 1951. Infection due to chromobacteria. AMA Arch. Intern. Med. *88:* 461–466.

2. Leifson, E. 1956. Morphological and physiological characters of the genus *Chromobacterium*. J. Bacteriol. *71:* 393–400.

3. Sneath, P.H.A. 1956. Cultural and biochemical characteristics of the genus *Chromobacterium*. J. Gen. Microbiol. *15:* 70–98.

4. Opinion 16. 1958. Int. Bull. Bacteriol. Nomencl. Taxon. *8:* 151–152, 1958.

5. Sivendra, R. 1976. Unusual *Chromobacterium violaceum*: aerogenic strains. J. Clin. Microbiol. *3:* 70–71.

6. Sorensen, R.U., M.R. Jacobs, and S.B. Shurin. 1985. *Chromobacterium violaceum* adenitis acquired in the northern United States as a complication of chronic granulomatous disease. Pediatr. Infect. Dis. *4:* 701–702.

Comments

Common in tropics; soil and water organisms; sometimes causes serious pyogenic or septicemic infections of humans and other animals. Information on the cellular fatty acid composition of this species is presented in Table 4.37 and Figure 4.42 of the cellular fatty acid section.

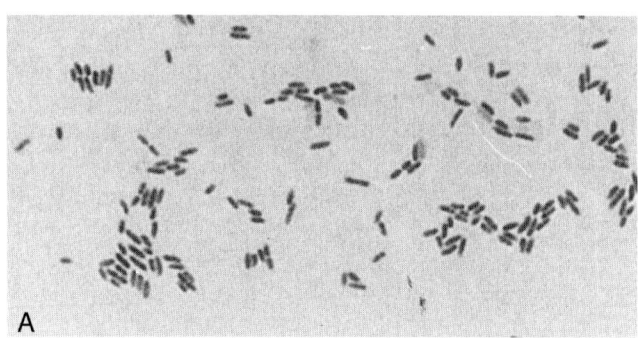

Chromobacterium violaceum strain KC531. Gram stain BAP 35° C 24 h x 1700.

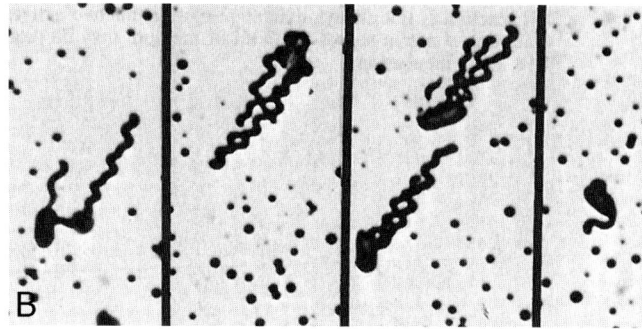

Chromobacterium violaceum strain F4417. Flagella stain HIA 25° C 24 h x 1700.

Chromobacterium violaceum

TEST PERFORMED	SIGN	% +	#+/T
	NUMBER OF STRAINS 37		
Morphology	mrs		
Motility; flagella	[m;p,L]		
Gas from glucose	–[a]	0	0/36
Action on blood	v	48 ß	14/29; 9 ly, 1 br
Fermentative or oxidative	[F]		
Carbohydrate base	F		
Acid from:			
D-Glucose	+	100	36/36
D-Xylose	–	0	0/36
D-Mannitol	[–]	0	0/36
Lactose	[–]	0	0/36
Sucrose	v	20 (6)	4, 3w (2)/35
Maltose	[–]	0 (3)	(1)/36
Catalase	+	97	30, 5w/36
Oxidase	v	67	19, 3w/33
Growth on:			
MacConkey	[+]	100	35, 1w/36
SS	[+ or (+)]	71 (23)	25 (8)/35
Simmons citrate	v	68 (9)	23 (3)/34
Urea, Christensen's	v	5 (14)	2 (5)/36
Nitrate reduction	+	97	34/35
Indole	v[b]	21	7/33
TSI slant, acid	–	8	3/35
TSI butt, acid	+	94	33/35
H$_2$S (TSI butt)	–	0	0/34
H$_2$S (Pb ac paper)	v	41	3, 11w/34
MR	v	37	11/30
VP	–	0	0/31
Gelatin hydrolysis[c]	v	86	8, 11w/22
Pigment:			
Insoluble violet	[+]	91	34/37
Litmus milk	pep	97	34/35
Growth at:			
25°C	+	100	34/34
35°C	+	100	34/34
42°C	v	85	29/34
Esculin hydrolysis	–	5[d]	2/37
Lysine decarboxylase	[–]	0	0/22
Arginine dihydrolase	[+]	100	22/22
Ornithine decarboxylase	[–]	0	0/25
Nutrient broth, 0% NaCl	[+]	100	13/13
Nutrient broth, 6% NaCl	–	8	1/13

[a]Gas producing strains have been described.
[b]Nonpigmented colonies may be indole-positive.
[c]Incubation of 7–14 days.
[d]The two esculin-positive strains were not pigmented.

Chryseomonas luteola

Gram-negative short to medium length, slightly thick rod, sometimes paired; motile (polar tuft of three or more flagella); cells often appear spindle-shaped and vacuolated; smooth and rough, wrinkled, adherent colonies may occur; produces a yellow growth pigment. Aerobic; grows on MacConkey agar; oxidase-negative; uses glucose, xylose, maltose, usually mannitol (often weakly), sometimes lactose and occasionally sucrose; produces arginine dihydrolase but not lysine or ornithine decarboxylase; hydrolyzes esculin; usually reduces nitrate to nitrite.

Sources (34 strains)

Blood (9%), urine (9%), lung (9%), cerebrospinal fluid (6%), eye (6%), wound (6%), other (31%), unknown (24%).

Reference strain

KC1808 ← JCM 3352 = ATCC 43273, type strain, from wound.

Literature

1. Pickett, M.J., and M.M. Pedersen. 1970. Characterization of saccharolytic nonfermentative bacteria associated with man. Can. J. Microbiol. *16:* 351–362.

2. Gilardi, G.L., S. Hirschl, and M. Mandel. 1975. Characteristics of yellow-pigmented nonfermentative bacilli (Groups VE-1 and VE-2) encountered in clinical bacteriology. J. Clin. Microbiol. *1:* 384–389.

3. Kodama, K., N. Kimura, and K. Komagata. 1985. Two new species of *Pseudomonas: P. oryzihabitans* isolated from rice paddy and clinical specimens and *P. luteola* isolated from clinical specimens. Int. J. Syst. Bacteriol. *35:* 467–474.

4. Holmes, B., A.G. Steigerwalt, R.E. Weaver, and D.J. Brenner. 1987. *Chryseomonas luteola* comb. nov. and *Flavimonas oryzihabitans* gen. nov., comb. nov., *Pseudomonas*-like species from human clinical specimens and formerly known, respectively, as Groups Ve-1 and Ve-2. Int. J. Syst. Bacteriol. *37:* 245–250.

Comments

Formerly designated CDC Group Ve-1. Other synonyms include *Chryseomonas polytricha* and *Pseudomonas luteola*. Cellular and colonial morphologies similar to *Flavimonas oryzihabitans* (formerly CDC Group Ve-2). Information on the cellular fatty acid composition of this species is presented in Table 4.38 and Figure 4.43 of the cellular fatty acid section.

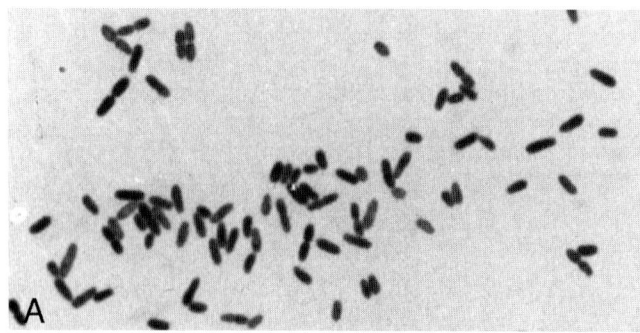

Chryseomonas luteola strain D4029(1). Gram stain BAP 35° C 24 h x 1700.

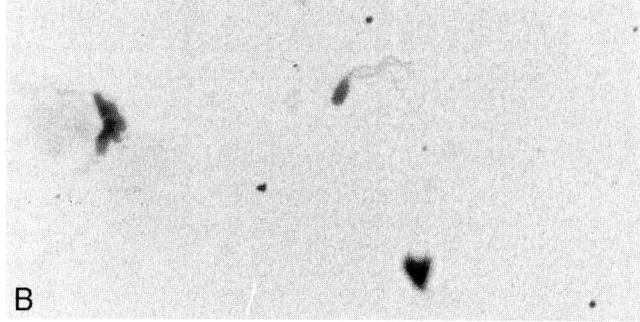

Chryseomonas luteola strain D4029(1). Flagella stain HIA 35° C 24 h x 1700.

Chryseomonas luteola

TEST PERFORMED	SIGN	% +	#+/T
	NUMBER OF STRAINS 34		
Morphology	mrs		
Motility; flagella	[m;p, >2]		
Action on blood	v	41 ly	13/32; 5 gr, 3 LG, 1 ß
Fermentative or oxidative	[O]		
Carbohydrate base	OF		
Acid from:			
D-Glucose	[+]	100	34/34
D-Xylose	+	100	34/34
D-Mannitol	[+ or (+)]	76 (18)	13, 12w (3), (3w)/33
Lactose	v	3 (24)	1w (8)/34
Sucrose	v	12	4/34
Maltose	+	100	33, 1w/34
Catalase	+	100	28, 6w/34
Oxidase	[−][a]	0	0/34
Growth on:			
MacConkey	[+]	100	34/34
SS	v	68	22, 1w/34
Cetrimide	−	0	0/34
Simmons citrate	+	100	34/34
Urea, Christensen's	v	26 (38)	9 (13)/34; 4 pk
Nitrate reduction	v	62	20, 1w/34
Gas from nitrate	−	0	0/34
Indole	−	0	0/34
TSI slant, acid	−	0	0/34
TSI butt, acid	−	0	0/34
H_2S (TSI butt)	−	0	0/34
H_2S (Pb ac paper)	v	12	2, 2w/34
Gelatin hydrolysis[b]	v	61	19/31
Litmus milk	v	44 k	15/34; 6 IR, 5 pep, 4 dirty
Pigment:			
Insoluble	[yel]	97	33/34
Growth at:			
25°C	+	100	32/32
35°C	+	100	32/32
42°C	+	94	30/32
Esculin hydrolysis	[+]	100	34/34
Lysine decarboxylase	[−]	0	0/25
Arginine dihydrolase	[+]	100	22, 3w/25
Ornithine decarboxylase	−	0	0/25
Nutrient broth, 0% NaCl	+	100	34/34
Nutrient broth, 6% NaCl	v	74 (3)	25 (1)/34

[a] A weak and late oxidase reaction was observed with 4 strains using the Kovacs' technique.
[b] Incubation of 7–14 days.

Comamonas acidovorans

Gram-negative medium to large rod, motile (polar tuft of three or more flagella); aerobic; grows on MacConkey agar and usually on SS agar; oxidase-positive; produces a weakly acid or neutral reaction in mannitol and fructose; alkaline in glucose, xylose, lactose, sucrose, and maltose; usually alkalinizes acetamide and tartrate; reduces nitrate without gas formation; urease-negative; does not produce H_2S on TSI agar.

Sources (72 strains)

Urine (19%), blood (18%), intravenous tubing (8%), water (6%), sputum (6%), other (35%), unknown (8%).

Reference strain

KC1069 ← R.Y. Stanier, RYS14 = ATCC 15668, type strain, from soil.

Literature

1. den Dooren de Jong, L.E. 1926. Bijdrage tot de kennis van het mineralisatieproces, Nijgh and van Ditmar Uitgevers-Mij, Rotterdam, pp. 1–200.

2. Stanier, R.Y., N.J. Palleroni, and M. Doudoroff. 1966. The aerobic pseudomonads: a taxonomic study. J. Gen. Microbiol. 43: 159–271.

3. Tamaoka, J., Duk-Mo Ha, and K. Komagata. 1987. Reclassification of *Pseudomonas acidovorans* den Doorn de Jong 1926 and *Pseudomonas testosteroni* Marcus and Talalay 1956 as *Comamonas acidovorans* comb. nov. and *Comamonas testosteroni* comb. nov. with an emended description of the genus *Comamonas*. Int. J. Syst. Bacteriol. 37: 52–59.

4. De Vos, P., K. Kersters, E. Falsen, B. Pot, M. Gillis, P. Segers, and J. De Ley. 1985. *Comamonas* Davis and Park 1962 gen. nov., nom. rev. emend., and *Comamonas terrigena* Hugh 1962 sp. nov., nom. rev. Int. J. Syst. Bacteriol. 35: 443–453.

5. Gilardi, G.L. 1991. *Pseudomonas* and related genera. In A. Balows, et al. (Eds.), Manual of Clinical Microbiology, 5th ed., American Society for Microbiology, Washington, pp. 429–441.

6. Horowitz, H., S. Gilroy, S. Feinstein, and G. Gilardi. 1990. Endocarditis associated with *Comamonas acidovorans*. J. Clin. Microbiol. 28: 143–145.

Comments

Formerly designated *Pseudomonas acidovorans*, this species, along with *C. testosteroni*, has been reclassified in the recently revived genus *Comamonas* (3). Cells are predominantly straight, occasionally curved; indole is not produced, but some strains cause a vivid yellow reaction in the test medium with xylene extraction and the addition of Ehrlich's reagent. Hugh and Welch reported that some strains produce anthranilic acid and kynurenine in tryptone broth, and that these and other tryptone derivatives cause Kovacs' reagent to become orange (5). A case of endocarditis caused by a mannitol-negative strain of *C. acidovorans* has been reported (6). Information on the cellular fatty acid composition of this species is presented in Table 4.9 and Figure 4.14 of the cellular fatty acid section.

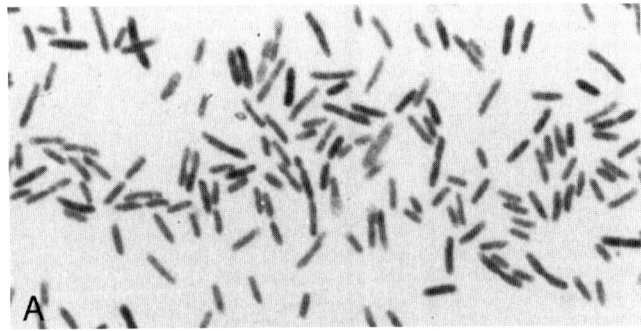

Comamonas acidovorans strain KC1069. Gram stain BAP 35° C 24 h x 1700.

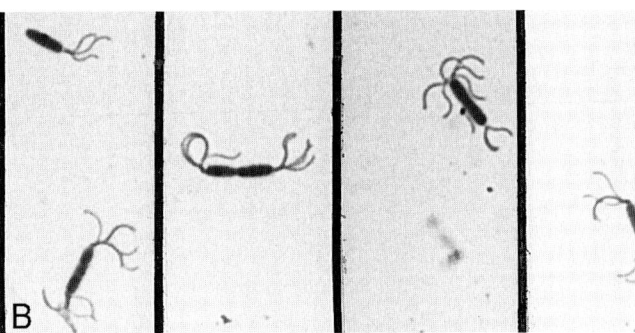

Comamonas acidovorans strain KC1069. Flagella stain BAP 35° C 24 h x 1700.

Comamonas acidovorans

TEST PERFORMED	SIGN	% +	#+/T
NUMBER OF STRAINS 69			
Morphology	mrs		
Motility; flagella	[m;p, >2]		
Action on blood	v	35 ly	23/65; 14 gr, 8 br, 5 LG, 3 al
Fermentative or oxidative	[n-o]		
Carbohydrate base	OF		
Acid from:			
D-Glucose	[−]	0	0/69
D-Xylose	[−]	0	0/69
D-Mannitol	[+]	100	69w/69
Lactose	−	0	0/69
Sucrose	−	0	0/69
Maltose	−	0	0/69
Fructose	+	100	11w/11
Catalase	+	100	69/69
Oxidase	[+]	100	59, 9w/68
Growth on:			
MacConkey	[+]	100	69/69
SS	v	63 (20)	44 (14)/69
Cetrimide	−	4 (3)	3 (2w)/69
Simmons citrate	+	94 (4)	65 (3)/69
Urea, Christensen's	[−]	(4)	(3)/69
Nitrate reduction	+	99	68/69
Gas from nitrate	[−]	0	0/69
Indole	−	0	0/69
TSI slant, acid	−	0	0/69
TSI butt, acid	−	0	0/69
H_2S (TSI butt)	−	0	0/69
H_2S (Pb ac paper)	v	57	3, 36w/69
Gelatin hydrolysis[a]	v	11	4/38
Litmus milk	k	97	67/69
Pigment:			
Fluorescent	v	26	18/69
Soluble	v	44 yel-ta	19/43
Growth at:			
25°C	+	100	69/69
35°C	+	100	69/69
42°C	v	29 (10)	20 (7)/69
Esculin hydrolysis	−	0	0/64
Lysine decarboxylase	−	0	0/23
Arginine dihydrolase	−	0	0/23
Ornithine decarboxylase	−	0	0/23
Nutrient broth, 0% NaCl	+	100	62/62
Nutrient broth, 6% NaCl	−	6	4/62
Alkalinization of:			
Acetamide	v	80 (8)	21 (2)/26
Serine	−	4	1w/26
Tartrate	v	88	23/26

[a] Incubation of 7–14 days.

Comamonas terrigena

Description of type strain: Gram-negative medium to long rod with curved filaments; motile (polar tuft of three or more flagella); aerobic; grows weak and late on MacConkey agar; does not grow on SS or cetrimide agars; does not use carbohydrates; oxidase-positive; reduces nitrate (without gas); does not alkalinize acetamide, serine, tartrate, or sodium acetate.

Source (type strain)

Hay infusion, United States.

Reference strain

KC1391 ← NCIB 8193 ← R. Hugh 147 = ATCC 8461, type strain, from hay.

Literature

1. Hugh, R. 1962. *Comamonas terrigena* comb. nov., with proposal of a neotype and request for an opinion. Int. Bull. Bacteriol. Nomencl. Taxon. *12:* 33–35.

2. Hugh, R. 1965. A comparison of *Pseudomonas testosteroni* and *Comamonas terrigena*. Int. Bull. Bacteriol. Nomencl. Taxon. *15:* 125–132.

3. Stanier, R.Y., N.J. Palleroni, and M. Doudoroff. 1966. The aerobic pseudomonads: a taxonomic study. J. Gen. Microbiol. *43:* 159–271.

4. De Vos, P., K. Kersters, E. Falsen, B. Pot, M. Gillis, P. Segers, and J. De Ley. 1985. *Comamonas* Davis and Park 1962 gen. nov., nom. rev. emend., and *Comamonas terrigena* Hugh 1962 sp. nov., nom. rev. Int. J. Syst. Bacteriol. *35:* 443–453.

5. Willems, A., J. De Ley, M. Gillis, and K. Kersters. 1991. *Comamonadaceae*, a new family encompassing the acidovorans rRNA complex, including *Variovorax paradoxus* gen. nov., comb. nov. for *Alcaligenes paradoxus* (Davis 1969). Int. J. Syst. Bacteriol. *41:* 445–450.

6. Gilardi, G.L. 1991. *Pseudomonas* and related genera. *In* A. Balows, et al. (Eds.), Manual of Clinical Microbiology, 5th ed., American Society for Microbiology, Washington, pp. 429–441.

Comments

The taxon *C. terrigena* recently has been revived and includes bacteria formerly designated *"Variovorax neocistes,"* and *"V. cyclosites."* Biochemical differentiation between *C. testosteroni* and *C. terrigena* is problematic, but preliminary studies in our laboratory with type strains of both species suggest that citrate and sodium acetate utilization may be useful tests. *C. terrigena* is the only *Comamonas* species that requires methionine and nicotinamide for growth. Cellular fatty acid analysis is also useful in differentiating these species. Information on the cellular fatty acid composition of this species is presented in Table 4.34 and Figure 4.39 of the cellular fatty acid section.

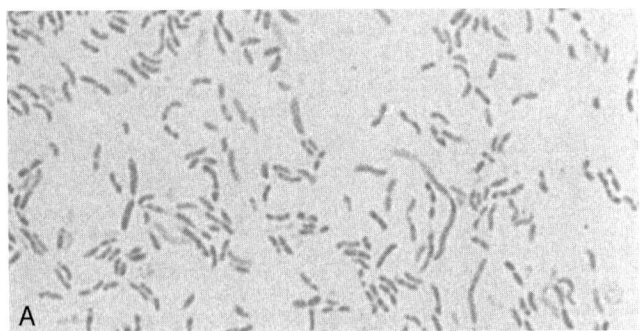

Comamonas terrigena strain KC1391. Gram stain BAP 35° C 24 h x 1700.

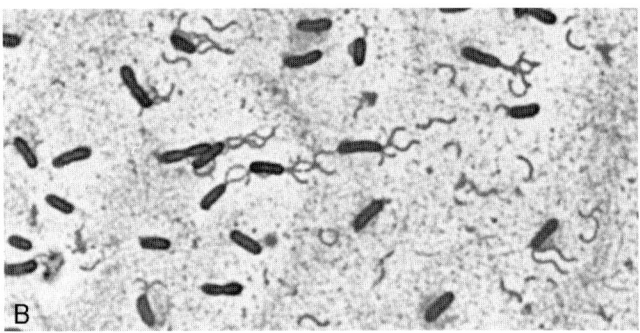

Comamonas terrigena strain KC1391. Flagella stain (Ryu method) TGY 25° C 24 h x 1700.

Comamonas testosteroni

Description of type strain: Gram-negative medium to long rod; motile (polar tuft of three or more flagella); aerobic; grows on MacConkey agar, most strains do not grow on SS agar; does not use carbohydrates; oxidase-positive, usually reduces nitrate (without gas); utilizes sodium acetate but not acetamide; does not produce H_2S on TSI agar.

Reference strain

KC1071 ← ATCC 11996, type strain, from soil.

Literature

1. Hugh, R. 1965. A comparison of *Pseudomonas testosteroni* and *Comamonas terrigena*. Int. Bull. Bacteriol. Nomencl. Taxon. *15:* 125–132.

2. Stanier, R.Y., N.J. Palleroni, and M. Doudoroff. 1966. The aerobic pseudomonads: a taxonomic study. J. Gen. Microbiol. *43:* 159–271.

3. Tamaoka, J., Duk-Mo Ha, and K. Komagata. 1987. Reclassification of *Pseudomonas acidovorans* den Doorn de Jong 1926 and *Pseudomonas testosteroni* Marcus and Talalay 1956 as *Comamonas acidovorans* comb. nov. and *Comamonas testosteroni* comb. nov. with an emended description of the genus *Comamonas*. Int. J. Syst. Bacteriol. 37: 52–59.

4. Gilardi, G.L. 1991. *Pseudomonas* and related genera. *In* A. Balows, et al. (Eds.), Manual of Clinical Microbiology, 5th ed., American Society for Microbiology, Washington, pp. 429–441.

5. De Vos, P., K. Kersters, E. Falsen, B. Pot, M. Gillis, P. Segers, and J. De Ley. 1985. *Comamonas* Davis and Park 1962 gen. nov., nom. rev. emend., and *Comamonas terrigena* Hugh 1962 sp. nov., nom. rev. Int. J. Syst. Bacteriol. *35:* 443–453.

Comments

Formerly designated *Pseudomonas testosteroni,* this species, along with *C. acidovorans,* has been reclassified in the recently revived genus *Comamonas*. Indole is not produced. It is reported that this organism uses testosterone (hence the name), but some pseudomonads and some fungi may also do so. Biochemical differentiation between *C. testosteroni* and *C. terrigena* is problematic, but preliminary studies in our laboratory with type strains of both species suggest that citrate and sodium acetate utilization may be useful tests. Cellular fatty acid analysis is also useful in differentiating these species. Information on the cellular fatty acid composition of this species is presented in Table 4.9 and Figure 4.14 of the cellular fatty acid section.

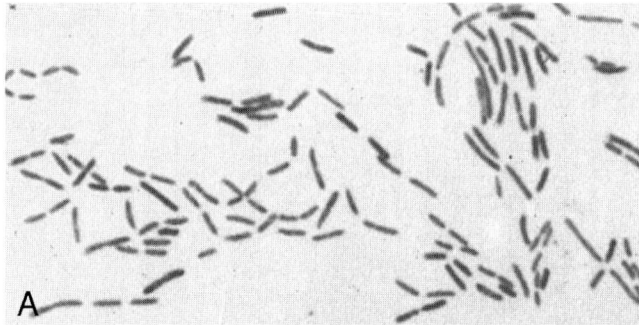

Comamonas testosteroni strain KC1071. Gram stain BAP 35° C 24 h x 1700.

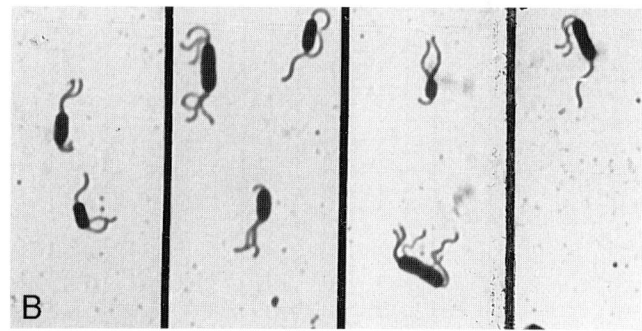

Comamonas testosteroni strain KC1071. Flagella stain FB 25° C 24 h x 1700.

Comamonas testosteroni and *Comamonas terrigena*

TEST PERFORMED	*Comamonas testosteroni* TYPE STRAIN SIGN	*Comamonas terrigena* TYPE STRAIN SIGN	*Comamonas* species NUMBER OF STRAINS 28[a]		
			SIGN	%+	#+/T
Morphology	[pr,r]	[lr,fc]	[pr,r]		
Motility; flagella	[m;p, >2]	[m;p, >2]	[m;p, >2]		
Action on blood	-	-	v	32 ly	9/28: 4 LG, 2 gr, 1 br
Fermentative or oxidative	[n-o]	[n-o]	[n-o]		
Carbohydrate base	OF	OF	OF		
Acid from:					
D-Glucose	[-]	[-]	[-]	0	0/28
D-Xylose	[-]	[-]	[-]	0	0/28
D-Mannitol	[-]	[-]	[-]	0	0/28
Lactose	-	-	-	0	0/28
Sucrose	-	-	-	0	0/28
Maltose	-	-	-	0	0/28
Fructose	[-]	[-]	[-]	6[b]	1w/16
Catalase	+	+	+	96	21, 6w/28
Oxidase	[+]	[+]	[+]	100	26, 2w/28
Growth on:					
MacConkey	[+]	[(+)]	[+]	96 (4)	27 (1)/28
SS	-	-	v	21 (11)	6 (3)/28
Cetrimide	-	-	-	(4)	(1)/28
Simmons citrate	[+]	[-]	[v]	47 (7)	13 (2)/28
Urea, Christensen's	[-]	[+]	[v]	7 (14)	2 (4)/28
Nitrate reduction	+	[+]	[+]	96	27/28
Gas from nitrate	[-]	[-]	[-]	0	0/28
Indole	-	-	-	0	0/28
TSI slant, acid	-	-	-	0	0/28
TSI butt, acid	-	-	-	0	0/28
H$_2$S (TSI butt)	-	-	-	0	0/28
H$_2$S (Pb ac paper)	-	+w	v	61	14, 3w/28
Gelatin hydrolysis[c]	-	-	-	0	0/18
Litmus milk	k	k	v	86 k	24/28
Pigment:					
Soluble	-	tan	v	27 yel-br	6/22
Growth at:					
25°C	+	+	+	100	25/25
35°C	+	+w	+	100	25/25

42°C	+w	-	v	17/25
Esculin hydrolysis	-	-	-	0/28
Lysine decarboxylase	-	-	-	0/17
Arginine dihydrolase	-	-	-	0/17
Ornithine decarboxylase	-	-	-	0/17
Nutrient broth, 0% NaCl	+	+	+	28/28
Nutrient broth, 6% NaCl	-	-	v	4/28
Alkalinization of:				
Acetamide	[-]	[-]	[-]	1w/18
Serine	[-]	[-]	[v]	3w/18
Tartrate	[-]	[-]	[-]	0/18
Sodium acetate	[+]	[-]		

[a]Sources: sputum (18%), peritoneal fluid (18%), urine (11%), urinary catheter (7%), appendix (7%), stool (7%), other (14%), unknown (18%). These strains were listed as *Pseudomonas testosteroni* in the first edition of this manual. They are phenotypically similar to both *C. testosteroni* and *C. terrigena*. The characteristics needed to differentiate these two species have not been determined.
[b]Fructose reaction weak.
[c]Incubation of 7–14 days.

DF-3

Gram-negative small coccoid to short rod; nonmotile; facultatively anaerobic; fastidious; ferments glucose, xylose, lactose, sucrose, and maltose; usually produces indole weakly; catalase- and oxidase-negative; produces no growth on MacConkey agar; does not reduce nitrate; hydrolyzes esculin; produces acid on slant and in butt of TSI agar.

Sources (21 strains)

Blood (26%), wound (15%), urine (9%), peritoneal fluid (6%), umbilicus (6%), abscess (4%), abdominal (4%), stool (4%), genital (4%), other (22%).

Reference strain

D7608, from shunt.

Literature

1. Pickett, M.J., D.G. Hollis, and E.J. Bottone. 1991. Miscellaneous gram-negative bacteria. In A. Balows, et al. (Eds.), Manual of Clinical Microbiology, 5th ed., American Society for Microbiology, Washington, pp. 410–428.

2. Gill, V.E., B. Travis, and D.Y. Williams. 1991. Clinical and microbiological observations on CDC group DF-3, a gram negative coccobacillus. J. Clin. Microbiol. *29:* 1589–1592.

3. Blum, R.N., C.D. Berry, M.G. Phillips, D.L. Hamlos, and E.W. Koneman. 1992. Clinical illness associated with isolation of dysgonic fermenter 3 from stool samples. J. Clin. Microbiol. *30:* 396–400.

4. Wallace, P.L., D.G. Hollis, R.E. Weaver, and C.W. Moss. 1989. Characterization of CDC group DF-3 by cellular fatty acid analysis. J. Clin. Microbiol. *27:* 735–737.

5. Bernard K., C. Cooper, S. Tessier, and E.P. Ewan. 1991. Use of chemotaxonomy as an aid to differentiate among *Capnocytophaga* species, CDC Group DF-3, and aerotolerant strains of *Leptotrichia buccalis*. J. Clin. Microbiol. *29:* 2263–2265.

Comments

Usually two drops of serum must be added to each carbohydrate tube for reaction, which is frequently delayed. Cultures have a sweet and bitter odor. Propionic, lactic, and succinic acids are the major end products of glucose fermentation (5). Information on the cellular fatty acid composition of this group is presented in Table 4.28 and Figure 4.33 of the cellular fatty acid section.

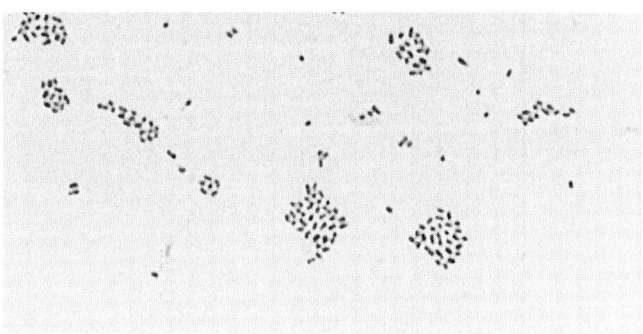

DF-3 strain D7608. Gram stain BAP 35° C 24 h x 1700.

DF-3

TEST PERFORMED	SIGN	% +	#+/T
Number of strains			21
Morphology	cc,srs		
Motility	nm		
Gas from glucose	-	0	0/21
Action on blood	[-]	10 al	2/21; 1 β
Fermentative or oxidative	[F]		
Carbohydrate base	F[a]		
Acid from:			
D-Glucose	[+ or (+)]	86 (14)	18 (3)/21
D-Xylose	[+ or (+)]	86 (14)	18 (3)/21
D-Mannitol	-	0	0/21
Lactose	[+ or (+)]	52 (43)	11 (9)/21
Sucrose	+ or (+)	62 (33)	13 (7)/21
Maltose	[+ or (+)]	81 (19)	17 (4)/21
Catalase	[-]	0	0/21
Oxidase	[-]	0	0/21
Growth on:			
MacConkey	[-]	0	0/21
SS	-	0	0/21
Simmons citrate	-	0	0/21
Urea, Christensen's	-	0 (4)	(1)/21
Nitrate reduction	[-]	0	0/21
Indole	[v]	71 (14)	15 (3)/21
TSI slant, acid	[+]	95	20/21
TSI butt, acid	[+]	100	21/21
H_2S (TSI butt)	-	0	0/21
H_2S (Pb ac paper)	+	90	19/21
Gelatin hydrolysis[b]	[-]	0	0/5
Litmus milk	v	76 A	16/21; 3 IR
Growth at:			
25°C	v	50	10/20
35°C	v	85	17/20
42°C	v	20	4/20
Esculin hydrolysis	[+]	100	21/21
Lysine decarboxylase	-	0	0/4
Arginine dihydrolase	-	0	0/4
Ornithine decarboxylase	-	0	0/4
Nutrient broth, 0% NaCl	v	38	8/21
Nutrient broth, 6% NaCl	-	10	2/21

[a]Rabbit serum (1–2 drops/3 ml of medium) may be required; reactions may be obtained within 4 h by using the rapid sugar test.
[b]Incubation of 7–14 days.

DF-3-like

Gram-negative small coccoid to short rod; nonmotile; produces a β-like hemolysis on rabbit blood agar; facultatively anaerobic; fastidious; ferments glucose, lactose, and maltose, but not xylose or sucrose; indole-positive; oxidase-negative; does not grow on MacConkey agar; does not reduce nitrate.

Sources (7 strains)

Wound (finger, hand, or foot) (4), blood (3).

Reference strain

E8526, from blood.

Literature

1. Daneshvar, M.I., D.G. Hollis, and C.W. Moss. 1991. Chemical characterization of clinical isolates which are similar to CDC Group DF-3 bacteria. J. Clin. Microbiol. *29*: 2351–2353.

Comments

Similar to DF-3. Differential characteristics include the failure to ferment xylose and sucrose. Often peptonizes litmus milk and hydrolyzes gelatin. Some strains produce catalase. Information on the cellular fatty acid composition of this group is presented in Table 4.29 and Figure 4.34 of the cellular fatty acid section.

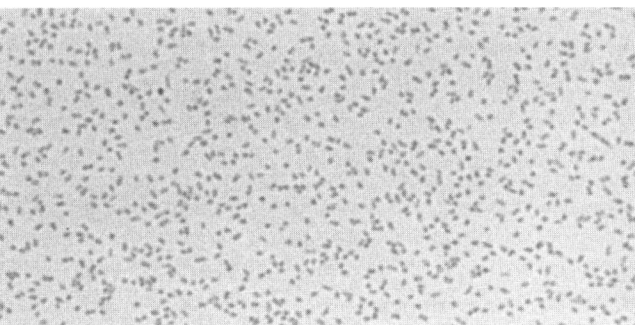

DF-3-like strain E8526. Gram stain BAP 35° C 24 h x 1700.

DF-3-like

TEST PERFORMED	SIGN	% +	#+/T
	NUMBER OF STRAINS 7		
Morphology	cc,srs		
Motility	[nm]		
Action on blood	[ß]	100	7/7
Fermentative or oxidative	[F]		
Carbohydrate base	F^a		
Acid from:			
D-Glucose	[+ or (+)]	43 (57)	3 (3), (1w)/7
D-Xylose	[−]	0	0/7
D-Mannitol	−	0	0/7
Lactose	+ or (+)	57 (43)	4 (2), (1w)/7
Sucrose	[−]	0	0/7
Maltose	+ or (+)	57 (43)	1, 3w, (1), (2w)/7
Catalase	v	29	2/7
Oxidase	[−]	0	0/7
Growth on:			
MacConkey	[−]	0	0/7
SS	−	0	0/7
Simmons citrate	−	0	0/7
Urea, Christensen's	−	0	0/7
Nitrate reduction	−	0	0/7
Gas from nitrate	−	0	0/7
Indole	[+]	100	7/7
TSI slant, acid	v	29	2/7
TSI butt, acid	v	29	2/7
H_2S (TSI butt)	−	0	0/7
H_2S (Pb ac paper)	v	86	6/7
Gelatin hydrolysis[b]	v	43	3/7
Litmus milk	v	71 pep	5/7
Growth at:			
25°C	v	14	1/7
35°C	v	57	4/7
42°C	v	14	1/7
Esculin hydrolysis	v	57 (14)	4 (1)/7
Lysine decarboxylase	−	0	0/2
Arginine dihydrolase	−	0	0/2
Ornithine decarboxylase	−	0	0/2
Nutrient broth, 0% NaCl	v	17	1/6
Nutrient broth, 6% NaCl	v	17	1/6

[a]Rabbit serum (1–2 drops/3 ml of medium) may be required.
[b]Incubation of 7–14 days.

EF-4a

Gram-negative coccoid to short rod, which stains deeply; nonmotile; ferments glucose only without gas; oxidase- and catalase-positive; sometimes grows on MacConkey agar; does not hydrolyze urea or produce indole; reduces nitrate, accompanied frequently by nitrite reduction with the formation of gas; most strains hydrolyze arginine and digest gelatin.

Sources (97 strains)

Dog bite (32%); wound (22%); cat bite (6%); gum or tonsil of dog or cat (5%); hand (3%); leg (3%); finger (3%); canine wound (3%); cat pulmonary (2%); eye (2%); skin (2%); dog (2%); wallaby mandible, pet monkey mass, and dog teeth (1% each); other (10%); unknown (2%).

Reference strain

D6567, from finger wound.

Literature

1. Tatum, H.W., W.H. Ewing, and R.E. Weaver. 1974. Miscellaneous Gram-negative bacteria. *In* E.H. Lennette, E.H. Spaulding, and J.P. Truant (Eds.), Manual of Clinical Microbiology, 2nd ed., American Society for Microbiology, Washington, pp. 289–291.

2. Holmes, B., and M.S. Ahmed. 1981. Group EF-4: A *Pasteurella*-like organism. *In* M. Kilian, W. Frederiksen, and E.L. Biberstein (Eds.), *Haemophilus, Pasteurella,* and *Actinobacillus,* Academic Press, London, pp. 161–174.

3. Holmes, B., M. Costas, and A.C. Wood. 1990. Numerical analysis of electrophoretic protein patterns of Group EF-4 bacteria, predominately from dog-bite wounds of humans. J. Appl. Bacteriol. *68:* 81–91.

Comments

Of 27 substrates, only glucose is fermented. Cultures smell like popcorn. Mostly associated with dog bites. Also associated with acute pneumonia in dogs and cats. Information on the cellular fatty acid composition of this group is presented in Table 4.12A and Figure 4.17 of the cellular fatty acid section.

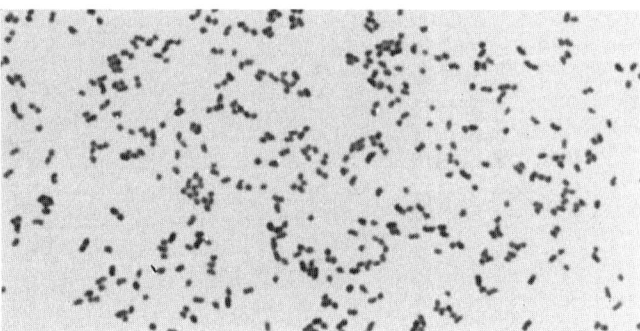

EF-4a strain D6567. Gram stain BAP 35° C 24 h x 1700.

EF-4a[a]

TEST PERFORMED	SIGN	% +	#+/T
	NUMBER OF STRAINS 97		
Morphology	cc,srs		
Motility	[nm]		
Gas form glucose	[−]	0	0/97
Action on blood	v	34 ly	30/87; 13 gr, 8 al, 1 ß
Fermentative or oxidative	[F]		
Carbohydrate base	F		
Acid from:			
D-Glucose	+	100	97/97
D-Xylose	−	0	0/97
D-Mannitol	−	0	0/97
Lactose	[−]	0	0/97
Sucrose	−	0	0/97
Maltose	[−]	0	0/97
Catalase	[+]	100	97/97
Oxidase	[+]	100	97/97
Growth on:			
MacConkey	[v]	42(8)	17, 23w (8)/96
SS	−	1	1/96
Simmons citrate	−	3 (1)	3 (1)/97
Urea, Christensen's	[−]	0	0/97
Nitrate reduction	[+]	97[b]	94/97
Gas from nitrate	[v]	62	60/97
Indole	[−]	0	0/97
TSI slant, acid	−	3	2, 1w/96
TSI butt, acid	v	73	54, 16w/96
H$_2$S (TSI butt)	−	0	0/96
H$_2$S (Pb ac paper)	+	95 (2)	77, 15w (2)/97
Gelatin hydrolysis[c]	v	60	21/35
Litmus milk	−	2(1) k	2 (1)/93
Growth at:			
25°C	v	89	43, 39w/92
35°C	+	100	92/92
42°C	v	70	22, 42w/92
Esculin hydrolysis	−	0	0/97
Lysine decarboxylase	−	0	0/57
Arginine dihydrolase	[v]	77 (2)	43, 1w (1)/57
Ornithine decarboxylase	−	0	0/57
Nutrient broth, 0% NaCl	+	94 (2)	73, 12w (2)/90
Nutrient broth, 6% NaCl	−	6 (2)	3, 2w (2)/90

[a]Yellow to amber pigment, soluble or insoluble, produced by some strains.
[b]Thirteen strains (13%) completely reduced nitrate and nitrite without gas formation.
[c]Incubation of 7–14 days.

EF-4b

Gram-negative coccoid to short rod, which stains deeply; nonmotile; oxidizes glucose only; oxidase- and catalase-positive; may grow on MacConkey agar; does not hydrolyze urea or produce indole; reduces nitrate (and sometimes nitrite) without producing gas; does not hydrolyze arginine; usually does not digest gelatin.

Sources (34 strains)

Dog bite (29%), cat bite (21%), cat scratch (9%), hand (12%), wound (6%), arm (6%), finger (6%), other (8%), unknown (3%).

Reference strain

A2617, from cat bite.

Literature

1. Tatum, H.W., W.H. Ewing, and R.E. Weaver. 1974. Miscellaneous Gram-negative bacteria. *In* E.H. Lennette, E.H. Spaulding, and J.P. Truant (Eds.), Manual of Clinical Microbiology, 2nd. ed. American Society for Microbiology, Washington, pp. 289–291.

2. Holmes, B., and M.S. Ahmed. 1981. Group EF-4: A *Pasteurella*-like organism. *In* M. Kilian, W. Frederiksen, and E.L. Biberstein (Eds.), *Haemophilus, Pasteurella,* and *Actinobacillus,* Academic Press, London, pp. 161–174.

3. Holmes, B., M. Costas, and A.C. Wood. 1990. Numerical analysis of electrophoretic protein patterns of Group EF-4 bacteria, predominantly from dog-bite wounds of humans. J. Appl. Bacteriol. *68:* 81–91.

Comments

This group, with EF-4a, originally was designated EF-4. Group EF-4b does not ferment glucose, does not hydrolyze arginine, and does not produce gas from nitrate during reduction. More of EF-4b strains (than EF-4a) are associated with cats. Cultures smell like popcorn. Information on the cellular fatty acid composition of this group is presented in Table 4.12A and Figure 4.17 of the cellular fatty acid section.

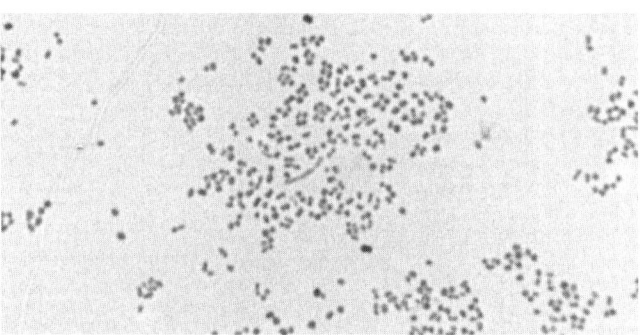

EF-4b strain A2617. Gram stain BAP 35° C 24 h x 1700.

EF-4b

TEST PERFORMED	SIGN	% +	#+/T
Morphology	cc, srs		
Motility	[nm]		
Action on blood	v	41 ly	13/32; 4 al, 1 gr, 1 ß
Fermentative or oxidative	[O]		
Carbohydrate base	OF		
Acid from:			
D-Glucose	+ or (+)	70 (26)	4, 12w (6)/23
D-Xylose	−	0	0/23
D-Mannitol	−	0	0/23
Lactose	[−]	0	0/23
Sucrose	−	0	0/23
Maltose	[−]	0	0/23
Catalase	[+]	100	34/34
Oxidase	+	100	34/34
Growth on:			
MacConkey	[v]	65 (6)	16, 6w (2)/34
SS	−	0	0/34
Simmons citrate	v	14 (6)	4, 1w (2)/34
Urea, Christensen's	[−]	0	0/34
Nitrate reduction	[+][a]	97	32, 1w/34
Gas from nitrate	[−]	0	0/34
Indole	[−]	0	0/34
TSI slant, acid	−	0	0/34
TSI butt, acid	−	6	2w/34
H_2S (TSI butt)	−	0	0/34
H_2S (Pb ac paper)	v	88	19, 11w/34
Gelatin hydrolysis[b]	−	9	3/33
Litmus milk	−	6k	2/34
Growth at:			
25°C	v	88	19, 9w/32
35°C	+	100	29, 4w/33
42°C	v	69	5, 15w/29
Esculin hydrolysis	−	0	0/34
Lysine decarboxylase	−	0	0/16
Arginine dihydrolase	[−]	0	0/16
Ornithine decarboxylase	−	0	0/16
Nutrient broth, 0% NaCl	+ or (+)	89 (7)	18, 6w (2)/27
Nutrient broth, 6% NaCl	−	0	0/27

NUMBER OF STRAINS 34

[a]Sixteen strains reduced nitrate and nitrite without gas formation.
[b]Incubation of 7–14 days.

Eikenella corrodens

Gram-negative slender straight rod, nonmotile; fastidious; does not produce acid from carbohydrates; oxidase- and nitrate-positive (no gas); no growth on MacConkey agar; catalase- and urease-negative; ornithine and usually lysine decarboxylase-positive.

Sources (506 strains)

Wound (15%), abscess (14%), sputum (6%), abdominal cavity (5%), peritoneal fluid (4%), throat (4%), appendix (4%), blood (4%), pleural fluid (4%), sinus (2%), pus (2%), eye (2%), bronchi (2%), other (26%), unknown (6%).

Reference strain

KC972 ← Henriksen, 333-54-55 = ATCC 23834, type strain, from sputum.

Literature

1. Eiken, M. 1958. Studies on an anaerobic rod-shaped Gram-negative microorganism: *Bacteroides corrodens* N. sp. Acta Pathol. Microbiol. Scand. *43:* 404–416.

2. Jackson, F.L., Y.E. Goodman, F.R. Bel, P.C. Wong, and R.L.S. Whitehouse. 1971. Taxonomic status of facultative and strictly anaerobic "corroding bacilli" that have been classified as *Bacteroides corrodens*. J. Med. Microbiol. *4:* 171–184.

3. Marsden, H.B., and W.A. Hyde. 1971. Isolation of *Bacteroides corrodens* from infections in children. J. Clin. Pathol. *24:* 117–119.

4. Jackson, F.L., and Y.E. Goodman. 1972. Transfer of the facultatively anaerobic organism *Bacteroides corrodens* Eiken to a new genus, *Eikenella*. Int. J. Syst. Bacteriol. *22:* 73–77.

5. Jackson, F.L., and Y. Goodman. 1984. Genus *Eikenella* Jackson and Goodman 1972, 74.[AL] *In* N.R. Krieg and J.G. Holt (Eds.), Bergey's Manual of Systematic Bacteriology, Vol. 1, Williams & Wilkins, Baltimore, pp. 591–597.

Comments

Growth enhanced by CO_2; usually pits or corrodes agar; colonial variation may occur; distinctive odor (hypochlorite-like), growth better in peptone nitrate broth than in infusion nitrate broth; frequently shows a slight yellowish growth at 3-7 days, particularly when accumulated on a loop. Information on the cellular fatty acid composition of this species is presented in Table 4.12A and Figure 4.17 of the cellular fatty acid section.

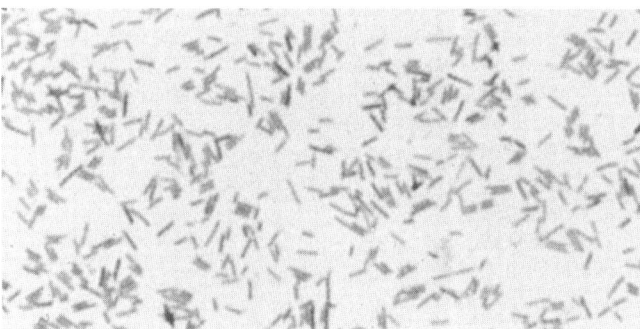

Eikenella corrodens strain KC972. Gram stain BAP 35° C 24 h x 1700.

Eikenella corrodens[a]

TEST PERFORMED	SIGN	% +	#+/T
	NUMBER OF STRAINS 506		
Morphology	mrs		
Motility; flagella	[nm]		
Action on blood	v	17 ly	67/400; 24 gr, 4 al, 2 ß
Fermentative or oxidative	[n-o]		
Carbohydrate base	OF		
Acid from:			
D-Glucose	[−]	0	0/67
D-Xylose	[−]	0	0/67
D-Mannitol	−	0	0/67
Lactose	−	0	0/67
Sucrose	−	0	0/67
Maltose	−	0	0/67
Catalase	[−]	8	20, 20w/482
Oxidase	[+]	100	502/502
Growth on:			
MacConkey	[−]	0	2/484
SS	−	0	0/481
Simmons citrate	−	0	0/484
Urea, Christensen's	[−]	0	0/450; 4 pk
Nitrate reduction	[+]	99	502/504
Gas from nitrate	[−]	0	0/504
Indole	−	0	0/489
TSI slant, acid	−	0	0/446
TSI butt, acid	−	0	0/446
H_2S (TSI butt)	−	0	0/479
H_2S (Pb ac paper)	v	73	131, 225w/489
Gelatin hydrolysis[b]	−	0	0/258
Litmus milk	−	1 k	4/484; 3 IR
Pigment	v	(70) sl yel	(160)/227
Growth at:			
25°C	v	15	50/337
35°C	+	99	339/342
42°C	v	52	174/337
Esculin hydrolysis	−	0	0/456
Lysine decarboxylase	v	82	49/59
Arginine dihydrolase	−	0	0/59
Ornithine decarboxylase	[+]	98	58/59
Nutrient broth, 0% NaCl	v	71	284/399
Nutrient broth, 6% NaCl	−	1	5/427

[a]Candle jar (CO_2) required or enhances growth.
[b]Incubation of 7–14 days.

EO-2

Gram-negative coccoid to short rod, frequently appearing vacuolated or peripherally stained ("O"-shaped), in pairs and short chains or packets; nonmotile; aerobic; oxidizes glucose, xylose, and lactose (sometimes delayed); oxidase-positive; reduces nitrate (occasionally with gas); does not produce indole; *Psychrobacter immobilis* transformation-negative.

Sources (11 strains)

Wound (3), eye (2), blood (1), cerebrospinal fluid (1), abdominal dialysate (1), biliary tract shunt (1), knee blister (1), toe (1).

Reference strain

G778, from wound.

Literature

1. Moss, C.W., P.L. Wallace, D.G. Hollis, and R.E. Weaver. 1988. Cultural and chemical characterization of CDC Groups EO-2, M-5, and M-6, *Moraxella* (*Moraxella*) species, *Oligella urethralis*, *Acinetobacter* species, and *Psychrobacter immobilis*. J. Clin. Microbiol. 26: 484–492.
2. Hudson, M.J., D.G. Hollis, R.E. Weaver, and C.G. Galvis. 1987. Relationship of CDC group EO-2 and *Psychrobacter immobilis*. J. Clin. Microbiol. 25: 1907–1910.
3. Pickett, M.J., D.G. Hollis, and E.J. Bottone. 1991. Miscellaneous gram-negative bacteria. In A. Balows, et al. (Eds.), Manual of Clinical Microbiology, 5th ed., American Society for Microbiology, Washington, pp. 410–428.

Comments

Growth frequently is mucoid. Biochemical reactions are similar to those of both saccharolytic *Acinetobacter* species and *Psychrobacter immobilis* strains; a positive oxidase reaction and a negative *Acinetobacter* transformation test distinguish this group from the *Acinetobacter* species. The EO-2 group can be differentiated from saccharolytic *P. immobilis* strains by a negative *P. immobilis* transformation test, cellular fatty acid analysis, and usually the presence of "O" forms on the Gram stain. Information on the cellular fatty acid composition of this group is presented in Table 4.30 and Figure 4.35 of the cellular fatty acid section.

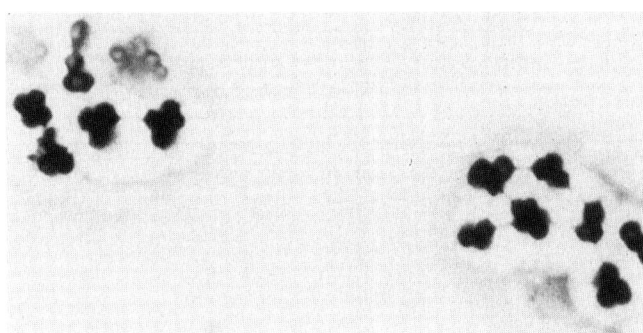

EO-2 strain D5834. Gram stain BAP 35° C 24 h x 1700.

EO-2[a]

	NUMBER OF STRAINS 11		
TEST PERFORMED	SIGN	% +	#+/T
Morphology	cc		
Motility; flagella	[nm]		
Action on blood	−	9 ly	1/11
Fermentative or oxidative	[O]		
Carbohydrate base	OF		
Acid from:			
D-Glucose	[+]	91 (9)	10 (1)/11
D-Xylose	[+]	91 (9)	10 (1)/11
D-Mannitol	−	0	0/11
Lactose	+ or (+)	45 (55)	5 (6)/11
Sucrose	[−]	0	0/11
Maltose	−	(9)	(1w)/11
Catalase	v	82	8, 1w/11
Oxidase	[+]	100	11/11
Growth on:			
MacConkey	v	64 (18)	7 (2)/11
SS	−	0	0/11
Cetrimide	−	0	0/11
Simmons citrate	+ or (+)	64 (36)	7 (4)/11
Urea, Christensen's	+ or (+)	36 (55)	4 (5), (1w)/11
Nitrate reduction	[+]	100	11/11
Gas from nitrate	v	18	2/11
Indole	[−]	0	0/11
TSI slant, acid	−	0	0/11
TSI butt, acid	−	0	0/11
H_2S (TSI butt)	−	0	0/11
H_2S (Pb ac paper)	v	64 (9)	5, 2w (1w)/11
Gelatin hydrolysis[b]	−	0	0/11
Litmus milk	v	82 k	9/11; 1 IR
Pigment	v	55 yel sol	6/11; 3 br
Growth at:			
25°C	v	73	8/11
35°C	[+]	100	11/11
42°C	v	36	4/11
Esculin hydrolysis	−	0	0/11
Nutrient broth, 0% NaCl	v	64	7/11
Nutrient broth, 6% NaCl	v	36	4/11
Psychrobacter transformation	[−]	0	0/1

[a]Colonies frequently mucoid.
[b]Incubation of 7–14 days.

EO-3

Gram-negative, coccoid to short thick rod; nonmotile; aerobic; oxidizes glucose and xylose, mannitol and lactose (usually weak or late), but not sucrose; grows on MacConkey agar; oxidase-positive; does not reduce nitrate; produces a yellow insoluble growth pigment; *Psychrobacter immobilis* transformation-negative.

Sources (7 strains)

Blood (3), eye (1), forearm (1), stitch abscess (1), vagina (1).

Reference strain

E7655, from vagina.

Literature

1. Moss, C.W., P.L. Wallace, D.G. Hollis, and R.E. Weaver. 1988. Cultural and chemical characterization of CDC Groups EO-2, M-5, and M-6, *Moraxella* (*Moraxella*) species, *Oligella urethralis*, *Acinetobacter* species, and *Psychrobacter immobilis*. J. Clin. Microbiol. 26: 484–492.
2. Pickett, M.J., D.G. Hollis, and E.J. Bottone. 1991. Miscellaneous gram-negative bacteria. *In* A. Balows, et al. (Eds.), Manual of Clinical Microbiology, 5th ed., American Society for Microbiology, Washington, pp. 410–428.

Comments

In the previous edition of this manual, these organisms were included in CDC Group EO-2. They can be differentiated from Group EO-2, as presently constituted, by their fatty acid profile, production of a yellow insoluble growth pigment, failure to reduce nitrate, and failure to show "O"-forms on Gram stain. Information on the cellular fatty acid composition of this group is presented in Table 4.31 and Figure 4.36 of the cellular fatty acid section.

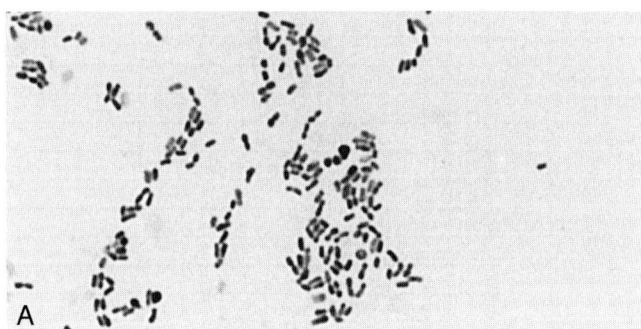

EO-3 strain E7655. Gram stain BAP 35° C 24 h x 1700.

EO-3

TEST PERFORMED	SIGN	% +	#+/T
NUMBER OF STRAINS 7			
Morphology	cc		
Motility	[nm]		
Action on blood	−	0	0/7
Fermentative or oxidative	[O]		
Carbohydrate base	OF		
Acid from:			
D-Glucose	[+]	100	7/7
D-Xylose	+	100	7/7
D-Mannitol	[+ or (+)]	57 (43)	4w (3w)/7
Lactose	+ or (+)	71 (29)	4, 1w (2)/7
Sucrose	−	0	0/7
Maltose	v	14 (14)	1w (1)/7
Catalase	+	100	7/7
Oxidase	[+]	100	7/7
Growth on:			
MacConkey	+	100	6, 1w/7
SS	−	0	0/7
Cetrimide	−	0	0/7
Simmons citrate	+ or (+)	29 (71)	2 (5)/7
Urea, Christensen's	+ or (+)	14 (86)	1 (6)/7
Nitrate reduction	[−]	0	0/7
Gas from nitrate	−	0	0/7
Indole	−	0	0/7
TSI slant, acid	−	0	0/7
TSI butt, acid	−	0	0/7
H_2S (TSI butt)	−	0	0/7
H_2S (Pb ac paper)	+	100	7/7
Gelatin hydrolysis[a]	−	0	0/7
Litmus milk	v	43 szf	3/7; 2 pep
Pigment:			
Insoluble	[yel]	100	7/7
Growth at:			
25°C	+	100	7/7
35°C	+	100	7/7
42°C	v	14	1/7
Esculin hydrolysis	−	0	0/7
Nutrient broth, 0% NaCl	v	71	5/7
Nutrient broth, 6% NaCl	v	43	3/7
Psychrobacter transformation	[−]	0	0/7

[a]Incubation of 7–14 days.

Flavimonas oryzihabitans

Gram-negative rod; motile (1–2 polar flagella); produces a yellow growth pigment; aerobic; oxidase-negative; grows on MacConkey agar; uses glucose, xylose, mannitol, maltose, and sometimes lactose and sucrose; does not hydrolyze esculin; arginine dihydrolase-variable, lysine and ornithine decarboxylase usually negative; does not usually reduce nitrate.

Sources (37 strains)

Urine (14%), eye (13%), wound (11%), blood (11%), skin (8%), sputum (5%), other (24%), unknown (14%).

Reference strain

KC1807 ← JCM 2952 = ATCC 43272, type strain, from a rice paddy, Japan.

Literature

1. Pickett, M.J., and M.M. Pedersen. 1970. Characterization of saccharolytic nonfermentative bacteria associated with man. Can. J. Microbiol. *16:* 351–362.

2. Gilardi, G.L., S. Hirschl, and M. Mandel. 1975. Characteristics of yellow-pigmented nonfermentative bacilli (Groups VE-1 and VE-2) encountered in clinical bacteriology. J. Clin. Microbiol. *1:* 384–389.

3. Kodama, K., N. Kimura, and K. Komagata. 1985. Two new species of *Pseudomonas*: *P. oryzihabitans* isolated from rice paddy and clinical specimens and *P. luteola* isolated from clinical specimens. Int. J. Syst. Bacteriol. *35:* 467–474.

4. Holmes, B., A.G. Steigerwalt, R.E. Weaver, and D.J. Brenner. 1987. *Chryseomonas luteola* comb. nov. and *Flavimonas oryzihabitans* gen. nov., comb. nov., *Pseudomonas*-like species from human clinical specimens and formerly known, respectively, as Groups Ve-1 and Ve-2. Int. J. Syst. Bacteriol. *37:* 245–250.

5. Hawkins, R.E., R.A. Moriarty, D.E. Lewis, and E.C. Oldfield. 1991. Serious infections involving the CDC group Ve bacteria *Chryseomonas luteola* and *Flavimonas oryzihabitans*. Rev. Infect. Dis. *13:* 257-260.

Comments

Formerly designated CDC Group Ve-2 and *Pseudomonas oryzihabitans*. Cellular and colonial morphologies similar to those of *Chryseomonas luteola* (formerly CDC Group Ve-1). Information on the cellular fatty acid composition of this species is presented in Table 4.38 and Figure 4.43 of the cellular fatty acid section.

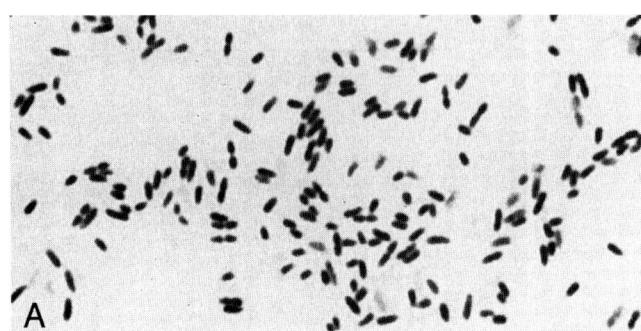

Flavimonas oryzihabitans strain D7021. Gram stain BAP 35° C 24 h x 1700.

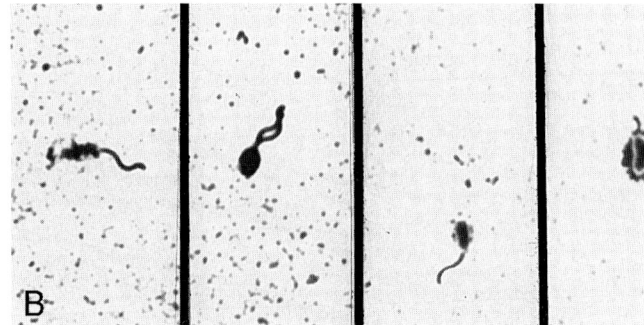

Flavimonas oryzihabitans strain D7021. Flagella stain BAP 35° C 24 h x 1700.

Flavimonas oryzihabitans

TEST PERFORMED	SIGN	%+	#+/T
NUMBER OF STRAINS 36			
Morphology	srs		
Motility; flagella	[m;p,1-2]		
Action on blood	v	43 ly	12/28; 3 LG, 1 ß
Fermentative or oxidative	[O]		
Carbohydrate base	OF		
Acid from:			
D-Glucose	[+]	100	36/36
D-Xylose	+	100	35, 1w/36
D-Mannitol	[+]	100	29, 7w/36
Lactose	v	14 (22)	2, 3w (8)/36
Sucrose	v	25	9/36
Maltose	+	97 (3)	30, 5w (1)/36
Catalase	+	94	34/36
Oxidase	[−][a]	0	0/36
Growth on:			
MacConkey	[+]	100	36/36
SS	v	22	7/28
Cetrimide	v	25 (28)	9w (10)/36
Simmons citrate	+	97	35/36
Urea, Christensen's	v	77	27/35
Nitrate reduction	[−]	6	2/36
Gas from nitrate	−	0	0/36
Indole	−	0	0/35
TSI slant, acid	−	0	0/36
TSI butt, acid	−	0	0/36
H$_2$S (TSI butt)	−	0	0/36
H$_2$S (Pb ac paper)	+	97	35w/36
Gelatin hydrolysis[b]	v	17	4/23
Litmus milk	v	57 k	20/35
Pigment:			
Insoluble	[yel]	100	35/35
Growth at:			
25°C	+	100	33/33
35°C	+	100	33/33
42°C	v	33	11/33
Esculin hydrolysis	[−]	0	0/34
Lysine decarboxylase	[−]	7	2/30
Arginine dihydrolase	v	14	4/30
Ornithine decarboxylase	−	3	1/30
Nutrient broth, 0% NaCl	+	100	32/32
Nutrient broth, 6% NaCl	v	62	20/32

[a] A few strains were weakly positive by the Kovacs' method.
[b] Incubation of 7–14 days.

Flavobacterium breve and Similar Strains

Gram-negative short to long rod, nonmotile; oxidizes glucose, maltose, and usually starch, not mannitol or trehalose; ONPG-negative; grows on MacConkey agar; indole- and oxidase-positive; nitrate- and urease-negative; proteolytic; produces a light yellow pigment; does not hydrolyze esculin.

Sources (7 strains)

Cerebrospinal fluid (1), bronchial fluid (1), hand (1), toe wound (1), vaginal discharge (1), unknown (2).

Reference strain

KC1428 ← NCTC 11099 = ATCC 43319, type strain, from bronchial fluid.

Literature

1. Holmes, B., J.J.S. Snell, and S.P. Lapage. 1978. Revised description, from clinical strains, of *Flavobacterium breve* (Lustig) Bergey, et al. 1923 and proposal of the neotype strain. Int. J. Syst. Bacteriol. *28:* 201–208.

2. Holmes, B., R.J. Owen, and T.A. McMeekin. 1984. Genus *Flavobacterium* Bergey et al. 1923, 97.[AL] *In* N.R. Krieg and J.G. Holt (Eds), Bergey's Manual of Systematic Bacteriology, Vol. 1, Williams & Wilkins, Baltimore, pp. 353–361.

3. Dees, S.B., C.W. Moss, D.G. Hollis, and R.E. Weaver. 1986. Chemical characterization of *Flavobacterium odoratum*, *Flavobacterium breve*, and *Flavobacterium*-like groups IIe, IIh, and IIf. J. Clin. Microbiol. *23:* 267–273.

4. Ursing, J., and B. Bruun. 1991. Genotypic heterogeneity of *Flavobacterium* group IIb and *Flavobacterium breve* demonstrated by DNA-DNA hybridization. A.P.M.I.S. *99:* 780–786.

Comments

Recent genetic studies indicate that *Flavobacterium breve* is a heterogeneous mix of multiple genomospecies. Therefore, we have listed the biochemical reactions of the type strain of *F. breve* with other phenotypically similar strains. The pale yellow growth pigment produced by these organisms is helpful in differentiating them from the biochemically similar, but more deeply pigmented, *Flavobacterium* species IIb. Information on the cellular fatty acid composition of this group is presented in Table 4.42 and Figure 4.47 of the cellular fatty acid section.

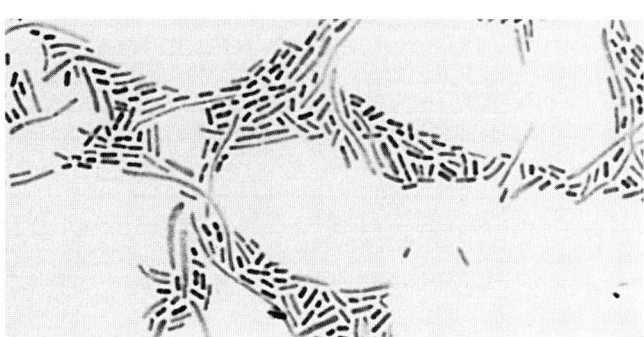

Flavobacterium breve strain KC1428. Gram stain BAP 35° C 24 h x 1700.

Flavobacterium breve and Similar Strains

TEST PERFORMED	TYPE STRAIN SIGN	*F. breve*-like NUMBER OF STRAINS 6		
		SIGN	%+	#+/T
Morphology	s-lr	s-lr, fc		
Motility	[nm]	[nm]		
Action on blood	ly	v	40 ly	2/5; 1 lav, 1 LG
Fermentative or oxidative	[O]	[O]		
Carbohydrate base	OF	OF		
Acid from:				
D-Glucose	[+]	[+ or (+)]	83 (17)	3 (1), 2 w/6
D-Xylose	−	−	0	0/6
D-Mannitol	[−]	[−]	0	0/6
Lactose	−	−	0	0/6
Sucrose	−	−	0	0/6
Maltose	+	+ or (+)	83 (17)	3 (1), 2w/6
Starch	+	v	75	3/4
Trehalose	[−]	[−]	0	0/3
ONPG	[−]	[−]	0	0/5
Catalase	+	+	100	6/6
Oxidase	[+]	[+]	100	6/6
Growth on:				
MacConkey	[+]	[+]	100	6/6
SS	−	−	0	0/6
Simmons citrate	−	−	0	0/6
Urea, Christensen's	−	−	0	0/6
Nitrate reduction	−	−	0	0/6
Indole	[+]	[+]	100	4, 2w/6
TSI slant, acid	−	−	0	0/6
TSI butt, acid	−	−	0	0/6
H_2S (TSI butt)	−	−	0	0/6
H_2S (Pb ac paper)	+	+	100	6/6
Gelatin hydrolysis[a]	+	+	100	6/6
Litmus milk	IR	v	50 IR	3/6, 2 pep
Pigment:				
Soluble	(br)	v	(50 br)	(3)/6
Insoluble	yel	v	66 sl yel	4/6
Growth at:				
25°C	+	+	100	6/6
35°C	+	+	100	6/6
42°C	−	−	0	0/6
Esculin hydrolysis	[−]	[−]	0	0/6
Lysine decarboxylase	−	−	0	0/2
Arginine dihydrolase	−	−	0	0/2
Ornithine decarboxylase	−	−	0	0/2
Nutrient broth, 0% NaCl	+	+	100	6/6
Nutrient broth, 6% NaCl	−	−	0	0/6

[a]Incubation of 7–14 days.

Flavobacterium species (IIb), including *F. gleum* and *F. indologenes*

Gram-negative slightly pleomorphic rod (cells appear thin to very thin in the central region, with thicker ends, the so-called "II forms"), motility not demonstrated; grows well aerobically; all strains use glucose, maltose, and trehalose, not mannitol; usually starch hydrolysis-positive; oxidase-positive; proteolytic, weakly indole-positive; produces a yellow-orange growth pigment.

Sources (155 strains)

Blood (16%), sputum (10%), throat (9%), urine (8%), lung (6%), spinal fluid (4%), brain (4%), nose (2%), trachea (2%), other (29%), unknown (10%).

Reference strain

B7755, from a water fountain.

Type strains

F. indologenes: 3716 = GIFU 1347 = ATCC 29897, type strain, from trachea.
F. gleum: F93 = NCTC 11432 = ATCC 35910, type strain, from vaginal swab.

Literature

1. Holmes, B., R.J. Owen, A.G. Steigerwalt, and D.J. Brenner. 1984. *Flavobacterium gleum*, a new species found in human clinical specimens. Int. J. Syst. Bacteriol. *34*: 21–25.
2. Yabuuchi, E., Y. Hashimoto, E. Takayuki, Y. Ido, and N. Takeuchi. 1990. Genotypic and phenotypic differentiation of *Flavobacterium indologenes* Yabuuchi et al. 1983 from *Flavobacterium gleum* Holmes et al 1984. Microbiol. Immunol. *34*: 73–76.
3. Yabuuchi, E., T. Kaneko, I. Yano, C.W. Moss, and N. Miyoshi. 1983. *Sphingobacterium* gen. nov., *Sphingobacterium spiritivorum* comb. nov., *Sphingobacterium multivorum* comb. nov., *Sphingobacterium mizutae* sp. nov., and *Flavobacterium indologenes* sp. nov.: glucose-nonfermenting gram-negative rods in CDC groups IIk-2 and IIb. Int. J. Syst. Bacteriol. *33*: 580–598.
4. King, E.O. 1959. Studies on a group of previously unclassified bacteria associated with meningitis in infants. Am. J. Clin. Pathol. *31*: 241–247.
5. Ursing, J., and B. Bruun. 1991. Genotypic heterogeneity of *Flavobacterium* group IIb and *Flavobacterium breve* demonstrated by DNA-DNA hybridization. A.P.M.I.S. *99*: 780–786.

Comments

Flavobacterium sp. (IIb) is heterogeneous, both genetically and phenotypically. Although two genomospecies within *F.* sp. (IIb) (*F. indologenes* and *F. gleum*) have been described, it will not be clear whether these species can be easily differentiated from each other and other genomospecies within the IIb group, until further DNA-DNA relatedness studies are completed. At this time, we have chosen not to subdivide the IIb strains and remove the known strains of *F. indologenes* and *F. gleum*. For comparative purposes, the characteristics of the type strains of *F. indologenes*, *F. gleum*, and 155 strains of *F.* sp. (IIb) are presented. Information on the cellular fatty acid composition of these groups is presented in Table 4.41 and Figure 4.46 of the cellular fatty acid section.

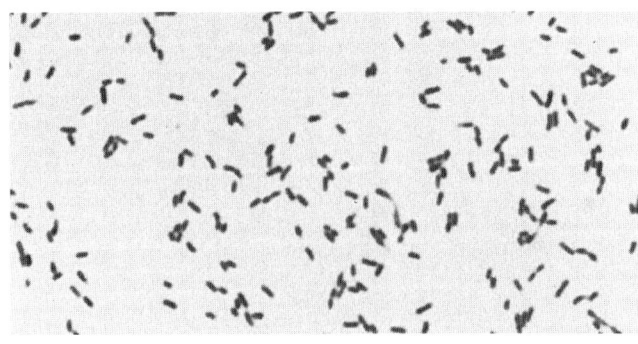

Flavobacterium species (IIb) strain B7755. Gram stain BAP 35° C 24 h x 1700.

Flavobacterium indologenes, *Flavobacterium gleum*, and *Flavobacterium species* (IIb)

TEST PERFORMED	*Flavobacterium indologenes* TYPE STRAIN SIGN	*Flavobacterium gleum* TYPE STRAIN SIGN	*Flavobacterium species* (IIb) NUMBER OF STRAINS 155 SIGN	% +	# +/T
Morphology	mrs,II	mrs,II	mrs,II		
Motility; flagella	[nm]	[nm]	[nm]		
Action on blood	-	-	v	26 LG	36/141; 35 ly, 32 ß, 22 gr
Fermentative or oxidative	[O]	[O]	[O]		
Carbohydrate base	OF	OF	OF		
Acid from:					
D-Glucose	(+)	(+)	[+]	92 (6)	82, 58w (10)/153
D-Xylose	-	(+)	v	30 (1)	2, 44w (1)/153
D-Mannitol	[-^{a}]	[-]	[-]	10	6, 10w/153
Lactose	-	-	-	0	0/153
Sucrose	-	-	v	13 (1)	14, 16w (2)/153
Maltose	(+)	(+)	+	92 (6)	85, 55w (9)/153
Starch	(+)	-	+	100	10, 4w/14
Trehalose	[(+)]	[(+)]	[+]	100	14/14
ONPG	-		v	57	8/14
Catalase	+	+	+	99	149, 5w/155
Oxidase	[+]	[+]	[+]	96	133, 15w/154
Growth on:					
MacConkey	+^{a}	+	v	54 (9)	52, 32w (14)/155
SS	-	-	-	0	0/155
Simmons citrate	+	+^{b}	-	2 (1)	4 (2)/155
Urea, Christensen's	-	(+)	v	14 (28)	15, 7w (43)/154
Nitrate reduction	-^{a}	+	v	22	34/155
Gas from nitrate	-^{a}	-	-	0	0/155
Nitrite reduction	-	+	v	20	31/155
Indole	[+]	[+]	[+]	98	109, 43w/155
TSI slant, acid	-	-	-	1	1w/155
TSI butt, acid	-	-	-	5 (5)	1, 7w (7)/153
H$_2$S (TSI butt)	-	-	-	1	1, 1w/155
H$_2$S (Pb ac paper)	+	+	+	99	135, 18w/155
Gelatin hydrolysis^c	+	+	v	78	120/153
Litmus milk	[pep]	[pep]	[pep]	93	144/155
Pigment:					
Insoluble	[yel]	[yel]	[yel]	99	154/155

(*continued*)

Flavobacterium indologenes, *Flavobacterium gleum*, and *Flavobacterium species* (IIb) *(continued)*

	Flavobacterium indologenes TYPE STRAIN	*Flavobacterium gleum* TYPE STRAIN	*Flavobacterium species* (IIb) NUMBER OF STRAINS 155	
Soluble		[br]	11 yel-br	17/155
Growth at:				
25°C	+	+	100	155/155
35°C	+	+	100	155/155
42°C	-	+	42	17, 48w/155
Esculin hydrolysis	+	+	70	105, 4w/149
Nutrient broth, 0% NaCl	+	+	100	143/143
Nutrient broth, 6% NaCl	-	-	0	0/142

[a] In the original description of *F. indologenes* (3), 30% of strains (4/13) oxidized mannitol, 46% of strains (6/13) grew on MacConkey agar, and 38% of strains (5/13) reduced nitrate to gas.
[b] The original description of *F. gleum* (1) lists the type strain as citrate negative.
[c] Incubation of 7–14 days.

Flavobacterium meningosepticum and Similar Strains

Gram-negative slightly pleomorphic rod [cells often appear thin to very thin in the central region, with thicker ends, "II-forms", so-called because these forms occurred in King's Group II organisms, which included *Flavobacterium meningosepticum* and *Flavobacterium* species (IIb)]; motility not demonstrated in motility medium; however, Webster and Hugh (1) demonstrated flagella and reported indirect evidence of motility; grows well aerobically; uses glucose, mannitol, maltose, and lactose (the latter, slowly); starch hydrolysis-negative; ONPG-positive, weakly indole-positive, oxidase-positive, nitrate-negative; usually grows on MacConkey agar; pigment, if present, only slightly yellow; proteolytic; hydrolyzes esculin.

Sources (148 strains)

Cerebrospinal fluid (24%), blood (19%), urine (7%), sputum (7%), trachea (3%), eye (3%), nose (2%), other (18%), unknown (17%).

Reference strain

KC1378 ← ATCC 13253, type strain, from spinal fluid of newborn.

Literature

1. Webster, J.A., and R. Hugh. 1979. *Flavobacterium aquatile* and *Flavobacterium meningosepticum*: glucose nonfermenters with similar flagellar morphologies. Int. J. Syst. Bacteriol. *29:* 333–337.

2. King, E.O. 1959. Studies on a group of previously unclassified bacteria associated with meningitis in infants. Am. J. Clin. Pathol. *31:* 241–247.

3. Holmes, B., R.J. Owen, and T.A. McMeekin. 1984. Genus *Flavobacterium* Bergey, Harrison, Breed, Hammer, and Huntoon 1923, 97.[AL] *In* N.R. Krieg and J.G. Holt (Eds.), Bergey's Manual of Systematic Bacteriology, Vol. 1, Williams & Wilkins, Baltimore, pp. 353–360.

4. Ursing, J., and B. Bruun. 1987. Genetic heterogeneity of *Flavobacterium meningosepticum* demonstrated by DNA-DNA hybridization. Acta Pathol. Microbiol. Immunol. Scand. Sect. B. *95:* 33–39.

Comments

Appears as an oxidizer at 48-h incubation; usually acidifies the butt of TSI in 3–7 days. Pathogenic for humans, especially premature infants; associated with meningitis and septicemia. Ongoing molecular taxonomic studies at the CDC and elsewhere indicate that *F. meningosepticum* represents a genetically heterogeneous group. In light of these findings, we present our reactions for the type strain of *F. meningosepticum* and a tabulation of other *F. meningosepticum*-like strains that we have studied. Information on the cellular fatty acid composition of this group is presented in Table 4.41 and Figure 4.46 of the cellular fatty acid section.

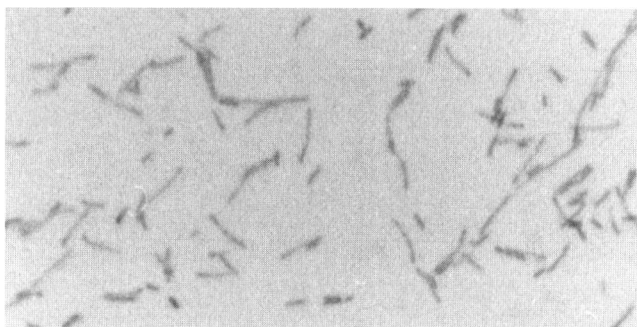

Flavobacterium meningosepticum strain KC1378.
Gram stain BAP 35° C 24 h x 1700.

Flavobacterium meningosepticum and Similar Strains

	TYPE STRAIN	*Flavobacterium meningosepticum*-like NUMBER OF STRAINS 148		
TEST PERFORMED	SIGN	SIGN	%+	#+/T
Morphology	lrs,II	lrs,II		
Motility	[nm]	[nm]a		
Action on blood	LG	v	75 LG	103/138; 22 ly, 4 al
Fermentative or oxidative	[O]	[O]		
Carbohydrate base	OF	OF		
Acid from:				
D-Glucose	[+]	[+]	95 (4)	123, 12w (6)/142
D-Xylose	−	−	2 (1)	1, 2w (2)/142
D-Mannitol	[+]	[+]	91 (8)	101, 30w(11)/143
Lactose	+	v	42 (15)	34, 23w(20)/134
Sucrose	−	−	0	0/143
Maltose	+	+	93 (7)	122, 9w (10)/141
Starch		−	0	0/15
Trehalose	+	+	100	15/15
ONPG	+	+	100	15/15
Catalase	+	+	100	146/146
Oxidase	[+]	[+]	99	147/148
Growth on:				
MacConkey	(+)	[+ or (+)]	89 (3)	130 (5)/146
SS	−	−	1	2/146
Simmons citrate	−	v	9 (3)	10, 3w (4)/146
Urea, Christensen's	−	−	3 (5)	4 (8)/147
Nitrate reduction	[−]	[−]	0	0/146
Indole	[+]	[+]	100	55, 45w/100
TSI slant, acid	−	−	0	0/148
TSI butt, acid	−	−	(3)	(5)/148
H_2S (TSI butt)	−	−	3	4w/148
H_2S (Pb ac paper)	+	+	98	145/148
Gelatin hydrolysisb	+	+	91	111/121
Litmus milk	[pep]	[pep]	98	141/144
Pigment:c				
Soluble		v	(37) ta-br	(61)/106
Growth at:				
25°C	+	+	100	144/144
35°C	+	+	100	144/144
42°C	+	v	45	65/143
Esculin hydrolysis	[+]	[+]	99	140/141
Nutrient broth, 0% NaCl	+	+	100	135/135
Nutrient broth, 6% NaCl	−	−	7	10/135

aPolar and lateral flagella have been demonstrated on some strains.
bIncubation of 7–14 days.
cSome strains exhibit slight yellow growth pigment

Flavobacterium mizutaii

Gram-negative rod; nonmotile; aerobic; oxidizes glucose, lactose, sucrose, maltose and xylose (xylose reaction usually delayed); does not grow on MacConkey agar, nitrate-negative, indole usually negative, although a very weak pink color may be observed in the xylene layer.

Sources (6 strains)

Blood (3), cerebrospinal fluid (2), wound (1).

Reference strain

KC1725 ← ATCC 33299, type strain, from ventricular fluid of fetus.

Literature

1. Yabuuchi, E., T. Kaneko, I. Yano, C.W. Moss, and M. Miyoshi. 1983. *Sphingobacterium* gen. nov., *Sphingobacterium spiritivorum* comb. nov., *Sphingobacterium multivorum* comb. nov., *Sphingobacterium mizutae* sp. nov., and *Flavobacterium indologenes* sp. nov.: glucose-nonfermenting gram-negative rods in CDC groups IIK-2 and IIb. Int. J. Syst. Bacteriol. *33:* 580–598.

2. Holmes, B., R.E. Weaver, A.G. Steigerwalt, and D.J. Brenner. 1988. A taxonomic study of *Flavobacterium spiritivorum* and *Sphingobacterium mizutae:* proposal of *Flavobacterium yabuuchiae* sp. nov. and *Flavobacterium mizutaii* comb. nov. Int. J. Syst. Bacteriol. *38:* 348–353.

Comments

F. mizutaii can be differentiated from *Flavobacterium multivorum* by its failure to grow on MacConkey agar. It has been proposed to include *F. mizutaii, F. multivorum,* and *F. spiritivorum* within the new genus *Sphingobacterium* (1). Information on the cellular fatty acid composition of this species is presented in Table 4.44 and Figure 4.49 of the cellular fatty acid section.

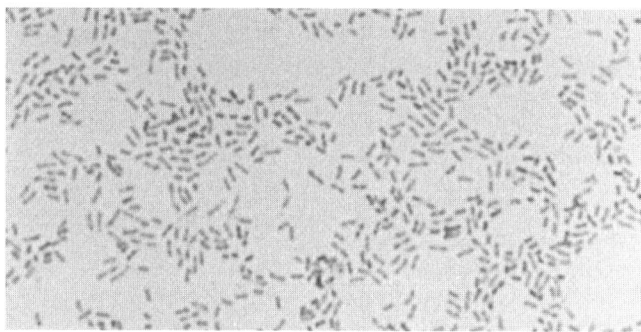

Flavobacterium mizutaii strain KC1725. Gram stain BAP 35° C 24 h x 1700.

Flavobacterium mizutaii

TEST PERFORMED	SIGN	%+	#+/T
	NUMBER OF STRAINS 6		
Morphology	srs, II		
Motility	[nm]		
Action on blood	v	17 ly	1/6; 1 lav
Fermentative or oxidative	[O]		
Carbohydrate base	OF		
Acid from:			
D-Glucose	[+ or (+)]	67 (33)	4 (2)/6
D-Xylose	[(+)]	(100)	(2) (4w)/6
D-Mannitol	[-]	0	0/6
Lactose	+	100	6/6
Sucrose	+ or (+)	50 (50)	2 (1), 1w (2w)/6
Maltose	+ or (+)	50 (50)	2 (3), 1w/6
Catalase	+	100	6/6
Oxidase	+	100	6/6
Growth on:			
MacConkey	[-]	0	0/6
SS	-	0	0/6
Cetrimide	-	0	0/6
Simmons citrate	-	0	0/6
Urea, Christensen's	-	0	0/6
Nitrate reduction	-	0	0/6
Nitrite reduction	+[a]	100	1/1
Indole	[-][b]	0	0/6
TSI slant, acid	v	17	1/6
TSI butt, acid	v	17	1w/6
H$_2$S (TSI butt)	-	0	0/6
H$_2$S (Pb ac paper)	+	100	6/6
Gelatin hydrolysis[c]	-	0	0/6
Litmus milk	v	50 IR	3/6; (1 Aw)
Pigment:			
Insoluble[d]	v	33 yel	2/6
Soluble	v	33 ta sol	2/6
Growth at:			
25°C	+	100	6/6
35°C	+	100	6/6
42°C	-	0	0/6
Esculin hydrolysis	+	100	6/6
Lysine decarboxylase	-	0	0/4
Arginine dihydrolase	v	25	1/4
Ornithine decarboxylase	-	0	0/4
Nutrient broth, 0% NaCl	+	100	6/6
Nutrient broth, 6% NaCl	-	0	0/6

[a] With the type strain a partial reaction is observed at 2 days in 0.1% nitrite, and a complete reaction is observed in 0.01% nitrite.
[b] Very weak pink color develops in the xylene layer.
[c] Incubation of 7–14 days. The rapid gelatin test was positive in 4 of 4 strains tested.
[d] Yellow growth pigment was observed at room temperature incubation.

Flavobacterium multivorum

Gram-negative coccoid to short rod; nonmotile; aerobic; grows on MacConkey agar; oxidizes glucose, xylose, lactose, sucrose, and maltose, not mannitol; hydrolyzes urea; usually does not reduce nitrate; oxidase-positive; indole-negative; hydrolyzes esculin; lysine- and arginine-negative.

Sources (22 strains)

Blood (36%), urine (14%), cerebrospinal fluid (9%), other (41%).

Reference strain

B5533 = NCTC 11343 = ATCC 33613, type strain, from spleen.

Literature

1. Holmes, B., R.J. Owen, and R.E. Weaver. 1981. *Flavobacterium multivorum*, a new species isolated from human clinical specimens and previously known as Group IIk, biotype 2. Int. J. Syst. Bacteriol. *31:* 21–34.

2. Tatum, H.W., W.H. Ewing, and R.E. Weaver. 1974. Miscellaneous gram-negative bacteria. *In* E.H. Lennette, E.H. Spaulding, and J.P. Truant (Eds.), Manual of Clinical Microbiology, 2nd. ed., American Society for Microbiology, Washington, pp. 270–294.

3. Holmes, B., R.J. Owen, and T.A. McMeekin. 1984. Genus *Flavobacterium* Bergey, Harrison, Breed, Hammer, and Huntoon 1923, 97.[AL] *In* N.R. Krieg and J.G. Holt (Eds.), Bergey's Manual of Systematic Bacteriology, Vol. 1, Williams & Wilkins, Baltimore, pp. 353–360.

4. Yabuuchi, E., T. Kaneko, I. Yano, C.W. Moss, and N. Miyoshi. 1983. *Sphingobacterium* gen. nov., *Sphingobacterium spiritivorum* comb. nov., *Sphingobacterium multivorum* comb. nov., *Sphingobacterium mizutae* sp. nov., and *Flavobacterium indologenes* sp. nov.: glucose-nonfermenting gram-negative rods in CDC groups IIK-2 and IIb. Int. J. Syst. Bacteriol. 33: 580–598.

Comments

Formerly called IIk-2. It has been proposed that *F. multivorum* be placed within the new genus *Sphingobacterium* (4). Information on the cellular fatty acid composition of this species is presented in Table 4.44 and Figure 4.49 of the cellular fatty acid section.

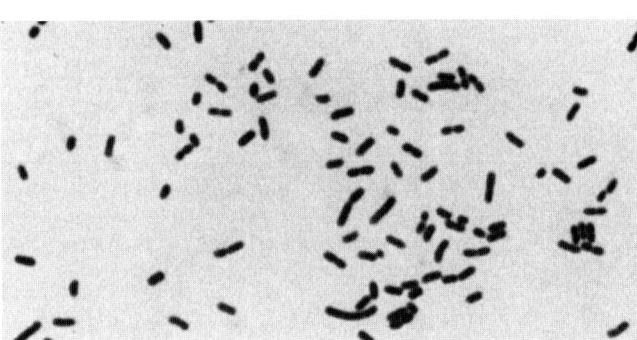

Flavobacterium multivorum strain B9570. Gram stain BAP 35° C 24 h x 1700.

Flavobacterium multivorum

TEST PERFORMED	SIGN	%+	#+/T
Morphology	srs		
Motility	[nm]		
Action on blood	v	14 gr	3/21; 3 lav, 2 LG, 1 ly, 1 al (stab)
Fermentative or oxidative	[O]		
Carbohydrate base	OF		
Acid from:			
D-Glucose	[+]	100	22/22
D-Xylose	+	100	22/22
D-Mannitol	[−]	0	0/22
Lactose	+	100	22/22
Sucrose	+	100	22/22
Maltose	[+]	100	22/22
Catalase	+	100	16, 5w/21
Oxidase	[+]	100	22/22
Growth on:			
MacConkey	[+]	100	19, 3w/22
SS	−	0	0/22
Cetrimide	−	0	0/22
Simmons citrate	−	0	0/22
Urea, Christensen's	[+]	95	21/22
Nitrate reduction	−	0	0/21
Indole	[−]	0	0/22
TSI slant, acid	v	55 (5)	10, 2w (1)/22
TSI butt, acid	v	5 (76)	1 (16)/21
H$_2$S (TSI butt)	−	0	0/21
H$_2$S (Pb ac paper)	+or(+)	86 (5)	19 (1)/22
Gelatin hydrolysis[a]	−	0	0/18
Litmus milk	v	(10) Aw	(2)/20; 1 k, 1 Aw
Pigment:			
Soluble	−	5 ta	1/21
Insoluble	v	57 sl yel	12/21
Growth at:			
25°C	+	100	20/20
35°C	+	100	20/20
42°C	−	0	0/20
Esculin hydrolysis	[+]	100	22/22
Lysine decarboxylase	[−]	0	0/4
Arginine dihydrolase	[−]	0	0/4
Ornithine decarboxylase	−	0	0/4
Nutrient broth, 0% NaCl	+	100	20/20
Nutrient broth, 6% NaCl	v	25	5/20

[a]Incubation of 7–14 days.

Flavobacterium odoratum

Gram-negative straight to slightly curved rod, nonmotile; aerobic; does not use carbohydrates; oxidase- and strongly urease-positive, indole- and nitrate-negative; usually grows on MacConkey agar; proteolytic; usually produces a yellow growth pigment; H_2S-negative on TSI agar.

Sources (74 strains)

Urine (62%), foot (5%), leg (5%), animal (5%), toe wound (4%), other wounds (4%), sputum (3%), other (11%).

Reference strain

KC1416 ← NCTC 11036 = ATCC 4651, type strain.

Literature

1. Stutzer, M., and A. Kwaschnina. 1929. In Aussaaten aus den Fazes des Menschen gelbe Kolonien bildende Bakterien (Gattung *Flavobacterium* u.a.). Zentralbl. Bakteriol. Parasitenkd. Infektionskr. Hyg. Abt. Orig. *102:* 113.

2. Holmes, B., J.J.S. Snell, and S.P. Lapage. 1977. Revised description, from clinical isolates, of *Flavobacterium odoratum* Stutzer and Kwaschnina 1929, and designation of the neotype strain. Int. J. Syst. Bacteriol. *27:* 330–336.

3. Holmes, B., R.J. Owen, and T.A. McMeekin. 1984. Genus *Flavobacterium* Bergey, Harrison, Breed, Hammer, and Huntoon 1923, 97.^AL In N.R. Krieg and J.G. Holt (Eds.), Bergey's Manual of Systematic Bacteriology, Vol. 1, Williams & Wilkins, Baltimore, pp. 353–360.

4. Dees, S.B., G.M. Carlone, D. Hollis, and C.W. Moss. 1985. Chemical and phenotypic characteristics of *Flavobacterium thalpophilum* compared with those of other *Flavobacterium* and *Sphingobacterium* species. Int. J. Syst. Bacteriol. *35:* 16–22.

Comments

Formerly called M-4f and IVf. Usually exhibits fruity odor; colonies are dull and spreading. The type strain reduces nitrite with gas formation and does not grow on MacConkey agar. Information on the cellular fatty acid composition of this species is presented in Table 4.41 and Figure 4.46 of the cellular fatty acid section.

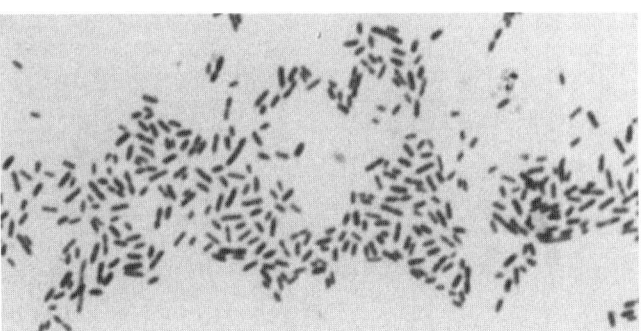

Flavobacterium odoratum strain D7703. Gram stain BAP 35° C 24 h x 1700.

Flavobacterium odoratum[a]

TEST PERFORMED	SIGN	% +	#+/T
Morphology	plr		
Motility	[nm]		
Action on blood	v	77 LG	57/74; 4 lav, 4 gr, 2 ly, 2 ß
Fermentative or oxidative	[n-o]		
Carbohydrate base	OF		
Acid from:			
D-Glucose	[−]	0	0/74
D-Xylose	[−]	0	0/74
D-Mannitol	−	0	0/74
Lactose	−	0	0/74
Sucrose	−	0	0/74
Maltose	−	0	0/74
Catalase	+	100	74/74
Oxidase	[+]	99	73/74
Growth on:			
MacConkey	+	91 (5)	67 (4)/74
SS	v	30 (11)	22 (8)/74
Simmons citrate	−	0	0/74
Urea, Christensen's	[+]	100	74/74
Nitrate reduction	[−]	0	0/74
Nitrite reduction	v	83	20/24
Indole	[−]	0	0/74
TSI slant, acid	−	0	0/74
TSI butt, acid	−	0	0/74
H_2S (TSI butt)	[−]	0	0/74
H_2S (Pb ac paper)	v	16	12/74
Gelatin hydrolysis[b]	[+]	96	53/55
Litmus milk	pep	93	69/74; 5 IR, 1 k
Pigment:			
Insoluble	v	85 yel	63/74
Soluble	v	28 ta-br	21/74
Growth at:			
25°C	+	100	74/74
35°C	+	100	74/74
42°C	v	31	23/74
Esculin hydrolysis	−	0	0/70
Lysine decarboxylase	−	0	0/11
Arginine dihydrolase	−	(9)	(1)/11
Ornithine decarboxylase	−	0	0/11
Nutrient broth, 0% NaCl	+	100	66/66
Nutrient broth, 6% NaCl	v	20 (5)	13 (3)/66

[a]The type strain reduces nitrite with gas formation and does not grow on MacConkey agar.
[b]Incubation of 7–14 days.

Flavobacterium spiritivorum

Gram-negative rod, nonmotile; aerobic; oxidizes glucose, mannitol, and other carbohydrates; oxidase-positive, indole-negative, urease-positive (sometimes delayed), and nitrate-negative; may grow on MacConkey agar.

Sources (12 strains)

Urine (3), vagina (2), and one each from nephrostomy tube, peritoneal fluid, sputum, bone marrow, uterus, humidifier, and mediastinal fluid.

Reference strain

KC1803 ← NCTC 11386 = ATCC 33861, type strain, from uterus.

Literature

1. Holmes, B., R.J. Owen, and D.G. Hollis. 1982. *Flavobacterium spiritivorum*, a new species isolated from human clinical specimens. Int. J. Syst. Bacteriol. *32:* 157–165.

2. Yabuuchi, E., T. Kaneko, I. Yano, C.W. Moss, and N. Miyoshi. 1983. *Sphingobacterium* gen. nov., *Sphingobacterium spiritivorum* comb. nov., *Sphingobacterium multivorum* comb. nov., *Sphingobacterium mizutae* sp. nov., and *Flavobacterium indologenes* sp. nov.: glucose-nonfermenting gram-negative rods in CDC groups IIk-2 and IIb. Int. J. Syst. Bacteriol. *33:* 580–598.

3. Holmes, B., R.E. Weaver, A.G. Steigerwalt, and D.J. Brenner. 1988. A taxonomic study of *Flavobacterium spiritivorum* and *Sphingobacterium mizutae*: proposal of *Flavobacterium yabuuchiae* sp. nov. and *Flavobacterium mizutaii* comb. nov. Int. J. Syst. Bacteriol. *38:* 348–353.

Comments

The ability of *F. spiritivorum* and *F. yabuuchiae* to produce acid from mannitol distinguishes them from *F. multivorum*, *F. mizutaii*, and *F. thalpophilum*. A positive acid phosphatase test is reported to differentiate *F. spiritivorum* from *F. yabuuchiae* (3). It has been proposed that *F. spiritivorum* be placed within the new genus *Sphingobacterium* (2). Information on the cellular fatty acid composition of this species is presented in Table 4.44 and Figure 4.49 of the cellular fatty acid section.

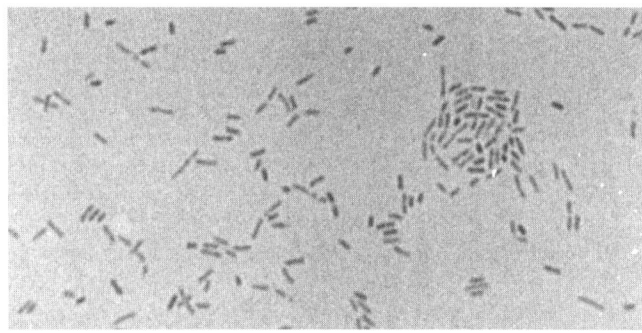

Flavobacterium spiritivorum strain KC1803. Gram stain BAP 35° C 24 h x 1700.

Flavobacterium spiritivorum

TEST PERFORMED	SIGN	% +	#+/T
	NUMBER OF STRAINS 13		
Morphology	srs		
Motility	[nm]		
Action on blood	v	15 LG	2/13; 1 ly
Fermentative or oxidative	O		
Carbohydrate base	OF		
Acid from:			
D-Glucose	[+]	100	13/13
D-Xylose	+	92 (8)	11, 1w (1)/13
D-Mannitol	[+]	100	13/13
Lactose	+	92 (8)	12 (1)/13
Sucrose	+	100	11, 2w/13
Maltose	+	92 (8)	12 (1)/13
Catalase	+	100	13/13
Oxidase	[+]	100	13/13
Growth on:			
MacConkey	[v]	(46)	(6)/13
SS	−	0	0/13
Cetrimide	−	0	0/13
Simmons citrate	−	0	0/13
Urea, Christensen's	[+ or (+)]	62 (38)	8 (5)/13
Nitrate reduction	−	0	0/13
Indole	[−]	0	0/13
TSI slant, acid	−	0	0/13
TSI butt, acid	−	0	0/13
H_2S (TSI butt)	−	0	0/11
H_2S (Pb ac paper)	v	56	5/9
Gelatin hydrolysis[a]	v	15	2/13
Litmus milk	−	0	0/11
Pigment:			
Insoluble	v	54 pale yel	7/13
Growth at:			
25°C	+	100	11/11
35°C	+	100	11/11
42°C	−	9	1/11
Esculin hydrolysis	+	100	13/13
Lysine decarboxylase	−	0	0/9
Arginine dihydrolase	v	25	2/8
Ornithine decarboxylase	−	0	0/9
Nutrient broth, 0% NaCl	+	100	13/13
Nutrient broth, 6% NaCl	−	0	0/13

[a]Incubation of 7–14 days.

Flavobacterium thalpophilum

Gram-negative uniformly staining rod with parallel sides and rounded ends; nonmotile; aerobic; oxidizes glucose, xylose, lactose, sucrose, maltose, and adonitol, but not mannitol; oxidase-positive, grows on MacConkey agar; urease- and nitrate-positive; grows at 42° C.

Sources (10 strains)

Wound (5), blood (2), one each from abscess, eye, and abdominal incision.

Reference strain

KC1806 ← NCTC 11429 = ATCC 43320, type strain, from wound swab.

Literature

1. Holmes, B., D.G. Hollis, A.G. Steigerwalt, M.J. Pickett, and D.J. Brenner. 1983. *Flavobacterium thalpophilum*, a new species recovered from human clinical material. Int. J. Syst. Bacteriol. *33:* 677–682.

2. Dees, S.B., G.M. Carlone, D. Hollis, and C.W. Moss. 1985. Chemical and phenotypic characteristics of *Flavobacterium thalpophilum* compared with those of other *Flavobacterium* and *Sphingobacterium* species. Int. J. Syst. Bacteriol. *35:* 16–22.

3. Takeuchi, M., and A. Yokota. 1992. Proposals of *Sphingobacterium faecium* sp. nov., *Sphingobacterium piscium* sp. nov., *Sphingobacterium heparinum* comb. nov., *Sphingobacterium thalpophilum* comb. nov., and two genospecies of the genus *Sphingobacterium*, and the synonymy of *Flavobacterium yabuuchiae* and *Sphingobacterium spiritivorum*. J. Gen. Appl. Microbiol. *38:* 465-482.

Comments

A positive nitrate test differentiates *F. thalpophilum* from the biochemically similar *F. multivorum* and *F. mizutaii*. Growth at 42° C also differentiates *F. thalpophilum* from most other flavobacteria. The chemical characteristics of *F. thalpophilum* indicate a closer taxonomic relationship of this organism to the species proposed to be moved to *Sphingobacterium* (*F. multivorum*, *F. spiritivorum*, and *F. mizutaii*) than to the other *Flavobacterium* species (2). Information on the cellular fatty acid composition of this species is presented in Table 4.44 and Figure 4.49 of the cellular fatty acid section.

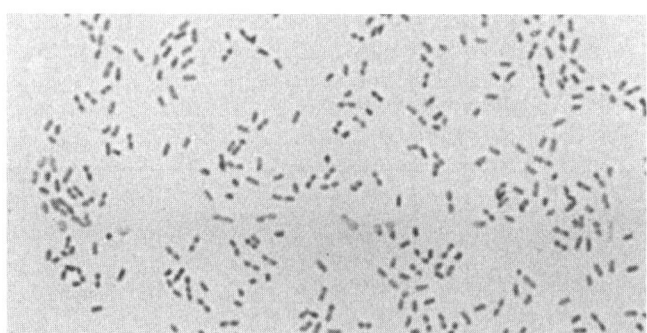

Flavobacterium thalpophilum strain KC1806. Gram stain BAP 35° C 24 h x 1700.

Flavobacterium thalpophilum

TEST PERFORMED	SIGN	% +	#+/T
NUMBER OF STRAINS 10			
Morphology	srs,II		
Motility	[nm]		
Action on blood	v	10 lav	1/10; 1 lys
Fermentative or oxidative	[O]		
Carbohydrate base	OF		
Acid from:			
D-Glucose	+	100	10/10
D-Xylose	+	100	10/10
D-Mannitol	[-]	0	0/10
Lactose	+	100	9, 1w/10
Sucrose	+	100	10/10
Maltose	+	100	10/10
Catalase	+	100	10/10
Oxidase	+	100	10/10
Growth on:			
MacConkey	[+]	100	10/10
SS	-	0	0/10
Cetrimide	-	0	0/10
Simmons citrate	-	0	0/10
Urea, Christensen's	+	90 (10)	9(1)/10
Nitrate reduction	[+]	100	10/10
Gas from nitrate	-	0	0/10
Indole	[-]	0	0/10
TSI slant, acid	[+]	100	9,1w/10
TSI butt, acid	v	10 (70)	1 (3), (4w)/10
H_2S (TSI butt)	-	0	0/10
H_2S (Pb ac paper)	+	100	10/10
Gelatin hydrolysis[a]	v	40	4/10
Litmus milk	v	20 pep	2/10; 2 Aw, 1 IR
Pigment:			
Insoluble	v	50 pale yel	5/10
Growth at:			
25°C	+	100	10/10
35°C	+	100	10/10
42°C	[+]	100	10/10
Esculin hydrolysis	+	100	10/10
Lysine decarboxylase	-	0	0/8
Arginine dihydrolase	-	0	0/8
Ornithine decarboxylase	-	0	0/8
Nutrient broth, 0% NaCl	+	100	10/10
Nutrient broth, 6% NaCl	-	10	1/10

[a]Incubation of 7–14 days.

Flavobacterium yabuuchiae

Gram-negative rod, nonmotile; aerobic; oxidizes glucose, mannitol, and other carbohydrates; oxidase-positive, indole- and nitrate-negative; may grow on MacConkey agar.

Sources (2 strains)

Peritoneal fluid and sputum.

Reference strain

D7529 ← NCTC 12113 = ATCC 49272, type strain, from sputum.

Literature

1. Holmes, B., R.E. Weaver, A.G. Steigerwalt, and D.J. Brenner. 1988. A taxonomic study of *Flavobacterium spiritivorum* and *Sphingobacterium mizutae*: proposal of *Flavobacterium yabuuchiae* sp. nov. and *Flavobacterium mizutaii* comb. nov. Int. J. Syst. Bacteriol. *38*: 348–353.

2. Pickett, M.J., D.G. Hollis, and E.J. Bottone. 1991. Miscellaneous gram-negative bacteria. *In* A. Balows et al. (Eds.), Manual of Clinical Microbiology, 5th ed., American Society for Microbiology, Washington, pp. 410–428.

3. Takeuchi, M., and A. Yokota. 1992. Proposals of *Sphingobacterium faecium* sp. nov., *Sphingobacterium piscium* sp. nov., *Sphingobacterium heparinum* comb. nov., *Sphingobacterium thalpophilum* comb. nov., and two genospecies of the genus *Sphingobacterium*, and the synonymy of *Flavobacterium yabuuchiae* and *Sphingobacterium spiritivorum*. J. Gen. Appl. Microbiol. *38*: 465–482.

Comments

The ability of *F. yabuuchiae* and *F. spiritivorum* to produce acid from mannitol distinguishes them from *F. multivorum*, *F. mizutaii*, and *F. thalpophilum*. A negative acid phosphatase test differentiates *F. yabuuchiae* from *F. spiritivorum* (1).

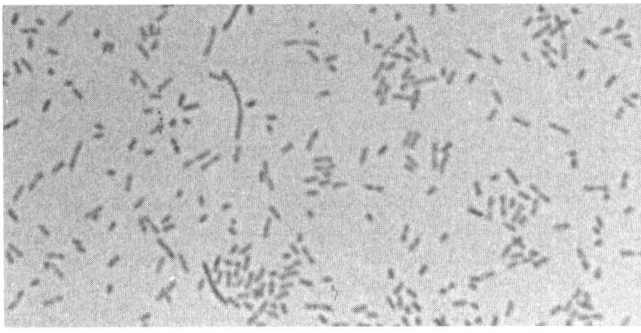

Flavobacterium yabuuchiae strain D7529. Gram stain BAP 35° C 24 h x 1700.

Flavobacterium yabuuchiae

TEST PERFORMED	SIGN	%+	#+/T
NUMBER OF STRAINS 2			
Morphology	srs		
Motility	[nm]		
Action on blood	−	0	0/2
Fermentative or oxidative	[O]		
Carbohydrate base	OF		
Acid from:			
D-Glucose	+	100	2/2
D-Xylose	+	100	2/2
D-Mannitol	[+]	100	2/2
Lactose	+	100	2/2
Sucrose	+	100	2/2
Maltose	+	100	2/2
Catalase	+	100	2/2
Oxidase	+	100	2/2
Growth on:			
MacConkey	v	(50)	(1)/2
SS	−	0	0/2
Cetrimide	−	0	0/2
Simmons citrate	−	0	0/2
Urea, Christensen's	v	50	1/2
Nitrate reduction	−	0	0/2
Gas from nitrate	−	0	0/2
Indole	[−]	0	0/2
TSI slant, acid	−	0	0/2
TSI butt, acid	−	0	0/2
H_2S (TSI butt)	−	0	0/2
H_2S (Pb ac paper)	−	0	0/2
Gelatin hydrolysis[a]	−	0	0/2
Litmus milk	−	0	0/2
Pigment:			
Soluble	v	50 br sol	1/2
Insoluble	v	50 sl yel	1/2
Growth at:			
25°C	+	100	2/2
35°C	+	100	2/2
42°C	−	0	0/2
Esculin hydrolysis	+	100	2/2
Lysine decarboxylase	−	0	0/2
Arginine dihydrolase	v	50	1/2
Ornithine decarboxylase	−	0	0/2
Nutrient broth, 0% NaCl	+	100	2/2
Nutrient broth, 6% NaCl	−	0	0/2

[a]Incubation of 7–14 days.

Francisella philomiragia

Gram-negative, very small coccoid form, nonmotile; halophilic; produces acid from glucose, sucrose, and maltose, but not xylose or mannitol (reactions may be weak and late); usually does not grow on MacConkey agar; oxidase-positive (Kovacs' modification); nitrate- and urease-negative; produces H_2S in triple sugar iron agar (reaction usually weak and delayed).

Sources (16 strains)

Blood (9 strains); lung biopsy (2); cerebrospinal fluid, pleural fluid, peritoneal fluid, unknown (2).

Reference strain

D2204 ← ATCC 25015, type strain, from lung biopsy.

Literature

1. Hollis, D.G., R.E. Weaver, A.G. Steigerwalt, J.D. Wenger, C.W. Moss, and D.J. Brenner. 1989. *Francisella philomiragia* comb. nov. (formerly *Yersinia philomiragia*) and *Francisella tularensis* biogroup novicida (formerly *Francisella novicida*) associated with human disease. J. Clin. Microbiol. *27:* 1601–1608.

2. Seger, R.A., D.G. Hollis, R.E. Weaver, and W.H. Hitzig. 1982. Chronic granulomatous disease: fatal septicemia caused by an unnamed gram-negative bacterium. J. Clin. Microbiol. *16:* 821–825.

3. Jensen, W.I., C.R. Owen, and W.J. Jellison. 1969. *Yersinia philomiragia* sp. n., a new member of the *Pasteurella* group of bacteria, naturally pathogenic for the muskrat (*Ondatra zibethica*). J. Bacteriol. *100:* 1237–1241.

Comments

Formerly designated *Yersinia philomiragia*, this organism has been isolated from water, muskrats, and humans. Less fastidious and apparently less virulent than *Francisella tularensis*. Carbohydrate reactions are enhanced in Difco OF media, presumably due to the higher NaCl content in the Difco formulation than in the King formulation. Information on the cellular fatty acid composition of this species is presented in Table 4.45 and Figure 4.50 of the cellular fatty acid section.

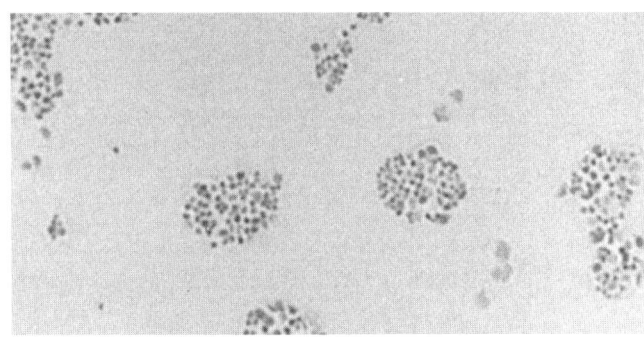

Francisella philomiragia strain D2204. Gram stain CA 35° C 24 h x 1700.

Francisella philomiragia

TEST PERFORMED	SIGN	%+	#+/T
Morphology	[rv]		
Motility; flagella	[nm]		
Action on blood			
Fermentative or oxidative	[O]		
Carbohydrate base	OF[a]		
Acid from:			
D-Glucose	[+ or (+)]	63 (37)	10w (6w)/16
D-Xylose	[−]	0	0/16
D-Mannitol	−	0	0/16
Lactose	−	0	0/16
Sucrose	[+ or (+)]	63 (37)	10w (6w)/16
Maltose	[+ or (+)]	63 (37)	10w (6w)/16
Catalase	+	94	7, 8w/16
Oxidase[b]	[+]	100	16/16
Growth on:			
MacConkey	v	19 (13)	3w (2w)/16
Simmons citrate	−	0	0/16
Urea, Christensen's	−	0	0/16
Nitrate reduction	−	0	0/16
Indole	−	0	0/16
TSI slant, acid	−	0	0/16
TSI butt, acid	−	0	0/16
H_2S (TSI butt or slant)	[+ or (+)]	56 (44)	9w (7w)/16
H_2S (Pb ac paper)	+	100	16/16
Gelatin hydrolysis[c]	v	75	(12)/16
Litmus milk	−	0	0/16
Growth at:			
25°C	+	100	16/16
35°C	+	100	16/16
42°C	v	19	3/16
Esculin hydrolysis	−	0	0/16
Lysine decarboxylase	−	0	0/16
Arginine dihydrolase	−	0	0/16
Ornithine decarboxylase	−	0	0/16
Nutrient broth, 0% NaCl	[−]	0	0/16
Nutrient broth, 6% NaCl	[+ or (+)]	88 (6)	14 (1)/16

[a]Reactions performed in Difco OF medium.
[b]Kovacs' modification.
[c]Incubation of 7–14 days.

Francisella tularensis biogroups tularensis, palaearctica, and novicida

Gram-negative, very small coccoid form, nonmotile; fastidious, requires or is stimulated by cystine or cysteine; glucose-cysteine-peptone agar with blood, chocolate agar with IsoVitaleX, and Thayer-Martin medium support growth (the latter medium is useful for isolating *F. tularensis* from a mixed bacterial population); heart infusion agar with rabbit blood and charcoal-yeast extract agar containing ketoglutarate enable some strains to grow; slide agglutination and/or direct fluorescent antibody tests are used to confirm identification. Carbohydrate reactions are determined on cysteine agar slants. Oxidase-negative; does not grow on MacConkey agar; urease- and nitrate-negative.

Sources (115 strains)

F. tularensis biogroup novicida (3 strains): water sample, cervical lymph node, blood.

F. tularensis biogroup palaearctica (43 strains): dog tick (21%), lymph node (16%), blood (16%), cervical lymph node (9%), mass (medial sternal, neck) (5%), pleural fluid (7%), abscess (5%), other (rabbit blood, arm ulcer, axillary node, cat bite, thumb, thoracentesis, pustule head, lung) (19%), unknown (2%).

F. tularensis biogroup tularensis (69 strains): blood (28%), lymph node (7%), lesion (7%), finger (10%), wound (6%), rabbit (6%), pleural fluid (4%), bone marrow (3%), pustule around tick bite (3%), abscess (buttock, axilla) (3%), insect bite (3%), other (thoracentesis fluid, postauricular node, periauricular swab, conjunctivitis, neck mass, wrist, axilla node, endometrial, cerebrospinal fluid, sputum) (14%), unknown (6%).

Reference strains

F. tularensis biogroup tularensis: KC619 ← Biegeleisen, P–35.
F. tularensis biogroup novicida: KC666 ← Utah 112, type strain, from water sample.

Literature

1. Hollis, D.G., R.E. Weaver, A.G. Steigerwalt, J.D. Wenger, C.W. Moss, and D.J. Brenner. 1989. *Francisella philomiragia* comb. nov. (formerly *Yersinia philomiragia*) and *Francisella tularensis* biogroup novicida (formerly *Francisella novicida*) associated with human disease. J. Clin. Microbiol. 27: 1601–1608.

2. Sandstrom, A., A. Sjostedt, M. Forsman, N.V. Pavlovich, and B.N. Nishankin. 1992. Characterization and classification of strains of *Francisella tularensis* isolated in the central Asian focus of the Soviet Union and in Japan. J. Clin. Microbiol. 30: 172–175.

3. Dorofe'ev, K.A. 1947. Classification of the causative agent of tularemia. Symp. Res. Works Inst. Epidemiol. Mikrobiol. Chita. 1: 170–180.

4. Olsutiev, N.G., O.S. Emelyanova, and T.N. Dunaeva. 1959. Comparative studies of strains of *B. tularense* in the Old and New World and their taxonomy. J. Hyg. Epidemiol. Microbiol. Immunol. (Prague). 3: 138–149.

Comments

Very infectious in the laboratory; causes tularemia in humans; usually is transmitted from wild animals to humans. Acid production from glycerol and sucrose differentiates *F. tularensis* biogroup tularensis from the less virulent *F. tularensis* biogroup palaearctica and *F. tularensis* biogroup novicida. The very small size of *Francisella tularensis* stained with the Gram stain should help point to its identity. Information on the cellular fatty acid composition of this group is presented in Table 4.45 and Figure 4.50 of the cellular fatty acid section.

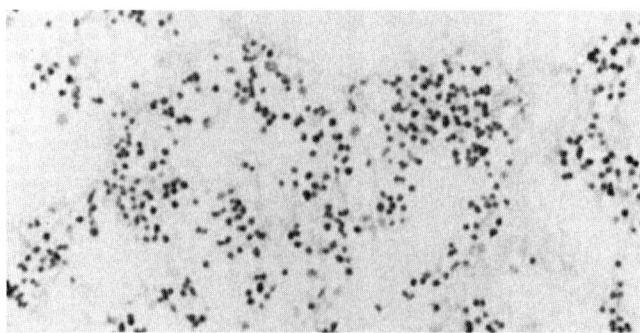

Francisella tularensis strain KC619. Gram stain CA 35° C 24 h x 1700.

Francisella tularensis biogroups tularensis, palaearctica, and novicida

TEST PERFORMED	Biogroup tularensis NUMBER OF STRAINS 69			Biogroup palaearctica NUMBER OF STRAINS 43			Biogroup novicida NUMBER OF STRAINS 3		
	SIGN	%+	#+/T	SIGN	%+	#+/T	SIGN	%+	#+/T
Morphology	[rv]			[rv]			[rv]		
Motility, flagella	nm			nm			nm		
Carbohydrate base	CA			CA			CA		
Acid from:									
D-Glucose	+	92 (8)	55, 8w, (4), (2w)/69	+ or (+)	79 (21)	14, 20w, (4), (5w)/43	(+)	(100)	(3w)/3
Sucrose	[-]	0	0/69	[-]	0	0/43	[+]	100	3w/3
Maltose							v	66	2w/3
Glycerol	[+ or (+)]	71 (29)	30, 19w, (12), (8w)/69	[-]	0	0/43	[-]	0	0/3
Oxidase	[-]	0	0/69	[-]	0	0/43	[-]	0	0/3
Growth on MacConkey	-	6	4w/69	-	0	0/43	v	(66)	(2w)/3
Urea, Christensen's	[-]	0	0/69	[-]	0	0/43	[-]	0	0/3
Nitrate reduction	[-]	0	0/69	[-]	0	0/43	[-]	0	0/3
TSI slant, acid	-	0	0/69	-	0	0/43	-	0	0/3
TSI butt, acid	-	0	0/69	-	0	0/43	-	0	0/3
H₂S (TSI butt)	-	0	0/69	-	0	0/43	-	0	0/3
Gelatin hydrolysis[a]	-	0	0/69	-	0	0/43	-	0	0/3
Nutrient broth, 0% NaCl	-	10	7/69	-	0	0/43	-	0	0/3
Nutrient broth, 6% NaCl	[-]	0	0/69	[-]	0	0/43	[v]	66	2/3

[a] Incubation of 7–14 days.

SECTION 3: DESCRIPTION OF SPECIES

Gardnerella vaginalis[a]

Gram-negative or Gram-variable, delicate "diphtheroid" rod, nonmotile; fastidious, but X and V factors are not required; ferments glucose, maltose, starch, and sometimes xylose, fructose, and sucrose, but not mannitol; serum usually is added to the carbohydrate tubes, and growth generally occurs only in positive tubes; in 2–7 days a wide (sometimes double) zone of incomplete β-like hemolysis usually is seen on heart infusion agar with 5% rabbit blood (this action may occur before colonies are visible); incubation in CO_2 or candle jar is required. Oxidase- and catalase-negative; no growth occurs on MacConkey agar.

Sources (126 strains)

Vagina (23%), urine (19%), blood (19%), cervix (8%), miscellaneous urogenital (penis, placenta, Bartholin gland, labium lesion, ovarian abscess, Skene's gland) (6%), amniotic fluid (3%), uterus (3%), urethra (2%), other (stool, arthrocentesis fluid, wound, cerebrospinal fluid, abscess, pilonidal cyst, throat, breast) (8%), unknown (9%).

Reference strains

KC1412 ← Skaggs, 594 = ATCC 1408, type strain, dark colony variant, from vaginal secretion.
KC1413 ← Skaggs, 594 = ATCC 1408, type strain, gold colony variant, from vaginal secretion.

Literature

1. Zinneman, K., and G.C. Turner. 1963. The taxonomic position of *"Haemophilus vaginalis"* (*Corynebacterium vaginale*). J. Pathol. Bacteriol. *85:* 213–219.

2. Dunkelberg, W.E., Jr., and I. McVeigh. 1969. Growth requirements of *Haemophilus vaginalis*. Antonie van Leeuwenhoek J. Microbiol. Serol. *35:* 129–145.

3. Dunkelberg, W.E., R.S. Skaggs, and D.S. Kellogg. 1970. A study and new description of *Corynebacterium vaginale* (*Haemophilus vaginalis*). Am. J. Clin. Pathol. *53:* 370–377.

4. Criswell, B.S., H.H. Marston, W.A. Stenback, S.H. Black, and H.L. Gardner. 1971. *Haemophilus vaginalis* 594, a gram-negative organism? Can. J. Microbiol. *17:* 868–869.

5. Malone, B.H., M. Schreiber, N.J. Schneider, and L.V. Holdeman. 1975. Obligately anaerobic strains of *Corynebacterium vaginale* (*Haemophilus vaginalis*). J. Clin. Microbiol. *2:* 272–275.

6. Greenwood, J.R., and M.J. Pickett. 1979. Salient features of *Haemophilus vaginalis*. J. Clin. Microbiol. *9:* 200–204.

7. Greenwood, J.R., and M.J. Pickett. 1980. Transfer of *Haemophilus vaginalis* Gardner and Dukes to a new genus, *Gardnerella: G. vaginalis* (Gardner and Dukes) comb. nov. Int. J. Syst. Bacteriol. *30:* 170–178.

8. Speigel, C.A., R. Amsel, D. Eschenbach, F. Schoenknecht, and K.K. Holmes. 1980. Anaerobic bacteria in nonspecific vaginitis. N. Engl. J. Med. *303:* 601–607.

Comments

G. vaginalis formerly was named *Haemophilus vaginalis* and *Corynebacterium vaginale*. This organism has been reported also to hydrolyze hippurate and to hemolyze human (not sheep) blood. A cause of "nonspecific" vaginitis and urethritis; in the discharge, the epithelial cells may be covered with masses of bacteria, "clue cells."

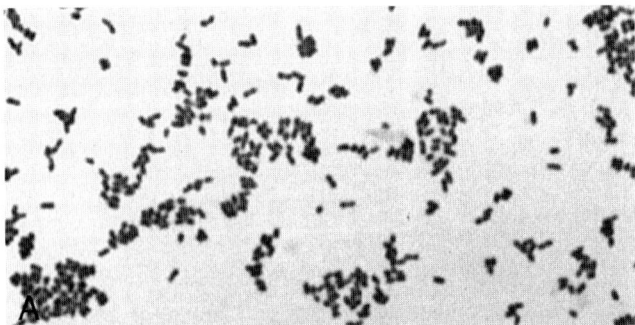

Gardnerella vaginalis strain KC1413. Gram stain BAP 35° C 24 h x 1700.

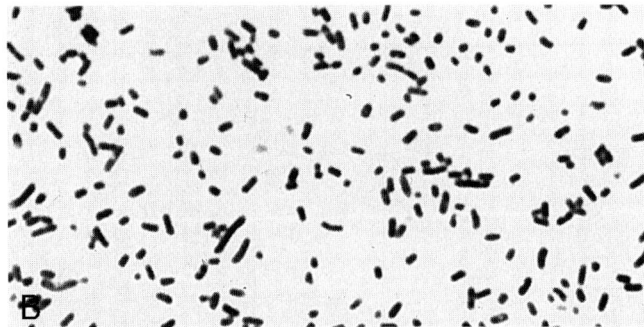

Gardnerella vaginalis strain KC1412. Gram stain BAP 35° C 24 h x 1700.

---

[a] Identification of *G. vaginalis* isolates at CDC now is done by the Sexually Transmitted Diseases Program, National Center for Infectious Diseases.

Gardnerella vaginalis

TEST PERFORMED	SIGN	%+	#+/T
Morphology	srs (Gram variable)		
Motility	nm		
Gas from glucose	−	0	0/122
Action on blood	[inc β]	93	98/105; 3 al
Fermentative or oxidative	F[a]		
Carbohydrate base	F[b]		
Acid from:			
D-Glucose	[+ or (+)]	81 (13)	93, 6w (16)/122
D-Xylose	v	36 (9)	40, 2w (10)/116
D-Mannitol	[−]	0	0/116
Lactose	−	(1)	(1)/114
Sucrose	v	11 (4)	11, 2w (5)/114
Maltose	[+ or (+)]	87 (13)	109 (16)/125
Fructose	v	60 (23)	53, 13w (25)/100
Catalase	[−]	2	2w/81
Oxidase	[−]	6	4w/73
Growth on:			
MacConkey	[−]	5	2w/42
SS	−	0	0/43
Simmons citrate	−	0	0/41
Urea, Christensen's	−	(5)	(2)/43
Nitrate reduction	−	0	0/85
Indole	−	0	0/61
TSI slant, acid	v	14	2, 4w/44; 28 ngr
TSI butt, acid	v	14	2, 4w/44; 28 ngr
H_2S (TSI butt)	−	0	0/16
H_2S (Pb ac paper)	v	31	5/16
Gelatin hydrolysis[c]	−	0	0/13
Litmus milk	v	32 (16) A	4, 8w (6)/38
Growth at:			
25°C	−	5	1/20
35°C	+	100	20/20
42°C	v	14	3/21
Esculin hydrolysis	−	0	0/37
Nutrient broth, 0% NaCl	v	24	8/33
Nutrient broth, 6% NaCl	−	0	0/36

[a]TSI reaction not always indicative of fermentative activity.
[b]Rabbit serum (1–2 drops/3 ml of medium) may be required.
[c]Incubation of 7–14 days.

Gilardi rod group 1

Gram-negative medium to wide, sometimes pleomorphic rod, nonmotile; asaccharolytic; oxidase-positive; grows on MacConkey agar; nitrate-, nitrite-, and urease-negative; phenylalanine deaminase-positive.

Sources (15 strains)

Blood (4), foot wound (3), leg wound (2), arm wound (2), amputation site, ankle, urine, and oral lesion (1 each).

Reference strain

F6511 ← ATCC 51249.

Literature

1. Moss, C.W., M.I. Daneshvar, and D.G. Hollis. 1993. Biochemical characteristics and fatty acid composition of Gilardi rod group 1 bacteria. J. Clin. Microbiol. *31:* 689–691.

2. Gilardi, G.L. 1990. Identification of glucose-nonfermenting gram-negative rods. Department of Laboratories, North General Hospital, New York.

Comments

Failure to reduce nitrite differentiates Gilardi rod group 1 from the biochemically similar *Neisseria weaveri*. Information on the cellular fatty acid composition of this group is presented in Table 4.46 and Figure 4.51 of the cellular fatty acid section.

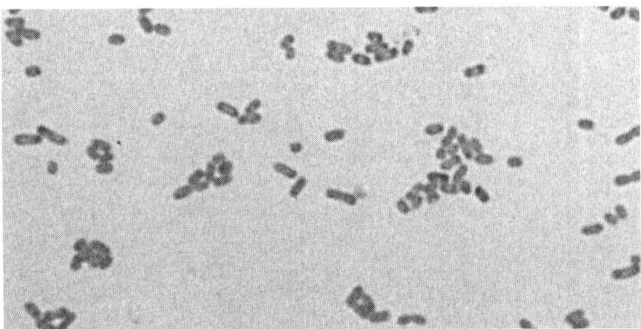

Gilardi rod group 1 strain F6511. Gram stain BAP 35° C 24 h x 1700.

Gilardi rod group 1

TEST PERFORMED	SIGN	% +	#+/T
Morphology	[sbr]		
Motility; flagella	nm		
Action on blood	v	20 gr + br	3/15; 3 gr, 2 lav, 1 ly
Fermentative or oxidative	n-o		
Carbohydrate base	OF		
Acid from:			
D-Glucose	[-]	0	0/15
D-Xylose	-	0	0/15
D-Mannitol	-	0	0/15
Lactose	-	0	0/15
Sucrose	-	0	0/15
Maltose	-	0	0/15
Catalase	+	100	15/15
Oxidase	+	100	15/15
Growth on:			
MacConkey	[+]	93	14/15
SS	v	80	12/15
Cetrimide	-	0	0/15
Simmons citrate	-	0	0/15
Urea, Christensen's	[-]	0	0/15
Nitrate reduction	[-]	0	0/15
Nitrite reduction	[-]	0	0/15
Indole	-	0	0/15
TSI slant, acid	-	0	0/15
TSI butt, acid	-	0	0/15
H_2S (TSI butt)	-	0	0/15
H_2S (Pb ac paper)	v	87	13/15
Gelatin hydrolysis[a]	-	0	0/15
Litmus milk	v	20 k	3/15
Pigment:			
Soluble	v	67 amb	10/15; 3 yel-br, 2 tan, 1 yel-tan
Insoluble	v	60 amb	9/15; 1 caramel, 1 apricot
Growth at:			
25°C	+	100	15/15
35°C	+	100	15/15
42°C	v	80	7, 5w/15
Esculin hydrolysis	-	0	0/15
Nutrient broth, 0% NaCl	+	100	15/15
Nutrient broth, 6% NaCl	v	7 (13)	1 (2)/15
Phenylalanine deaminase	[+]	100	15/15

[a]Incubation of 7–14 days.

Haemophilus aphrophilus

Gram-negative small rod or filament, nonmotile; fastidious, requires or is enhanced by a candle-jar atmosphere; ferments glucose with slight gas formation, lactose, sucrose, and maltose; catalase-negative, nitrate-positive; indole-negative; esculin hydrolysis-negative.

Sources (203 strains)

Blood (25%); brain abscess (8%); other abscesses (6%); pleural fluid (4%); wound (4%); mouth (4%); eye (4%); sputum (3%); sinus (2%); mandible (2%); brain (2%); cerebrospinal fluid (2%); bone fragment, throat, trachea, nose, cyst, neck, tooth, skin, lung, pus, bronchus, abdominal drainage, urogenital, and lymph node (1% each); other (11%); unknown (5%).

Reference strain

KC425 ← Margaret Pittman, 5908.

Literature

1. Khairat, O. 1940. Endocarditis due to a new species of *Haemophilus*. J. Pathol. Bacteriol. *50:* 497–505.

2. King, E.O., and H.W. Tatum. 1962. *Actinobacillus actinomycetemcomitans* and *Haemophilus aphrophilus*. J. Infect. Dis. *111:* 85–94.

3. Sneath, P.H.A., and V.B.D. Skerman. 1966. A new list of type and reference strains of bacteria. Int. J. Syst. Bacteriol. *16:* 1–113.

4. Page, M.I., and E.O. King. 1966. Infection due to *Actinobacillus actinomycetemcomitans* and *Haemophilus aphrophilus*. N. Engl. J. Med. *275:* 181–188.

5. Sutter, V.L., and S.M. Finegold. 1970. *Haemophilus* infections: clinical and bacteriologic studies. Ann. NY Acad. Sci. *174:* 468–487.

Comments

Usually granular growth in broth, with colonies adhering to sides of tube. Originally described as requiring X factor; however, none of the strains received in our laboratory or the reference strain have required X factor. Genus status is questionable. After repeated subcultures, *H. aphrophilus* strains may become elongated and may exhibit some curved filaments. We have studied 15 *H. aphrophilus*-like isolates that differ from *H. aphrophilus* only in the lactose reaction. Information on the cellular fatty acid composition of this species is presented in Table 4.4B and Figure 4.9 of the cellular fatty acid section.

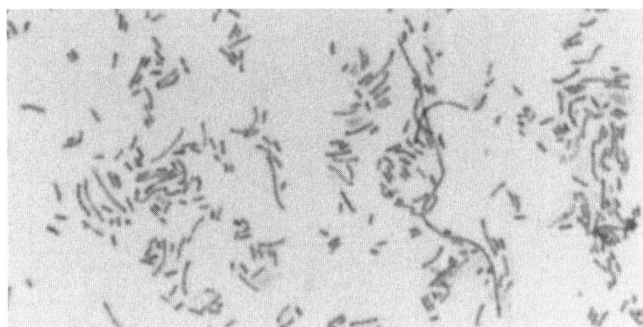

Haemophilus aphrophilus strain KC425. Gram stain BAP 35° C 24 h x 1700.

Haemophilus aphrophilus

TEST PERFORMED	SIGN	%+	#+/T
	NUMBER OF STRAINS 203		
Morphology	[rv or plr]		
Motility; flagella	nm		
Gas form glucose	[+][a]	98	168/172
Action on blood	v	41 al	56/135; 15 gr, 1 LG
Fermentative or oxidative	[F]		
Carbohydrate base	F		
Acid from:			
D-Glucose	[+]	98 (2)	187, 12w (4)/203
D-Xylose	-	0	0/202
D-Mannitol	-	0	0/202
Lactose	[+]	96 (4)	180, 13w (9)/202
Sucrose	+	99 (1)	191, 9w (3)/203
Maltose	[+]	99 (1)	190, 10w (3)/203
Catalase	[-]	5	1, 10w/203
Oxidase	v	28	46, 9w/199
Growth on:			
MacConkey	v	19	5, 28/171
SS	-	0	0/169
Simmons citrate	-	0	0/170
Urea, Christensen's	-	0	0/150
Nitrate reduction	[+]	100	202/202
Gas from nitrate	-	0	0/202
Indole	[-]	0	0/192
TSI slant, acid	+	100	166, 4w/170
TSI butt, acid	+	100	163, 5w/168
H_2S (TSI butt)	-	0	0/169
H_2S (Pb ac paper)	+	100	126, 36w/162
MR	v	12	2, 9w/91
VP	-	0	0/63
Gelatin hydrolysis[b]	-	0	0/121
Litmus milk	v	30 (7) A	53 (12)/179; 12 IR
Growth at:			
25°C	v	55	86/156
35°C	+	99	157/158
42°C	v	26	40/155
Esculin hydrolysis	[-]	0	0/200
Lysine decarboxylase	-	0	0/9
Arginine dihydrolase	-	0	0/9
Ornithine decarboxylase	-	0	0/9
Nutrient broth, 0% NaCl	v	56	91/162
Nutrient broth, 6% NaCl	-	1	1/165

[a]Volume of gas is frequently small.
[b]Incubation of 7–14 days.

Haemophilus ducreyi[a]

Gram-negative slender rod or coccoid, sometimes occurring in pairs, chains, or filaments, nonmotile; requires X factor; grows in air or in air with increased CO_2 (growth is enhanced by a moist atmosphere); shows α-hemolysis in stabs at 48 h; most strains reduce nitrate; alkaline phosphatase positive; indole- and catalase-negative; porphyrin test-negative; oxidase-positive. (For further biochemical reactions see Chart 45 entitled "Species and biotypes of X- or V-factor dependent *Haemophilus* isolated from humans and two animal species." p. 218.)

Sources (31 strains)

Genital lesion (55%), chancroid ulcer (32%), penile lesion (10%), infected lymph node (3%).

Reference strain

KC1449 ← G. Hammond ← Pasteur Institute A77.

Literature

1. Kilian, M. 1976. A taxonomic study of the genus *Haemophilus*, with the proposal of a new species. J. Gen. Microbiol. *93:* 9–62.

2. Hammond, G.W., C.J. Lian, J.C. Witt, and A.R. Ronald. 1978. Comparison of specimen collection and laboratory techniques for isolation of *Haemophilus ducreyi*. J. Clin. Microbiol. 7: 39–43.

3. Morse, S.A. 1989. Chancroid and *Haemophilus ducreyi*. Clin. Microbiol. Rev. *2:* 137–157.

Comments

Growth usually is difficult to remove from agar with an inoculating needle; the colonies can be pushed intact across the surface of the plate. Overall appearance may be that of a mixed culture. Difficulties in isolating this organism have been reported. See Reference 3 for a review of isolation strategies. Of 29 *H. ducreyi* isolates, 19 were β-lactamase positive. Information on the cellular fatty acid composition of this species is presented in Table 4.4B and Figure 4.9 of the cellular fatty acid section.

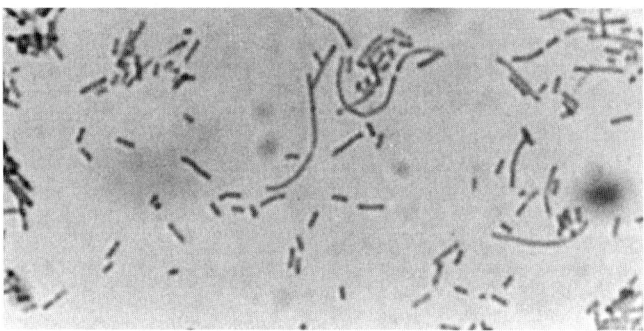

Haemophilus ducreyi strain KC1449. Gram stain BAP 35° C 24 h x 1700.

---

[a]Identification of *H. ducreyi* isolates at CDC is now done by the Sexually Transmitted Diseases Program, National Center for Infectious Diseases.

"Haemophilus felis"

Gram-negative coccobacillus, nonmotile; requires V factor and CO_2 for growth; oxidase-negative; catalase-positive; uses glucose and lactose; urease-, indole-, and ornithine decarboxylase-negative. (For further biochemical reactions see Chart 45 entitled "Species and biotypes of X- or V-factor dependent *Haemophilus* isolated from humans and two animal species," p. 218.)

Source

Feline respiratory tract.

Reference strain

G2295 ← TI189, from Thomas Inzana.

Literature

1. Inzana, T.J., J.L. Johnson, et al. 1992. Isolation and characterization of a newly identified *Haemophilus* species from cats: *"Haemophilus felis"*. J. Clin. Microbiol. *30:* 2108–2112.

Comments

Biochemically similar to *H. paraphrophilus*. Commensal colonizer of the nasopharynx of cats. Not known to cause disease in humans.

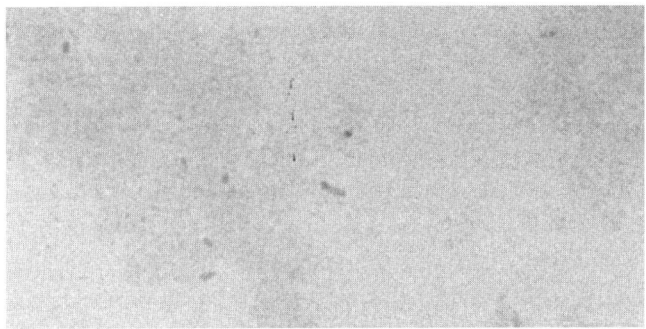

"Haemophilus felis" strain G2295. Gram stain BAP 35° C 24 h x 1700.

Haemophilus haemoglobinophilus

Gram-negative small pleomorphic rod; sometimes forms filaments; nonmotile; requires X factor only; grows in air or with increased CO_2; nonhemolytic; uses glucose, xylose, mannitol, and sucrose; oxidase-positive; indole- and nitrate-positive; urease- and ornithine-negative; alkaline phosphatase-negative (all other *Haemophilus* species are positive, with the exception of *H. piscium*). (For further biochemical reactions see Chart 45 entitled "Species and biotypes of X- or V-factor dependent *Haemophilus* isolated from humans and two animal species," p. 218.)

Sources

One from dog prepuce and one from source unknown.

Reference strain

KC613 ← NCTC 1659 = ATCC 19416, type strain, from dog prepuce.

Literature

1. Kilian, M. 1976. A taxonomic study of the genus *Haemophilus*, with the proposal of a new species. J. Gen. Microbiol. *93:* 9–62.

2. Biberstein, E.L. 1990. *Haemophilus* and *Taylorella*. In G.R. Carter and J.R. Cole (Eds.), Diagnostic Procedures in Veterinary Bacteriology and Mycology, 5th ed., Academic Press, New York, pp. 151–164.

3. MacLachlan, G.K., and G.F. Hopkins. 1978. Early death in pups to *Haemophilus haemoglobinophilus* (*canis*) infection. Vet. Rec. *103:* 409–410.

Comments

Usually isolated from prepuce or vagina of dog and considered a commensal. These organisms have also been referred to as *H. canis*. We have a group of at least 21 *Haemophilus*-like organisms that exhibited a heavy mucoid growth and required X factor (but not V factor) after one passage from blood agar. Most of these cultures were from sputum. We are not certain of their relationship to *H. haemoglobinophilus*.

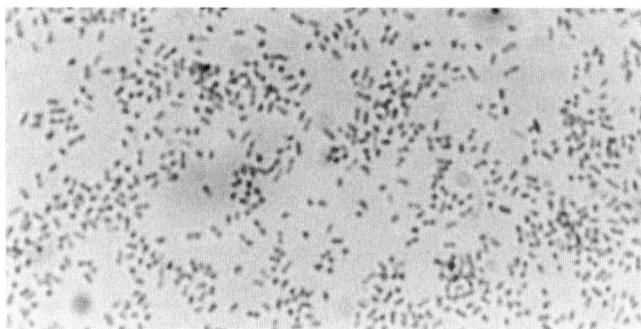

Haemophilus haemoglobinophilus strain KC613.
Gram stain BAP 35° C 24 h x 1700.

Haemophilus haemolyticus

Gram-negative coccoid rod; may form tangled filaments; nonmotile; requires X and V factors; grows in air or in air with increased CO_2; uses glucose, usually with gas production; usually urease-, catalase-, oxidase-, and nitrate-positive; indole-variable; ornithine-negative; hemolytic on horse and rabbit blood. (For further biochemical reactions see Chart 45 entitled "Species and biotypes of X- or V-factor dependent *Haemophilus* isolated from humans and two animal species," p. 218.)

Sources (18 strains)

Sputum (28%), throat (22%), blood (17%), other (33%).

Reference strain

D8243, from blood.

Literature

1. Stillman, E.G., and J.M. Bourn. 1920. Biological study of the hemophilic bacilli. J. Exp. Med. *32:* 665–682.

2. Kilian, M. 1976. A taxonomic study of the genus *Haemophilus,* with the proposal of a new species. J. Gen. Microbiol. *93:* 9–62.

3. Kilian, M. 1992. *Haemophilus. In* A. Balows, et al. (Eds.), Manual of Clinical Microbiology, 5th ed., American Society for Microbiology, Washington, pp. 463–470.

Comments

Commensal organism found in the human oropharynx.

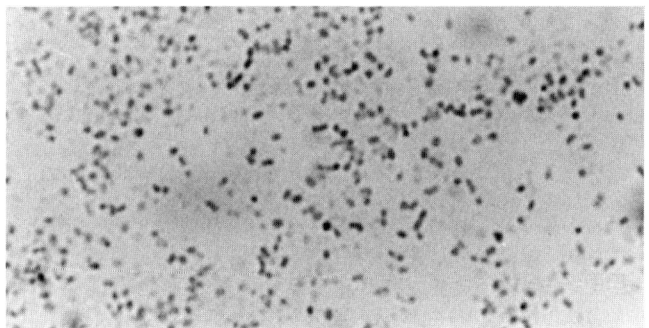

Haemophilus haemolyticus strain D8243. Gram stain BAP 35° C 24 h x 1700.

Haemophilus influenzae

Gram-negative small coccoid to slightly pleomorphic rod, sometimes with tendency to filament formation; nonmotile; some strains are encapsulated; requires X and V factors; grows in air or in air with increased CO_2; uses glucose and usually xylose; oxidase- and ornithine-variable; usually nitrate- and urease-positive; and most strains indole-positive. (For further biochemical reactions see Chart 45 entitled "Species and biotypes of X- or V-factor dependent *Haemophilus* isolated from humans and two animal species," p. 218.)

Sources (154 strains)

Blood (25%), cerebrospinal fluid (22%), eye (10%), sputum (9%), throat (7%), nose and nasopharynx (5%), bronchial washing (3%), cervix and vagina (3%), lung (2%), amniotic fluid (1%), transtracheal aspirate (1%), cyst (1%), other (6%), unknown (5%)

Reference strain

KC1318 ← ATCC 19418 = NCTC 4560.

Literature

1. Winslow, C.-E.A., J. Broadhurst, R.E. Buchanan, C. Krumweide, Jr., L.A. Rogers, and G.H. Smith. 1917. The families and genera of the bacteria. Preliminary report of the committee of the Society of American Bacteriologists of characterization and classification of bacterial types. J. Bacteriol. *2*: 505–566.
2. Sneath, P.H.A., and V.B.D. Skerman. 1966. A new list of type and reference strains of bacteria. Int. J. Syst. Bacteriol. *16*: 1–113.
3. Kilian, M. 1976. A taxonomic study of the genus *Haemophilus*, with the proposal of a new species. J. Gen. Microbiol. *93*: 9–62.
4. Oberhofer, T.R., and A.E. Back. 1979. Biotypes of *Haemophilus* encountered in clinical laboratories. J. Clin. Microbiol. *10*: 168–174.
5. Kilian, M. 1992. *Haemophilus. In* A. Balows, et al. (Eds.), Manual of Clinical Microbiology, 5th ed., American Society for Microbiology, Washington, pp. 463–470.

Comments

Eight biotypes, based upon urease, ornithine decarboxylase, and indole production, have been described (5). Six serological types (a–f) based upon capsular carbohydrate antigens are also recognized; most isolates from cerebrospinal fluid and blood are serotype b; usually strains from the respiratory tract are not serotype-specific. Information on the cellular fatty acid composition of this species is presented in Table 4.4B and Figure 4.9 of the cellular fatty acid section.

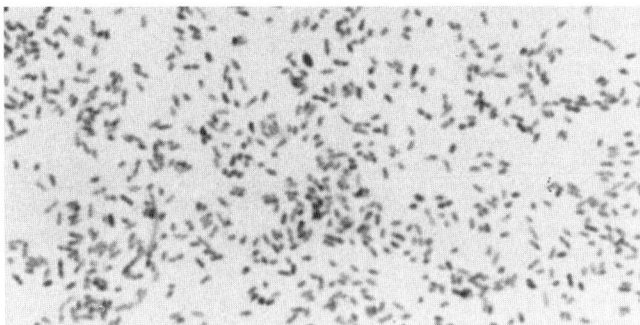

Haemophilus influenzae strain KC1318. Gram stain BAP 35° C 24 h x 1700.

Haemophilus influenzae biogroup aegyptius *(H. aegyptius)*

Gram-negative small to medium rod, sometimes filamentous, nonmotile; requires X and V factors; grows in air with increased CO_2; uses glucose but not xylose; oxidase-variable; indole- and ornithine-negative; primary isolates usually agglutinate human erythrocytes. (For further biochemical reactions see footnote "[a]" of Chart 45 entitled "Species and biotypes of X- or V-factor dependent *Haemophilus* isolated from humans and two animal species," p. 218.)

Source

Conjunctiva.

Reference strain

KC1018 ← ATCC 11116 = NCTC 8502. Suggested "working type culture."

Literature

1. Pittman, M., and D.J. Davis. 1950. Identification of the Koch-Weeks bacillus (*Haemophilus aegyptius*). J. Bacteriol. *59:* 413–426

2. Mazloum, H.A., M. Kilian, Z.M. Mohammed, and M.D. Said. 1982. Differentiation of *Haemophilus aegyptius* and *Haemophilus influenzae*. Acta Pathol. Microbiol. Immunol. Scand. Sect. B *90:* 109–112.

3. Brenner, D.J., L.W. Mayer, et al. 1988. Biochemical, genetic, and epidemiologic characterization of *Haemophilus influenzae* biogroup aegyptius (*Haemophilus aegyptius*) strains associated with Brazilian purpuric fever. J. Clin. Microbiol. 26: 1524–1534.

4. Kilian, M. 1976. A taxonomic study of the genus *Haemophilus*, with the proposal of a new species. J. Gen. Microbiol. *93:* 9–62.

Comments

These organisms are related to *H. influenzae* at the species level by DNA hybridization and share a common biochemical profile with *H. influenzae* biotype III. Phenotypic tests that have been described to differentiate these organisms from *H. influenzae* biotype III include distinct bacillary morphology, troleandromycin sensitivity, failure to grow on tryptic soy agar in the presence of X and V factors, and failure to produce acid from xylose (2); however, these tests are not absolute. These organisms commonly cause acute or subacute infectious conjunctivitis. Certain clones have been identified as the etiologic agent of Brazilian purpuric fever, a highly fatal systemic infection (3). Common name: Koch-Weeks bacillus. Information on the cellular fatty acid composition of this group is presented in Table 4.4B and Figure 4.9 of the cellular fatty acid section.

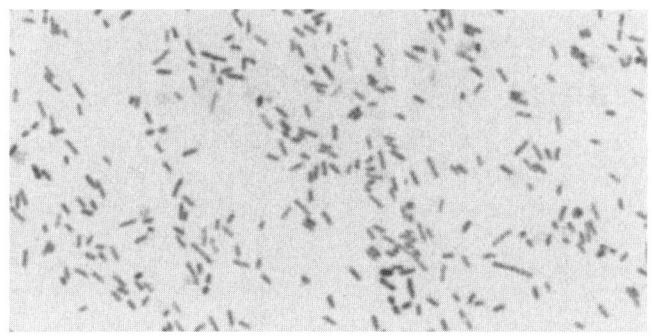

Haemophilus influenzae biogroup aegyptius strain KC1018. Gram stain BAP 35° C 24 h x 1700.

Haemophilus parahaemolyticus

Gram-negative thin rod, sometimes with tendency to filament formation; nonmotile; requires V but not X factor; β-hemolytic on blood agar; uses glucose but not lactose; urease-positive, indole- and ornithine decarboxylase-negative. (For further biochemical reactions see Chart 45 entitled "Species and biotypes of X- or V-factor dependent *Haemophilus* isolated from humans and two animal species," p. 218.)

Sources (54 strains)

Throat (59%), sputum (17%), spinal fluid (4%), other (broncheal alveolar lavage, nasal, eye, tonsil, ulcer) (9%), unknown (11%).

Reference strain

KC421 ← Margaret Pittman, 761.

Literature

1. Kilian, M. 1992. *Haemophilus*. In A. Balows, et al. (Eds.), Manual of Clinical Microbiology, 5th ed., American Society of Microbiology, Washington, pp. 463–470.

2. Kilian, M., and E.L. Biberstein. 1984. Genus II. *Haemophilus* Winslow, Broadhurst, Buchanan, Krumweide, Rogers and Smith 1917, 561.[AL] In N.R. Krieg and J.G. Holt (Eds.), Bergey's Manual of Systematic Bacteriology, Vol. 1, Williams & Wilkins, Baltimore, pp. 558–569.

3. Albritton, W.L. 1982. Infections due to *Haemophilus* species other than *H. influenzae*. Ann. Rev. Microbiol. *36:* 199–206.

Comments

β-hemolysis on blood agar differentiates *H. parahaemolyticus* from *H. parainfluenzae* biotype III. Information on the cellular fatty acid composition of this species is presented in Table 4.4B and Figure 4.9 of the cellular fatty acid section.

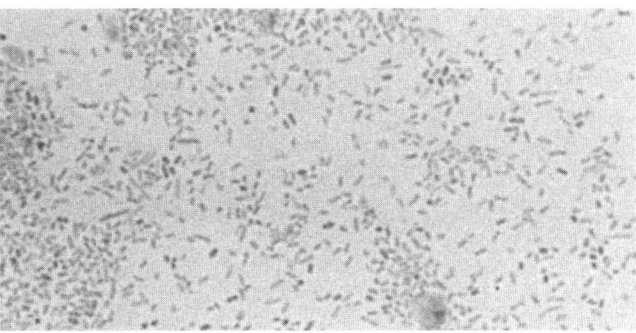

Haemophilus parahaemolyticus strain KC421. Gram stain BAP 35° C 24 h x 1700.

Haemophilus parainfluenzae

Gram-negative small pleomorphic rod or long filament; nonmotile; requires V factor only; synthesizes porphyrins from D-amino-levulinic acid; aerobic; growth not improved by increased CO_2 except for some biotype III strains; uses glucose, often with gas formation, and sucrose; most strains produce β-galactosidase; oxidase-variable; indole-negative and nitrate-positive. (For further biochemical reactions see Chart 45 entitled "Species and biotypes of X- or V-factor dependent *Haemophilus* isolated from humans and two animal species," p. 218.)

Sources (41 strains)

Blood (24%), throat (22%), sputum (15%), eye (5%), nasopharynx (5%), urine (5%), other (24%).

Reference strain

KC363 ← ATCC 7901 ← Margaret Pittman, 429.

Literature

1. Public Health Rep *55:* 915, 1940.
2. Gullekson, E.H., and M. Dumoff. 1966. *Haemophilus parainfluenzae* meningitis in a newborn. J. Am. Med. Assoc. *198:* 1221.
3. Kilian, M. 1992. *Haemophilus*. In A. Balows, et al. (Eds.), Manual of Clinical Microbiology, 5th ed., American Society for Microbiology, Washington, pp. 463–470.
4. Oberhofer, T.R., and A.E. Back. 1979. Biotypes of *Haemophilus* encountered in clinical laboratories. J. Clin. Microbiol. *10:* 168–174.

Comments

Seven biotypes, based on urease, ornithine decarboxylase, and indole production, currently are recognized (3). *H. paraphrohaemolyticus* is considered to be a CO_2-dependent variant of biotype III. Oberhofer and Back (4) described a *H. parainfluenzae* biotype that has the same characteristics as Kilian's *H. segnis*, i.e., V factor-requiring, ornithine decarboxylase-, urease-, and indole-negative. The taxonomic status of these strains is currently unresolved. If they are found to be *H. parainfluenzae,* Kilian (3) has suggested that they be designated biotype V. Currently, there is no *H. parainfluenzae* biotype V because the original description of this biotype was published previously as biotype IV (3). Information on the cellular fatty acid composition of this species is presented in Table 4.4B and Figure 4.9 of the cellular fatty acid section.

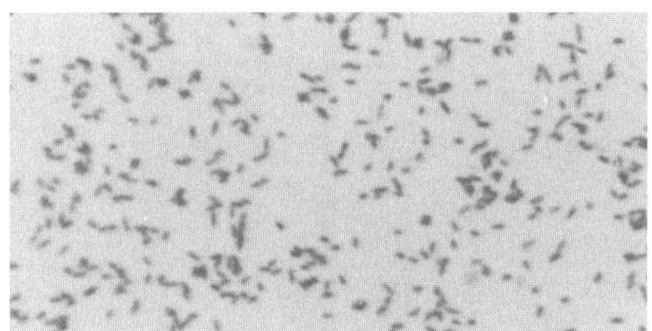

Haemophilus parainfluenzae strain KC363. Gram stain BAP 35° C 24 h x 1700.

Haemophilus paraphrophilus

Gram-negative small rod or filament; nonmotile; requires V factor only; improvement of growth in CO_2 is variable; uses glucose and lactose; urease-, indole-, and ornithine-negative; nitrate reduction is positive. (For further biochemical reactions see Chart 45 entitled "Species and biotypes of X- or V-factor dependent *Haemophilus* isolated from humans and two animal species," p. 218.)

Sources (16 strains)

Blood (44%); vagina (13%); and one strain (6% each) each of cheek abscess, liver abscess, ear, pleural fluid, neck wound, sinus, and dental abscess.

Reference strain

KC977 ← NCTC 10556, from parietal abscess.

Literature

1. Zinneman, K., K.B. Rogers, J. Frazer, and J.M.H. Boyce. 1968. A new V-dependent *Haemophilus* species preferring increased CO_2 tension for growth and named *Haemophilus paraphrophilus*, nov. sp. J. Pathol. Bacteriol. *96:* 413–419.

2. Kilian, M., and E.L. Biberstein. 1984. Genus II. *Haemophilus* Winslow, Broadhurst, Buchanan, Krumweide, Rogers, and Smith 1917, 561.[AL] In N.R. Krieg and J.G. Holt (Eds.), Bergey's Manual of Systematic Bacteriology, Vol. 1, Williams & Wilkins, Baltimore, pp. 558–569.

3. Kilian, M. 1992. *Haemophilus*. In A. Balows, et al. (Eds.), Manual of Clinical Microbiology, 5th ed., American Society for Microbiology, Washington, pp. 463–470.

Comment

Information on the cellular fatty acid composition of this species is presented in Table 4.4B and Figure 4.9 of the cellular fatty acid section.

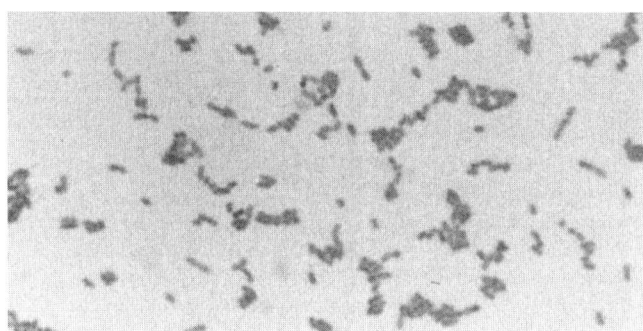

Haemophilus paraphrophilus strain KC977. Gram stain BAP 35° C 24 h x 1700.

Haemophilus segnis

Gram-negative small rod or filament; nonmotile; requires V factor; nonhemolytic; uses glucose, usually weakly; oxidase-, lactose-, urease-, indole-, and ornithine-negative. (For further biochemical reactions, see Chart 45 entitled "Species and biotypes of X- or V-factor dependent *Haemophilus* isolated from humans and two animal species," p. 218.)

Literature

1. Kilian, M. 1976. A taxonomic study of the genus *Haemophilus*, with the proposal of a new species. J. Gen. Microbiol. *93:* 9–62.

2. Oberhofer, T.R., and A.E. Back. 1979. Biotypes of *Haemophilus* encountered in clinical laboratories. J. Clin. Microbiol. *10:* 168–174.

3. Kilian, M. 1992. *Haemophilus. In* A. Balows, et al. (Eds.), Manual of Clinical Microbiology, 5th ed., American Society for Microbiology, Washington, pp. 463–470.

Comments

M. Kilian proposed *H. segnis* as a new species. The strains studied were isolated primarily from the human oral cavity. We have not studied a reference strain, but we have received 40 strains with characteristics similar to those described for the species. Sources of these strains were blood (9 strains); eye (4 strains); chest/pleural fluid (3 strains); lung (3 strains); finger abscess (3 strains); peritoneal fluid, mandibular abscess, and wound (2 strains each); and one strain each from sputum, throat, brain aspirate, abscess, hand wound, neck, brain abscess, gastric fluid, toe wound, intrauterine device, breast cyst, and pilodontal cyst. Information on the cellular fatty acid composition of this species is presented in Table 4.4B and Figure 4.9 of the cellular fatty acid section.

Biotype IV of *H. parainfluenzae,* as proposed by Oberhofer and Back (2), has the same characteristics as Kilian's *H. segnis.*

Kingella denitrificans

Gram-negative coccoid or short rod occurring in pairs or short chains; nonmotile; facultatively anaerobic; produces acid fermentatively from glucose (but not from maltose as stated in 1976 Reference 2; oxidase-positive, catalase-negative; usually reduces nitrate and nitrite with gas formation; does not grow on MacConkey agar.

Sources (60 strains)

Throat (68%), urogenital (7%), chimpanzee throat (7%), mouth (3%), sputum (3%), nasopharynx (3%), other (5%) (1 strain each: epiglottis, mandibular abscess, blood), unknown (3%).

Reference strain

C1080, from throat.

Literature

1. Hollis, D.G., G.L. Wiggins, and R.E. Weaver. 1972. An unclassified gram-negative rod isolated from the pharynx on Thayer-Martin medium (selective agar). Appl. Microbiol. 24: 772–777.

2. Snell, J.J.S., and S.P. Lapage. 1976. Transfer of some saccharolytic *Moraxella* species to *Kingella* Henriksen and Bøvre 1976, with descriptions of *Kingella indologenes*, sp. nov. and *Kingella denitrificans* sp. nov. Int. J. Syst. Bacteriol. 26: 451–458.

3. Minamoto, G.Y., and E.M. Sordillo. 1992. *Kingella denitrificans* as a cause of granulomatous disease in a patient with AIDS.

Comments

Formerly designated TM-1. Frequently pits the agar. Grows on Thayer-Martin selective medium; nonpitting colonies are similar to those of *Neisseria gonorrhoeae*. Information on the cellular fatty acid composition of this species is presented in Table 4.53 and Figure 4.58 of the cellular fatty acid section.

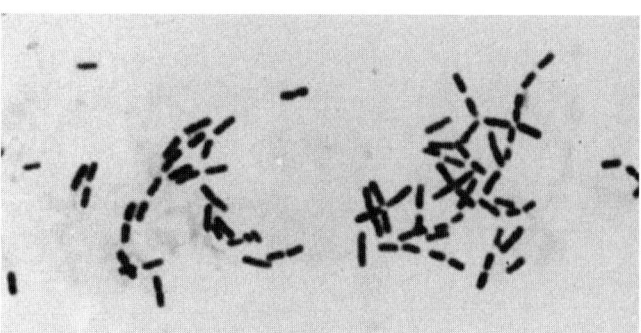

Kingella denitrificans strain C1080. Gram stain BAP 35° C 24 h x 1700.

Kingella denitrificans

TEST PERFORMED	SIGN	% +	#+/T
Morphology	cc,pr,chs		
Motility; flagella	[nm]		
Action on blood	−	0	0/38
Fermentative or oxidative	[F]a		
Carbohydrate base	Fb		
Acid from:			
D-Glucose	[+ or (+)]	30 (62)	7, 11w (37)/60
D-Xylose	−	0	0/60
D-Mannitol	−	0	0/60
Lactose	−	0	0/60
Sucrose	−	0	0/60
Maltose	[−]	0	0/60
Catalase	[−]	10	5,1w/60
Oxidase	[+]	100	60/60
Growth on:			
MacConkey	[−]	0	0/50
SS	−	0	0/50
Simmons citrate	−	0	0/50
Urea, Christensen's	−	0	0/29
Nitrate reduction	[+]	93	53/57
Gas from nitrate	[v]	88	50/57
Indole	−	0	0/57
TSI slant, acid	−	2	1w/52
TSI butt, acid	−	0	0/52
H$_2$S (TSI butt)	−	0	0/44
H$_2$S (Pb ac paper)	v	73	22, 10w/44
Gelatin hydrolysisc	−	0	0/12
Litmus milk	−	6 IR	2/47; 1 k
Pigment:			
Insoluble	−	2 yel	1/60
Soluble	−	2 yel	1/60
Growth at:			
25°C	v	40	19w/48
35°C	+	100	48w/48
42°C	v	48	23w/48
Esculin hydrolysis	−	0	0/48
Nutrient broth, 0% NaCl	v	78 (4)	9, 26w (2)/45
Nutrient broth, 6% NaCl	−	0	0/45

NUMBER OF STRAINS 60

aTSI reaction not always indicative of fermentative activity.
bRabbit serum (1–2 drops/3 ml of medium) may be required.
cIncubation of 7–14 days.

Kingella kingae

Gram-negative coccoid to short rod occurring in pairs and chains; nonmotile; produces acid fermentatively from glucose and maltose, not sucrose; oxidase-positive, catalase-negative, indole-negative, usually nitrate-negative; proteolytic; usually β-hemolytic.

Sources (137 strains)

Blood (54%), bone or joint fluid (21%), throat (11%), other (11%), unknown (3%).

Reference strain

A1702, from knee joint.

Literature

1. Henriksen, S.D., and K. Bøvre. 1968. *Moraxella kingii* spec. nov., a haemolytic, saccharolytic species of the genus *Moraxella*. J. Gen. Microbiol. *51:* 377–385.

2. Snell, J.J.S., and S.P. Lapage. 1976. Transfer of some saccharolytic *Moraxella* species to *Kingella* Henriksen and Bøvre 1976, with descriptions of *Kingella indologenes,* sp. nov. and *Kingella denitrificans* sp. nov. Int. J. Syst. Bacteriol. *26:* 451–458.

3. Morrison, V.A., and K.F. Wagner. 1989. Clinical manifestations of *Kingella kingae* infections: case report and review. Rev. Infect. Dis. *11:* 776–782.

4. Graham, D.R., J.D. Band, C. Thornsberry, D.G. Hollis, and R.E. Weaver. 1990. Infections caused by *Moraxella, Moraxella urethralis, Moraxella*-like groups M-5 and M-6, and *Kingella kingae* in the United States, 1953-1980. Rev. Infect. Dis. *12:* 423–431.

Comments

Formerly *Moraxella kingae* and *M. kingii.* May be confused with β-hemolytic *Streptococcus.* Shows some tendency to resist decolorization. Maltose reaction frequently occurs earlier and is stronger than that of glucose. Of 32 patients of known age, 22 (69%) were 5 years of age or younger. Information on the cellular fatty acid composition of this species is presented in Table 4.53 and Figure 4.58 of the cellular fatty acid section.

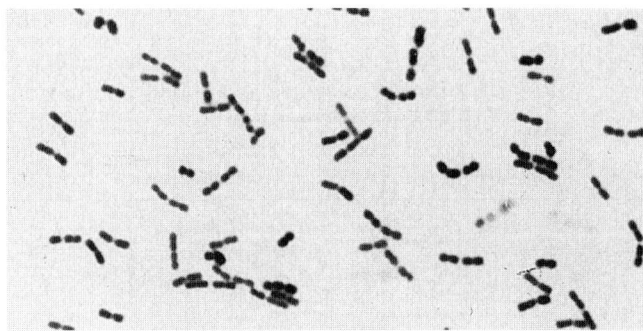

Kingella kingae strain A1702. Gram stain BAP 35° C 24 h × 1700.

Kingella kingae

TEST PERFORMED	SIGN	%+	#+/T
Morphology	cc,pr,chs		
Motility; flagella	[nm]		
Action on blood	v	81 β	77/96; 1 al, 2 ly
Fermentative or oxidative	F[a]		
Carbohydrate base	F[b]		
Acid from:			
D-Glucose	[+ or (+)]	41 (54)	47 (71), 9w (3w)/136
D-Xylose	−	0	0/136
D-Mannitol	−	0	0/136
Lactose	−	0	0/136
Sucrose	[−]	0	0/136
Maltose	[+ or (+)]	71 (29)	86 (39), 10w/136
Catalase	[−]	0	0/135
Oxidase	[+]	100	135/135
Growth on:			
MacConkey	−	9 (1)	8 (1), 4w/137
SS	−	0	0/135
Simmons citrate	−	0	0/132
Urea, Christensen's	−	0	0/108
Nitrate reduction	−	3 (1)	2 (1), 1w/125
Gas from nitrate	−	0	0/95
Nitrite reduction	v[c]	50	9/18
Indole	−	0	0/123
TSI slant, acid	v	21	21, 7w/128
TSI butt, acid	v	11	11, 3w/128
H$_2$S (TSI butt)	−	0	0/131
H$_2$S (Pb ac paper)	v	60	61/101
Gelatin hydrolysis[d]	−	1	1/98
Litmus milk	v	82 pep	111/135; 1 k
Pigment	−	2 br-yel sol	2/104
Growth at:			
25°C	v	34	39/116
35°C	+	98	122/124
42°C	−	10	11/113
Esculin hydrolysis	−	0	0/123
Lysine decarboxylase	−	0	0/8
Arginine dihydrolase	−	0	0/8
Ornithine decarboxylase	−	0	0/8
Nutrient broth, 0% NaCl	v	60	80/134
Nutrient broth, 6% NaCl	−	9	11/123

[a]TSI reaction is not always indicative of fermentative activity.
[b]Rabbit serum (1–2 drops/3 ml of medium) is required.
[c]Some strains reduced 0.01% nitrite but not 0.1% nitrite; some of these strains failed to grow in the higher nitrite concentration.
[d]Incubation of 7–14 days.

Leptotrichia buccalis

Gram-negative long, straight, or slightly curved rod with tapered or curved ends; nonmotile; anaerobic, but may grow aerobically in a candle-jar atmosphere; oxidase- and catalase-negative; does not grow on MacConkey agar; ferments glucose, sucrose, and maltose.

Sources (9 strains)

Blood (8), Unknown (1).

Reference strain

KC1930 ← ATCC 14201, type strain, from supragingival calculus.

Literature

1. Hofstad, T. 1984. Genus III *Leptotrichia* Trevisan 1879, 138.[AL] *In* N.R. Krieg and J.G. Holt (Eds.), Bergey's Manual of Systematic Bacteriology, Vol. 1, Williams & Wilkins, Baltimore, pp. 637–641.

2. Bernard, K., C. Cooper, S. Tessier, and E.P. Ewan. 1991. Use of chemotaxonomy as an aid to differentiate among *Capnocytophaga* species, CDC Group DF-3, and aerotolerant strains of *Leptotrichia buccalis*. J. Clin. Microbiol. *29:* 2263–2265.

3. Weinberger, M., T. Wu, M. Rubin, V.J. Gill, and P.A. Pizzo. 1991. *Leptotrichia buccalis* bacteremia in patients with cancer: report of four cases and review. Rev. Infect. Dis. *13:* 201–206.

Comments

Cells from young cultures may stain partially Gram-positive. Natural inhabitant of the oral cavity of humans. Lactic acid is the major end product of glucose fermentation. End product and cellular fatty acid profile analyses are useful in differentiating *L. buccalis* from *Capnocytophaga* sp. (2). Lactic acid is the major fermentation endproduct of *L. buccalis* and succinic acid is the major fermentation endproduct of *Capnocytophaga* species.

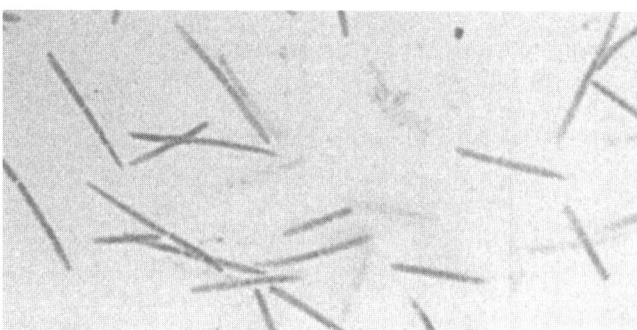

Leptotrichia buccalis strain KC1930. Gram stain BAP 35° C 24 h x 1700.

Leptotrichia buccalis

TEST PERFORMED	SIGN	%+	#+/T
Morphology	[ltrt]		
Motility; flagella	–		
Action on blood	–	0	0/9
Fermentative or oxidative	[F]		
Carbohydrate base	F[a]		
Acid from:			
D-Glucose	[+ or (+)]	88 (12)	7 (1), (1w)/9
D-Xylose	–	0	0/9
D-Mannitol	–	0	0/9
Lactose	v	44 (11)	3 (1), 1w/9
Sucrose	[+ or (+)]	44 (56)	4, (3), (2w)/9
Maltose	[+ or (+)]	67 (33)	5 (1), 1w (2w)/9
Catalase	[–]	0	0/9
Oxidase	[–]	0	0/9
Growth on:			
MacConkey	[–]	0	0/8
SS	–	0	0/7
Simmons citrate	–	0	0/7
Urea, Christensen's	–	0	0/9
Nitrate reduction	–	0	0/8
Indole	–	0	0/9
TSI slant, acid	v	11 (33)	1w (3w)/9
TSI butt, acid	v	11 (33)	1w (3w)/9
H_2S (TSI butt)	–	0	0/9
H_2S (Pb ac paper)	v	11	1/9
Gelatin hydrolysis[b]	–	0	0/8
Litmus milk	v	44 A	4/9; 3 cl, 1 pep, 1 IR
Pigment	–	0	0/6
Growth at:			
25°C	v	17	1/6
35°C	+	100	6/6
42°C	v	17	1/6
Esculin hydrolysis	v	89	8/9
Lysine decarboxylase	–	0	0/1
Arginine dihydrolase	–	0	0/1
Ornithine decarboxylase	–	0	0/1
Nutrient broth, 0% NaCl	–	0	0/5
Nutrient broth, 6% NaCl	–	0	0/5

[a]Carbohydrate reactions may require addition of 1–2 drops of normal rabbit serum/3 ml of enteric fermentation broth.
[b]Incubation of 7–14 days.

Methylobacterium species

Gram-negative pleomorphic vacuolated rod with large intracellular fat bodies; motile (single polar flagellum); aerobic; oxidizes xylose, usually methanol, and sometimes glucose (weakly), but not lactose; oxidase-positive; indole-negative; pink to coral growth pigment.

Sources (90 strains)

Blood (24%), bronchus (9%), sputum (6%), nebulizer (6%), cerebrospinal fluid (3%), eye (3%), water (3%), urine (2%), bone (2%), chest fluid (2%), exudate (2%), other (25%), unknown (13%).

Reference strains

KC1496 ← ATCC 29983, *M. mesophilicum* type strain, from a leaf.
KC1896 ← ATCC 43645, *M. extorquens* type strain, from soil.
KC1876 ← ATCC 27329, *M. radiotolerans* type strain, from rice.
KC1897 ← ATCC 43883, *M. zatmanii* type strain, from a fermenter.
KC1875 ← ATCC 14821, *M. rhodinum* type strain, from *Alnus* rhizosphere.

Literature

1. Stocks, P.K., and C.S. McCleskey. 1964. Identity of the pink pigmented methanol-oxidizing bacteria as *Vibrio extorquens*. J. Bacteriol. *88:* 1065–1070.
2. Austin, B., and M. Goodfellow. 1979. *Pseudomonas mesophilica,* a new species of pink bacteria isolated from leaf surfaces. Int. J. Syst. Bacteriol. *29:* 373–378.
3. Green, P.N., and I.J. Bousfield. 1982. A taxonomic study of some gram-negative facultatively methylotrophic bacteria. J. Gen. Microbiol. *128:* 623–638.
4. Green, P.N., and I.J. Bousfield. 1983. Emendation of *Methylobacterium* Patt, Cole, and Hanson 1976; *Methylobacterium rhodinum* (Heumann 1962) comb. nov. corrig.; *Methylobacterium radiotolerans* (Ito and Iizuka 1971) comb. nov. corrig.; and *Methylobacterium mesophilicum* (Austin and Goodfellow 1979) comb. nov. Int. J. Syst. Bacteriol. *33:* 875–877.
5. Bousfeld, I.J., and P.N. Green. 1985. Reclassification of bacteria of the genus *Protomonas* Urakami and Komagata 1984 in the genus *Methylobacterium* (Patt, Cole, and Hanson) emend. Green and Bousfeld 1983. Int. J. Syst. Bacteriol. *35:* 209.
6. Hood, D.W., C.S. Dow, and P.N. Green. 1987. DNA:DNA hybridization studies on the pink-pigmented facultative methylotrophs. J. Gen. Microbiol. *133:* 709–720.
7. Green, P.N., I.J. Bousfeld, and D. Hood. 1988. Three new *Methylobacterium* species: *M. rhodesianum* sp. nov., *M. zatmanii* sp. nov. and *M. fujisawaense* sp. nov. Int. J. Syst. Bacteriol. *38:* 124–127.

Comments

Methylobacterium mesophilicum and *M. extorquens* formerly have been called *Pseudomonas mesophilica* and *Vibrio extorquens,* respectively. DNA:DNA relatedness studies by Hood, et al. (6) indicate that *M. mesophilicum* and *M. extorquens* are separate species; however, it is not clear that they can be differentiated from each other or from other described species of *Methylobacterium* by phenotypic characteristics. Other described *Methylobacterium* species include *M. organophilum, M. rhodesianum,* and *M. fujisawaense.* In the absence of DNA relatedness studies, therefore, we currently limit identifications to the genus level. Pale-staining Gram-positive members may be observed. May prefer media without blood. May be a cause of systemic illness as a secondary invader. Has been found as a contaminant of laboratory media and is called the "Red Phantom." Information on the cellular fatty acid composition of this group is presented in Table 4.54 and Figure 4.59 of the cellular fatty acid section.

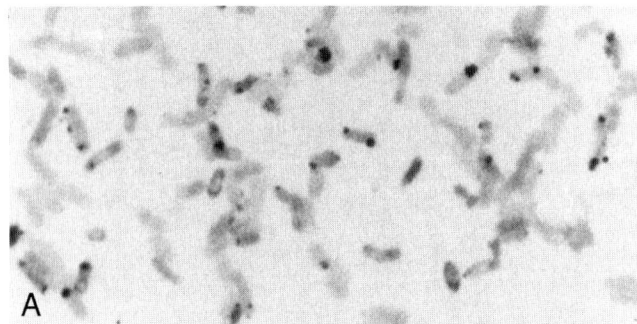

Methylobacterium sp. strain D1912. Gram stain HIA
35° C 24 h x 1700.

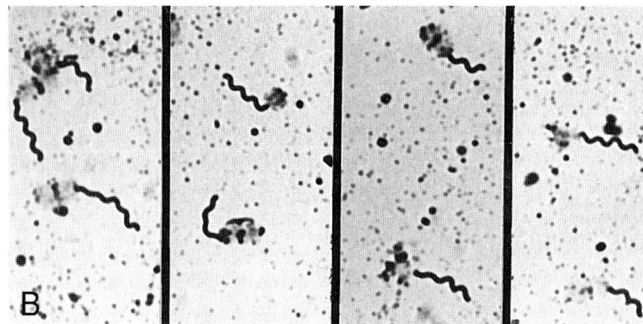

Methylobacterium sp. strain B4139. Flagella stain FB
25° C 24 h x 1700.

Methylobacterium species

TEST PERFORMED	M. mesophilicum KC1496 SIGN	M. extorquens KC1896 SIGN	M. radiotolerans KC1876 SIGN	M. zatmanii KC1897 SIGN	M. rhodinum KC1875 SIGN	PHENOTYPICALLY SIMILAR STRAINS 90 STRAINS SIGN	% +	# +/T
Morphology	[vps-mr]	[vps-mr]	[vps-mr]	[vps-mr]	[vps-mr]	[vps-mr]		
Motility; flagella	[m;p,1]	[m;p,1]	[m;p,1]	[m;p,1]	[m;p,1]	[m;p,1]		
Action on blood	-	-	-	-	-	-	8 ly	4/47; 1 al
Fermentative or oxidative	[n-o]	[n-o]	[O]	[n-o]	[n-o]	[n-o or O]		
Carbohydrate base	OF	OF	OF	OF	OF	OF		
Acid from:								
D-Glucose	-	-	(+w)	-	-	v	40	4, 32w/90
D-Xylose	+w	(+w)	(+w)	+w	(+w)	[+]	94	51, 34w/90
D-Mannitol	-	-	-	-	-	-	2	2w/90
Lactose	-	-	-	-	-	[-]	0	0/90
Sucrose	-	-	-	-	-	-	0	0/90
Maltose	-	-	-	-	-	-	2	2w/90
Catalase	+	+	+	+	+	+	100	82, 6w/88
Oxidase	+	+	+	+	+	[+]	96	72, 3w/78
Growth on:								
MacConkey	+	+	-	-	-	v	15	7, 6w/89
SS	-	-	-	-	-	-	0	0/89
Simmons citrate	(+)	-	-	-	-	-	2 (3)	2 (3)/89
Urea, Christensen's	(+)	(+)	(+)	(+)	(+)	v	29 (26)	23 (20)/78
Nitrate reduction	-	-	+	-	-	v	25	19, 3w/88
Gas from nitrate	-	-	-	-	-	-	0	0/88
Indole	-	-	-	-	-	[-]	0	0/85
TSI slant, acid	-	-	-	-	-	-	0	0/83
TSI butt, acid	-	-	-	-	-	-	0	0/75
H$_2$S (TSI butt)	-	-	-	-	-	-	0	0/84
H$_2$S (Pb ac paper)	+	+	+	+	+	v	47	9, 31w/85
Gelatin hydrolysis[a]	-	-	-	-	-	-	0	0/57
Litmus milk	(k)	kw	-	(k)	-	v	49 k	43/87
Pigment	[pk-col]	[pk-col]	[pk-col]	[pk-col]	[pk-col]	[pk-col]	99	89/90
Growth at:								
25°C	+	+	+	+	+	+	98	81/83
35°C	+	-	+	+	-	+	93	76/82
42°C	-	-	-	-	-	v	12	9/73
Esculin hydrolysis	-	-	-	-	-	-	0	0/81
Nutrient broth, 0% NaCl	+	+	+	+	+	+	93	83/89
Nutrient broth, 6% NaCl	+	-	-	-	-	-	0	0/79
Sodium acetate	-	+	+	(+)	+	+ or (+)	10 (80)	1w (8w)/10
Acetamide	-	-	-	-	-	-	6 (2)	2 (1), 1w/51
Serine	-	-	-	-	-	-	(10)	(5w)/51
Tartrate	-	+	+	+	-	v	12 (14)	6 (7)/51

[a] Incubation of 7–14 days.

Moraxella atlantae

Gram-negative coccoid to medium-length thick rod (may tend to retain crystal violet stain), nonmotile; usually does not grow in OF medium or use carbohydrates; grows on MacConkey agar, otherwise fastidious; aerobic; strongly oxidase-positive, nitrate-negative, urease-negative.

Sources (73 strains)

Blood (59%), cerebrospinal fluid (8%), sputum (4%), leg wound (3%), dialysis fluid (3%), abscess (9%), other (19%) (1 each: ulcer, wound, urine, urethra, hamster, skin, back, gall bladder, ear, eye, bronchial wash, cyst), unknown (4%).

Reference strain

5118 = ATCC 29525, type strain, from blood.

Literature

1. Bøvre, K., J.E. Fuglesang, N. Hagen, E. Jansen, and L.O. Frøholm. 1976. *Moraxella atlantae* sp. nov. and its distinction from *Moraxella phenylpyruvica*. Int. J. Syst. Bacteriol. 26: 511–521.

2. Buchman, A.L., and M.J. Pickett. 1991. *Moraxella atlantae* bacteraemia in a patient with systemic lupus erythematosis. J. Infect. 23: 197–199.

3. Graham, D.R., J.D. Band, C. Thornsberry, D.G. Hollis, and R.E. Weaver. 1990. Infections caused by *Moraxella* species, *Moraxella urethralis*, *Moraxella*-like groups M-5 and M-6, and *Kingella kingae* in the United States, 1953–1980. Rev. Infect. Dis. 12: 423–431.

Comments

Formerly CDC Group M-3. Usually two colony variants can be observed, one convex with entire edges, the other flatter, with irregular edges and a tendency to spread; may pit the agar. Information on the cellular fatty acid composition of this species is presented in Table 4.55 and Figure 4.60 of the cellular fatty acid section.

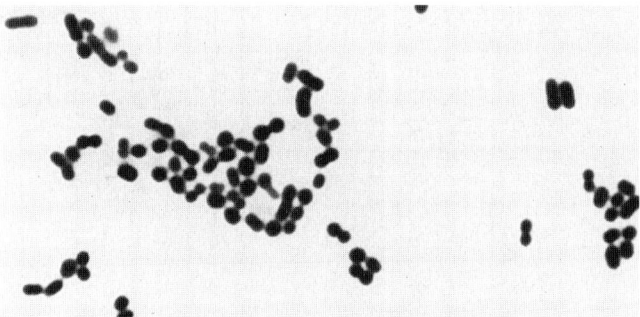

Moraxella atlantae strain B7333. Gram stain BAP 35° C 24 h x 1700.

Moraxella atlantae

TEST PERFORMED	SIGN	% +	#+/T
NUMBER OF STRAINS 73			
Morphology	sbr,cc		
Motility; flagella	[nm]		
Action on blood	–	3 al	2/63; 1 ly, 1 LG
Fermentative or oxidative	[n-o]		
Carbohydrate base	OF		
Acid from:			
D-Glucose	[–][a]	0	0/59
D-Xylose	–	0	0/59
D-Mannitol	–	0	0/59
Lactose	–	0	0/59
Sucrose	–	0	0/59
Maltose	–	0	0/59
Catalase	+	95	48, 21w/73
Oxidase	[+]	100	73/73
Growth on:			
MacConkey	[+ or (+)]	80 (20)	52 (11), 6w (4w)/73
SS	–	0	0/73
Cetrimide	–	0	0/73
Simmons citrate	–	0	0/73
Urea, Christensen's	[–]	0	0/73
Nitrate reduction	[–]	0	0/73
Gas from nitrate	–	0	0/73
Nitrite reduction	–	3	1/38
Indole	–	0	0/73
TSI slant, acid	–	0	0/73
TSI butt, acid	–	0	0/73
H_2S (TSI butt)	–	0	0/56
H_2S (Pb ac paper)	v	61	24, 10w/56
Gelatin hydrolysis[b]	–	0	0/61
Litmus milk	–	3 IR	2/68
Pigment	–	0	0/40
Growth at:			
25°C	v	51	37/72
35°C	+	99	71/72
42°C	v	46	33/72
Esculin hydrolysis	–	0	0/56
Nutrient broth, 0% NaCl	–	0	0/56
Nutrient broth, 6% NaCl	–	0	0/56
Phenylalanine deaminase	–	0	0/44
Penicillin sensitivity[c]	+	100	54/54

[a] Usually does not grow in OF medium.
[b] Incubation of 7–14 days.
[c] Based on results obtained by streaking a blood agar plate with growth from an 18- to 36- h culture and then placing a 10-Unit penicillin disk on the streaked area. A positive reaction is indicated by the appearance of a zone of inhibition.

Moraxella bovis

Gram-negative coccoid to filamentous thick rod (may tend to retain crystal violet stain); nonmotile; fastidious; aerobic; asaccharolytic, hemolytic, strongly oxidase-positive; does not grow on MacConkey agar; nitrate-negative, urease-negative, gelatin-positive (serum may need to be added).

Sources (7 strains)

Bovine eye (5), bovine nose (1), unknown (1).

Reference strain

KC416(1) ← ATCC 17947 = NCTC 8561, smooth form.

Literature

1. Hughes, D.E., and G.W. Pugh. 1970. Isolation and description of a *Moraxella* from horses with conjunctivitis. Am. J. Vet. Res. *31:* 457–462.

2. Henriksen, S.D. 1973. *Moraxella, Acinetobacter,* and the *Mimeae.* Bacteriol. Rev. *37:* 522–561.

3. Inzana, T.J. 1990. Miscellaneous glucose-nonfermenting gram-negative bacteria. *In* G.R. Carter and J.R. Cole (Eds.), Diagnostic Procedures in Veterinary Bacteriology and Mycology, 5th ed., Academic Press, New York, pp. 165–176.

Comments

Penicillin-sensitive. Causes "pink-eye" in cattle. A nonhemolytic variant of this organism is associated with conjunctivitis in horses and has been referred to as *M. equi* (3). Information on the cellular fatty acid composition of this species is presented in Table 4.56 and Figure 4.61 of the cellular fatty acid section.

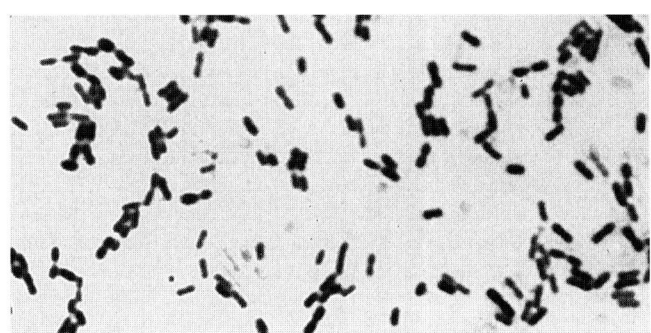

Moraxella bovis strain KC416(1). Gram stain BAP 35° C 24 h x 1700.

Moraxella bovis

TEST PERFORMED	SIGN	% +	#+/T
	NUMBER OF STRAINS 7		
Morphology	cc,sbr		
Motility; flagella	[nm]		
Action on blood	[β]	100	7/7
Fermentative or oxidative	n-o		
Carbohydrate base	OF		
Acid from:			
D-Glucose	[-]	0	0/7
D-Xylose	-	0	0/2
D-Mannitol	-	0	0/2
Lactose	-	0	0/6
Sucrose	-	0	0/2
Maltose	-	0	0/3
Catalase	v	14	1w/7
Oxidase	[+]	100	7/7
Growth on:			
MacConkey	[-]	0	0/7
SS	-	0	0/7
Simmons citrate	-	0	0/7
Urea, Christensen's	[-]	0	0/7
Phenylalanine deaminase	-	0	0/1
Nitrate reduction	v	14	1/7
Indole	-	0	0/7
TSI slant, acid	-	0	0/7
TSI butt, acid	-	0	0/7
H_2S (TSI butt)	-	0	0/6
H_2S (Pb ac paper)	+	100	4, 2w/6
Gelatin hydrolysis[a]	[+]	100	7/7
Litmus milk	pep	100	7/7
Growth at:			
25°C	+	100	7/7
35°C	+	100	7/7
42°C	-	0	0/7
Penicillin sensitivity[b]	+	100	2/2

[a]Incubation of 7–14 days.
[b]Based on results obtained by streaking a blood agar plate with growth from an 18- to 36- h culture and then placing a 10-Unit penicillin disk on the streaked area. A positive reaction is indicated by the appearance of a zone of inhibition.

Moraxella (Branhamella) catarrhalis

Gram-negative coccus; nonmotile; asaccharolytic; catalase-positive, strongly oxidase-positive; usually reduces nitrate and nitrite without gas; usually grows on nutrient agar at 35° and 25° C; grows only rarely on MacConkey agar; amylosucrase-negative.

Sources (74 strains)

Eye (26%), sputum (7%), blood (7%), ear (5%), throat (5%), nose (4%), other (15%), unknown (31%).

Reference strain

KC917 ← S. Branham, N-9 = ATCC 25240.

Literature

1. Henriksen, S.D., and K. Bøvre. 1968. The taxonomy of the genera *Moraxella* and *Neisseria*. J. Gen. Microbiol. *51:* 387–392.

2. Bøvre, K. 1984. Genus II. *Moraxella* Lwoff 1939, 173 emend. Henriksen and Bøvre 1968, 391.[AL] *In* N.R. Krieg and J.G. Holt (Eds.), Bergey's Manual of Systematic Bacteriology, Vol. 1, Williams & Wilkins, Baltimore, pp. 296–303.

3. Doern, G.V. 1986. *Branhamella catarrhalis*—an emerging human pathogen. Diagn. Microbiol. Infect. Dis. *4:* 191–201.

4. Rossau, R., A. Van Landschoot, M. Gillis, and J. De Ley. 1991. Taxonomy of *Moraxellaceae* fam. nov., a new bacterial family to accommodate the genera *Moraxella, Acinetobacter,* and *Psychrobacter* and related organisms. Int. J. Syst. Bacteriol. *41:* 310–319.

5. Vaneechoutte, M., G. Verschraegen, G. Claeys, and P. Flamen. 1988. Rapid identification of *Branhamella catarrhalis* with 4-methylumbelliferyl butyrate. J. Clin. Microbiol. *26:* 1227–1228.

6. Janda, W.M., and P. Ruther. 1989. B.CAT CONFIRM, a rapid test for confirmation of *Branhamella catarrhalis*. J. Clin. Microbiol. *27:* 1130–1131.

Comments

Formerly *Neisseria catarrhalis*. This organism is now classified in the subgenus *Branhamella* of the genus *Moraxella* with the name *Moraxella (Branhamella) catarrhalis* (2). Recent DNA-rRNA hybridization studies have led to the proposal that the subgenus be abandoned and that this group be considered an unequivocal *Moraxella* species (4). Usually, no growth pigment occurs on Loeffler blood serum medium. A few strains that have been identified as *M. catarrhalis* have grown on Thayer-Martin medium. Deoxyribonuclease- and butyrate esterase-positive. Information on the cellular fatty acid composition of this species is presented in Table 4.57 and Figure 4.62 of the cellular fatty acid section.

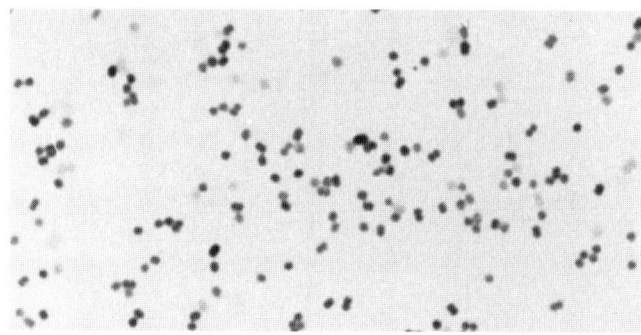

Moraxella (Branhamella) catarrhalis strain KC917. Gram stain BAP 35° C 24 h x 1700.

Moraxella (Branhamella) catarrhalis[a]

TEST PERFORMED	SIGN	% +	#+/T
	NUMBER OF STRAINS 74		
Morphology	[cd]		
Motility; flagella	[nm]		
Action on blood	v	32 ly	7/32
Fermentative or oxidative	n-o		
Carbohydrate base	CTA		
Acid from:			
D-Glucose	[-]	0	0/69
D-Xylose	-	0	0/15
D-Mannitol	-	0	0/66
Lactose	-	0	0/66
Sucrose	-	0	0/70
Maltose	[-]	0	0/69
Catalase	+	100	37/37
Oxidase	[+]	100	73/73
Growth on:			
MacConkey	-	5	2/39
SS	-	0	0/39
Simmons citrate	-	0	0/38
Urea, Christensen's	-	0	0/40
Phenylalanine deaminase	v	68	19, 5w/34
Nitrate reduction	[+]	92	68/74
Gas from nitrate	-	0	0/74
Nitrite reduction	v	86	32/37
Indole	-	0	0/37
TSI slant, acid	-	0	0/29
TSI butt, acid	-	0	0/29
H$_2$S (TSI butt)	-	0	0/31
H$_2$S (Pb ac paper)	v	73	3, 21w/33
Gelatin hydrolysis[b]	-	0	0/31
Litmus milk	-	6k	2w/36; 2 IR
Pigment	v	25 sl yel	15 sl yel/60
Growth at:			
25°C	v	85	28/33
35°C	+	97	30/31
42°C	v	23	5/22
Esculin hydrolysis	-	0	0/14
Nutrient broth, 0% NaCl	v	47	8/17

[a]See chart 46 "Differentiation *Moraxella* (*Branhamella*) *catarrhalis* and *Neisseria* species with coccoid morphology " (page 220) for distinguishing *M. catarrhalis* and *Neisseria* species.
[b]Incubation of 7–14 days.

Moraxella lacunata

Gram-negative lanceolate coccoid to medium length rod of varying length in pairs or short chains; nonmotile; aerobic; asaccharolytic; strongly oxidase-positive, nitrate-positive; sodium acetate-negative; urease-negative; digests Loeffler blood serum agar (obtained commercially); usually no growth in nutrient broth.

Sources (66 strains)

Eye (74%), blood (14%), other (12%).

Reference strain

KC784 ← ATCC, 17967, neotype strain, from eye.

Literature

1. Lwoff, A. 1939. Revision et démembrement des *Hemophilae* le genre *Moraxella* nov. gen. Ann. Inst. Pasteur, Paris *62:* 168–176.

2. Henriksen, S.D. 1969. Proposal of a neotype strain for *Moraxella lacunata*. Int. J. Syst. Bacteriol. *19:* 263.

3. Judicial Commission. 1971. Opinion 41. Conservation of the generic name *Moraxella* Lwoff. Int. J. Syst. Bacteriol. *21:* 106.

4. Henriksen, S.D. 1973. *Moraxella, Acinetobacter,* and the *Mimeae*. Bacteriol. Rev. *37:* 522–561.

5. Moss, C.W., P.L. Wallace, D.G. Hollis, and R.E. Weaver. 1988. Cultural and chemical characterization of CDC groups EO-2, M-5, and M-6, *Moraxella (Moraxella)* species, *Oligella urethralis, Acinetobacter* species, and *Psychrobacter immobilis*. J. Clin. Microbiol. *26:* 484–492.

6. Rossau, R., A. Van Landschoot, M. Gillis, and J. De Ley. 1991. Taxonomy of *Moraxellaceae* fam. nov., a new bacterial family to accommodate the genera *Moraxella, Acinetobacter,* and *Psychrobacter* and related organisms. Int. J. Syst. Bacteriol. *41:* 310–319.

7. Tønjum, T., D.A. Caugant, and K. Bøvre. 1992. Differentiation of *Moraxella nonliquefaciens, M. lacunata,* and *M. bovis* by using multilocus enzyme electrophoresis and hybridization with pilin-specific probes. J. Clin. Microbiol. *30:* 3099–3107.

Comments

Moraxella lacunata is the type species; synonyms are *Moraxella liquefaciens* and *Diplobacillus moraxaxenfeld*. Penicillin-sensitive. Associated with conjunctivitis. Two biogroups, based on colony morphology and fatty acid profile, have been recognized (5). Information on the cellular fatty acid composition of this group is presented in Table 4.56 and Figure 4.61 of the cellular fatty acid section.

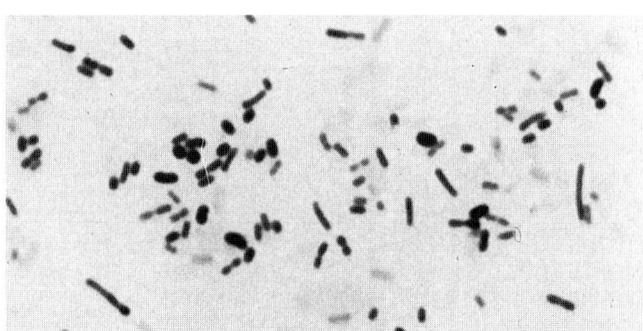

Moraxella lacunata strain KC1376. Gram stain BAP 35° C 24 h x 1700.

Moraxella lacunata

TEST PERFORMED	SIGN	% +	#+/T
NUMBER OF STRAINS 66			
Morphology	cc,mr		
Motility; flagella	[nm]		
Action on blood	v	24 al	14/59; 6 ly, 1 gr
Fermentative or oxidative	n-o		
Carbohydrate base	OF[a]		
Acid from:			
D-Glucose	[-]	0	0/46
D-Xylose	-	0	0/46
D-Mannitol	-	0	0/46
Lactose	-	0	0/46
Sucrose	-	0	0/46
Maltose	-	0	0/46
Catalase	+	100	46, 16w/62
Oxidase	[+]	100	66/66
Growth on:			
MacConkey	-	2	1w/66
SS	-	0	0/66
Simmons citrate	-	0	0/65
Urea, Christensen's	[-]	0	0/54
Nitrate reduction	[+]	98	63/64
Gas from nitrate	-	0	0/64
Nitrite reduction	-	0	0/2
Indole	-	0	0/58
TSI slant, acid	-	0	0/45
TSI butt, acid	-	0	0/45
H_2S (TSI butt)	-	0	0/45
H_2S (Pb ac paper)	v	34	6, 10w/47
Gelatin hydrolysis[b]	v	42	27/64
Litmus milk	v	48 pep	30/63; 1 k
Loeffler slant, digestion	[+]	100	65/65
Pigment	-	2 br-sol	1/66
Growth at:			
25°C	v	33	18/54
35°C	v	73	41/56
42°C	-	0	0/54
Esculin hydrolysis	-	0	0/148
Nutrient broth, 0% NaCl	[-]	5	2, 1w/57
Nutrient broth, 6% NaCl	-	2	1/57
Phenylalanine deaminase	v	17	5/30
Alkalinization of:			
Sodium acetate	[-]	0	0/25
Acetamide	-	0	0/15
Serine	-	0	0/15
Tartrate	-	0	0/15
Penicillin sensitivity[c]	[+]	95	38/40

[a]Reactions neutral (does not grow or grows poorly) in King's OF medium.
[b]Incubation of 7–14 days.
[c]Based on results obtained by streaking a blood agar plate with growth from an 18- to 36- h culture and then placing a 10-unit penicillin disk on the streaked area; a positive reaction is indicated by the appearance of a zone of inhibition.

Moraxella lincolnii

Gram-negative coccus-like to plump rod; often occurs in pairs; may form short chains; nonmotile; oxidase-positive; asaccharolytic; does not grow on MacConkey agar; urease-, indole-, and nitrate-negative; alkaline phosphatase-negative.

Source (type strain)

Human nasopharynx.

Reference strain

KC1956 ← CCUG 9405 (Culture Collection, University of Göteborg, Göteborg, Sweden), type strain, nasopharynx.

Literature

1. Vandamme, P., M. Gillis, M. Vancanneyt, B. Hoste, K. Kersters, and E. Falsen. 1993. *Moraxella lincolnii*, sp. nov., isolated from the human respiratory tract, and reevaluation of the taxonomic position of *Moraxella osloensis*. Int. J. Syst. Bacteriol. *43:* 474–481.

Comments

Grows best at 30° C; prefers Chocolate II (BBL) agar over rabbit blood agar or BCYE agar. Although the type strain is nitrite-negative, nitrite-positive strains have been described (1). The cellular fatty acid composition of this species is presented in Table 4.56 and Figure 4.61 of the cellular fatty acid section.

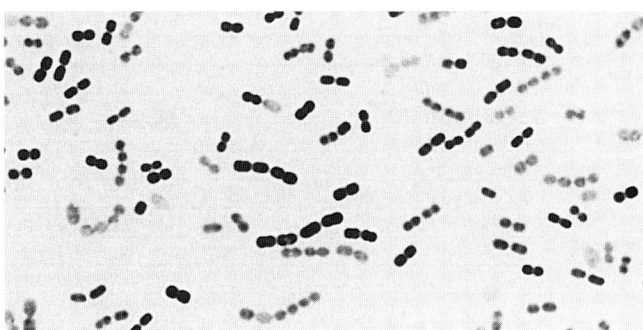

Moraxella lincolnii strain KC1956. Gram stain BAP 35° C 24 h x 1700.

Moraxella lincolnii

TYPE STRAIN	
TEST PERFORMED	SIGN
Morphology	[cc,ch]
Motility; flagella	[nm]
Action on blood	–
Fermentative or oxidative	[n-o]
Carbohydrate base	OF
Acid from:	
D-Glucose	[–]
D-Xylose	–
D-Mannitol	–
Lactose	–
Sucrose	–
Maltose	–
Catalase	+
Oxidase	[+]
Growth on:	
MacConkey	–
SS	–
Cetrimide	–
Simmons citrate	–
Urea, Christensen's	–
Nitrate reduction	[–]
Gas from nitrate	–
Nitrite reduction	–[a]
Indole	–
TSI slant, acid	–[b]
TSI butt, acid	–[b]
H_2S (TSI butt)	
H_2S (Pb ac paper)	
Gelatin hydrolysis[c]	–
Litmus milk	–
Pigment	–
Growth at:	
25°C	+
35°C	+
42°C	–
Esculin hydrolysis	–
Nutrient broth, 0% NaCl	[–]
Nutrient broth, 6% NaCl	–
Alkalinization of:	
Acetamide	–
Serine	–
Tartrate	–
Sodium acetate	[–]
Alkaline phosphatase	–

[a]Nitrite positive strains have been described.
[b]No growth was detected.
[c]Incubation of 7–14 days.

Moraxella nonliquefaciens

Gram-negative, deeply staining coccoid to filament-length thick rod (may tend to retain crystal violet stain), in pairs and short chains; nonmotile; aerobic; asaccharolytic; strongly oxidase-positive, usually MacConkey-negative; nitrate-positive, urease-negative, indole- and sodium acetate-negative; no digestion of Loeffler blood serum agar (obtained commercially).

Sources (243 strains)

Nose (21%), throat (12%), eye (11%), nasopharynx (9%), sputum (8%), blood (6%), other (33%).

Reference strains

KC129 ← Henriksen.
KC770 ← ATCC 17953 (NCTC 7784).

Literature

1. Lwoff, A. 1939. Revision et démembrement des *Hemophilae* le genre *Moraxella* nov. gen. Ann. Inst. Pasteur, Paris 62: 168–176.
2. Bøvre, K., and S.D. Henriksen. 1967. A new *Moraxella* species, *Moraxella osloensis,* and a revised description of *Moraxella nonliquefaciens.* Int. J. Syst. Bacteriol. *17:* 127–135.
3. Henriksen, S.D. 1973. *Moraxella, Acinetobacter,* and the *Mimeae.* Bacteriol. Rev. *37:* 522–561.
4. Graham, D.R., J.D. Band, C. Thornsberry, D.G. Hollis, and R.E. Weaver. 1990. Infections caused by *Moraxella, Moraxella urethralis, Moraxella*-like groups M-5 and M-6, and *Kingella kingae* in the United States, 1953-1980. Rev. Infect. Dis. *12:* 423–431.

Comments

Usually penicillin-sensitive. *Moraxella osloensis* transformation test-negative. Does not grow (or grows poorly) in King's OF medium, on MacConkey agar, and in nutrient broth. Information on the cellular fatty acid composition of this species is presented in Table 4.56 and Figure 4.61 of the cellular fatty acid section.

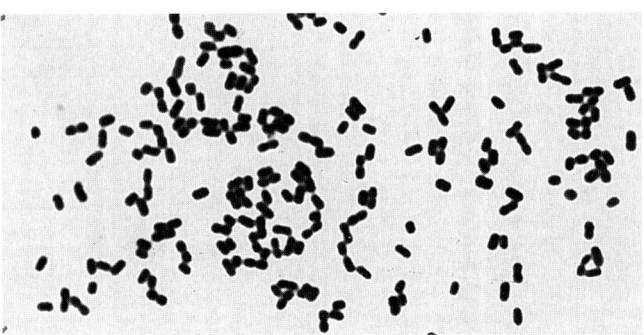

Moraxella nonliquefaciens strain KC129. Gram stain BAP 35° C 24 h x 1700.

Moraxella nonliquefaciens

TEST PERFORMED	SIGN	% +	#+/T
Morphology	cc,sbr		
Motility; flagella	[nm]		
Action on blood	v	10 ly	21/220; 11 al, 3 gr, 2 LG
Fermentative or oxidative	n-o		
Carbohydrate base	OF[a]		
Acid from:			
D-Glucose	[-]	0	0/142
D-Xylose	-	0	0/142
D-Mannitol	-	0	0/144
Lactose	-	0	0/144
Sucrose	-	0	0/143
Maltose	-	0	0/142
Catalase	+	95	204/214
Oxidase	[+]	100	237/237
Growth on:			
MacConkey	[-]	8 (2)	5, 15w (4)/239
SS	-	0	0/239
Simmons citrate	-	0	0/237
Urea, Christensen's	[-]	0	0/201
Nitrate reduction	[+]	95	225/237
Gas from nitrate	-	0	0/237
Indole	-	0	0/237
TSI slant, acid	-	0	0/227
TSI butt, acid	-	0	0/229
H_2S (TSI butt)	-	0	0/230
H_2S (Pb ac paper)	v	83	102, 89w/231
Gelatin hydrolysis[b]	-	0	0/224
Loeffler slant, digestion	[-]	0	0/104
Litmus milk	-	2 (1) k	5, 1w (3)/233
Pigment	-	3 yel sol	4/117
Growth at:			
25°C	+	93	204/219
35°C	v	88	196/223
42°C	v	15	34/224
Esculin hydrolysis	-	0	0/170
Nutrient broth, 0% NaCl	v	22	39/175
Nutrient broth, 6% NaCl	-	0	0/175
Penicillin sensitivity[c]	+	99	136/137
Alkalinization of sodium acetate	[-]	0	0/16

[a]Reactions neutral (does not grow or grows poorly) in King's OF medium.
[b]Incubation of 7–14 days.
[c]Based on results obtained by streaking a blood agar plate with growth from an 18- to 36- h culture and then placing a 10-unit penicillin disk on the streaked area. A positive reaction is indicated by the presence of a zone of inhibition.

Moraxella osloensis

Gram-negative deeply staining coccoid form to thick rod occurring in pairs and short chains, with filaments sometimes observed; may tend to retain crystal violet stain; nonmotile; aerobic; asaccharolytic; strongly oxidase-positive; indole-negative; nitrite-negative; sodium acetate-positive; urease-negative; grows in nutrient broth.

Sources (163 strains)

Blood (17%), genital (8%), throat (7%), cerebrospinal fluid (7%), urine (7%), wound (5%), pyogenic lesion (5%), sputum (3%), chest cavity (3%), eye (2%), nasopharyngeal (2%), rectal (2%), other (20%), unknown (12%).

Reference strain

KC1088 ← ATCC 19976 = NCTC 10465, type strain, from cerebrospinal fluid.

Literature

1. Bøvre, K., and S.D. Henriksen. 1967. A new *Moraxella* species, *Moraxella osloensis,* and a revised description of *Moraxella nonliquefaciens*. Int. J. Syst. Bacteriol. 17: 127–135.

2. Baumann, P., M. Doudoroff, and R.Y. Stanier. 1968. A study of the *Moraxella* group. II. Oxidative-negative species (genus *Acinetobacter*). J. Bacteriol. 95: 1520–1541.

3. Henriksen, S.D. 1973. *Moraxella, Acinetobacter,* and the *Mimeae*. Bacteriol. Rev. 37: 522–561.

4. Graham, D.R., J.D. Band, C. Thornsberry, D.G. Hollis, and R.E. Weaver. 1990. Infections caused by *Moraxella, Moraxella urethralis, Moraxella*-like groups M-5 and M-6, and *Kingella kingae* in the United States, 1953–1980. Rev. Infect. Dis. 12: 423–431.

5. Vandamme, P., M. Gillis, M. Vancanneyt, B. Hoste, K. Kersters, and E. Falsen. 1993. *Moraxella lincolnii,* sp. nov., isolated from the human respiratory tract, and reevaluation of the taxonomic position of *Moraxella osloensis*. Int. J. Syst. Bacteriol. 43: 474–481.

Comments

Usually penicillin sensitive. *Moraxella osloensis* transformation test recommended to confirm identification. An occasional strain may appear "fuzzy" in motility medium, suggesting motility; however, no flagella could be demonstrated. This taxon includes many of the strains formerly called *Mima polymorpha* variety oxidans. Information on the cellular fatty acid composition of this species is presented in Table 4.58 and Figure 4.63 of the cellular fatty acid section.

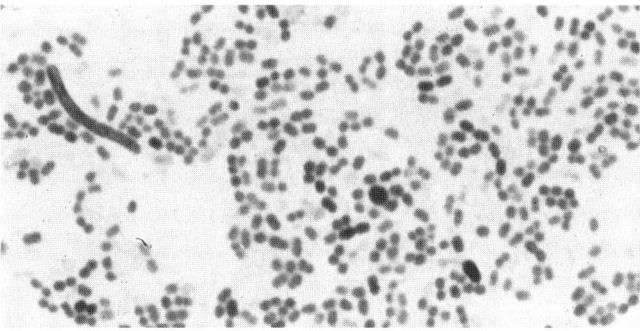

Moraxella osloensis strain KC1088. Gram stain BAP 35° C 24 h x 1700.

Moraxella osloensis

TEST PERFORMED	SIGN	% +	#+/T
Morphology	cc,sbr		
Motility; flagella	[nm]		
Action on blood	v	7 al	11/147; 10 ly, 8 gr, 1 ß
Fermentative or oxidative	n-o		
Carbohydrate base	OF		
Acid from:			
D-Glucose	[-]	0	0/156
D-Xylose	-	0	0/156
D-Mannitol	-	0	0/154
Lactose	-	0	0/154
Sucrose	-	0	0/154
Maltose	-	0	0/154
Catalase	+	95	151/159
Oxidase	[+]	100	158/158
Growth on:			
MacConkey	v	70	111/158
SS	-	0	0/156
Simmons citrate	-	0	0/156
Urea, Christensen's	[-]	0	0/158
Nitrate reduction	v	24	38/159
Gas from nitrate	-	0	0/159
Nitrite reduction	-	0	0/9
Indole	-	0	0/158
TSI slant, acid	-	0	0/159
TSI butt, acid	-	0	0/159
H_2S (TSI butt)	-	0	0/149
H_2S (Pb ac paper)	v	74	114/153
Gelatin hydrolysis[a]	-	0	0/120
Litmus milk	-	9 k	14/155; 2 IR
Pigment	-	6 yel-br sol	10/56
Growth at:			
25°C	+	96	152/158
35°C	+	98	154/157
42°C	v	51	80/157
Esculin hydrolysis	-	0	0/158
Nutrient broth, 0% NaCl	[+]	98	155/158
Nutrient broth, 6% NaCl	v	12	19/158
Penicillin sensitivity[b]	+	92	80/87
Phenylalanine deaminase	v	14	4/29
Alkalinization of sodium acetate	[+]	100	16/16

[a]Incubation of 7–14 days.
[b]Based on results obtained by streaking a blood plate with growth from an 18- to 36- h culture and then placing a 10-unit penicillin disk on the streaked area. A positive reaction is indicated by the appearance of a zone of inhibition.

Moraxella phenylpyruvica

Gram-negative coccoid to filamentous thick rod (may tend to retain crystal violet stain); nonmotile; fastidious; aerobic; asaccharolytic; strongly oxidase-positive; catalase-positive; urease-positive; deaminates phenylalanine to phenylpyruvic acid; sodium acetate-negative, nitrite-negative.

Sources (52 strains)

Urine (17%), urogenital tract (15%), blood (13%), cerebrospinal fluid (8%), superficial lesion (6%), peritoneal fluid (6%), fowl (4%), pyogenic lesion (4%), ear (4%), other (19%), unknown (4%).

Reference strain

2863 = ATCC 23333, type strain, from blood.

Literature

1. Bøvre, K., and Henriksen, S.D. 1967. A revised description of *Moraxella polymorpha* Flamm 1957, with a proposal of a new name, *Moraxella phenylpyruvica,* for this species. Int. J. Syst. Bacteriol. *17:* 343–360.

2. Judicial Commission. 1971. Opinion 42. Conservation of the specific epithet *"phenylpyruvica"* in the name *Moraxella phenylpyruvica* Bøvre and Henriksen. Int. J. Syst. Bacteriol. *21:* 107.

3. Snell, J.J., L.R. Hill, and S.P. Lapage. 1972. Identification and characterization of *Moraxella phenylpyruvica.* J. Clin. Pathol. *25:* 959–965.

4. Graham, D.R., J.D. Band, C. Thornsberry, D.G. Hollis, and R.E. Weaver. 1990. Infections caused by *Moraxella, Moraxella urethralis, Moraxella*-like groups M-5 and M-6, and *Kingella kingae* in the United States, 1953-1980. Rev. Infect. Dis. *12:* 423–431.

5. Batchelor, B.I., R.J. Brindle, G.F. Gilks, and J.B. Selkon. 1992. Biochemical mis-identification of *Brucella melitensis* and subsequent laboratory-acquired infections. J. Hosp. Infect. *22:* 159–162.

6. Rossau, R., A. Van Landschoot, M. Gillis, and J. De Ley. 1991. Taxonomy of *Moraxellaceae* fam. nov., a new bacterial family to accommodate the genera *Moraxella, Acinetobacter,* and *Psychrobacter* and related organisms. Int. J. Syst. Bacteriol. *41:* 310–319.

Comments

Misidentification of *Brucella melitensis* as *M. phenylpyruvica* by commercial identification systems has been reported (5). Cellular morphology is useful in differentiating between these agents. We have also studied approximately 50 *M. phenylpyruvica*-like isolates that share the same biochemical profile with *M. phenylpyruvica,* but produce an incompatible cellular fatty acid profile. *M. phenylpyruvica* and some asaccharolytic *Psychrobacter immobilis* strains share phenotypic characteristics; however, cellular fatty acid analysis and the transformation test are helpful in the differentiation of these organisms. Recent molecular taxonomic studies suggest that *M. phenylpyruvica* should be removed from the genus *Moraxella* (6). Information on the cellular fatty acid composition of this species is presented in Table 4.59 and Figure 4.64 of the cellular fatty acid section.

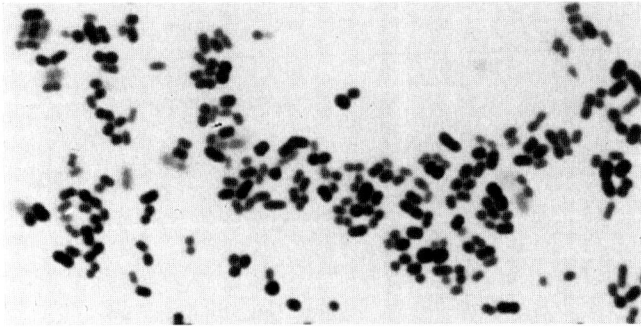

Moraxella phenylpyruvica strain 2863. Gram stain BAP 35° C 24 h x 1700.

Moraxella phenylpyruvica

TEST PERFORMED	SIGN	% +	#+/T
Morphology	[cc,br]		
Motility; flagella	[nm]		
Action on blood	v	16 ly	8/46; 6 al, 2 gr, 1 LG
Fermentative or oxidative	[n-o]		
Carbohydrate base	OF		
Acid from:			
D-Glucose	[-]	0	0/30
D-Xylose	-	0	0/30
D-Mannitol	-	0	0/30
Lactose	-	0	0/30
Sucrose	-	0	0/30
Maltose	-	0	0/30
Catalase	+	90	30, 14w/49
Oxidase	[+]	100	50/50
Growth on:			
MacConkey	v	80 (6)	40 (3)/50
SS	-	0	0/50
Simmons citrate	-	0	0/50
Urea, Christensen's	[+]	100	50/50
Nitrate reduction	v	68	34/50
Gas from nitrate	-	0	0/50
Nitrite reduction	-	0	0/3
Indole	-	0	0/48
TSI slant, acid	-	0	0/50
TSI butt, acid	-	0	0/50
H_2S (TSI butt)	-	0	0/47
H_2S (Pb ac paper)	v	47	16, 7w/49
Gelatin hydrolysis[a]	-	0	0/43
Litmus milk	v	22 (20) k	10, 2w (9)/46
Pigment	-	8 yel-br sol	4/50
Growth at:			
25°C	v	85	41/48
35°C	+	100	48/48
42°C	v	29	14/48
Esculin hydrolysis	-	0	0/44
Nutrient broth, 0% NaCl	v	53	21/40
Nutrient broth, 6% NaCl	v	19	8/43
Penicillin sensitivity[b]	v	73	19/26
Alkalinization of sodium acetate	v	43	5, 4w/21
Phenylalanine deaminase	[+]	97	38/39

[a] Incubation of 7–14 days.

[b] Based on results obtained by streaking a blood agar plate with growth from an 18- to 36- h culture and then placing a 10-unit penicillin disk on the streaked area. A positive reaction is indicated by the appearance of a zone of inhibition.

Neisseria canis

Gram-negative coccus; nonmotile; oxidase-positive; light growth on MacConkey agar; catalase-positive; nitrate-positive; urease-negative; produces acid from glucose in CTA base (weak and late) but not lactose, sucrose, or maltose; produces a yellow growth pigment.

Source (Reference strain)

Dog pharynx.

Reference strain

KC811 ← ATCC 14687, type strain, from dog pharynx.

Literature

1. Berger, U. 1962. Ueber das Vorkommen von Neisserien bei einigen. Tieren. Z. Hyg. *148:* 445–457.
2. Hoke, C., and N.A. Vedros. 1982. Characterization of atypical aerobic gram-negative cocci isolated from humans. J. Clin. Microbiol. *15:* 906–914.
3. Guibourdenche, M., T. Lambert, and J.Y. Riou. 1989. Isolation of *Neisseria canis* in mixed culture from a patient after a cat bite. J. Clin. Microbiol. *27:* 1673–1674.
4. Vedros, N.A. 1984. Genus I. *Neisseria* Trevisan 1885, 105[AL]. *In* N.R. Krieg and J.G. Holt (Eds.), Bergey's Manual of Systematic Bacteriology, Vol. 1, Williams & Wilkins, Baltimore, pp. 290–296.

Comments

Isolated from throats of dogs and cats. Associated with bite wound infections. Biochemical and cellular fatty acid profiles are similar to EF-4b, but these organisms can be differentiated by cellular morphology. A weakly acidic glucose reaction was also found by Guibourdenche et al. (3), but others, (2, 4) report the reaction to be negative. Information on the cellular fatty acid composition of this species is presented in Table 4.12B and Figure 4.17 of the cellular fatty acid section.

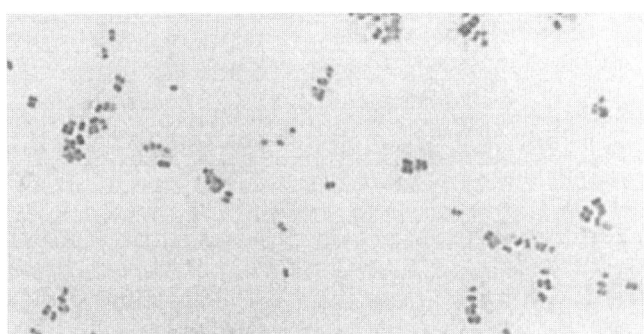

Neisseria canis strain KC811. Gram stain BAP 35° C 24 h x 1700.

Neisseria canis

TEST PERFORMED	SIGN
Morphology	[cd]
Motility; flagella	[nm]
Action on blood	−
Fermentative or oxidative	O
Carbohydrate base	CTA
Acid from:	
D-Glucose	(+)
Fructose	−
D-Mannitol	−
Lactose	−
Sucrose	−
Maltose	−
Catalase	[+]
Oxidase	[+]
Growth on:	
MacConkey	(+)
SS	−
Simmons citrate	−
Urea, Christensen's	[−]
Nitrate reduction	[+]
Gas from nitrate	−
Indole	−
TSI slant, acid	−
TSI butt, acid	−
H$_2$S (TSI butt)	−
H$_2$S (Pb ac paper)	+
Gelatin hydrolysis[a]	−
Litmus milk	−
Pigment	[yel]
Growth at:	
25°C	
35°C	+
42°C	
Esculin hydrolysis	−
Nutrient broth, 0% NaCl	+
Nutrient broth, 6% NaCl	−

[a]Incubation of 7–14 days.

Neisseria cinerea

Gram-negative coccus; may be observed in pairs or scattered clusters; nonmotile; oxidase-positive; asaccharolytic; nitrate-negative; nitrite-positive; amylosucrase-negative; susceptible to colistin (10 µg disc).

Sources (58 strains)

Eye (23%), blood (15%), sputum (13%), throat (9%), other (cerebrospinal fluid, cervix, urethra, urine, rectal abscess, hip wound) (22%), unknown (18%).

Reference strain

KC796 ← ATCC 14685, type strain, from nasopharynx.

Literature

1. Vedros, N.A. 1984. Genus I. *Neisseria* Trevisan 1885, 105.^AL *In* N.R. Krieg and J.G. Holt (Eds.), Bergey's Manual of Systematic Bacteriology, Vol. 1, Williams & Wilkins, Baltimore, pp. 290–296.

2. Knapp, J.S., P.A. Totten, M.H. Mulks, and B.H. Minshew. 1984. Characterization of *Neisseria cinerea*: a nonpathogenic species isolated on Martin-Lewis medium selective for pathogenic *Neisseria* spp. J. Clin. Microbiol. *19*: 63–67.

3. Southern, P.M., Jr., and A.E. Kutchner. 1987. Bacteremia due to *Neisseria cinerea*: report of two cases. Diagn. Microbiol. Infect. Dis. 7: 143–147.

4. Knapp, J.S., and E.W. Hook, III. 1988. Prevalence and persistance of *Neisseria cinerea* and other *Neisseria* spp. in adults. J. Clin. Microbiol. *26*: 896–900.

5. Denison, M.R., S. Pearlman, and R. Andersen. 1988. Misidentification of *Neisseria* species in a neonate with conjunctivitis. Pediatrics *81*: 877–878.

6. Boyce, J.M., and E.B. Mitchell, Jr. 1985. Difficulties in differentiating *Neisseria cinerea* from *Neisseria gonorrhoeae* in rapid systems used for identifying pathogenic *Neisseria* species. J. Clin. Microbiol. *22*: 731–734.

Comments

Confusion with *N. gonorrhoeae* may occur because *N. cinerea* has been isolated on Martin-Lewis medium (2) and has given *N. gonorrhoeae*-like reactions in some rapid identification systems (6). Gas production from nitrite may be difficult to demonstrate because many strains will not grow in 0.1% nitrite and the volume of gas generated from 0.01% nitrite is often too small to detect in a Durham tube. We have studied eight strains that share all of the characteristics of *N. cinerea*, except that they fail to reduce nitrite. Information on the cellular fatty acid composition of this species is presented in Table 4.12B and Figure 4.17 of the cellular fatty acid section.

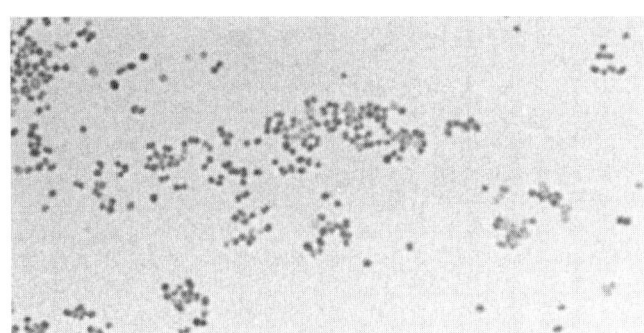

Neisseria cinerea strain KC796. Gram stain BAP 35° C 24 h x 1700.

Neisseria cinerea

TEST PERFORMED	SIGN	%+	#+/T
NUMBER OF STRAINS 58			
Morphology	cd		
Motility; flagella	[nm]		
Action on blood	-	0	0/20
Fermentative or oxidative	n-o		
Carbohydrate base	CTA		
Acid from:			
D-Glucose	[-]	0	0/58
D-Mannitol	-	0	0/58
Lactose	-	0	0/58
Sucrose	-	0	0/58
Maltose	-	0	0/58
Fructose	-	0	0/58
Catalase	+	100	7, 2w/9
Oxidase	[+]	100	58/58
Growth on:			
MacConkey	-	0	0/2
SS	-	0	0/2
Thayer-Martin	[v]	20	10/50
Simmons citrate	-	0	0/5
Urea, Christensen's	-	0	0/4
Nitrate reduction	[-]	0	0/58
Nitrite reduction	+	100	54/54
Gas from nitrite[a]	v	22	12/54
Indole	-	0	0/1
TSI slant, acid	-	0	0/8
TSI butt, acid	-	0	0/8
H_2S (TSI butt)	-	0	0/2
H_2S (Pb ac paper)	v	33	1w/3
Pigment:[b]			
Insoluble	v	57 lt yel	33/58
Growth at:[c]			
25°C	v	74	43/58
35°C	[+]	91	53/58
Amylosucrase	[-]	0	0/58

[a]Only a small volume of gas is produced in 0.01% nitrite. Many strains have not grown in 0.1% nitrite.
[b]Grown on Loeffler's agar.
[c]Grown on nutrient agar.

Neisseria elongata subsp. *elongata*

Gram-negative rod, often occurring as diplobacilli or in short chains; nonmotile; aerobic; no acid produced from carbohydrates; strongly oxidase-positive; variable growth on MacConkey agar (when growth occurs, it is weak and delayed); nitrate-negative and nitrite-positive; catalase-negative; sparse growth on Thayer-Martin medium.

Sources (15 strains)

Blood (5 strains), human bite wound (2 strains), other (one strain each: bronchial wash, eustachian tube, chest drainage, lung tissue, nasopharynx, sputum, eye, and urine).

Reference strain

KC1898 ← ATCC 25295, type strain, from human nasopharynx.

Literature

1. Bøvre, K., and E. Holten. 1970. *Neisseria elongata* sp. nov., a rod shaped member of the genus *Neisseria*. Reevaluation of cell shape as a criterion in classification. J. Gen. Microbiol. *60:* 67–75.

2. Vedros, N.A. 1984. Genus I. *Neisseria* Trevisan 1885, 105.[AL] *In* N.R. Krieg and J.G. Holt (Eds.), Bergey's Manual of Systematic Bacteriology, Vol. 1, Williams & Wilkins, Baltimore, pp. 290–296.

3. Grant, P.E., D.J. Brenner, A.G. Steigerwalt, D.G. Hollis, and R.E. Weaver. 1990. *Neisseria elongata* subsp. *nitroreducens* subsp. nov., formerly CDC group M-6, a gram-negative bacterium associated with endocarditis. J. Clin. Microbiol. *28:* 2591–2596.

Comments

Failure to acidify glucose and produce catalase differentiates *N. elongata* subsp. *elongata* from *N. elongata* subsp. *glycolytica*. Failure to acidify glucose and to reduce nitrate differentiates *N. elongata* subsp. *elongata* from *N. elongata* subsp. *nitroreducens*. Although described as positive for the production of gas from nitrite (2), this character is difficult to demonstrate using a Durham tube. Information on the cellular fatty acid composition of this group is presented in Table 4.12B and Figure 4.17 of the cellular fatty acid section.

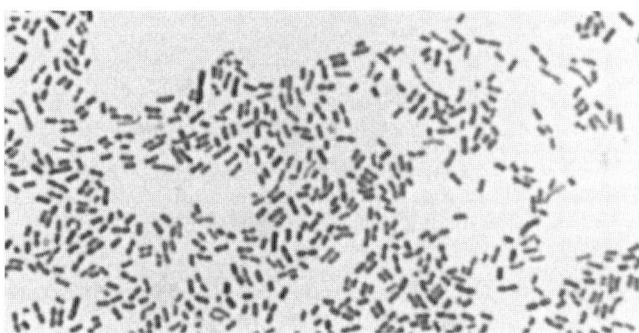

Neissseria elongata subsp. *elongata* strain KC1898.
Gram stain BAP 35° C 24 h x 1700.

Neisseria elongata subsp. *elongata*

TEST PERFORMED	SIGN	% +	#+/T
Morphology	[cc,srs]		
Motility; flagella	[nm]		
Action on blood	–	0	0/14
Fermentative or oxidative	n-o		
Carbohydrate base	OF		
Acid from:			
D-Glucose	[–]	0	0/15
D-Xylose	–	0	0/15
D-Mannitol	–	0	0/15
Lactose	–	0	0/15
Sucrose	–	0	0/15
Maltose	–	0	0/15
Catalase	[–]	0	0/15
Oxidase	[+]	100	15/15
Growth on:			
MacConkey	v	13 (53)	2w (8w)/15
SS	–	0	0/15
Cetrimide	–	0	0/15
Simmons citrate	–	0	0/15
Urea, Christensen's	–	0	0/15
Nitrate reduction	[–]	0	0/15
Nitrite reduction	[+]	92	11/12
Indole	–	0	0/15
TSI slant, acid	–	0	0/15
TSI butt, acid	–	0	0/15
H_2S (TSI butt)	–	0	0/15
H_2S (Pb ac paper)	v	67	8, 2w/15
Gelatin hydrolysis[a]	–	0	0/15
Litmus milk	–	0	0/15
Pigment	v	13 yel-sol	2/15; 1 tan-sol
Growth at:			
25°C	v	67	10/15
35°C	+	100	15/15
42°C	v	27	4/15
Esculin hydrolysis	–	0	0/15
Nutrient broth, 0% NaCl	+	100	14/14
Nutrient broth, 6% NaCl	–	0	0/14
Alkalinization of:			
Acetamide	–	0	0/12
Serine	–	0	0/12
Tartrate	–	0	0/12
Sodium acetate	v	36 (9)	4 (1)/11

[a]Incubation of 7–14 days.

Neisseria elongata subsp. *glycolytica*

Gram-negative rod, often occurring as diplobacilli or in short chains; nonmotile; aerobic; weak acid produced from glucose; strongly oxidase-positive; weak and/or delayed growth on MacConkey agar; nitrate-negative and nitrite-positive; catalase-positive.

Sources (2 strains)

Throat swab and sputum.

Reference strain

KC1899 ← ATCC 25315, type strain, from human throat swab.

Literature

1. Bøvre, K., and E. Holten. 1970. *Neisseria elongata* sp. nov., a rod shaped member of the genus *Neisseria*. Reevaluation of cell shape as a criterion in classification. J. Gen. Microbiol. *60:* 67–75.

2. Vedros, N.A. 1984. Genus I. *Neisseria* Trevisan 1885, 105.[AL] In N.R. Krieg and J.G. Holt (Eds.), Bergey's Manual of Systematic Bacteriology, Vol. 1, Williams & Wilkins, Baltimore, pp. 290–296.

3. Grant, P.E., D.J. Brenner, A.G. Steigerwalt, D.G. Hollis, and R.E. Weaver. 1990. *Neisseria elongata* subsp. *nitroreducens* subsp. nov., formerly CDC group M-6, a gram-negative bacterium associated with endocarditis. J. Clin. Microbiol. *28:* 2591–2596.

4. Henriksen S.D., and E. Holten. 1976. *Neisseria elongata* subsp. *glycolytica* new subspecies. Int. J. Syst. Bacteriol. *26:* 479–481.

5. Bøvre, K., L.O. Froholm, S.D. Henriksen, and E. Holten. 1977. Relationship of *Neisseria elongata* subsp. *glycolytica* to other members of the family *Neisseriaceae*. Acta Pathol. Microbiol. Scand. Sect. B Microbiol. *85:* 18–26.

Comments

Weak acidification of glucose and catalase production differentiates *N. elongata* subsp. *glycolytica* from *N. elongata* subsp. *elongata*. Catalase production and failure to reduce nitrate distinguishes this organism from *N. elongata* subsp. *nitroreducens*. Acid production from glucose is difficult to demonstrate, requiring the use of the rapid sugar test. Although described as positive for the production of gas from nitrite (2), this character is difficult to demonstrate using a Durham tube. Information on the cellular fatty acid composition of this group is presented in Table 4.12B and Figure 4.17 of the cellular fatty acid section.

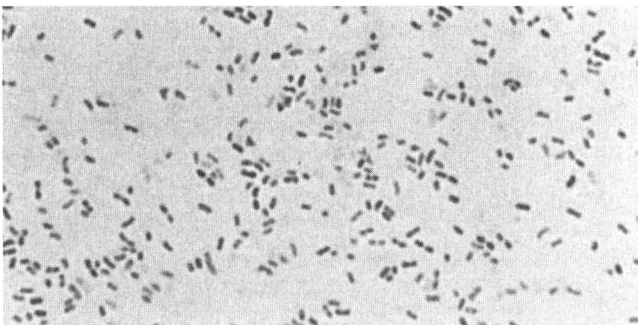

Neisseria elongata subsp. *glycolytica* strain KC1899.
Gram stain BAP 35° C 24 h x 1700.

Neisseria elongata subsp. *glycolytica*

TEST PERFORMED	SIGN	% +	#+/T
Morphology	[cc,srs]		
Motility; flagella	[nm]		
Action on blood	−	0	0/2
Fermentative or oxidative	[O]		
Carbohydrate base	OF		
Acid from:			
D-Glucose[a]	[−]	0	0/2
D-Xylose	−	0	0/2
D-Mannitol	−	0	0/2
Lactose	−	0	0/2
Sucrose	−	0	0/2
Maltose	−	0	0/2
Catalase	[+]	100	2/2
Oxidase	[+]	100	2/2
Growth on:			
MacConkey	[+ or (+)]	50 (50)	1 (1w)/2
SS	−	0	0/2
Cetrimide	−	0	0/2
Simmons citrate	−	0	0/2
Urea, Christensen's	−	0	0/2
Nitrate reduction	[−]	0	0/2
Nitrite reduction	v	50	1/2
Indole	−	0	0/2
TSI slant, acid	−	0	0/2
TSI butt, acid	−	0	0/2
H_2S (TSI butt)	−	0	0/2
H_2S (Pb ac paper)	+	100	2/2
Gelatin hydrolysis[b]	−	0	0/2
Litmus milk	−	0	0/2
Pigment	v	50 yel-sol	1/2
Growth at:			
25°C	+	100	2/2
35°C	+	100	2/2
42°C	−	0	0/2
Esculin hydrolysis	−	0	0/2
Nutrient broth, 0% NaCl	+	100	2/2
Nutrient broth, 6% NaCl	−	0	0/2
Alkalinization of:			
Acetamide	−	0	0/2
Serine	−	0	0/2
Tartrate	−	0	0/2
Sodium acetate	+	100	1/1

NUMBER OF STRAINS 2

[a]Acid production may be detected in rapid sugar test base.
[b]Incubation of 7–14 days.

Neisseria elongata subsp. nitroreducens

Gram-negative rod; nonmotile; aerobic; strongly oxidase-positive; catalase-, urease-, and indole-negative; weakly positive nitrate reaction in routine nitrate broth (peptone base); complete reduction of nitrate and nitrite without gas formation in special nitrate broth (heart infusion broth base).

Sources (95 strains)

Throat (20%); sputum (21%); blood (27%); wound (13%); peritoneal fluid (4%); other, including appendix exudate, urine, ear, stool, pleural fluid, pericardial fluid, chest drainage, lung tissue, corneal ulcer, and cervical exudate (15%)

Reference strain

B1019 = ATCC 49377, type strain, from blood.

Literature

1. Bøvre, K., and E. Holten. 1970. *Neisseria elongata* sp. nov., a rod shaped member of the genus *Neisseria*. Reevaluation of cell shape as a criterion in classification. J. Gen. Microbiol. *60:* 67–75.

2. Grant, P.E., D.J. Brenner, A.G. Steigerwalt, D.G. Hollis, and R.E. Weaver. 1990. *Neisseria elongata* subsp. *nitroreducens* subsp. nov., formerly CDC group M-6, a gram-negative bacterium associated with endocarditis. J. Clin. Microbiol. *28:* 2591–2596.

3. Wong, J.D., and J.M. Janda. 1992. Association of an important *Neisseria* species, *Neisseria elongata* subsp. *nitroreducens*, with bacteremia, endocarditis, and osteomyelitis. J. Clin. Microbiol. *30:* 719–720.

Comments

Formerly M-6. Usually sensitive to penicillin. These organisms differ from *Moraxella* species in that they are catalase-negative and reduce nitrate to completion. They differ from other subspecies of *N. elongata* by reducing nitrate. Has emerged in recent years as a cause of endocarditis. Information on the cellular fatty acid composition of this group is presented in Table 4.12B and Figure 4.17 of the cellular fatty acid section.

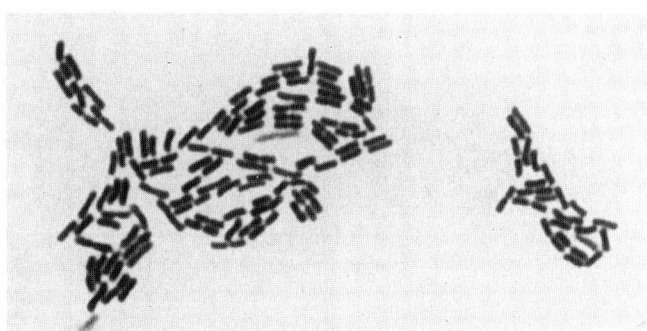

Neisseria elongata subsp. *nitroreducens* strain D3205. Gram stain BAP 35° C 24 h x 1700.

Neisseria elongata subsp. *nitroreducens*

TEST PERFORMED	SIGN	% +	#+/T
NUMBER OF STRAINS 26			
Morphology	[cc,srs]		
Motility; flagella	[nm]		
Action on blood	−	0	0/26
Fermentative or oxidative	n-o or O		
Carbohydrate base	OF		
Acid from:			
D-Glucose	v	23	6w/26
D-Xylose	−	0	0/26
D-Mannitol	−	0	0/26
Lactose	−	0	0/26
Sucrose	−	0	0/26
Maltose	−	0	0/26
Catalase	[−]	0	0/26
Oxidase	[+]	100	26/26
Growth on:			
MacConkey	v	19 (35)	5, (9w)/26
SS	−	0	0/26
Simmons citrate	−	0	0/26
Urea, Christensen's	−	0	0/26
Nitrate reduction	[+]	100	26/26
Gas from nitrate	−	0	0/26
Nitrite reduction	[+]	100	26/26
Indole	−	0	0/26
TSI slant, acid	−	0	0/26
TSI butt, acid	−	0	0/26
H_2S (TSI butt)	−	0	0/26
H_2S (Pb ac paper)	+	100	26/26
Gelatin hydrolysis[a]	−	0	0/26
Litmus milk	−	0	0/26
Pigment	v	15 ta-yel-sol	4/26
Growth at:			
25°C	v	62	16/26
35°C	+	100	26/26
42°C	v	23	6/26
Esculin hydrolysis	−	0	0/26
Lysine decarboxylase	−	0	0/26
Arginine dihydrolase	−	0	0/26
Ornithine decarboxylase	−	0	0/26
Nutrient broth, 0% NaCl	v	88	23/26
Nutrient broth, 6% NaCl	−	4	1/26
Alkalinization of:			
Sodium acetate	v	85	22/26
Acetamide	−	0	0/26
Serine	−	0	0/26
Tartrate	−	0	0/26
Phenylalanine deaminase	−	0	0/26

[a]Incubation of 7–14 days.

Neisseria flavescens

Gram-negative coccus occuring singly and in pairs with adjacent sides flattened; nonmotile; grows in air without increased CO_2; produces no acid from carbohydrates; strongly oxidase-positive; nitrate-negative, nitrite-positive; usually grows on nutrient agar at 35° C or 25° C; usually produces yellow growth pigment on Loeffler blood serum agar (obtained commercially); produces polysaccharide on medium with 5% sucrose (amylosucrase test); will not grow on Thayer-Martin medium.

Sources (10 strains)

Trachea (1), lymph node (1), cerebrospinal fluid (1), throat (1), unknown (6).

Reference strain

KC651 ← ATCC 13120, type strain, from cerebrospinal fluid.

Literature

1. Branham, S. 1930. A new meningococcus-like organism (*Neisseria flavescens* n. sp.) from epidemic meningitis. US Public Health Serv. Public Health. Rep. *45:* 845–849.
2. Henriksen, S.D., and K. Bøvre. 1968. The taxonomy of the genera *Moraxella* and *Neisseria*. J. Gen. Microbiol. *51:* 387–392.
3. Lewis, V.J., R.E. Weaver, and D.G. Hollis. 1968. Fatty acid composition of *Neisseria* species as determined by gas chromatography. J. Bacteriol. *96:* 1–5.
4. Faur, Y.C., M.H. Weisburd, and M.E. Wilson. 1975. Carbohydrate fermentation plate medium for confirmation of *Neisseria* species. J. Clin. Microbiol. *1:* 294–297.
5. Vedros, N.A. 1984. Genus I. *Neisseria* Trevisan 1885, 105.[AL] *In* N.R. Krieg and J.G. Holt (Eds.), Bergey's Manual of Systematic Bacteriology, Vol. 1, Williams & Wilkins, Baltimore, pp. 290–296.
6. Szabo, S., J.P. Lieberman, and Y.A. Lue. 1990. Unusual pathogens in narcotic-associated endocarditis. Rev. Infect. Dis. *12:* 412–415.

Comments

Isolations from patients with meningitis and septicemia have been reported. Production of gas from the reduction of nitrite has been described (5). May cause narcotic-associated endocarditis.

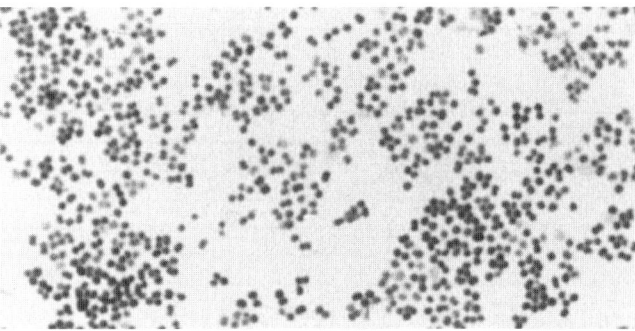

Neisseria flavescens strain KC651. Gram stain BAP 35° C 24 h x 1700.

Neisseria flavescens

TEST PERFORMED	SIGN	% +	#+/T
Morphology	[cd]		
Motility; flagella	[nm]		
Action on blood	v	33 ly	1/3
Fermentative or oxidative	n-o		
Carbohydrate base	CTA		
Acid from:			
D-Glucose	[-]	0	0/10
D-Mannitol	-	0	0/10
Lactose	-	0	0/10
Sucrose	-	0	0/10
Maltose	-	0	0/10
Fructose	-	0	0/10
Catalase	+	100	3/3
Oxidase	[+]	100	10/10
Growth on:			
MacConkey	v	67	1w/3
SS	-	0	0/3
Simmons citrate	-	0	0/3
Urea, Christensen's	-	0	0/3
Nitrate reduction	[-]	0	0/10
Nitrite reduction	+	100	2/2
TSI slant, acid	-	0	0/2
TSI butt, acid	-	0	0/2
H_2S (TSI butt)	-	0	0/2
H_2S (Pb ac paper)	+	100	2/2
Gelatin hydrolysis[a]	-	0	0/3
Litmus milk	v	50 k	1w/2
Pigment on Loeffler	[yel]	90	9/10
Growth on nutrient agar:			
25°C	v	86	6/7
35°C	[+]	100	7/7
Amylosucrase	[+]	100	2/2
Growth on Thayer-Martin	[-]	0	0/2

[a]Incubation of 7–14 days.

Neisseria gonorrhoeae

Gram-negative coccus, occurring singly and in pairs with adjacent sides flattened; nonmotile; grows in air with increased CO_2; usually uses glucose, but not maltose; strongly oxidase-positive; nitrate-negative; does not grow on nutrient agar (without NaCl) either at 35° C or 25° C; most strains (83/86) grow well on Thayer-Martin selective medium; superoxol-positive (5).

Sources (204 strains)

Cervix (17%), vagina (16%), blood (12%), urethra (11%), joint fluid (7%), eye (6%), urine (4%), other genital sources (5%), throat (2%), cerebrospinal fluid (1%), rectum (1%), other (8%), unknown (10%).

Reference strain

KC1321 ← ATCC 19424, type strain.

Literature

1. Thayer, J.D., and M.B. Moore, Jr. 1964. Gonorrhoea: present knowledge, research, and control efforts. Med. Clin. North Am. *48:* 755–765.
2. Sneath, P.H.A., and S.T. Cowan. 1958. An electro-taxonomic survey of bacteria. J. Gen. Microbiol. *19:* 551–565.
3. Henriksen, S.D., and K. Bøvre. 1968. The taxonomy of the genera *Moraxella* and *Neisseria*. J. Gen. Microbiol. *51:* 387–392.
4. Morello, J.A., W.M. Janda, and G.V. Doern. 1991. Chapter 30 *Neisseria* and *Branhamella. In* A. Balows, et al. (Eds.), Manual of Clinical Microbiology, 5th ed., American Society for Microbiology, Washington, pp. 258–269.
5. Knapp, J.S. 1988. Historical perspectives and identification of *Neisseria* and related species. Clin. Microbiol. Rev. *1:* 415–431.

Comments

Causes gonorrhea. In recent years, penicillin-resistant strains have emerged. Alternative methods for identification include fluorescent antibody, rapid enzymatic, and DNA probe techniques (4). Identification of *N. gonorrhoeae* at CDC is performed by the Sexually Transmitted Diseases Program. Information on the cellular fatty acid composition of this species is presented in Table 4.12B and Figure 4.17 of the cellular fatty acid section.

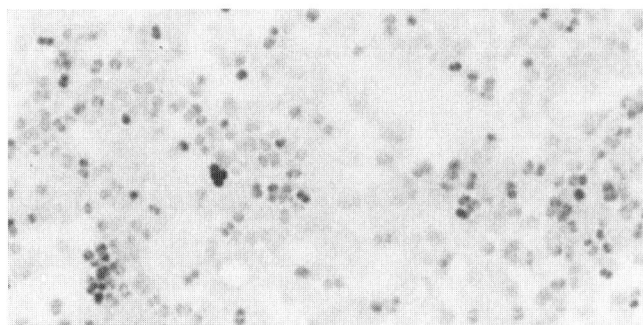

Neisseria gonorrhoeae strain KC1321. Gram stain BAP 35° C 24 h x 1700.

Neisseria gonorrhoeae

TEST PERFORMED	SIGN	% +	#+/T
Morphology	[cd]		
Motility; flagella	[nm]		
Action on blood	−	0	0/197
Fermentative or oxidative	O		
Carbohydrate base	CTA		
Acid from:			
D-Glucose	[+][a]	90 (3)	127, 51w (6)/197
D-Mannitol	−	0	0/197
Lactose	−	0	0/197
Sucrose	−	0	0/197
Maltose	[−]	0	0/197
Fructose	−	0	0/19
ONPG	−	0	0/19
Oxidase	[+]	100	197/197
Growth on:			
MacConkey	−	0	0/7
Nitrate reduction	[−]	0	0/69
Nitrite reduction (0.1%)	−	0	0/10
Nitrite reduction (0.01%)	v	78	7/9
Pigment on Loeffler	−	10 sl yel	1/10
Growth on nutrient agar at:			
25°C	−	0	0/149
35°C	[−]	0	0/149
Amylosucrase	−	0	0/14
Growth on Thayer-Martin	[+]	96	83/86

[a]Occasional negative reaction if caps on CTA tubes are not tightened before incubation.

Neisseria lactamica

Gram-negative coccus occurring singly and in pairs with adjacent sides flattened; nonmotile; grows in air with increased CO_2; uses glucose, maltose, and lactose; ONPG-positive; strongly oxidase-positive; nitrate-negative, nitrite-positive; usually grows on nutrient agar (without NaCl) at 35° C, few strains grow at 25° C; grows well on Thayer-Martin selective medium.

Sources (287 strains)

Throat (63%), nasopharynx (7%), trachea (2%), lung (1%), sputum (1%), vagina (1%), other (1%), unknown (24%).

Reference strain

A7515 = ATCC 23970, type strain, from nasopharynx.

Literature

1. Hollis, D.G., G.L. Wiggins, and R.E. Weaver. 1969. *Neisseria lactamicus* sp. n., a lactose-fermenting species resembling *Neisseria meningitidis*. Appl. Microbiol. *17:* 71–77.
2. Hollis, D.G. 1973. Sources of *Neisseria lactamicus*. Lancet. *i:* 1010.
3. Pykett, A.H. 1973. Isolation of *Neisseria lactamicus* from the nasopharynx. J. Clin. Pathol. *26:* 399–400.
4. Lauer, B.A., and C.E. Fisher. 1976. *Neisseria lactamica* meningitis. Am. J. Dis. Child. *130:* 198–199.
5. Vedros, N.A. 1984. Genus I. *Neisseria* Trevisan 1885, 105.^AL *In* N.R. Krieg and J.G. Holt (Eds.), Bergey's Manual of Systematic Bacteriology, Vol. 1, Williams & Wilkins, Baltimore, pp. 290–296.
6. Orden, B., and M.A. Amérigo. 1991. Acute otitis media caused by *Neisseria lactamica*. Eur. J. Clin. Microbiol. Infect. Dis. *10:* 986–987.

Comments

Most strains are serologically rough. Rarely isolated from blood or cerebrospinal fluid.

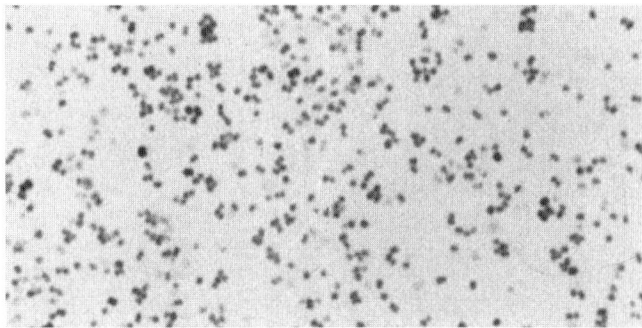

Neisseria lactamica strain A7515. Gram stain BAP 35° C 24 h x 1700.

Neisseria lactamica

TEST PERFORMED	SIGN	% +	#+/T
NUMBER OF STRAINS 287			
Morphology	[cd]		
Motility; flagella	[nm]		
Action on blood	−	0	0/29
Fermentative or oxidative	F		
Carbohydrate base	CTA		
Acid from:			
D-Glucose	[+]	96 (3)	267, 2w (8)/279
D-Mannitol	−	0	0/269
Lactose	[+]	95 (5)	261, 5w (13)/279
Sucrose	−	0	0/274
Maltose	[+]	98 (2)	270, 3w (5)/280
Fructose	−	0	0/258
ONPG	[+]	100	49, 1w/50
Catalase	+	100	29, 1w/30
Oxidase	[+]	100	278/278
Growth on:			
MacConkey	−	0	0/33
SS	−	0	0/33
Simmons citrate	−	0	0/29
Urea, Christensen's	−	0	0/27
Nitrate reduction	[−]	0	0/219
Nitrite reduction	+	100	8/8
TSI slant, acid	v	30	8/27
TSI butt, acid	v	19	4,1/26
H_2S (TSI butt)	−	0	0/32
H_2S (Pb ac paper)	v	88	11, 17w/32
Pigment on Loeffler (Insoluble)	v	38 sl yel	15/39
Phenylalanine deaminase	−	0	0/4
Growth on nutrient agar at:			
25°C	v	22	18, 40w (1)/262
35°C	v	74	166/223
Amylosucrase	−	0	0/3
Growth on Thayer-Martin	[+]	100	23/23

Neisseria meningitidis

Gram-negative coccus occurring singly and in pairs with adjacent sides flattened; nonmotile; grows in air with increased CO_2; uses glucose and maltose (some strains do not produce acid from one or both), but not lactose and sucrose; strongly oxidase-positive; nitrate-negative, nitrite-variable; no growth on nutrient agar (without NaCl) at 25° C, but a few carrier-source strains grow at 35° C on this medium; grows well on Thayer-Martin selective medium; amylosucrase-negative.

Sources (375 strains)

Cerebrospinal fluid (34%), blood (17%), throat (10%), sputum (4%), nasopharynx (3%), eye (1%), trachea (1%), other (3%), unknown (27%).

Reference strain

KC785 ← S. Branham, M 1027 = ATCC 13077, type strain, from cerebrospinal fluid.

Literature

1. Murray, E.G.D. 1929. The meningococcus. Med. Res. Counc. Spec. Rep. Ser. *124:* 7–142.
2. Lewis, V.J., R.E. Weaver, and D.G. Hollis. 1968. Fatty acid composition of *Neisseria* species as determined by gas chromatography. J. Bacteriol. *96:* 1–5.
3. Judicial Commission. 1970. Opinion 35. Conservation of the specific epithet *meningitidis* in the scientific name of the meningococcus. Int. J. Syst. Bacteriol. *20:* 13–14.
4. Beck, A., J.L. Fluker, and D.J. Platt. 1974. *Neisseria meningitidis* in urogenital infection. Br. J. Vener. Dis. *50:* 367–369.
5. Morello, J.A., W.H. Janda, and G.V. Doern. 1991. Chapter 30 *Neisseria* and *Branhamella*. *In* A. Balows et al. (Eds.), Manual of Clinical Microbiology, 5th ed., American Society for Microbiology, Washington, pp. 258–269.

Comments

Thirteen serogroups, based on capsular or outer membrane protein antigens, are recognized (A, B, C, D, H, I, K, L, W135, X, Y, Z, and Z'). Groups A, B, C, Y, and W135 are most commonly implicated in systemic disease. A primary cause of meningitis and septicemia. Information on the cellular fatty acid composition of this species is presented in Table 4.12C and Figure 4.17 of the cellular fatty acid section.

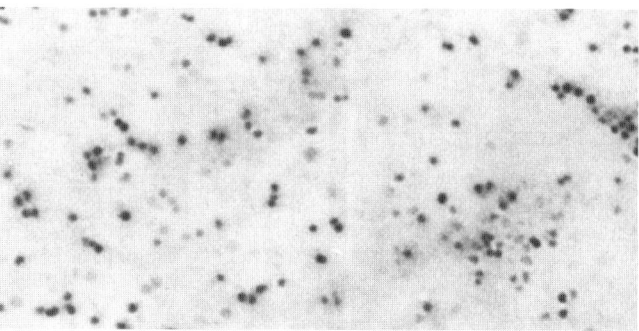

Neisseria meningitidis strain KC785. Gram stain BAP 35° C 24 h x 1700.

Neisseria meningitidis

TEST PERFORMED	SIGN	%+	#+/T
Morphology	[cd]		
Motility; flagella	[nm]		
Action on blood	–	2 ly	2/101
Fermentative or oxidative	O		
Carbohydrate base	CTA		
Acid from:			
D-Glucose	[+][a]	99	247, 5w/255
D-Mannitol	–	0	0/236
Lactose	[–]	0	0/255
Sucrose	[–]	0	0/238
Maltose	[+]	99	248, 4w/254
Fructose	–	0	0/237
ONPG	[–]	0	0/44
Oxidase	[+]	100	354/354
Growth on:			
MacConkey	–	0	0/47
Nitrate reduction	–	0	0/47
Nitrite reduction	v	29	2/7
Pigment on Loeffler	–	7 sl yel	6/82
Growth on nutrient agar at:			
25°C	–	0	0/207
35°C	[–]	1	2/209
Amylosucrase	–	0	0/17
Growth on Thayer-Martin	[+]	100	45/45

NUMBER OF STRAINS 375

[a]Occasional negative reaction if caps on CTA are not tightened before incubation.

Neisseria mucosa

Gram-negative coccus occurring singly and in pairs with adjacent sides flattened; nonmotile; grows in air without increased CO_2, also anaerobically; uses glucose (usually), sucrose, maltose, and fructose but not lactose; strongly oxidase-positive; reduces nitrate and nitrite with gas formation; usually grows on nutrient agar (without NaCl) at 35° and 25° C, usually does not grow on Thayer-Martin selective medium; produces polysaccharide on medium with 5% sucrose (amylosucrase test).

Sources (30 strains)

Blood (30%), eye (7%), cerebrospinal fluid (7%), wound (7%), urine (7%), sputum (7%), other (13%), unknown (22%).

Reference strain

C2629, from blood.

Literature

1. Véron, M., P. Thibault, and L. Secord. 1961. *Neisseria mucosa* (*Diplococcus mucosus* Lingelsheim). II. Etude antigénique et classification. Ann. Inst. Pasteur (Paris) *100:* 166–179.

2. Vedros, N.A. 1984. Genus I. *Neisseria* Trevisan 1885, 105[AL]. *In* N.R. Krieg and J.G. Holt (Eds.), Bergey's Manual of Systematic Bacteriology, Vol. 1, Williams & Wilkins, Baltimore, pp. 290–296.

3. Ingram, R.J.H., B. Cornere, and R.B. Ellis-Pegler. 1991. Endocarditis due to *Neisseria mucosa:* two cases and review. Clin. Infect. Dis. *15:* 321–324.

Comments

Occasionally pathogenic for humans. Associated with endocarditis. Information on the cellular fatty acid composition of this species is presented in Table 4.12C and Figure 4.17 of the cellular fatty acid section.

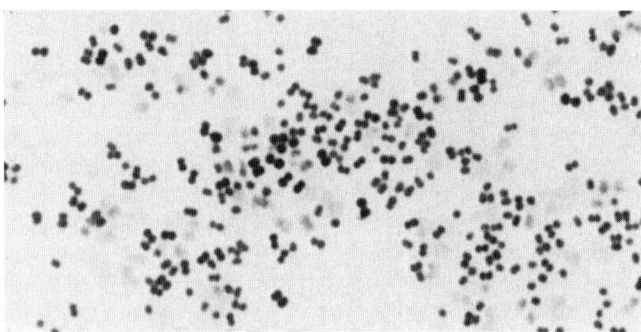

Neisseria mucosa strain C2629. Gram stain BAP 35° C 24 h x 1700.

Neisseria mucosa

TEST PERFORMED	SIGN	%+	#+/T
Morphology	[cd]		
Motility; flagella	[nm]		
Action on blood	v	25 ly	5/20; 2al
Fermentative or oxidative	F		
Carbohydrate base	CTA		
Acid from:			
D-Glucose	[+ or (+)]	83 (10)	25 (3)/30
D-Mannitol	–	0	0/30
Lactose	[–]	0	0/30
Sucrose	[+]	90 (10)	27 (3)/30
Maltose	+	90 (10)	27 (3)/30
Fructose	[+]	90 (10)	27 (3)/30
Catalase	+	100	8, 1w/9
Oxidase	[+]	100	29/29
Growth on:			
MacConkey	v	57 (3)	2, 2w (2)/7
SS	–	0	0/8
Thayer-Martin	[v]	15	4w/27
Simmons citrate	–	0	0/8
Urea, Christensen's	–	0	0/6
Nitrate reduction	[+]	100	29/29
Gas from nitrate	[+]	100	29/29
Nitrite reduction	[+]	100	29/29
TSI slant, acid	+	100	8/8
TSI butt, acid	v	88	7//8
H_2S (TSI butt)	–	0	0/8
H_2S (Pb ac paper)	+	100	7, 1w/8
Gelatin hydrolysis[a]	–	0	0/7
Litmus milk	–	0	0/7
Pigment on Loeffler	v	50 sl yel	10/20
Growth on nutrient agar at:			
25°C	+	96	24/25
35°C	[+]	100	16/16
Esculin hydrolysis	–	0	0/5
Amylosucrase	+	100	2/2

NUMBER OF STRAINS 30

[a]Incubation of 7–14 days.

Neisseria polysaccharea

Gram-negative cocci in pairs and tetrads; nonmotile; oxidase-positive; produces acid from glucose and maltose, but not fructose, mannitol, or lactose; nitrate-negative; amylosucrase-positive (produces a large amount of extracellular polysaccharide on media containing 1–5% sucrose); grows on Thayer-Martin selective medium.

Source (reference strain)

Child's throat.

Reference strain

KC1900 ← J. Knapp N-462 = ATCC 43768, type strain, from throat.

Literature

1. Riou, J.Y., and M. Guibourdenche. 1987. *Neisseria polysaccharea* sp. nov. Int. J. Syst. Bacteriol. *37:* 163–165.
2. Knapp, J.S. 1988. Historical perspectives and identification of *Neisseria* and related species. Clin. Microbiol. Rev. *1:* 415–431.

Comments

Some strains produce acid from sucrose (1). Sucrose-negative strains can be differentiated from *N. meningitidis* by the amylosucrase test. Information on the cellular fatty acid composition of this species is presented in Table 4.12C and Figure 4.17 of the cellular fatty acid section.

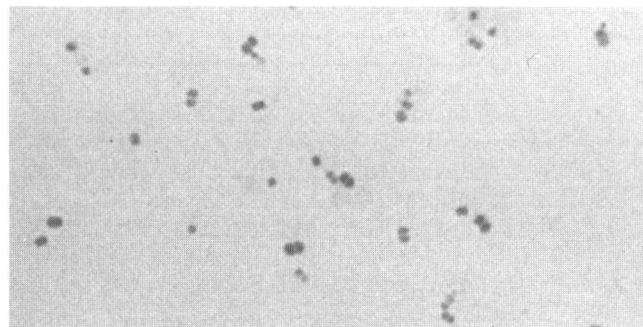

Neisseria polysaccharea strain KC1900. Gram stain BAP 35° C 24 h x 1700.

Neisseria polysaccharea

TEST PERFORMED	SIGN
Morphology	[cd]
Motility; flagella	[nm]
Action on blood	–
Fermentative or oxidative	O
Carbohydrate base	CTA
Acid from:	
D-Glucose	[+]
D-Mannitol	–
Lactose	[–]
Sucrose	(+w)
Maltose	[+]
Fructose	–
Catalase	+
Oxidase	[+]
Nitrate reduction	–
Nitrite reduction[a]	+
Pigment on Loeffler	sl yel
Growth on nutrient agar at:	
25°C	(+)
35°C	[+]
Growth on Thayer-Martin agar	[+]
Amylosucrase	[+]

[a]Nitrite, 0.01%. Negative reaction in 0.1% nitrite.

Neisseria sicca

Gram-negative coccus occurring singly and in pairs with adjacent sides flattened; nonmotile; grows in air without increased CO_2, also anaerobically; uses glucose (usually), sucrose, maltose, and fructose, but not lactose; strongly oxidase-positive, nitrate-negative, nitrite-positive; usually grows on nutrient agar (without NaCl) at 35° C and 25° C; generally does not grow on Thayer-Martin selective medium; produces polysaccharide on medium with 5% sucrose (amylosucrase test); no growth pigment on Loeffler blood serum agar; colonies usually dull, dry, wrinkled, and adherent to agar.

Sources (43 strains)

Throat (19%), sputum (9%), urine (7%), blood (7%), eye (5%), other (12%), unknown (41%).

Reference strain

KC181 ← S. Branham, N5.

Literature

1. Bergey, D.H., F.C. Harrison, R.S. Breed, B.W. Hammer, and F.M. Huntoon. 1923. Bergey's Manual of Determinative Bacteriology, 1st ed. Williams & Wilkins, Baltimore.

2. Hoke, C., and N. Vedros. 1982. Taxonomy of the *Neisseriae*: deoxyribonucleic acid base composition, interspecific transformation, and deoxyribonucleic acid hybridization. Int. J. Syst. Bacteriol. *32*: 57–66.

3. Vedros, N.A. 1984. Genus I. *Neisseria* Trevisan 1885, 105.[AL] In N.R. Krieg and J.G. Holt (Eds.), Bergey's Manual of Systematic Bacteriology, Vol. 1, Williams & Wilkins, Baltimore, pp. 290–296.

4. Heiddal, S., J.T. Sverrisson, F.E. Yngvason, N. Cariglia, and K.G. Kristinsson. 1993. Native-valve endocarditis due to *Neisseria sicca*: case report and review. Clin. Infect. Dis. *16*: 667–670.

Comments

Unable to differentiate with certainty the nonpigmented *N. sicca* and the nonpigmented strains (on Loeffler agar) of *N. subflava* biovar perflava. Genetic studies indicate a close relationship of *N. sicca* to *N. subflava* (2). Information on the cellular fatty acid composition of this species is presented in Table 4.12C and Figure 4.17 of the cellular fatty acid section.

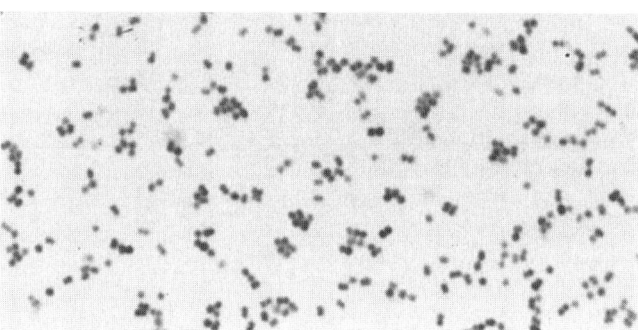

Neisseria sicca strain KC181. Gram stain BAP 35° C 24 h x 1700.

Neisseria sicca

TEST PERFORMED	SIGN	% +	#+/T
Morphology	[cd]		
Motility; flagella	[nm]		
Action on blood	v	22 ly	4/18; 3ß, 1 al
Fermentative or oxidative	F		
Carbohydrate base	CTA		
Acid from:			
D-Glucose	[+ or (+)]	78 (14)	24, 5w (5)/37
D-Mannitol	-	0	0/39
Lactose	[-]	0	0/38
Sucrose	[+]	92 (8)	32, 4w (3)/39
Maltose	[+]	95 (5)	37 (2)/39
Fructose	[+]	100	36, 2w/38
ONPG	-	0	0/3
Catalase	v	80	15, 1w/20
Oxidase	[+]	100	41/41
Growth on:			
MacConkey	v	68	7, 8w/22
SS	-	0	0/22
Simmons citrate	-	0	0/22
Urea, Christensen's	-	0	0/22
Nitrate reduction	[-]	0	0/37
Nitrite reduction	+	100	2/2
TSI slant, acid	+	100	22/22
TSI butt, acid	+ or (+)	82 (9)	16, 2w (2)/22
H_2S (TSI butt)	-	0	0/22
H_2S (Pb ac paper)	+	100	21, 1w/22
Gelatin hydrolysis[a]	-	0	0/15
Litmus milk	-	0	0/16
Pigment on Loeffler	-	0	0/22
Growth on nutrient agar at:			
25°C	+	100	18/18
35°C	[+]	90 (10)	18 (2)/20
Esculin hydrolysis	-	0	0/14
Amylosucrase	+	100	3/3
Growth on Thayer-Martin	[-]	0	0/5

Number of strains 43

[a] Incubation of 7–14 days.

Neisseria subflava

Gram-negative coccus occurring singly and in pairs with adjacent sides flattened; nonmotile; grows in air without increased CO_2, also anaerobically; uses glucose (usually) and maltose, sometimes sucrose and fructose, lactose-negative; strongly oxidase-positive; nitrate-negative; nitrite-positive; usually grows on nutrient agar (without NaCl) at 35° C and 25° C (generally does not grow on Thayer-Martin selective medium); usually produces yellow growth pigment on Loeffler blood serum agar (commercially obtained); may produce polysaccharide on medium with 5% sucrose (amylosucrase test).

Sources (151 strains)

Throat (28%), blood (9%), cerebrospinal fluid (7%), sputum (6%), nasopharynx (4%), urogenital (4%), eye (2%), other (14%), unknown (26%).

Reference strain

KC808 ← ATCC 14221.

Literature

1. Trevisan, V. 1889. I. Generi e le Specie delle Battiericee. Zanaboni e Gabuzzi, Milano.

2. Lewin, R.A., and W.I. Hughes. 1966. *Neisseria subflava* as a cause of meningitis and septicemia in children. J. Am. Med. Assoc. *195:* 821–823.

3. Hoke, C., and N. Vedros. 1982. Taxonomy of the *Neisseriae*: deoxyribonucleic acid base composition, interspecific transformation, and deoxyribonucleic acid hybridization. Int. J. Syst. Bacteriol. *32:* 57–66.

4. Vedros, N.A. 1984. Genus I. *Neisseria* Trevisan 1885, 105.[AL] *In* N.R. Krieg and J.G. Holt (Eds.), Bergey's Manual of Systematic Bacteriology, Vol. 1, Williams & Wilkins, Baltimore, pp. 290–296.

5. Demmler, G.J., R.S. Couch, and L.H. Taber. 1985. *Neisseria subflava* bacteremia and meningitis in a child: report of a case and review of the literature. Pediatr. Infect. Dis. *4:* 286–288.

Comments

Bergey's Manual of Systematic Bacteriology combined the former species *N. subflava*, *N. flava*, and *N. perflava* into one species, *N. subflava* (4). Information on the cellular fatty acid composition of this species is presented in Table 4.12C and Figure 4.17 of the cellular fatty acid section.

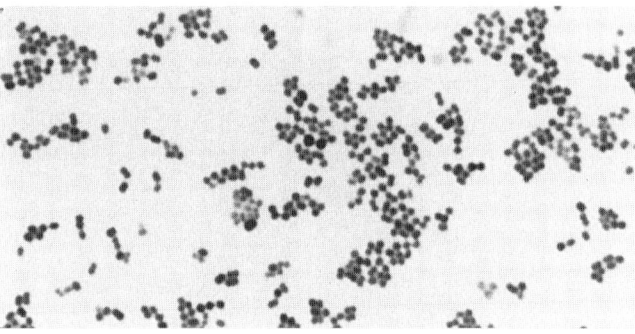

Neisseria subflava strain KC808. Gram stain BAP 35° C 24 h x 1700.

Neisseria subflava

TEST PERFORMED	SIGN	% +	#+/T
	NUMBER OF STRAINS 153		
Morphology	[cd]		
Motility; flagella	[nm]		
Action on blood	v	56 ly	28/50; 14 ß, 6 al, 1 inc ß, 1 gr
Fermentative or oxidative	F		
Carbohydrate base	CTA		
Acid from:			
D-Glucose	v	68 (6)	71, 25w (8)/141
D-Mannitol	-	0	0/143
Lactose	[-]	0	0/141
Sucrose	[v]	56 (5)	63, 19w (7)/145
Maltose	[+]	99 (1)	132, 10w (2)/144
Fructose	[v]	77 (5)	87, 23w (7)/143
ONPG	-	0	0/15
Catalase	v	80	47/59
Oxidase	[+]	100	150/150
Growth on:			
MacConkey	v	47 (2)	12, 13w (1)/53
SS	-	0	0/53
Simmons citrate	-	0	0/53
Urea, Christensen's	-	0	0/51
Nitrate reduction	[-]	0	0/144
Nitrite reduction	+	100	94/94
TSI slant, acid	v	68	34, 3w/54
TSI butt, acid	v	56 (4)	23, 7w (2)/54
H_2S (TSI butt)	-	0	0/54
H_2S (Pb ac paper)	+	100	45, 9w/54
Gelatin hydrolysis[a]	-	0	0/35
Litmus milk	v	14 IR	2/14
Pigment on Loeffler	[yel]	99	146/147
Growth on nutrient agar at:			
25°C	+	95	82/86
35°C	[+]	95	112/118
Phenylalanine deaminase	v	32	7/22
Amylosucrase	v	73	8/11
Growth on Thayer-Martin	[-]	8	3/39

[a]Incubation of 7–14 days.

Neisseria weaveri

Gram-negative thin to medium-width rod; nonmotile; aerobic; does not use carbohydrates; strongly oxidase-positive; urease-negative, nitrate- and indole-negative.

Sources (132 strains)

Dog bite (60%), hand wound (14%), unspecified wound (9%), dog's mouth (5%), leg wound (5%), cheek or face wound (4%), other (3%).

Reference strain

8142 = ATCC 51223, type strain, from dog bite.

Literature

1. Tatum, H.W., W.H. Ewing, and R.E. Weaver. 1974. Miscellaneous gram-negative bacteria. *In* E.H. Lennette, E.H. Spaulding, and J.P. Truant (Eds.), Manual of Clinical Microbiology, 2nd ed., American Society for Microbiology, Washington, pp. 270–294.
2. Bailie, W.E., E.C. Stowe, and A.M. Schmitt. 1978. Aerobic bacterial flora of oral and nasal fluids of canines with reference to bacteria associated with bites. J. Clin. Microbiol. 7: 223–231.
3. Moss, C.W., P.L. Wallace, D.G. Hollis, and R.E. Weaver. 1988. Cultural and chemical characterization of CDC groups EO-2, M-5, and M-6, *Moraxella* species, *Oligella urethralis*, *Acinetobacter* species, and *Psychrobacter immobilis*. J. Clin. Microbiol. 26: 484–492.
4. Andersen, B.M., A.G. Steigerwalt, S.P. O'Connor, D.G. Hollis, R.S. Weyant, R.E. Weaver, and D.J. Brenner. 1993. *Neisseria weaveri* sp. nov., formerly CDC group M-5, a gram-negative bacterium associated with dog bite wounds. J. Clin. Microbiol. 31: 2456–2466.
5. Holmes, B., M. Costas, S.L.W. On, P. Vandamme, E. Falsen, and K. Kersters. 1993. *Neisseria weaveri* sp. nov. (formerly CDC Group M-5), from dog bite wounds of humans. Int. J. Syst. Bacteriol. 43: 687–693.

Comments

Formerly M-5. Usually sensitive to penicillin; the *Moraxella osloensis* transformation test is negative. Growth on MacConkey agar is variable, and is often weak and/or delayed when noted. Associated with animal-bite wounds. Information on the cellular fatty acid composition of this species is presented in Table 4.12C and Figure 4.17 of the cellular fatty acid section.

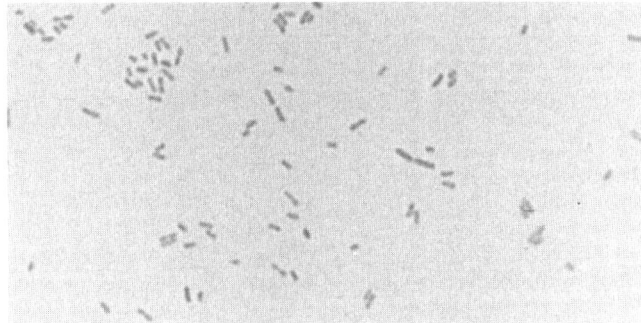

Neisseria weaveri strain 8142. Gram stain BAP 35° C 24 h x 1700.

Neisseria weaveri

TEST PERFORMED	SIGN	% +	#+/T
Morphology	[mrs]		
Motility; flagella	[nm]		
Action on blood	v	39 ly	50/128; 25 gr, 7 LG, 6 al, 1 lav, 1 br
Fermentative or oxidative	n-o		
Carbohydrate base	OF		
Acid from:			
D-Glucose	[-]	0	0/132
D-Xylose	[-]	0	0/132
D-Mannitol	-	0	0/132
Lactose	-	0	0/132
Sucrose	-	0	0/132
Maltose	-	0	0/132
Catalase	[+]	100	132/132
Oxidase	[+]	100	132/132
Growth on:			
MacConkey	v	27 (18)	23 (6), 12w (18w)/132
SS	-	0	0/132
Cetrimide	-	0	0/132
Simmons citrate	-	0	0/132
Urea, Christensen's	-	0	0/132
Nitrate reduction	[-]	0	0/132
Gas from nitrate	-	0	0/132
Nitrite reduction	[+]	100	41/41
Indole	-	0	0/132
TSI slant, acid	-	0	0/132
TSI butt, acid	-	0	0/132
H_2S (TSI butt)	-	0	0/132
H_2S (Pb ac paper)	v	86	71, 42w/131
Gelatin hydrolysis[a]	-	0	0/118
Litmus milk	-	6 k	8/126, 2 IR
Pigment	v	24 yel-ta sol	32/132
Growth at:			
25°C	+	94	121/129
35°C	+	100	129/129
42°C	v	41	53/128
Esculin hydrolysis	-	0	0/124
Lysine decarboxylase	-	0	0/45
Arginine dihydrolase	-	0	0/45
Ornithine decarboxylase	-	0	0/45
Nutrient broth, 0% NaCl	v	85	106/125
Nutrient broth, 6% NaCl	v	18	23/125
Phenylalanine deaminase[b]	v	71	71/100
Alkalinization of:			
Sodium acetate	-	4	3/78
Acetamide	-	2	1/63
Serine	-	0	0/63
Tartrate	-	0	0/63
Penicillin sensitivity	+	97	98/101

[a]Incubation of 7–14 days.
[b]Reaction is enhanced when using strains grown on trypticase soy agar with 5% sheep blood.

NO-1

Gram-negative rod; nonmotile; asaccharolytic; oxidase-negative; usually does not grow on MacConkey agar; catalase- and nitrate-positive; *Acinetobacter* transformation-negative.

Sources (22 strains)

Dog bite wound (hand, face, arm, and wrist) (70%), cat bite wound (15%), other (unspecified foot and hand wound) (15%).

Reference strain

A2795 = ATCC 51247.

Literature

1. Hollis, D.G., C.W. Moss, M.I. Daneshvar, L. Meadows, J. Jordan, and B. Hill. 1993. Characterization of Centers for Disease Control Group NO-1, a fastidious, nonoxidative, gram-negative organism associated with dog and cat bites. J. Clin. Microbiol. *31:* 746-748.

Comments

Associated with dog or cat bite wounds. Similar to fastidious *Acinetobacter* species, but can be differentiated by cellular fatty acid analysis and the *Acinetobacter* transformation assay. Although a positive nitrate reaction suggests identification as group NO-1, a small percentage (<10%) of *Acinetobacter* species also reduce nitrate. Information on the cellular fatty acid composition of this group is presented in Table 4.32 and Figure 4.37 of the cellular fatty acid section.

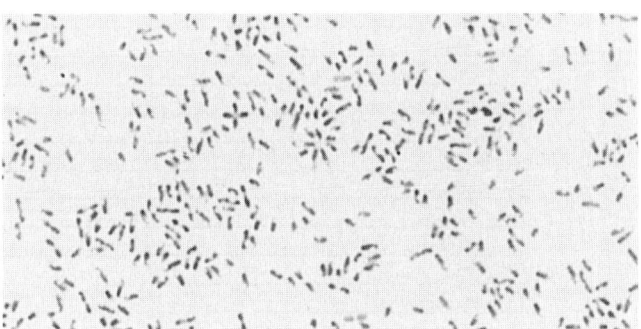

NO-1 strain A2795. Gram stain BAP 35° C 24 h x 1700.

NO-1

	NUMBER OF STRAINS 22		
TEST PERFORMED	SIGN	%+	#+/T
Morphology	rs		
Motility; flagella	[nm]		
Action on blood	v	30 ly	6/20
Fermentative or oxidative	[n-o]		
Carbohydrate base	OF		
Acid from:			
D-Glucose	[-]	0	0/20
D-Xylose	-	0	0/20
D-Mannitol	-	0	0/20
Lactose	-	0	0/20
Sucrose	-	0	0/20
Maltose	-	0	0/20
Catalase	+	100	19/19
Oxidase	[-]	5	1w/22
Growth on:			
MacConkey	[v]	5 (15)	1w (1) (2w)/20
SS	-	0	0/20
Simmons citrate	-	0	0/20
Urea, Christensen's	-	5	1w/20
Nitrate reduction	[+]	100	20/20
Gas from nitrate	-	0	0/20
Indole	-	0	0/20
TSI slant, acid	-	0	0/20
TSI butt, acid	-	0	0/20
H_2S (TSI butt)	-	0	0/20
H_2S (Pb ac paper)	v	50	1, 9w/20
Gelatin hydrolysis[a]	-	0	0/20
Litmus milk	v	(20 k)	(4)/20
Pigment	-	0	0/20
Growth at:			
25°C	v	20	4/20
35°C	+	100	20/20
42°C	v	15	3/20
Esculin hydrolysis	-	0	0/20
Lysine decarboxylase	-	0	0/7
Arginine dihydrolase	-	0	0/7
Ornithine decarboxylase	-	0	0/7
Nutrient broth, 0% NaCl	v	10 (5)	2 (1)/20
Nutrient broth, 6% NaCl	-	0	0/20
Acinetobacter transformation	[-]	0	0/17

[a] Incubation of 7–14 days.

Bordetella holmesii (formerly NO-2)

Gram-negative small coccoid and short rod, rarely medium wide long rod; grows on MacConkey agar, otherwise fastidious and slow growing; oxidase-negative; nonmotile; does not use carbohydrates; indole-, urease-, nitrate-, and nitrite-negative; catalase-variable; produces brown soluble pigment; *Acinetobacter* transformation-negative.

Source (13 strains)

Blood (100%).

Reference strain

F5101 = ATCC 51541, from blood.

Literature

1. Weyant, R.S., D.G. Hollis, R.E. Weaver, et al. 1995. *Bordetella holmesii* sp. nov., a new gram-negative species associated with septicemia. J. Clin. Microbiol. *33:* 1–7.

Comments

At least four of the 13 strains were from multiple blood cultures, three of these from splenectomized patients. Predominately isolated from male patients. Cellular fatty acid composition for this group is presented in Figure 4.21 and Table 4.16 of the cellular fatty acid section.

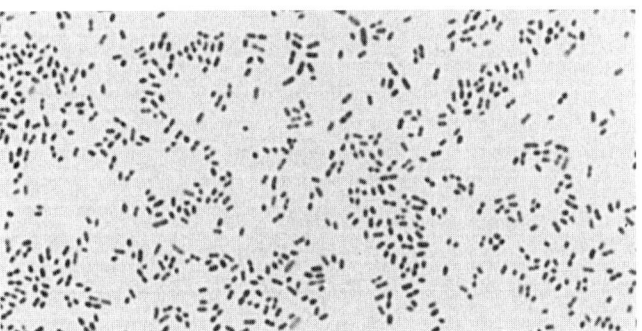

NO-2 strain F5101. Gram stain BAP 35° C 24 h x 1700.

Bordetella holmesii (formerly NO-2)

TEST PERFORMED	SIGN	% +	#+/T
	NUMBER OF STRAINS 13		
Morphology	cc,srs,occas.lr		
Motility; flagella	[nm]		
Action on blood	v	38 gr	5/13; 2 br, 1 ly
Fermentative or oxidative	[n-o]		
Carbohydrate base	OF		
Acid from:			
D-Glucose	[-]	0	0/13
D-Xylose	-	0	0/13
D-Mannitol	-	0	0/13
Lactose	-	0	0/13
Sucrose	-	0	0/13
Maltose	-	0	0/13
Catalase	v	38	5w/13
Oxidase	[-]	0	0/13
Growth on:			
MacConkey	[+ or (+)][a]	77(23)	10(3)/13
SS	-	0	0/13
Cetrimide	-	0	0/13
Simmons citrate	-	0	0/13
Urea, Christensen's	[-]	0	0/13
Nitrate reduction	[-]	0	0/13
Gas from nitrate	-	0	0/13
Nitrite reduction	-	0	0/4
Indole	-	0	0/13
TSI slant, acid	-	0	0/13
TSI butt, acid	-	0	0/13
H_2S (TSI butt)	-	0	0/13
H_2S (Pb ac paper)	v	62	8w/13
Gelatin hydrolysis[b]	-	0	0/13
Litmus milk	v	31 k	4/13; 1 IR
Pigment	[br sol]	100	13/13
Growth at:			
25°C	v	67	8/12
35°C	+	100	13/13
42°C	-	0	0/12
Esculin hydrolysis	-	0	0/12
Lysine decarboxylase	-	0	0/4
Arginine dihydrolylase	-	0	0/4
Ornithine decarboxylase	-	0	0/4
Nutrient broth, 0% NaCl	v	46(15)	6(2)/13
Nutrient broth, 6% NaCl	-	0	0/13
Acinetobacter transformation	-	0	0/13

[a]Light growth.
[b]Incubation of 7–14 days.

Ochrobactrum anthropi

Gram-negative rod; motile (peritrichous; frequently individual cells have only a single long flagellum either polar, subpolar, or lateral with a fairly short wavelength); oxidizes glucose, xylose, mannitol, sucrose, and maltose (these reactions may be delayed), but not lactose; grows on MacConkey and SS agars; oxidase- and urease-positive; reduces nitrate with gas formation; indole-negative.

Sources (14 strains)

Blood (21%), cervix (14%), throat (14%), other (ear, bronchogenic cyst, sputum, incision, bladder wound, rectum, urine) (51%).

Reference strains

A8409 = NCTC 12169 = ATCC 49187, type strain, from bronchogenic cyst.

Literature

1. Riley, P.S., and R.E. Weaver. 1977. Comparison of thirty-seven strains of Vd-3 bacteria with *Agrobacterium radiobacter:* morphological and physiological observations. J. Clin. Microbiol. *5:* 172–177.

2. Holmes, B., M. Popoff, M. Kiredjian, and K. Kersters. 1988. *Ochrobactrum anthropi* gen. nov., sp. nov. from human clinical specimens and previously known as group Vd. Int. J. Syst. Bacteriol. *38:* 406–416.

3. Cieslak, R.J., M.L. Robb, C.J. Drabick, and G.W. Fischer. 1992. Catheter-associated sepsis caused by *Ochrobactrum anthropi:* report of a case and review of related nonfermentative bacteria. Clin. Infect. Dis. *14:* 902–907.

4. Holmes, B., C.W. Moss, and M.I. Daneshvar. 1993. Cellular fatty acid compositions of "*Achromobacter*" groups B and E". J. Clin. Microbiol. *31:* 1007–1008.

Comments

Formerly Vd. This group originally was composed of organisms designated as groups Vd-1 (mannitol- and sucrose-negative) and Vd-2 (mannitol- and sucrose-positive) and *Achromobacter* sp. biotypes 1 and 2. This organism, unlike *Agrobacterium radiobacter*, does not oxidize lactose. Differentiation of esculin-positive strains from "*Achromobacter*" groups B and E requires cellular fatty acid analysis. Information on the cellular fatty acid composition of this organism is presented in Table 4.60 and Figure 4.65 of the cellular fatty acid section.

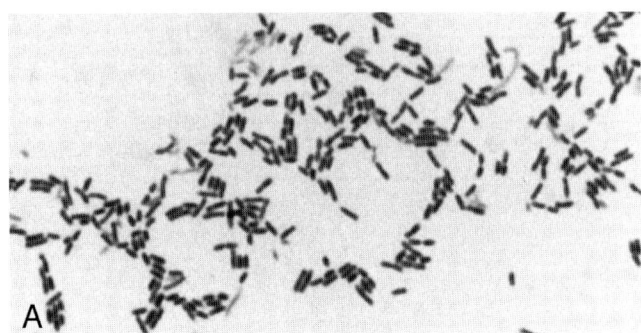

Ochrobactrum anthropi strain D2701. Gram stain BAP 35° C 24 h x 1700.

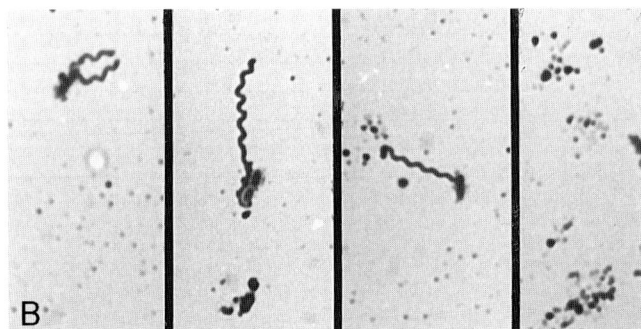

Ochrobactrum anthropi strain D2701. Flagella stain FB 25° C 24 h x 1700.

Ochrobactrum anthropi

TEST PERFORMED	SIGN	% +	#+/T
Morphology	mrs		
Motility; flagella	[m;pe]a		
Action on blood	v	43 ly	6/14; 4 LG, 1 gr
Fermentative or oxidative	[O]		
Carbohydrate base	OF		
Acid from:			
D-Glucose	[+]	93 (7)	6, 7w (1)/14
D-Xylose	+	100	12, 2w/14
D-Mannitol	v	43 (14)	3 (1), 3w (1w)/14
Lactose	[−]	0	0/14
Sucrose	v	50	2, 5w/14
Maltose	v	64	3, 6w/14
Catalase	+	100	14/14
Oxidase	[+]	100	14/14
Growth on:			
MacConkey	[+]	100	14/14
SS	[+]	100	13, 1w/14
Cetrimide	−	0	0/14
Simmons citrate	v	64	9/14
Urea, Christensen's	[+]	100	14/14
Nitrate reduction	v	86	12/14
Gas from nitrate	v	43	6/14
Indole	[−]	0	0/14
TSI slant, acid	−	0	0/14
TSI butt, acid	−	0	0/14
H_2S (TSI butt)	v	43	4, 2w/14
H_2S (Pb ac paper)	+	100	14/14
Gelatin hydrolysisb	−	0	0/12
Litmus milk	k	93	13/14
Pigment	v	21 yel-br sol	3/14
Growth at:			
25°C	+	100	14/14
35°C	+	100	14/14
42°C	v	64	9/14
Esculin hydrolysis	v	29 (7)	4 (1)/14
Lysine decarboxylase	−	0	0/7
Arginine dihydrolase	v	71	5/7
Ornithine decarboxylase	−	0	0/7
Nutrient broth, 0% NaCl	+	100	10/10
Nutrient broth, 6% NaCl	v	60	6/10
3-Ketolactonate	[−]	0	0/3

Number of strains: 14

aFrequently individual cells have only a single flagellum, either polar, subpolar, or lateral.
bIncubation of 7–14 days.

OFBA-1

Gram-negative medium to long rod; motile (1–2 polar flagella); oxidizes glucose, xylose, mannitol, lactose, sucrose, and maltose; produces an acidic reaction in OF medium blank; grows on MacConkey and SS agars; oxidase-positive; reduces nitrate to gas.

Sources (6 strains)

Blood (1), leg ulcer (2), abdominal wound (1), bronchial wash (1), catheter tunnel, continuous ambulatory peritoneal dialysis (1).

Reference strain

G6901 ← A. von Graevenitz, V09-1635.

Literature

1. von Graevenitz, A., G.E. Pfyffer, M.J. Pickett, R.E. Weaver, and J. Wüst. 1993. Isolation of an unclassified non-fermentative gram-negative rod from a patient on continuous ambulatory peritoneal dialysis. Eur. J. Clin. Microbiol. Infect. Dis. *12:* 568–570.

Comments

When grown in fermentative broth base without carbohydrate, the reference strain produced a combination of oxaloacetic, succinic, and acetic acids. Information on the cellular fatty acid composition of this group is presented in Table 4.33 and Figure 4.38 of the cellular fatty acid section.

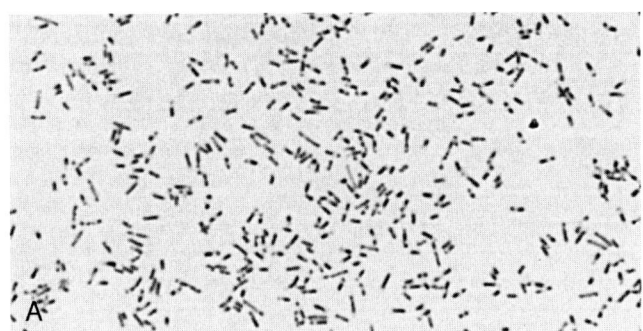

OFBA-1 strain G6901. Gram stain BAP 35° C 24 h x 1700.

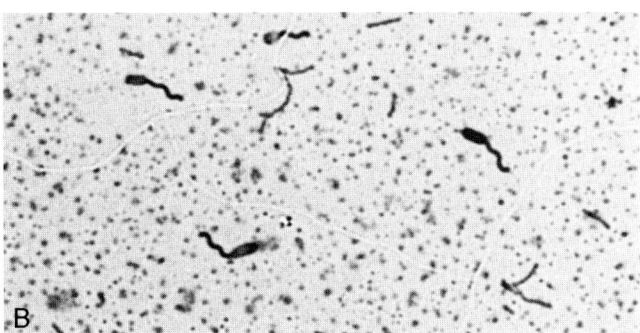

OFBA-1 strain G6901. Flagella stain (Ryu method) TGY 25° C 24 h x 1700.

OFBA-1

TEST PERFORMED	SIGN	%+	#+/T
Morphology	mrs		
Motility; flagella	[m;p,1-2]		
Action on blood	β	100	5, 1w/6
Fermentative or oxidative	[unknown]		
Carbohydrate base	OF		
Acid from:			
D-Glucose	[+]	100	6/6
D-Xylose	+	100	6/6
D-Mannitol	+ or (+)	67 (33)	4 (2)/6
Lactose	+ or (+)	67 (33)	3, 1w (2)/6
Sucrose	+ or (+)	67 (33)	3, 1w (2)/6
Maltose	+ or (+)	67 (33)	3, 1w (2)/6
Blank	[+ or (+)]	67 (33)	3, 1w (2)/6
Catalase	+	100	6/6
Oxidase	[+]	100	6/6
Growth on:			
MacConkey	[+]	100	6/6
SS	+	100	6/6
Cetrimide	+	100	4, 2w/6
Simmons citrate	v	33 (33)	2 (1) (1w)/6
Urea, Christensen's	v	33 (17)	2 (1)/6
Nitrate reduction	+	100	6/6
Gas from nitrate	[+]	100	6/6
Indole	-	0	0/6
TSI slant, acid	+	100	6/6
TSI butt, acid	+ or (+)	33 (67)	2 (3) (1w)/6
H_2S (TSI butt)	-	0	0/6
H_2S (Pb ac paper)	v	50	2, 1w/6
Gelatin hydrolysis[a]	v	50	3/6
Litmus milk	v	50 pep	3/6; 2 szf
Pigment	[-]	0	0/6
Growth at:			
25°C	+	100	5, 1w/6
35°C	+	100	6/6
42°C	+	100	6/6
Esculin hydrolysis	-	0	0/6
Lysine decarboxylase	-	0	0/5
Arginine dihydrolase	+	100	5/5
Ornithine decarboxylase	-	0	0/5
Nutrient broth, 0% NaCl	+	100	4/4
Nutrient broth, 6% NaCl	v	75	3/4
Alkalinization of:			
Acetamide	-	0	0/5
Serine	-	0	0/5
Tartrate	-	0	0/5

Number of strains 6

[a] Incubation of 7–14 days.

Oligella ureolytica

Small coccoid gram-negative rod; motile (peritrichous, often with one long polar and only a few lateral flagella); aerobic; does not use carbohydrates; oxidase-positive; frequently grows on MacConkey agar, but not on SS agar; strongly urease-positive; nitrate and nitrite reduced, usually with some gas production; no H_2S formed on TSI agar.

Sources (37 strains)

Urine (92%), blood (3%), unknown (5%).

Reference strains

C379 = ATCC 43534, type strain, from urine.

Literature

1. Rossau, R., K. Kersters, E. Falsen, E. Jantzen, P. Segers, A. Union, L. Nehls, and J. De Ley. 1987. *Oligella*, a new genus including *Oligella urethralis* comb. nov. (formerly *Moraxella urethralis*) and *Oligella ureolytica* sp. nov. (formerly CDC group IVe): relationship to *Taylorella equigenitalis* and related taxa. Int. J. Syst. Bacteriol. 37: 198–210.

Comments

Formerly CDC group IVe. Most frequent source has been urine, particularly from males. Motility may be delayed, or difficult to demonstrate. Information on the cellular fatty acid composition of this species is presented in Table 4.21 and Figure 4.26 of the cellular fatty acid section.

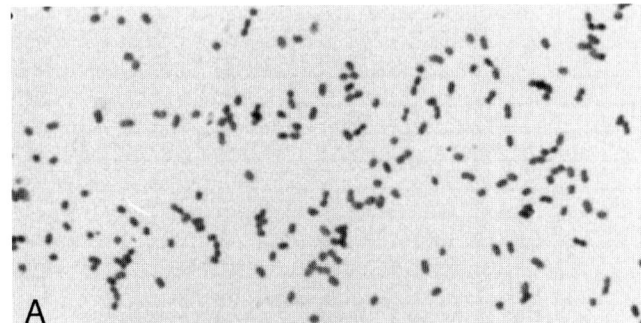

Oligella ureolytica strain D3926. Gram stain BAP 35° C 24 h x 1700.

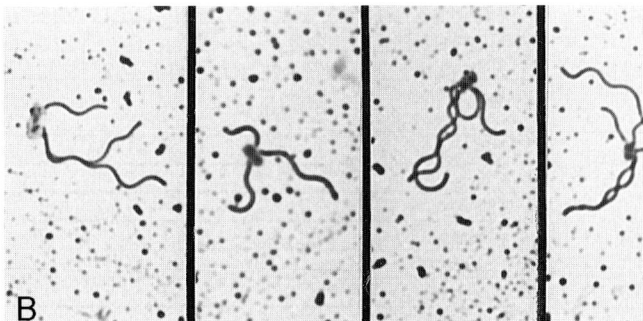

Oligella ureolytica strain D3926. Flagella stain TGY 25° C 24 h x 1700.

Oligella ureolytica

TEST PERFORMED	SIGN	%+	#+/T
	NUMBER OF STRAINS 37		
Morphology	srs		
Motility; flagella	[m;pe]a		
Action on blood	v	71 ly	5/7; 1 gr, 1 br
Fermentative or oxidative	[n-o]		
Carbohydrate base	OF		
Acid from:			
D-Glucose	[-]	0	0/37
D-Xylose	[-]	0	0/37
D-Mannitol	-	0	0/37
Lactose	-	0	0/37
Sucrose	-	0	0/37
Maltose	-	0	0/37
Catalase	+	100	37/37
Oxidase	[+]	100	37/37
Growth on:			
MacConkey	[v]	62 (27)	23 (10)/37
SS	[-]	5	2/37
Cetrimide	-	0 (3)	(1)/37
Simmons citrate	v	14 (16)	5 (6)/37
Urea, Christensen's	[+]	97	36/37
Nitrate reduction	[+]	100	37/37
Gas from nitrate	[v]	60	22/37
Indole	-	0	0/37
TSI slant, acid	-	0	0/36
TSI butt, acid	-	0	0/36
H_2S (TSI butt)	[-]	0	0/36
H_2S (Pb ac paper)	v	38	14/37
Gelatin hydrolysisb	-	0	0/33
Litmus milk	v	65 k	24/37
Pigment	-	3 yel sol	1/37
Growth at:			
25°C	v	67	22/33
35°C	v	88	29/33
42°C	v	18	6/33
Esculin hydrolysis	-	0	0/32
Nutrient broth, 0% NaCl	v	19 (3)	7 (1)/37
Nutrient broth, 6% NaCl	v	15 (5)	5 (2)/37

[a]Motility may be delayed or difficult to demonstrate.
[b]Incubation of 7–14 days.

Oligella urethralis

Gram-negative small coccoid form and rod (may tend to retain crystal violet stain); nonmotile; aerobic; asaccharolytic; strongly oxidase-positive; grows on MacConkey agar; nitrate-negative; nitrite-positive with gas formation; deaminates phenylalanine to phenylpyruvic acid; frequently citrate-positive, urease-negative.

Sources (22 strains)

Urinary tract (68%), ear (14%), other (18%).

Reference strain

KC1290 ← J. L. Mitchell, Brooks AFB ← H. Lautrop, WM20.

Literature

1. Lautrop, H., K. Bøvre, and W. Frederiksen. 1970. *Moraxella*-like microorganism isolated from the genitourinary tract of man. Acta. Pathol. Microbiol. Scand. Sect. B *78:* 255–256.

2. Riley, P.S., D.G. Hollis, and R.E. Weaver. 1974. Characterization and differentiation of 59 strains of *Moraxella urethralis* from clinical specimens. Appl. Microbiol. *28:* 355–358.

3. Henriksen, S.D. 1973. *Moraxella, Acinetobacter,* and the *Mimeae*. Bacteriol. Rev. *37:* 522–561.

4. Rossau, R., K. Kersters, E. Falsen, E. Jantzen, P. Segers, A. Union, L. Nehls, and J. De Ley. 1987. *Oligella*, a new genus including *Oligella urethralis* comb. nov. (formerly *Moraxella urethralis*) and *Oligella ureolytica* sp. nov. (formerly CDC group IVe): relationship to *Taylorella equigenitalis* and related taxa. Int. J. Syst. Bacteriol. *37:* 198–210.

5. Graham, D.R., J.D. Band, C. Thornsberry, D.G. Hollis, and R.E. Weaver. 1990. Infections caused by *Moraxella, Moraxella urethralis, Moraxella*-like groups M-5 and M-6, and *Kingella kingae* in the United States, 1953-1990. Rev. Infect. Dis. *12:* 423–431.

6. Moss, C.W., P.L. Wallace, D.G. Hollis, and R.E. Weaver. 1988. Cultural and chemical characterization of CDC groups EO-2, M-5, and M-6, *Moraxella (Moraxella)* species, *Oligella urethralis, Acinetobacter* species, and *Psychrobacter immobilis*. J. Clin. Microbiol. *26:* 484–492.

Comments

Formerly called *Moraxella urethralis*. Colonies nearly opaque. This taxon also includes some of the strains formerly called *Mima polymorpha* variety oxidans. Information on the cellular fatty acid composition of this species is presented in Table 4.21 and Figure 4.26 of the cellular fatty acid section.

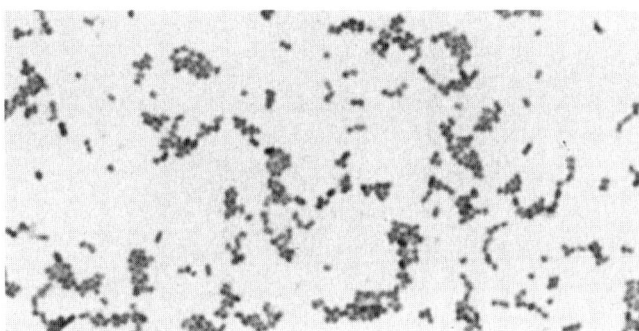

Oligella urethralis strain KC1290. Gram stain BAP 35° C 24 h x 1700.

Oligella urethralis

TEST PERFORMED	SIGN	%+	#+/T
NUMBER OF STRAINS 22			
Morphology	[cc]		
Motility; flagella	[nm]		
Action on blood	v	18 ly	4/22; 1 gr
Fermentative or oxidative	n-o		
Carbohydrate base	OF		
Acid from:			
D-Glucose	[-]	0	0/22
D-Xylose	[-]	0	0/22
D-Mannitol	-	0	0/22
Lactose	-	0	0/22
Sucrose	-	0	0/22
Maltose	-	0	0/22
Catalase	+	100	22/22
Oxidase	[+]	100	22/22
Growth on:			
MacConkey	[+]	96	21/22
SS	-	9	2w/22
Simmons citrate	v	46	10/22
Urea, Christensen's	[-]	0	0/22
Phenylalanine deaminase	[+]	100	20/20
Nitrate reduction	[-]	0	0/22
Nitrite reduction	[+]	100	21/21
Indole	-	0	0/22
TSI slant, acid	-	0	0/22
TSI butt, acid	-	0	0/22
H$_2$S (TSI butt)	-	0	0/22
H$_2$S (Pb ac paper)	-	9	1, 1w/22
Gelatin hydrolysis[a]	-	0	0/14
Litmus milk	v	36 k	8/22
Pigment	-	4 amb sol	1/22
Growth at:			
25°C	v	50	11/22
35°C	+	100	22/22
42°C	v	59	13/22
Esculin hydrolysis	-	0	0/20
Nutrient broth, 0% NaCl	+	96	21/22
Nutrient broth, 6% NaCl	v	59	13/22
Penicillin sensitivity	+	100	13/13

[a]Incubation of 7–14 days.

O-1

Gram-negative rod; motile (1–2 polar with occasionally lateral flagella, may be difficult to demonstrate); oxidizes glucose (frequently weakly and delayed); usually oxidase-positive; variable growth on MacConkey agar; nitrate-, urease-, and indole-negative, produces a yellow growth pigment.

Sources (62 strains)
Blood (55%), cerebrospinal fluid (6%), pleural fluid (6%), wound (5%), cervix (5%), other (vagina, eye, aortic valve, IV fluid, lymph node, platelets for transfusion, sternum, scapula, finger, bone marrow, peritoneal fluid, allergenic extract, water bath, unknown) (23%).

Reference strain
G8807, from blood.

Literature
None

Comment
Cellular fatty acid profile is consistent with that of *Sphingomonas paucimobilis*; however, the lack of acid prodution from xylose, lactose, sucrose, and maltose differentiate this group from *S. paucimobilis*.

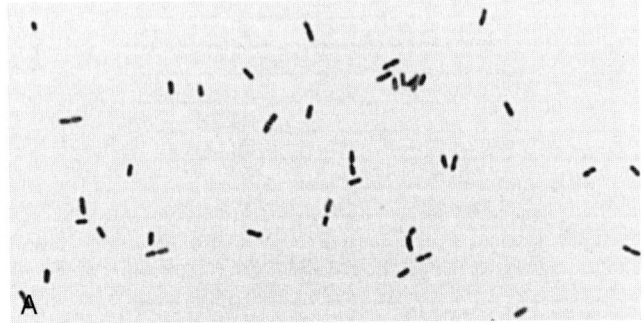

O-1 strain G8807. Gram stain BAP 35° C 24 h x 1700.

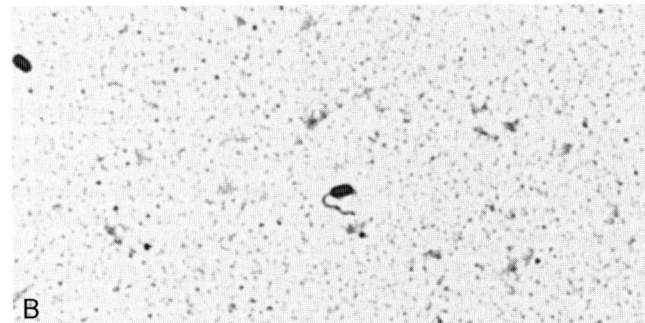

O-1 strain G8807. Flagella stain (Ryu method) TGY 25° C 24 h x 1700.

O-1

	NUMBER OF STRAINS 62		
TEST PERFORMED	SIGN	%+	#+/T
Morphology	rs		
Motility; flagella	[m;p,1-2]a		
Action on blood	v	8 LG	5/62; 5 lav, 3 gr
Fermentative or oxidative	[O]		
Carbohydrate base	OF		
Acid from:			
D-Glucose	[+ or (+)]	69 (31)	15 (6), 28w (13w)/62
D-Xylose	-	0	0/62
D-Mannitol	[-]	0	0/62
Lactose	-	0	0/62
Sucrose	[-]	0	0/62
Maltose	[-]	0	0/62
Catalase	+	98	56, 5w/62
Oxidase	v	77	47/61
Growth on:			
MacConkey	v	6 (40)	2 (25), 2w/62
SS	-	0	0/62
Cetrimide	-	0	0/62
Simmons citrate	-	0	0/62
Urea, Christensen's	-	2	1/62
Nitrate reduction	-	0	0/62
Nitrite reduction	-	0	0/8
Indole	-	0	0/62
TSI slant, acid	-	0	0/62
TSI butt, acid	-	0	0/62
H_2S (TSI butt)	-	0	0/61
H_2S (Pb ac paper)	+	93	57/61
Gelatin hydrolysisb	v	25	10/40
Litmus milk	-	5 IR	3/62
Pigment:			
Insoluble	[yel]	100	62/62
Soluble	v	74 br-ta	46/62
Growth at:			
25°C	+	90	56/62
35°C	+	100	62/62
42°C	v	24	15/62
Esculin hydrolysis	[+]	93 (2)	56 (1)/60
Lysine decarboxylase	-	0	0/23
Arginine dihydrolase	-	4	1/23
Ornithine decarboxylase	-	0	0/23
Nutrient broth, 0% NaCl	v	34	21/61
Nutrient broth, 6% NaCl	-	0	0/61

aMotility is often difficult to demonstrate.
bIncubation of 7–14 days.

O-2

Gram-negative slightly pleomorphic rod (cells sometimes appear thin in the central region, with thickened ends, the so-called "II forms"); variable motility (difficult to detect when present); oxidizes sucrose, maltose, and usually glucose, but not mannitol, xylose, or lactose; oxidase-positive; usually does not grow on MacConkey agar; produces a yellow growth pigment.

Sources (66 strains)

Urine (21%), blood (18%), environmental (ocean water, whirlpool water, hospital equipment) (12%), respiratory (throat, bronchial washing, sinus, lung, tracheal aspirate, pleural fluid) (11%), eye (6%), cerebrospinal fluid (3%), other (26%), unknown (3%).

Reference strain

G2149, from human bone.

Literature

None

Comment

Fatty acid profile analysis, performed on six of these strains, indicates that this group is heterogeneous.

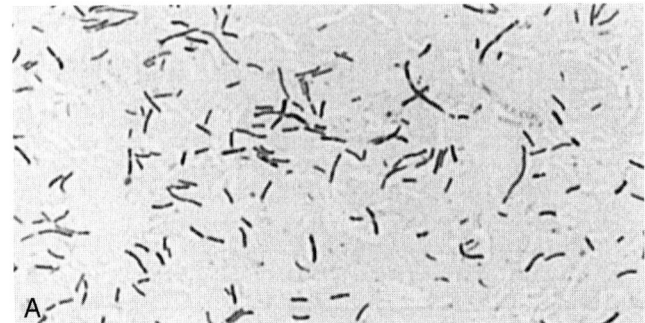

O-2 strain G2149. Gram stain BAP 35° C 24 h x 1700.

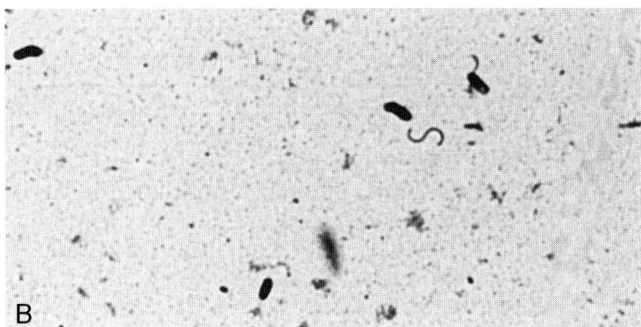

O-2 strain G2149. Flagella stain (Ryu method) TGY 25° C 24 h x 1700.

O-2

	NUMBER OF STRAINS 66		
TEST PERFORMED	SIGN	% +	#+/T
Morphology	srs,II		
Motility; flagella	[v;[a] p,1-2 or p,L]	20	13/66; p,1-2 or p,L
Action on blood	v	14 ß	6/44; 5 ly, 5 gr, 4 LG, 2 lav, 1 al
Fermentative or oxidative	[O or n-o]		
Carbohydrate base	OF		
Acid from:			
D-Glucose	v	73 (11)	25 (2), 23w (5w)/66
D-Xylose	−	2	1w/66
D-Mannitol	−	2	1w/66
Lactose	−	2	1w/66
Sucrose	[+ or (+)]	64 (36)	37 (5), 20w (4w)/66
Maltose	[+ or (+)]	71 (27)	46 (1), 14w (4w)/66
Catalase	+	91	56, 4w/66
Oxidase	+	97	63, 1w/66
Growth on:			
MacConkey	−	5 (5)	2 (2), 1w (1w)/66
SS	−	0	0/66
Cetrimide	−	0	0/54
Simmons citrate	−	0	0/65
Urea, Christensen's	v	12	8w/66
Nitrate reduction	v	15	10/66
Gas from nitrate	−	0	0/66
Indole	−	0	0/66
TSI slant, acid	v	18	4, 8w/66
TSI butt, acid	v	20	2, 11w/66
H_2S (TSI butt)	−	0	0/66
H_2S (Pb ac paper)	+	91	53, 7w/66
Gelatin hydrolysis[b]	v	38	18/47
Litmus milk	v	48 pep	31/65; 4 k, 3 IR
Pigment:			
Insoluble	[yel-or]	100	66/66
Soluble	v	18 ta-br	12/66; 1 yel
Growth at:			
25°C	v	89	58/65
35°C	+	100	65/65
42°C	v	31	20/65
Esculin hydrolysis	v	64	39/61
Lysine decarboxylase	−	0	0/18
Arginine dihydrolase	v	22	4/18
Ornithine decarboxylase	−	6	1/18
Nutrient broth, 0% NaCl	+	92	60/65
Nutrient broth, 6% NaCl	v	22	14/65

[a] Motility was detected by wet preparation only with 2 strains, by flagella stain only with 7 strains, and in OF medium only in 1 strain.

[b] Incubation of 7–14 days.

Pasteurella aerogenes

Gram-negative coccobacillus; nonmotile; ferments glucose with gas, sucrose, maltose, and usually xylose (some strains received by CDC have fermented lactose weakly after 48 h); grows on MacConkey agar; oxidase- and urease-positive (oxidase often weak); nitrate-positive; indole-negative; usually ornithine decarboxylase-positive.

Sources (16 strains)

Swine (intestine, placenta, lung) (44%), cattle (25%), swine bite (19%), other (12%).

Reference strain

C1485, from pig intestine.

Literature

1. McAllister, H.A., and G.R. Carter. 1974. An aerogenic *Pasteurella*-like organism recovered from swine. Am. J. Vet. Res. *35:* 917–922.

2. Carter, G.R. 1984. Genus 1 *Pasteurella*. *In* N.R. Krieg and J.G. Holt (Eds.), Bergey's Manual of Systematic Bacteriology, Vol. 1, Williams & Wilkins, Baltimore, pp. 552–558.

3. Mutters, R., P. Ihm, S. Pohl, W. Frederiksen, and W. Mannheim. 1985. Reclassification of the genus *Pasteurella* Trevisan 1887 on the basis of deoxyribonucleic acid homology, with proposals for the new species *Pasteurella dagmatis, Pasteurella stomatis, Pasteurella anatis,* and *Pasteurella langaa.* Int. J. Syst. Bacteriol. *35:* 309–322.

Comments

Usually associated with swine. The taxonomic position of this species in the genus *Pasteurella* has been questioned (3). Information on the cellular fatty acid composition of this species is presented in Table 4.4C and Figure 4.9 of the cellular fatty acid section.

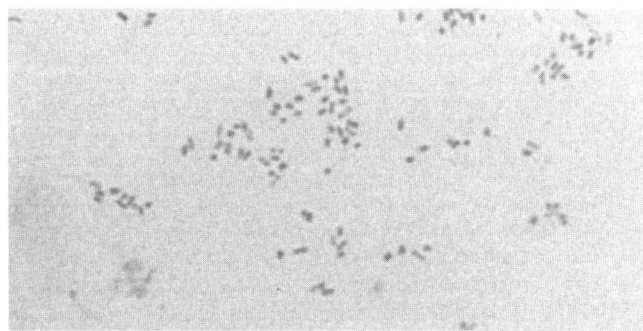

Pasteurella aerogenes strain C1485. Gram stain BAP 35° C 24 h x 1700.

Pasteurella aerogenes

TEST PERFORMED	SIGN	% +	#+/T
Morphology	cc,srs		
Motility; flagella	[nm]		
Gas from glucose	[+]	100	16/16
Action on blood	v	52 al	9/16; 1 ly
Fermentative or oxidative	[F]		
Carbohydrate base	F		
Acid from:			
D-Glucose	+	100	16/16
D-Xylose	v	81	13/16
D-Mannitol	−	6	1/16
Lactose	v	19 (38)	3w (6)/16
Sucrose	+	94	15/16
Maltose	+	100	16/16
Catalase	+	100	15, 1w/16
Oxidase	[+]	100	10, 6w/16
Growth on:			
MacConkey	[+]	100	15, 1w/16
SS	−	6	1w/16
Simmons citrate	−	0	0/16
Urea, Christensen's	[+]	100	16/16
Nitrate reduction	[+]	100	16/16
Gas from nitrate	−	0	0/16
Indole	[−]	0	0/16
TSI slant, acid	+	100	16/16
TSI butt, acid	−	0	0/16
H_2S (TSI butt)	−	0	0/16
H_2S (Pb ac paper)	+	100	16/16
MR	−	0	0/14
VP	v	29	4w/14
Gelatin hydrolysis[a]	−	0	0/14
Litmus milk	v	56A	1, 8w/16
Growth at:			
25°C	+	94	13, 2w/16
35°C	+	100	16/16
42°C	+	94	12, 3w/16
Esculin hydrolysis	−	0	0/16
Lysine decarboxylase	−	0	0/16
Arginine dihydrolase	−	0	0/16
Ornithine decarboxylase	v	88	14/16
Nutrient broth, 0% NaCl	+	100	14, 2w/16
Nutrient broth, 6% NaCl	−	0	0/16

[a]Incubation of 7–14 days.

Pasteurella bettyae

Gram-negative rod; nonmotile; fastidious; ferments glucose with a small amount of gas but not mannitol or sucrose; growth on MacConkey agar variable; oxidase-variable; catalase- and urease-negative; indole- and nitrate-positive.

Sources (88 strains)

Genitourinary, including Bartholin cyst, amniotic fluid, placenta, cervix, lochia, penis, scrotum, perineum, urethra, vagina, and labia (57%); urine (11%); blood (10%); other, including finger, appendix, toe, leg, buttock, gastric secretion, eye, axilla, and graft site (22%).

Reference strain

C7682, from blood.

Literature

1. King, E.O. 1964. The identification of unusual pathogenic gram negative bacteria. Communicable Disease Center, Atlanta.
2. Baddur, L.M., et al. 1989. CDC Group HB-5 as a cause of genitourinary infections in adults. J. Clin. Microbiol. 27: 801–805.
3. Sneath, P.H.A., and M. Stevens. 1990. *Actinobacillus rossi* sp. nov., *Actinobacillus seminis* sp. nov., nom. rev., *Pasteurella bettii* sp. nov., *Pasteurella lymphangitidis* sp. nov., *Pasteurella mairi* sp. nov., and *Pasterella trehalosi* sp. nov. Int. J. Syst. Bacteriol. 40: 148–153.
4. Sneath, P.H.A. 1992. Correction of orthography of epithets in *Pasteurella* and some problems with recommendations on Latinization. Int. J. Syst. Bacteriol. 42: 658–659.

Comments

Formerly HB-5. Most isolates are from genitourinary tract specimens, particularly those of females. Three of the eight blood isolates that we have studied were obtained from neonates. Named in honor of Elizabeth (Betty) O. King. Information on the cellular fatty acid composition of this species is presented in Table 4.4C and Figure 4.9 of the cellular fatty acid section.

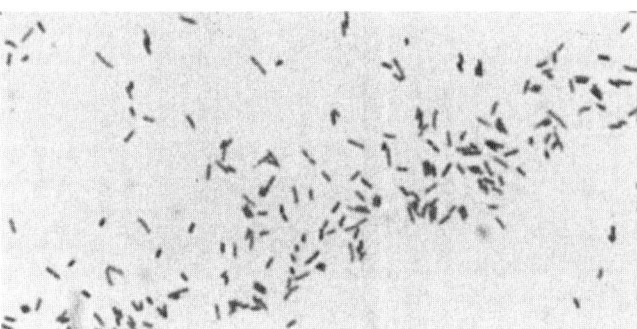

Pasteurella bettyae strain C7682. Gram stain BAP 35° C 24 h x 1700.

Pasteurella bettyae

TEST PERFORMED	SIGN	% +	#+/T
	NUMBER OF STRAINS 88		
Morphology	tsrs		
Motility; flagella	[nm]		
Gas from glucose	[+][a]	92	81/88
Action on blood	v	24 al	19/80; 5 ly, 3 gr, 1 lav
Fermentative or oxidative	[F]		
Carbohydrate base	F		
Acid from:			
D-Glucose	[+]	100	88/88
D-Xylose	−	0	0/88
D-Mannitol	[−]	0	0/88
Lactose	−	0	0/88
Sucrose	[−]	0	0/88
Maltose	−	0	0/88
Fructose	+	100	38/38
Catalase	−	1	1w/81
Oxidase	v	61	31w, 23/88
Growth on:			
MacConkey	v	20 (24)	17 (21)/86
SS	−	0	0/86
Simmons citrate	−	0	0/44
Urea, Christensen's	[−]	0	0/85
Nitrate reduction	[+]	100	88/88
Gas from nitrate	−	2	2/88
Indole	[+]	100	40, 48w/88
TSI slant, acid	+	100	72, 11w/83
TSI butt, acid	[+]	100	76, 7w/83
H_2S (TSI butt)	−	0	0/83
H_2S (Pb ac paper)	+	94	70, 8w/83
MR	−	7	2w/29
VP	−	0	0/19
Gelatin hydrolysis[b]	−	0	0/60
Litmus milk	−	5	4 IR/84
Pigment	−	0	0/49
Growth at:			
25°C	v	79	66/84
35°C	+	96	81/84
42°C	v	13	11/84
Esculin hydrolysis	−	0	0/84
Lysine decarboxylase	−	0	0/10
Arginine dihydrolase	−	0	0/10
Ornithine decarboxylase	−	0	0/11
Nutrient broth, 0% NaCl	v	83	33, 35w/82
Nutrient broth, 6% NaCl	−	10	6, 2w/82

[a] Volume of gas is frequently small.
[b] Incubation of 7–14 days.

Pasteurella canis

Gram-negative small rod and coccoid form; nonmotile; ferments glucose and sucrose, but not maltose, xylose, or mannitol; oxidase-positive; no growth on MacConkey agar; nitrate-positive; ornithine-positive.

Sources (31 strains)

Hand, leg, or finger wound (20 strains); eye (4 strains); urine (2 strains); skin (2 strains); cerebrospinal fluid (1 strain); sputum (1 strain); unknown (1 strain).

Reference strain

G2320, from hand wound.

Literature

1. Mutters, R., P. Ihm, S. Pohl, W. Frederiksen, and W. Mannheim. 1985. Reclassification of the genus *Pasteurella* Trevisan 1887 on the basis of deoxyribonucleic acid homology, with proposals for the new species *Pasteurella dagmatis*, *Pasteurella canis*, *Pasteurella stomatis*, *Pasteurella anatis*, and *Pasteurella langaa*. Int. J. Syst. Bacteriol. 35: 309–322.

2. Escande, F., and C. Lion. 1993. Epidemiology of human infections by *Pasteurella* and related groups in France. Zbl. Bakt. 279: 131–139.

3. Holst, E., J. Rollof, L. Larsson, and J.P. Nielsen. 1992. Characterization and distribution of *Pasteurella* species recovered from infected humans. J. Clin. Microbiol. 30: 2984–2987.

Comments

Associated with animal bite wounds. Two biotypes have been described (1). Biotype 1 (indole-positive) is found in oral cavities of dogs, biotype 2 (indole-negative) has been isolated from calves. Information on the cellular fatty acid composition of this species is presented in Table 4.4C and Figure 4.9 of the cellular fatty acid section.

Pasteurella canis strain G2320. Gram stain BAP 35° C 24 h x 1700.

Pasteurella canis

TEST PERFORMED	SIGN	%+	#+/T
Morphology	srs		
Motility; flagella	[nm]		
Gas from glucose	−	0	0/31
Action on blood	v	29 al	6/21; 2 ly, 3 gr
Fermentative or oxidative	[F]		
Carbohydrate base	F		
Acid from:			
D-Glucose	+	100	28, 3w/31
D-Xylose	−	0	0/31
D-Mannitol	−	0	0/31
Lactose	[−]	0	0/31
Sucrose	[+]	100	28, 3w/31
Maltose	[−]	0	0/31
Catalase	+	96	17, 6w/24
Oxidase	[+]	92	19, 3w/24
Growth on:			
MacConkey	[−]	0	0/24
SS	−	0	0/24
Simmons citrate	−	0	0/24
Urea, Christensen's	−	0	0/24
Nitrate reduction	+	96	23/24
Gas from nitrate	−	4	1/24
Indole	+	96	23/24
TSI slant, acid	+	96	20, 3w/24
TSI butt, acid	+	96	19, 4w/24
H_2S (TSI butt)	−	0	0/24
H_2S (Pb ac paper)	+	92	17, 5w/24
Gelatin hydrolysis[a]	−	0	0/18
Litmus milk	v	14 IR	3/21; 2 a
Growth at:			
25°C	v	62	13/21
35°C	+	100	20, 1w/21
42°C	−	0	0/21
Esculin hydrolysis	−	0	0/22
Lysine decarboxylase	−	7	2/30
Arginine dihydrolase	−	0	0/30
Ornithine decarboxylase	[+]	100	31/31
Nutrient broth, 0% NaCl	v	25	6/24
Nutrient broth, 6% NaCl	v	25	6/24

[a]Incubation of 7–14 days.

Pasteurella dagmatis

Gram-negative rod; nonmotile; ferments glucose, sucrose and maltose (sometimes produces slight gas); xylose- and mannitol-negative; no growth on MacConkey agar; oxidase- and indole-positive (oxidase often weak); urease-positive; reduces nitrate; ornithine decarboxylase-negative.

Sources (129 strains)

Wound (arm, hand, leg, foot, toe, head, face, unspecified) (32%), dog bite (29%), cat (mouth, lung, liver, unspecified) (7%), throat (5%), dog (jaw, unspecified) (5%), blood (3%), other (17%), unknown (2%).

Reference strain

KC558 ← Carpenter, 305-62.

Literature

1. Gump, D.W., and R.A. Holden. 1972. Endocarditis caused by a new species of *Pasteurella*. Ann. Intern. Med. *76:* 275–278.

2. Mutters, R., P. Ihm, S. Pohl, W. Frederiksen, and W. Mannheim. 1985. Reclassification of the genus *Pasteurella* Trevisan 1887 on the basis of deoxyribonucleic acid homology, with proposals for the new species *Pasteurella dagmatis, Pasteurella canis, Pasteurella stomatis, Pasteurella anatis,* and *Pasteurella langaa*. Int. J. Syst. Bacteriol. 35: 309–322.

3. Escande, F., and C. Lion. 1993. Epidemiology of human infections by *Pasteurella* and related groups in France. Zbl. Bakt. *279:* 131–139.

4. Holst, E., J. Rollof, L. Larsson, and J.P. Nielsen. 1992. Characterization and distribution of *Pasteurella* species recovered from infected humans. J. Clin. Microbiol. *30:* 2984–2987.

Comments

Formerly called "Pasteurella gas", *Pasteurella* sp. "new species 1," and *Pasteurella pneumotropica* type Henriksen. Thioglycollate broth with glucose and maltose may be useful for detecting gas production, which is often difficult to demonstrate, even with the hot-needle technique (J. Infect. Dis. *3:* 85-94, 1962). Similar to Bisgaard's taxon 16, except for the urease reaction. Isolated from dog and cat bite wound infections. Information on the cellular fatty acid composition of this species is presented in Table 4.4C and Figure 4.9 of the cellular fatty acid section.

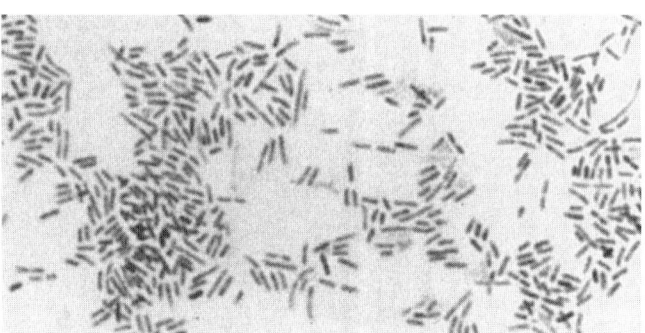

Pasteurella dagmatis strain KC558. Gram stain BAP 35° C 24 h x 1700.

Pasteurella dagmatis

TEST PERFORMED	SIGN	% +	#+/T
Morphology	srs		
Motility; flagella	[nm]		
Gas from glucose	v	25[a]	33/129
Action on blood	v	43 al	49/114; 3 ly, 2 gr
Fermentative or oxidative	[F]		
Carbohydrate base	F		
Acid from:			
D-Glucose	+	100	125, 4w/129
D-Xylose	[−]	0	0/129
D-Mannitol	[−]	0	0/129
Lactose	−	3	3, 1w/129
Sucrose	+	100	125, 4w/129
Maltose	[+]	100	120, 9w/129
Catalase	+	99	109, 15w/125
Oxidase	[+]	98	101, 24w/127
Growth on:			
MacConkey	[−]	1 (2)	1 (2w)/127
SS	−	0	0/127
Simmons citrate	−	0	0/127
Urea, Christensen's	[+]	95 (5)	120, 1w (6)/127
Nitrate reduction	[+]	100	127/127
Gas from nitrate	−	0	0/127
Indole	[+]	100	126, 3w/129
TSI slant, acid	+	100	127/127
TSI butt, acid	+	100	127/127
H_2S (TSI butt)	−	0	0/127
H_2S (Pb ac paper)	+	98 (1)	104, 20w (1w)/127
Gelatin hydrolysis[b]	v	19	18/95
Litmus milk	−	4 IR	5/118; 2 A, 1 k
Growth at:			
25°C	v	87	86, 23w/125
35°C	+	97	116, 5w/125
42°C	v	29	19, 17w/125
Esculin hydrolysis	−	0	0/126
Lysine decarboxylase	−	0	0/63
Arginine dihydrolase	−	0	0/63
Ornithine decarboxylase	−	0	0/63
Nutrient broth, 0% NaCl	v	68 (4)	55, 13w (4)/104
Nutrient broth, 6% NaCl	−	5	3, 2w/104

[a]Gas was observed directly in Durham tubes for 21 strains, and was observed by testing a thioglycollate culture with a hot needle for 21 strains.
[b]Incubation of 7–14 days.

Pasteurella gallinarum

Gram-negative rod and coccoid form; nonmotile; ferments glucose, sucrose, maltose, and sometimes xylose; catalase- and oxidase-positive (often weak); urease- and indole-negative; nitrate-positive.

Sources (10 strains)

Avian (6), lung (1), unknown (3).

Reference strain

KC316 ← W. J. Hall P787 = ATCC 13361, type strain, from chicken upper respiratory tract.

Literature

1. Hall, W.J., K.L. Heddleston, D.H. Legenhausen, and R.W. Hughes. 1955. Studies on pasteurellosis. I. A new species of *Pasteurella* encountered in chronic fowl cholera. Am. J. Vet. Res. *16:* 598–604.
2. Carter, G.R. 1984. Genus 1 *Pasteurella. In* N.R. Krieg and J.G. Holt (Eds.), Bergey's Manual of Systematic Bacteriology, Vol. 1, Williams & Wilkins, Baltimore, pp. 552–558.
3. Mutters, R., P. Ihm, S. Pohl, W. Frederiksen, and W. Mannheim. 1985. Reclassification of the genus *Pasteurella* Trevisan 1887 on the basis of deoxyribonucleic acid homology, with proposals for the new species *Pasteurella dagmatis, Pasteurella stomatis, Pasteurella anatis,* and *Pasteurella langaa*. Int. J. Syst. Bacteriol. *35:* 309–322.

Comment

Usually avian-associated.

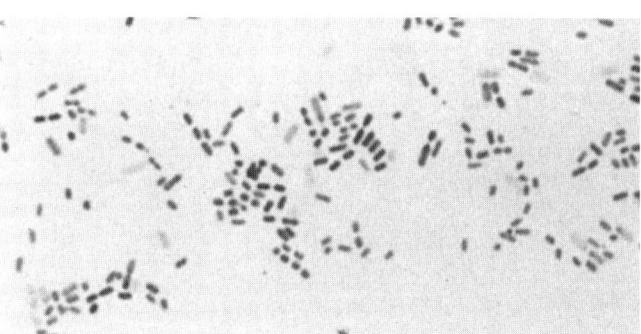

Pasteurella gallinarum strain KC316. Gram stain BAP 35° C 24 h x 1700.

Pasteurella gallinarum

TEST PERFORMED	SIGN	% +	#+/T
Morphology	srs		
Motility; flagella	[nm]		
Gas from glucose	−	0	0/10
Action on blood	v	30 al	3/10
Fermentative or oxidative	[F]		
Carbohydrate base	F		
Acid from:			
D-Glucose	+	90 (10)	9 (1)/10
D-Xylose	v	33	3/9
D-Mannitol	−	0	0/10
Lactose	−	0	0/10
Sucrose	+	100	9, 1w/10
Maltose	+	100	9, 1w/10
Catalase	[+]	100	9, 1w/10
Oxidase	[+]	90	4, 5w/10
Growth on:			
MacConkey	v	20 (10)	2 (1)/10
SS	−	0	0/10
Simmons citrate	−	0	0/10
Urea, Christensen's	[−]	0	0/10
Nitrate reduction	[+]	100	10/10
Gas from nitrate	−	0	0/10
Indole	[−]	0	0/10
TSI slant, acid	+	100	8, 2w/10
TSI butt, acid	+	100	8, 2w/10
H_2S (TSI butt)	−	0	0/10
H_2S (Pb ac paper)	+	90	6, 3w/10
MR	−	0	0/5
VP	v	50	2, 2w/8
Gelatin hydrolysis[a]	−	0	0/10
Litmus milk	−	10	1w/10
Growth at:			
25°C	v	88	7/8
35°C	+	100	8/8
42°C	v	67	4/6
Esculin hydrolysis	−	0	0/9
Lysine decarboxylase	−	0	0/8
Arginine dihydrolase	−	0	0/8
Ornithine decarboxylase	v	25	2/8
Nutrient broth, 0% NaCl	v	25	2w/8
Nutrient broth, 6% NaCl	−	0	0/9

[a]Incubation of 7–14 days.

Pasteurella haemolytica sensu stricto

Gram-negative rod and coccoid form; nonmotile; ferments glucose, xylose, mannitol, and maltose, but not trehalose; β-hemolytic; usually grows on MacConkey agar; catalase-positive; oxidase-positive; urease- and indole-negative; nitrate-positive; lysine decarboxylase-negative; arginine dihydrolase-negative.

Sources (28 strains)

Bovine (brain, lung, liver, stomach, neck) (33%), ovine (lung, pleural fluid, brain) (17%), other animals (chinchilla, hamster, swine) (24%), human (finger, throat, lung) (13%), unknown (13%).

Reference strains

KC627 ← E.L. Biberstein H21 = ATCC 33372, from sheep lung.
KC629 ← E.L. Biberstein H29 = ATCC 33366, from sheep liver.

Literature

1. Newsome, I.E., and F. Cross. 1932. Some bipolar organisms found in pneumonia of sheep. J. Am. Vet. Med. Assoc. *80:* 711–719.

2. Sneath, P.H.A., and M. Stevens. 1990. *Actinobacillus rossi* sp. nov., *Actinobacillus seminis* sp. nov., nom. rev., *Pasteurella bettii* sp. nov., *Pasteurella lymphangitidis* sp. nov., *Pasteurella mairi* sp. nov., and *Pasteurella trehalosi* sp. nov. Int. J. Syst. Bacteriol. *40:* 148–153.

3. Sneath, P.H.A., and M. Stevens. 1985. A numerical taxonomic study of *Actinobacillus, Pasteurella,* and *Yersinia.* J. Gen. Microbiol. *131:* 2711–2738.

4. Carter, G.R. 1984. Genus I. Pasteurella Trevisan 1887, 94[AL] (Nom. cons. Opin. 13, Jud. Comm. 1954, 153). *In* N.R. Krieg and J.G. Holt (Eds.), Bergey's Manual of Systematic Bacteriology, Vol. 1, Williams & Wilkins, Baltimore, pp. 552–557.

Comments

Dies out quickly in culture. Formerly *Pasteurella haemolytica* biovar A. DNA relatedness studies suggest that this group should be transferred to the genus *Actinobacillus* (4).

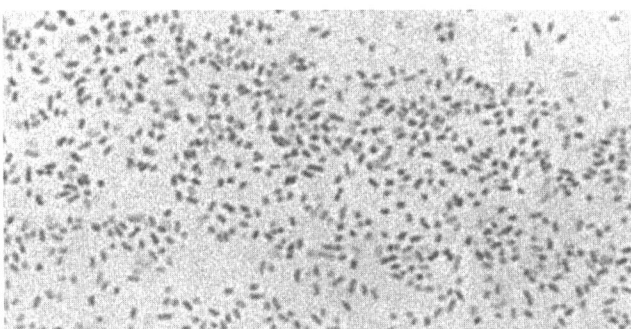

Pasteurella haemolytica strain KC627. Gram stain BAP 35° C 24 h x 1700.

Pasteurella haemolytica sensu stricto

TEST PERFORMED	2 REFERENCE STRAINS		28 PHENOTYPICALLY SIMILAR STRAINS		
	KC627	KC629	SIGN	%+	#+/T
Morphology	srs	srs	srs		
Motility; flagella	[nm]	[nm]	[nm]		
Gas from glucose	−	−	−	0	0/28
Action on blood	β	gr	v	75 β	21/28; 2 ly, 3 al, 1 gr
Fermentative or oxidative	[F]	[F]	[F]		
Carbohydrate base	F	F	F		
Acid from:					
D-Glucose	+	+	+	96 (4)	26, 1w (1)/28
D-Xylose	+	+w	[+]	96 (4)	26, 1w (1)/28
D-Mannitol	+	+w	[+]	96 (4)	26, 1w (1)/28
Lactose	(+w)	−	v	7 (43)	2, (5) (7w)/28
Sucrose	+	+w	+	96 (4)	25, 2w (1)/28
Maltose	+w	+w	+	93 (7)	24, 2w (2)/28
L-Arabinose[a]	−	−	v	23	3/13
D-Galactose	+	+w	[+]	92	11, 1w/13
Trehalose	[−]	[−]	[−]	0	0/13
Catalase	+	+	+	96	22/23
Oxidase	[+]	[+]	[+]	96	24/25
Growth on:					
MacConkey	+	+	v	64 (14)	18 (3) (1w)/28
SS	−	−	−	0	0/28
Simmons citrate	−	−	−	0	0/27
Urea, Christensen's	[−]	[−]	[−]	0	0/28
Nitrate reduction	+	+	+	100	28/28
Gas from nitrate	−	−	−	0	0/28
Indole	−	−	−	0	0/28
TSI slant, acid	+	+	+	100	22/22
TSI butt, acid	+	+	+	100	22/22
H$_2$S (TSI butt)	−	−	−	0	0/22
H$_2$S (Pb ac paper)	+	+	+	100	20, 2w/22
Gelatin hydrolysis[b]	−	−	−	0	0/24
Litmus milk			v	19 A	5/27; 1 k, 1 IR
Growth at:					
25°C	+	+	v	77	13, 4w/22
35°C	+	+	+	100	19, 3w/22
42°C	+w	−	v	14	1, 2w/22
Esculin hydrolysis	−	−	−	0	0/21
Lysine decarboxylase	−	−	−	0	0/6
Arginine dihydrolase	−	−	−	0	0/6
Ornithine decarboxylase	−	−	−	0	0/6
Nutrient broth, 0% NaCl	+	+	+	94	15/16
Nutrient broth, 6% NaCl	−	−	−	6	1/16

[a] >90% of *P. haemolytica* strains in reference 2 were positive.
[b] Incubation of 7–14 days.

Pasteurella multocida

Gram-negative small coccoid rod; nonmotile; ferments glucose, mannitol, and sucrose without gas; no growth on MacConkey agar; oxidase- and indole-positive (oxidase often weak); urease-negative; nitrate-positive; ornithine decarboxylase-positive.

Sources (225 strains)

Sputum (28%), wound (15%), blood (13%), other (dog bite, dog or cat gums or tonsils, cat bite, abscess, hand, throat, bronchi, cat scratch, tissue, exudate) (29%), cerebrospinal fluid (6%), unknown (9%).

Reference strain

D5515, from cerebrospinal fluid.

Literature

1. Rosenbusch, C.T., and I.A. Merchant. 1939. A study of the hemorrhagic septicemia *Pasteurellaceae*. J. Bacteriol. *37:* 69–89.

2. Torphy, D.E., and C.G. Ray. 1969. *Pasteurella multocida* in dog and cat bite infections. Pediatrics *43:* 295–297.

3. Mutters, R., P. Ihm, S. Pohl, W. Frederiksen, and W. Mannheim. 1985. Reclassification of the genus *Pasteurella* Trevisan 1887 on the basis of deoxyribonucleic acid homology, with proposals for the new species *Pasteurella dagmatis*, *Pasteurella canis*, *Pasteurella stomatis*, *Pasteurella anatis*, and *Pasteurella langaa*. Int. J. Syst. Bacteriol. *35:* 309–322.

4. Escande, F., and C. Lion. 1993. Epidemiology of human infections by *Pasteurella* and related groups in France. Zbl. Bakt. *279:* 131–139.

Comments

Associated with animals and animal bites. Strains from human respiratory sources usually are mucoid. Three subspecies have been described (3): *P. multocida* subsp. *multocida* (sorbitol-positive, dulcitol-negative); *P. multocida* subsp. *septica* (sorbitol-negative, dulcitol-negative); and *P. multocida* subsp. *gallicida* (sorbitol-positive, dulcitol-positive). Information on the cellular fatty acid composition of this group is presented in Table 4.4C and Figure 4.9 of the cellular fatty acid section.

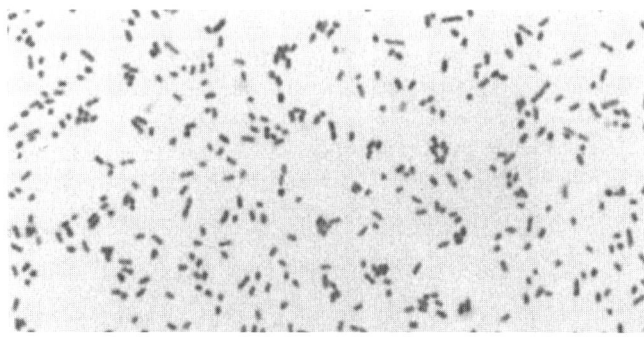

Pasteurella multocida strain D5515. Gram stain BAP 35° C 24 h x 1700.

Pasteurella multocida[a]

TEST PERFORMED	SIGN	% +	#+/T
Morphology	srs		
Motility; flagella	[nm]		
Gas from glucose	[-]	0	0/225
Action on blood	v	29 al	52/180; 5 ly, 4 gr, 3 ß, 1 br, 1 lav
Fermentative or oxidative	[F]		
Carbohydrate base	F		
Acid from:			
D-Glucose	+	98 (2)	221 (4)/225
D-Xylose	v	81 (4)	182 (9)/225
D-Mannitol	+	97 (3)	218 (6) (1w)/225
Lactose	-	8 (<1)	17 (1)/225
Sucrose	+	96 (3)	216 (7)/225
Maltose	[-]	<1	1w/225
Catalase	+	98	198, 17w/224
Oxidase	+	96	153, 58w/220
Growth on:			
MacConkey	[-]	1 (1)	2w (2)/223
SS	-	0	0/223
Simmons citrate	-	0	0/221
Urea, Christensen's	[-]	0	0/219
Nitrate reduction	[+]	99	220/223
Gas from nitrate	-	0	0/223
Indole	[+]	98	214, 2w/220
TSI slant, acid	+	99	220/223
TSI butt, acid	+	92	206/223
H_2S (TSI butt)	-	0	0/216
H_2S (Pb ac paper)	v	59	49, 78w/216
MR	-	3	3/107
VP	v	26	7, 4w/43
Gelatin hydrolysis[b]	-	0	0/161
Litmus milk	-	6 A	13/219; 4 IR
Growth at:			
25°C	v	82	174/211
35°C	+	100	219/219
42°C	v	29	61/209
Esculin hydrolysis	-	0	0/215
Lysine decarboxylase	-	3	3/115
Arginine dihydrolase	-	3	3/208
Ornithine decarboxylase	[+]	93	111/119
Nutrient broth, 0% NaCl	v	79 (1)	174 (2)/219
Nutrient broth, 6% NaCl	v	15	32/217

[a]Three subspecies have been described (3): *P. multocida* subsp. *multocida* (sorbitol-positive, dulcitol-negative); *P. multocida* subsp. *septica* (sorbitol-negative, dulcitol-negative); and *P. multocida* subsp. *gallicida* (sorbitol-positive, dulcitol-positive).
[b]Incubation of 7–14 days.

Pasteurella pneumotropica

Gram-negative rod; nonmotile; ferments (without gas) glucose, xylose, sucrose, maltose, and occasionally lactose, mannitol rarely; may grow on MacConkey agar; oxidase-, urease-, and indole- positive (oxidase often weak); nitrate-positive; ornithine decarboxylase-positive.

Sources (107 strains)

Mouse (58%), rat (10%), other small animals (12%), human (cellulitis, throat, rabbit bite, lymph node, pleural fluid) (15%), other (10%), unknown (5%).

Reference strain

KC315 ← ATCC 12555.

Literature

1. Jawetz, E. 1950. A pneumotropic pasteurella of laboratory animals. I. Bacteriological and serological characteristics of the organism. J. Infect. Dis. *86:* 172–183.

2. Mutters, R., P. Ihm, S. Pohl, W. Frederiksen, and W. Mannheim. 1985. Reclassification of the genus *Pasteurella* Trevisan 1887 on the basis of deoxyribonucleic acid homology, with proposals for the new species *Pasteurella dagmatis, Pasteurella canis, Pasteurella stomatis, Pasteurella anatis,* and *Pasteurella langaa.* Int. J. Syst. Bacteriol. *35:* 309–322.

3. Ryll, M., R. Mutters, and W. Mannheim. 1991. The genetic classification of the *Pasteurella pneumotropica* complex. Berl. Muench. Tieraerztl. Wochenschr. *104:* 243–245.

Comments

Frequently isolated from mice and rats. A recent study (3) indicates that isolates previously identified as *P. pneumotropica* represent at least two species, one of which should be placed in a new genus. Information on the cellular fatty acid composition of this species is presented in Table 4.4C and Figure 4.9 of the cellular fatty acid section.

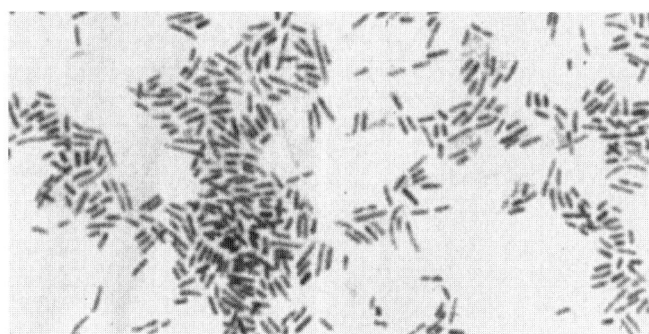

Pasteurella pneumotropica strain KC315. Gram stain BAP 35° C 24 h x 1700.

Pasteurella pneumotropica

TEST PERFORMED	SIGN	% +	#+/T
		NUMBER OF STRAINS 107	
Morphology	srs		
Motility; flagella	[nm]		
Gas from glucose	[-]	0	0/107
Action on blood	v	32 al	33/104; 3 ly, 2 gr, 1 LG
Fermentative or oxidative	[F]		
Carbohydrate base	F		
Acid from:			
D-Glucose	+	97 (3)	99, 5w (3)/107
D-Xylose	[+ or (+)]	76 (19)	77, 4w (20)/107
D-Mannitol	[-]	2 (1)	2 (1)/107
Lactose	v	14 (39)	9, 6w (42)/107
Sucrose	+	97 (3)	99, 5w (3)/107
Maltose	+	97 (3)	99, 5w (3)/107
Catalase	+	100	106/106
Oxidase	[+]	99	76, 29w/106
Growth on:			
MacConkey	[v]	36 (17)	38 (18)/107
SS	-	0	0/107
Simmons citrate	-	0	0/107
Urea, Christensen's	[+][a]	95 (1)	102 (1)/107
Nitrate reduction	[+]	100	107/107
Gas from nitrate	-	0	0/107
Indole	[+]	90	96/107
TSI slant, acid	+	100	102/102
TSI butt, acid	+	97	99/102
H_2S (TSI butt)	-	0	0/103
H_2S (Pb ac paper)	+	100	89, 18w/107
MR	-	3	2w/78
VP	v	13	8, 3w/82
Gelatin hydrolysis[b]	-	0	0/92
Litmus milk	-	9 A	5/104; 3 IR, 2 kw
Growth at:			
25°C	v	74	75/101
35°C	+	98	99/101
42°C	-	8	5/64
Esculin hydrolysis	-	0	0/64
Lysine decarboxylase	v	33	9/27
Arginine dihydrolase	-	0	0/27
Ornithine decarboxylase	[+]	100	27/27
Nutrient broth, 0% NaCl	v	84	37/44
Nutrient broth, 6% NaCl	-	2	1/43

[a]May require a drop of serum on slant or a heavy inoculum.
[b]Incubation of 7–14 days.

Pasteurella stomatis

Gram-negative rod; nonmotile; ferments glucose and sucrose, not xylose, mannitol, or maltose; does not grow on MacConkey agar; oxidase-, catalase-, and indole-positive; urease- and ornithine decarboxylase-negative.

Sources (8 strains)

Face wound, cat bite wound on forearm, pleural fluid, right palm wound, blood, and head wound (1 strain each); unknown (2 strains).

Reference strain

G1504 = NCTC 11623 = ATCC 43327, type strain, from dog throat.

Literature

1. Mutters, R., P. Ihm, S. Pohl, W. Frederiksen, and W. Mannheim. 1985. Reclassification of the genus *Pasteurella* Trevisan 1887 on the basis of deoxyribonucleic acid homology, with proposals for the new species *Pasteurella dagmatis*, *Pasteurella canis*, *Pasteurella stomatis*, *Pasteurella anatis*, and *Pasteurella langaa*. Int. J. Syst. Bacteriol. *35:* 309–322.
2. Holst, E., J. Rollof, L. Larsson, and J.P. Nielsen. 1992. Characterization and distribution of *Pasteurella* species recovered from infected humans. J. Clin. Microbiol. *30:* 2984–2987.
3. Pouëdras, P., P.Y. Donnio, Y.L. Tulzo, and J.L. Avril. 1993. *Pasteurella stomatis* infection following a dog bite. Eur. J. Clin. Microbiol. Infect. Dis. *12:* 65.
4. Frederiksen, W. 1993. Ecology and significance of *Pasteurellaceae* in man—an update. Zbl. Bakt. *279:* 27–34.

Comments

Found in the throat of dogs and cats. Associated with bite wound infection. Similar to *P. canis* except for the ornithine decarboxylase reaction. Similar to Bisgaard's taxon 16 except for the negative maltose reaction. Information on the cellular fatty acid composition of this species is presented in Table 4.4C and Figure 4.9 of the cellular fatty acid section.

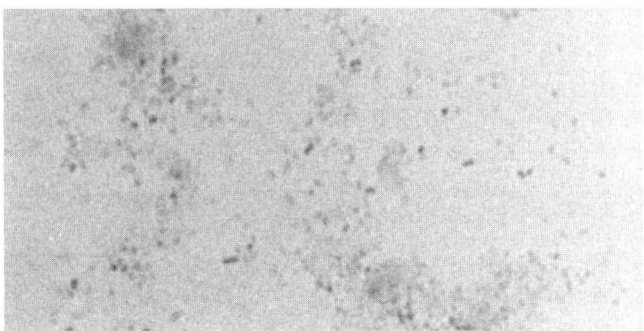

Pasteurella stomatis strain G1504. Gram stain BAP 35° C 24 h x 1700.

Pasteurella stomatis

TEST PERFORMED	SIGN	% +	#+/T
Morphology	srs		
Motility; flagella	[nm]		
Action on blood	v	38 al	3/8; 1 ly
Fermentative or oxidative	[F]		
Carbohydrate base	F		
Acid from:			
D-Glucose	+	100	6, 2w/8
D-Xylose	[-]	0	0/8
D-Mannitol	-	0	0/8
Lactose	-	0	0/8
Sucrose	[+]	100	6, 2w/8
Maltose	[-]	0	0/8
Catalase	+	100	7, 1w/8
Oxidase	[+]	100	7, 1w/8
Growth on:			
MacConkey	[-]	0	0/8
SS	-	0	0/8
Simmons citrate	-	0	0/8
Urea, Christensen's	[-]	0	0/8
Nitrate reduction	+	100	8/8
Gas from nitrate	-	0	0/8
Indole	[+]	100	8/8
TSI slant, acid	+	100	8/8
TSI butt, acid	+	100	7, 1w/8
H_2S (TSI butt)	-	0	0/8
H_2S (Pb ac paper)	+	100	6, 2w/8
Gelatin hydrolysis[a]	-	0	0/7
Litmus milk	v	38 IR	3/8; 1 A, 1 Aw
Growth at:			
25°C	v	38	3/8
35°C	+	100	8/8
42°C	v	25	2/8
Esculin hydrolysis	-	0	0/8
Lysine decarboxylase	-	0	0/8
Arginine dihydrolase	-	0	0/8
Ornithine decarboxylase	[-]	0	0/8
Nutrient broth, 0% NaCl	v	(25)	(1) (1w)/8
Nutrient broth, 6% NaCl	v	13	1/8

[a]Incubation of 7–14 days.

Pasteurella trehalosi

Gram-negative rod and coccoid form; nonmotile; ferments glucose, mannitol, maltose, and trehalose, not arabinose, xylose, or galactose; catalase-negative or weakly positive; usually grows on MacConkey agar; oxidase-positive; urease- and indole-negative; nitrate-positive; ornithine decarboxylase-negative.

Sources (11 strains)

Bovine (lung, trachea, liver, blood, fetus) (8 strains), other animals (1 strain each, ovine lung, ovine spleen, caprine lymph node).

Reference strain

5247, from bovine lung.

Literature

1. Sneath, P.H.A., and M. Stevens. 1990. *Actinobacillus rossii* sp. nov., *Actinobacillus seminis* sp. nov., nom. rev., *Pasteurella bettii* sp. nov., *Pasteurella lymphangitidis* sp. nov., *Pasteurella mairi* sp. nov., and *Pasteurella trehalosi* sp. nov. Int. J. Syst. Bacteriol. 40: 148–153.

2. Sneath, P.H.A., and M. Stevens. 1985. A numerical taxonomic study of *Actinobacillus, Pasteurella,* and *Yersinia.* J. Gen. Microbiol. 131: 2711–2738.

3. Carter, G.R. 1984. Genus I. *Pasteurella* Trevisan 1887, 94[AL] (Nom. cons. Opin. 13, Jud. Comm. 1954, 153). *In* N.R. Krieg and J.G. Holt (Eds.), Bergey's Manual of Systematic Bacteriology, Vol. 1, Williams & Wilkins, Baltimore, pp. 552–557.

Comment

Formerly biovar T of *Pasteurella haemolytica*.

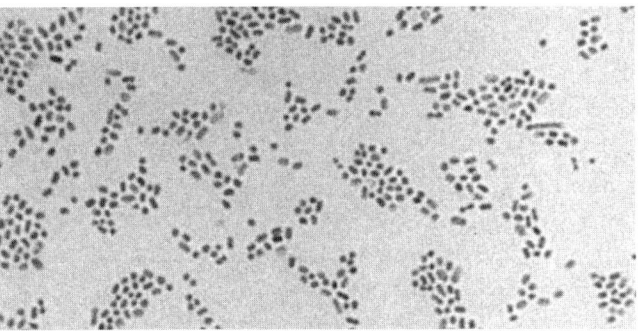

Pasteurella trehalosi strain 5247. Gram stain BAP 35° C 24 h x 1700.

Pasteurella trehalosi

TEST PERFORMED	SIGN	% +	#+/T
Morphology	srs		
Motility; flagella	[nm]		
Action on blood	v	27 ß	3/11; 1 al
Fermentative or oxidative	[F]		
Carbohydrate base	F		
Acid from:			
D-Glucose	+	100	8, 3w/11
D-Xylose	-	0	0/11
D-Mannitol	[+ or (+)]	82 (18)	6, 4w (1w)/11
Lactose	-	0	0/11
Sucrose	+	100	8, 3w/11
Maltose	+ or (+)	82 (18)	8, 2w (1w)/11
L-Arabinose	[-]	0	0/11
D-Galactose	[-]	0	0/11
Trehalose	[+]	100	8, 3w/11
Catalase	v	27	3w/11
Oxidase	+	100	8, 3w/11
Growth on:			
MacConkey	[+]	100	11/11
SS	-	0	0/11
Simmons citrate	-	0	0/11
Urea, Christensen's	[-]	0	0/11
Nitrate reduction	+	100	11/11
Gas from nitrate	-	0	0/11
Indole	[-]	0	0/11
TSI slant, acid	+	100	11/11
TSI butt, acid	+	100	11/11
H_2S (TSI butt)	-	0	0/11
H_2S (Pb ac paper)	+	100	9, 2w/11
Gelatin hydrolysis[a]	v	36	4/11
Litmus milk	-	9 A	1/11
Pigment	-	0	0/11
Growth at:			
25°C	v	55	3, 3w/11
35°C	+	100	11/11
42°C	v	27	2, 1w/11
Esculin hydrolysis	v	64	7/11
Lysine decarboxylase	-	0	0/7
Arginine dihydrolase	-	0	0/7
Ornithine decarboxylase	[-]	0	0/7
Nutrient broth, 0% NaCl	v	18	2w/11
Nutrient broth, 6% NaCl	-	0	0/11

[a]Incubation of 7–14 days.

Pseudomonas aeruginosa

Gram-negative rod; motile (single, occasionally two, polar flagella); aerobic; uses glucose, xylose, and usually mannitol, not sucrose or maltose; grows on MacConkey, SS, and cetrimide agars; oxidase- and arginine dihydrolase-positive, lysine decarboxylase-negative; reduces nitrate to gas; indole-negative; grows at 42° C; acetamide- and 2-ketogluconate-positive; proteolytic; may produce pyocyanin (blue), pyoverdin (yellow-green) (fluoresces with a Woods lamp), pyorubrin (red), and pyomelanin (brown) water-soluble pigments.

Sources (201 strains)

Urine (29%), sputum (18%), environmental (8%), ear (5%), other (28%), unknown (12%).

Reference strains

KC1072 ← R. Y. Stanier, 45.
KC1073 ← R. Y. Stanier, 52.

Literature

1. Migula, W. 1900. System der Bakterien, Vol. 2. G. Fischer, Jena.

2. Stanier, R.Y., N.J. Palleroni, and M. Doudoroff. 1966. The aerobic pseudomonads; a taxonomic study. J. Gen. Microbiol. *43:* 159–271.

3. Judicial Commission. 1970. Opinion 36. Designation of strain ATCC 10145 as the neotype strain of *Pseudomonas aeruginosa* (Schroeter) Migula. Int. J. Syst. Bacteriol. *20:* 15.

4. Gilardi, G.L. 1991. *Pseudomonas* and related genera. In A. Balows, et al. (Eds.), Manual of Clinical Microbiology, 5th ed., American Society for Microbiology, Washington, pp. 429–441.

5. Gilligan, P.H. 1991. Microbiology of airway disease in patients with cystic fibrosis. Clin. Microbiol. Rev. *4:* 35–51.

Comments

Growth frequently exhibits a metallic sheen; mucoid strains occur and frequently may be atypical biochemically. To our knowledge, *P. aeruginosa* is the only species of *Pseudomonas* that produces pyocyanin. Information on the cellular fatty acid composition of this species is presented in Table 4.37 and Figure 4.42 of the cellular fatty acid section.

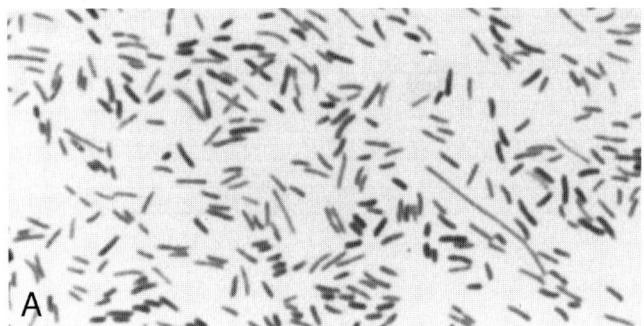

Pseudomonas aeruginosa strain KC1072. Gram stain BAP 35° C 24 h x 1700.

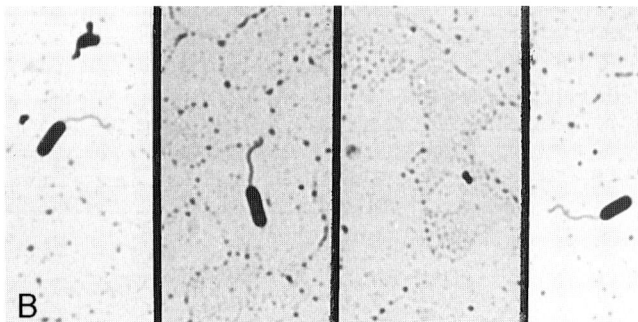

Pseudomonas aeruginosa strain KC1072. Flagella stain BAP 35° C 24 h x 1700.

Pseudomonas aeruginosa

	NUMBER OF STRAINS 201		
TEST PERFORMED	SIGN	%+	#+/T
Morphology	mrs		
Motility; flagella	[m;p,1-2]		
Action on blood	v	42 LG	70/168; 33 ß, 32 ly, 17 gr, 9 br
Fermentative or oxidative	[O]		
Carbohydrate base	OF		
Acid from:			
D-Glucose	[+]	97 (1)	195 (2)/201
D-Xylose	+	90 (1)	182 (2)/201
D-Mannitol	v	70 (3)	141 (6)/201
Lactose	−	<1 (<1)	1 (1)/201
Sucrose	[−]	0	0/201
Maltose	−	(<1)	(1)/201
Catalase	+	100	167/167
Oxidase	[+]	99	163/165
Growth on:			
MacConkey	[+]	100	166/166
SS	+	96	158/165
Cetrimide	[+]	94 (2)	156 (4)/165
Simmons citrate	+	95 (1)	158 (1)/166
Urea, Christensen's	v	48 (9)	75, 3w (15)/161
Nitrate reduction	+	98	163/167
Gas from nitrate	[+]	93	156/167
Indole	−	0	0/168
TSI slant, acid	−	0	0/164
TSI butt, acid	−	0	0/164
H$_2$S (TSI butt)	−	0	0/168
H$_2$S (Pb ac paper)	[−]	4	2, 2w/168
Gelatin hydrolysis[a]	[v]	82	111/135
Litmus milk	[v]	89 pep	153/172; 2 IR, 2 k, 2 c
Pigment:			
Pyoverdin	[v]	65	131/201
Pyocyanin	[v]	46	92/201
Pyorubrin	v	25	22/85
Other	v	23 pyomelanin	47/201
Growth at:			
25°C	+	100	201/201
35°C	+	100	201/201
42°C	[+]	100	201/201
Esculin hydrolysis	−	0	0/156
Lysine decarboxylase	[−]	0	0/15
Arginine dihydrolase	[+]	100	15/15
Ornithine decarboxylase	−	0	0/15
Nutrient broth, 0% NaCl	+	100	148/148
Nutrient broth, 6% NaCl	v	65	96/147
Alkalinization of acetamide	[+]	100	70, 1w/71
2-Ketogluconate	+	96	27/28

[a]Incubation of 7–14 days.

Pseudomonas alcaligenes

Gram-negative rod or filament; motile (1–2 polar flagella); aerobic; grows on MacConkey agar; does not use carbohydrates; oxidase-positive, urease-negative, usually does not grow at 42° C; sometimes reduces nitrate, without gas.

Sources (26 strains)

Urine (15%), blood (12%), sputum (12%), ear (8%), tissue culture (8%), other (41%), unknown (4%).

Reference strain

KC1075 ← ATCC 14909, type strain, from water in a swimming pool.

Literature

1. Monias, B.L. 1928. Classification of *Bacterium alcaligenes, pyocyaneum,* and *fluorescens.* J. Infect. Dis. *43:* 330–334.
2. Hugh, R., and P. Ikari. 1964. The proposed neotype strain of *Pseudomonas alcaligenes* Monias (1928). Int. Bull. Bacteriol. Nomencl. Taxon. *14:* 103–107.
3. Stanier, R.Y., N.J. Palleroni, and M. Doudoroff. 1966. The aerobic pseudomonads: a taxonomic study. J. Gen. Microbiol. *43:* 243–247.
4. Gilardi, G.L. 1991. *Pseudomonas* and related genera. *In* A. Balows, et al. (Eds.), Manual of Clinical Microbiology, 5th ed., American Society for Microbiology, Washington, pp. 429–441.

Comment

Information on the cellular fatty acid composition of this species is presented in Table 4.9 and Figure 4.14 of the cellular fatty acid section.

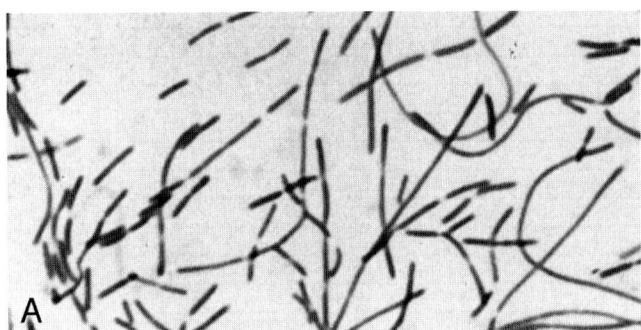

Pseudomonas alcaligenes strain KC1075. Gram stain BAP 35° C 24 h x 1700.

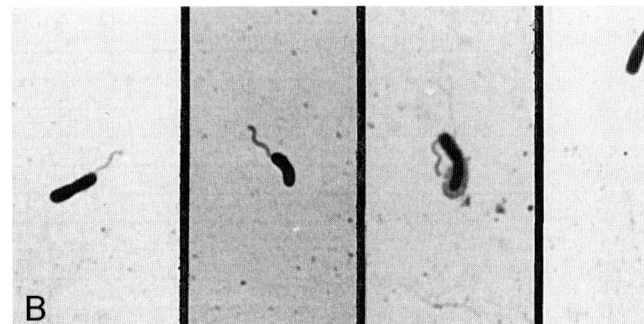

Pseudomonas alcaligenes strain KC1075. Flagella stain BAP 35° C 24 h x 1700.

Pseudomonas alcaligenes

TEST PERFORMED	SIGN	%+	#+/T
	NUMBER OF STRAINS 26		
Morphology	mrs		
Motility; flagella	[m;p,1-2]		
Action on blood	v	31 ly	8/26; 3 LG, 2 gr, 1 al, 1 ß
Fermentative or oxidative	[n-o]		
Carbohydrate base	OF		
Acid from:			
D-Glucose	[-]	0	0/26
D-Xylose	[-]	0	0/26
D-Mannitol	-	0	0/26
Lactose	-	0	0/26
Sucrose	-	0	0/26
Maltose	[-]	0	0/26
Fructose	[-]	0	0/12
Catalase	+	92	24/26
Oxidase	[+]	96	25/26
Growth on:			
MacConkey	[+]	96	25/26
SS	v	38 (8)	10 (2)/26
Cetrimide	v	15 (4)	4 (1)/26
Simmons citrate	v	57 (8)	15 (2)/26
Urea, Christensen's	[-]	0	0/25
Nitrate reduction	v	54	14/26
Gas from nitrate	[-]	0	0/26
Indole	-	0	0/26
TSI slant, acid	-	0	0/26
TSI butt, acid	-	0	0/26
H_2S (TSI butt)	-	0	0/26
H_2S (Pb ac paper)	v	65	17/26
Gelatin hydrolysis[a]	-	0	0/18
Litmus milk	v	46 k	12/26; 4 IR
Pigment:			
Soluble	v	32 ta	12/19; sl yel
Growth at:			
25°C	+	100	26/26
35°C	+	100	26/26
42°C	-	0	0/26
Esculin hydrolysis	-	0	0/26
Lysine decarboxylase	-	0	0/8
Arginine dihydrolase	v	12	1/8
Ornithine decarboxylase	-	0	0/8
Nutrient broth, 0% NaCl	+	95	18/19
Nutrient broth, 6% NaCl	v	41	9/22

[a]Incubation of 7–14 days.

Pseudomonas cepacia

Gram-negative straight or slightly curved rod; motile (polar tuft of three or more flagella); oxidizes glucose, xylose, mannitol, lactose, and maltose; usually lysine decarboxylase-positive, arginine dihydrolase-negative; grows on MacConkey agar; sometimes reduces nitrate, without gas.

Sources (159 strains)

Urine (31%), blood (16%), cerebrospinal fluid (8%), bronchial washing (7%), detergicide (6%), sputum (6%), abscess, exudate, and wound (6%), other (9%), unknown (11%).

Reference strain

KC1371 ← ATCC 25416, type strain, from onion.

Literature

1. Burkholder, W.H. 1950. Sour skin, a bacterial rot of onion bulbs. Phytopathology *40:* 115–-117.

2. Daily, R.H., and E.J. Benner. 1968. Necrotizing pneumonitis due to the pseudomonad "eugonic oxidizer group 1." N. Engl. J. Med. *279:* 361–362.

3. Jonsson, V. 1970. Proposal of a new species *Pseudomonas kingii*. Int. J. Syst. Bacteriol. *20:* 255–257.

4. Ballard, R.W., N.J. Palleroni, M. Doudoroff, R.Y. Stanier, and M. Mandel. 1970. Taxonomy of the aerobic pseudomonads: *Pseudomonas cepacia, P. marginata, P. alliicola*, and *P. caryophylli*. J. Gen. Microbiol. *60:* 199–214.

5. Bassett, D.C. 1970. Wound infection with *Pseudomonas multivorans*. Lancet *1:* 1188–1191.

6. Sinsabaugh, H.H., and G.W. Howard, Jr. 1975. Emendation of the description of *Pseudomonas cepacia* Burkholder (synonyms: *Pseudomonas multivorans* Stanier et al., *Pseudomonas kingae* Jonsson; EO-1 Group). Int. J. Syst. Bacteriol. *25:* 187–201.

7. Goldmann, D.A., and J.D. Klinger. 1986. *Pseudomonas cepacia:* biology, mechanisms of virulence, epidemiology. J. Pediatr. *108:* 806–812.

8. Gilligan, P.H. 1991. Microbiology of airway disease in patients with cystic fibrosis. Clin. Microbiol. Rev. *4:* 35–51.

9. Yabuuchi, E., et al. 1992. Proposal of *Burkholderia* gen. nov. and transfer of seven species of the genus *Pseudomonas* homology group II to the new genus, with the type species *Burkholderia cepacia* (Palleroni and Holmes 1981) comb. nov. Microbiol. Immunol. *36:* 1251–1275.

Comments

Synonyms: *P. multivorans, P. kingii*, EO-1 (eugonic oxidizer). Surface growth dies out quickly; survives "disinfection" with quaternary ammonium compounds. Some strains produce a yellow growth pigment on TSI agar and Kligler's iron agar, but not on media that do not contain iron; others produce a chartreuse growth pigment on common laboratory media; most strains exhibit no growth pigment. Since the early 1980s, it has emerged as a potential pathogen in patients with cystic fibrosis. It recently has been proposed that these organisms, along with other *Pseudomonas* homology group II species, be placed in the new genus *Burkholderia* (9). Information on the cellular fatty acid composition of this species is presented in Table 4.62 and Figure 4.67 of the cellular fatty acid section.

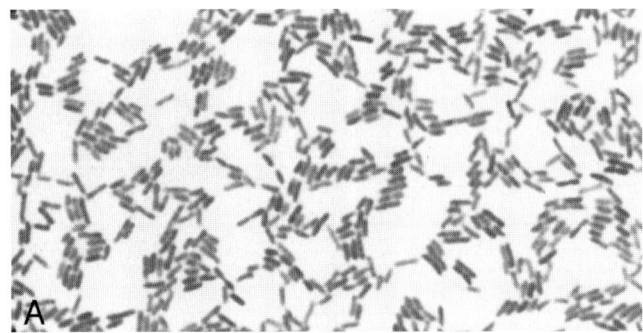
Pseudomonas cepacia strain KC1371. Gram stain BAP 35° C 24 h x 1700.

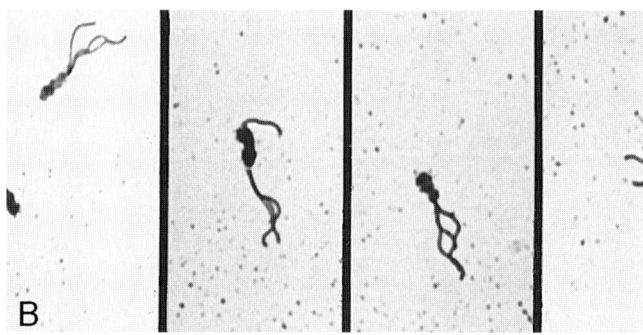
Pseudomonas cepacia strain KC1371. Flagella stain FB 25° C 24 h x 1700.

Pseudomonas cepacia

TEST PERFORMED	SIGN	% +	#+/T
	NUMBER OF STRAINS 159		
Morphology	mrs		
Motility; flagella	[m:p, >2]		
Action on blood	v	47 ly	65/139; 30 LG, 16 gr, 17 lav, 9 al
Fermentative or oxidative	[O]		
Carbohydrate base	OF		
Acid from:			
D-Glucose	[+]	100	159/159
D-Xylose	+	100	159/159
D-Mannitol	[+]	100	159/159
Lactose	[+]	99 (1)	156, 2w (1)/159
Sucrose	v	86 (1)	137 (1)/159
Maltose	[+]	99 (1)	157 (2)/159
Catalase	+	99	109, 44w/155
Oxidase	v	86	53, 80w/154
Growth on:			
MacConkey	[+]	100	155/155
SS	v	6 (18)	9 (27)/154
Cetrimide	v	44 (22)	69 (34)/155
Simmons citrate	+	94 (5)	146 (8)/156
Urea, Christensen's	v	60 (18)	93 (28)/155
Nitrate reduction	v	57	87/152
Gas from nitrate	[-]	0	0/152
Indole	-	0	0/152
TSI slant, acid	-	1	2/157
TSI butt, acid	-	0	0/157
H_2S (TSI butt)	-	0	0/140
H_2S (Pb ac paper)	[-]	0	0/141
Gelatin hydrolysis[a]	v	20	24/118
Litmus milk	pep	94	132/140; 6 IR, 1 k
Pigment:			
Insoluble	v	60 yel (TSI)[b]	95/159
Growth at:			
25°C	+	98	142/145
35°C	+	100	145/145
42°C	v	83	120/145
Esculin hydrolysis	v	63 (6)	85 (8)/135
Lysine decarboxylase	[v]	80	32/40
Arginine dihydrolase	[-]	0	0/40
Ornithine decarboxylase	v	48	19/40
Nutrient broth, 0% NaCl	+	100	132/132
Nutrient broth, 6% NaCl	-	7	9/135

[a]Incubation of 7–14 days.

[b]Sixty percent of the *P. cepacia* strains produced a yellow pigment, which occurs only on iron-containing media (e.g., TSI). This usually is not evident before 42–48 h. Sometimes a chartreuse pigment occurs on ordinary media.

"Pseudomonas denitrificans"

Gram-negative medium to long rod; motile (single polar flagellum); oxidizes glucose, xylose, and maltose; grows on MacConkey, cetrimide, and usually SS agar; oxidase-positive; reduces nitrate, usually with gas formation; arginine dihydrolase-positive.

Sources (29 strains)

Blood (24%), sputum (24%), peritoneal fluid (10%), environmental (IV solution, water, culture media) (10%), other (18%), unknown (14%).

Reference strain

KC1439 ← ATCC 13867.

Literature

1. Bergey, D.H., F.C. Harrison, R.S. Breed, B.W. Hammer, and F.M. Huntoon (Eds.). 1923. Bergey's Manual of Determinative Bacteriology, 1st ed., Williams & Wilkins, Baltimore.

2. Doudoroff, M., R. Contopoulou, R. Kunisawa, and N.J. Palleroni. 1974. Taxonomic validity of *Pseudomonas denitrificans* (Christensen) Bergey et al. Request for an opinion. Int. J. Syst. Bacteriol. 24: 294–300.

3. Judicial Commission of the International Committee of Systematic Bacteriology. 1982. Opinion 54. Rejection of the species name *Pseudomonas denitrificans* (Christensen) Bergey et al. 1923. Int. J. Syst. Bacteriol. 32: 466.

Comments

The name "*Pseudomonas denitrificans*" currently has no standing in the literature. The strains tabulated below are similar to strain KC1439, which was originally described as *P. denitrificans*. The taxonomic position of these strains is unsettled. Information on the cellular fatty acid composition of this species is presented in Table 4.37 and Figure 4.42 of the cellular fatty acid section.

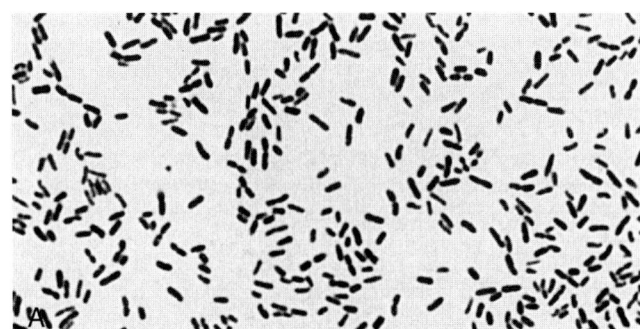 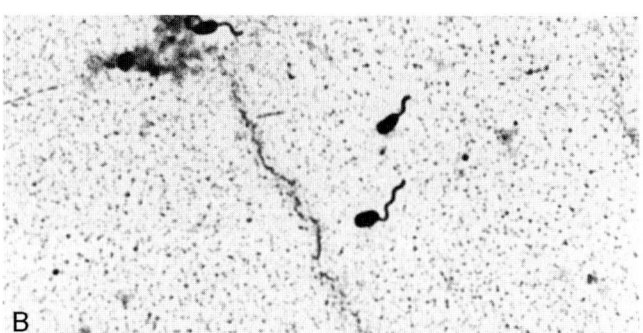

"*Pseudomonas denitrificans*" strain KC1439. Gram stain BAP 35° C 24 h x 1700.

"*Pseudomonas denitrificans*" strain KC1439. Flagella stain (Ryu method) TGY 25° C 24 h x 1700

"Pseudomonas denitrificans"

TEST PERFORMED	REFERENCE STRAIN SIGN	28 PHENOTYPICALLY SIMILAR STRAINS		
		SIGN	%+	#+/T
Morphology	mrs	mrs		
Motility; flagella	[m;p,1-2]	[m;p,1-2]		
Action on blood	ly	v	31 ly	8/26; 6 LG, 5 gr, 1 lav
Fermentative or oxidative	O	O		
Carbohydrate base	OF	OF		
Acid from:				
D-Glucose	[+]	[+]	100	28/28
D-Xylose	(+)	+ or (+)	43 (57)	8 (8), 4w (8w)/28
D-Mannitol	–	–	0	0/28
Lactose	+	+	96 (4)	26, 1w (1w)/28
Sucrose	–	–	0	0/28
Maltose	+	+	100	28/28
Catalase	+	+	100	28/28
Oxidase	+	+	100	28/28
Growth on:				
MacConkey	[+]	[+]	100	28/28
SS	[+]	[+ or (+)]	89 (4)	24 (1), 1w/28
Cetrimide	+	+	100	28/28
Simmons citrate	+	+	96	27/28
Urea, Christensen's	–	v	25 (18)	6, 1w (5w)/28
Nitrate reduction[a]	[+]	[+]	100	28/28
Gas from nitrate	[+]	[v]	71	20/28
Indole	–	–	0	0/28
TSI slant, acid	–	–	0	0/28
TSI butt, acid	–	–	0	0/28
H_2S (TSI butt)	–	–	7	2/28
H_2S (Pb ac paper)	+	+	100	28/28
Gelatin hydrolysis[b]	–	–	0	0/25
Litmus milk	k	v	82 k	23/28; 2 IR
Pigment:				
Soluble	tan	v	44 yel-ta	12/27; 5 br
Insoluble	–	v	44 yel-ta	12/27
Growth at:				
25°C	+	+	93	26/28
35°C	+	+	100	28/28
42°C	+	v	39	11/28
Esculin hydrolysis	–	–	0	0/28
Lysine decarboxylase	–	–	0	0/24
Arginine dihydrolase	[+]	[+]	100	24/24
Ornithine decarboxylase	–	–	0	0/24
Nutrient broth, 0% NaCl	+	+	100	28/28
Nutrient broth, 6% NaCl	+	v	14	4/28
Alkalinization of:				
Acetamide	–	–	0	0/10
Serine	(+)	v	20 (50)	2 (5)/10
Tartrate	–	–	0	0/10

[a]Nitrate was reduced through nitrite to completion.
[b]Incubation of 7–14 days.

Pseudomonas diminuta

Gram-negative short to long rod; motile (single polar flagellum with distinctive, very short wavelength); aerobic; grows on MacConkey agar; usually does not use carbohydrates, although some strains oxidize glucose; produces oxidase; rarely reduces nitrate to nitrite; indole negative.

Sources (68 strains)

Blood (29%), cerebrospinal fluid (10%), urine (10%), wound (7%), other (32%), unknown (12%).

Reference strains

KC679 ← ATCC 11568, type strain.
D7054 (1), from toe wound.

Literature

1. Leifson, E., and R. Hugh. 1954. A new type of polar monotrichous flagellation. J. Gen. Microbiol. *10:* 68–70.
2. Ballard, R.W., M. Doudoroff, R.Y. Stanier, and M. Mandel. 1968. Taxonomy of the aerobic pseudomonads: *Pseudomonas diminuta* and *P. vesiculare*. J. Gen. Microbiol. *53:* 349–361.
3. Kaltenbach, C.M., C.W. Moss, and R.E. Weaver. 1975. Cultural and biochemical characteristics and fatty acid composition of *Pseudomonas diminuta* and *Pseudomonas vesiculare*. J. Clin. Microbiol. *1:* 339–344.
4. Gilardi, G.L. 1991. *Pseudomonas* and related genera. *In* A. Balows, et al. (Eds.), Manual of Clinical Microbiology, 5th ed., American Society for Microbiology, Washington, pp. 429–441.

Comments

This organism distinctively produces a brown, water-soluble pigment on heart infusion agar with tyrosine. Information on the cellular fatty acid composition of this species is presented in Table 4.63 and Figure 4.68 of the cellular fatty acid section.

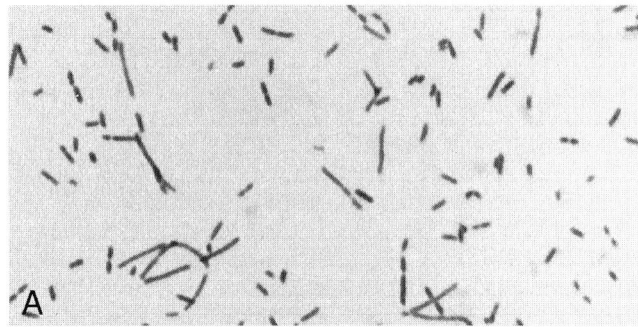

Pseudomonas diminuta strain KC679. Gram stain BAP 35° C 24 h x 1700.

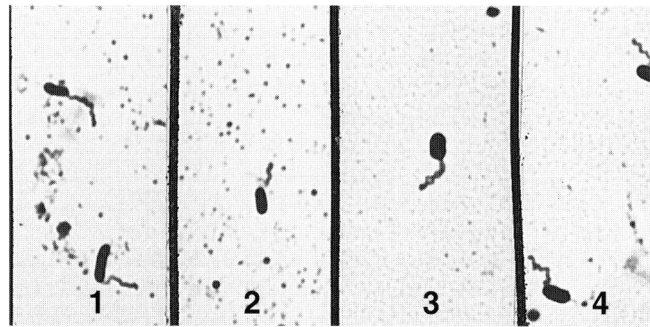

Pseudomonas diminuta strains KC679 (plates 1 and 2) and D7054 (plates 3 and 4). Flagella stain BAP 35° C 24 h x 1700.

Pseudomonas diminuta

TEST PERFORMED	SIGN	% +	# +/T
Morphology	rs		
Motility; flagella	[m;p,1-2]a		
Action on blood	v	69 LG or gr	38/55; 6 br, 6 ly, 5 lav
Fermentative or oxidative	[n-o or O]		
Carbohydrate base	OF		
Acid from:			
D-Glucose	[v]	21 (9)	13 (6)/68
D-Xylose	[-]	0	0/68
D-Mannitol	-	0	0/68
Lactose	-	0	0/68
Sucrose	-	0	0/68
Maltose	[-]	0	0/68
Catalase	+	98	67/68
Oxidase	[+]	100	68/68
Growth on:			
MacConkey	[+]	97 (3)	66 (2)/68
SS	-	1 (1)	1 (1)/68
Cetrimide	-	(1)	(1)/68
Simmons citrate	-	1	1w/68
Urea, Christensen's	v	(13)	(9)/68
Nitrate reduction	[-]	3	2/68
Gas from nitrate	[-]	0	0/68
Indole	[-]	0	0/68
TSI slant, acid	-	0	0/68
TSI butt, acid	-	0	0/68
H_2S (TSI butt)	-	0	0/68
H_2S (Pb ac paper)	v	34	23/68
Gelatin hydrolysisb	v	68	46/68
Litmus milk	v	35 k	17/48; 15 IR, 9 pep
Pigment:			
Soluble	[br-ta]	96	64/67; 2 yel, 1 or
Growth at:			
25°C	+	100	68/68
35°C	+	100	68/68
42°C	v	38	26/68
Esculin hydrolysis	-	5 (1)	3 (1)/68
Lysine decarboxylase	-	0	0/6
Arginine dihydrolase	-	0	0/6
Ornithine decarboxylase	-	0	0/6
Nutrient broth, 0% NaCl	+	100	58/58
Nutrient broth, 6% NaCl	v	21 (2)	12 (1)/56

NUMBER OF STRAINS 68

aFlagella have a short wavelength with low amplitude.
bIncubation of 7–14 days.

Pseudomonas fluorescens

Gram-negative short to long rod; motile (polar tuft of three or more flagella); aerobic; uses glucose, xylose, often sucrose, and sometimes other carbohydrates; aerobic; grows on MacConkey and SS agars; oxidase-positive; arginine dihydrolase-positive; lysine decarboxylase-negative; liquefies gelatin at 25° C; usually does not reduce nitrate; indole-negative; usually prefers to grow at 25° C, grows poorly at 35° C, does not grow at 42° C; produces a fluorescent yellow green pigment (pyoverdin), does not produce pyocyanin; acetamide-negative.

Sources (155 strains)

Sputum (30%), blood (9%), urine (8%), cerebrospinal fluid (6%), other (40%), unknown (7%).

Reference strain

KC678 ← ATCC 13525 (type strain).

Literature

1. Migula, W. 1895. Bakteriaceae. *In* Engler und Prantl, Pflanzenfamilien. W. Engelmann, Leipzig, Teil 1.
2. Rhodes, M.E. 1959. The characterization of *Pseudomonas fluorescens*. J. Gen. Microbiol. *21:* 221–263.
3. Judicial Commission. 1970. Opinion 37. Designation of strain ATCC 13525 as the neotype strain of *Pseudomonas fluorescens* Migula. Int. J. Syst. Bacteriol. *20:* 18.
4. Palleroni, N.J. 1984. Genus I. *Pseudomonas* Migula 1894, 237[AL] (Nom. cons. Opin. 5, Jud. Comm. 1952, 237). *In* N.R. Krieg and J.G. Holt (Eds.), Bergey's Manual of Systematic Bacteriology, Vol. 1, Williams & Wilkins, Baltimore, pp. 141–213
5. Gilardi, G.L. 1991. *Pseudomonas* and related genera. *In* A. Balows, et al. (Eds.), Manual of Clinical Microbiology, 5th ed., American Society for Microbiology, Washington, pp. 429–441.
6. Gibb, A.P., et al. 1992. Bacterial growth in blood for transfusion. Lancet *340:* 1222–1223.

Comments

Associated with post-transfusion septicemia. Survives and grows at 5° C. Five biovars have been described (4). Information on the cellular fatty acid composition of this species is presented in Table 4.37 and Figure 4.42 of the cellular fatty acid section.

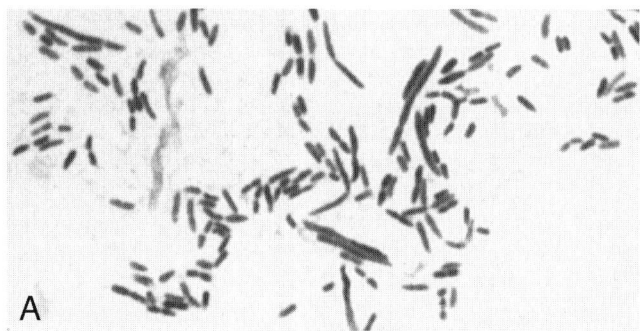

Pseudomonas fluorescens strain KC678. Gram stain BAP 35° C 24 h x 1700.

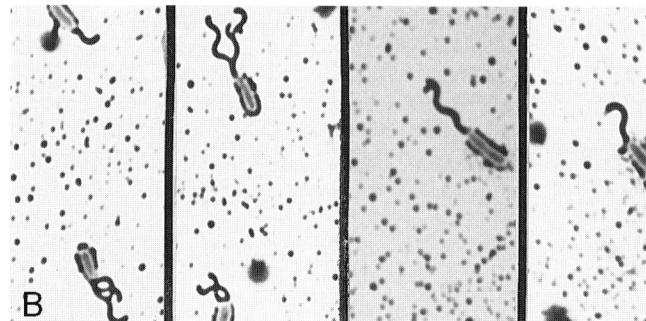

Pseudomonas fluorescens strain KC678. Flagella stain FB 25° C 24 h x 1700.

Pseudomonas fluorescens

NUMBER OF STRAINS 155

TEST PERFORMED	SIGN	% +	#+/T
Morphology	mrs		
Motility; flagella	[m;p,>2]		
Action on blood	v	50 ly	60/119; 16 gr, 6 LG, 3 lav
Fermentative or oxidative	[O]		
Carbohydrate base	OF		
Acid from:			
D-Glucose	[+]	100	155/155
D-Xylose	+	100	155/155
D-Mannitol	v	53 (2)	80 (3)/151
Lactose	v	24 (3)	38 (5)/155
Sucrose	[v]	48	75/155
Maltose	[-]	2	3/155
Catalase	+	99 (1)	149 (1)/150
Oxidase	[+]	97	148, 2w/154
Growth on:			
MacConkey	[+]	100	152/152
SS	[+ or (+)]	86 (7)	130 (11)/152
Cetrimide	v	89	135/151
Simmons citrate	+	93 (3)	142 (5)/152
Urea, Christensen's	v	21 (31)	32 (47)/151
Nitrate reduction	v	19	29/151
Gas from nitrate	[-]	3	5/151
Indole	[-]	0	0/155
TSI slant, acid	−	0	0/145
TSI butt, acid	−	0	0/145
H$_2$S (TSI butt)	−	0	0/155
H$_2$S (Pb ac paper)	v	12	5, 14w/152
Gelatin hydrolysis[a]	[+]	100	75/75
Litmus milk	[pep]	95	146/154; 7 k, 1 A
Pigment:			
Pyoverdin	[+]	96	141/147
Pyocyanin	[-]	0	0/147
Pyorubrin	−	0	0/147
Other	−	4 br sol	6w/147; 3 yel sol
Growth at:			
5°C	+	95 (1)	136 (1)/143
25°C	+	100	153/153
35°C	+	90	138/153
42°C	[-]	0	0/153
Esculin hydrolysis	−	0	0/138
Lysine decarboxylase	[-]	0	0/37
Arginine dihydrolase	[+]	97	36/37
Ornithine decarboxylase	−	0	0/37
Nutrient broth, 0% NaCl	+	99	147/149
Nutrient broth, 6% NaCl	v	43	64/149
Alkalinization of acetamide	v	6 (12)	1 (2)/17

[a] Incubation of 7–14 days.

Pseudomonas gladioli

Gram-negative rod; motile (>2 polar flagella, but frequently more than 1–2 flagella/cell cannot be demonstrated); aerobic; oxidizes glucose, xylose, mannitol, dulcitol, and inositol, but not maltose or erythritol; grows on MacConkey agar; oxidase-negative (some strains weakly positive by Kovacs' method); peptonizes litmus milk; urease-positive; does not produce lysine decarboxylase or arginine dihydrolase.

Sources (58 strains)

Sputum (31%), blood (21%), cerebrospinal fluid (7%), throat (3%), leg (3%), other (23%), unknown (12%).

Reference strains

KC1370 ← ATCC 10248, type strain.
KC1159 = ATCC 19302, neopathotype strain.

Literature

1. McCulloch, L. 1924. A leaf and corn disease of gladioli caused by *Bacterium marginatum*. J. Agric. Res. *29:* 159–177.

2. Ballard, R.E., N.J. Palleroni, M. Doudoroff, and R.Y. Stanier. 1970. Taxonomy of the aerobic pseudomonads: *Pseudomonas cepacia, P. marginata, P. alliicola,* and *P. caryophylli*. J. Gen. Microbiol. *60:* 199–214.

3. Hildebrand, D.C., Palleroni, N.J., and M. Doudoroff. 1973. Synonomy of *Pseudomonas gladioli* Severini 1913 and *Pseudomonas marginata* (McCulloch 1921) Stapp 1928. Int. J. Syst. Bacteriol. *23:* 433–437.

4. Palleroni, N.J. 1984. Genus I. *Pseudomonas* Migula 1894, 237[AL] (Nom. cons. Opin. 5, Jud. Comm. 1952, 237). *In* N.R. Krieg and J.G. Holt (Eds.), Bergey's Manual of Systematic Bacteriology, Vol. 1, Williams & Wilkins, Baltimore, pp. 141–213.

5. Christenson, J.C., D.F. Welch, G. Mukwaya, M.J. Muszynski, R.E. Weaver, and D.J. Brenner. 1989. Recovery of *Pseudomonas gladioli* from respiratory tract specimens of patients with cystic fibrosis. J. Clin. Microbiol. *27:* 270–273.

6. Yabuuchi, E., et al. 1992. Proposal of *Burkholderia* gen. nov. and transfer of seven species of the genus *Pseudomonas* homology group II to the new genus, with the type species *Burkholderia cepacia* (Palleroni and Holmes 1981) comb. nov. Microbiol. Immunol. *36:* 1251–1275.

Comments

Formerly *P. marginata*. Reference strain KC1370 oxidized lactose weakly in 8–21 days when removed from the 35° C incubator after 7 days and held at ambient temperature (25–28° C). *P. gladioli* is described in *Bergey's Manual of Systematic Bacteriology* as motile, with polar multitrichous flagellation, and oxidase-positive (4). Both of these characters may be difficult to demonstrate using the methods described in this manual. Recently, it has been proposed that this species be placed in the new genus *Burkholderia* (6). In recent years, it has been recovered from respiratory tract specimens of patients with cystic fibrosis and may be confused with *P. cepacia*. Information on the cellular fatty acid composition of this species is presented in Table 4.62 and Figure 4.67 of the cellular fatty acid section.

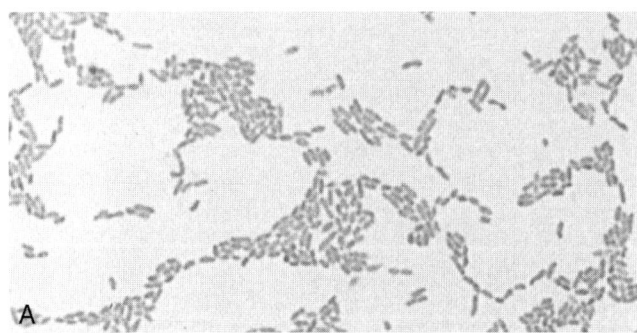

Pseudomonas gladioli strain KC1370. Gram stain BAP 35° C 24 h x 1700.

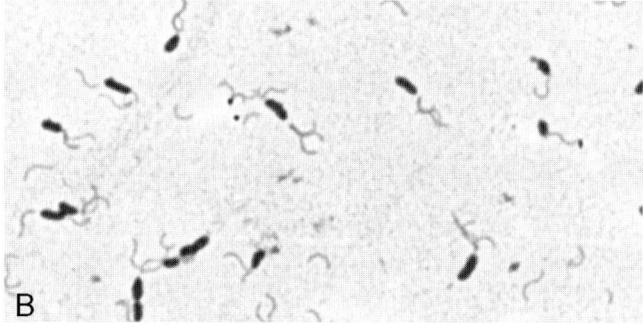

Pseudomonas gladioli strain KC1370. Flagella stain (Ryu method) TGY 25° C 24 h x 1700.

Pseudomonas gladioli

NUMBER OF STRAINS 58

TEST PERFORMED	SIGN	% +	# +/T
Morphology	mrs		
Motility; flagella	[m;p, >2][a]		
Action on blood	v	27 gr	15/56; 6 ly, 4 LG, 2 lav, 2 β
Fermentative or oxidative	[O]		
Carbohydrate base	OF		
Acid from:			
D-Glucose	[+]	98 (2)	56, 1w (1)/58
D-Xylose	+	98 (2)	55, 2w (1)/58
D-Mannitol	[+]	91 (9)	48, 5w (5)/58
Lactose	v	9 (28)	2, 3w (4) (12w)/58
Sucrose	−	0	0/58
Maltose	[−]	0	0/58
Dulcitol	[+ or (+)]	78 (22)	15, 6w (5) (1w)/27
Inositol	[+]	96 (4)	23, 2w (1w)/26
Erythritol	[−]	0	0/26
Catalase	+	98	52, 5w/58
Oxidase	v[b]	47	20, 7w/58
Growth on:			
MacConkey	[+]	97 (3)	54, 2w (2w)/58
SS	−	3 (1)	2, (1w)/58
Cetrimide	−	3 (1)	1, 1w (1w)/58
Simmons citrate	+	93 (5)	53, 1w (3)/58
Urea, Christensen's	v	30 (28)	14, 3w (10) (6w)/58
Nitrate reduction	v	43	25/58
Gas from nitrate	−	0	0/58
Indole	−	0	0/58
TSI slant, acid	−	0	0/58
TSI butt, acid	−	0	0/58
H_2S (TSI butt)	−	0	0/58
H_2S (Pb ac paper)	v	31 (5)	7, 11w (3w)/58
Gelatin hydrolysis[c]	v	12	6/52
Litmus milk	v	71 pep	35, 5w/56; 7 IR, 4 k
Pigment	v	41 yel sol	24/58
Growth at:			
25°C	+	97	55, 1w/58
35°C	+	100	57, 1w/58
42°C	−	9	4, 1w/58
Esculin hydrolysis	−	0	0/58
Lysine decarboxylase	[−]	0	0/45
Arginine dihydrolase	[−]	2	1/45
Ornithine decarboxylase	−	0	0/45
Nutrient broth, 0% NaCl	+	100	58/58
Nutrient broth, 6% NaCl	−	10	6/58

[a]Frequently difficult to demonstrate more than 1–2 polar flagella.
[b]Most strains studied were only positive with the Kovacs' method. Reactions were frequently weak.
[c]Incubation of 7–14 days.

Pseudomonas-like group 2

Gram-negative; motile (>2 polar flagella); aerobic; oxidizes glucose, xylose, mannitol and lactose, but not dulcitol or inositol; grows on MacConkey agar; oxidase-positive; urease-positive; indole-negative.

Sources (11 strains)

Blood (6 strains), sputum (2 strains), boil and lymph node (1 strain each), unknown (1 strain).

Reference strain

A9800, from sputum.

Literature

1. Dees, S.B., D.G. Hollis, R.E. Weaver, and C.W. Moss. 1983. Cellular fatty acid composition of *Pseudomonas marginata* and closely associated bacteria. J. Clin. Microbiol. *18:* 1073–1078.

Comments

Similar to *Pseudomonas gladioli* (formerly *P. marginata*). Differential tests include dulcitol, inositol, and erythritol oxidation, along with cellular fatty acid analysis. Information on the cellular fatty acid composition of this group is presented in Table 4.37 and Figure 4.42 of the cellular fatty acid section.

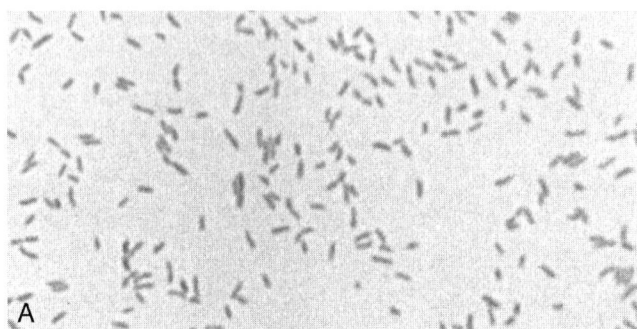 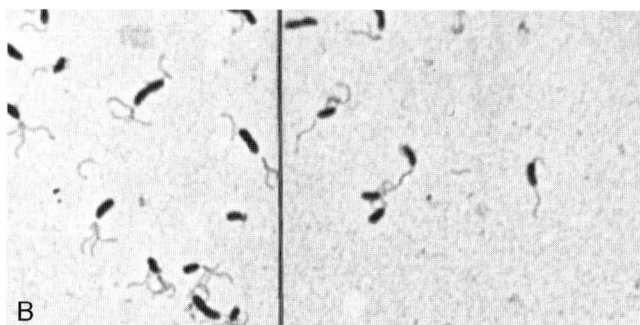

Pseudomonas-like group 2 strain A9800. Gram stain BAP 35° C 24 h x 1700.

Pseudomonas-like group 2 strain A9800. Flagella stain (Ryu method) TGY 25° C 24 h x 1700.

Pseudomonas-like group 2

	NUMBER OF STRAINS 11		
TEST PERFORMED	SIGN	% +	#+/T
Morphology	mrs		
Motility; flagella	[m;p, >2]		
Action on blood	v	20 ly	2/10; 1 lav, 1 β
Fermentative or oxidative	[O]		
Carbohydrate base	OF		
Acid from:			
D-Glucose	[+]	100	11/11
D-Xylose	+	100	11/11
D-Mannitol	[+]	100	11/11
Lactose	[+]	100	1, 10w/11
Sucrose	−	0	0/11
Maltose	[−]	0	0/11
Dulcitol	[−]	0	0/11
Inositol	[−]	0	0/11
Erythritol	[v]	73	6, 2w/11
Catalase	+	100	4, 7w/11
Oxidase	[+]	100	11/11
Growth on:			
MacConkey	+	100	11/11
SS	−	0	0/11
Cetrimide	−	0	0/11
Simmons citrate	+	100	11/11
Urea, Christensen's	+	91 (9)	10 (1)/11
Nitrate reduction	v	18	2/11
Gas from nitrate	−	0	0/11
Indole	−	0	0/11
TSI slant, acid	−	0	0/11
TSI butt, acid	−	0	0/11
H_2S (TSI butt)	−	0	0/11
H_2S (Pb ac paper)	v	89	6, 2w/9
Gelatin hydrolysis[a]	−	0	0/11
Litmus milk	v	73 k	8/11; 3 IR
Pigment	−	0	0/11
Growth at:			
25°C	v	80	8/10
35°C	+	100	10/10
42°C	v	20	2/10
Esculin hydrolysis	−	0	0/11
Lysine decarboxylase	−	0	0/11
Arginine dihydrolase	v	30	3/10
Ornithine decarboxylase	−	0	0/10
Nutrient broth, 0% NaCl	+	100	10/10
Nutrient broth, 6% NaCl	−	0	0/10

[a]Incubation of 7–14 days.

Pseudomonas mallei

Gram-negative rod; nonmotile; aerobic; oxidizes glucose and sometimes (delayed) xylose, mannitol, lactose, and maltose, but not sucrose; growth on MacConkey agar is variable; usually oxidase-negative; reduces nitrate to nitrite without gas; produces arginine dihydrolase, not lysine decarboxylase; does not use citrate (Simmons), digest gelatin, or grow at 42° C.

Sources (8 strains)

Equine abscess (1), unknown (7).

Reference strain

KC234 ← ATCC 23344, type strain, from human infection.

Literature

1. Redfearn, M.S., N.J. Palleroni, and R.Y. Stanier. 1966. A comparative study of *Pseudomonas pseudomallei* and *Bacillus mallei*. J. Gen. Microbiol. *43:* 293–313.

2. Rogul, M., J.J. Brendle, D.K. Haapala, and A.D. Alexander. 1970. Nucleic acid similarities among *Pseudomonas multivorans* and *Actinobacillus mallei*. J. Bacteriol. *101:* 827–835.

3. Palleroni, N.J. 1984. Genus I. *Pseudomonas* Migula 1894, 237[AL] (Nom. cons. Opin. 5, Jud. Comm. 1952, 237). *In* N.R. Krieg and J.G. Holt (Eds.), Bergey's Manual of Systematic Bacteriology, Vol. 1, Williams & Wilkins, Baltimore, pp. 141–213.

4. Yabuuchi, E., et al. 1992. Proposal of *Burkholderia* gen. nov. and transfer of seven species of the genus *Pseudomonas* homology group II to the new genus, with the type species *Burkholderia cepacia* (Palleroni and Holmes 1981) comb. nov. Microbiol. Immunol. *36:* 1251–1275.

Comments

Etiologic agent of glanders, primarily a natural equine disease. Considered to be too hazardous for routine laboratory study. It recently has been proposed that these organisms, along with other *Pseudomonas* homology group II species, be placed in the new genus *Burkholderia* (4). Information on the cellular fatty acid composition of this species is presented in Table 4.62 and Figure 4.67 of the cellular fatty acid section.

Pseudomonas mallei

TEST PERFORMED	SIGN	% +	#+/T
Morphology	cc		
Motility; flagella	[nm]		
Action on blood	ly	100	8/8
Fermentative or oxidative	[O]		
Carbohydrate base	OF		
Acid from:			
D-Glucose	[+]	100	7, 1w/8
D-Xylose	v	12 (50)[a]	1w (4)/8
D-Mannitol	v	62 (14)[a]	5w (1)/8
Lactose	v	12 (62)[a]	1w (5)/8
Sucrose	[-]	0	0/8
Maltose	v	(75)[a]	(6)/8
Catalase	+	100	8/8
Oxidase	v	25	2/8
Growth on:			
MacConkey	v	88	7w/8
SS	-	0	0/8
Cetrimide	-	0	0/8
Simmons citrate	[-]	0	0/8
Urea, Christensen's	v	12	1/8
Nitrate reduction	[+]	100	8/8
Gas from nitrate	[-]	0	0/8
Indole	-	0	0/8
TSI slant, acid	-	0	0/8
TSI butt, acid	-	0	0/8
H_2S (TSI butt)	-	0	0/8
H_2S (Pb ac paper)	+	100	8/8
Gelatin hydrolysis[b]	-	0	0/8
Litmus milk	-	0	0/8
Pigment:			
Soluble	-	0	0/8
Insoluble	-	0	0/8
Growth at:			
25°C	+	100	8/8
35°C	+	100	8/8
42°C	[-]	0	0/8
Esculin hydrolysis	-	0	0/8
Lysine decarboxylase	[-]	0	0/8
Arginine dihydrolase	[+]	100	8/8
Ornithine decarboxylase	-	0	0/8
Nutrient broth, 0% NaCl	+	100	8/8
Nutrient broth, 6% NaCl	-	0	0/8

[a]Some delayed carbohydrate reactions required 7–14 days of incubation.
[b]Incubation of 7–14 days.

Pseudomonas mendocina

Gram-negative short to medium length slender rod; motile (single polar flagellum); aerobic; oxidizes glucose and xylose only; does not hydrolyze starch; grows on MacConkey and SS agars; oxidase-positive; reduces nitrate to gas; produces arginine dihydrolase, but not lysine decarboxylase; 2-ketogluconate-negative; acetamide-negative; H_2S detected on lead acetate paper suspended above TSI agar slant; nonproteolytic.

Sources (4 strains)

Water (2), soil (1), blood (1).

Reference strains

KC1140 ← N. J. Palleroni, CH20 (ATCC 25412).
KC1218 ← ATCC 25411 (Palleroni CH50), type strain.
KC1220 ← ATCC 25413 (Palleroni CH120).

Literature

1. Stanier, R.Y., N.J. Palleroni, and M. Doudoroff. 1966. The aerobic pseudomonads: a taxonomic study. J. Gen. Microbiol. *43:* 159–271.

2. Palleroni, N.J., M. Doudoroff, and R.Y. Stanier. 1970. Taxonomy of the aerobic pseudomonads: the properties of the *Pseudomonas stutzeri* group. J. Gen. Microbiol. *60:* 215–231.

3. Palleroni, N.J. 1984. Genus I. *Pseudomonas* Migula 1894, 237[AL] (Nom. cons. Opin. 5, Jud. Comm. 1952, 237). *In* N.R. Krieg and J.G. Holt (Eds.), Bergey's Manual of Systematic Bacteriology, Vol. 1, Williams & Wilkins, Baltimore, pp. 141–213.

4. Aragone, M.D.R., D.M. Maurizi, L.O. Clara, J.L.N. Estrada, and A. Ascione. 1992. *Pseudomonas mendocina,* an environmental bacterium isolated from a patient with human infective endocarditis. J. Clin. Microbiol. *30:* 1583–1584.

Comments

P. mendocina is similar to nonpigmented strains of *P. aeruginosa,* except that the latter is usually H_2S-negative (lead acetate paper method) and acetamide, gelatin, and 2-ketogluconate-positive. Information on the cellular fatty acid composition of this species is presented in Table 4.36 and Figure 4.41 of the cellular fatty acid section.

A small group of similar organisms which oxidize mannitol were included in previously published "Round-Table" charts as Vb-2. Such strains were isolated from a variety of sources (two from inline transducer and one each from urine, vagina, wound, clavicle node, serous otitis, and exudate). These strains are not included in this manual. Further studies needed to establish their taxonomy have not been done.

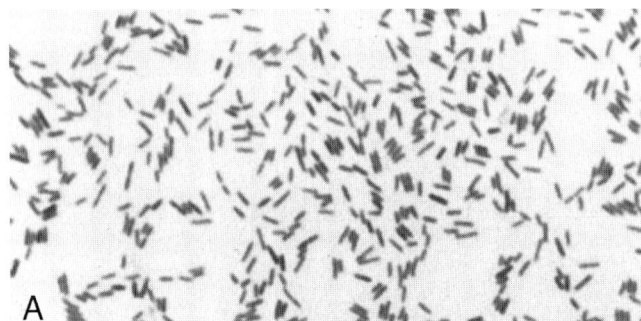

Pseudomonas mendocina strain KC1140. Gram stain BAP 35° C 24 h x 1700.

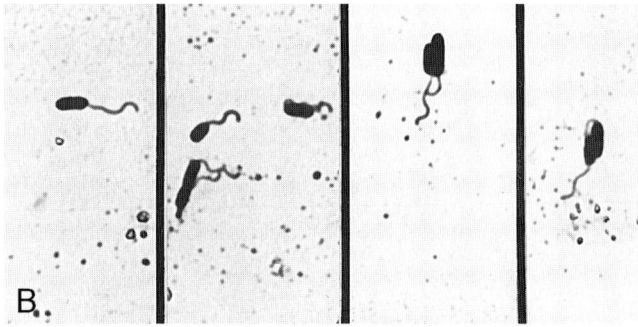

Pseudomonas mendocina strain KC1140. Flagella stain BAP 35° C 24 h x 1700.

Pseudomonas mendocina

TEST PERFORMED	SIGN	% +	#+/T
	NUMBER OF STRAINS 4		
Morphology	mrs		
Motility; flagella	[m;p,1-2]		
Action on blood	ly	100	2/2
Fermentative or oxidative	[O]		
Carbohydrate base	OF		
Acid from:			
D-Glucose	[+]	100	4/4
D-Xylose	+ or (+)	75 (25)	3, (1)/4
D-Mannitol	-	0	0/4
Lactose	[-]	0	0/4
Sucrose	-	0	0/4
Maltose	[-]	0	0/4
Starch	-	0	0/3
Catalase	+	100	2, 2w/4
Oxidase	[+]	100	4/4
Growth on:			
MacConkey	[+]	100	4/4
SS	[+]	100	4/4
Cetrimide	+ or (+)	75 (25)	3 (1w)/4
Simmons citrate	+	100	4/4
Urea, Christensen's	v	50	2/4
Nitrate reduction	+	100	4/4
Gas from nitrate	[+]	100	4/4
Indole	-	0	0/4
TSI slant, acid	-	0	0/4
TSI butt, acid	-	0	0/4
H_2S (TSI butt)	-	0	0/4
H_2S (Pb ac paper)	[+]	100	4/4
Gelatin hydrolysis[a]	[-]	0	0/4
Litmus milk	k or (k)	25 (75)	1 (3)/4
Pigment	sl yel ins	100	4/4
Growth at:			
25°C	+	100	4/4
35°C	+	100	4/4
42°C	+	100	4/4
Esculin hydrolysis	-	0	0/4
Lysine decarboxylase	[-]	0	0/4
Arginine dihydrolase	[+]	100	4/4
Ornithine decarboxylase	-	0	0/4
Nutrient broth, 0% NaCl	+	100	4/4
Nutrient broth, 6% NaCl	+	100	2/2
Alkalinization of acetamide	[-]	0	0/4
2-Ketogluconate	[-]	0	0/3

[a]Incubation of 7–14 days.

Pseudomonas pertucinogena

Gram-negative medium length rod; motile (single polar flagellum); oxidizes glucose and xylose (weak and late), not mannitol or sucrose; grows on MacConkey agar; oxidase-positive; citrate-, urease-, nitrate-, and arginine dihydrolase-negative.

Sources (2 strains)

Unknown, possibly from human respiratory tract, see Reference 1.

Reference strains

KC1467 ← E. Yabuuchi, 1319 ← ATCC 190, type strain.
KC1468 ← E. Yabuuchi, 1320 ← ATCC 6627.

Literature

1. Kawai, Y., and E. Yabuuchi. 1975. *Pseudomonas pertucinogena* sp. nov., an organism previously misidentified as *Bordetella pertussis*. Int. J. Syst. Bacteriol. 25: 317–323.

2. Palleroni, N.J. 1984. Genus I. *Pseudomonas* Migula 1894, 237[AL] (Nom. cons. Opin. 5, Jud. Comm. 1952, 237). *In* N.R. Krieg and J.G. Holt (Eds.), Bergey's Manual of Systematic Bacteriology, Vol. 1, Williams & Wilkins, Baltimore, pp. 141–213.

Comments

Colonial morphology mimics that of rough strains of *Bordetella pertussis* but motility and flagellar morphology distinguish them from *B. pertussis*. Information on the cellular fatty acid composition of this species is presented in Table 4.36 and Figure 4.41 of the cellular fatty acid section.

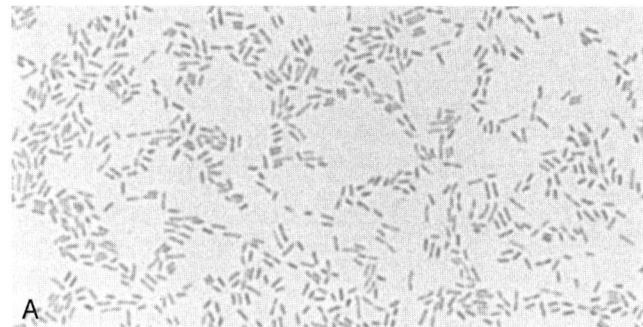

Pseudomonas pertucinogena strain KC1467. Gram stain BAP 35° C 24 h x 1700.

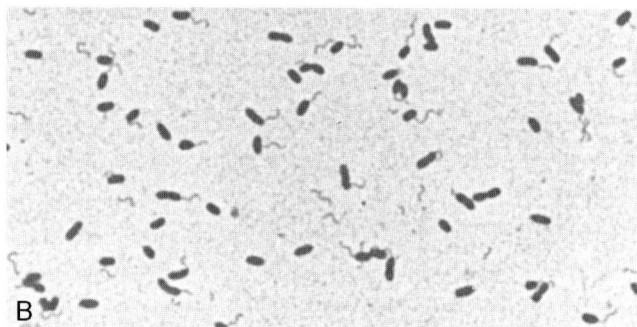

Pseudomonas pertucinogena strain KC1467. Flagella stain (Ryu method) TGY 25° C 24 h x 1700.

Pseudomonas pertucinogena

TEST PERFORMED	SIGN	% +	#+/T
Morphology	mrs		
Motility; flagella	[m;p,1-2]		
Action on blood	–	0	0/2
Fermentative or oxidative	[O]		
Carbohydrate base	OF		
Acid from:			
D-Glucose	+ or (+)	50 (50)	1w (1w)/2
D-Xylose	[(+)]	(100)	(2w)/2
D-Mannitol	–	0	0/2
Lactose	–	0	0/2
Sucrose	–	0	0/2
Maltose	–	0	0/2
Catalase	+	100	2/2
Oxidase	+	100	2/2
Growth on:			
MacConkey	[+]	100	2/2
SS	–	0	0/2
Cetrimide	–	0	0/2
Simmons citrate	–	0	0/2
Urea, Christensen's	–	0	0/2
Nitrate reduction	[–]	0	0/2
Gas from nitrate	–	0	0/2
Indole	–	0	0/2
TSI slant, acid	–	0	0/2
TSI butt, acid	–	0	0/2
H_2S (TSI butt)	–	0	0/2
H_2S (Pb ac paper)	+	100	2/2
Gelatin hydrolysis[a]	–	0	0/2
Litmus milk	–	0	0/2
Pigment	–	0	0/2
Growth at:			
25°C	+	100	2/2
35°C	+	100	2/2
42°C	+	100	2/2
Esculin hydrolysis	–	0	0/2
Lysine decarboxylase	–	0	0/2
Arginine dihydrolase	–	0	0/2
Ornithine decarboxylase	–	0	0/2
Nutrient broth, 0% NaCl	+	100	2/2
Nutrient broth, 6% NaCl	+	100	2/2

[a]Incubation of 7–14 days.

Pseudomonas pickettii

Gram-negative short to medium length rod; motile (single polar flagellum); aerobic; oxidizes glucose and xylose; grows on MacConkey agar; oxidase-positive; citrate-positive; urease-positive; reduces nitrate with gas formation (may be difficult to detect); lysine decarboxylase- and usually arginine dihydrolase-negative; grows in nutrient broth.

Sources Biovar 1 (70 strains)

Blood (11%); water (10%); sputum (10%); urine (10%); cerebrospinal fluid (7%); eye (6%); exudate, dialysis, lymph node, cervix, and trachea (3% each); other (25%); unknown (6%).

Sources Biovar 2 (54 strains)

Urine (22%), nasopharynx (20%), abscess and wound (18%), blood (13%), other (12%), unknown (15%).

Reference strains

Biovar 1
C6712, from ear exudate.
D3927, from conjunctiva.

Biovar 2
KC1295 ← ATCC 27511, type strain.

Literature

1. Ralston, E., N.J. Palleroni, and M. Doudoroff. 1973. *Pseudomonas pickettii*, a new species of clinical origin related to *Pseudomonas solanacearum*. Int. J. Syst. Bacteriol. 23: 15–19.

2. Riley, P.S., and R.E. Weaver. 1975. Recognition of *Pseudomonas pickettii* in the clinical laboratory: biochemical characterization of 62 strains. J. Clin. Microbiol. 1: 61–64.

3. Fass, R.J., and J. Barnishan. 1975. Acute meningitis due to a *Pseudomonas*-like group Va-1 bacillus. Ann. Intern. Med. 84: 51–52.

4. Pickett, M.J., and J.R. Greenwood. 1980. A study of the Va-1 group of pseudomonads and its relationship to *Pseudomonas pickettii*. J. Gen. Microbiol. 120: 439–446.

5. Palleroni, N.J. 1984. Genus I. *Pseudomonas* Migula 1894, 237^{AL} (Nom. cons. Opin. 5, Jud. Comm. 1952, 237). *In* N.R. Krieg and J.G. Holt (Eds.), Bergey's Manual of Systematic Bacteriology, Vol. 1, Williams & Wilkins, Baltimore, pp. 141–213.

6. Yabuuchi, E., et al. 1992. Proposal of *Burkholderia* gen. nov. and transfer of seven species of the genus *Pseudomonas* homology group II to the new genus, with the type species *Burkholderia cepacia* (Palleroni and Holmes 1981) comb. nov. Microbiol. Immunol. 36: 1251–1275.

7. Raveh, D., A. Simhon, Z. Gimmon, T. Sacks, and M. Shapiro. 1993. Infections caused by *Pseudomonas pickettii* in association with permanent indwelling intravenous devices: four cases and a review. Clin. Infect. Dis. 17: 877–880.

Comments

Motility weak or delayed, not readily detectable. Gas production during denitrification also may not be detected readily unless Stanier's semiaerobic nitrate medium is used. Biovar 1 differs from Biovar 2 by the oxidation of lactose and maltose. It has recently been proposed to move these organisms to the new genus *Burkholderia* (6). Information on the cellular fatty acid composition of this species is presented in Table 4.10 and Figure 4.15 of the cellular fatty acid section.

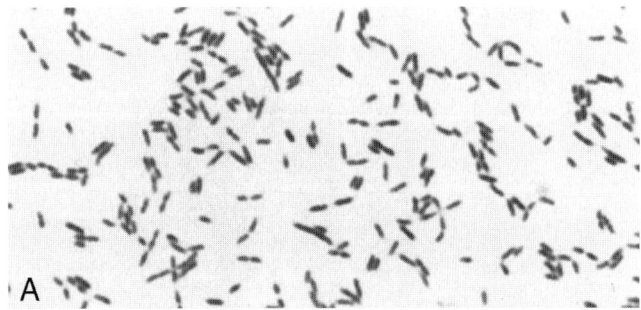

Pseudomonas pickettii strain KC1295. Gram stain BAP 35° C 24 h x 1700.

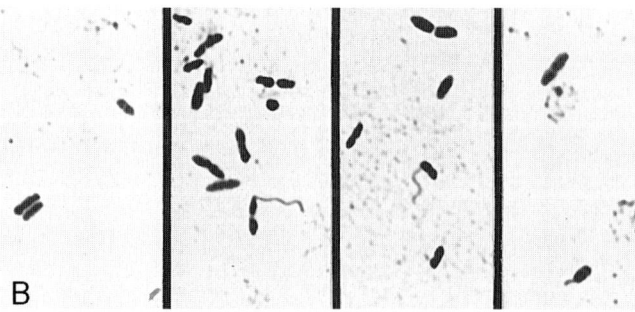

Pseudomonas pickettii strain KC1295. Flagella stain FB 25° C 24 h x 1700.

Pseudomonas pickettii

TEST PERFORMED	Biovar 1			Biovar 2		
	SIGN	% +	#+/T	SIGN	% +	#+/T
		NUMBER OF STRAINS 70			NUMBER OF STRAINS 54	
Morphology	mrs			mrs		
Motility; flagella	[m;p,1-2][a]			[m;p,1-2][a]		
Action on blood	v	56 ly	14/25: 5 LG, 4 al, 2 gr	v	21 ly	6/29: 5 al, 4 LG, 2 gr
Fermentative or oxidative	[O]			[O]		
Carbohydrate base	OF			OF		
Acid from:						
D-Glucose	[+]	100	70/70	[+]	100	54/54
D-Xylose	[+]	100	70/70	[+]	100	54/54
D-Mannitol	-	0	0/70	-	0	0/54
Lactose	[+]	100	70/70	[-]	0	0/54
Sucrose	-	0	0/70	-	0	0/54
Maltose	[+]	100	69/69	[-]	0	0/54
Catalase	+	100	70w/70	+	100	54/54
Oxidase	[+]	100	70/70	[+]	100	54/54
Growth on:						
MacConkey	[+]	99 (1)	69 (1)/70	[+]	100	54/54
SS	-	0	0/70	-	0	0/54
Cetrimide	-	1	1/70	-	0	0/54
Simmons citrate	+	99	69/70	+	100	54/54
Urea, Christensen's	[+]	100	70/70	[+]	100	54/54
Nitrate reduction	+	100	70/70	+	100	54/54
Gas from nitrate[b]	[v]	86	60/70	[+]	100	54/54
Indole	-	0	0/70	-	0	0/54
TSI slant, acid	-	0	0/70	-	0	0/54
TSI butt, acid	-	0	0/70	-	0	0/54
H$_2$S (TSI butt)	-	0	0/70	-	0	0/54
H$_2$S (Pb ac paper)	v	67	47/70	v	38	11/29
Gelatin hydrolysis[c]	v	12	7/59	v	33	12/36
Litmus milk	v	83 k	40/48: 2 pep	k	90	26/29
Pigment:						
Soluble	v	36 yel-ta	25/70		0	0/54
Growth at:						
25°C	+	100	69/69	+	100	54/54
35°C	+	100	69/69	+	100	54/54
42°C	v	83	57/69	+	94	51/54
Esculin hydrolysis	-	0	0/70	-	0	0/54
Lysine decarboxylase	[-]	0	0/53	[-]	0	0/54
Arginine dihydrolase	[-]	6	3/53	[-]	0	0/54
Ornithine decarboxylase	-	0	0/53	-	0	0/54
Nutrient broth, 0% NaCl	+	100	70/70	+	100	29/29
Nutrient broth, 6% NaCl	-	4	3/70	-	3	1/29

[a]Weakly motile, may be difficult to demonstrate.
[b]May be negative if not tested in semiaerobic nitrate medium.
[c]Incubation of 7–14 days.

Pseudomonas pseudoalcaligenes

Gram-negative medium length rod; motile (1–2 polar flagella); aerobic; grows on MacConkey agar; oxidizes fructose, not glucose or maltose; oxidase-positive; grows at 42° C; urease-negative; usually reduces nitrate without gas; does not produce H_2S on TSI agar.

Sources (34 strains)

Sputum (12%), blood (9%), urine (9%), stool (6%), wound (6%), hospital environment (6%), river water (6%), other (30%), unknown (6%).

Reference strain

KC946 ← Pickett, K-491.

Literature

1. Stanier, R.Y., N.J. Palleroni, and M. Doudoroff. 1966. The aerobic pseudomonads: a taxonomic study. J. Gen. Microbiol. *43:* 159–271.

2. Palleroni, N.J. 1984. Genus I. *Pseudomonas* Migula 1894, 237[AL] (Nom. cons. Opin. 5, Jud. Comm. 1952, 237). *In* N.R. Krieg and J.G. Holt (Eds.), Bergey's Manual of Systematic Bacteriology, Vol. 1, Williams & Wilkins, Baltimore, pp. 141–213.

3. Gavini, F., B. Holmes, D. Izard, A. Beji, A. Bernigaud, and E. Jakubczak. 1989. Numerical taxonomy of *Pseudomonas alcaligenes, P. pseudoalcaligenes, P. mendocina, P. stutzeri,* and related bacteria. Int. J. Syst. Bacteriol. *39:* 135–144.

Comments

Oxidation of fructose and growth at 42° C are useful in differentiating *P. pseudoalcaligenes* from *P. alcaligenes,* although a few 42° C growth-positive *P. alcaligenes* strains have been described (3). This group is genetically heterogeneous (1). Information on the cellular fatty acid composition of this species is presented in Table 4.9 and Figure 4.14 of the cellular fatty acid section.

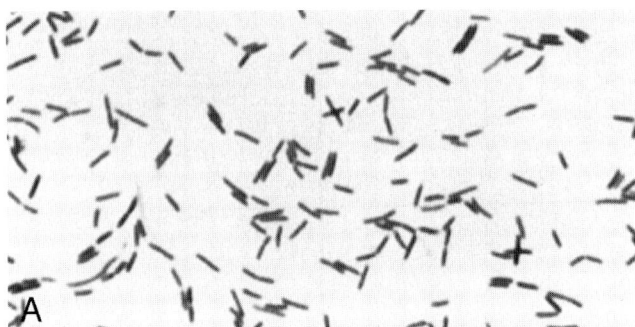

Pseudomonas pseudoalcaligenes strain KC946. Gram stain BAP 35° C 24 h x 1700.

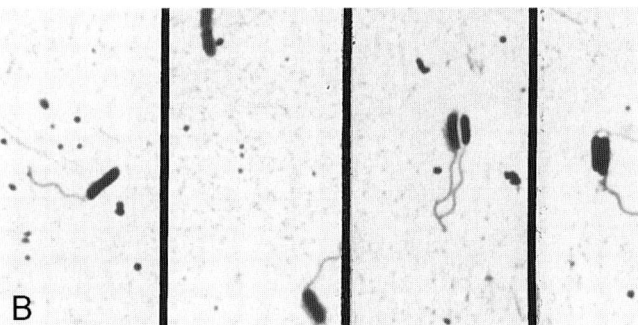

Pseudomonas pseudoalcaligenes strain KC946. Flagella stain BAP 35° C 24 h x 1700.

Pseudomonas pseudoalcaligenes

TEST PERFORMED	REFERENCE STRAIN SIGN	34 PHENOTYPICALLY SIMILAR STRAINS		
		SIGN	%+	#+/T
Morphology	mrs	mrs		
Motility; flagella	[1 polar]	[m;p,1-2]		
Action on blood	sl LG	v	65 ly	17/26; 3 LG, 3 gr, 2 br, 1 al
Fermentative or oxidative	[n-o]	[n-o]		
Carbohydrate base	OF	OF		
Acid from:				
D-Glucose	[-]	[-]	9	3w/34
D-Xylose	+w	v	18 (12)	6w (4w)/34
D-Mannitol	-	-	0	0/34
Lactose	-	-	0	0/34
Sucrose	-	-	0	0/34
Maltose	[-]	[-]	0	0/34
Fructose	[+]	[+ or (+)]	79 (21)	17 (1) 10w (6w)/34
Catalase	+w	+	97	29, 4w/34
Oxidase	[+]	[+]	100	34/34
Growth on:				
MacConkey	[+]	[+]	100	34/34
SS	-	+	90 (3)	29, 2w (1)/34
Cetrimide	+	v	56 (18)	15, 4w (6)/34
Simmons citrate	-	v	26 (9)	9 (3)/34
Urea, Christensen's	[-]	[-]	3 (6)	1 (2w)/34
Nitrate reduction	+	+	100	34/34
Gas from nitrate	[-]	[-]	0	0/34
Indole	-	-	0	0/34
TSI slant, acid	-	-	0	0/34
TSI butt, acid	-	-	0	0/34
H$_2$S (TSI butt)	[-]	[-]	0	0/34
H$_2$S (Pb ac paper)	+	+	97	31, 2w/34
Gelatin hydrolysis[a]	-	-	0	0/32
Litmus milk	k	v	38 k	13/34; 1 IR, 1 pep
Pigment:				
Soluble	-	v	18 yel-br	6/34; 2 tan
Growth at:				
25°C	+	+	100	34/34
35°C	+	+	100	34/34
42°C	[+w]	[+]	94	32/34
Esculin hydrolysis	-	-	0	0/34
Lysine decarboxylase	-	-	0	0/18
Arginine dihydrolase	+	+	78	14/18
Ornithine decarboxylase	-	-	0	0/18
Nutrient broth, 0% NaCl	+	+	100	34/34
Nutrient broth, 6% NaCl	+	v	62 (6)	21 (2)/34

[a]Incubation of 7–14 days.

Pseudomonas pseudomallei

Gram-negative rod, motile (polar tufts of three or more flagella); aerobic; oxidizes glucose, xylose, mannitol, lactose, maltose, and sometimes sucrose; grows on MacConkey agar; oxidase-positive; hydrolyzes arginine, but does not decarboxylate lysine; proteolytic; reduces nitrate, usually with formation of gas. Usually shows characteristic white opaque growth with a sheen by 48 h; at first, colonies are smooth and convex, but slowly they become umbonate with an uneven, wrinkled surface. In a young culture at 35° C, most cells will show bipolar staining; after 48 h, most cells are oval to round, and only the periphery stains; such cells may be mistaken for endospores.

Sources (70 strains)

Sputum (21%), wound (14%), animal isolate (13%), blood (10%), abscess or drainage (6%), lung (4%), skin (3%), other (including one prostatic fluid) (15%), unknown (14%).

Reference strains

KC871-1 ← WRAMIR, 295 MP-G.
KC872 ← WRAMIR, 3 MP-H.

Literature

1. Lysenko, O. 1961. *Pseudomonas*—an attempt at a general classification. J. Gen. Microbiol. 25: 379–408.
2. Redfearn, M.S., N.J. Palleroni, and R.Y. Stanier. 1966. A comparative study of *Pseudomonas pseudomalleii* and *Bacillus mallei*. J. Gen. Microbiol. 43: 293–313.
3. Rogul, M., J.J. Brendle, D.K. Haapala, and A.D. Alexander. 1970. Nucleic acid similarities among *Pseudomonas pseudomallei*, *Pseudomonas multivorans*, and *Actinobacillus mallei*. J. Bacteriol. 101: 827–835.
4. Zierdt, C.H., and H.H. Marsh. 1971. Identification of *Pseudomonas pseudo mallei*. Am. J. Clin. Pathol. 55: 596–603.
5. Palleroni, N.J. 1984. Genus I. *Pseudomonas* Migula 1894, 237[AL] (Nom. cons. Opin. 5, Jud. Comm. 1952, 237). *In* N.R. Krieg and J.G. Holt (Eds.), Bergey's Manual of Systematic Bacteriology, Vol. 1, Williams & Wilkins, Baltimore, pp. 141–213.
6. Yabuuchi, E., et al. 1992. Proposal of *Burkholderia* gen. nov. and transfer of seven species of the genus *Pseudomonas* homology group II to the new genus, with the type species *Burkholderia cepacia* (Palleroni and Holmes 1981) comb. nov. Microbiol. Immunol. 36: 1251–1275.

Comments

Causes melioidosis, an endemic glanders-like disease. Most infections are acquired in the Far East. Melioidosis was reported in American military personnel during World War II and during the Vietnam conflict. Laboratory infections have been reported. Frequently this organism has a characteristic musty, earthy odor; it is resistant to colistin. The amount of gas formed in the reduction of nitrate may be small or may be demonstrated only at 25° C. *P. stutzeri* may be confused with *P. pseudomallei*; however, *P. stutzeri* usually develops a slightly yellow growth pigment, does not oxidize lactose, is rarely proteolytic, and is motile by a single polar flagellum. Also, *P. cepacia* may be confused with *P. pseudomallei* when gas is not formed readily from nitrate; however, *P. cepacia* is lysine-positive and arginine-negative. It recently has been proposed that *P. pseudomallei* be placed in the new genus *Burkholderia* (6). Information on the cellular fatty acid composition of this species is presented in Table 4.62 and Figure 4.67 of the cellular fatty acid section.

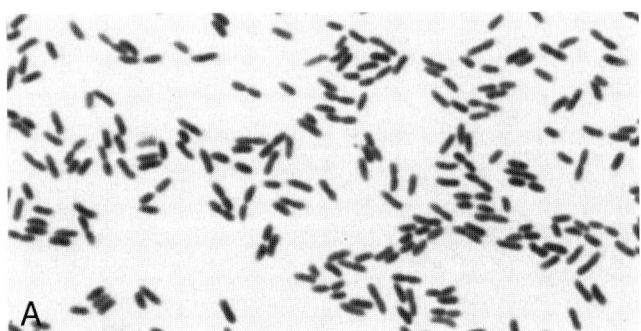

Pseudomonas pseudomallei strain KC871-1. Gram stain BAP 35° C 24 h x 1700.

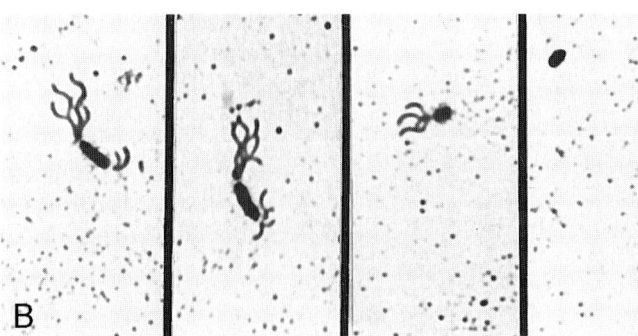

Pseudomonas pseudomallei strain KC871-1. Flagella stain BAP 35° C 24 h x 1700.

Pseudomonas pseudomallei

TEST PERFORMED	SIGN	% +	#+/T
	NUMBER OF STRAINS 70		
Morphology	mrs		
Motility; flagella	[m;p, >2]		
Action on blood	v	40 gr	23/57; 17 ly, 11 LG, 6 lav
Fermentative or oxidative	[O]		
Carbohydrate base	OF		
Acid from:			
D-Glucose	[+]	100	70/70
D-Xylose	+ or (+)	86 (14)	58, 2w (10)/70
D-Mannitol	+	94 (6)	66 (4)/70
Lactose	[+]	99 (1)	69 (1)/70
Sucrose	v	66 (4)	42, 4w (3)/70
Maltose	[+]	99 (1)	69 (1)/70
Catalase	+	100	58, 1w/59
Oxidase	[+]	100	55, 6w/61
Growth on:			
MacConkey	[+]	100	62/62
SS	v	8 (31)	5 (19)/61
Cetrimide	-	7(3)	5 (2)/68
Simmons citrate	v	77 (4)	53 (3)/69
Urea, Christensen's	v	13 (8)	9 (6)/69
Nitrate reduction	[+]	100	70/70
Gas from nitrate	[+][a]	100	70/70
Indole	-	0	0/60
TSI slant, acid	v	72	49/68
TSI butt, acid	-	0	0/68
H_2S (TSI butt)	-	0	0/70
H_2S (Pb ac paper)	v	26	18w/69
Gelatin hydrolysis[b]	v	79	46/58
Litmus milk	pep	96	67/70; 2 IR
Pigment:			
Insoluble	v	51 white	28/54; 21 yel
Growth at:			
25°C	+	100	55/55
35°C	+	100	55/55
42°C	+	100	55/55
Esculin hydrolysis	v	59	34/58
Lysine decarboxylase	[-]	0	0/8
Arginine dihydrolase	[+]	100	8/8
Ornithine decarboxylase	-	0	0/8
Nutrient broth, 0% NaCl	+	100	61/61
Nutrient broth, 6% NaCl	v	12	7/61

[a]The volume of gas may be small. Gas may not be detected unless incubated at 25° C.
[b]Incubation of 7–14 days.

Pseudomonas putida

Gram-negative medium to slightly long rod; motile (polar tuft of three or more flagella); aerobic; oxidizes glucose and xylose, but not sucrose; grows on MacConkey agar; oxidase-positive; usually produces pyoverdin, a fluorescent yellow-green pigment; does not grow at 42° C; does not reduce nitrate; does not liquefy gelatin at 25° C; lysine decarboxylase-negative; arginine dihydrolase-positive; acetamide-negative.

Sources (99 strains)

Urine (18%), blood (7%), cosmetics (7%), sputum (7%), lung (6%), cerebrospinal fluid (5%), throat (5%), other (33%), unknown (12%).

Reference strain

KC1074 = ATCC 12633 ← Stanier, 90; type strain.

Literature

1. Blazevic, D.J., M.H. Koepke, and J.M. Matsen. 1973. Incidence and identification of *Pseudomonas fluorescens* and *Pseudomonas putida* in the clinical laboratory. Appl. Microbiol. 25: 107–110.

2. Palleroni, N.J. 1984. Genus I. *Pseudomonas* Migula 1894, 237[AL] (Nom. cons. Opin. 5, Jud. Comm. 1952, 237). In N.R. Krieg and J.G. Holt (Eds.), Bergey's Manual of Systematic Bacteriology, Vol. 1, Williams & Wilkins, Baltimore, pp. 141–213.

3. Anaissie, E., V. Fainstein, et al. 1987. *Pseudomonas putida* newly recognized pathogen in patients with cancer. Am. J. Med. 82: 1191–1194.

Comments

Similar to *P. fluorescens*, except for lack of gelatinase and certain other proteolytic enzymes. Two biovars, A and B, have been described (2). Differentiation of these biovars requires the analysis of L-tryptophan, L-kynurenine, and anthranilate utilization. In addition to those tabluated, we have studied approximately 30 strains that resemble *P. putida*, except that they grow at 42° C.

Information on the cellular fatty acid composition of this group is presented in Table 4.37 and Figure 4.42 of the cellular fatty acid section. These strains form two similar cellular fatty acid groups (*P. putida* A and *P. putida* 1) that are differentiated by the presence or absence of 16:1ω7t. The type strain produces the *P. putida* A profile. The taxonomic significance of the *P. putida* 1 group has not been investigated.

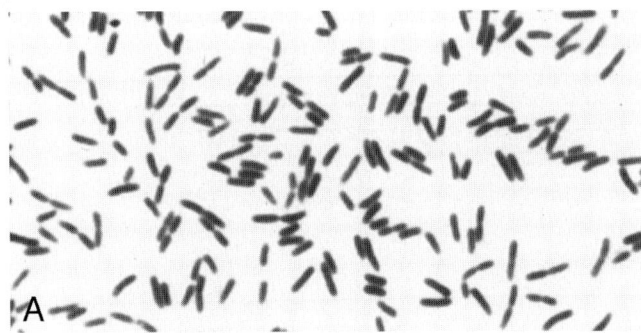

Pseudomonas putida strain KC1074. Gram stain BAP 35° C 24 h x 1700.

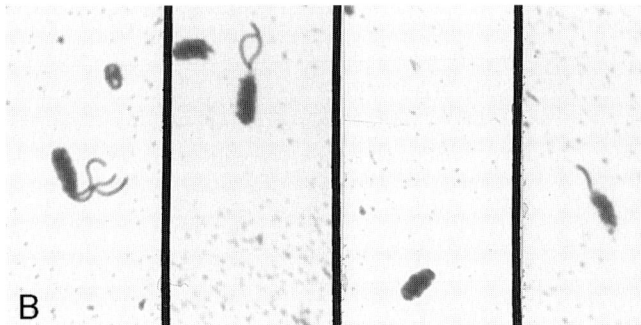

Pseudomonas putida strain KC1074. Flagella stain BAP 35° C 24 h x 1700.

Pseudomonas putida

TEST PERFORMED	SIGN	% +	#+/T
	NUMBER OF STRAINS 16		
Morphology	mrs		
Motility; flagella	[m;p>2]		
Action on blood	v	38 gr	6/16; 4 gr & ly, 4 ly
Fermentative or oxidative	[O]		
Carbohydrate base	OF		
Acid from:			
D-Glucose	[+]	100	16/16
D-Xylose	+	100	16/16
D-Mannitol	v	25	3, 1w/16
Lactose	v	25 (13)	4 (2)/16
Sucrose	[-]	0	0/16
Maltose	[v]	31	3, 2w/16
Catalase	+	100	16/16
Oxidase	[+]	100	16/16
Growth on:			
MacConkey	[+]	100	16/16
SS	+	100	16/16
Cetrimide	v	81 (6)	13 (1)/16
Simmons citrate	+	94 (6)	15 (1)/16
Urea, Christensen's	v	13 (44)	2 (7)/16
Nitrate reduction	[-]	0	0/16
Gas from nitrate	-	0	0/16
Indole	-	0	0/16
TSI slant, acid	-	0	0/16
TSI butt, acid	-	0	0/16
H_2S (TSI butt)	-	0	0/16
H_2S (Pb ac paper)	-	6	(1w)/16
Gelatin hydrolysis[a]	[-]	0	0/16
Litmus milk	v	62 k	10/16
Pigment:			
Pyoverdin	[+]	93	15/16
Pyocyanin	[-]	0	0/16
Pyorubrin	-	0	0/16
Growth at:			
25°C	+	100	16/16
35°C	+	100	16/16
42°C	[-]	0	0/16
Esculin hydrolysis	-	0	0/16
Lysine decarboxylase	[-]	0	0/16
Arginine dihydrolase	[+]	100	16/16
Ornithine decarboxylase	-	0	0/16
Nutrient broth, 0% NaCl	+	100	16/16
Nutrient broth, 6% NaCl	+	100	16/16
Acetamide alkalinization	[-]	0	0/13

[a]Incubation at 25° C for 7–14 days

Pseudomonas sp. CDC Group 1

Gram-negative straight to slightly curved rod; motile (1–2 polar flagella); aerobic; usually grows on MacConkey agar; does not utilize carbohydrates; oxidase-positive; urease-negative; reduces both nitrate and nitrite to gas; produces no H_2S in TSI agar.

Sources (31 strains)

Urine (26%), cerebrospinal fluid (13%), blood (10%), throat (10%), environment (6%), 1 strain each from sputum, lung, ear, toenail exudate, knee fluid, cul-de-sac pericardium, leg lesion, and horse uterus (29%), unknown (6%).

Reference strain

G8475.

Literature

1. Gilardi, G.L. 1991. *Pseudomonas* and related genera. *In* A. Balows, et al. (Eds.), Manual of Clinical Microbiology, 5th ed., American Society for Microbiology, Washington, pp. 429–441.

Comments

Prior to 1978, this was a group of pseudomonads that was referred to as *"Pseudomonas denitrificans"* in our laboratory. Reduction of nitrate to gas differentiates these organisms from *P. alcaligenes*. Information on the cellular fatty acid composition of this group is presented in Table 4.64 and Figure 4.69 of the cellular fatty acid section.

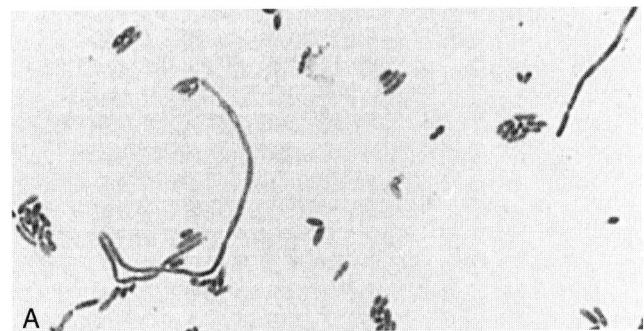

Pseudomonas sp. CDC group 1 strain G8475. Gram stain BAP 35° C 24 h x 1700.

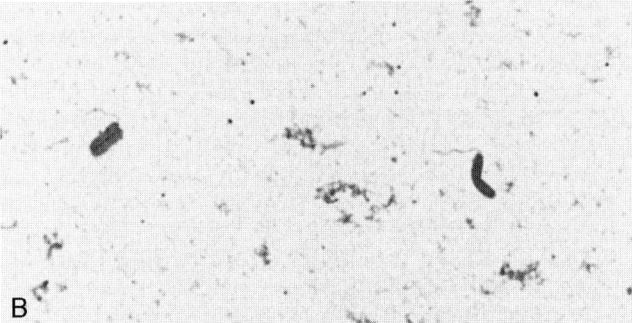

Pseudomonas sp. CDC group 1 strain G8475. Flagella stain (Ryu method) TGY 25° C 24 h x 1700.

Pseudomonas sp. CDC Group 1

	NUMBER OF STRAINS 31		
TEST PERFORMED	SIGN	%+	#+/T
Morphology	mrc		
Motility; flagella	[m;p,1-2]		
Action on blood	v	33 ly or gr	10/30; 2 al, 1 LG
Fermentative or oxidative	[n-o]		
Carbohydrate base	OF		
Acid from:			
D-Glucose	[-]	0	0/31
D-Xylose	[-]	0	0/31
D-Mannitol	-	0	0/31
Lactose	-	0	0/31
Sucrose	-	0	0/31
Maltose	[-]	0	0/31
Catalase	v	81	19, 6w/31
Oxidase	[+]	100	30, 1w/31
Growth on:			
MacConkey	[+]	97 (3)	30 (1)/31
SS	v	30 (6)	9 (2)/31
Cetrimide	v	13 (6)	4 (2)/31
Simmons citrate	v	42 (6)	13 (2)/31
Urea, Christensen's	[-]	3 (7)	1 (2)/30
Nitrate reduction	+	100	31/31
Gas from nitrate	[+]	100	31/31
Indole	-	0	0/31
TSI slant, acid	-	0	0/31
TSI butt, acid	-	0	0/31
H_2S (TSI butt)	-	0	0/31
H_2S (Pb ac paper)	v	68	10, 11w/31
Gelatin hydrolysis[a]	-	4	1/26
Litmus milk	v	39 k	12/31
Pigment:			
Soluble	v	60 yel-br	12/20
Growth at:			
25°C	+	97	30/31
35°C	+	97	30/31
42°C	v	48	15/31
Esculin hydrolysis	-	0	0/28
Lysine decarboxylase	-	0	0/3
Arginine dihydrolase	v	33	1/3
Ornithine decarboxylase	-	0	0/3
Nutrient broth, 0% NaCl	+	93	26/28
Nutrient broth, 6% NaCl	v	14	4/28

[a]Incubation of 7–14 days.

Pseudomonas stutzeri

Gram-negative medium to long rod; motile (single polar flagellum); aerobic; oxidizes glucose, xylose, maltose, and (usually) mannitol, not lactose; hydrolyzes starch; grows on MacConkey agar; oxidase-positive; reduces nitrate to gas; indole-negative; does not grow on cetrimide agar; no H_2S on TSI agar; nonproteolytic; arginine dihydrolase- and lysine decarboxylase-negative.

Sources (28 strains)

Wound (14%); urogenital (14%); blood (7%); fecal material (7%); nasopharynx (7%); cerebrospinal fluid, heel fluid, tooth socket, bronchial wash (1 each) (14%); other (nonhuman) (12%); unknown (25%).

Reference strain

KC1077 ← ATCC 17587 (Stanier's 220).

Literature

1. von Graevenitz, A. 1965. *Pseudomonas stutzeri* isolated from clinical specimens. Am. J. Clin. Pathol. 43: 357–360.
2. Stanier, R.Y., N.J. Palleroni, and M. Doudoroff. 1966. The aerobic pseudomonads: a taxonomic study. J. Gen. Microbiol. 43: 159–271.
3. Palleroni, N.J., M. Doudoroff, and R.Y. Stanier. 1970. Taxonomy of the aerobic pseudomonads: the properties of the *Pseudomonas stutzeri* group. J. Gen. Microbiol. 60: 215–231.
4. Holmes, B. 1986. Identification and distribution of *Pseudomonas stutzeri* in clinical material. J. Appl. Bacteriol. 60: 401–411.
5. Gavini, F., B. Holmes, D. Izard, A. Beji, A. Bernigaud, and E. Jakubczak. 1989. Numerical taxonomy of *Pseudomonas alcaligenes, P. pseudoalcaligenes, P. mendocina, P. stutzeri*, and related bacteria. Int. J. Syst. Bacteriol. 39: 135–144.
6. Palleroni, N.J. 1984. Genus 1. *Pseudomonas* Migula 1894, 237[AL] (Nom. cons. Opin. 5, Jud. Comm. 1952, 237). In N.R. Krieg and J.G. Holt (Eds.), Bergey's Manual of Systematic Bacteriology, Vol. 1, Williams & Wilkins, Baltimore, pp. 141–199.

Comments

Formerly called Vb-1. A nonfluorescent pseudomonad; produces smooth and rough, slightly yellow, wrinkled colonies. *Pseudomonas stutzeri* is described as being arginine dihydrolase-negative in *Bergey's Manual of Systematic Bacteriology* (6), and we have included only arginine dihydrolase-negative strains in our tabulation. We continue to place arginine dihydrolase-positive strains in group Vb-3, although Gavini et al. (5) and Holmes (4) have identified arginine dihydrolase-positive strains as *P. stutzeri*. DNA-relatedness studies are needed to resolve the relationship of *P. stutzeri* and Vb-3. Information on the cellular fatty acid composition of this species is presented in Table 4.36 and Figure 4.41 of the cellular fatty acid section.

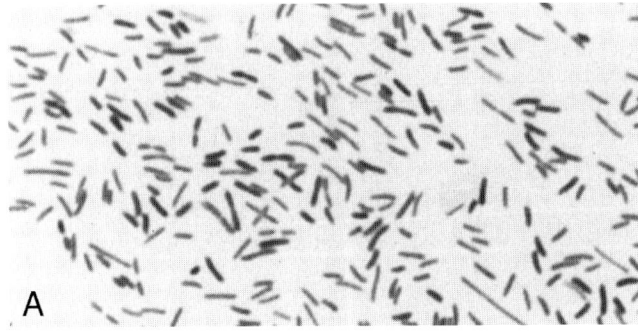

Pseudomonas stutzeri strain KC1077. Gram stain BAP 35° C 24 h x 1700.

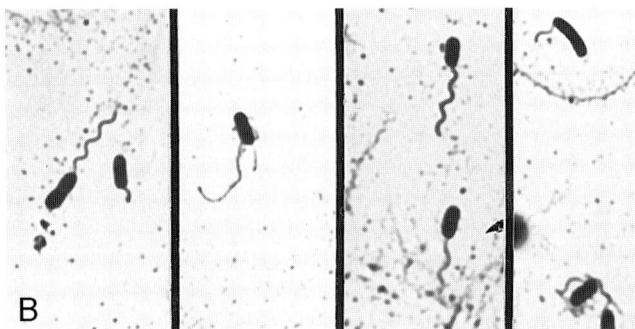

Pseudomonas stutzeri strain KC1077. Flagella stain BAP 35° C 24 h x 1700.

Pseudomonas stutzeri

	NUMBER OF STRAINS 28		
TEST PERFORMED	SIGN	% +	#+/T
Morphology	mrs		
Motility; flagella	[m;p,1-2]		
Action on blood	v	26 ly	7/27; 6 gr, 3 LG, 3 br
Fermentative or oxidative	[O]		
Carbohydrate base	OF		
Acid from:			
D-Glucose	[+]	96 (4)	25, 2w (1)/28
D-Xylose	+	93 (7)	25, 1w (2)/28
D-Mannitol	+ or (+)	89 (4)	10, 15w (1)/28
Lactose	[-]	0	0/28
Sucrose	-	0	0/28
Maltose	[+]	100	13, 15w/28
Starch	+	100	5/5
Catalase	+	100	25, 3w/28
Oxidase	[+]	100	28/28
Growth on:			
MacConkey	[+]	100	28/28
SS	v	54	13, 2w/28
Cetrimide	-	4	1w/28
Simmons citrate	+ or (+)	82 (14)	23 (4)/28
Urea, Christensen's	v	33 (22)	9 (6)/27
Nitrate reduction	+	100	28/28
Gas from nitrate	[+]	100	28/28
Indole	-	0	0/28
TSI slant, acid	-	0	0/28
TSI butt, acid	-	0	0/28
H_2S (TSI butt)	-	0	0/28
H_2S (Pb ac paper)	v	36	4, 6w/28
Gelatin hydrolysis[a]	[-]	0	0/22
Litmus milk	v	57 k	16/28; 4 IR, 1 pep
Pigment	v	86 sl yel ins	24/28
Growth at:			
25°C	+	100	26/26
35°C	+	100	26/26
42°C	v	69	16, 2w/26
Esculin hydrolysis	-	0	0/27
Lysine decarboxylase	[-]	0	0/28
Arginine dihydrolase	[-]	0	0/28
Ornithine decarboxylase	-	0	0/28
Nutrient broth, 0% NaCl	+	96	24/25
Nutrient broth, 6% NaCl	+ or (+)	80 (16)	19, 1w (4)/25
Alkalinization of acetamide	-	0	0/3
2-Ketogluconate	-	0	0/5

[a]Incubation of 7–14 days.

"Pseudomonas thomasii"

Gram-negative short to medium length rod; motile (single polar flagellum); aerobic; oxidizes glucose, xylose, mannitol, lactose, and maltose, but not sucrose; grows on MacConkey agar; oxidase-positive; usually does not reduce nitrate; arginine dihydrolase-negative and lysine and ornithine decarboxylases-negative.

Sources (31 strains)

Blood (26%), sputum (13%), abscess (10%), urine (6%), other (39%), unknown (6%).

Reference strains

C3408 ← I. Phillips, 73-9288.
C1700, from blood.

Literature

1. Phillips, I., S. Eykyn, and M. Kaber. 1972. Outbreak of hospital infection caused by contaminated autoclave fluids. Lancet *1*: 1258–1260.
2. King, A., B. Holmes, I. Phillips, and S. Lapage. 1979. A taxonomic study of clinical isolates of *Pseudomonas pickettii*, "*P. thomasii*" and "Group IVd" bacteria. J. Gen. Microbiol. *114*: 137–147.
3. Palleroni, N.J. 1984. Genus I. *Pseudomonas* Migula 1894, 237[AL] (Nom. cons. Opin. 5, Jud. Comm. 1952, 237). *In* N.R. Krieg and J.G. Holt (Eds.), Bergey's Manual of Systematic Bacteriology, Vol. 1, Williams & Wilkins, Baltimore, pp. 141–213.
4. Costas, M., B. Holmes, L.L. Sloss, and S. Heard. 1990. Investigation of a pseudo-outbreak of "*Pseudomonas thomasii*" in a special-care baby unit by numerical analysis of SDS-PAGE protein analysis. Epidemiol. Infect. *105*: 127–137.

Comments

This group of cultures is similar to a culture received from Ian Phillips, London, England, designated "*Pseudomonas thomasii*." The taxonomic location of these organisms is unsettled. Numerical taxonomy studies indicate "*P. thomasii*" may be a biovar of *P. pickettii* (2). However, protein profile analysis indicates that these organisms represent a different species (4). Information on the cellular fatty acid composition of this species is presented in Table 4.10 and Figure 4.15 of the cellular fatty acid section.

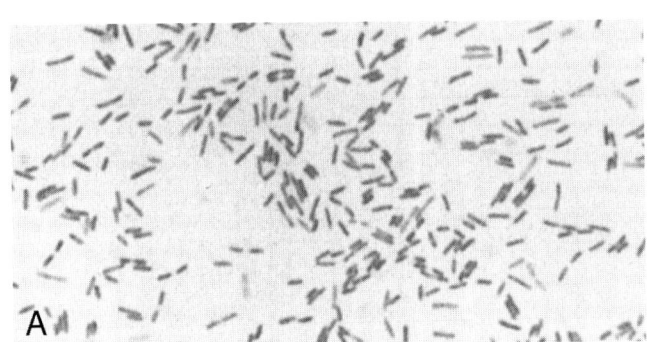

"*Pseudomonas thomasii*" strain C1700. Gram stain BAP 35° C 24 h x 1700.

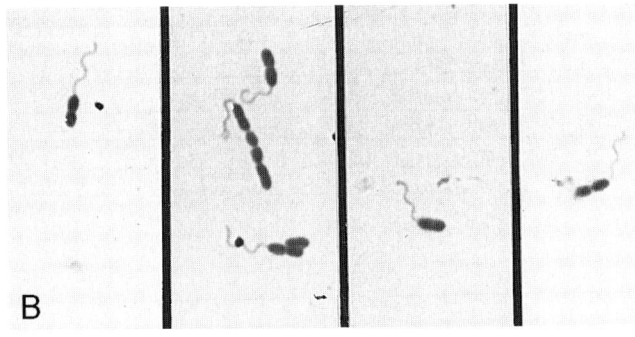

"*Pseudomonas thomasii*" strain C1700. Flagella stain BAP 35° C 24 h x 1700.

"Pseudomonas thomasii"

NUMBER OF STRAINS 31

TEST PERFORMED	SIGN	% +	#+/T
Morphology	mrs		
Motility; flagella	[m;p,1-2]		
Action on blood	v	24 gr	7/29; 6 ly, 2 LG, 1 inc β
Fermentative or oxidative	[O]		
Carbohydrate base	OF		
Acid from:			
D-Glucose	[+]	100	31/31
D-Xylose	+	100	31/31
D-Mannitol	[+]	100	3, 28w/31
Lactose	[+]	100	31/31
Sucrose	[−]	0	0/31
Maltose	[+]	100	31/31
Catalase	v	87	5, 22w/31
Oxidase	[+]	100	21, 10w/31
Growth on:			
MacConkey	[+]	100	31/31
SS	−	0	0/31
Cetrimide	−	(3)	(1w)/31
Simmons citrate	+	100	31/31
Urea, Christensen's	+ or (+)	81 (19)	25 (6)/31
Nitrate reduction	v	13	4/31
Gas from nitrate	[−]	0	0/31
Indole	−	0	0/31
TSI slant, acid	−	0	0/31
TSI butt, acid	−	0	0/31
H_2S (TSI butt)	−	0	0/31
H_2S (Pb ac paper)	v	23	1, 8w/31
Gelatin hydrolysis[a]	v	30	8/27
Litmus milk	v	52 k	15/29; 5 IR, 2 pep
Pigment:			
Soluble	v	33 yel-ta	10/31
Growth at:			
25°C	+	97	17, 13w/31
35°C	+	100	31/31
42°C	v	84	21, 5w/31
Esculin hydrolysis	−	0	0/31
Lysine decarboxylase	[−]	0	0/31
Arginine dihydrolase	[−]	3	1/31
Ornithine decarboxylase	−	0	0/31
Nutrient broth, 0% NaCl	+	100	31/31
Nutrient broth, 6% NaCl	−	0	0/31

[a] Incubation of 7–14 days.

Pseudomonas vesicularis

Gram-negative medium to long, slender rod; motile (single polar flagellum with distinctively short wavelength); aerobic; usually grows on MacConkey agar; oxidizes glucose, maltose, and sometimes xylose; produces oxidase; does not usually reduce nitrate to nitrite.

Sources (94 strains)

Blood (30%), cerebrospinal fluid (7%), urine (7%), eye (7%), pleural fluid (4%), peritoneal fluid (4%), endocervical (3%), wound (3%), ear (2%), blister (2%), environmental (2%), other (23%), unknown (6%).

Reference strain

KC1099 ← H. Lautrop, AB102, from whooping cough plate.

Literature

1. Galarneault, T.P., and E. Leifson. 1964. *Pseudomonas vesiculare* (Busing et al.) nov. comb. Int. Bull. Bacteriol. Nomencl. Taxon. *14:* 165–168.
2. Ballard, R.W., M. Doudoroff, R.Y. Stanier, and M. Mandel. 1968. Taxonomy of the aerobic Pseumonads: *Pseudomonas diminuta* and *P. vesiculare*. J. Gen. Microbiol. *53:* 349–361.
3. Kaltenbach, C.M., C.W. Moss, and R.E. Weaver. 1975. Cultural and biochemical characteristics and fatty acid composition of *Pseudomonas diminuta* and *Pseudomonas vesiculare*. J. Clin. Microbiol. *1:* 339–344.
4. Gilardi, G.L. 1991. *Pseudomonas* and related genera. *In* A. Balows, et al. (Eds.), Manual of Clinical Microbiology, 5th ed., American Society for Microbiology, Washington, pp. 429–441.

Comments

Formerly called "*Pseudomonas vesiculare*." Sixty percent of strains produced a tan-brown, water-soluble pigment on heart infusion agar with tyrosine, and 52% of the strains produced a yellow or orange growth pigment. In addition to the strains tabulated, we have also studied numerous strains that phenotypically resemble *P. vesicularis*, but produce a fatty acid profile resembling *Sphingomonas paucimobilis*. Information on the cellular fatty acid composition of this species is presented in Table 4.63 and Figure 4.68 of the cellular fatty acid section.

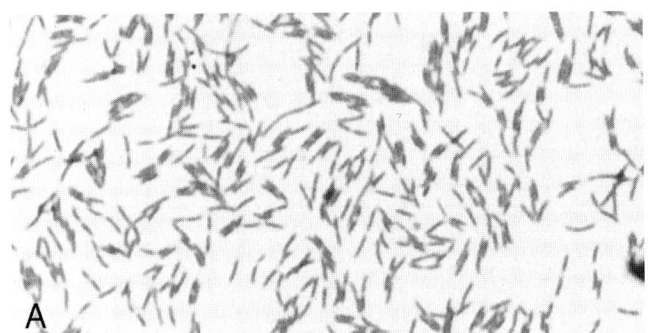

Pseudomonas vesicularis strain D3834. Gram stain BAP 35° C 24 h x 1700.

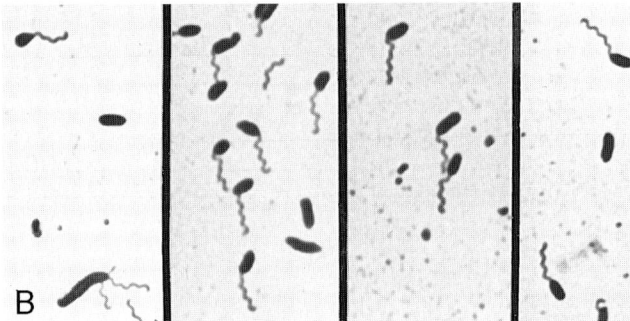

Pseudomonas vesicularis strain D3834. Flagella stain FB 25° C 24 h x 1700.

Pseudomonas vesicularis

TEST PERFORMED	SIGN	% +	#+/T
Morphology	mrs		
Motility; flagella	[m;p,1-2]^a		
Action on blood	v	7 LG	7/94; 6 ly, 2 gr, 1 lav, 1 br
Fermentative or oxidative	[O]		
Carbohydrate base	OF		
Acid from:			
D-Glucose	[+ or (+)]	87 (12)	77, 5w (3) (8w)/94
D-Xylose	[v]	27 (9)	24, 1w (3) (5w)/94
D-Mannitol	−	0	0/94
Lactose	−	0	0/94
Sucrose	−	0	0/94
Maltose	[+]	94 (6)	85, 3w (2) (4w)/94
Catalase	v	83	47, 30w/93
Oxidase	[+]	98	88, 3w/93
Growth on:			
MacConkey	[v]	43 (23)	37, 3w (18) (3w)/93
SS	−	1	1/93
Cetrimide	−	0	0/89
Simmons citrate	−	1 (1)	1 (1w)/93
Urea, Christensen's	−	2 (5)	2 (5)/93
Nitrate reduction	[−]	5	5/93
Gas from nitrate	[−]	0	0/93
Indole	−	0	0/93
TSI slant, acid	−	0	0/93
TSI butt, acid	−	0	0/93
H_2S (TSI butt)	−	0^b	0/93^b
H_2S (Pb ac paper)	v	49 (6)	21, 25w (1) (5w)/93
Gelatin hydrolysis^c	v	25	23/93
Litmus milk	v	17 IR	15/87; 7 k, 7 pep
Pigment:			
Insoluble	v	52 yel-or	48/93; 1 co
Soluble	v	60 br-ta	56/93; 6 amb, 4 yel
Growth at:			
25°C	+	97	90/93
35°C	+	97	90/93
42°C	v	19	18/93
Esculin hydrolysis	[+ or (+)]	88 (6)	82 (6)/93
Lysine decarboxylase	−	0	0/45
Arginine dihydrolase	−	7	3/45
Ornithine decarboxylase	−	0	0/45
Nutrient broth, 0% NaCl	+	95	88/93
Nutrient broth, 6% NaCl	v	23 (2)	21 (2)/93

^aFlagella have a short wavelength with low amplitude.
^bBlack color was observed around the stab in the TSI butt at 7 days of incubation for one strain.
^cIncubation of 7–14 days.

Psychrobacter immobilis

Asaccharolytic strains

Gram-negative coccobacillus; psychrotrophic; nonmotile; aerobic; does not use carbohydrates; oxidase-positive; variable growth on MacConkey agar; does not produce indole; *Psychrobacter immobilis* transformation-positive.

Sources (5 strains)

Left hip prosthesis (1), catheter site (1), upper thigh wound (1), unknown (2).

Reference strain

KC1840 ← ATCC 17955.

Saccharolytic strains

Saccharolytic *P. immobilis* strains share all of the characteristics of the asaccharolytic strains listed above except that glucose, xylose, and lactose, but not sucrose and maltose, are oxidized.

Sources (7 strains)

Blood (3), vagina (1), unknown (3).

Reference strain

KC1837 ← ATCC 43116, type strain.

Literature

1. Juni, E., and G.A. Heym. 1986. *Psychrobacter immobilis* gen. nov., sp. nov.: genospecies composed of gram-negative, aerobic, oxidase-positive coccobacilli. Int. J. Syst. Bacteriol. *36:* 388–391.
2. Lloyd-Puryear, M., D. Wallace, T. Baldwin, and D.G. Hollis. 1991. Meningitis caused by *Psychrobacter immobilis* in an infant. J. Clin. Microbiol. *29:* 2041–2042.
3. Moss, C.W., P.L. Wallace, D.G. Hollis, and R.E. Weaver. 1988. Cultural and chemical characterization of CDC Groups EO-2, M-5, and M-6, *Moraxella* (*Moraxella*) species, *Oligella urethralis, Acinetobacter species,* and *Psychrobacter immobilis.* J. Clin. Microbiol. *26:* 484–492.
4. Pickett, J.M., D.G. Hollis, and E.J. Bottone. 1991. Miscellaneous gram-negative bacteria. *In* A. Balows, et al. (Eds.), Manual of Clinical Microbiology, 5th ed., American Society for Microbiology, Washington, pp. 410–428.

Comments

Saccharolytic strains of *P. immobilis* were included in group EO-2 in the previous edition of this manual. Asaccharolytic strains are phenotypically similar to *Moraxella phenylpyruvica*. Characteristics that are useful for differentiating *P. immobilis* from groups EO-2 and EO-3 and from *M. phenylpyruvica* are a positive *P. immobilis* transformation test, cellular fatty acid profile, and optimal growth temperatures <35° C. Many strains produce a rose-like odor similar to that associated with phenylethyl alcohol blood agar. Information on the cellular fatty acid composition of this group is presented in Table 4.65 and Figure 4.70 of the cellular fatty acid section.

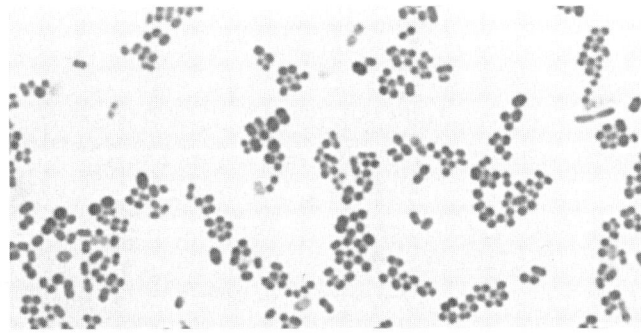

Psychrobacter immobilis strain KC1840. Gram stain BAP 35° C 24 h x 1700.

Asaccharolytic *Psychrobacter immobilis*

TEST PERFORMED	SIGN	% +	# +/T
	NUMBER OF STRAINS 5		
Morphology	cc		
Motility	[nm]		
Action on blood	−	0	0/5
Fermentative or oxidative	[n-o]		
Carbohydrate base	OF		
Acid from:			
D-Glucose	[−]	0	0/5
D-Xylose	−	0	0/5
D-Mannitol	−	0	0/5
Lactose	−	0	0/5
Sucrose	−	0	0/5
Maltose	−	0	0/5
Catalase	+	100	5/5
Oxidase	[+]	100	5/5
Growth on:			
MacConkey	v	40	2/5
SS	−	0	0/5
Cetrimide	−	0	0/5
Simmons citrate	v	20	1/5
Urea, Christensen's	v	(20)	(1)/5
Nitrate reduction	v	40 (20)	2, (1)/5
Gas from nitrate	−	0	0/5
Indole	[−]	0	0/5
TSI slant, acid	−	0	0/5
TSI butt, acid	−	0	0/5
H$_2$S (TSI butt)	−	0	0/5
H$_2$S (Pb ac paper)	v	20 (20)	1 (1)/5
Gelatin hydrolysis[a]	−	0	0/5
Litmus milk	−	0	0/5
Pigment	[−]	0	0/5
Growth at:			
25°C	[+][b]	100	5/5
35°C	[v]	40	1, 1w/5
42°C	v	20	1/5
Esculin hydrolysis	−	0	0/5
Nutrient broth, 0% NaCl	+	100	5/5
Nutrient broth, 6% NaCl	v	60	3/5
Psychrobacter transformation	[+]	100	5/5

[a]Incubation of 7–14 days.
[b]Usually prefers 25° C.

Saccharolytic *Psychrobacter immobilis*

TEST PERFORMED	SIGN	% +	#+/T
NUMBER OF STRAINS 7			
Morphology	cc		
Motility	[nm]		
Action on blood	v	29 ly	2/7; 1 gr
Fermentative or oxidative	[O]		
Carbohydrate base	OF		
Acid from:			
D-Glucose	[+ or (+)]	57 (43)	4 (3)/7
D-Xylose	[+ or (+)]	57 (43)	4 (3)/7
D-Mannitol	−	0	0/7
Lactose	[+ or (+)]	57 (43)	4 (3)/7
Sucrose	−	0	0/7
Maltose	−	0	0/7
Catalase	+	100	7/7
Oxidase	[+]	100	7/7
Growth on:			
MacConkey	+	100	6, 1w/7
SS	−	0	0/7
Cetrimide	−	0	0/7
Simmons citrate	−	0	0/7
Urea, Christensen's	v	(43)	(3)/7
Nitrate reduction	v	86	6/7
Gas from nitrate	−	0	0/7
Indole	−	0	0/7
TSI slant, acid	−	0	0/7
TSI butt, acid	−	0	0/7
H_2S (TSI butt)	−	0	0/7
H_2S (Pb ac paper)	v	43 (14)	2, 1w (1w)/7
Gelatin hydrolysis[a]	−	0	0/7
Litmus milk	v	14 k	1/7
Pigment	v	43 tan-sol	3/7; 1 yel insoluble
Growth at:			
25°C	[+][b]	100	7/7
35°C	[v]	57	4/7
42°C	−	0	0/7
Esculin hydrolysis	−	0	0/7
Lysine decarboxylase	−	0	0/7
Arginine dihydrolase	v	29	2/7
Ornithine decarboxylase	−	0	0/7
Nutrient broth, 0% NaCl	+ or (+)	43 (57)	3 (4)/7
Nutrient broth, 6% NaCl	+	100	7/7
Psychrobacter transformation	[+]	100	7/7

[a]Incubation of 7–14 days.
[b]Usually prefers 25° C.

[1]Table Footnotes

Riemerella anatipestifer

Gram-negative rod; nonmotile; aerobic; oxidizes glucose and maltose weakly; lactose-negative; does not grow on MacConkey agar; oxidase-positive, nitrate-negative; indole-negative; proteolytic; characteristically produces a brown, tan, or yellow water-soluble pigment.

Sources (5 strains)

Avian (4), unknown (1).

Reference strain

KC1631 ← ATCC 11845, type strain, from duck blood.

Literature

1. Hauduroy, P., et al. 1953. Dictionnaire des Bactéries Pathogènes, 28th ed., Paris.

2. Hendrickson, J.M., and K.F. Hilbert. 1932. A new and serious septicemic disease of young ducks with a description of the causative organism. *Pfeifferella anatipestifer*. N.S. Cornell Vet. 22: 239-252.

3. Bøvre, K. 1984. Genus II. *Moraxella* Lwoff 1939. In N.R. Krieg and J.G. Holt (Eds.), Bergey's Manual of Systematic Bacteriology, Vol. 1, Williams & Wilkins, Baltimore, pp. 296–303.

4. Piechulla, K., S. Pohl, and W. Mannheim. 1986. Phenotypic and genetic relationships of so-called *Moraxella (Pasteurella) anatipestifer* to the *Flavobacterium/Cytophaga* group. Vet. Microbiol. 11: 261–270.

5. Segers, P., W. Mannheim, M. Vancanneyt, K. De Brandt, K.-H. Hinz, K. Kersters, and P. Vandamme. 1993. *Riemerella anatipestifer* gen. nov., comb. nov., the causative agent of septicemia anserum exsudativa, and its phylogenetic affiliation within the *Flavobacterium-Cytophaga* rRNA homology group. Int. J. Syst. Bacteriol. 43: 768–776.

Comments

Isolated from ducks and ducklings, geese, turkeys, and waterfowl with septicemic disease. Other names that have been used to designate this species in the past include *Pfeifferella anatipestifer*, *Pasteurella anatipestifer*, and *Moraxella anatipestifer*. It recently has been proposed that these organisms be placed in the *Flavobacterium/Cytophaga* rRNA group. Information on the cellular fatty acid composition of this species is presented in Table 4.66 and Figure 4.71 of the cellular fatty acid section.

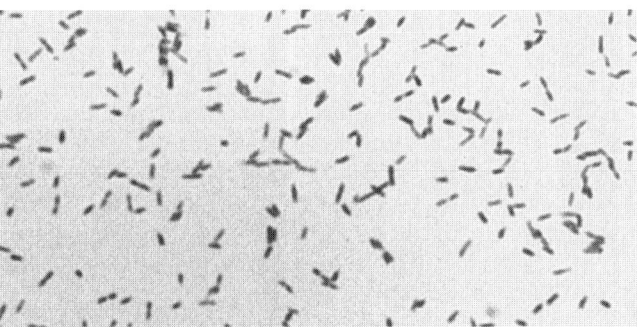

Riemerella anatipestifer strain KC534. Gram stain BAP 35° C 24 h x 1700.

Riemerella anatipestifer

TEST PERFORMED	SIGN	% +	# +/T
		NUMBER OF STRAINS 5	
Morphology	srs		
Motility; flagella	[nm]		
Action on blood	v	60 LG	3/5, 1 lav, 1 ly
Fermentative or oxidative	[O]		
Carbohydrate base	OF		
Acid from:			
D-Glucose	[+ or (+)]	60 (40)	3w (2w)/5
D-Xylose	-	0	0/5
D-Mannitol	-	0	0/5
Lactose	[-]	0	0/5
Sucrose	-	0	0/5
Maltose	[+ or (+)]	40 (60)	2w (3w)/5
Catalase	+	100	5/5
Oxidase	+	100	5/5
Growth on:			
MacConkey	-	0	0/5
SS	-	0	0/5
Cetrimide	-	0	0/5
Simmons citrate	-	0	0/5
Urea, Christensen's	v	40 (20)	2 (1)/5
Nitrate reduction	[-]	0	0/5
Gas from nitrate	-	0	0/5
Indole	[-]	0	0/5
TSI slant, acid	-	0	0/5
TSI butt, acid	-	0	0/5
H_2S (TSI butt)	-	0	0/5
H_2S (Pb ac paper)	+	100	4, 1w/5
Gelatin hydrolysis[a]	+	100	3/3
Litmus milk	v	60 pep	3/5
Pigment	v	80 ta-yel sol	4/5, 1 br sol
Growth at:			
25°C	+	100	5/5
35°C	+	100	5/5
42°C	-	0	0/5
Esculin hydrolysis	-	0	0/5

[a]Incubation of 7–14 days.

Roseomonas species

Gram-negative, plump coccoid rod, appears as single cells, pairs, or short chains; motility varies among the species; oxidase-variable; catalase- and urease-positive; produces a pink growth pigment. These strains were formerly designated CDC Pink Coccoid groups I through IV (1). Biochemical differentiation of the species and genomospecies of *Roseomonas* follows in Table 3.139. Data for this Table were obtained from Rihs et al. (2).

Literature

1. Wallace, P.L., D.G. Hollis, R.E. Weaver, and C.W. Moss. 1990. Biochemical and chemical characterization of pink-pigmented oxidative bacteria. J. Clin. Microbiol. *28:* 689–693.

2. Rihs, J.D., D.J. Brenner, R.E. Weaver, A.G. Steigerwalt, D.G. Hollis, and V.L. Yu. 1993. *Roseomonas,* a new genus associated with bacteremia and other human infections. J. Clin. Microbiol. *31:* 3275–3283.

Comments

These organisms may be confused with *Methylobacterium* species. Distinguishing characteristics include cellular fatty acid profiles and the ability of *Roseomonas* species to oxidize methanol.

Table 3.139 Biochemical differentiation of the species and genomospecies of *Roseomonas* [data from Rihs, et al. (2)]

TEST PERFORMED	R. gilardii	R. cervicalis	Number of strains positive/number of strains tested			
			R. fauriae	Genomospecies 4	Genomospecies 5	Genomospecies 6
Esculin hydrolysis	0/23	0/7	5/5	0/3	0/3	1/1
Oxidation of:						
L-Arabinose	15/23	4/7	5/5	2/3	1/3	0/1
D-Galactose	16/23	0/7	5/5	2/3	1/3	0/1
Glycerol	23/23	0/7	5/5	0/3	1/3[a]	1/1
D-Mannose	1/23	0/7	0/3	3/3	0/3	0/1
D-Xylose	11/22	4/7	5/5	3/3	2/3	0/1
Citrate	23/23	6/7	3/5	0/3	1/3[a]	1/1
Motility	8/23	7/7	5/5	1/3	0/3	1/1

[a]None of the three strains of genomospecies 5 was positive for both oxidation of glycerol and utilization of citrate.

Roseomonas cervicalis

Gram-negative plump coccoid rod, appears as single cells, pairs or short chains; motile (single polar flagellum); variable oxidizer of xylose; no acid produced from glucose, mannitol, or lactose; grows on MacConkey agar; oxidase- and urease-positive; produces a pink growth pigment.

Sources (7 strains)

Female urogenital (3 strains), eye (2 strains), unknown (2 strains).

Reference strain

E7107 = ATCC 49957, type strain, from cervix.

Literature

1. Gilardi, G.L., and Y.C. Faur. 1984. *Pseudomonas mesophilica* and an unnamed taxon, clinical isolates of pink-pigmented oxidative bacteria. J. Clin. Microbiol. *20:* 626–629.

2. Wallace, P.L., D.G. Hollis, R.E. Weaver, and C.W. Moss. 1990. Biochemical and chemical characterization of pink-pigmented oxidative bacteria. J. Clin. Microbiol. *28:* 689–693.

3. Rihs, J.D., D.J. Brenner, R.E. Weaver, A.G. Steigerwalt, D.G. Hollis, and V.L. Yu. 1993. *Roseomonas,* a new genus associated with bacteremia and other human infections. J. Clin. Microbiol. *31:* 3275–3283.

Comments

The pink coccoid (pc) groups of Wallace et al. (2) are *Roseomonas* species, but the pc groups do not completely correlate with the various genomospecies of *Roseomonas*. *R. cervicalis* contains some of the strains formerly designated pc groups I and II. Information on the cellular fatty acid composition of this species is presented in Table 4.60 and Figure 4.65 of the cellular fatty acid section.

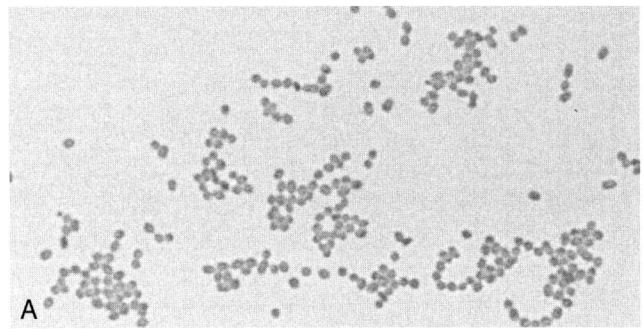

Roseomonas cervicalis strain E7107. Gram stain BAP 35° C 24 h x 1700.

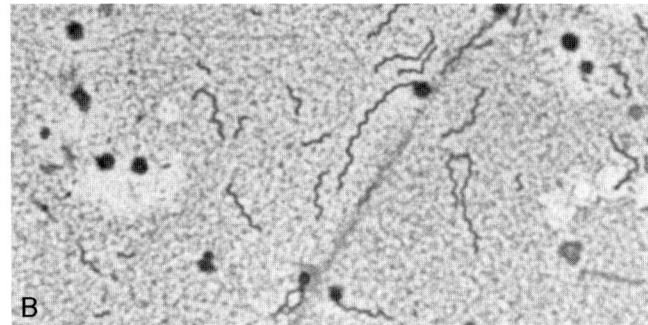

Roseomonas cervicalis strain E7107. Flagella stain (Ryu method) TGY 25° C 24 h x 1700.

Roseomonas cervicalis

TEST PERFORMED	SIGN	%+	#+/T
	NUMBER OF STRAINS 7		
Morphology	cc		
Motility; flagella	[m;p,1-2]		
Action on blood	v	14 ß	1/7
Fermentative or oxidative	[n-o]		
Carbohydrate base	OF		
Acid from:			
D-Glucose	[−]	0	0/7
D-Xylose	v	43	2, 1w/7
D-Mannitol	[−]	0	0/7
Lactose	−	0	0/7
Sucrose	−	0	0/7
Maltose	−	0	0/7
Catalase	+	100	7/7
Oxidase	[+]	100	7/7
Growth on:			
MacConkey	+	100	7/7
SS	−	0	0/7
Cetrimide	−	0	0/7
Simmons citrate	+ or (+)	86 (14)	5, 1w (1)/7
Urea, Christensen's	+ or (+)	86 (14)	6 (1)/7
Nitrate reduction	−	0	0/7
Gas from nitrate	−	0	0/7
Indole	−	0	0/7
TSI slant, acid	−	0	0/7
TSI butt, acid	−	0	0/7
H$_2$S (TSI butt)	−	0	0/7
H$_2$S (Pb ac paper)	+	100	7/7
Gelatin hydrolysis[a]	−	0	0/7
Litmus milk	v	86 k	6/7
Pigment	[pk]	100	7/7
Growth at:			
25°C	+	100	7/7
35°C	+	100	7/7
42°C	+	100	7/7
Esculin hydrolysis	[−]	0	0/7
Nutrient broth, 0% NaCl	+	100	7/7
Nutrient broth, 6% NaCl	−	0	0/7

[a]Incubation of 7–14 days.

Roseomonas fauriae

Gram-negative, coccobacillus and rod that appears singly or in pairs; motile (single polar flagellum); oxidizes xylose, but not mannitol, lactose, or sucrose; variable oxidation of glucose; grows on MacConkey agar; oxidase-, urease-, esculin-, and nitrate-positive; produces a pink growth pigment.

Sources (5 strains)

Blood (2 strains); hand, knee wound, and sputum (1 strain each).

Reference strain

C610 = ATCC 49958, type strain, from hand wound.

Literature

1. Gilardi, G.L., and Y.C. Faur. 1984. *Pseudomonas mesophilica* and an unnamed taxon, clinical isolates of pink-pigmented oxidative bacteria. J. Clin. Microbiol. *20:* 626–629.

2. Wallace, P.L., D.G. Hollis, R.E. Weaver, and C.W. Moss. 1990. Biochemical and chemical characterization of pink-pigmented oxidative bacteria. J. Clin. Microbiol. *28:* 689–693.

3. Rihs, J.D., D.J. Brenner, R.E. Weaver, A.G. Steigerwalt, D.G. Hollis, and V.L. Yu. 1993. *Roseomonas*, a new genus associated with bacteremia and other human infections. J. Clin. Microbiol. *31:* 3275–3283.

Comments

The pink coccoid (pc) groups of Wallace et al. (2) are *Roseomonas* species, but the pc groups do not completely correlate with the various genomospecies of *Roseomonas*. *R. fauriae* strains were previously in pc group IV. Information on the cellular fatty acid composition of this species is presented in Table 4.54 and Figure 4.59 of the cellular fatty acid section.

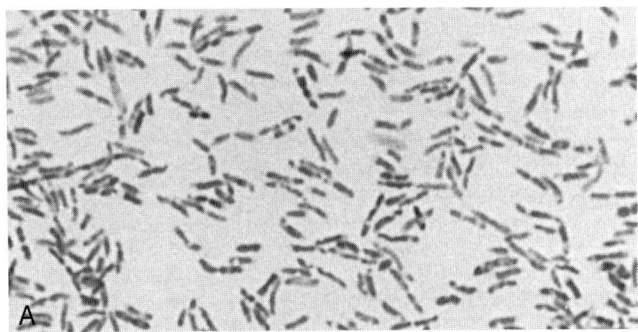

Roseomonas fauriae strain C610. Gram stain BAP 35° C 24 h x 1700.

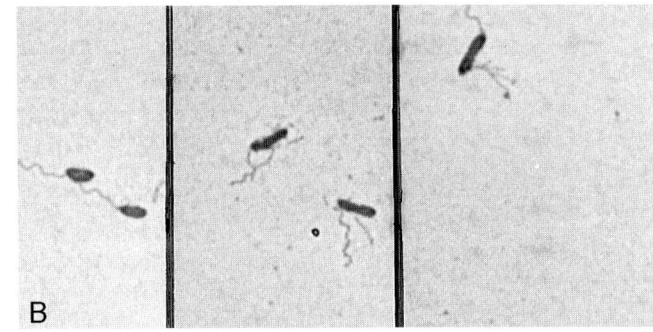

Roseomonas fauriae strain C610. Flagella stain (Ryu method) TGY 25° C 24 h x 1700.

Roseomonas fauriae

TEST PERFORMED	SIGN	%+	#+/T
NUMBER OF STRAINS 5			
Morphology	cc,srs		
Motility; flagella	[m;p,1-2]		
Action on blood	v	25 ly	1/4
Fermentative or oxidative	[n-o or O]		
Carbohydrate base	OF		
Acid from:			
D-Glucose	v	20	1w/5
D-Xylose	[+ or (+)]	80 (20)	2, 2w (1)/5
D-Mannitol	–	0	0/5
Lactose	–	0	0/5
Sucrose	–	0	0/5
Maltose	–	0	0/5
Catalase	+	100	5/5
Oxidase	[+]	100	5/5
Growth on:			
MacConkey	+ or (+)	60 (40)	2, 1w (1) (1w)/5
SS	v	20	1w/5
Cetrimide	–	0	0/5
Simmons citrate	v	60 (20)	2, 1w (1)/5
Urea, Christensen's	+	100	5/5
Nitrate reduction	[+]	100	5/5
Gas from nitrate	v	20	1/5
Indole	–	0	0/5
TSI slant, acid	–	0	0/5
TSI butt, acid	–	0	0/5
H_2S (TSI butt)	–	0	0/5
H_2S (Pb ac paper)	+	100	4, 1w/5
Gelatin hydrolysis[a]		0	0/5
Litmus milk	k	100	4, 1w/5
Pigment	[pk]	100	5/5
Growth at:			
25°C	+	100	5/5
35°C	+	100	5/5
42°C	+	100	5/5
Esculin hydrolysis	[+]	100	5/5
Nutrient broth, 0% NaCl	+	100	5/5
Nutrient broth, 6% NaCl	v	20	1/5

[a]Incubation of 7–14 days.

Roseomonas genomospecies 4

Gram-negative plump coccoid rod, appears as single cells, pairs, or short chains; motile with a single polar flagellum (motility may be difficult to demonstrate with some strains); oxidizes xylose, but not glucose, lactose, or mannitol; grows on MacConkey agar; oxidase-, urease-, and nitrate-positive; citrate-negative; produces a pink growth pigment.

Sources (3 strains)

Ear, wound, and cervix.

Reference strain

E7832 = ATCC 49959, type strain, from an ear.

Literature

1. Gilardi, G.L., and Y.C. Faur. 1984. *Pseudomonas mesophilica* and an unnamed taxon, clinical isolates of pink pigmented oxidative bacteria. J. Clin. Microbiol. *20:* 626–629.

2. Wallace, P.L., D.G. Hollis, R.E. Weaver, and C.W. Moss. 1990. Biochemical and chemical characterization of pink-pigmented oxidative bacteria. J. Clin. Microbiol. *28:* 689–693.

3. Rihs, J.D., D.J. Brenner, R.E. Weaver, A.G. Steigerwalt, D.G. Hollis, and V.L. Yu. 1993. *Roseomonas,* a new genus associated with bacteremia and other human infections. J. Clin. Microbiol. *31:* 3275–3283.

Comments

The pink coccoid (pc) groups of Wallace et al. (2) are *Roseomonas* species, but the pc groups do not completely correlate with the various genomospecies of *Roseomonas*. All known strains of genomospecies 4 previously were designated pc group I; however, pc group I also contains strains of *R. cervicalis* and *Roseomonas* genomospecies 5. Information on the cellular fatty acid composition of this group is presented in Table 4.60 and Figure 4.65 of the cellular fatty acid section.

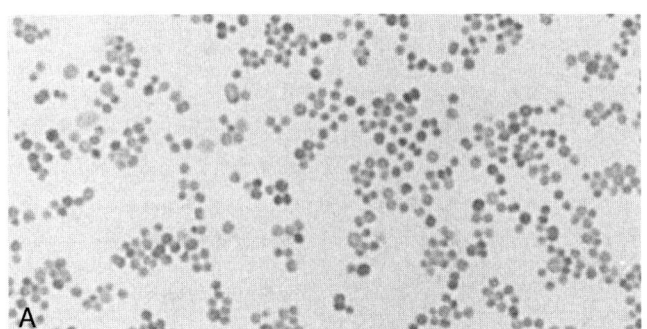

Roseomonas genomospecies 4 strain E7832. Gram stain BAP 35° C 24 h x 1700.

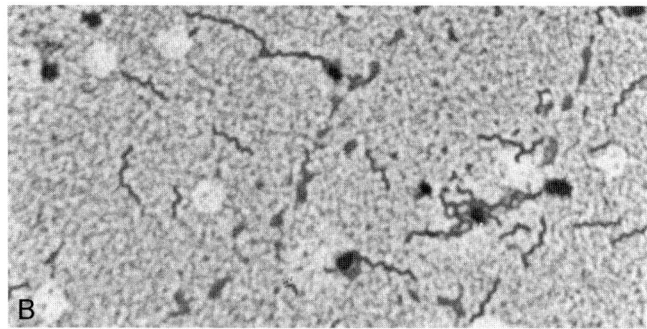

Roseomonas genomospecies 4 strain E7832. Flagella stain (Ryu method) TGY 25° C 24 h x 1700.

Roseomonas genomospecies 4

TEST PERFORMED	SIGN	%+	#+/T
	NUMBER OF STRAINS 3		
Morphology	cc		
Motility; flagella	v[a]	67	2/3
Action on blood	v	33 lav	1/3
Fermentative or oxidative	[n-o]		
Carbohydrate base	OF		
Acid from:			
D-Glucose	[-]	0	0/3
D-Xylose	[+]	100	3/3
D-Mannitol	[-]	0	0/3
Lactose	-	0	0/3
Sucrose	-	0	0/3
Maltose	-	0	0/3
Catalase	+	100	3/3
Oxidase	[+]	100	3/3
Growth on:			
MacConkey	+	100	3/3
SS	-	0	0/3
Cetrimide	-	0	0/3
Simmons citrate	[-]	0	0/3
Urea, Christensen's	+ or (+)	67 (33)	2 (1)/3
Nitrate reduction	[+]	100	3/3
Gas from nitrate	-	0	0/3
Indole	-	0	0/3
TSI slant, acid	-	0	0/3
TSI butt, acid	-	0	0/3
H_2S (TSI butt)	-	0	0/3
H_2S (Pb ac paper)	+	100	1, 2w/3
Gelatin hydrolysis[b]	-	0	0/2
Litmus milk	k	100	3/3
Pigment:			
Insoluble	[pk]	100	3/3
Growth at:			
25°C	+	100	3/3
35°C	+	100	3/3
42°C	+	100	3/3
Esculin hydrolysis	[-]	0	0/3
Nutrient broth, 0% NaCl	+	100	3/3
Nutrient broth, 6% NaCl	v	33	1/3

[a]Motile strains demonstrated 1–2 polar flagella.
[b]Incubation of 7–14 days.

Roseomonas genomospecies 5

Gram-negative plump coccoid rod, appears as single cells, pairs, or short chains; nonmotile; variable oxidation of xylose, but not glucose or mannitol; grows on MacConkey agar; oxidase- and urease-positive; esculin- and nitrate-negative; produces a pink growth pigment.

Sources (3 strains)

Bone, breast, and blood.

Reference strain

F4700 = ATCC 49960, type strain, from bone.

Literature

1. Odugbemi, T., C. Nwofor, and K.T. Joiner. 1988. Isolation of an unidentified pink-pigmented bacterium in a clinical specimen. J. Clin. Microbiol. *26:* 1072–1073.

2. Wallace, P.L., D.G. Hollis, R.E. Weaver, and C.W. Moss. 1990. Biochemical and chemical characterization of pink-pigmented oxidative bacteria. J. Clin. Microbiol. *28:* 689–693.

3. Rihs, J.D., D.J. Brenner, R.E. Weaver, A.G. Steigerwalt, D.G. Hollis, and V.L. Yu. 1993. *Roseomonas*, a new genus associated with bacteremia and other human infections. J. Clin. Microbiol. *31:* 3275–3283.

Comments

The pink coccoid (pc) groups of Wallace et al. (2) are *Roseomonas* species, but the pc groups do not completely correlate with the various genomospecies of *Roseomonas*. Strains in genomospecies 5 were formerly designated pc groups I or II. The citrate reaction and oxidation of glycerol will differentiate the described strains of genomospecies 5 from *R. gilardii*. All three strains of genomospecies 5 are negative for one or both reactions (3). Information on the cellular fatty acid composition of this group is presented in Table 4.60 and Figure 4.65 of the cellular fatty acid section.

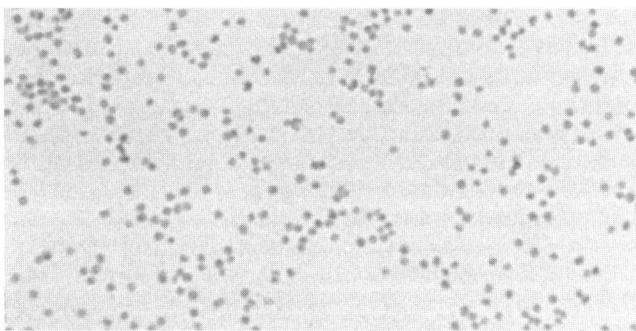

Roseomonas genomospecies 5 strain F4700. Gram stain BAP 35° C 24 h x 1700.

Roseomonas genomospecies 5

TEST PERFORMED	SIGN	%+	#+/T
	NUMBER OF STRAINS 3		
Morphology	cc		
Motility; flagella	[nm]		
Action on blood	-		
Fermentative or oxidative	[n-o]		
Carbohydrate base	OF		
Acid from:			
D-Glucose	[-]	0	0/3
D-Xylose	v	67	2w/3
D-Mannitol	[-]	0	0/3
Lactose	-	0	0/3
Sucrose	-	0	0/3
Maltose	-	0	0/3
Catalase	+	100	3/3
Oxidase	+	100	3/3
Growth on:			
MacConkey	+ or (+)	67 (33)	2, (1w)/3
SS	-	0	0/3
Cetrimide	-	0	0/3
Simmons citrate	[v]	33	1/3
Urea, Christensen's	+	100	3/3
Nitrate reduction	-	0	0/3
Gas from nitrate	-	0	0/3
Indole	-	0	0/3
TSI slant, acid	-	0	0/3
TSI butt, acid	-	0	0/3
H_2S (TSI butt)	-	0	0/3
H_2S (Pb ac paper)	+	100	3/3
Gelatin hydrolysis[a]	-	0	0/3
Litmus milk	v	67 k	2/3
Pigment	[pk]	100	3/3
Growth at:			
25°C	+	100	3/3
35°C	+	100	3/3
42°C	v	67	2/3
Esculin hydrolysis	[-]	0	0/3
Nutrient broth, 0% NaCl	+	100	3/3
Nutrient broth, 6% NaCl	-	0	0/3

[a]Incubation of 7–14 days.

Roseomonas genomospecies 6

Gram-negative coccobacillus and rod that appears singly or in pairs; motile (single polar flagellum); does not oxidize glucose, xylose, or mannitol; grows weakly on MacConkey agar; citrate-, urease-, nitrate-, and esculin-positive; produces a pink growth pigment.

Source (reference strain)

Breast incision.

Reference strain

F4626 = ATCC 49961, type strain, from breast incision.

Literature

1. Wallace, P.L., D.G. Hollis, R.E. Weaver, and C.W. Moss. 1990. Biochemical and chemical characterization of pink-pigmented oxidative bacteria. J. Clin. Microbiol. 28: 689–693.

2. Rihs, J.D., D.J. Brenner, R.E. Weaver, A.G. Steigerwalt, D.G. Hollis, and V.L. Yu. 1993. *Roseomonas*, a new genus associated with bacteremia and other human infections. J. Clin. Microbiol. 31: 3275–3283.

Comments

The pink coccoid (pc) groups of Wallace et al. (1) are *Roseomonas* species, but the pc groups do not completely correlate with the various genomospecies of *Roseomonas*. The only known strain of genomospecies 6 belongs to pc group IV, which also contains the known strains of genomospecies 3. Negative xylose, arabinose, and galactose oxidation reactions have been reported as useful in differentiating this organism from *R. fauriae* (2). Information on the cellular fatty acid composition of this group is presented in Table 4.54 and Figure 4.59 of the cellular fatty acid section.

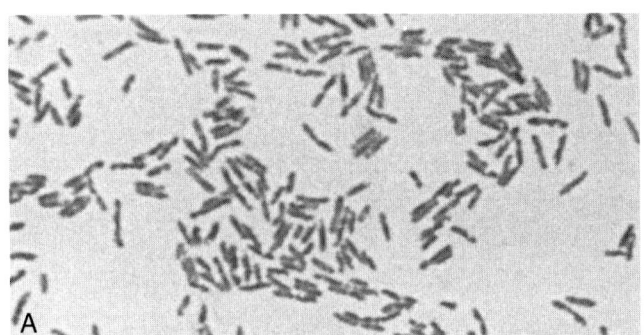

Roseomonas genomospecies 6 strain F4626. Gram stain BAP 35° C 24 h x 1700.

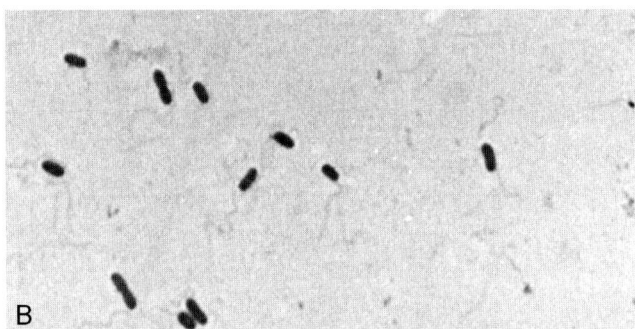

Roseomonas genomospecies 6 strain F4626. Flagella stain (Ryu method) TGY 25° C 24 h x 1700.

Roseomonas genomospecies 6

TEST PERFORMED	REFERENCE STRAIN SIGN
Morphology	cc,srs
Motility; flagella	[m;p,1-2]
Action on blood	−
Fermentative or oxidative	[n-o]
Carbohydrate base	OF
Acid from:	
D-Glucose	[−]
D-Xylose	−
D-Mannitol	[−]
Lactose	−
Sucrose	−
Maltose	−
Catalase	+
Oxidase	+
Growth on:	
MacConkey	(+w)
SS	−
Cetrimide	−
Simmons citrate	(+)
Urea, Christensen's	+
Nitrate reduction	+
Gas from nitrate	−
Indole	−
TSI slant, acid	−
TSI butt, acid	−
H_2S (TSI butt)	−
H_2S (Pb ac paper)	+
Gelatin hydrolysis[a]	−
Litmus milk	k
Pigment: Insoluble	[pk]
Growth at:	
25°C	+
35°C	+
42°C	+
Esculin hydrolysis	[+]
Nutrient broth, 0% NaCl	+
Nutrient broth, 6% NaCl	−

[a]Incubation of 7–14 days.

Roseomonas gilardii

Gram-negative plump coccoid rod, appears as single cells, pairs or short chains; the type strain is motile with single polar or subpolar flagella (motility may be difficult to demonstrate with some strains); oxidizes glycerol; variable oxidizer of glucose, xylose, and mannitol; does not oxidize lactose, maltose, or sucrose; usually grows on MacConkey agar; citrate- and urease-positive; produces a pink growth pigment.

Sources (21 strains)

Blood (37%), water (19%), eye (10%), kidney and hip wound (1 strain each (5% each)), unknown (24%).

Reference strain

G5300 ← J.D. Rihs 5424 = ATCC 49956, type strain, from potable water.

Literature

1. Gilardi, G.L., and Y.C. Faur. 1984. *Pseudomonas mesophilica* and an unnamed taxon, clinical isolates of pink-pigmented oxidative bacteria. J. Clin. Microbiol. *20:* 626–629.

2. Wallace, P.L., D.G. Hollis, R.E. Weaver, and C.W. Moss. 1990. Biochemical and chemical characterization of pink-pigmented oxidative bacteria. J. Clin. Microbiol. *28:* 689–693.

3. Rihs, J.D., D.J. Brenner, R.E. Weaver, A.G. Steigerwalt, D.G. Hollis, and V.L. Yu. 1993. *Roseomonas*, a new genus associated with bacteremia and other human infections. J. Clin. Microbiol. *31:* 3275–3283.

Comments

The pink coccoid (pc) groups of Wallace et al. (2) are *Roseomonas* species, but the pc groups do not completely correlate with the various genomospecies of *Roseomonas*. *R. gilardii* contains strains of pink coccoid groups II and III. Information on the cellular fatty acid composition of this species is presented in Table 4.60 and Figure 4.65 of the cellular fatty acid section.

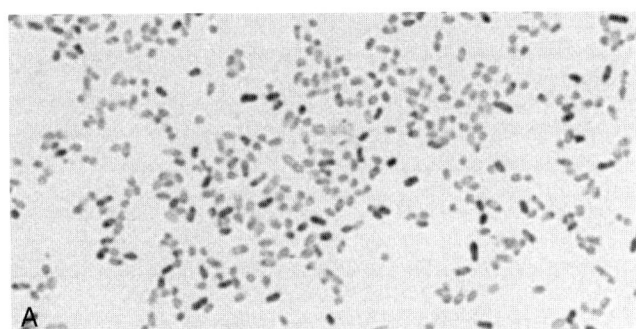
Roseomonas gilardii strain G5300. Gram stain BAP 35° C 24 h x 1700.

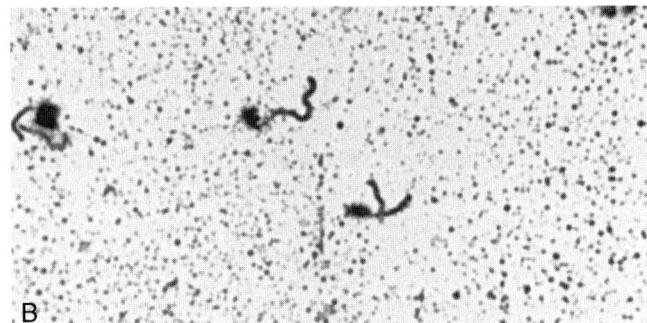
Roseomonas gilardii strain G5300. Flagella stain (Ryu method) TGY 25° C 24 h x 1700.

Roseomonas gilardii

TEST PERFORMED	SIGN	% +	#+/T
	NUMBER OF STRAINS 21		
Morphology	cc		
Motility; flagella	v[a]	33	7/21
Action on blood	−	5 ly	1/21
Fermentative or oxidative	[n-o or O]		
Carbohydrate base	OF		
Acid from:			
D-Glucose	v	(43)	(4) (5w)/21
D-Xylose	v	19 (57)	4w (1) (11)/21
D-Mannitol	[v]	14 (38)	3w (6) (2w)/21
Lactose	−	0	0/21
Sucrose	−	0	0/21
Maltose	−	0	0/21
Catalase	+	100	21/21
Oxidase	v	52	7, 6w/21
Growth on:			
MacConkey	+ or (+)	43 (52)	9 (8) (3w)/21
SS	−	0	0/21
Cetrimide	−	0	0/21
Simmons citrate	[+]	100	21/21
Urea, Christensen's	+ or (+)	71 (29)	15 (6)/21
Nitrate reduction	−	5	1/21
Gas from nitrate	−	0	0/21
Indole	−	0	0/21
TSI slant, acid	−	0	0/21
TSI butt, acid	−	0	0/21
H$_2$S (TSI butt)	−	0	0/21
H$_2$S (Pb ac paper)	+	100	20, 1w/21
Gelatin hydrolysis[b]	−	0	0/21
Litmus milk	k or (k)	76 (19)	14, 2w (4)/21
Pigment: Insoluble	[pk]	100	21/21
Growth at:			
25°C	+	90	18, 1w/21
35°C	+	95	20/21
42°C	v	67	8, 6w/21
Esculin hydrolysis	[−]	0	0/21
Nutrient broth, 0% NaCl	+	100	21/21
Nutrient broth, 6% NaCl	v	24	2, 3w/21

[a]Motility was more easily demonstrated in OF medium than in motility medium. Motile strains demonstrated either 1–2 polar/subpolar flagella, or detached flagella.

[b]Incubation of 7–14 days.

Shewanella putrefaciens

Gram-negative short to long rod or filament; motile (1–2 polar flagella); aerobic; grows on MacConkey agar; produces oxidase and ornithine decarboxylase; produces H_2S in TSI agar; reduces indicator in litmus milk overnight; also reduces nitrate to nitrite without gas.

Two biotypes are recognized at CDC: biotype 1 oxidizes glucose, sucrose, and maltose, and grows heavily in nutrient broth without sodium chloride after overnight incubation; biotype 2 does not oxidize glucose and grows poorly or not at all in nutrient broth without sodium chloride (prefers or requires NaCl), and may not grow in OF basal medium (King) (there is no NaCl in this medium), although it grows well in Difco OF basal medium, in which it produces an alkaline reaction.

Sources Biotype 1 (24 strains)

Stool (17%), throat (12%), hand (8%), water (8%), other (38%), unknown (17%).

Sources Biotype 2 (26 strains)

Wound (19%); blood, ear, stool, sea water, sputum, water (8% each); other (19%); unknown (14%).

Reference strains

KC988 ← ATCC 8073.
D3845, from foot.

Literature

1. von Graevenitz, A., and G. Simon. 1970. Potentially pathogenic, nonfermentative, H_2S-producing gram-negative rod (Ib). Appl. Microbiol. *19:* 176.

2. Riley, P.S., H.W. Tatum, and R.E. Weaver. 1972. *Pseudomonas putrefaciens* isolates from clinical specimens. Appl. Microbiol. *24:* 798–800.

3. Owen, R.J., R.M. Legros, and S.P. Lapage. 1978. Base composition, size and sequence similarities of genome deoxyribonucleic acids from clinical isolates of *Pseudomonas putrefaciens*. J. Gen. Microbiol. *104:* 127–138.

4. van Landschoot, A., and J. De Ley. 1983. Intra- and intergeneric similarities of the rRNA cistrons of *Alteromonas*, *Marinomonas* (gen. nov.) and some other gram-negative bacteria. J. Gen. Microbiol. *129:* 3057–3074.

5. Lee, J.V., D.M. Gibson, and J.M. Shewan. 1977. A numerical taxonomic study of some pseudomonas-like marine bacteria. J. Gen. Microbiol. *98:* 439–451.

6. Palleroni, N.J. 1984. Genus I. *Pseudomonas* Migula 1894, 247.^AL *In* N.R. Krieg and J.G. Holt (Eds.), Bergey's Manual of Systematic Bacteriology, Vol. 1, Williams & Wilkins, Baltimore, pp. 141–199.

7. MacDonell, M.T., and R.R. Colwell. 1985. Phylogeny of the *Vibrionaceae*, and recommendation for two genera, *Listonella* and *Shewanella*. Syst. Appl. Microbiol. *6:* 171–182.

8. Moule, A.L., and S.G. Wilkinson. 1987. Polar lipids, fatty acids, and isoprenoid quinones of *Alteromonas putrefaciens* (*Shewanella putrefaciens*). Syst. Appl. Microbiol. *9:* 192–198.

9. Gilardi, G.L. 1991. *Pseudomonas* and related genera. *In* A. Balows et al. (Eds.), Manual of Clinical Microbiology, 5th ed., American Society for Microbiology, Washington, pp. 429–441.

Comments

The taxonomic classification of these bacteria is unsettled. Formerly CDC groups Ib-1 and Ib-2, these organisms are designated *Pseudomonas putrefaciens* species incertae sedis in *Bergey's Manual of Systematic Bacteriology* (6). It has been proposed that they be placed in the genus *Alteromonas* and more recently in the new genus *Shewanella* (5, 7, 8). This species is biochemically and genetically heterogeneous (2, 3). At CDC, two biotypes are recognized, based upon the requirement of NaCl for growth, oxidation of sucrose and maltose, and the ability to grow on SS agar (NaCl+/SS+/Suc−/Mal− and NaCl−/SS−/Suc+/Mal+). Recently, Gilardi has described a third biotype that is negative for all of these characters (NaCl−/SS− /Suc−/Mal−) (9). Information on the cellular fatty acid composition of this group is presented in Table 4.67 and Figure 4.72 of the cellular fatty acid section.

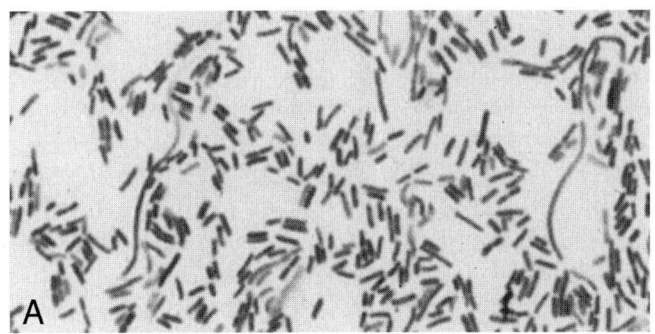

Shewanella putrefaciens strain KC988. Gram stain BAP 35° C 24 h x 1700.

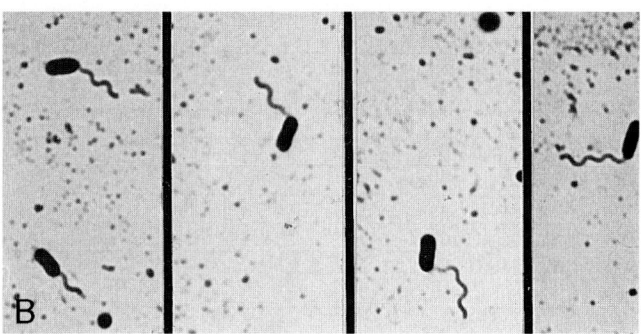

Shewanella putrefaciens strain KC988. Flagella stain FB 25° C 24 h x 1700.

Shewanella putrefaciens (biotype 1)

TEST PERFORMED	SIGN	% +	# +/T
	NUMBER OF STRAINS 24		
Morphology	mrs		
Motility; flagella	[m;p,1-2]		
Action on blood	v	75 LG	18/24; 2 ly, 2 al, 1 br, 1 β
Fermentative or oxidative	[O or n-o]		
Carbohydrate base	OF		
Acid from:			
D-Glucose	[v]	17 (33)	4 (8)/24
D-Xylose	[-]	0	0/24
D-Mannitol	-	0	0/24
Lactose	-	0	0/24
Sucrose	[+]	96 (4)	23 (1)/24
Maltose	+	92 (8)	22 (2)/24
Catalase	+	100	24/24
Oxidase	[+]	100	24/24
Growth on:			
MacConkey	[+]	100	24/24
SS	-	(8)	(2)/24
Cetrimide	-	4	1/24
Simmons citrate	-	4 (4)	1 (1)/24
Urea, Christensen's	v	4 (8)	1 (2)/24
Nitrate reduction	[+]	100	24/24
Gas from nitrate	[-]	0	0/24
Indole	-	0	0/24
TSI slant, acid	-	0	0/24
TSI butt, acid	-	0	0/24
H_2S (TSI butt)	[+]	96	23/24
H_2S (Pb ac paper)	+	100	24/24
Gelatin hydrolysis[a]	v	65	13/20
Litmus milk	v	61 pep	14/23; 9 IR
Pigment:			
Soluble	v	71 br-ta	17/24
Growth at:			
25°C	+	100	24/24
35°C	+	96	23/24
42°C	v	38	9/24
Esculin hydrolysis	-	0	0/24
Lysine decarboxylase	-	0	0/24
Arginine dihydrolase	-	0	0/24
Ornithine decarboxylase	[+]	100	24/24
Nutrient broth, 0% NaCl	[+]	100	23/23
Nutrient broth, 6% NaCl	v	43	10/23

[a]Incubation of 7–14 days.

Shewanella putrefaciens (biotype 2)

	NUMBER OF STRAINS 26		
TEST PERFORMED	SIGN	%+	#+/T
Morphology	mrs		
Motility; flagella	[m;p,1-2]		
Action on blood	LG	92	24/26
Fermentative or oxidative	[n-o]		
Carbohydrate base	OF		
Acid from:			
D-Glucose	[-]	0	0/26
D-Xylose	[-]	0	0/26
D-Mannitol	-	0	0/26
Lactose	-	0	0/26
Sucrose	-	0	0/26
Maltose	-	0	0/26
Catalase	+	100	26/26
Oxidase	[+]	100	26/26
Growth on:			
MacConkey	[+]	100	26/26
SS	+	92 (4)	24 (1)/26
Cetrimide	-	8	2/26
Simmons citrate	-	8	2/26
Urea, Christensen's	v	42	11/26
Nitrate reduction	[+]	100	26/26
Gas from nitrate	[-]	0	0/26
Indole	-	0	0/26
TSI slant, acid	-	0	0/26
TSI butt, acid	-	0	0/26
H_2S (TSI butt)	[+]	100	26/26
H_2S (Pb ac paper)	+	100	26/26
Gelatin hydrolysis[a]	+	100	23/23
Litmus milk	v	50 IR & pep	13/26; 4 IR, 3 pep
Pigment:			
Soluble	br-yel	100	26/26
Growth at:			
25°C	+	100	26/26
35°C	+	100	26/26
42°C	v	23	6/26
Esculin hydrolysis	-	0	0/25
Lysine decarboxylase	-	0	0/3
Arginine dihydrolase	-	0	0/3
Ornithine decarboxylase	[+]	100	3/3
Nutrient broth, 0% NaCl	[-]	0	0/26
Nutrient broth, 6% NaCl	[+]	100	26/26

[a]Incubation of 7–14 days.

Simonsiella species

Gram-negative disc-shaped cell that forms a multicellular filament with the long axis of each individual cell perpendicular to the long axis of the filament; motile by gliding of the entire filament in the direction of the long axis; aerobic; produces acid from glucose and sucrose, not from lactose or xylose; oxidase-positive; no growth on MacConkey agar; indole- and urease-negative.

Sources (3 strains)

Throat (2 strains), abscess of face.

Reference strain

KC1456 ← ATCC 15533, *S. crassa* type strain, from sheep saliva.

Literature

1. Steed, P.D.M. 1962. *Simonsiellaceae* fam. nov. with characterization of *Simsonsiella crassa* and *Alysiella filiformis*. J. Gen. Microbiol. 29: 615–624.
2. Whitehouse, R.L.S., H. Jackson, M.C. Jackson, and M.M. Ramji. 1987. Isolation of *Simonsiella* sp. from a neonate. J. Clin. Microbiol. 25: 522–525.
3. Larkin, J.M. 1984. Genus I. *Simonsiella* Schmid in Simons 1922, 504.[AL] In J.T. Staley, M.P. Bryant, N. Pfennig, and J.G. Holt (Eds.), Bergey's Manual of Systematic Bacteriology, Vol. 3, Williams & Wilkins, Baltimore, pp. 2107–2110.

Comments

The unique "roll of coins" morphology of these organisms differentiates them from other Gram-negative clinical isolates. *Bergey's Manual of Systematic Bacteriology* (3) describes three: *Simonsiella* species *S. muelleri*, *S. crassa*, and *S. steedae*; associated with humans, dogs, and sheep, respectively. Found in the oral cavity of humans and other mammals. Information on the cellular fatty acid composition of this group is presented in Table 4.53 and Figure 4.58 of the cellular fatty acid section.

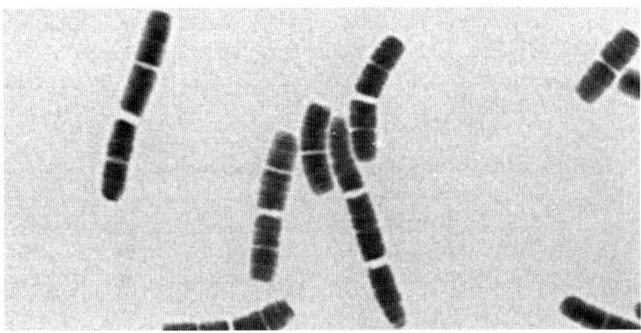

Simonsiella crassa strain KC1456. Gram stain BAP 35° C 24 h x 1700.

Simonsiella species

TEST PERFORMED	S. crassa REFERENCE STRAIN SIGN	3 PHENOTYPICALLY SIMILAR STRAINS		
		SIGN	%+	#+/T
Morphology	[dskfil][a]	[dskfil]		
Motility; flagella	-[b]	-[b]		
Action on blood	lys	v	67 ß	2/3
Fermentative or oxidative				
Carbohydrate base	F[c]	F		
Acid from:				
D-Glucose	[+w]	[+]	100	2, 1w/3
D-Xylose	[-]	[-]	0	0/3
D-Mannitol	-	-	0	0/3
Lactose	-	-	0	0/3
Sucrose	+w	-	0	0/3
Maltose	-	+	100	3/3
Catalase	-	+	100	2w, 1/3
Oxidase	+	+	100	3/3
Growth on:				
MacConkey	-	-	0	0/3
SS	-	-	0	0/3
Cetrimide	-	-	0	0/3
Simmons citrate	-	-	0	0/3
Urea, Christensen's	-	-	0	0/3
Nitrate reduction	+	v	67	2/3
Gas from nitrate	+	v	67	2/3
Indole	-	-	0	0/3
TSI slant, acid	-	-	0	0/3
TSI butt, acid	-	-	0	0/3
H_2S (TSI butt)	-	-	0	0/3
H_2S (Pb ac paper)	-	+	100	3/3
Gelatin hydrolysis[d]	-	-	0	0/2
Litmus milk	a	-	0	0/2
Pigment	-	-	0	0/3
Growth at:				
25°C	-	v	67	1, 1w/3
35°C	+	+	100	2, 1w/3
42°C	-	-	0	0/3
Esculin hydrolysis	-	-	0	0/3
Nutrient broth, 0% NaCl	+	+	100	3/3
Nutrient broth, 6% NaCl	+	-	0	0/3

[a]Multicellular filaments made up of disc-like cells in a "roll of coins" arrangement.
[b]Gliding motility of multicellular filaments is observed. Individual cells do not express flagella.
[c]Acid production from carbohydrates was not observed in OF or TSI agar media.
[d]Incubation of 7–14 days.

Sphingomonas paucimobilis and *Sphingomonas parapaucimobilis*

Gram-negative medium, sometimes a long rod; motility is difficult to demonstrate, especially at 35° C (single, occasionally two polar flagella); produces a yellow insoluble growth pigment; aerobic; does not usually grow on MacConkey agar; uses glucose, xylose, lactose, sucrose, and maltose, not mannitol; oxidase-variable; usually does not produce urease nor reduce nitrate; lysine decarboxylase-, arginine dihydrolase-, and 3-ketolactonate-negative.

Sources of 134 phenotypically similar strains

Blood (24%); environmental contaminant (9%); humidifier (8%); water (7%); urine (7%); wound (7%); cerebrospinal fluid (6%); ear, eye, throat, abdomen, vagina, dialysis fluid (2% each); other (16%); unknown (4%).

Reference strain: *S. paucimobilis*

KC1426 ← NCTC 11030 = ATCC 29837, type strain, from hospital respirator.

Reference strains: *S. parapaucimobilis*

KC1945 ← CCUG 27291 = GIFU 11387, JCM 7510, type strain, from urine.
KC1952 ← CCUG 27292 = GIFU 2135, JCM 7512, from vaginal swab.

Literature

1. Holmes, B., R.J. Owen, A. Evans, H. Malnick, and W.R. Wilcox. 1977. *Pseudomonas paucimobilis*, a new species isolated from human clinical specimens, the hospital environment, and other sources. Int. J. Syst. Bacteriol. 27: 133–146.

2. Morrison, A.J., and J. Shulman. 1986. Community-acquired bloodstream infection caused by *Pseudomonas paucimobilis*: case report and review of the literature. J. Clin. Microbiol. 24: 853–885.

3. Reina, J., A. Bassa, I. Llompart, D. Portela, and N. Borrell. 1991. Infections with *Pseudomonas paucimobilis*: report of four cases and review. Rev. Infect. Dis. 13: 1072–1076.

4. Yabuuchi, E., I. Yano, H. Oyaizu, Y. Hashimoto, T. Ezaki, and H. Yamamoto. 1990. Proposal for *Sphingomonas paucimobilis* gen. nov. and comb. nov., *Sphingomonas parapaucimobilis* sp. nov., *Sphingomonas yanoikuyae* sp. nov., *Sphingomonas adhaesiva* sp. nov., *Sphingomonas capsulata* comb. nov., and two genospecies of the genus *Sphingomonas*. Microbiol. Immunol. 34: 99–119.

5. Holmes, B., and P. Roberts. 1981. The classification, identification and nomenclature of agrobacteria. J. Appl. Bacteriol. 50: 443–467.

6. Swann, R.A., S.J. Foulkes, B. Holmes, J.B. Young, R.G. Mitchell, and S.T. Reeders. 1985. "*Agrobacterium* yellow group" and *Pseudomonas paucimobilis* causing peritonitis in patients receiving continuous ambulatory peritoneal dialysis. J. Clin. Pathol. 38: 1293–1299.

7. Owen, R.J., and P.J.H. Jackman. 1982. The similarities between *Pseudomonas paucimobilis* and allied bacteria derived from analysis of deoxyribonucleic acids and electrophoretic protein patterns. J. Gen. Microbiol. 128: 2945–2954.

Comments

Yabuuchi et al. (4) proposed placing *Pseudomonas paucimobilis* into the new genus *Sphingomonas*. Cellular fatty acid analysis and negative reactions for oxidation of glycerol and rhamnose differentiate *S. paucimobilis* from *S. parapaucimobilis*. These two species are being described together because, except for the reference strains, the CDC collection has not been tested for the characteristics that differentiate them. A negative 3-ketolactonate reaction differentiates *P. paucimobilis* from the 3-ketolactonate-positive unclassified "*Agrobacterium* yellow group" of Holmes and Roberts (5) and Swann et al. (6). A negative reaction for oxidation of glycerol differentiates *P. paucimobilis* from the phenotypically similar "*P. azotocolligans*" (7). CDC has received several strains that are similar to these organisms, except they do not oxidize lactose. DNA-relatedness studies are needed to determine if the lactose-negative strains are a different species. Information on the cellular fatty acid composition of *S. paucimobilis* is presented in Table 4.61 and Figure 4.66 of the cellular fatty acid section. Information on the cellular fatty acid composition of *S. parapaucimobilis* is presented in Table 4.68 and Figure 4.73 of the cellular fatty acid section.

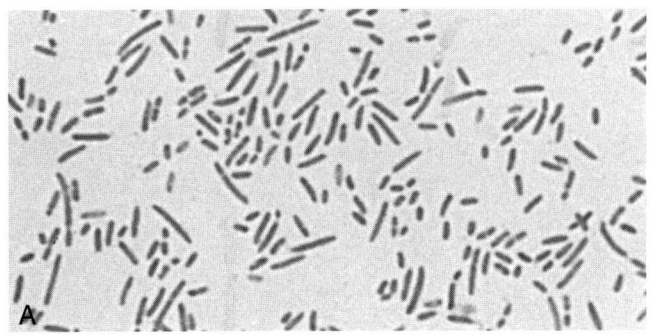

Sphingomonas parapaucimobilis strain KC1945. Gram stain BAP 35° C 24 h x 1700.

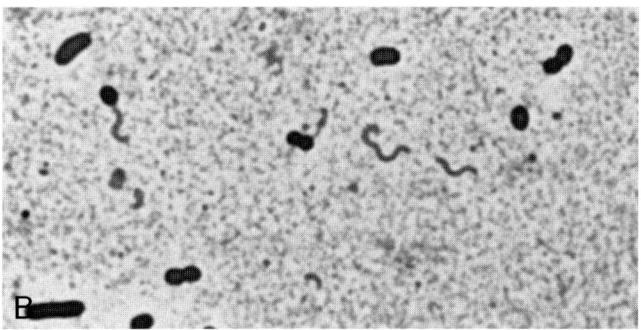

Sphingomonas parapaucimobilis strain KC1945. Flagella stain (Ryu method) TGY 25° C 24 h x 1700.

Sphingomonas paucimobilis and *Sphingomonas parapaucimobilis*

TEST PERFORMED	S. paucimobilis TYPE STRAIN SIGN	S. parapaucimobilis REFERENCE STRAINS SIGN	S. parapaucimobilis REFERENCE STRAINS #+/T	134 PHENOTYPICALLY SIMILAR STRAINS SIGN	134 PHENOTYPICALLY SIMILAR STRAINS %+	134 PHENOTYPICALLY SIMILAR STRAINS #+/T
Morphology	mrs	mrs		mrs		
Motility; flagella	[m;p,1-2][a]	[m;p,1-2][a]		[m;p,1-2][a]		
Action on blood	ly	v	1 ly/2	v	32 gr	13/40
Fermentative or oxidative	[O]	[O]		[O]		
Carbohydrate base	OF	OF		OF		
Acid from:						
D-Glucose	+w	[(+)]	(2)/2	[+]	93 (7)	103, 22w (9)/134
D-Xylose	+	+	2/2	+	96 (4)	129 (5)/134
D-Mannitol	[-]	[-]	0/2	[-]	0	0/134
Lactose	[+]	[+ or (+)]	1 (1)/2	[+]	93 (7)	116, 8w (10)/134
Sucrose	+	+	2/2	+	93 (7)	116, 9w (9)/134
Maltose	[+]	[+ or (+)]	1 (1)/2	[+]	97 (3)	125, 5w (4)/134
Glycerol	[-]	[+]	2/2			
L-Rhamnose	[-]	[+]	2/2			
Catalase	+	+	2/2	+	95	127/134
Oxidase	+w	v	1w/2	v	75	72, 28w/134
Growth on:						
MacConkey	-	-	0/2	v	10 (13)	14 (17)/134
SS	-	-	0/2	-	0	0/134
Cetrimide	-	-	0/2	-	0	0/134
Simmons citrate	-	-	0/2	-	6 (1)	8 (1)/134
Urea, Christensen's	[-]	-	0/1	[-]	6 (3)	8 (4)/133
Nitrate reduction	-	-	0/2	-	3	3, 1w/134
Gas from nitrate	-	-	0/2	-	0	0/134
Indole	-	-	0/2	-	0	0/134
TSI slant, acid	-	+w	1w/1	v	20 (3)	14 (2)/70
TSI butt, acid	-	-	0/1	v	6 (31)	8 (41)/134
H₂S (TSI butt)	-	-	0/2	-	0	0/132
H₂S (Pb ac paper)	[-]	[+]	2/2	v	76	100/132
Gelatin hydrolysis[b]	-	-	0/2	-	2	3/118
Litmus milk	k/IR	(IR)	(2)/2	v	60 IR	24/40; 6 k, 5 A, 4 pep
Pigment:						
Insoluble	[yel]	[yel]	2/2	[yel]	95	121/128; 3 or
Soluble	(wk tan)	tan-br	2/2	v	14 tan-br	18/128; 5 yel
Growth at:						
25°C	+	+	2/2	+	100	129/129
35°C	+	+	2/2	+	100	129/129

42°C	+w	–	0/2	v	24	31/129
Esculin hydrolysis	+	+	1/1	+	91	108/119
Lysine decarboxylase				[–]	0	0/26
Arginine dihydrolase				[–]	8	2/26
Ornithine decarboxylase				–	0	0/26
Nutrient broth, 0% NaCl	+	+	2/2	+	93	124/133
Nutrient broth, 6% NaCl	–	v	1/2	–	4	5/133
3-Ketolactonate	[–]	[–]	0/2	[–]	0	0/20

[a]Usually appears nonmotile in motility medium, motility may be demonstrated by the wet mount method.
[b]Incubation of 7–14 days.

Streptobacillus moniliformis

Gram-negative pleomorphic rod, frequently in chains and tangled filaments with bulbous or *"Monilia"*-like swellings; nonmotile; usually requires 20% serum or ascitic fluid or 10% whole blood; facultatively anaerobic; ferments glucose and maltose weakly, but not mannitol or lactose; no growth on MacConkey agar; catalase- and oxidase-negative; does not reduce nitrate.

Sources (48 strains)

Blood (64%), synovial fluid (9%), knee joint (9%), other (rat bite, peacock lung, shoulder, wrist, hip, elbow, mitral valve, joint fluid) (16%), unknown (2%).

Reference strains

KC199 ← R. Wickelhausen, strain Vancouver.
D7626, from blood.

Literature

1. Levaditi, C., S. Nicolau, and P. Poincloux. 1925. Sur le rôle étiologique de *Streptobacillus moniliformis* (nov. spec.) dars l'érythème polymorph aigu septicémique. CR Hebd. Séances Acad. Sci. (Paris) *180:* 1188–1190.

2. Robinson, L. 1963. *Streptobacillus moniliformis* infections. *In* A.H. Harris and M.B. Coleman (Eds.), Diagnostic procedures and reagents, 4th ed., American Public Health Association, New York, pp. 642–651.

3. Savage, N. 1984. Genus *Streptobacillus* Levaditi, Nicolau and Pioncloux 1925, 1188[AL]. *In* N.R. Krieg and J.G. Holt (Eds.), Bergey's Manual of Systematic Bacteriology, Vol. 1, Williams & Wilkins, Baltimore, pp. 598–600.

4. Azimi, P. 1990. Pets can be dangerous. Pediatr. Infect. Dis. J. *9:* 670–684.

5. Piot, P. 1991. *Gardnerella, Streptobacillus, Spirillum*, and *Calymmatobacterium*. *In* A. Balows et al. (Eds.), Manual of Clinical Microbiology, 5th ed., American Society for Microbiology, Washington, pp. 483–487.

6. Rupp, M.E. 1991. *Streptobacillus moniliformis* endocarditis: case report and review. Clin. Infect. Dis. *14:* 769–772.

Comments

Grows in broth producing discrete "fluff-ball-like" colonies. A causative agent of rat-bite fever in humans.

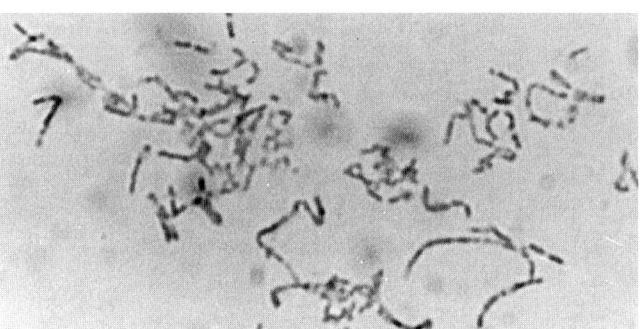

Streptobacillus moniliformis strain D7626. Gram stain thioglycolate broth 35° C 24 h x 1700.

Streptobacillus moniliformis

TEST PERFORMED	SIGN	% +	# +/T
	NUMBER OF STRAINS 48		
Morphology	[plr,ch]		
Motility; flagella	nm		
Action on blood	v	12 al	3/25; 1 ß
Fermentative or oxidative	F		
Carbohydrate base	F[a]		
Acid from:			
D-Glucose	[+ or (+)]	79 (21)	12 (3), 26w (7w)/48
D-Xylose	-	(4)	(1) (1w)/48
D-Mannitol	-	0	0/48
Lactose	-	0	0/48
Sucrose	-	0	0/48
Maltose	[+ or (+)]	81 (19)	18 (4), 21w (5w)/48
Catalase	-	2	1/41
Oxidase	-	3	1/36
Growth on:			
MacConkey	[-]	0	0/32
SS	-	0	0/28
Simmons citrate	-	0	0/28
Urea, Christensen's	-	0	0/41
Nitrate reduction	[-]	0	0/43
Indole	-	0	0/43
TSI slant, acid	v	29 (12)	2 (1), 10w (4w)/42
TSI butt, acid	v	26 (7)	2 (1), 9w (2w)/42
H_2S (TSI butt)	-	7	1, 1w/28
H_2S (Pb ac paper)	v	50	13, 1w/28
Gelatin hydrolysis[b]	-	0	0/19
Litmus milk	-	8	2/26
Growth at:			
25°C	v	40	3, 9w/30
35°C	v	87	22, 4w/30
42°C	v	20	3, 3w/30
Esculin hydrolysis	v	47 (9)	13 (3), 2w/32
Arginine dihydrolase	+	100	3/3
Nutrient broth, 0% NaCl	-	7	2/27
Nutrient broth, 6% NaCl	-	4	1/27

[a]Usually requires 20% serum or ascitic fluid.
[b]Incubation of 7–14 days.

Suttonella indologenes

Gram-negative plump rod that may stain irregularly; nonmotile; facultatively anaerobic; produces acid fermentatively from glucose, sucrose, and maltose, but not mannitol or xylose; catalase-negative, oxidase- and indole-positive; no growth on MacConkey agar; nitrate- and nitrite-negative.

Source (6 strains)

Eye (5), unknown (1).

Reference strain

KC1142 ← NCTC 10717, type strain, from conjunctiva.

Literature

1. Snell, J.J.S., and S.P. Lapage. 1976. Transfer of some saccharolytic *Moraxella* species to *Kingella* Henriksen and Bøvre 1976, with descriptions of *Kingella indologenes* sp. nov. and *Kingella denitrificans* sp. nov. Int. J. Syst. Bacteriol. 26: 451–458.
2. Jenny, D.B., P.W. Lenendre, and G. Iverson. 1987. Endocarditis caused by *Kingella indologenes*. Rev. Infect. Dis. 9: 787–788.
3. Dewhirst, F.E., B.J. Paster, and S. La Fontaine. 1990. Transfer of *Kingella indologenes* (Snell and Lapage 1976) to the genus *Suttonella* gen. nov. as *Suttonella indologenes* comb. nov.; transfer of *Bacteroides nodosus* (Beveridge 1941) to the genus *Dichelobacter* gen. nov. and *Dichelobacter nodosus* comb. nov.; and assignment of the genera *Cardiobacterium*, *Dichelobacter*, and *Suttonella* to *Cardiobacteriaceae* fam. nov. in the gamma division of *Proteobacteria* on the basis of 16S rRNA sequence comparisons. Int. J. Syst. Bacteriol. 40: 426–433.

Comments

Previously, *Kingella indologenes*. Pitting of agar medium or spreading edges of colonies may occur, especially in freshly isolated strains. Strains of this organism have been referred to as "Bijsterveld/Sutton" strains.

The rare mannitol-negative strains of *Cardiobacterium hominis* would be difficult to differentiate biochemically from *S. indologenes*. Snell and Lapage (1) reported that casein digestion, phosphatase production, and Tween 20 and Tween 40 hydrolysis are negative for *C. hominis*, but positive for *S. indologenes*. These species also differ in their cellular fatty acid profiles. Information on the cellular fatty acid composition of this species is presented in Table 4.12A and Figure 4.17 of the cellular fatty acid section.

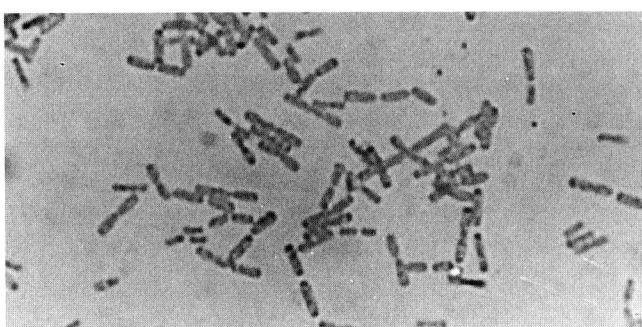

Suttonella indologenes strain KC1142. Gram stain BAP 35° C 24 h x 1700.

Suttonella indologenes

TEST PERFORMED	SIGN	%+	#+/T
	NUMBER OF STRAINS 6		
Morphology	vbr		
Motility; flagella	[nm]		
Action on blood	v	17 al	1/6
Fermentative or oxidative	F		
Carbohydrate base	F		
Acid from:			
D-Glucose	[+]	100	4, 2w/6
D-Xylose	−	0	0/6
D-Mannitol	[−]	0	0/6
Lactose	−	0	0/6
Sucrose	[+]	100	4, 2w/6
Maltose	[+ or (+)]	17 (83)	1w (5w)/6
Catalase	v	17	1 tr/6
Oxidase	[+]	100	6/6
Growth on:			
MacConkey	v	17	(1)/6
SS	−	0	0/6
Cetrimide	−	0	0/2
Simmons citrate	−	0	0/6
Urea, Christensen's	−	0	0/6
Nitrate reduction	−	0	0/6
Indole	[+]	100	6/6
TSI slant, acid	+	100	5, 1w/6
TSI butt, acid	v	33 (50)	2 (3)/6
H_2S (TSI butt)	−	0	0/6
H_2S (Pb ac paper)	+	100	6/6
Gelatin hydrolysis[a]	−	0	0/5
Litmus milk	−	0	0/4
Pigment	−	0	0/6
Growth at:			
25°C	v	50	3/6
35°C	+	100	6/6
42°C	−	0	0/6
Esculin hydrolysis	−	0	0/5
Lysine decarboxylase	−	0	0/1
Arginine dihydrolase	−	0	0/1
Ornithine decarboxylase	−	0	0/1
Nutrient broth, 0% NaCl	−	0	0/6
Nutrient broth, 6% NaCl	v	17	1/6
Tween-40 hydrolysis	+	100	3/3
Alkaline phosphatase	+	100	3/3
Casein hydrolysis	+	100	3/3

[a] Incubation of 7–14 days.

Taylorella equigenitalis

Gram-negative small rod, nonmotile; microaerophilic; does not require X or V factors, but growth is enhanced by X factor; glucose not used; urease-, indole-, and ornithine-negative.

Source

Cervical swab of mare.

Reference strain

KC1457← NCTC 11184 = ATCC 35865, from equine cervical swab.

Literature

1. Taylor, C.E.D., R.O. Rosenthal, D.F.J. Brown, S.P. Lapage, L.R. Hill, and R.M. Legros. 1977. The causative organism of contagious equine metritis 1977: proposal for a new species to be known as *Haemophilus equigenitalis*. Equine Vet. J. *10:* 136–144.
2. Sugimoto, C., Y. Isayama, R. Sakazaki, and S. Kuramouchi. 1983. Transfer of *Haemophilus equigenitalis* Taylor et al. 1978 to the genus *Taylorella* gen. nov. as *Taylorella equigenitalis* comb. nov. Curr. Microbiol. *9:* 155–162.
3. Biberstein, E.L. 1990. *Haemophilus* and *Taylorella*. *In* G.R. Carter and J.R. Cole (Eds.), Diagnostic Procedures in Veterinary Bacteriology and Mycology, Academic Press, New York, pp. 151–164.

Comments

The causative organism of contagious equine metritis. This organism was originally placed in the genus *Haemophilus*, but has since been transferred to the new genus *Taylorella* (2). Human infections associated with this species have not, to the best of our knowledge, been described.

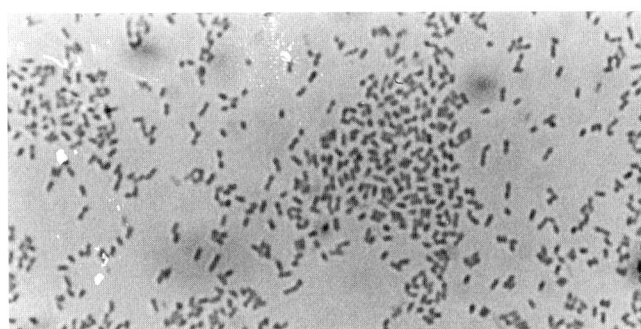

Taylorella equigenitalis strain KC1457. Gram stain BAP 35° C 24 h x 1700.

Taylorella equigenitalis

TEST PERFORMED	REFERENCE STRAIN SIGN
Morphology	srs
Motility; flagella	nm
Action on blood	−
Fermentative or oxidative	[n-o]
Carbohydrate base	OF
Acid from:	
D-Glucose	[−]
D-Xylose	−
D-Mannitol	−
Lactose	−
Sucrose	−
Maltose	−
Catalase	+
Oxidase	+
Growth on:	
MacConkey[a]	(+)
SS	−
Cetrimide	−
Simmons citrate	−
Urea, Christensen's	[−]
Nitrate reduction	−
Indole	[−]
TSI slant, acid	−
TSI butt, acid	−
H_2S (TSI butt)	−
H_2S (Pb ac paper)	−
Gelatin hydrolysis[b]	−
Litmus milk	−
Pigment	−
Growth at:	
25°C	+
35°C	+
42°C	−
Esculin hydrolysis	−
Lysine decarboxylase	−
Arginine dihydrolase	−
Ornithine decarboxylase	[−]
Nutrient broth, 0% NaCl	−
Nutrient broth, 6% NaCl	+

[a]Very light growth observed at 7 days of incubation.
[b]Incubation of 7–14 days.

Weeksella virosa

Gram-negative slightly pleomorphic rod (cells sometimes appear thin to very thin in the central region, with thickened ends, the so-called "II forms"); motility not demonstrated; grows well aerobically; does not use carbohydrates; MacConkey-negative, oxidase-positive; produces indole weakly; does not break down urea or nitrate; proteolytic.

Sources (87 strains)

Urine (45%), vagina (15%), cervix (15%), other (17%), unknown (8%).

Reference strain

9751 = NCTC 11634 = ATCC 43766, type strain, from urine.

Literature

1. Owen, R.J., and J.J.S. Snell. 1973. Comparison of group IIf with *Flavobacterium* and *Moraxella*. Antonie V. Leeuwenhoek *39*: 473–480.

2. Tatum, H.W., W.H. Ewing, and R.E. Weaver. 1974. Miscellaneous gram-negative bacteria. *In* E.H. Lennette, E.H. Spaulding, and J.P. Truant (Eds.), Manual of Clinical Microbiology, 2nd ed., American Society for Microbiology, Washington, pp. 270–294.

3. Holmes, B., A.G. Steigerwalt, R.E. Weaver, and D.J. Brenner. 1986. *Weeksella virosa* gen. nov., sp. nov. (formerly group IIf), found in human clinical specimens. Syst. Appl. Microbiol. *8:* 185–190.

Comments

Formerly group IIf. Isolated from genitourinary tract, primarily female. Grows on Thayer-Martin selective medium. Produces very heavy confluent growth on heart infusion agar in 24 h; cells are encapsulated. Information on the cellular fatty acid composition of this species is presented in Table 4.41 and Figure 4.46 of the cellular fatty acid section.

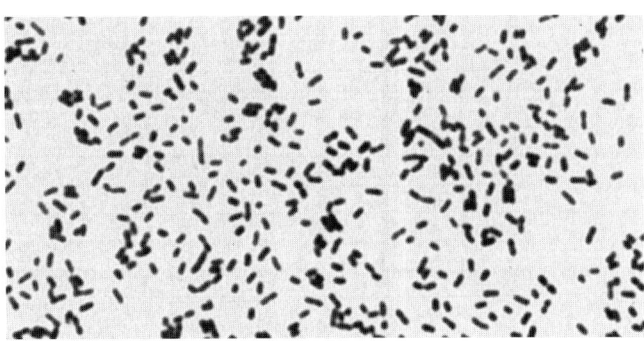

Weeksella virosa strain B4353. Gram stain BAP 35° C 24 h x 1700.

Weeksella virosa

TEST PERFORMED	SIGN	%+	#+/T
NUMBER OF STRAINS 87			
Morphology	srs,II		
Motility; flagella	[nm]		
Action on blood	v	76 LG	62/81; 7 lav, 3 gr, 2 br
Fermentative or oxidative	[n-o]		
Carbohydrate base	OF		
Acid from:			
D-Glucose	[-]	0	0/87
D-Xylose	[-]	0	0/87
D-Mannitol	-	0	0/87
Lactose	-	0	0/87
Sucrose	-	0	0/87
Maltose	-	0	0/87
Catalase	+	98	74, 11w/87
Oxidase	[+]	100	86, 1w/87
Growth on:			
MacConkey	[-]	(10)	(9)/87
SS	-	0	0/87
Simmons citrate	[-]	0	0/87
Urea, Christensen's	[-][a]	0	0/86; (27 pk)
Nitrate reduction	-	0	0/87
Gas from nitrate	-	0	0/87
Indole	[+]	100	51, 36w/87
TSI slant, acid	-	0	0/87
TSI butt, acid	-	0	0/87
H_2S (TSI butt)	-	0	0/87
H_2S (Pb ac paper)	+	95	83/87
Gelatin hydrolysis[b]	[+]	100	79/79
Litmus milk	v	39 pep	34/87; 15 IR, 3 k
Pigment:			
Insoluble	-	0	0/87
Soluble	br-ta	98	85/87
Growth at:			
25°C	v	58	45/77
35°C	+	100	77/77
42°C	v	70	54/77
Esculin hydrolysis	-	0	0/87
Lysine decarboxylase	-	0	0/2
Arginine dihydrolase	-	0	0/2
Ornithine decarboxylase	-	0	0/2
Nutrient broth, 0% NaCl	+	99	85/86
Nutrient broth, 6% NaCl	-	7	6/86

[a]In 3–7 days, 27 of 86 strains produced pink reactions that might be interpreted as weakly positive.
[b]Incubation of 7–14 days.

Weeksella zoohelcum

Gram-negative medium to long rod, may exhibit "II forms"; motility not demonstrated; does not use carbohydrates; usually does not grow or change indicator in OF carbohydrate media; oxidase-positive; no growth on MacConkey agar; indole- and urease-positive; nitrate not reduced.

Sources (41 strains)

Dog bite (34%), cat bite and scratch (15%), abscess and wound (10%), canine isolation (10%), sputum (5%), cerebrospinal fluid (5%), other (16%), unknown (5%).

Reference strain

D658 = NCTC 11660 = ATCC 43767, type strain, from sputum.

Literature

1. Tatum, H.W., W.H. Ewing, and R.E. Weaver. 1974. Miscellaneous gram-negative bacteria. *In* E.H. Lennette, E.H. Spaulding, and J.P. Truant (Eds.), Manual of Clinical Microbiology, 2nd ed., American Society for Microbiology, Washington, pp. 270–294.

2. Bailie, W.E., E.C. Stone, and A.M. Schmitt. 1978. Aerobic bacterial flora of oral and nasal fluids of canines with reference to bacteria associated with bites. J. Clin. Microbiol. 7: 223–231.

3. Holmes, B., A.G. Steigerwalt, R.E. Weaver, and D.J. Brenner. 1986. *Weeksella zoohelcum* sp. nov. (formerly group IIj) from human clinical specimens. Syst. Appl. Microbiol. 8: 191–196.

Comments

Formerly group IIj. Growth on rabbit blood agar is usually sticky. Associated with animal bite wounds. Information on the cellular fatty acid composition of this species is presented in Table 4.41 and Figure 4.46 of the cellular fatty acid section.

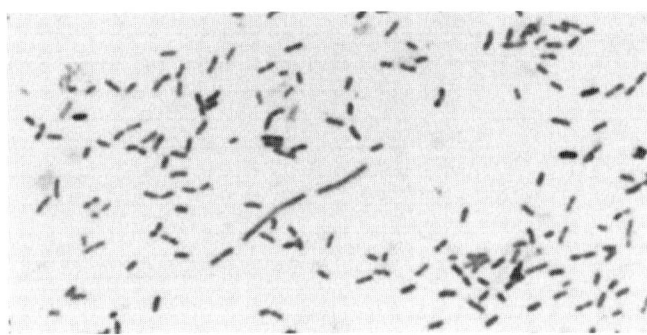

Weeksella zoohelcum strain B2704. Gram stain BAP 35° C 24 h x 1700.

Weeksella zoohelcum

TEST PERFORMED	SIGN	%+	#+/T
NUMBER OF STRAINS 41			
Morphology	s-lr,II		
Motility; flagella	[nm]		
Action on blood	v	36 LG	15/41; 4 al, 3 gr, 2 lav, 1 ß
Fermentative or oxidative	[n-o]		
Carbohydrate base	OF		
Acid from:			
D-Glucose	[-]	0	0/41
D-Xylose	[-]	0	0/41
D-Mannitol	-	0	0/41
Lactose	-	0	0/41
Sucrose	-	0	0/41
Maltose	-	0	0/41
Catalase	+	100	39, 2w/41
Oxidase	[+]	100	39, 1w/40
Growth on:			
MacConkey	[-]	2	1/41
SS	-	0	0/41
Simmons citrate	-	0	0/40
Urea, Christensen's	[+]	100	40/40
Nitrate reduction	[-]	0	0/40
Gas from nitrate	-	0	0/40
Indole	[+]	98	31, 9w/41
TSI slant, acid	-	0	0/41
TSI butt, acid	-	0	0/41
H_2S (TSI butt)	-	0	0/41
H_2S (Pb ac paper)	v	59	21, 2w/39
Gelatin hydrolysis[a]	[+]	98	39/40
Litmus milk	v	18 pep	7/40; 3 k, 1 IR
Pigment:			
Insoluble	-	0	0/39
Soluble	ta-yel	100	39/39
Growth at:			
25°C	v	30	12/40
35°C	+	95	38/40
42°C	-	10	4/40
Esculin hydrolysis	-	0	0/15
Lysine decarboxylase	-	0	0/2
Arginine dihydrolase	+	100	2/2
Ornithine decarboxylase	-	0	0/2
Nutrient broth, 0% NaCl	v	15	6/40
Nutrient broth, 6% NaCl	-	0	0/40

[a] Incubation of 7–14 days.

Xanthomonas maltophilia

Gram-negative short to medium rod; motile (polar tuft of three or more flagella); aerobic; oxidizes glucose (frequently weak or delayed) and maltose, but not mannitol; grows on MacConkey agar; catalase-positive, oxidase-negative; lysine decarboxylase-positive; hydrolyzes gelatin; sometimes reduces nitrate, but without gas; does not produce H_2S in TSI agar.

Sources (228 strains)

Blood (14%), sputum (14%), urine (10%), wound (10%), throat (5%), other (37%), unknown (10%).

Reference strains

KC1078 ← ATCC 13637, type strain, from oropharynx.
C3507, from stool of mouse.

Literature

1. Hugh, R., and E. Ryschenkow. 1961. *Pseudomonas maltophilia* an *Alcaligenes*-like species. J. Gen. Microbiol. *26:* 123–132.

2. Stanier, R.Y., N.J. Palleroni, and M. Doudoroff. 1966. The aerobic pseudomonads: a taxonomic study. J. Gen. Microbiol. *43:* 159–271.

3. Swings, J., P. De Vos, M. Van den Mooter, and J. De Ley. 1983. Transfer of *Pseudomonas maltophilia* Hugh 1981 to the genus *Xanthomonas* as *Xanthomonas maltophilia* (Hugh 1981) comb. nov. Int. J. Syst. Bacteriol. *33:* 409–413.

4. Khardori, N., L. Elting, E. Wong, B. Shable, and G.P. Bodey. 1990. Nosocomial infections due to *Xanthomonas maltophilia* (*Pseudomonas maltophilia*) in patients with cancer. Rev. Infect. Dis. *12:* 997–1003.

5. Palleroni, N.J., and J.F. Bradbury. 1993. *Stenotrophomonas*, a new bacterial genus for *Xanthomonas maltophilia* (Hugh 1980) Swings et al. 1983. Int. J. Syst. Bacteriol. *43:* 606–609.

Comments

The taxonomy of this species is unsettled. Originally described as *Pseudomonas maltophilia*, these organisms were transferred to the genus *Xanthomonas* in 1983 and recently a transfer to the new genus *Stenotrophomonas* has been proposed (5). Produces a lavender-green color on rabbit blood agar at 18–24 h, and a distinctive brown water-soluble pigment on heart infusion agar with tyrosine. Information on the cellular fatty acid composition of this species is presented in Table 4.75 and Figure 4.80 of the cellular fatty acid section.

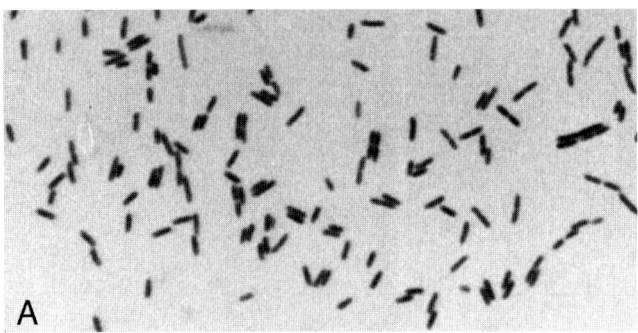

Xanthomonas maltophilia strain KC1078. Gram stain BAP 35° C 24 h x 1700.

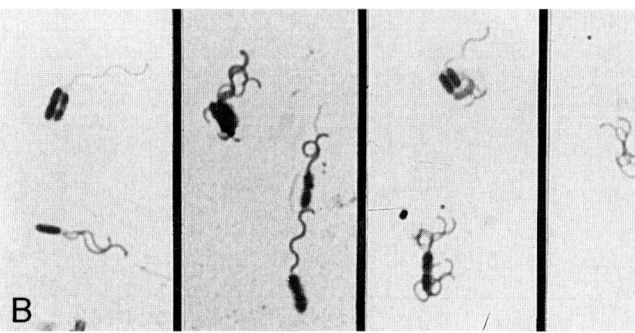

Xanthomonas maltophilia strain C3507. Flagella stain FB 25° C 24 h x 1700.

Xanthomonas maltophilia

TEST PERFORMED	SIGN	% +	#+/T
	NUMBER OF STRAINS 228		
Morphology	mrs		
Motility; flagella	[m;p, > 2]		
Action on blood	v	81 LG	158/194; 27 lav, 15 ly, 6 gr, 3 ß
Fermentative or oxidative	[O]		
Carbohydrate base	OF		
Acid from:			
D-Glucose	[+ or (+)]	85 (5)	6, 56w (4)/73
D-Xylose	v	35 (1)	2, 24w (1)/73
D-Mannitol	[−]	0	0/43
Lactose	v	60 (1)	44w (1)/73
Sucrose	v	63 (1)	46w (1)/73
Maltose	[+][a]	100	67, 6w/73
Catalase	[+]	100	226/226
Oxidase	[−][b]	0	0/73
Growth on:			
MacConkey	[+]	100	228/228
SS	v	22 (21)	51 (49)/228
Cetrimide	−	2 (7)	5 (17)/227
Simmons citrate	v	34 (12)	77 (28)/226
Urea, Christensen's	v	3 (12)	6 (27)/227
Nitrate reduction	v	39	88/226
Gas from nitrate	[−]	0	0/226
Indole	−	0	0/226
TSI slant, acid	−	0	0/226
TSI butt, acid	−	0	0/226
H_2S (TSI butt)	[−]	0	0/228
H_2S (Pb ac paper)	+	95	217/228
Gelatin hydrolysis[c]	[+]	93	178/191
Litmus milk	pep	96	214/224
Pigment:			
Soluble	br-ta	98	223/228
Growth at:			
25°C	+	99	224/225
35°C	+	99	224/225
42°C	v	48	108/225
Esculin hydrolysis	v	39	87/224
Lysine decarboxylase	[+]	93	52/56
Arginine dihydrolase	−	0	0/56
Ornithine decarboxylase	−	0	0/56
Nutrient broth, 0% NaCl	+	100	225/225
Nutrient broth, 6% NaCl	v	22	49/225

[a] Acid reaction in maltose usually stronger than acid reactions in the other carbohydrates listed.
[b] Fourteen of 73 strains that were negative when the reagent was applied directly to growth on a blood agar plate were positive when tested by Kovacs' method; ten of 14 strains were recorded as only weakly positive.
[c] Incubation of 7–14 days.

Ic

Gram-negative slender short to long rod; motile with 1–2 polar flagella; grows on MacConkey, SS, and usually cetrimide agars; oxidizes glucose and maltose; reduces nitrate to nitrite without gas formation; H_2S (Pb acetate paper)-positive, usually strong; arginine dihydrolase positive; urease- and citrate-variable; esculin and gelatin hydrolysis-negative.

Sources (34 strains)

Urine (21%); sputum (15%); blood (9%); bronchial wash (6%); wound (6%); bile or bile tube (6%); one strain each from: throat, infected tooth, cold steamer, humerus, dialysis, leg ulcer, arm, sinus, peritoneal fluid, bile exudate, and cervix (32%); unknown (7%).

Reference strain

B4637, source unknown.

Literature

None

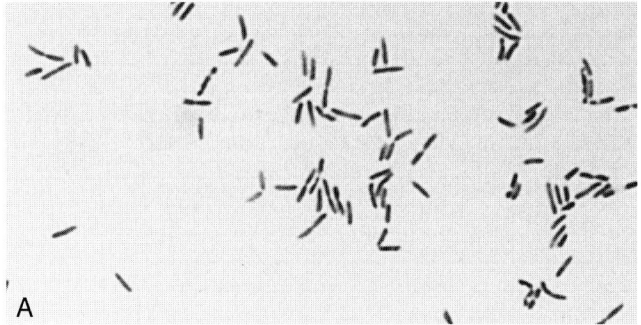

Ic strain B4637. Gram stain BAP 35° C 24 h x 1700.

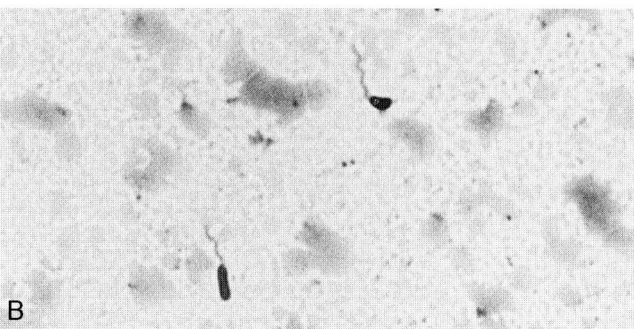

Ic strain B4637. Flagella stain (Ryu method) TGY 25° C 24 h x 1700.

Ic

TEST PERFORMED	SIGN	%+	#+/T
\\ NUMBER OF STRAINS 34			
Morphology	t,s-lr		
Motility; flagella	[m;p,1-2][a]		
Action on blood	v	42 ly	14/33, 4 LG, 4 lav, 2 gr, 1 lav br
Fermentative or oxidative	[O]		
Carbohydrate base	OF		
Acid from:			
D-Glucose	[+]	97	33/34
D-Xylose	−	0	0/34
D-Mannitol	−	0	0/34
Lactose	−	0	0/34
Sucrose	−	0	0/34
Maltose	[+]	100	34/34
Catalase	+	97	33/34
Oxidase	[+]	100	32, 2w/34
Growth on:			
MacConkey	[+]	100	34/34
SS	+	100	34/34
Cetrimide	[+]	94	32/34
Simmons citrate	v	41 (15)	14 (5)/34
Urea, Christensen's	v	18 (15)	6 (5)/34
Nitrate reduction	[+]	100	34/34
Gas from nitrate	−	0	0/34
Indole	−	0	0/34
TSI slant, acid	−	0	0/34
TSI butt, acid	−	0	0/34
H_2S (TSI butt)	−	3	1w/34
H_2S (Pb ac paper)	[+]	100	34/34
Gelatin hydrolysis[b]	−	0	0/26
Litmus milk	v	24 (3) k	4, 4w (1w)/33; 4 IR, 3 dirty, 1 muddy
Pigment:			
Soluble	v	32 tan-br	11/34; 3 yel, 1 gr
Insoluble	v	9 w pk	3/34; 3 golden, 1 buff, 1 w yel
Growth at:			
25°C	+	100	33/33
35°C	+	100	33/33
42°C	+	97	32/33
Esculin hydrolysis	−	0	0/31
Lysine decarboxylase	−	0	0/31
Arginine dihydrolylase	[+]	100	31/31
Ornithine decarboxylase	−	0	0/31
Nutrient broth, 0% NaCl	+	100	34/34
Nutrient broth, 6% NaCl	+	91	31/34
Alkalinization of acetamide	v	12 (6)	1, 1w (1)/17

[a]Medium wavelength.
[b]Incubation of 7–14 days.

SECTION 3: DESCRIPTION OF SPECIES

IIe

Gram-negative slightly pleomorphic rod (cells sometimes appear thin to very thin in the central region, with thicker ends, the so-called "II forms"); motility not demonstrated; grows well aerobically; uses glucose and maltose, but not mannitol and lactose; no growth on MacConkey agar; weakly indole-positive; nitrate not reduced and esculin not hydrolyzed; produces no insoluble pigment; does not hydrolyze gelatin.

Sources (30 strains)

Blood (17%), wound (17%), eye (13%), urine (10%), hand (7%), other (cerebrospinal fluid, chest fluid, peritoneal fluid, genital, ear, nasopharyngeal, lung tissue, meninges, leg) (33%), unknown (3%).

Reference strains

B5408, from blood.
KC198 ← ATCC 10811 ← J. M. Coffey, New York Dept Publ Hlth, 33444, from cervical gland of cow.

Literature

None

Comments

Groups IIe and IIh differ mainly in the esculin reaction. Information on the cellular fatty acid composition of this group is presented in Table 4.43 and Figure 4.48 of the cellular fatty acid section.

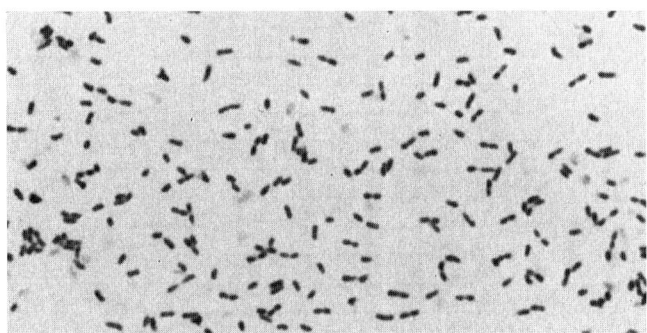

IIe strain KC198. Gram stain BAP 35° C 24 h x 1700.

IIe

	NUMBER OF STRAINS 30		
TEST PERFORMED	SIGN	%+	#+/T
Morphology	srs,II		
Motility; flagella	[nm]		
Action on blood	v	7 br	2/28; 2 LG, 2 gr, 1 al
Fermentative or oxidative	[O]		
Carbohydrate base	OF		
Acid from:			
D-Glucose	[+ or (+)]	83 (17)	21 (4), 4w (1w)/30
D-Xylose	−	0	0/30
D-Mannitol	[−]	0	0/30
Lactose	[−]	0	0/30
Sucrose	−	0	0/30
Maltose	[+]	97 (3)	26 (1), 3w/30
Catalase	+	100	27, 3w/30
Oxidase	+	100	30/30
Growth on:			
MacConkey	[−]	3	1/30
SS	−	0	0/30
Cetrimide	−	0	0/22
Simmons citrate	−	0	0/30
Urea, Christensen's	−	0	0/30
Nitrate reduction	[−]	0	0/30
Indole	[+]	100	30/30
TSI slant, acid	−	0	0/30
TSI butt, acid	−	0	0/30
H_2S (TSI butt)	−	0	0/30
H_2S (Pb ac paper)	v	87	26/30
Gelatin hydrolysis[a]	[−]	3	1/29
Litmus milk	v	14 IR	4/29; 1 k
Pigment:			
Insoluble	−	7 w-yel	2/29
Soluble	v	62 br-ta-yel	18/29
Growth at:			
25°C	+	90	26/29
35°C	+	100	29/29
42°C	−	0	0/29
Esculin hydrolysis	[−]	0	0/30
Lysine decarboxylase	−	0	0/2
Arginine dihydrolase	−	0	0/2
Ornithine decarboxylase	−	0	0/2
Nutrient broth, 0% NaCl	+	97	29/30
Nutrient broth, 6% NaCl	−	3	1/30

[a]Incubation of 7–14 days.

IIg

Gram-negative small coccoid to small rod forms, occasionally with medium to long filaments; motility not demonstrated; asaccharolytic; oxidase-positive; grows on MacConkey agar; indole-positive; nitrate- and esculin-negative; does not hydrolyze gelatin.

Sources (12 strains)

Wound (one strain each) scrotal, inguinal, buttock, leg, pus, throat, scrotum abscess, endometrial brushing, urethra, ear, cyst.

Reference strain

G8864, from leg wound.

Literature

None

Comments

Growth on MacConkey agar and lack of gelatin hydrolysis are useful in differentiating this group from *Weeksella virosa* (formerly IIf). Information on the cellular fatty acid composition of this group is presented in Table 4.35 and Figure 4.40 of the cellular fatty acid section.

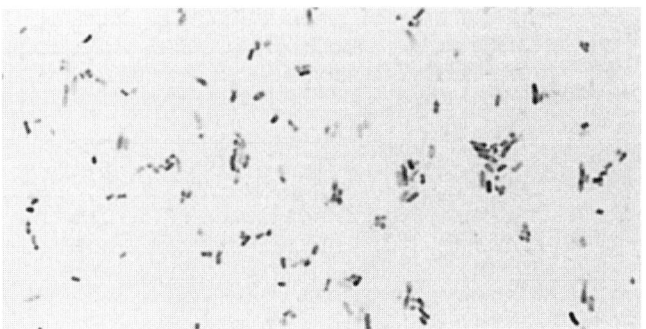

IIg strain G8864. Gram stain BAP 35° C 24 h x 1700.

IIg

TEST PERFORMED	SIGN	% +	#+/T
Morphology	cc,rv		
Motility; flagella	[nm]		
Action on blood	v	42 ly	5/12
Fermentative or oxidative	n-o		
Carbohydrate base	OF		
Acid from:			
D-Glucose	[-]	0	0/12
D-Xylose	-	0	0/12
D-Mannitol	-	0	0/12
Lactose	-	0	0/12
Sucrose	-	0	0/12
Maltose	[-]	0	0/12
Catalase	+	92	11/12
Oxidase	+	100	12/12
Growth on:			
MacConkey	[+]	100	12/12
SS	-	0	0/12
Cetrimide	-	0	0/12
Simmons citrate	-	8	1w/12
Urea, Christensen's	-	0	0/12
Nitrate reduction	[-]	0	0/12
Nitrite reduction	+	100	4/4
Indole	[+]	100	12/12
TSI slant, acid	-	0	0/12
TSI butt, acid	-	0	0/12
H_2S (TSI butt)	-	0	0/12
H_2S (Pb ac paper)	v	50	6w/12
Gelatin hydrolysis[a]	[-]	0	0/12
Litmus milk	v	18 k	2/11
Pigment:			
Insoluble	v	17 ta	2/12; 1 salmon
Soluble	v	17 br-yel	2/12
Growth at:			
25°C	+	100	10/10
35°C	+	100	11/11
42°C	+	90	9/10
Esculin hydrolysis	[-]	0	0/12
Nutrient broth, 0% NaCl	+	100	11/11
Nutrient broth, 6% NaCl	-	0	0/11

NUMBER OF STRAINS 12

[a]Incubation of 7–14 days.

IIh

Small Gram-negative slightly pleomorphic rod (cells sometimes appear thin to very thin in the central region, with thicker ends, the so-called "II forms"); motility not demonstrated; grows well aerobically; uses glucose and maltose, but not mannitol or lactose; MacConkey-negative, oxidase-positive; produces indole; hydrolyzes esculin; does not break down urea or nitrate; produces no growth pigment; nonproteolytic.

Sources (21 strains)

Animal (38%), cerebrospinal fluid (10%), ear (10%), other human (42%).

Reference strains

781, source unknown.
D3016, from guinea pig kidney.

Literature

1. Pickett, M.J., D.G. Hollis, and E.J. Bottone. 1991. Miscellaneous gram-negative bacteria. *In* A. Balows et al. (Eds.), Manual of Clinical Microbiology, 5th ed., American Society for Microbiology, Washington, pp. 410–428.

Comment

Groups IIh and IIe differ mainly in the esculin reaction. Information on the cellular fatty acid composition of this group is presented in Table 4.43 and Figure 4.48.

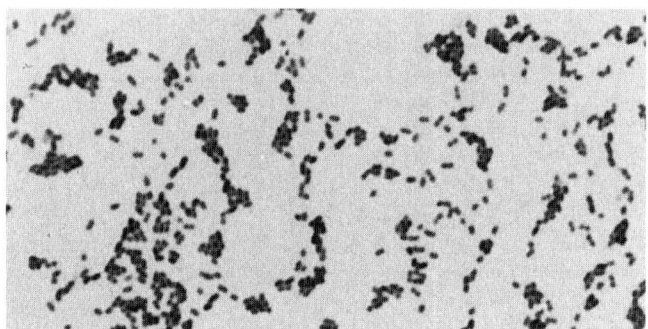

IIh strain 781. Gram stain BAP 35° C 24 h x 1700.

IIh

TEST PERFORMED	SIGN	NUMBER OF STRAINS 21 %+	#+/T
Morphology	srs,II		
Motility; flagella	[nm]		
Action on blood	v	32 LG	6/19; 4 gr, 3 al (stab)
Fermentative or oxidative	[O]		
Carbohydrate base	OF		
Acid from:			
D-Glucose	[+ or (+)]	85 (15)	17 (3)/20
D-Xylose	-	5	1w/20
D-Mannitol	[-]	0	0/20
Lactose	[-]	0	0/20
Sucrose	-	0	0/20
Maltose	[+]	95	19/20
Catalase	+	100	21/21
Oxidase	[+]	100	21/21
Growth on:			
MacConkey	[-]	0	0/21
SS	-	0	0/21
Simmons citrate	-	0	0/21
Urea, Christensen's	-	0	0/21
Nitrate reduction	[-]	0	0/21
Indole	[+]	100	18, 3w/21
TSI slant, acid	-	0	0/21
TSI butt, acid	-	5	1w/21
H_2S (TSI butt)	-	0	0/21
H_2S (Pb ac paper)	+	100	21/21
Gelatin hydrolysis[a]	[-]	7	1/15
Litmus milk	-	7 IR	1/15
Pigment:			
Insoluble	-	0	0/21
Soluble	br	100	19/19
Growth at:			
25°C	+	100	21/21
35°C	+	100	21/21
42°C	-	5	1/21
Esculin hydrolysis	[+]	100	21/21
Nutrient broth, 0% NaCl	v	86	18/21
Nutrient broth, 6% NaCl	-	5	1/21

[a]Incubation of 7–14 days.

IIi

Gram-negative slightly pleomorphic rod (cells sometimes appear thin to very thin in the central region, with thicker ends, the so-called "II forms"); motility not demonstrated; grows well aerobically; uses glucose, xylose, lactose, sucrose, and maltose, but not mannitol; reaction on TSI agar is alkaline slant, no change in the butt; no growth on MacConkey agar; oxidase-, catalase-, and indole-positive; nonproteolytic.

Sources (23 strains)

Blood (35%), urine (26%), wound (13%), other (13%), unknown (13%).

Reference strain

C9138, from leg wound.

Literature

None

Comments

Cellular fatty acid profile analysis suggests that these organisms are closely related to *Flavobacterium* (*Sphingobacterium*) species. Information on the cellular fatty acid composition of this group is presented in Table 4.44 and Figure 4.49 of the cellular fatty acid section.

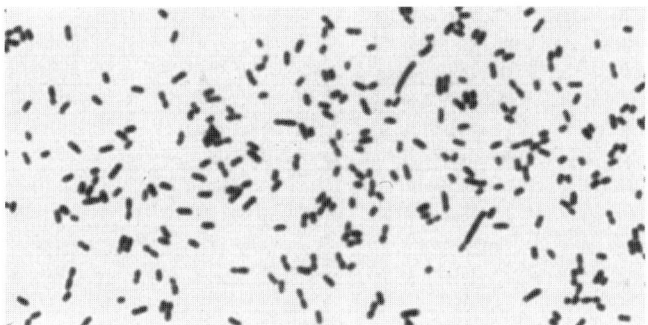

IIi strain C9138. Gram stain BAP 35° C 24 h x 1700.

IIi

	NUMBER OF STRAINS 23		
TEST PERFORMED	SIGN	%+	#+/T
Morphology	srs,II		
Motility; flagella	[nm]		
Action on blood	v	39 LG	9/23; 5 ly, 1 lav, 1 gr, 1 al
Fermentative or oxidative	[O]		
Carbohydrate base	OF		
Acid from:			
D-Glucose	[+]	91 (9)	21 (2)/23
D-Xylose	+ or (+)	87 (13)	16, 4w (3)/23
D-Mannitol	[-]	0	0/23
Lactose	+	91 (9)	21 (2)/23
Sucrose	[+]	91 (9)	21 (2)/23
Maltose	+	91 (9)	21 (2)/23
Catalase	[+]	100	23/23
Oxidase	[+]	100	21, 2w/23
Growth on:			
MacConkey	[-]	0	0/22
SS	-	0	0/22
Simmons citrate	-	0	0/22
Urea, Christensen's	v	14 (18)	3 (4)/22
Nitrate reduction	-	0	0/23
Indole	[+]	100	10, 13w/23
TSI slant, acid	-	0	0/23
TSI butt, acid	[-]	0	0/23
H_2S (TSI butt)	-	0	0/23
H_2S (Pb ac paper)	v	70	16/23
Gelatin hydrolysis[a]	[-]	0	0/16
Litmus milk	-	4 A	1/22
Pigment:			
Insoluble	v	22 yel	3, 2w/23
Soluble	v	48 yel-ta	11/23
Growth at:			
25°C	+	100	22/22
35°C	+	100	22/22
42°C	v	36	8/23
Esculin hydrolysis	+	96	22/23
Lysine decarboxylase	-	0	0/3
Arginine dihydrolase	-	0	0/3
Ornithine decarboxylase	-	0	0/3
Nutrient broth, 0% NaCl	+	100	23/23
Nutrient broth, 6% NaCl	-	9	2/23

[a]Incubation of 7–14 days.

IVc-2

Gram-negative short to medium rod, may stain irregularly; motile (peritrichous flagella); aerobic; grows on MacConkey agar, not on SS medium; does not use carbohydrates; oxidase-positive; urease test partially positive in 24 h; usually nitrate-negative (no gas if positive); no H_2S formed on TSI agar.

Sources (36 strains)

Water (19%), blood (14%), eye (8%), sputum (8%), urine (5%), throat (5%), tissue culture contaminant (5%), peritoneal fluid (5%), other (15%), unknown (16%).

Reference strains

A1036, from blood.
D3518, from peritoneal fluid.

Literature

1. Dan, M., S.A. Berger, D. Aderka, and Y. Levo. 1986. Septicemia caused by the gram-negative bacterium CDC IVc-2 in an immunocompromised human. J. Clin. Microbiol. 23: 803.

2. Rossau, R., K. Kersters, E. Falsen, E. Jantzen, P. Segers, A. Union, L. Nehls, and J. De Ley. 1987. *Oligella*, a new genus including *Oligella urethralis* comb. nov. (formerly *Moraxella urethralis*) and *Oligella ureolytica* sp. nov. (formerly CDC group IVe): relationship to *Taylorella equigenitalis* and related taxa. Int. J. Syst. Bacteriol. 37: 198–210.

3. Crowe, H.M., and S.M. Brecher. 1987. Nosocomial septicemia with CDC group IVc-2, an unusual gram-negative bacillus. J. Clin. Microbiol. 25: 2225–2226.

4. Zapardiel, J., G. Blum, C. Caramelo, R. Fernandez-Roblas, J.L. Rodriguez-Tudela, and F. Soriano. 1991. Peritonitis with CDC group IVc-2 bacteria in a patient on continuous ambulatory peritoneal dialysis. Eur. J. Clin. Microbiol. Infect. Dis. 10: 509–511.

Comments

Shown to be closely related to *Alcaligenes eutrophus*, an organism that phenotypically differs from IVc-2 by reducing nitrate and nitrite (0.01% medium). Closely resembles *Bordetella bronchiseptica* except for nitrate, SS, and tartrate reactions. Information on the cellular fatty acid composition of this group is presented in Table 4.10 and Figure 4.15 of the cellular fatty acid section.

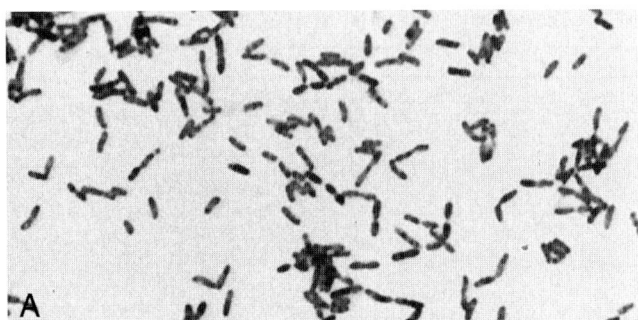

IVc-2 strain A1036. Gram stain BAP 35° C 24 h x 1700.

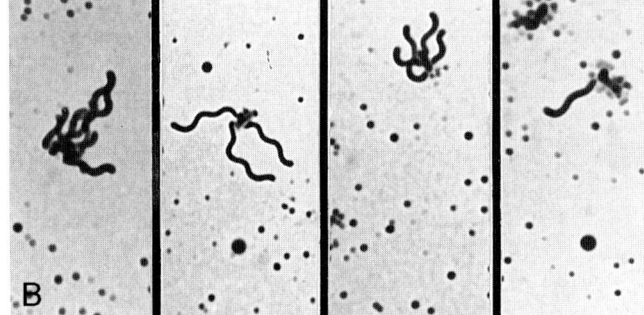

IVc-2 strain D3518. Flagella stain BF 25° C 24 h x 1700.

IVc-2

TEST PERFORMED	SIGN	%+	#+/T
	NUMBER OF STRAINS 36		
Morphology	srs		
Motility; flagella	[m;pe]		
Action on blood	v	53 ly	8/15; 4 gr, 2 br, 1 al
Fermentative or oxidative	[n-o]		
Carbohydrate base	OF		
Acid from:			
D-Glucose	[-]	0	0/36
D-Xylose	[-]	0	0/36
D-Mannitol	-	0	0/36
Lactose	-	0	0/36
Sucrose	-	0	0/36
Maltose	-	0	0/36
Catalase	+	100	36/36
Oxidase	[+]	100	36/36
Growth on:			
MacConkey	[+]	94 (6)	34 (2)/36
SS	[-]	3 (6)	1 (2)/36
Cetrimide	-	0 (3)	(1)/36
Simmons citrate	+	100	36/36
Urea, Christensen's	[+]	100	36/36
Nitrate reduction	[v]	11	4/36
Gas from nitrate	[-]	0	0/36
Indole	-	0	0/36
TSI slant, acid	-	0	0/36
TSI butt, acid	-	0	0/36
H_2S (TSI butt)	[-]	0	0/35
H_2S (Pb ac paper)	v	51	18/35
Gelatin hydrolysis[a]	-	0	0/30
Litmus milk	k	97	35/36
Pigment	v	22 yel-ta sol	8/36
Growth at:			
25°C	+	94	34/36
35°C	+	100	36/36
42°C	v	86	31/36
Esculin hydrolysis	-	0	0/30
Nutrient broth, 0% NaCl	+	100	36/36
Nutrient broth, 6% NaCl	v	11	4/36
Tartrate alkalinization	[+]	100	9/9

[a]Incubation of 7–14 days.

Vb-3

Gram-negative medium rod; motile (single, occasionally two polar flagella); aerobic; uses glucose, xylose, maltose, and (some strains) mannitol, but not lactose and sucrose; hydrolyzes starch; oxidase-positive; grows on MacConkey agar; reduces nitrate to gas; no H_2S produced on TSI agar; usually nonproteolytic; produces arginine dihydrolase, but not lysine decarboxylase.

Sources (65 strains)

Wound (12%), ear (11%), leg (8%), blood (8%), sputum (6%), eye (5%), cerebrospinal fluid (5%), urine (5%), knee abscess (3%), dialysis supply tank (3%), other (31%), unknown (3%).

Reference strain

D2986, from leg wound.

Literature

1. Gilardi, G.L. 1991. *Pseudomonas* and related genera. *In* A. Balows et al. (Eds.), Manual of Clinical Microbiology, 5th ed., American Society for Microbiology, Washington, pp. 429–441.

Comments

Very similar to *Pseudomonas stutzeri*, possibly a biotype (arginine dihydrolase production is positive with Vb-3 and negative with *P. stutzeri*); genetic studies are needed to determine the species status of this group. Information on the cellular fatty acid composition of this group is presented in Table 4.36 and Figure 4.41 of the cellular fatty acid section.

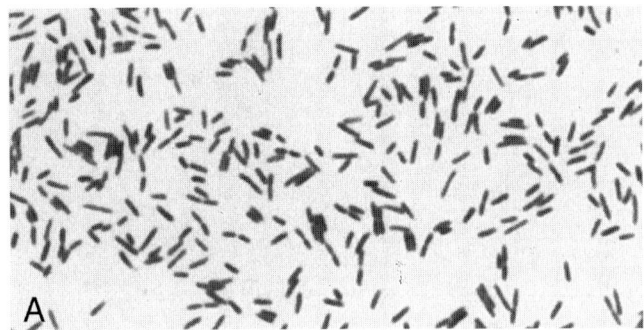

Vb-3 strain D2986. Gram stain BAP 35° C 24 h x 1700.

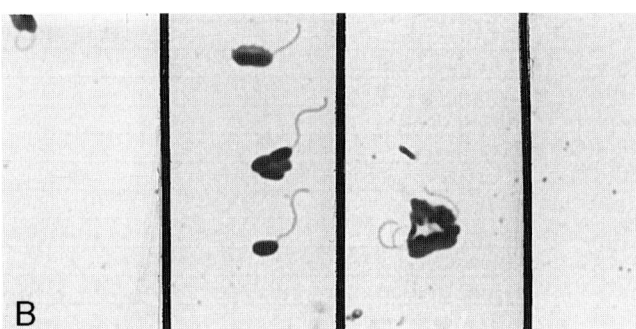

Vb-3 strain D2986. Flagella stain FB 25° C 24 h x 1700.

Vb-3

TEST PERFORMED	SIGN	%+	#+/T
\<td colspan=4\>NUMBER OF STRAINS 65			
Morphology	mrs		
Motility; flagella	[m;p,1-2]		
Action on blood	v	32 gr	20/63; 17 ly, 11 br & ly, 7 LG
Fermentative or oxidative	[O]		
Carbohydrate base	OF		
Acid from:			
D-Glucose	[+]	100	61, 4w/65
D-Xylose	+	97 (3)	61, 2w (2)/65
D-Mannitol	v	65 (9)	19, 23w (6)/65
Lactose	[-]	0	0/65
Sucrose	-	0	0/65
Maltose	[+]	95 (5)	42, 20w (3)/65
Starch	+	100	9, 1w/10
Catalase	+	98	61, 3w/65
Oxidase	[+]	100	62, 3w/65
Growth on:			
MacConkey	[+]	100	65/65
SS	+	100	64, 1w/65
Cetrimide	v	18 (2)	4, 8w (1)/65
Simmons citrate	+	92 (6)	60 (4)/65
Urea, Christensen's	v	12 (45)	7, 1w (29)/65
Nitrate reduction	+	100	65/65
Gas from nitrate	[+]	100	65/65
Indole	-	0	0/65
TSI slant, acid	-	0	0/65
TSI butt, acid	-	0	0/65
H_2S (TSI butt)	-	0	0/65
H_2S (Pb ac paper)	v	5 (14)	3 (9)/65
Gelatin hydrolysis[a]	[-]	4	2/45
Litmus milk	v	69 k	44/64; 14 IR, 2 pep
Pigment	yel ins	95	21, 41w/65; 7 yel-br sol
Growth at:			
25°C	+	98	64/65
35°C	+	100	65/65
42°C	-	0	0/65
Esculin hydrolysis	-	0	0/65
Lysine decarboxylase	[-]	0	0/65
Arginine dihydrolase	[+]	100	65/65
Ornithine decarboxylase	-	0	0/65
Nutrient broth, 0% NaCl	+	100	61/61
Nutrient broth, 6% NaCl	v	85 (3)	45, 7w (2)/61
Acetamide	-	0	0/8
2-Ketogluconate	-	10	1w/10

[a] Incubation of 7–14 days.

BACTERIAL IDENTIFICATION BY CELLULAR FATTY ACID ANALYSIS

INTRODUCTION

In 1965, the Analytical Chemistry Laboratory began extensive studies to develop gas-liquid chromatographic (GLC) methods and procedures for rapid and sensitive determination of the cellular fatty acid (CFA) composition of bacteria. The results of these studies, published in 1967, were the first detailed report to show the usefulness of CFA for identification and classification of microorganisms. Subsequent studies in this laboratory and others throughout the world have firmly established the value of CFAs as an important aid in rapid identification. Today, the CFA compositions of most of the named bacterial species have been determined. CFA data are also routinely included in the description of essentially all new or recently discovered organisms. For the last several years, this laboratory has routinely determined the CFA composition of named bacteria as well as unknown isolates submitted for identification to specialty laboratories within the Centers for Disease Control and Prevention (CDC). A large CFA data base has been established in our laboratory for these Gram-negative aerobic and facultative anaerobic bacteria identified by the CDC Special Bacteriology Reference Laboratory. A detailed description of the CFA composition of these organisms, together with methods and procedures for their determination are described in this manual.

Culture and Growth Conditions

Pure cultures are required for CFA analysis. After primary isolation, the culture is inoculated onto growth media and incubated under optimum conditions to produce good to heavy growth. The growth medium and the time and temperature of incubation for good growth differ among bacterial species and groups. Most nonfermentative Gram-negative bacteria, and other organisms isolated in the clinical laboratory, give good growth after 24- to 36-h incubation at 32–35° C on trypticase soy agar (TSA) or heart infusion agar (HIA) supplemented with 5% rabbit blood. Some organisms such as *Brucella, Capnocytophaga,* and *Campylobacter* often require 48- to 72-h incubation for good growth. Some organisms such as *Legionella* require special media supplements (i.e., L-cysteine and iron salts) for good growth. Thus, selection and use of standard media that produce good growth of the organisms under study are important considerations for comparable CFA data. Length of incubation is another important factor, since the CFA composition of an organism is most stable and reproducible at the stationary phase of growth. Physiological age of the cell rather than the exact time of incubation is the important parameter. For example, when examining many strains within a species or group, one or more strains may be slow growing and require 48 h to obtain equivalent growth to other strains incubated for 24 h. Cells of the slow growing strain(s) most likely will contain relatively larger amounts of unsaturated fatty acids than physiologically older cells. The fatty acid composition of the slow growing strain at 48 h (or until good growth) will be most similar to the other strains that show good growth at 24 h. Thus, familiarity with the growth characteristics of the organisms under study will lead to the most consistent and reproducible fatty acid results for all strains within the group. Obviously, all organisms within the study or comparison group should be grown on the same medium and under similar conditions of incubation.

Hydrolysis of Cells

The lipids of microorganisms are found in the cell wall/cell membrane fraction where the fatty acids are chemically bonded to proteins, carbohydrates, and other chemical entities. In general, the total amount of lipid is greater in Gram-negative bacteria than in Gram-positive organisms, and this factor must be considered in determining the amount of cells (growth) required for satisfactory fatty acid analysis. Most, but not all, of the cellular lipid can be removed by extraction with organic solvents such as chloroform, chloroform-methanol, or hexane. However, to determine total CFA composition, the cells (or extracted lipid from cells) must be hydrolyzed with acid or base to liberate the unit fatty acids. Examples of the various types of fatty acids found in bacteria are presented later in Table 4.1. The hydrolysis procedure is critical to understanding the final results, as acid hydrolysis degrades cyclopropane acids, and base hydrolysis fails to liberate all of the amide-linked hydroxy fatty acids.

We developed a simple and rapid base hydrolysis (saponification) procedure that gives accurate and reproducible results with either lyophilized or fresh whole cells. Later, this procedure is outlined in Figure 4.1 and detailed in Protocol 4.1. Growth from the surface of an agar plate or slant is removed by adding 0.5 – 1 ml of distilled water and gently scraping. The turbid cell suspension is placed in a screw-capped tube (13 × 100 mm) fitted with a Teflon-lined cap and saponified by heating at 100° C for 30 min after adding 1 ml of 15% NaOH in 50% aqueous methanol. The sample is cooled to ambient temperature, 1.5 ml of 25% HCl in methanol (methylation reagent) is added, and the contents (pH 2.0–3.0) are mixed. The sample is heated at 85° C for 15 min to form fatty acid methyl esters (FAME). The FAME sample is then cooled to room temperature, 1.5 ml of diethylether:hexane (1:1) is added, and the contents are mixed by shaking. The phases are allowed to separate by standing 1–3 min, and the lower (aqueous) layer is removed carefully with a Pasteur pipette and discarded (or saved to test for amide-linked hydroxy acids). One milliliter of 0.3 M phosphate buffer (pH 11.0) is added to the organic layer, the contents are mixed well by shaking, and the phases are allowed to separate by standing 3–5 min or by brief centrifugation.

About two-thirds of the top (organic) layer containing the FAME is removed to a clean test tube or sample vial for subsequent analysis by GLC. The capped FAME sample should be stored at –20° C if GLC analysis is not initiated within 3–4 h.

Amide-linked hydroxy acids, if present, are not totally released with this saponification procedure. These acids, which remain in the methanolic aqueous layer after the methylation and extraction steps in the above procedure, can be released completely by adding 1 ml of concentrated HCl to the aqueous layer and heating at 100° C for 1 h. The resulting FAME of these acids are extracted with 1:1 ether:hexane and can be analyzed separately by GLC or combined with the FAME from the regular saponification procedure. Most of the bacterial fatty acid data generated in this laboratory and in most other laboratories throughout the world have been obtained with the above saponification procedure without subsequent acid hydrolysis for amide-linked acids.

Gas Liquid Chromatography (GLC)

GLC has been used almost exclusively to analyze bacterial fatty acids because of the speed, sensitivity, and resolution of this technique. In our early studies, GLC analysis was done on 6 or 8 ft × 4-mm (inside diameter) glass columns packed with nonpolar (SE-30, OV-1, or OV-101) or a polar stationary phase (EGA or OV-17). GLC retention time (RT) comparisons with known standards on both polar and nonpolar phases were essential for tentative identification of the individual fatty acid components. Confirmation of identity required additional ancillary techniques such as hydrogenation, acetylation, and/or mass spectrometry.

In 1980, this laboratory described the use of fused silica capillary columns for GLC analysis of bacterial fatty acids. This column, which is typically 15–30 m in length with internal diameter of 0.2–0.4 mm, gives superior resolution and separating efficiency compared with packed columns. Most of the fatty acids known to occur in bacteria can be resolved (as methyl esters) with a 50 m nonpolar phase capillary column, and the accuracy of RT measurements with currently available gas chromatographs is routinely greater than 0.005 min. Therefore, precise RT matches of FAME peaks from the analytical sample with those of known standards under standard chromatographic conditions give a high probability of identification. The accuracy of identification by RT measurements is increased by hydrogenation of unsaturated acids and acetylation of hydroxy acids with subsequent GLC analysis under identical conditions as the original FAME sample. When hydrogenated, peaks of unsaturated acids will disappear from the chromatogram as these will be converted to their saturated homologs (i.e., C-16:1 to C-16:0) and the size of the saturated homolog peak (i.e., C-16:0) will be increased in proportion to the amount of unsaturated homolog (i.e., C-16:1) in the original sample. With acetylation, hydroxy acid methyl esters are converted to a more volatile derivative that elutes faster from a nonpolar column than the original hydroxy acid methyl ester. With these combinations of GLC analyses, almost all the fatty acids in a given microorganism can be identified by RT alone and accurately quantitated with a recording integrator. If available, other analytical techniques such as infrared spectroscopy, nuclear magnetic resonance spectroscopy, and mass spectroscopy are useful for confirmation and are required for identification of any significant component not identified by RT measurements. Some fatty acids known to occur in bacteria are not available commercially for use as standards, but the RTs of these relatively unusual acids (i.e., 2—OH-13:0, i-2—OH-15:0, a-17:1, etc.) can be determined by consulting the literature and then analyzing microorganisms known to contain these acids.

Trifluorocetylation and Hydrogenation of Fatty Acid Methyl Esters (FAME)

The procedures for trifluoroacetylation and hydrogenation of FAME are illustrated later in Figures 4.2 and 4.3, respectively, and detailed in Protocols 4.2 and 4.3. An example of the results obtained by trifluoroacetylation and hydrogenation are illustrated in Figures 4.4–4.6 with a strain of *Neisseria subflava*. The chromatogram in Figure 4.4 shows the peaks of the FAME of this organism processed with the saponification procedure outlined in Protocol 4.1. Tentative identification of peaks in the chromatogram was made on the basis of RT comparison to a known standard and were recorded on the chromatogram in their shorthand designation as illustrated in Table 4.1. To confirm the presence of the hydroxyl group in the three peaks tentatively identified as hydroxy acids (3—OH-12:0, 3—OH-14:0, 3—OH-16:0), the FAME sample was treated with trifluoroacetic anhydride as described in Protocol 4.2 and then reanalyzed by GLC on the same column under identical conditions as in Figure 4.4.

A chromatogram of the acetylated FAME of *N. subflava* is shown in Figure 4.5. By comparing chromatograms, it is clear that the peaks of hydroxy FAME at 6.2, 9.5, and 12.9 min in Figure 4.4 are not present in Figure 4.5, and that each of these peaks has been converted to a more volatile derivative that elutes at 5.2, 8.0, and 11.3 min, respectively, in Figure 4.5. These data, which give identical RT matches to known standards in which two functional groups (—COOH, —OH) have been derivatized, give strong support for positive identification of these three acids. It is also apparent that no peaks other than the three hydroxy acids were affected by the acetylation step as these peaks have the same RT in both chromatograms.

A chromatogram of the hydrogenated FAME of *N. subflava* is shown in Figure 4.6. All labeled peaks of unsaturated FAME in Figure 4.4 (14:1, 16:1ω7c, 18:2, 18:1ω9c, 18:1ω7c) are absent in the hydrogenated sample in Figure 4.6, as each unsaturated FAME has been converted to its saturated homolog. The increase in the amount of the saturated homolog is proportional to that amount of unsaturated component(s) present, as illustrated by comparing the relative peak heights of 14:0, 16:0, and 18:0 in Figure 4.4 with those in Figure 4.6. The area of the 18:0 peak in Figure 4.6 represents the total of all unsaturated 18-carbon FAME in Figure 4.4 (18:2, 18:1ω9c, 18:1ω7c) as well as the area of 18:0 in the original sample. The 14:1 and 16:1 peaks are not present in Figure 4.6, since these have been converted to 14:0 and 16:0, respectively. These data from hydrogenation give additional evidence for identification of unsaturated as well as saturated FAME. The fact that the hydroxy acid FAME were unaffected by hydrogenation confirmed that these acids were saturated. However, unsaturated hydroxy FAME occasionally are present in bacteria (i.e., *Pseudomonas cepacia*) and will require the combined analyses as FAME, acetylated FAME, and hydrogenated FAME for accurate identification of these acids. Organisms with relatively complex fatty acid profiles such as *Xanthomonas maltophilia* and *Flavobacterium meningosepticum* also require these procedures for accurate peak identification.

After the individual fatty acids are identified, the fatty acid composition of the unknown organism is compared with phenotypically related organisms whose fatty acid contents are known. Visual comparison of chromatograms are made on the basis of qualitative (RT match) and quantitative similarities or differences in fatty acids of the unknown to known organisms. The relative amount of each component is estimated visually by observing peak heights; accurate quantitative data, including total peak area and percentage of composition of each peak, are available from the recording integrator. Identification by visual comparison requires time, experience, and familiarity with the fatty acid composition of many bacterial genera and species. As the number of species for comparison grows, the difficulty of visual identification increases because of the large amount of data that must be correlated. Use of a microcomputer for microbial identification is a logical and important step for widespread use of CFA data outside research or specialty laboratories.

Qualitative and quantitative CFA data of various Gram-negative bacteria we have examined over the years are presented in the following chromatograms and tables. Organisms with similar CFA composition are listed together and generally cannot be distinguished from each other on the basis of CFA; however, most of the groups are easily differentiated from each other by either qualitative differences or large quantitative differences in their CFA composition. Organisms that we consider to have very similar fatty acid composition are listed together in the same table. The presence and relative amount (expressed as percentage (%) of total) of fatty acids of each organism within the group are listed in the table. The percentage (%) composition value for each acid is the mean from multiple analyses (two or more) of several known strains (10 or more strains if available) of each organism listed. The identity of all strains included in the CFA data was established with conventional cultural and biochemical tests; the identities of many strains were confirmed with results from DNA-relatedness studies.

FAME profiles that are representative of each organism within a CFA group are presented in the chromatogram. Each chromatogram was developed using a 15- × 0.2-mm (inside diameter) nonpolar stationary phase (DB-5) fused silica capillary column with a flame ionization detector. The column temperature was programed at 5° C/min from 170–270° C, then at 30° C/min to 310° C for 3 min, and then back to 170° C. The major fatty acid peaks in each chromatogram are identified and labeled according to their shorthand designation (Table 4.1); RT (in minutes) is listed on the horizontal line at the bottom of each chromatogram. The individual FAME peaks in a given sample will elute in the same order as shown in the chromatograms when analyzed on a nonpolar stationary phase column. Moreover, the separation (or resolution) between peaks in the chromatogram, as well as their relative position (or distance) from each other, are essentially identical on capillary columns with the same or similar nonpolar stationary phase (i.e., OV-1, OV-101, DB-5, etc.). Thus, it is possible to accurately compare GLC chromatograms developed in different laboratories by observing the relative position of peaks to each other rather than comparing **exact** RT values that are difficult to duplicate from one GLC column to another.

When visually comparing the CFA of the test (or unknown) organism to those given in the following chromatograms, it is helpful to use the following procedure. 1) Locate the largest peak in the chromatogram and identify it by obtaining an exact RT match with that of a known standard and/or with a known culture in which the peak has been accurately identified. 2) Identify (RT comparison) the next largest three or four peaks and compare their relative concentration to each other and to the largest peak. If the largest peak is off scale, then the relative amounts of the major acids are determined from the integrator data that gives the percentage (%) area (and RT) for each peak. 3) One should note any relatively unusual acids even though they may be present in small (1–2%) amounts. With these factors in mind, one can make a meaningful comparison as illustrated in the following example:

Example: The CFA profile of the organism under study showed a major peak (>60%) of 18:1ω7c, approximately equal amounts (6%) of 16:0, 17:0, 18:0, and Br-19:1 and small amounts (1–2%) of 3—OH-14:0, 3—OH-18:0, and 20:1ω9t. By comparing these data, one can determine that only 13 of the CFA groups listed have 18:1ω7c in concentrations greater than 47%. Four of these 13 CFA groups are easily eliminated from consideration by their absence of hydroxy acids and each of the remaining nine are eliminated by the absence of 3—OH-18:0. Moreover, the Br-19:1 acid is an unusual acid. It is present at concentrations greater than 6% in only two CFA groups, and one of these can be eliminated by the absence of hydroxy acids. Thus, on the basis of CFA composition, this unknown organism would be identified as an *Achromobacter* group B or group E (Figure 4.7, Table 4.2).

Accurate placement in the correct CFA group requires that both the qualitative and quantitative data from the test organism match with that of the correct CFA group. After some familiarization with the CFA data presented, one can use the procedure outlined in the example to place test organisms in their appropriate CFA groups. If the organism under study does not fit into any CFA group, it is possible that it may represent a new or previously undescribed organism. About 20% of the isolates examined in this laboratory do not fit any of our existing CFA groups. Thus, an important part of our ongoing research is to identify and classify these organisms through detailed biochemical, cultural, chemical, and DNA-relatedness studies. Recent examples are several *Afipia* species, which were first recognized as possible new organism on the basis of their unique CFA profiles.

List of Figures

Fig. 4.1. Saponification and methylation procedure for cellular fatty acid analysis of microorganisms (See Protocol 4.1).
Fig. 4.2. Procedure For trifluoroacetylation of fatty acid methyl esters (FAME) (see Protocol 4.2).
Fig. 4.3. Procedure for hydrogenation of fatty acid methyl esters (FAME) (see Protocol 4.3).
Fig. 4.4. Gas chromatogram of the cellular fatty acids of *Neisseria subflava* analyzed with a fused silica capillary column.
Fig. 4.5. Gas chromatogram of acetylated cellular fatty acids of *Neisseria subflava* analyzed with a fused silica capillary column under identical conditions as in Figure 4.4.
Fig. 4.6. Gas chromatogram of hydrogenated cellular fatty acids of *Neisseria subflava* analyzed with a fused silica capillary column under identical conditions as in Figures 4.4 and 4.5.
Fig. 4.7. Gas chromatogram of the cellular fatty acids of *Achromobacter* groups B and E (see Table 4.2).
Fig. 4.8. Gas chromatogram of the cellular fatty acids of *Acinetobacter* (see Table 4.3).
Fig. 4.9. Gas chromatogram of the cellular fatty acids of *Actinobacillus*, *Haemophilus*, and *Pasteurella* (see Tables 4.4A, 4.4B, and 4.4C).
Fig. 4.10. Gas chromatogram of the cellular fatty acids of *Aeromonas hydrophila* and *Aeromonas salmonicida* (see Table 4.5).
Fig. 4.11. Gas chromatogram of the cellular fatty acids of *Afipia broomeae*, *Afipia clevelandensis*, *Afipia felis*, *Afipia* genomospecies 1, *Afipia* genomospecies 2, and *Afipia* genomospecies 3 (see Table 4.6).
Fig. 4.12. Gas chromatogram of the cellular fatty acids of *Agrobacterium radiobacter* biovar 1 (see Table 4.7).
Fig. 4.13. Gas chromatogram of the cellular fatty acids of *Alcaligenes faecalis*, *Alcaligenes xylosoxidans* subsp. *denitrificans* group 1, *Alcaligenes xylosoxidans* subsp. *xylosoxidans*, and *Bordetella bronchiseptica* (see Table 4.8).
Fig. 4.14. Gas chromatogram of the cellular fatty acids of *Alcaligenes piechaudii*, *Comamonas acidovorans*, *Comamonas testosteroni*, *Pseudomonas alcaligenes*, and *Pseudomonas pseudoalcaligenes* (see Table 4.9).
Fig. 4.15. Gas chromatogram of the cellular fatty acids of *Alcaligenes*-like Group 1, IVc-2, *Pseudomonas pickettii*, and "*Pseudomonas thomasii*" (see Table 4.10).
Fig. 4.16. Gas chromatogram of the cellular fatty acids of *Arcobacter butzleri*, *Arcobacter cryaerophilus*, and *Arcobacter skirrowii* (see Table 4.11).
Fig. 4.17. Gas chromatogram of the cellular fatty acids of *Arcobacter nitrofigilis*, EF-4a and EF-4b, *Eikenella corrodens*, *Suttonella indologenes*, and *Neisseria* spp. (see Tables 4.12A, 4.12B, and 4.12C).
Fig. 4.18. Gas chromatogram of the cellular fatty acids of *Balneatrix alpica* (see Table 4.13).
Fig. 4.19. Gas chromatogram of the cellular fatty acids of *Bartonella bacilliformis* (see Table 4.14).
Fig. 4.20. Gas chromatogram of the cellular fatty acids of *Bartonella elizabethae*, *Bartonella henselae*, *Bartonella quintana*, and *Bartonella vinsonii* (see Table 4.15).
Fig. 4.21. Gas chromatogram of the cellular fatty acids of *Bordetella avium* and *Bordetella holmesii* (see Table 4.16).
Fig. 4.22. Gas chromatogram of the cellular fatty acids of *Bordetella parapertussis* (see Table 4.17).
Fig. 4.23. Gas chromatogram of the cellular fatty acids of *Bordetella pertussis* (see Table 4.18).
Fig. 4.24. Gas chromatogram of the cellular fatty acids of *Brucella abortus*, *Brucella melitensis*, *Brucella ovis*, and *Brucella suis* (see Table 4.19).
Fig. 4.25. Gas chromatogram of the cellular fatty acids of *Brucella canis* (see Table 4.20).
Fig. 4.26. Gas chromatogram of the cellular fatty acids of *Campylobacter coli*, *Campylobacter helveticus*, *Campylobacter jejuni*, *Campylobacter lari*, *Campylobacter upsaliensis*, *Oligella ureolytica*, and *Oligella urethralis* (see Table 4.21).
Fig. 4.27. Gas chromatogram of the cellular fatty acids of *Campylobacter concisus*, *Campylobacter mucosalis*, and *Campylobacter* group DG (see Table 4.22).
Fig. 4.28. Gas chromatogram of the cellular fatty acids of *Campylobacter fetus* subsp. *fetus* and *Campylobacter fetus* subsp. *venerealis* (see Table 4.23).
Fig. 4.29. Gas chromatogram of the cellular fatty acids of *Campylobacter hyointestinalis* (see Table 4.24).
Fig. 4.30. Gas chromatogram of the cellular fatty acids of *Campylobacter sputorum* subsp. *bubulus*, *Campylobacter sputorum* subsp. *faecalis*, and *Campylobacter sputorum* subsp. *sputorum* (see Table 4.25).
Fig. 4.31. Gas chromatogram of the cellular fatty acids of *Capnocytophaga canimorsus*, *Capnocytophaga cynodegmi*, *Capnocytophaga gingivalis*, and *Capnocytophaga ochracea* (see Table 4.26).

Fig. 4.32.   Gas chromatogram of the cellular fatty acids of *Cardiobacterium hominis* (see Table 4.27).
Fig. 4.33.   Gas chromatogram of the cellular fatty acids of DF-3 (see Table 4.28).
Fig. 4.34.   Gas chromatogram of the cellular fatty acids of DF-3-like (see Table 4.29).
Fig. 4.35.   Gas chromatogram of the cellular fatty acids of EO-2 (see Table 4.30).
Fig. 4.36.   Gas chromatogram of the cellular fatty acids of EO-3 (see Table 4.31).
Fig. 4.37.   Gas chromatogram of the cellular fatty acids of NO-1 (see Table 4.32).
Fig. 4.38.   Gas chromatogram of the cellular fatty acids of OFBA-1 (see Table 4.33).
Fig. 4.39.   Gas chromatogram of the cellular fatty acids of *Acidovorax delafieldii*, WO-1, WO-1A, and *Comamonas terrigena* (see Table 4.34).
Fig. 4.40.   Gas chromatogram of the cellular fatty acids of IIg (see Table 4.35).
Fig. 4.41.   Gas chromatogram of the cellular fatty acids of Vb-3, *Pseudomonas mendocina*, *Pseudomonas pertucinogena*, and *Pseudomonas stutzeri* (see Table 4.36).
Fig. 4.42.   Gas chromatogram of the cellular fatty acids of *Chromobacterium violaceum*, *Pseudomonas aeruginosa*, "*Pseudomonas denitrificans*," *Pseudomonas fluorescens*, *Pseudomonas*-like Group 2, and *Pseudomonas putida* (see Table 4.37).
Fig. 4.43.   Gas chromatogram of the cellular fatty acids of *Chryseomonas luteola*, *Flavimonas oryzihabitans*, and *Pseudomonas syringae* (see Table 4.38).
Fig. 4.44.   Gas chromatogram of the cellular fatty acids of *Cytophaga johnsonae* (see Table 4.39).
Fig. 4.45.   Gas chromatogram of the cellular fatty acids of *Escherichia coli*, *Salmonella dublin*, *Salmonella newport*, *Salmonella oranienburg*, *Salmonella typhi*, and *Salmonella zaiman* (see Table 4.40).
Fig. 4.46.   Gas chromatogram of the cellular fatty acids of *Flavobacterium balustinum*, *Flavobacterium gleum*, *Flavobacterium indologenes*, *Flavobacterium meningosepticum*, *Flavobacterium odoratum*, *Flavobacterium* species (IIb), *Weeksella virosa*, and *Weeksella zoohelcum* (see Table 4.41).
Fig. 4.47.   Gas chromatogram of the cellular fatty acids of *Flavobacterium breve* (see Table 4.42).
Fig. 4.48.   Gas chromatogram of the cellular fatty acids of IIc, IIe, and IIh (see Table 4.43).
Fig. 4.49.   Gas chromatogram of the cellular fatty acids of IIi, *Flavobacterium (Sphingobacterium) mizutaii*, *Flavobacterium (Sphingobacterium) multivorum*, *Flavobacterium (Sphingobacterium) spiritivorum*, and *Flavobacterium (Sphingobacterium) thalpophilum* (see Table 4.44).
Fig. 4.50.   Gas chromatogram of the cellular fatty acids of *Francisella philomiragia*, *Francisella tularensis* biogroup novicida, and *Francisella tularensis* biogroup tularensis (see Table 4.45).
Fig. 4.51.   Gas chromatogram of the cellular fatty acids of Gilardi rod group 1 (see Table 4.46).
Fig. 4.52.   Gas chromatogram of the cellular fatty acids of *Helicobacter cinaedi* and *Campylobacter* group CLO1B (see Table 4.47).
Fig. 4.53.   Gas chromatogram of the cellular fatty acids of *Helicobacter fennelliae* and *Campylobacter* CLO3 (see Table 4.48).
Fig. 4.54.   Gas chromatogram of the cellular fatty acids of *Helicobacter mustelae* (see Table 4.49).
Fig. 4.55.   Gas chromatogram of the cellular fatty acids of *Helicobacter pylori* (see Table 4.50).
Fig. 4.56.   Gas chromatogram of the cellular fatty acids of *Hydrogenophaga palleronii* (see Table 4.51).
Fig. 4.57.   Gas chromatogram of the cellular fatty acids of *Hydrogenophaga pseudoflava* (see Table 4.52).
Fig. 4.58.   Gas chromatogram of the cellular fatty acids of *Kingella denitrificans*, *Kingella kingae*, and *Simonsiella crassa* (see Table 4.53).
Fig. 4.59.   Gas chromatogram of the cellular fatty acids of *Methylobacterium extorquens*, *Methylobacterium mesophilicum*, *Methylobacterium zatmanii*, *Roseomonas fauriae*, and *Roseomonas* genomospecies 6 (see Table 4.54).
Fig. 4.60.   Gas chromatogram of the cellular fatty acids of *Moraxella atlantae* (see Table 4.55).
Fig. 4.61.   Gas chromatogram of the cellular fatty acids of *Moraxella bovis*, *Moraxella canis*, *Moraxella cuniculi*, *Moraxella lacunata* I, *Moraxella lacunata* II, *Moraxella lincolnii*, *Moraxella nonliquefaciens*, and *Moraxella ovis* (see Table 4.56).
Fig. 4.62.   Gas chromatogram of the cellular fatty acids of *Moraxella (Branhamella) catarrhalis* (see Table 4.57).
Fig. 4.63.   Gas chromatogram of the cellular fatty acids of *Moraxella osloensis* (see Table 4.58).
Fig. 4.64.   Gas chromatogram of the cellular fatty acids of *Moraxella phenylpyruvica* (see Table 4.59).
Fig. 4.65.   Gas chromatogram of the cellular fatty acids of *Ochrobactrum anthropi*, *Roseomonas cervicalis*, *Roseomonas gilardii*, and *Roseomonas* genomospecies 4 and 5 (see Table 4.60).
Fig. 4.66.   Gas chromatogram of the cellular fatty acids of "*Pseudomonas azotocolligans*", *Sphingomonas capsulata*, *Sphingomonas* genomospecies 2, *Sphingomonas paucimobilis*, and *Sphingomonas yanoikuyae* (see Table 4.61).

Fig. 4.67. Gas chromatogram of the cellular fatty acids of *Pseudomonas cepacia*, *Pseudomonas gladioli*, *Pseudomonas mallei*, and *Pseudomonas pseudomallei* (see Table 4.62).

Fig. 4.68. Gas chromatogram of the cellular fatty acids of *Pseudomonas diminuta* and *Pseudomonas vesicularis* (see Table 4.63).

Fig. 4.69. Gas chromatogram of the cellular fatty acids of *Pseudomonas* sp. CDC Group 1 (see Table 4.64).

Fig. 4.70. Gas chromatogram of the cellular fatty acids of *Psychrobacter immobilis* (see Table 4.65).

Fig. 4.71. Gas chromatogram of the cellular fatty acids of *Riemerella anatipestifer* (see Table 4.66).

Fig. 4.72. Gas chromatogram of the cellular fatty acids of *Shewanella putrefaciens* biotypes 1 and 2 (see Table 4.67).

Fig. 4.73. Gas chromatogram of the cellular fatty acids of *Sphingomonas adhaesiva* and *Sphingomonas parapaucimobilis* (see Table 4.68).

Fig. 4.74. Gas chromatogram of the cellular fatty acids of *Sporocytophaga myxococcoides* (see Table 4.69).

Fig. 4.75. Gas chromatogram of the cellular fatty acids of *Tatumella ptyseos* (see Table 4.70).

Fig. 4.76. Gas chromatogram of the cellular fatty acids of *Vibrio cholerae*, *Vibrio fluvialis*, and *Vibrio parahaemolyticus* (see Table 4.71).

Fig. 4.77. Gas chromatogram of the cellular fatty acids of *Vibrio furnissii* (see Table 4.72).

Fig. 4.78. Gas chromatogram of the cellular fatty acids of *Wolinella curva* and *Wolinella succinogenes* (see Table 4.73).

Fig. 4.79. Gas chromatogram of the cellular fatty acids of *Wolinella recta* (see Table 4.74).

Fig. 4.80. Gas chromatogram of the cellular fatty acids of *Xanthomonas beticola*, *Xanthomonas campestris*, *Xanthomonas maltophilia*, *Xanthomonas nakataecorchori*, and *Xanthomonas phaseoli* (see Table 4.75).

Fig. 4.81. Gas chromatogram of the cellular fatty acids of *Xenorhabdus luminescens* (see Table 4.76).

Fig. 4.82. Gas chromatogram of the cellular fatty acids of *Yersinia enterocolitica* (see Table 4.77).

List of Tables

Table 4.1. Names of some fatty acids found in bacteria.
Table 4.2. Cellular fatty acid composition of *Achromobacter* groups B and E (see Fig. 4.7).
Table 4.3. Cellular fatty acid composition of *Acinetobacter* (see Fig. 4.8).
Table 4.4A. Cellular fatty acid composition of *Actinobacillus* species (see Fig. 4.9).
Table 4.4B. Cellular fatty acid composition of *Haemophilus* species (see Fig. 4.9).
Table 4.4C. Cellular fatty acid composition of *Pasteurella* species (see Fig. 4.9).
Table 4.5. Cellular fatty acid composition of *Aeromonas hydrophila* and *Aeromonas salmonicida* (see Fig. 4.10).
Table 4.6. Cellular fatty acid composition of *Afipia* (see Fig. 4.11).
Table 4.7. Cellular fatty acid composition of *Agrobacterium radiobacter* biovar 1 (see Fig. 4.12).
Table 4.8. Cellular fatty acid composition of *Alcaligenes faecalis, Alcaligenes xylosoxidans* subsp. *denitrificans* 1, *Alcaligenes xylosoxidans* subsp. *xylosoxidans,* and *Bordetella bronchiseptica* (see Fig. 4.13).
Table 4.9. Cellular fatty acid composition of *Alcaligenes piechaudii, Comamonas acidovorans, Comamonas testosteroni, Pseudomonas alcaligenes,* and *Pseudomonas pseudoalcaligenes* (see Fig. 4.14).
Table 4.10. Cellular fatty acid composition of *Alcaligenes*-like Group 1, IVc-2, *Pseudomonas pickettii,* and *"Pseudomonas thomasii"* (see Fig. 4.15).
Table 4.11. Cellular fatty acid composition of *Arcobacter butzleri, Arcobacter cryaerophilus,* and *Arcobacter skirrowii* (see Fig. 4.16).
Table 4.12A. Cellular fatty acid composition of *Arcobacter nitrofigilis,* EF-4a, EF-4b, *Eikenella corrodens,* and *Suttonella indologenes* (see Fig. 4.17).
Table 4.12B. Cellular fatty acid composition of *Neisseria canis, Neisseria cinerea, Neisseria elongata* subsp. *elongata* and *glycolytica, Neisseria elongata* subsp. *nitroducens, Neisseria flava,* and *Neisseria gonorrhoeae* (see Fig. 4.17).
Table 4.12C. Cellular fatty acid composition of *Neisseria meningitidis, Neisseria mucosa, Neisseria polysaccharea, Neisseria sicca, Neisseria subflava,* and *Neisseria weaveri* (see Fig. 4.17).
Table 4.13. Cellular fatty acid composition of *Balneatrix alpica* (see Fig. 4.18).
Table 4.14. Cellular fatty acid composition of *Bartonella bacilliformis* (see Fig. 4.19).
Table 4.15. Cellular fatty acid composition of *Bartonella elizabethae, Bartonella henselae, Bartonella quintana,* and *Bartonella vinsonii* (see Fig. 4.20).
Table 4.16. Cellular fatty acid composition of *Bordetella avium* and *Bordetella holmesii* (see Fig. 4.21).
Table 4.17. Cellular fatty acid composition of *Bordetella parapertussis* (see Fig. 4.22).
Table 4.18. Cellular fatty acid composition of *Bordetella pertussis* (see Fig. 4.23).
Table 4.19. Cellular fatty acid composition of *Brucella abortus, Brucella melitensis, Brucella ovis,* and *Brucella suis* (see Fig. 4.24).
Table 4.20. Cellular fatty acid composition of *Brucella canis* (see Fig. 4.25).
Table 4.21. Cellular fatty acid composition of *Campylobacter coli, Campylobacter helveticus, Campylobacter jejuni, Campylobacter lari, Campylobacter upsaliensis, Oligella ureolytica,* and *Oligella urethralis* (see Fig. 4.26).
Table 4.22. Cellular fatty acid composition of *Campylobacter concisus, Campylobacter mucosalis,* and *Campylobacter* group DG (see Fig. 4.27).
Table 4.23. Cellular fatty acid composition of *Campylobacter fetus* subsp. *fetus* and *Campylobacter fetus* subsp. *venerealis* (see Fig. 4.28).
Table 4.24. Cellular fatty acid composition of *Campylobacter hyointestinalis* (see Fig. 4.29).
Table 4.25. Cellular fatty acid composition of *Campylobacter sputorum* subsp. *bubulus, Campylobacter sputorum* subsp. *faecalis,* and *Campylobacter sputorum* subsp. *sputorum* (see Fig. 4.30).
Table 4.26. Cellular fatty acid composition of *Capnocytophaga canimorsus, Capnocytophaga cynodegmi, Capnocytophaga gingivalis,* and *Capnocytophaga ochracea* (see Fig. 4.31).
Table 4.27. Cellular fatty acid composition of *Cardiobacterium hominis* (see Fig. 4.32).
Table 4.28. Cellular fatty acid composition of DF-3 (see Fig. 4.33).
Table 4.29. Cellular fatty acid composition of DF-3-like (see Fig. 4.34).
Table 4.30. Cellular fatty acid composition of EO-2 (see Fig. 4.35).
Table 4.31. Cellular fatty acid composition of EO-3 (see Fig. 4.36).
Table 4.32. Cellular fatty acid composition of NO-1 (see Fig. 4.37).

Table 4.33.	Cellular fatty acid composition of OFBA-1 (see Fig. 4.38).
Table 4.34.	Cellular fatty acid composition of *Acidovorax delafieldii*, WO-1, WO-1A, and *Comamonas terrigena* (see Fig. 4.39).
Table 4.35.	Cellular fatty acid composition of IIg (see Fig. 4.40).
Table 4.36.	Cellular fatty acid composition of Vb-3, *Pseudomonas mendocina, Pseudomonas pertucinogena,* and *Pseudomonas stutzeri* (see Fig. 4.41).
Table 4.37.	Cellular fatty acid composition of *Chromobacterium violaceum, Pseudomonas aeruginosa,* "*Pseudomonas denitrificans,*" *Pseudomonas fluorescens, Pseudomonas*-like Group 2, and *Pseudomonas putida* A and *Pseudomonas putida* 1 (see Fig. 4.42).
Table 4.38.	Cellular fatty acid composition of *Chryseomonas luteola, Flavimonas oryzihabitans,* and *Pseudomonas syringae* (see Fig. 4.43).
Table 4.39.	Cellular fatty acid composition of *Cytophaga johnsonae* (see Fig. 4.44).
Table 4.40.	Cellular fatty acid composition of *Escherichia coli, Salmonella dublin, Salmonella newport, Salmonella oranienburg, Salmonella typhi,* and *Salmonella zaiman* (see Fig. 4.45).
Table 4.41.	Cellular fatty acid composition of *Flavobacterium balustinum, Flavobacterium gleum, Flavobacterium indologenes, Flavobacterium meningosepticum, Flavobacterium odoratum, Flavobacterium* species (IIb); *Weeksella virosa,* and *Weeksella zoohelcum* (see Fig. 4.46).
Table 4.42.	Cellular fatty acid composition of *Flavobacterium breve* (see Fig. 4.47).
Table 4.43.	Cellular fatty acid composition of IIc, IIe, and IIh (see Fig. 4.48).
Table 4.44.	Cellular fatty acid composition of IIi, *Flavobacterium (Sphingobacterium) mizutaii, Flavobacterium (Sphingobacterium) multivorum, Flavobacterium (Sphingobacterium) spiritivorum,* and *Flavobacterium (Sphingobacterium) thalpophilum* (see Fig. 4.49).
Table 4.45.	Cellular fatty acid composition of *Francisella philomiragia, Francisella tularensis* biogroup novicida, and *Francisella tularensis* biogroup tularensis (see Fig. 4.50).
Table 4.46.	Cellular fatty acid composition of Gilardi rod group 1 (see Fig. 4.51).
Table 4.47.	Cellular fatty acid composition of *Helicobacter cinaedi* and *Campylobacter* CLO1B (see Fig. 4.52).
Table 4.48.	Cellular fatty acid composition of *Helicobacter fennelliae* and *Campylobacter* CLO3 (see Fig. 4.53).
Table 4.49.	Cellular fatty acid composition of *Helicobacter mustelae* (see Fig.4. 54).
Table 4.50.	Cellular fatty acid composition of *Helicobacter pylori* (see Fig. 4.55).
Table 4.51.	Cellular fatty acid composition of *Hydrogenophaga palleronii* (see Fig. 4.56).
Table 4.52.	Cellular fatty acid composition of *Hydrogenophaga pseudoflava* (see Fig. 4.57).
Table 4.53.	Cellular fatty acid composition of *Kingella denitrificans, Kingella kingae,* and *Simonsiella crassa* (see Fig. 4.58).
Table 4.54.	Cellular fatty acid composition of *Methylobacterium extorquens, Methylobacterium mesophilicum, Methylobacterium zatmanii, Roseomonas fauriae,* and *Roseomonas* genomospecies 6 (see Fig. 4.59).
Table 4.55.	Cellular fatty acid composition of *Moraxella atlantae* (see Fig. 4.60).
Table 4.56.	Cellular fatty acid composition of *Moraxella bovis, Moraxella canis, Moraxella cuniculi, Moraxella lacunata* I, *Moraxella lacunata* II, *Moraxella lincolnii, Moraxella nonliquefaciens,* and *Moraxella ovis* (see Fig. 4.61).
Table 4.57.	Cellular fatty acid composition of *Moraxella (Branhamella) catarrhalis* (see Fig. 4.62).
Table 4.58.	Cellular fatty acid composition of *Moraxella osloensis* (see Fig. 4.63).
Table 4.59.	Cellular fatty acid composition of *Moraxella phenylpyruvica* (see Fig. 4.64).
Table 4.60.	Cellular fatty acid composition of *Ochrobactrum anthropi, Roseomonas cervicalis, Roseomonas gilardii,* and *Roseomonas* genomospecies 4 and 5 (see Fig. 4.65).
Table 4.61.	Cellular fatty acid composition of "*Pseudomonas azotocolligans,*" *Sphingomonas capsulata, Sphingomonas* genospecies 2, *Sphingomonas paucimobilis,* and *Sphingomonas yanoikuyae* (see Fig. 4.66).
Table 4.62.	Cellular fatty acid composition of *Pseudomonas cepacia, Pseudomonas gladioli, Pseudomonas mallei,* and *Pseudomonas pseudomallei* (see Fig. 4.67).
Table 4.63.	Cellular fatty acid composition of *Pseudomonas diminuta* and *Pseudomonas vesicularis* (see Fig. 4.68).
Table 4.64.	Cellular fatty acid composition of *Pseudomonas* sp. CDC Group 1 (see Fig. 4.69).
Table 4.65.	Cellular fatty acid composition of *Psychrobacter immobilis* (see Fig. 4.70).
Table 4.66.	Cellular fatty acid composition of *Riemerella anatipestifer* (see Fig. 4.71).
Table 4.67.	Cellular fatty acid composition of *Shewanella putrefaciens* biotypes 1 and 2 (see Fig. 4.72).
Table 4.68.	Cellular fatty acid composition of *Sphingomonas adhaesiva* and *Sphingomonas parapaucimobilis* (see Fig. 4.73).
Table 4.69.	Cellular fatty acid composition of *Sporocytophaga myxococcoides* (see Fig. 4.74).

Table 4.70. Cellular fatty acid composition of *Tatumella ptyseos* (see Fig. 4.75).
Table 4.71. Cellular fatty acid composition of *Vibrio cholerae, Vibrio fluvialis,* and *Vibrio parahaemolyticus* (see Fig. 4.76).
Table 4.72. Cellular fatty acid composition of *Vibrio furnissii* (see Fig. 4.77).
Table 4.73. Cellular fatty acid composition of *Wolinella curva* and *Wolinella succinogenes* (see Fig. 4.78).
Table 4.74. Cellular fatty acid composition of *Wolinella recta* (see Fig. 4.79).
Table 4.75. Cellular fatty acid composition of *Xanthomonas beticola, Xanthomonas campestris, Xanthomonas maltophilia, Xanthomonas nakataecorchori,* and *Xanthomonas phaseoli* (see Fig. 4.80).
Table 4.76. Cellular fatty acid composition of *Xenorhabdus luminescens* (see Fig. 4.81).
Table 4.77. Cellular fatty acid composition of *Yersinia enterocolitica* (see Fig. 4.82).

Protocols

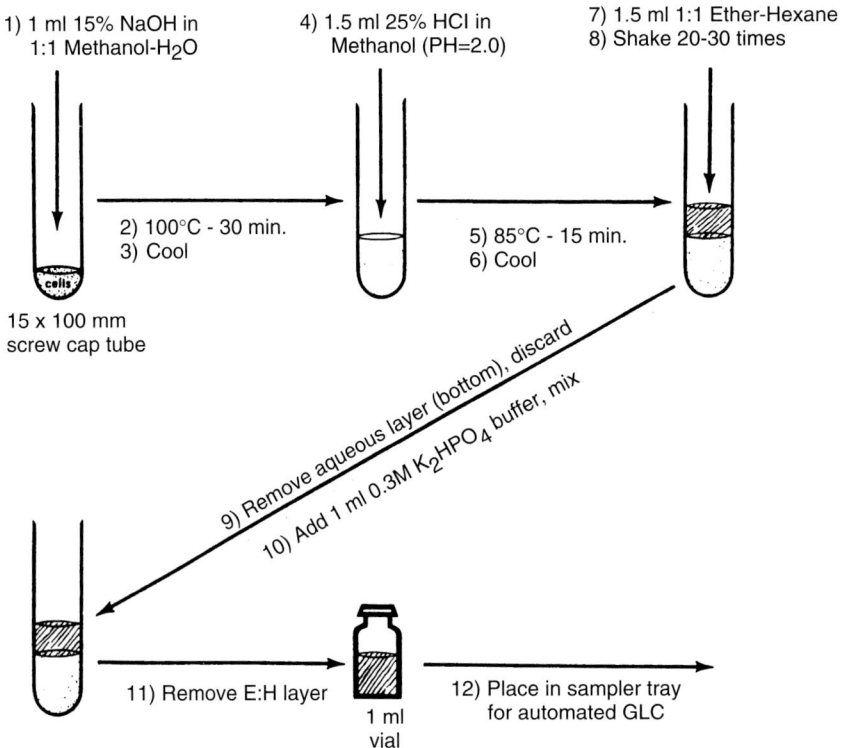

Figure 4.1. Saponification and methylation procedure for cellular fatty acid analysis of microorganisms.

Protocol 4.1   Saponification and methylation procedure for cellular fatty acid analysis of microorganisms.

After an appropriate incubation period to obtain good growth, cells are removed from an agar plate or slant using 0.5–1 ml of distilled water and gently scraped and transferred to a clean (hexane rinsed) 13- × 100-mm screw-capped tube. Cultures grown in liquid medium are centrifuged and washed with distilled water before proceeding with the saponification procedure.

1. Place 4–8 drops of the turbid bacterial cell suspension into a clean (hexane rinsed) 13- × 100-mm screw-capped test tube.

2. Add 1 ml of saponification reagent (15% NaOH in 50% aqueous methanol), seal the tube with a Teflon-lined cap, and mix the contents well.

3. Heat the mixture for 30 min in a boiling water bath. Occasionally, hairline cracks occur in test tubes that may cause the contents of the tube to boil away. Observe the sample for 2–3 min after heating begins, and if bubbles are filling the tube, remove it immediately and cool; then transfer the sample to another tube and continue heating.

4. After heating 30 min, cool the sample to room temperature, add 1.5 ml of methylation reagent (25% HCl in methanol), and mix well. Check the pH of the mixture with pH paper. If the pH is not 2.0–3.0, add additional methylation reagent and recheck pH.

5. Seal the tube with the Teflon-lined cap, heat in a 85° C water bath for 15 min, and then cool to room temperature.

6. Add 1.5 ml of a 1:1 mixture of diethylether:hexane, seal the tube with the Teflon-lined cap, and mix contents well by shaking vigorously (end to end) 20 to 30 times.

7. Allow layers to separate and remove the aqueous (bottom) layer to discard, using a disposable Pasteur pipette. (If an emulsion forms at step 6, add 2–3 drops of methanol and allow to stand 3–5 min to allow the two layers to separate clearly).

8. Add 1 ml of 0.3 M phosphate buffer (pH 11.0) to the organic layer (ether:hexane layer) remaining in the tube. Mix well by shaking 10–20 times and allow to stand 3–5 min. If the layers do not clearly separate, add 2–3 drops of methanol or centrifuge 3–5 min.

9. Using a disposable Pasteur pipette, remove most (i.e., two-thirds) of the organic layer (ether:hexane, top layer) that contains the FAME to a clean (hexane rinsed) test tube or sample vial and seal with a Teflon-lined cap or septum. ***Do not transfer any of the bottom layer (buffer) during this step.*** Store the capped FAME sample at −20° C if gas-liquid chromatography (GLC) is not initiated within 3–4 h.

10. Analyze by GLC.

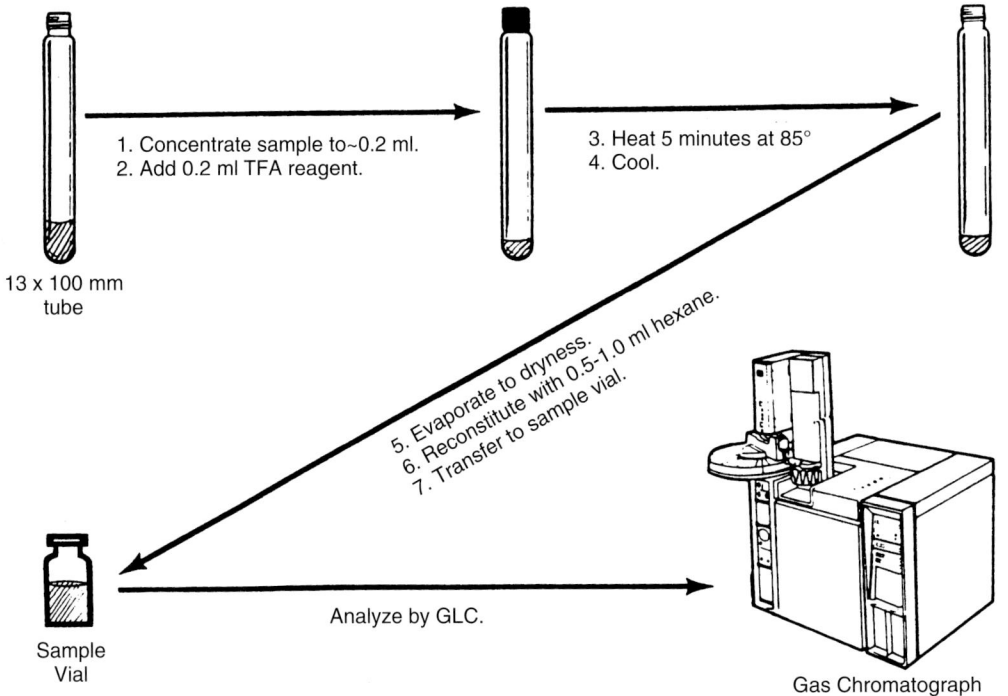

Figure 4.2. Procedure for trifluoroacetylation of fatty acid methyl esters (FAME).

Protocol 4.2 Trifluoroacetylation of fatty acid methyl esters (FAME)

1. Dry the FAME by adding 0.3–0.5 g of anhydrous Na_2SO_4 to the FAME sample dissolved in 0.5 ml of hexane and mix.

2. Centrifuge for 3–5 min at $1000 \times g$ and remove hexane to a clean 13- × 100-mm screw-capped test tube.

3. Rinse Na_2SO_4 with 0.5 ml of hexane, mix, centrifuge, and combine with first hexane aliquot. Concentrate to about 0.2 ml under nitrogen.

4. Add 0.2 ml of trifluoroacetic anhydride (TFAA) at room temperature, seal with a Teflon-lined screwcap and mix gently.

    NOTE: If TFAA is refrigerated, allow it to warm to room temperature before opening.

5. Heat the sample for 5 min at 85°C and then cool to room temperature.

6. Evaporate contents of the tube with nitrogen **just to dryness;** then reconstitute to 0.5–1 ml with hexane. Transfer to sample vial.

7. Analyze an aliquot of the acetylated FAME sample by GLC and compare the resulting chromatogram with that of the unacetylated FAME run under identical GLC conditions.

$$\underset{\text{Hydroxy acid ME}}{R\text{—}\underset{\underset{\text{OH}}{|}}{CH}\text{—}COOCH_3} + \underset{\text{TFAA}}{CF_3\text{—}\underset{\underset{O}{\|}}{C}\text{—}\underset{\underset{O}{\|}}{C}\text{—}CF_3} \rightarrow \underset{\text{Acetylated ME}}{R\text{—}\underset{\underset{O\text{-}C\text{-}CF_3}{|}}{CH}\text{—}COOCH_3} + \underset{\text{TFA Acid}}{CF_3COOH}$$

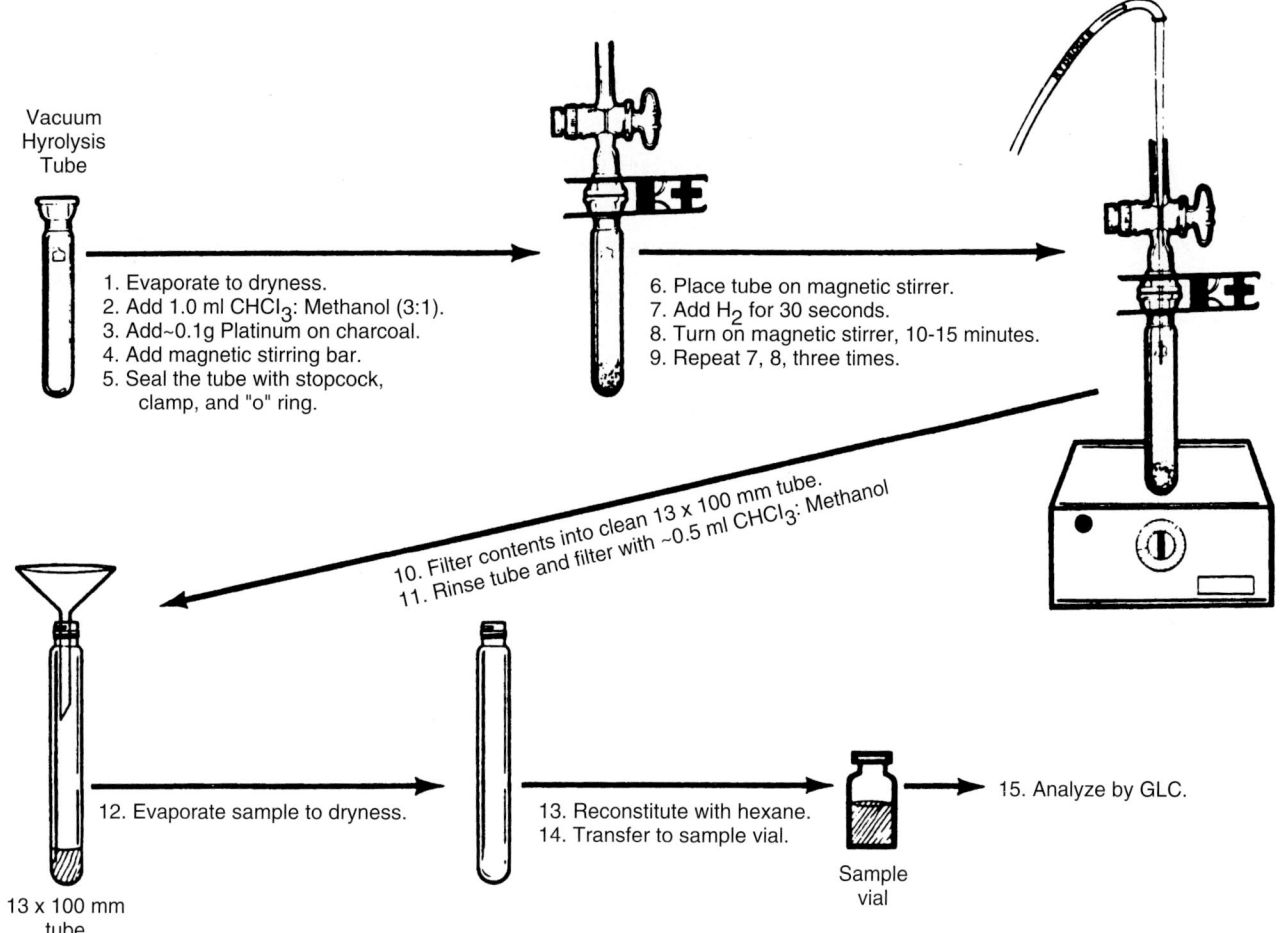

Figure 4.3. Procedure for hydrogenation of fatty acid methyl esters (FAME).

Protocol 4.3. Hydrogenation of fatty acid methyl esters (FAME).

1. Dry the FAME sample with Na_2SO_4 as described in steps 1–3 of Protocol 4.2. Transfer 0.2–0.3 ml of the FAME (in hexane) to a 10-ml vacuum hydrolysis tube and evaporate to dryness with nitrogen.

2. Add 1 ml of a 3:1 mixture of chloroform:methanol, 0.05–0.1 g of catalyst (5% platinum on powdered charcoal) and a small magnetic stirrer.

   NOTE: Chloroform may be carcinogenic.

3. Connect to the top part of vacuum tube with "O" ring and clamp and tighten.

4. Be sure that the opening in the tube is clear by gently inserting the tip of a 9-inch capillary Pasteur pipette into the opening and gently flush tube with nitrogen.

5. Gently and slowly add a small flow of hydrogen gas to the sample through the pipette for 0.5 min. Close stopcock while hydrogen is flowing but be sure the tip of the pipette is clear of stopcock hole before closing.

   *NOTE: All work with hydrogen gas should be done **with extreme care with very low flow volumes** to avoid small **explosions.***

6. Turn off hydrogen gas, turn on magnetic stirrer for 10–15 min and then turn stirrer off.

7. Carefully open stopcock and listen for release of pressure indicating that the tube was filled with hydrogen.

8. Repeat steps 5, 6, and 7 three more times.

9. After the last addition of hydrogen, open tube and filter contents through a chloroform-saturated glass fiber filter paper into a clean 13 × 100 mm tube.
10. Rinse reaction tube and filter paper with 0.5 ml of 3:1 chloroform:methanol and combine with solvent in the tube.
11. Evaporate hydrogenated FAME sample **just to dryness** with nitrogen gas and then reconstitute to 0.2–0.3 ml with hexane. Transfer to sample vial.
12. Analyze by GLC and compare the resulting chromatogram with that of the unhydrogenated FAME run under identical GLC conditions.

$$R-CH=CH-COOCH_3 + H_2 \xrightarrow{Pt} R-CH_2-CH_2-COOCH_3$$

Unsaturated FAME　　　　　　　　　Saturated FAME

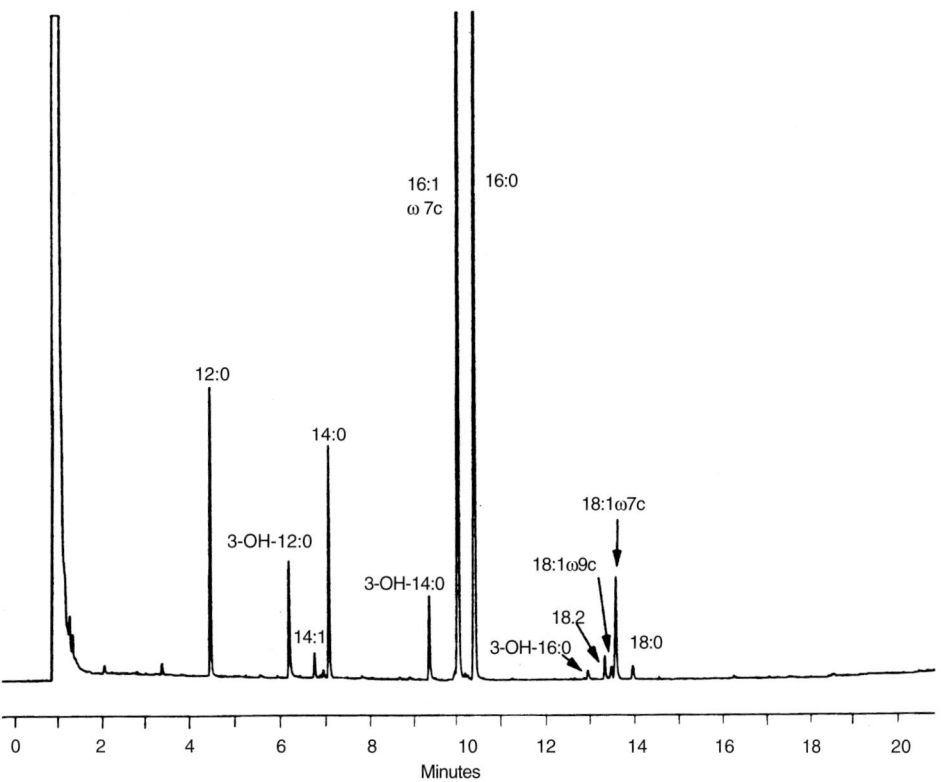

Figure 4.4. Gas chromatogram of the cellular fatty acids of *Neisseria subflava* analyzed with a fused silica capillary column.

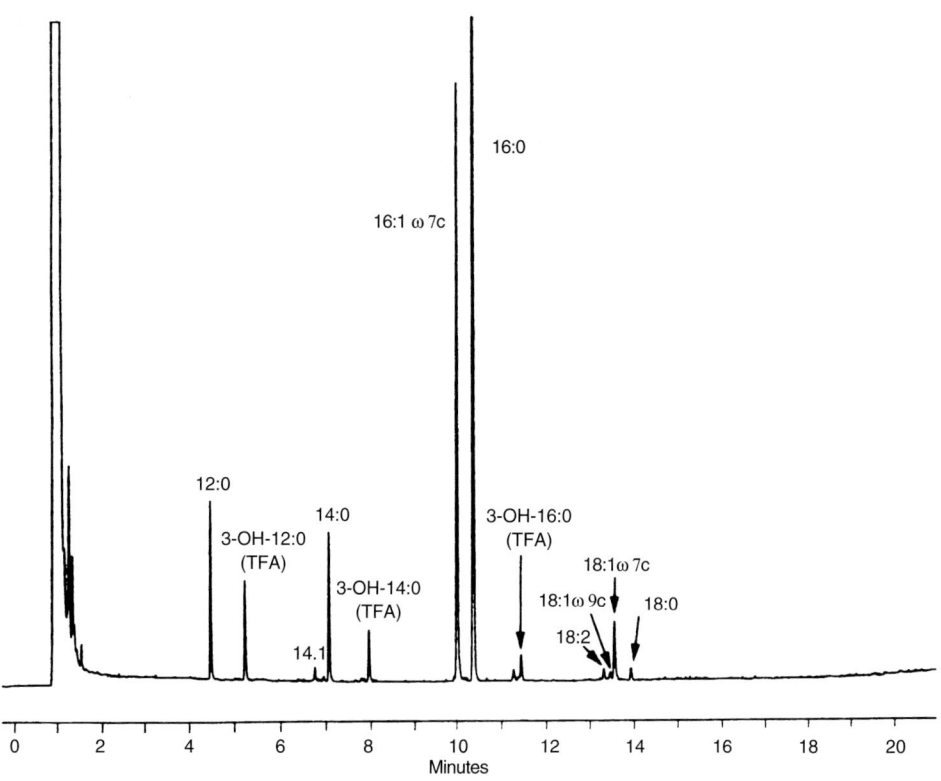

Figure 4.5. Gas chromatogram of acetylated cellular fatty acids of *Neisseria subflava* analyzed with a fused silica capillary column under identical conditions as in Figure 4.4.

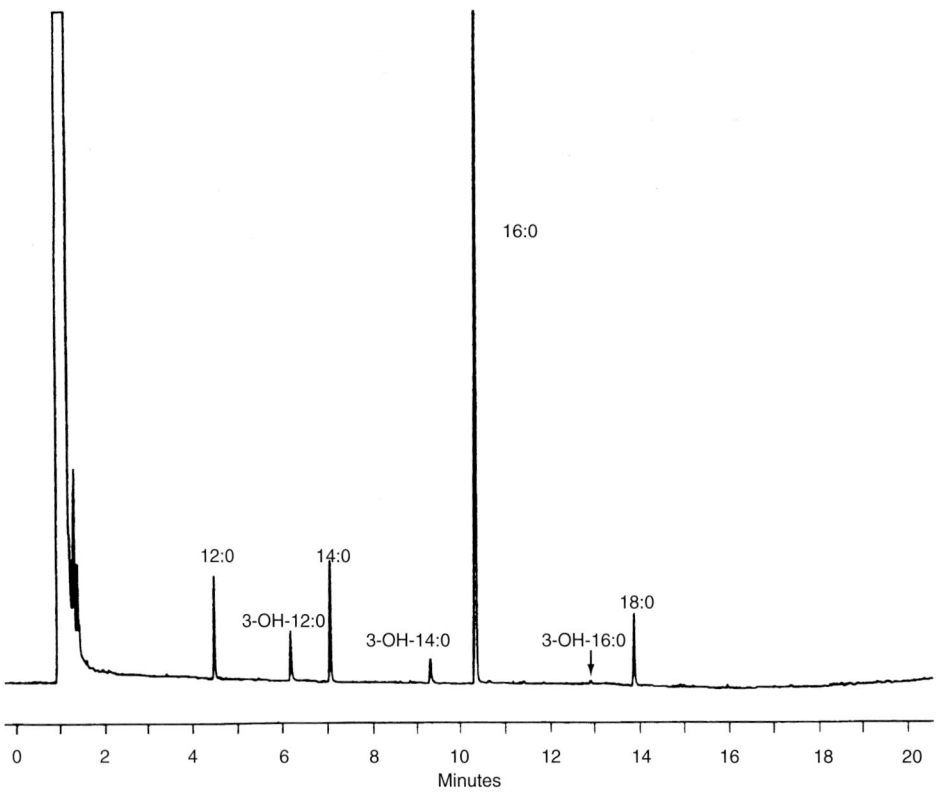

Figure 4.6. Gas chromatogram of hydrogenated cellular fatty acids of *Neisseria subflava* analyzed with a fused silica capillary column under identical conditions as in Figures 4.4 and 4.5.

Table 4.1. Names of some fatty acids found in bacteria

Types of Fatty Acid	Shorthand Formula	Systematic Name	Common Name
Saturated	10:0[a]	Decanoic	Capric
	11:0	Undecanoic	Undecylic
	12:0	Dodecanoic	Lauric
	13:0	Tridecanoic	Tridecylic
	14:0	Tetradecanoic	Myristic
	15:0	Pentadecanoic	Pentadecylic
	16:0	Hexadecanoic	Palmitic
	17:0	Heptadecanoic	Margaric
	18:0	Octadecanoic	Stearic
	19:0	Nonadecanoic	Nondecylic
	20:0	Eicosanoic	Arachidic
Unsaturated	14:1ω5c[b]	cis-9-Tetradecenoic	Myristoleic
	16:1ω7c	cis-9-Hexadecenoic	Palmitoleic
	18:1ω9c	cis-9-Octadecenoic	Oleic
	18:1ω9t	trans-9-Octadecenoic	Elaidic
	18:1ω7c	cis-11-Octadecenoic	cis-Vaccenic
Polyunsaturated	18:2	cis-9,12-Octadecadienoic	Linoleic
	20:4	5,8,11,14-Eicosatetrienoic	Arachidonic
Branched chain	i-15:0[c]	13-Methyltetradecanoic	Isopentadecanoic
	a-15:0[d]	12-Methyltetradecanoic	Anteisopentadecanoic
	i-16:0	14-Methylpentadecanoic	Isopalmitic
	a-16:0	13-Methylpentadecanoic	Anteisopalmitic
	10-CH$_3$-18:0	10-Methyloctadecanoic	Tuberculostearic (TBSA)
	Br-19:1	11-Methyloctadeca-12-enoic	
Branched chain unsaturated	i-17:1	15-Methylhexadecenoic	Isoheptadecenoic
	a-17:1	14-Methylheptadecenoic	Anteisoheptadecenoic
Hydroxy	2-OH-12:0[e]	2-Hydroxydodecanoic	α-Hydroxylauric
	3-OH-12:0	3-Hydroxydodecanoic	ß-Hydroxylauric
	2-OH-14:0	2-Hydroxytetradecanoic	α-Hydroxymyristic
	3-OH-14:0	3-Hydroxytetradecanoic	ß-Hydroxymyristic
Branched chain hydroxy	i-2-OH-15:0	2-Hydroxy-13-methyltetradecanoic	α-Hydroxyisopentadecanoic
	i-3-OH-17:0	3-Hydroxy-15-methylhexadecanoic	ß-Hydroxyisoheptadecanoic
Cyclopropane	17:0cyc[f]	Δ-cis-9,10-Methylenehexadecanoic	
	19:0cyc	Δ-cis-11,12-Methyleneoctadecanoic	Lactobacillic

[a]The number before the colon indicates the number of carbons, and the number after the colon is the number of double bonds.
[b]Indicates the position of the double bond counting from the hydrocarbon end of the carbon chain; c = cis isomer; t = trans isomer.
[c]i- indicates a methyl branch at the iso position.
[d]a- indicates a methyl branch at the anteiso position.
[e]OH- indicates a hydroxy group at the 2(α)- or 3(β)- position from the carboxyl end.
[f]cyc- indicates a cyclopropane ring structure; Δ- indicates position of cyc ring.

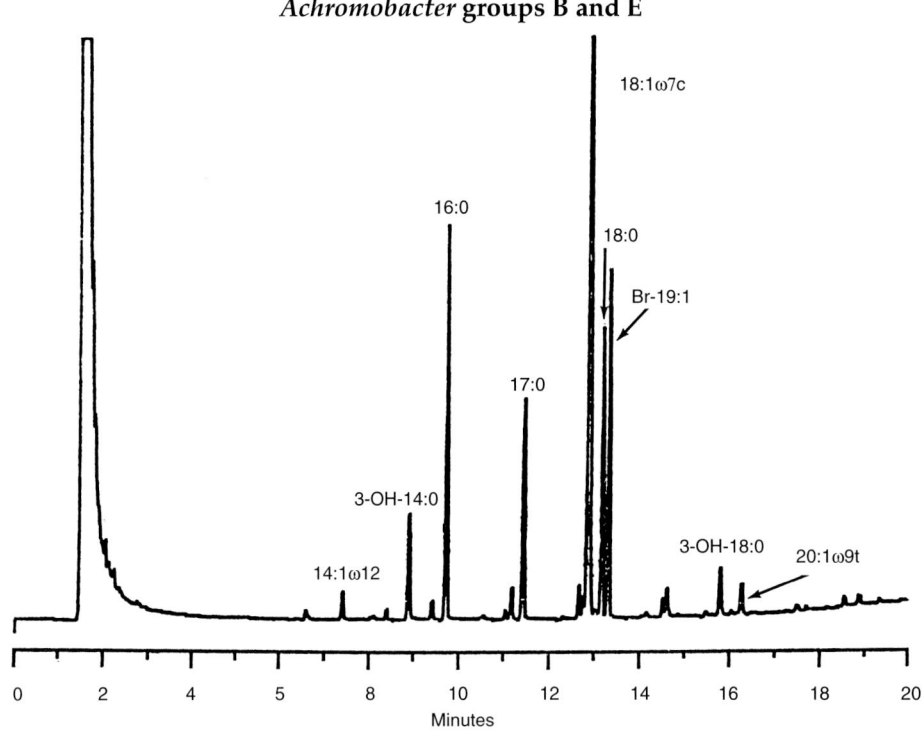

Figure 4.7. Gas chromatogram of the cellular fatty acids of *Achromobacter* groups B and E.

Comments

Achromobacter groups B and E have a unique CFA profile characterized by large amounts (>60%) of 18:1ω7c, 6% Br-19:1, 2% of 3—OH-14:0 and 1% of 3—OH-18:0. Groups B and E cannot be distinguished from each other by CFA composition. The CFA profile of groups B and E is most similar to group EO-2 except for the presence of Br-19:1 and the absence of 3—OH-10:0 and 12:1ω7.

Literature

1. Holmes, B., C.W. Moss, and M.I. Daneshvar. 1993. Cellular fatty acid composition of *Achromobacter* groups B and E. J. Clin. Microbiol. *31:* 1007–1008.

Table 4.2. Cellular fatty acid composition of *Achromobacter* groups B and E.

Fatty Acid	*Achromobacter* group B	*Achromobacter* group E
	%	%
14:1ω12	T[a]	–
3-OH-14:0	2	5
16:1ω7c	T	T
16:0	7	7
17:1ω6c	T	1
17:0	3	5
18:2	1	1
18:1ω9c	T	T
18:1ω7c	70	62
18:0	5	6
Br-19:1	4	7
19:0cyc$^{9-10}$	1	1
3-OH-18:0	1	1
20:1ω9t	1	1

[a]T = Trace.

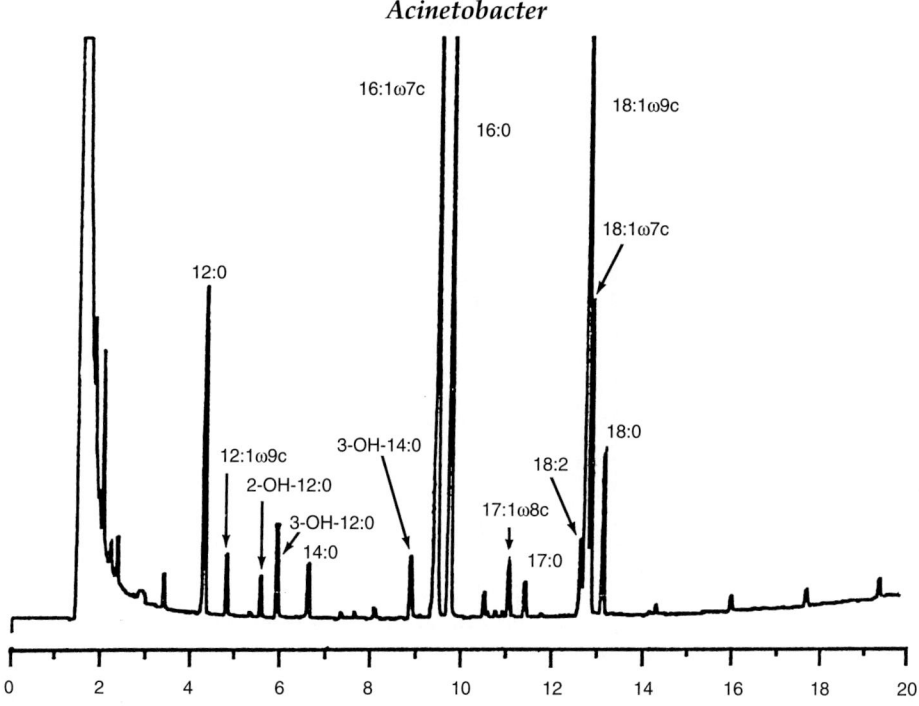

Figure 4.8. Gas chromatogram of the cellular fatty acids of *Acinetobacter*.

Comments

Major acids are 18:1ω9c, 16:1ω7c, and 16:0 with small amounts of 2—OH-12:0, 3—OH-12:0, 12:1ω9, 17:1ω8c, and 17:0. Cannot distinguish *Acinetobacter* species by CFA composition. CFA profile most similar to *Neisseria* except for the presence of 12:1ω9, 2—OH-12:0, 17:1ω8c, 17:0, and more 18:1ω9c (>25% vs. <5% in *Neisseria*).

Literature

1. Jantzen, E., K. Bryn, T. Bergan, and K. Bøvre. 1975. Gas chromatography of bacterial whole cell methanolysate. VII. Fatty acid composition of *Acinetobacter* in relation to the taxonomy of *Neisseriaceae*. Acta Pathol. Microbiol. Scand. Sect. B *83*: 569–580.

2. Moss, C.W., P.L. Wallace, D.G. Hollis, and R.E. Weaver. 1988. Cultural and chemical characterization of CDC groups EO-2, M-5, and M-6, *Moraxella (Moraxella)* species, *Oligella urethralis*, *Acinetobacter* species, and P*sychrobacter immobilis*. J. Clin. Microbiol. *26*: 484–492.

Table 4.3. Cellular fatty acid composition of *Acinetobacter*.

Fatty Acid	Organism *Acinetobacter* species %
10:0	T[a]
12:0	7
12:1ω9c	1
2-OH-12:0	2
3-OH-12:0	4
14:0	1
3-OH-14:0	1
15:0	1
16:1ω9c	1
16:1ω7c	16
16:0	20
17:1ω8c	4
17:0	2
18:2	2
18:1ω9c	36
18:1ω7c	2
18:0	1

[a]T = Trace.

Actinobacillus, *Haemophilus,* and *Pasteurella*

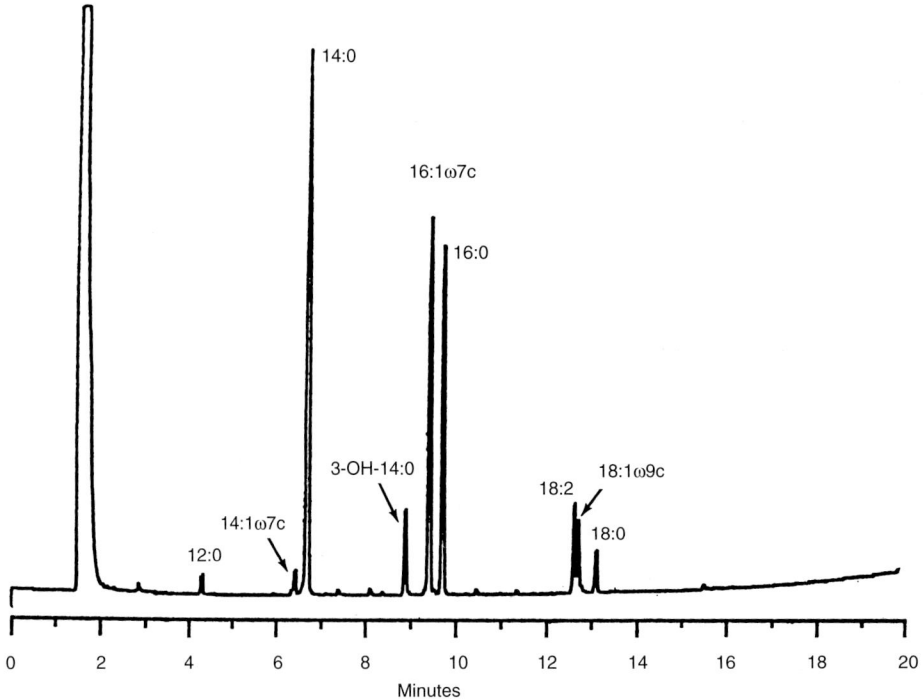

Figure 4.9. Gas chromatogram of the cellular fatty acids of *Actinobacillus, Haemophilus,* and *Pasteurella.*

Comments

Species of the genera *Actinobacillus, Haemophilus,* and *Pasteurella* have essentially identical CFA profiles that are characterized by large amounts (20–30% each) of 14:0, 16:1ω7c, and 16:0, with smaller amounts of 3—OH-14:0, 18:2, 18:0, 14:1ω7c, and 14:1ω12, and no cyclopropane acids. 18:1ω7c is absent (or present in only trace–4% amounts) in all species of the family *Pasteurellaceae,* except *Pasteurella testudinis* which contains 7%. *Helicobacter pylori* has 34% 14:0 and *"Campylobacter fecalis"* has 18% 14:0, but these two organisms and all other *Campylobacter* species are distinguished from *Actinobacillus/Haemophilus/Pasteurella* by the presence of large amounts of 18:1ω7c and/or 3—OH-16:0, and/or cyclopropane acids (17:0cyc or 19:0cyc).

Literature

1. Jantzen, E., B.P. Berdal, and T. Omland. 1981. Cellular fatty acid taxonomy of *Haemophilus, Pasteurella,* and *Actinobacillus. In* M. Kilian, W. Frederiksen, and E.L. Biberstein (Eds.), *Haemophilus, Pasteurella,* and *Actinobacillus,* Academic Press, London, pp. 197–204.

2. Schlater, L.K., D.J. Brenner, A.G. Steigerwalt, C.W. Moss, M.A. Lambert, and R.A. Packer. 1989. *Pasteurella caballi,* a new species from equine clinical specimens. J. Clin. Microbiol. 27: 2169–2174.

3. Brondz, I., I. Olsen, and M. Sjostrom. 1990. Multivariate analysis of quantitative chemical and enzymic characterization data in classification of *Actinobacillus, Haemophilus,* and *Pasteurella* spp. J. Gen. Microbiol. *136:* 507–513.

4. Engelhard, E., R.M. Kroppenstedt, R. Muttus, and W. Mannheim. 1991. Carbohydrate patterns, cellular lipoquinones, fatty acids, and phospholipids of the genus *Pasteurella* sensu stricto. Med. Microbiol. Immunol. *180:* 79–92.

Table 4.4A. Cellular fatty acid composition of *Actinobacillus*.

Fatty Acid	Organism[a]					
	A	B	C	D	E	F
	%	%	%	%	%	%
12:0	-	1	1	1	1	1
14:1ω7c	-	1	1	1	2	1
14:0	21	27	25	25	30	29
14:1ω12	1	1	1	1	1	1
3-OH-14:0	8	7	5	7	6	6
16:1ω7c	25	26	20	31	26	23
16:0	33	20	28	22	18	24
18:2	4	8	8	5	7	6
18:1ω9c	3	5	6	3	4	5
18:1ω7c	-	-	-	1	-	-
18:0	3	3	4	2	3	3
20:4	-	1	1	1	1	-

[a]A, *Actinobacillus actinomycetemcomitans*; B, *Actinobacillus equuli*; C, *Actinobacillus hominis*; D, *Actinobacillus lignieresii*; E, *Actinobacillus suis*; F, *Actinobacillus ureae*.

Table 4.4B. Cellular fatty acid composition of *Haemophilus*.

Fatty Acid	Organism[a]							
	A	B	C	D	E	F	G	H
	%	%	%	%	%	%	%	%
12:0	-	1	T[b]	-	1	-	1	-
14:1ω7c	T	T	1	-	1	-	-	-
14:0	20	22	30	20	29	15	25	21
14:1ω12	1	1	-	1	1	1	1	1
3-OH-14:0	7	7	6	6	8	6	8	9
16:1ω7c	33	25	40	28	31	12	28	19
16:0	34	34	22	35	22	29	29	30
18:2	1	4	-	2	3	18	3	9
18:1ω9c	T	3	-	1	2	12	2	7
18:1ω7c	1	-	-	4	-	-	-	-
18:0	2	2	T	3	2	6	2	4
20:4	-	-	-	-	-	1	-	-

[a]A, *Haemophilus biogroup aegyptius*; B, *Haemophilus aphrophilus*; C, *Haemophilus ducreyi*; D, *Haemophilus influenzae*; E, *Haemophilus parahaemolyticus*; F, *Haemophilus parainfluenzae*; G, *Haemophilus paraphrophilus*; H, *Haemophilus segnis*.
[b]T = Trace.

Table 4.4C. Cellular fatty acid composition of *Pasteurella*.

Fatty Acid	Organism[a]											
	A	B	C	D	E	F	G	H	I	J	K	L
	%	%	%	%	%	%	%	%	%	%	%	%
12:0	-	1	2	T[b]	1	-	T	T	T	T	1	1
14:1ω7c	1	1	2	T	1	-	T	T	T	1	T	1
14:0	22	27	30	21	26	19	20	20	20	25	21	17
14:1ω12	2	1	1	T	1	1	1	1	1	1	1	1
15:0	T	T	-	T	T	-	1	1	T	T	T	-
3-OH-14:0	7	6	9	5	7	7	5	5	5	5	7	7
16:1ω7c	31	27	28	28	30	36	30	30	30	31	28	29
16:0	28	25	16	29	23	25	31	32	32	31	25	31
18:2	3	5	5	7	6	5	4	4	4	1	8	3
18:1ω9c	2	4	4	4	3	4	3	2	3	2	4	2
18:1ω7c	1	-	-	1	T	1	1	1	1	1	1	7
18:0	2	2	2	3	2	2	3	3	3	2	4	1
20:4	T	T	T	1	1	1	1	1	1	-	1	T

[a]A, *Pasteurella aerogenes*; B, *Pasteurella bettyae*; C, *Pasteurella caballi*; D, *Pasteurella canis*; E, *Pasteurella dagmatis*; F, *Pasteurella langaa*; G, *Pasteurella multocida* subsp. *gallicida*; H, *Pasteurella multocida* subsp. *multocida*; I, *Pasteurella multocida* subsp. *septica*; J, *Pasteurella pneumotropica*; K, *Pasteurella stomatis*; L, *Pasteurella testudinis*.
[b]T = Trace.

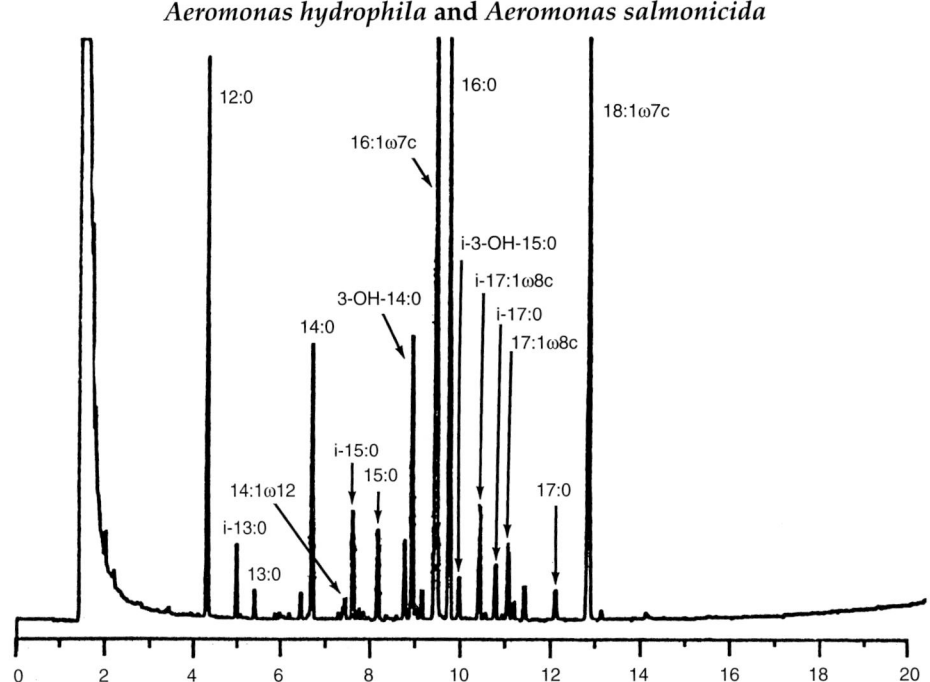

Figure 4.10. Gas chromatogram of the cellular fatty acids of *Aeromonas hydrophila* and *Aeromonas salmonicida*.

Comments

The major fatty acids of *Aeromonas* species are 16:1ω7c, 16:0, 18:1ω7c, 12:0, 3—OH-14:0, and 14:0. The presence of small amounts of 14:1ω12 and the relatively uncommon monounsaturated branched-chain acids (i-16:1ω11c, i-17:1ω8c) are useful chemical markers of this genus. *A. hydrophila* contains small amounts of i-13:0, i-15:0, i-16:0, and 17:1ω8c that are absent in *A. salmonicida*. These differences, if substantiated with results from additional strains, are sufficient for differentiation of these two species by their CFA composition. *Aeromonas* differs from *Vibrio* by the presence of i-17:1ω8c and the absence of 3—OH-12:0 and 16:1ω7t that are present in small amounts (trace–7%) in *Vibrio*.

Literature

1. Boe, B., and J. Gjerde. 1980. Fatty acid patterns in the classification of some representatives of the families *Enterobacteriaceae* and *Vibrionaceae*. J. Gen. Microbiol. *116*: 41–49.

2. Lambert, M.A., F.W. Hickman-Brenner, J.J. Farmer, III, and C.W. Moss. 1983. Differentiation of *Vibrionaceae* by their cellular fatty acid composition. Int. J. Syst. Bacteriol. *33*: 777–792.

3. Urdaci, M.C., M. Marchand, and P.A. Grimont. 1990. Characterization of 22 *Vibrio* species by gas chromatography analysis of their cellular fatty acids. Res. Microbiol. *141*: 437–452.

4. Hansen, W., J. Freney, M. Labbe, F. Renaud, E. Yourassowsky, and J. Fleurette. 1991. Gas-liquid chromatographic analysis of cellular fatty acid methyl esters in *Aeromonas* species. Zlb. Bakt. *275*: 1–10.

Table 4.5. Cellular fatty acid composition of *Aeromonas hydrophila* and *Aeromonas salmonicida*.

	Organism	
	Aeromonas hydrophila	*Aeromonas salmonicida*
Fatty Acid	%	%
12:0	6	6
i-13:0	1	–
14:0	4	4
14:1ω12	2	1
i-15:0	2	–
15:0	2	3
3-OH-14:0	6	9
i-16:1ω11c	1	1
i-16:0	1	–
16:1ω7c	32	40
16:0	22	22
i-3-OH-15:0	2	–
i-17:1ω8c	2	3
i-17:0	3	–
17:1ω8c	2	–
17:0	1	2
18:1ω7c	10	7

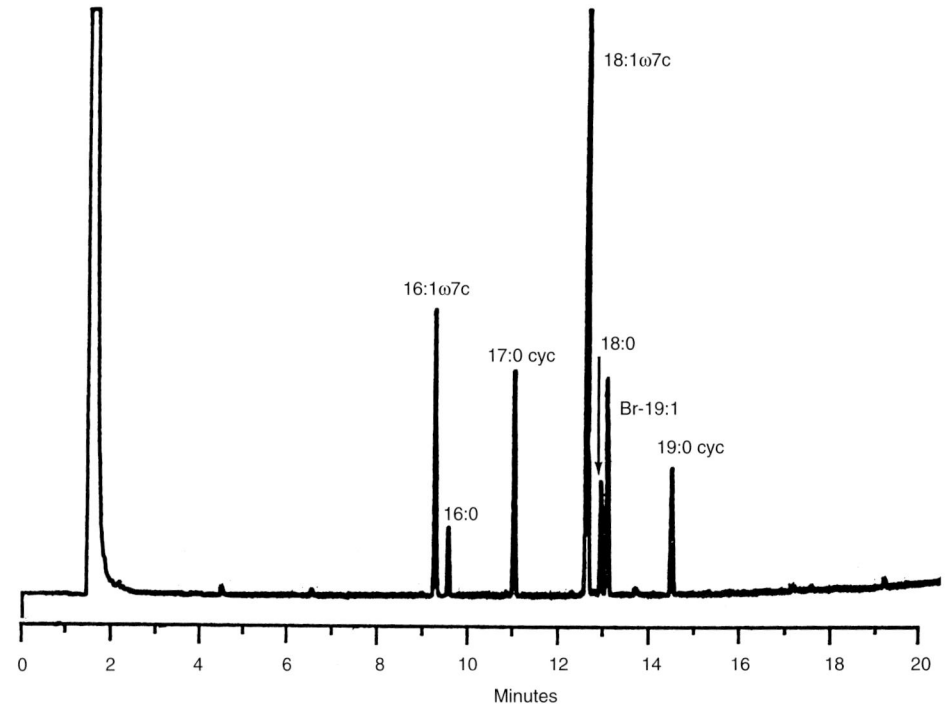

Figure 4.11. Gas chromatogram of the cellular fatty acids of *Afipia broomeae*, *Afipia clevelandensis*, *Afipia felis*, *Afipia* genomospecies 1, *Afipia* genomospecies 2, and *Afipia* genomospecies 3.

Comments

All members of the genus *Afipia* contain Br-19:1 and 18:1ω7c acids as major components with only small to trace amounts of hydroxy acids. The large amount (10–23%) of Br-19:1 is a unique chemical marker of *Afipia*, as it has been detected only in small amounts in some *Brucella* and *Pseudomonas* species at levels less than 4% and in *Achromobacter* groups B and E at levels less than 10%. All *Afipia* species contain 17:0cyc and/or 19:0cyc[11–12].

Literature

1. Moss, C.W., G. Holzer, P.L. Wallace, and D.G. Hollis. 1990. Cellular fatty acid compositions of an unidentified organism and a bacterium associated with cat scratch disease. J. Clin. Microbiol. *28:* 1071–1074.

2. Brenner, D.J., D.G. Hollis, C.W. Moss, C.K. English, G.S. Hall, J. Vincent, J. Radosevic, K.A. Birkness, W.F. Bibb, F.D. Quinn, B. Swaminathan, R.E. Weaver, M.W. Reeves, S.P. O'Connor, P.G. Hayes, F.C. Tenover, A.G. Steigerwalt, B.A. Perkins, M.I. Daneshvar, B.C. Hill, J.A. Washington, T.C. Woods, S.B. Hunter, T.L. Hadfield, G.W. Ajello, A.F. Kaufmann, D.J. Wear, and J.D. Wenger. 1991. Proposal of *Afipia* gen. nov., with *Afipia felis* sp. nov. (formerly the cat scratch disease bacillus), *Afipia clevelandensis* sp. nov. (formerly the Cleveland Clinic Foundation strain), *Afipia broomeae* sp. nov., and three unnamed genospecies. J. Clin. Microbiol. *29:* 2450–2460.

Table 4.6. Cellular fatty acid composition of *Afipia*.

Fatty Acid	Organism[a]					
	A	B	C	D	E	F
	%	%	%	%	%	%
14:1ω7	-	-	-	1	2	-
15:0	1	-	-	1	1	-
16:1ω7c	9	12	10	2	2	3
16:1ω5	-	-	-	3	8	-
16:0	4	4	4	6	9	12
17:1A	-	-	-	-	1	-
17:1ω8c	4	-	-	T[b]	1	-
17:1ω6c	3	-	-	3	-	-
17:0cyc	15	11	10	5	3	1
17:0	5	1	-	3	-	-
18:3	-	-	-	-	1	-
18:1ω7c	24	36	38	35	39	65
18:0	10	9	9	3	4	6
Br-19:1	18	22	15	22	18	10
19:0cyc$^{11-12}$	4	4	13	15	10	2
19:0	2	-	-	-	-	-
20:1ω9t	1	1	-	-	-	-

[a] A, *Afipia broomeae*; B, *Afipia clevelandensis*; C, *Afipia felis*; D, *Afipia* genomospecies 1; E, *Afipia* genomospecies 2; F, *Afipia* genomospecies 3.
[b] T = Trace.

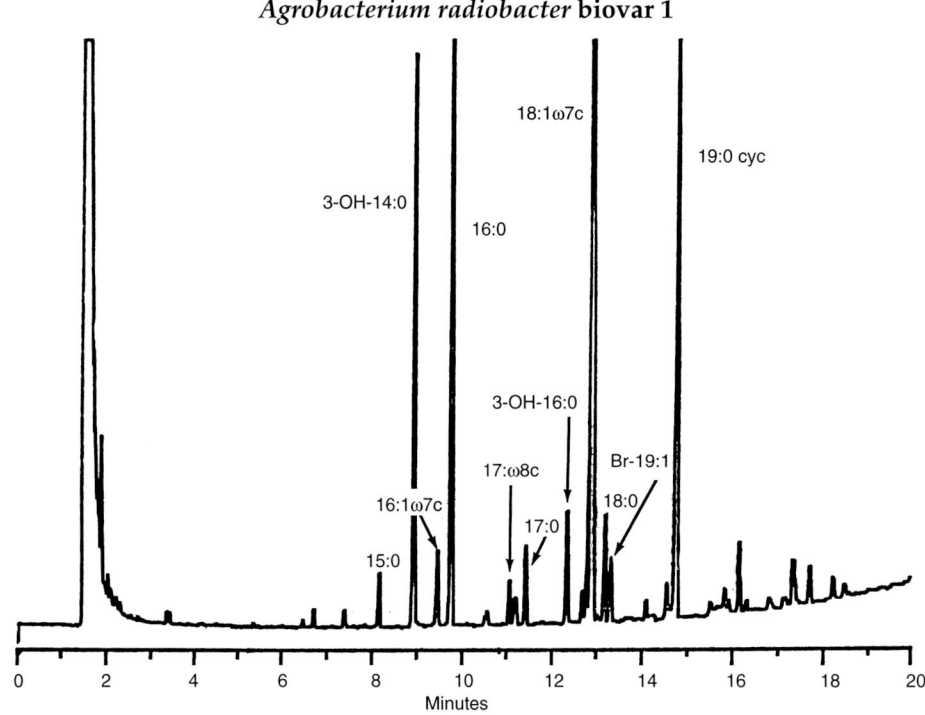

Figure 4.12. Gas chromatogram of the cellular fatty acids of *Agrobacterium radiobacter* biovar 1.

Comments

Agrobacterium radiobacter biovar 1 is characterized by the presence of >50% 18:1ω7c, 13% 19:0cyc[11–12], 7% 3—OH-14:0, 1% 3—OH-16:0, and trace of Br-19:1. This profile is somewhat similar to *Roseomonas cervicalis, R. gilardii, Roseomonas* genomospecies 4 and 5, *Pseudomonas diminuta, P. vesicularis,* and *Ochrobactrum anthropi;* but differs from these organisms by the presence of 3—OH-14:0 and the absence of 2—OH-18:1 and 2—OH-19:0cyc. *A. radiobacter* biovar 1 differs from *Campylobacter jejuni, C. coli, C. upsaliensis, C. lari,* and *Oligella* species by the absence of 14:0 and smaller amounts of 16:0 (18% versus >30%).

Literature

1. Moss, C.W., P.L. Wallace, D.G. Hollis, and R.E. Weaver. 1988. Cultural and chemical characterization of CDC groups EO-2, M-5, M-6, *Moraxella (Moraxella)* species, *Oligella urethralis, Acinetobacter* species, and *Psychrobacter immobilis.* J. Clin. Microbiol. *26:* 484–492.

2. Veys, A., W. Callewaert, E. Waelkens, and K.V.D. Abbeele. 1989. Application of gas-liquid chromatography to the routine identification of nonfermentative gram-negative bacteria in clinical specimens. J. Clin. Microbiol. *27:* 1538–1542.

3. Wallace, P.L., D.G. Hollis, R.E. Weaver, and C.W. Moss. 1990. Biochemical and chemical characterization of pink-pigmented oxidative bacteria. J. Clin. Microbiol. *28:* 689–693.

Table 4.7. Cellular fatty acid composition of *Agrobacterium radiobacter* biovar 1.

Fatty Acid	Organism *Agrobacterium radiobacter* biovar 1 %
3-OH-14:0	8
16:1ω7c	2
16:0	18
17:0cyc	1
3-OH-16:0	1
18:2	1
18:1ω9c	T[a]
18:1ω7c	51
18:0	1
Br-19:1	T
19:0cyc$^{11\text{-}12}$	13

[a]T = Trace.

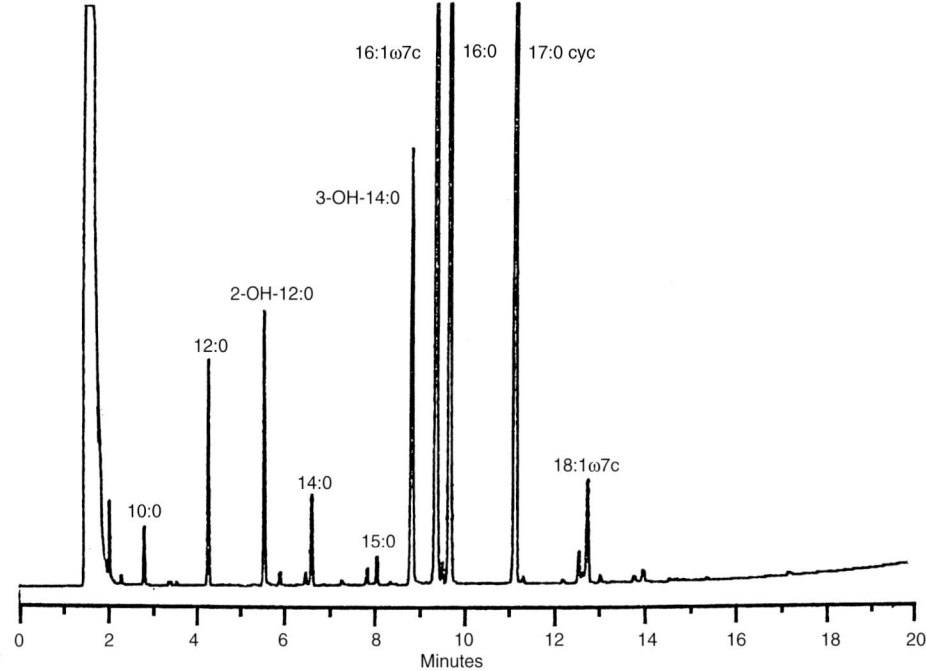

Figure 4.13. Gas chromatogram of the cellular fatty acids of *Alcaligenes faecalis, Alcaligenes xylosoxidans* subsp. *denitrificans* group 1, *Alcaligenes xylosoxidans* subsp. *xylosoxidans*, and *Bordetella bronchiseptica*.

Comments

The CFA compositions of species in this CFA group are characterized by 16:1ω7c, 16:0, and 17:0cyc as the major acids with 5–11% of 3—OH-14:O, 2–5% of 2—OH-12:0, and 2–7% of 18:1ω7c. 3—OH-10:0 is absent. *Alcaligenes xylosoxidans* subsp. *xylosoxidans* and *A. xylosoxidans* subsp. *denitrificans* group 1 contain 2—OH-14:0 that is absent in *A. faecalis* and *Bordetella bronchiseptica;* only *A. faecalis* contains more than trace amounts of 10:0. The possibility of further differentiation of these species by CFA analysis awaits testing of additional well documented strains of each species. This CFA group differs from *B. parapertussis* by the absence of 3—OH-10:0, larger amounts of 16:1ω7c (>18 versus 6%), and the presence of 12:0 and 18:1ω7c. This CFA group differs from *B. pertussis* by the presence of large amounts of 17:0cyc (14–23%) that is absent in *B. pertussis*.

Literature

1. Dees, S.B., and C.W. Moss. 1975. Cellular fatty acids of *Alcaligenes* and *Pseudomonas* species isolated from clinical specimens. J. Clin. Microbiol. *1:* 414–419.

2. Veys, A., W. Callewaert, E. Waelkens, and K.V.D. Abbeele. 1989. Application of gas-liquid chromatography to the routine identification of nonfermentative gram-negative bacteria in clinical specimens. J. Clin. Microbiol. *27:* 1538–1542.

3. Osterhout, G.J., V.H. Shull, and J.D. Dick. 1991. Identification of clinical isolates of gram-negative nonfermentative bacteria by an automated cellular fatty acid identification system. J. Clin. Microbiol. *29:* 1822–1830.

Table 4.8. Cellular fatty acid composition of *Alcaligenes faecalis*, *Alcaligenes xylosoxidans* subsp. *denitrificans* 1, *Alcaligenes xylosoxidans* subsp. *xylosoxidans*, and *Bordetella bronchiseptica*.

Fatty Acid	Organism[a]			
	A	B	C	D
	%	%	%	%
10:0	2	-	-	-
12:0	4	4	1	2
2-OH-12:0	5	4	2	4
14:0	2	4	4	6
15:0	1	T[b]	T	-
2-OH-14:0	-	3	3	-
3-OH-14:0	11	6	7	5
16:1ω7c	19	18	22	27
16:0	31	32	34	35
17:0cyc	19	23	14	14
17:0	-	1	T	-
2-OH-16:0	-	-	2	-
18:2	T	T	1	1
18:1ω7c	3	2	7	2
18:0	-	1	1	2
19:0cyc[11-12]	-	-	2	-

[a] A, *Alcaligenes faecalis*; B, *Alcaligenes xylosoxidans* subsp. *denitrificans* 1; C, *Alcaligenes xylosoxidans* subsp. *xylosoxidans*; D, *Bordetella bronchiseptica*.
[b] T = Trace.

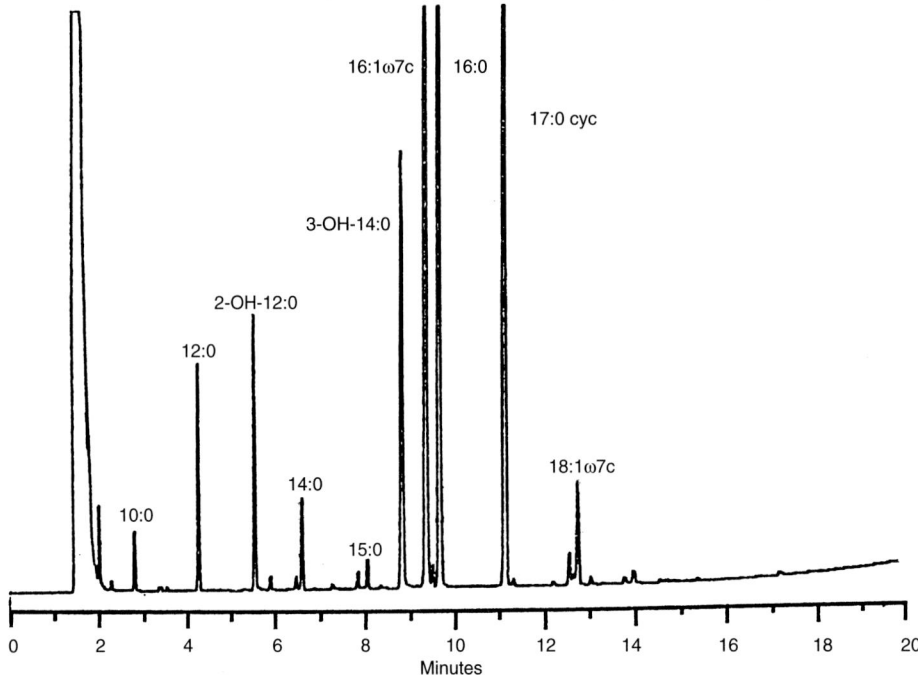

Figure 4.14. Gas chromatogram of the cellular fatty acids of *Alcaligenes piechaudii*, *Comamonas acidovorans*, *Comamonas testosteroni*, *Pseudomonas alcaligenes*, and *Pseudomonas pseudoalcaligenes*.

Comments

Organisms in this group have heterogeneous CFA profiles but are placed together because of the overlap among strains of species in the group. The relative amounts of some acids vary significantly among species as well as among strains within a given species: 16:1ω7c (14–31%), 16:0 (23–44%), 17:0cyc (4–23%), 18:1ω7c (5–37%), and 19:0cyc$^{11-12}$ (0–4%). Strains of *Pseudomonas alcaligenes* form at least six distinct CFA groups that differ from each other by the presence of and/or relative amounts of cyclopropane, hydroxy, and monounsaturated acids. Further differentiation of the group may be feasible as additional strains are tested. For example, *Comamonas acidovorans* has 3—OH-10:0 as the only hydroxy acid, *C. testosteroni* has 3—OH-10:0 and 2—OH-16:0, *P. pseudoalcaligenes* has 3—OH-10:0 and 3—OH-12:0, and *Alcaligenes piechaudii* has 2—OH-12:0, but no 3—OH-10:0. This CFA group differs from *Bordetella parapertussis* by the presence of 12:0 and 18:1ω7c, both of which are absent in *B. parapertussis*. Except for the presence of 17:0cyc and different ratios of 16:1 to 16:0, the overall CFA profile of organisms in this group is most similar to the CFA group that contains *C. terrigena*, *Acidovorax delafieldii*, and WO-1 and WO-1A. Some strains within this group have CFA profiles essentially identical to one of the species in the CFA group containing *A. faecalis*, *A. xylosoxidans* subsp. *denitrificans* group 1, *A. xylosoxidans* subsp. *xylosoxidans*, and *B. bronchiseptica*. Thus, additional strains must be examined to evaluate the usefulness of CFA for distinguishing among these organisms.

Literature

1. Dees, S.B., and C.W. Moss. 1975. Cellular fatty acids of *Alcaligenes* and *Pseudomonas* species isolated from clinical specimens. J. Clin. Microbiol. *1:* 414–419.

2. Moss, C.W., and S.B. Dees. 1976. Cellular fatty acids and metabolic products of *Pseudomonas* species obtained from clinical specimens. J. Clin. Microbiol. *4:* 492–502.

3. Veys, A., W. Callewaert, E. Waelkens, and K.V.D. Abbeele. 1989. Application of gas-liquid chromatography to the routine identification of nonfermentative gram-negative bacteria in clinical specimens. J. Clin. Microbiol. *27:* 1538–1542.

Table 4.9. Cellular fatty acid composition of *Alcaligenes piechaudii*, *Comamonas acidovorans*, *Comamonas testosteroni*, *Pseudomonas alcaligenes*, **and** *Pseudomonas pseudoalcaligenes*.

Fatty Acid	Organism[a]				
	A	B	C	D	E
	%	%	%	%	%
3-OH-10:0	-	3	4	2	3
12:0	1	2	3	3	6
2-OH-12:0	3	-	-	-	-
3-OH-12:0	-	-	-	-	1
14:0	5	1	2	2	1
15:0	-	1	3	5	1
3-OH-14:0	6	-	-	2	-
16:1ω7c	17	31	22	21	14
16:0	36	44	32	24	23
i-17:0	-	-	-	-	2
17:0cyc	23	7	6	7	4
17:0	T[b]	-	2	1	1
2-OH-16:0	T	-	1	1	-
18:2	-	-	-	T	-
18:1ω7c	5	11	23	22	37
18:0	2	T	T	2	1
19:0cyc$^{11\text{-}12}$	T	-	-	4	2

[a]A, *Alcaligenes piechaudii*; B, *Comamonas acidovorans*; C, *Comamonas testosteroni*; D, *Pseudomonas alcaligenes*; E, *Pseudomonas pseudoalcaligenes*.
[b]T = Trace.

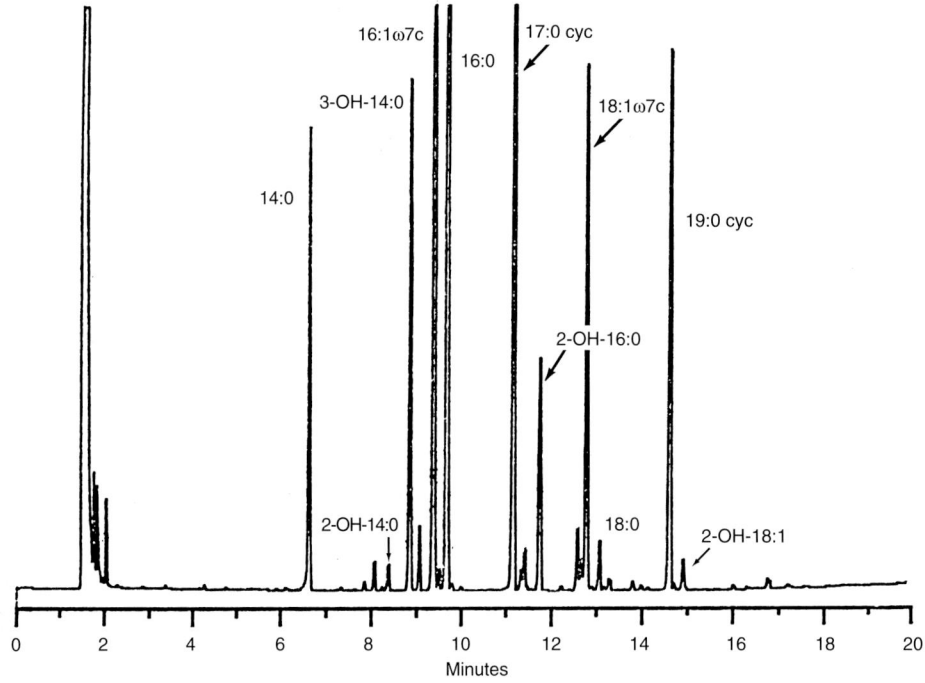

Figure 4.15. Gas chromatogram of the cellular fatty acids of *Alcaligenes*-like Group 1, IVc-2, *Pseudomonas pickettii*, and "*Pseudomonas thomasii*."

Comments

All species in this CFA group contain 16:1ω7c and 16:0 as major acids with smaller amounts (2–10%) of 3—OH-14:0 and 17:0cyc. Also, all species contain trace to 2% amounts of 2—OH-14:0, 2—OH-16:1, 2—OH-16:0, and 2—OH-18:1, which distinguishes this group from *Tatumella ptyseos* that does not contain these hydroxy acids. **No** 3—OH-10:0 is present in this CFA group compared to 2% or greater amounts in the CFA group containing *Alcaligenes piechaudii*, *Pseudomonas alcaligenes*, *P. pseudoalcaligenes*, *Comamonas testosteroni*, and *C. acidovorans*. Also, species in the latter CFA group do not contain 2—OH-14:0 and 2—OH-18:1 acids.

Literature

1. Dees, S.B., and C.W. Moss. 1975. Cellular fatty acids of *Alcaligenes* and *Pseudomonas* species isolated from clinical specimens. J. Clin. Microbiol. *1:* 414–419.

2. Veys, A., W. Callewaert, E. Waelkens, and K.V.D. Abbeele. 1989. Application of gas-liquid chromatography to the routine identification of nonfermentative gram-negative bacteria in clinical specimens. J. Clin. Microbiol. *27:* 1538–1542.

3. Osterhout, G.J., V.H. Shull, and J.D. Dick. 1991. Identification of clinical isolates of gram-negative nonfermentative bacteria by an automated cellular fatty acid identification system. J. Clin. Microbiol. *29:* 1822–1830.

Table 4.10. Cellular fatty acid composition of *Alcaligenes*-like Group 1, **IVc-2**, *Pseudomonas pickettii*, and "*Pseudomonas thomasii*".

Fatty Acid	Organism[a]			
	A	B	C	D
	%	%	%	%
14:0	6	6	5	5
15:0	T[b]	T	1	1
2-OH-14:0	1	T	1	1
3-OH-14:0	7	7	7	7
16:1ω7c	26	28	26	25
16:0	31	25	28	27
17:0cyc	9	10	4	5
17:0	T	T	1	1
2-OH-16:1	1	1	2	2
2-OH-16:0	3	4	1	2
18:2	1	T	T	T
18:1ω7c	11	14	19	19
18:0	1	1	1	1
19:0cyc$^{11\text{-}12}$	3	2	1	2
2-OH-18:1	1	1	2	1

[a]A, *Alcaligenes*-like Group 1; B, IVc-2; C, *Pseudomonas pickettii*; D, "*Pseudomonas thomasii*".
[b]T = Trace.

Arcobacter butzleri, Arcobacter cryaerophilus, and *Arcobacter skirrowii*

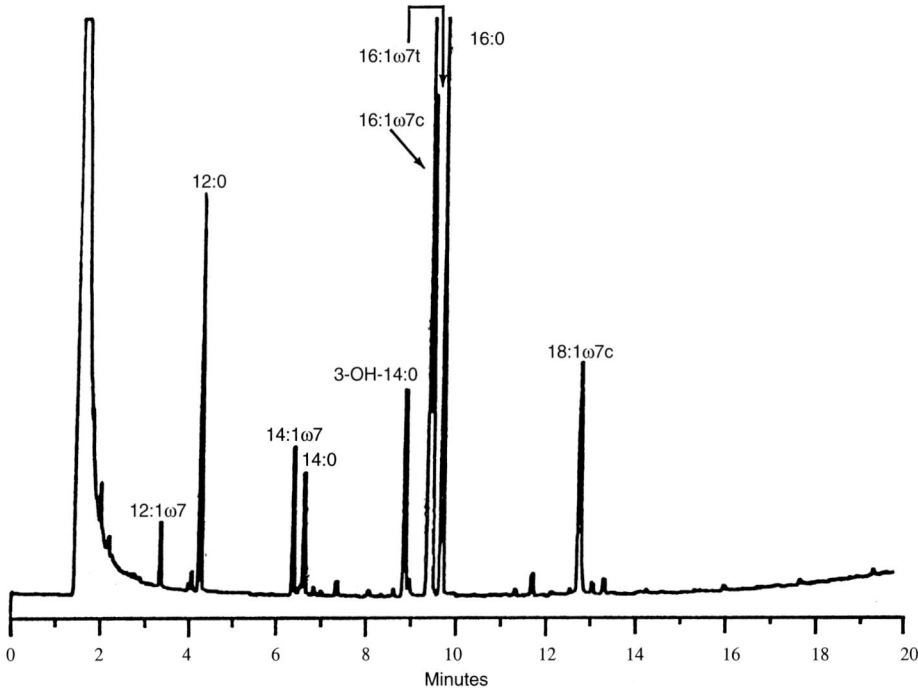

Figure 4.16. Gas chromatogram of the cellular fatty acids of *Arcobacter butzleri, Arcobacter cryaerophilus,* and *Arcobacter skirrowii.*

Comments

These three species have a unique CFA profile that is recognized by the presence of two relatively uncommon monounsaturated acids (4% of 14:1ω7 and 14% of 16:1ω7t). They also contain 16:1ω7c (25%), 16:0 (26%), 18:1ω7c (17%), 3—OH-14:0 (5%), and small amounts of 12:0, 14:0, and 14:1ω12. Vandamme et al. (3) reported the presence of an unidentified component (eluting just after 2—OH-14:0) in *Arcobacter skirrowii* that they considered useful for distinguishing this species. However, in our laboratory, this component was absent in 13 of 16 *A. skirrowii* strains and present in amounts less than 0.7% in the other three strains (some of the same *A. skirrowii* strains were examined in both laboratories). Vandamme et al. (3) also reported the absence of 16:1ω7t in some strains of *A. skirrowii,* but these same strains when examined in our laboratory had significant amounts (15–25%) of 16:1ω7t.

Literature

1. Lambert, M.A., C.M. Patton, T.J. Barrett, and C.W. Moss. 1987. Differentiation of *Campylobacter* and *Campylobacter*-like organisms by cellular fatty acids. J. Clin. Microbiol. 25: 706–713.

2. Moss, C.W., and M.A. Lambert-Fair. 1989. Location of double bonds in monounsaturated fatty acids of *Campylobacter cryaerophila* with dimethyl disulfide derivatives and combined gas chromatography-mass spectrometry. J. Clin. Microbiol. 27: 1467–1470.

3. Vandamme, P., M. Vancanneyt, B. Pot, L. Mels, B. Hoste, D. Dewettinck, L. Vlaes, C.V.D. Borre, R. Higgins, J. Hommez, K. Kersters, J.P. Butzler, and H. Goossens. 1992. Polyphasic taxonomic study of the amended genus *Arcobacter* with *Arcobacter butzleri* comb. nov. and *Arcobacter skirrowii* sp. nov., an aerotolerant bacterium isolated from veterinary specimens. Int. J. Syst. Bacteriol. 42: 344–356.

Table 4.11. Cellular fatty acid composition of *Arcobacter butzleri*, *Arcobacter cryaerophilus*, and *Arcobacter skirrowii*.

	Organism		
	Arcobacter butzleri	*Arcobacter cryaerophilus*	*Arcobacter skirrowii*
Fatty Acid	%	%	%
12:0	5	4	10
14:1ω7c	4	4	2
14:0	3	2	3
14:1ω12	1	T[a]	1
3-OH-14:0	6	5	8
16:1ω7c	23	25	21
16:1ω7t	15	14	14
16:0	27	26	26
18:1ω7c	15	17	15
18:0	1	1	1

[a] T = Trace.

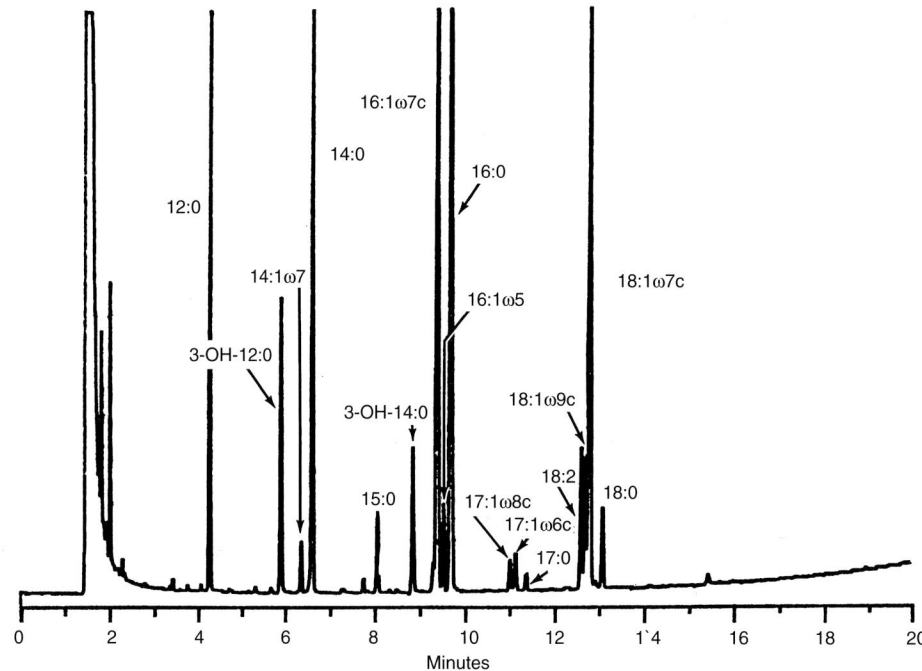

Figure 4.17. Gas chromatogram of the cellular fatty acids of *Arcobacter nitrofigilis*, EF-4a, EF-4b, *Eikenella corrodens*, *Suttonella indologenes*, and *Neisseria* spp.

Comments

Organisms in this CFA group are characterized by 16:1ω7c, 16:0, and 18:1ω7c as major fatty acids; moderate amounts of 12:0 and 14:0; and smaller amounts of 18:2, 18:1ω9c, and 18:0. All contain 2%–7% of 3—OH-12:0 and trace to 1% amounts of 3—OH-14:0; some contain trace to small amounts of 2—OH-16:0, but none contain 3—OH-10:0. The absence of 3—OH-10:0 distinguishes this CFA group from *Chromobacterium violaceum* and from the CFA group containing *Comamonas terrigena*, *Acidovorax delafieldii*, WO-1, and WO-1A. Some members of this CFA group contain trace to 2% amounts of 14:1ω7.

Literature

1. Moss, C.W., D.S. Kellogg, Jr., D.C. Farshy, M.A. Lambert, and J.D. Thayer. 1970. Cellular fatty acids of pathogenic *Neisseria*. J. Bacteriol. *104*: 63–68.

2. Lambert, M.A., D.G. Hollis, C.W. Moss, R.E. Weaver, and M.L. Thomas. 1971. Cellular fatty acids of nonpathogenic *Neisseria*. Can. J. Microbiol. *17*: 1491–1502.

3. Jantzen, E., K. Bryn, T. Bergan, and K. Bøvre. 1974. Gas chromatography of bacterial whole cell methanolysates. V. Fatty acid composition of *Neisseria* and *Moraxella*. Acta. Pathol. Microbiol. Immunol. *82*: 767–779.

4. Moss, C.W., P.L. Wallace, D.G. Hollis, and R.E. Weaver. 1988. Cultural and chemical characterization of CDC groups EO-2, M-5, and M-6, *Moraxella (Moraxella)* species, *Oligella urethralis*, *Acinetobacter* species, and *Psychrobacter immobilis*. J. Clin. Microbiol. *26*: 484–492.

5. Osterhout, G.J., V.H. Shull, and J.D. Dick. 1991. Identification of clinical isolates of gram-negative nonfermentative bacteria by an automated cellular fatty acid identification system. J. Clin. Microbiol. *29*: 1822–1830.

Table 4.12A. Cellular fatty acid composition of *Arcobacter nitrofigilis*, EF-4a, EF-4b, *Eikenella corrodens*, and *Suttonella indologenes*.

Fatty Acid	Organism[a]				
	A	B	C	D	E
	%	%	%	%	%
12:0	5	7	7	6	5
3-OH-12:0	3	4	4	5	2
14:1ω7	T[b]	1	–	–	–
14:0	5	8	4	2	11
15:0	–	2	–	–	T
3-OH-14:0	1	1	1	T	1
16:1ω7c	29	25	23	20	19
16:1ω5	–	1	1	–	–
16:0	36	27	30	29	26
17:1ω8c	–	1	–	–	–
17:1ω6c	–	1	–	–	–
2-OH-16:0	–	–	2	–	–
18:2	1	4	3	4	3
18:1ω9c	1	2	2	2	2
18:1ω7c	16	15	20	28	26
18:0	2	1	2	3	2

[a]A, *Arcobacter nitrofigilis*; B, EF-4a; C, EF-4b; D, *Eikenella corrodens*; E, *Suttonella indologenes*.
[b]T = Trace.

Table 4.12B. Cellular fatty acid composition of *Neisseria canis*, *Neisseria cinerea*, *Neisseria elongata* subsp. *elongata* and *glycolytica*, *Neisseria elongata* subsp. *nitroreducens*, *Neisseria flava*, and *Neisseria gonorrhoeae*.

Fatty Acid	Organism[a]					
	A	B	C	D	E	F
	%	%	%	%	%	%
12:0	5	8	7	6	10	8
3-OH-12:0	3	4	5	3	5	4
14:1ω7	–	1	–	–	–	1
14:0	5	5	5	4	8	4
15:0	–	2	1	2	T[b]	–
3-OH-14:0	2	1	1	1	1	1
16:1ω7c	25	33	27	25	30	35
16:1ω5	–	1	–	–	–	1
16:0	33	25	35	38	27	29
2-OH-16:0	1	–	–	–	–	–
18:2	2	2	4	4	4	2
18:1ω9c	1	1	2	3	2	1
18:1ω7c	20	12	11	11	11	12
18:0	2	1	1	2	1	1

[a]A, *Neisseria canis*; B, *Neisseria cinerea*; C, *Neisseria elongata* subsp. *elongata* and *glycolytica*; D, *Neisseria elongata* subsp. *nitroreducens*; E, *Neisseria flava*; F, *Neisseria gonorrhoeae*.
[b]T = Trace.

Table 4.12C. Cellular fatty acid composition of *Neisseria meningitidis, Neisseria mucosa, Neisseria polysaccharea, Neisseria sicca, Neisseria subflava,* and *Neisseria weaveri.*

Fatty Acid	Organism[a]					
	A	B	C	D	E	F
	%	%	%	%	%	%
12:0	8	12	6	8	7	7
3-OH-12:0	4	7	4	3	4	4
14:1ω7	-	-	T[b]	-	T	1
14:0	6	3	5	4	6	6
15:0	T	-	1	1	-	-
3-OH-14:0	1	1	1	1	1	1
16:1ω7c	31	32	29	33	32	26
16:1ω5	1	1	1	1	1	1
16:0	31	24	32	26	29	25
2-OH-16:0	-	2	-	1	-	-
18:2	1	3	3	2	2	2
18:1ω9c	4	2	3	2	4	1
18:1ω7c	11	12	11	14	11	23
18:0	1	1	3	1	1	1

[a] A, *Neisseria meningitidis*; B, *Neisseria mucosa*; C, *Neisseria polysaccharea*; D, *Neisseria sicca*; E, *Neisseria subflava*; F, *Neisseria weaveri*.
[b] T = Trace.

SECTION 4: CELLULAR FATTY ACID ANALYSIS

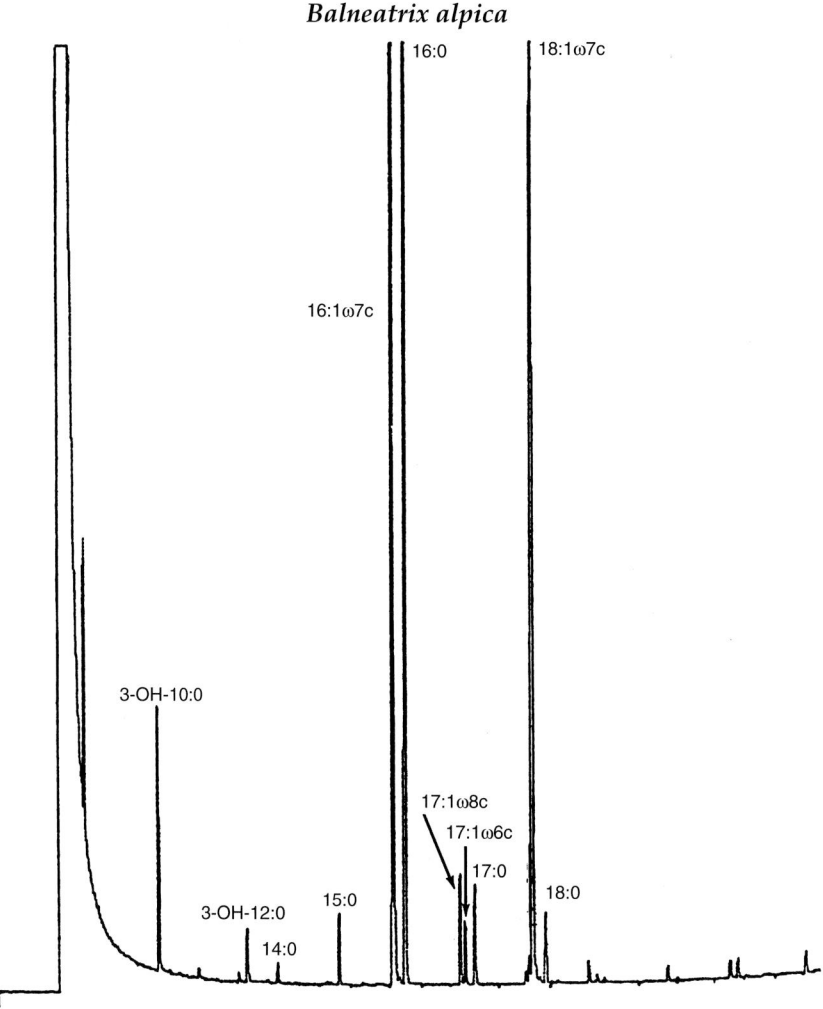

Figure 4.18. Gas chromatogram of the cellular fatty acids of *Balneatrix alpica*.

Comments

This organism contains large amounts of 16:1ω7c (42%), 16:0 (22%), and 18:1ω7c (24%); trace to small amounts of 14:0, 15:0, 17:1ω8c, 17:1ω6c, 17:0, 18:1ω9c, and 18:0; and two hydroxy acids, 3—OH-10:0 (4%) and 3—OH-12:0 (1%). This organism is most similar to the CFA group containing *Pseudomonas aeruginosa*, *Pseudomonas*-like Group 2, "*P. denitrificans*," *P. fluorescens*, *P. putida*, and *Chromobacterium violaceum* except for larger amounts of 16:1ω7c (42% versus 10–35%); presence of 1% each of 17:1ω8c, 17:1ω6c, and 17:0; and the absence of 12:0 (0% versus 2%), 2—OH-12:0 (0% versus 1–6%), and 17:0cyc (0% versus trace–12%).

Literature

1. Dauga, C., M. Gillis, P. Vandamme, E. Ageron, F. Grimont, K. Kersters, C. de Mahenge, Y. Peloux, and P.A.D. Grimont. 1993. *Balneatrix alpica* gen. nov., sp. nov., a bacterium associated with pneumonia and meningitis in a spa therapy centre. Res. Microbiol. *144:* 35–46.

Table 4.13. Cellular fatty acid composition of *Balneatrix alpica*.

Fatty Acid	Organism *Balneatrix alpica* %
3-OH-10:0	4
3-OH-12:0	1
14:0	T
15:0	1
16:1ω7c	42
16:0	22
17:1ω8c	1
17:1ω6c	1
17:0	1
18:1ω9c	T
18:1ω7c	24
18:0	1

[a]T = Trace.

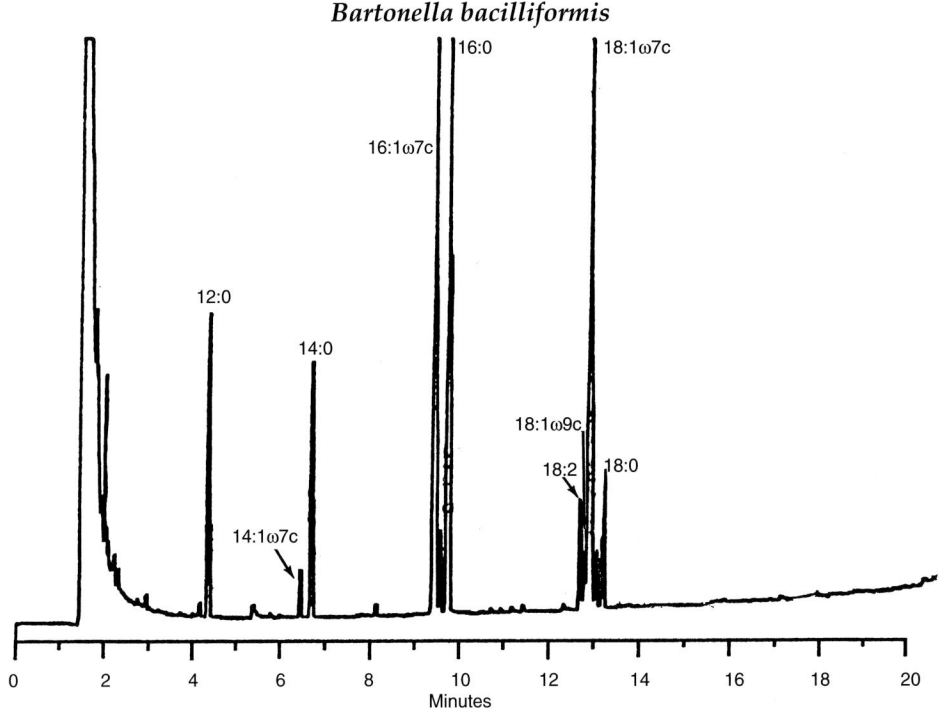

Figure 4.19. Gas chromatogram of the cellular fatty acids of *Bartonella bacilliformis*.

Comments

The CFA profile of *Bartonella bacilliformis* is characterized by the presence of 18:1ω7c as the major acid (49%), with significant amounts of 16:0 (19%) and 16:1ω7c (18%), and the **absence** of hydroxy acids. This profile differs from other *Bartonella* species by the presence of small amounts (1–4%) of 12:0, 14:0, and 16:1ω5; larger amounts of 16:1ω7c (18% versus 2%); absence of 17:0 (0% versus 1–21%); and less 18:0 (2% versus 8%). *B. bacilliformis* differ from *Brucella canis* by the presence of 12:0 (4%), 14:0 (2%), and smaller amounts of 18:1ω7c (49% versus 80%) and from *Flavimonas* by the **absence** of hydroxy acids.

Literature

1. Daly, J.S., M.G. Worthington, D.J. Brenner, C.W. Moss, D.G. Hollis, R.S. Weyant, A.G. Steigerwalt, R.E. Weaver, M.I. Daneshvar, and S.P. O'Connor. 1993. *Rochalimaea elizabethae* sp. nov. isolated from a patient with endocarditis. J. Clin. Microbiol. *31:* 872–881.

Table 4.14. Cellular fatty acid composition of *Bartonella bacilliformis*.

	Organism
	Bartonella bacilliformis
Fatty Acid	%
12:0	4
14:0	2
16:1ω7c	18
16:1ω5	1
16:0	19
18:2	2
18:1ω9c	1
18:1ω7c	49
18:0	2

Bartonella elizabethae, Bartonella henselae, Bartonella quintana, and *Bartonella vinsonii*

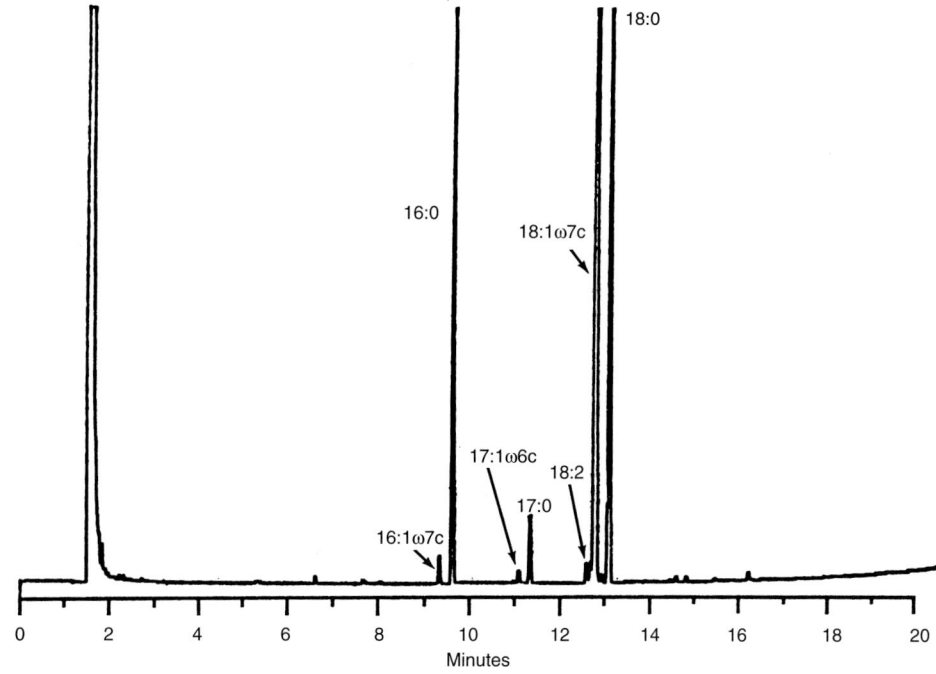

Figure 4.20. Gas chromatogram of the cellular fatty acids of *Bartonella elizabethae, Bartonella henselae, Bartonella quintana,* and *Bartonella vinsonii.*

Comments

All species of *Bartonella* (except *B. bacilliformis*) have a unique CFA profile characterized by large amounts of 18:1ω7c, 16:0, and 18:0, which accounts for 64%–97% of total acids. *B. henselae* and *B. quintana* have essentially identical CFA features, but both differ from *B. vinsonii* and *B. elizabethae* by smaller amounts of 17:0. *B. elizabethae* contains more 17:0 and less 16:0 and 18:1ω7c than does *B. vinsonii*, but these differences must be considered tentative until additional strains are discovered and tested. The CFA profile of *B. bacilliformis* differs from other *Bartonella* species by the presence of small amounts (1–4%) of 12:0, 14:0, and 16:1ω5; larger amounts of 16:1ω7c (18% versus 2%); the absence of 17:0 (0% versus 1–21%); and less 18:0 (2% versus 8%).

Literature

1. Daly, J.S., M.G. Worthington, D.J. Brenner, C.W. Moss, D.G. Hollis, R.S. Weyant, A.G. Steigerwalt, R.E. Weaver, M.I. Daneshvar, and S.P. O'Connor. 1993. *Rochalimaea elizabethae* sp. nov. isolated from a patient with endocarditis. J. Clin. Microbiol. *31:* 872–881.

Table 4.15. Cellular fatty acid composition of *Bartonella elizabethae, Bartonella henselae, Bartonella quintana,* and *Bartonella vinsonii.*

	Organism			
	Bartonella elizabethae	*Bartonella henselae*	*Bartonella quintana*	*Bartonella vinsonii*
Fatty Acid	%	%	%	%
15:0	2	-	-	2
16:1ω7c	1	1	1	2
16:0	13	18	17	20
17:1ω8c	1	-	-	-
17:1ω6c	8	-	-	3
17:0	21	2	1	9
18:2	1	1	T[a]	1
18:1ω9c	1	1	T	T
18:1ω7c	43	52	59	54
18:0	8	21	21	9

[a]T = Trace.

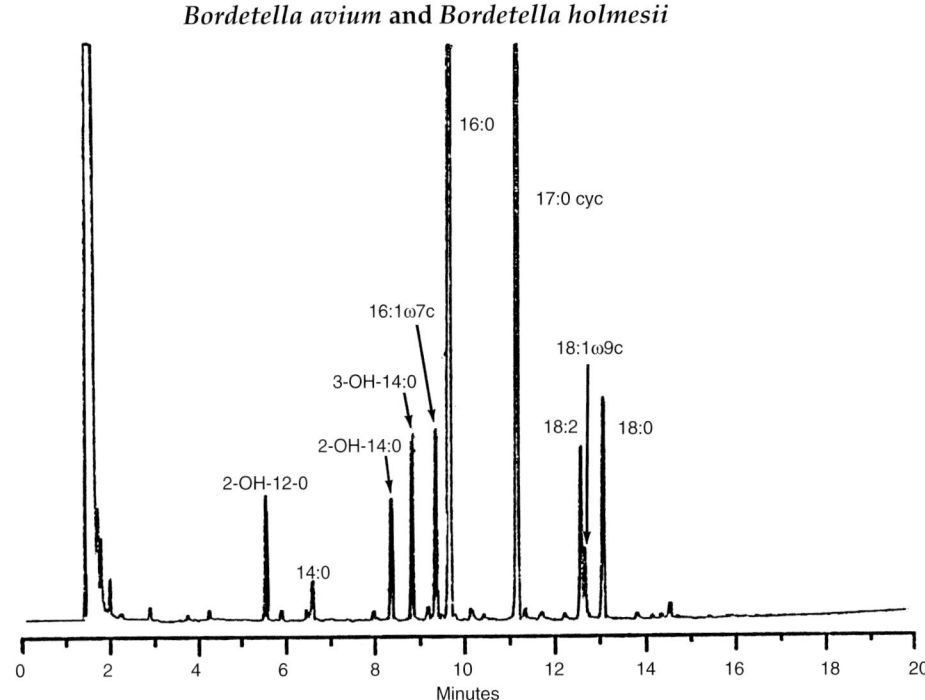

Figure 4.21. Gas chromatogram of the cellular fatty acids of *Bordetella avium* and *Bordetella holmesii* (NO-2).

Comments

The CFA profile of the organisms in this group is characterized by the presence of small amounts (2–5%) of 2—OH-12:0, 2—OH-14:0, and 3—OH-14:0; large amounts (30–41%) of 16:0 and 17:0cyc; and only trace to small amounts of unsaturated acids (16:1ω7c = 2%, 18:1ω9c = 1%). This profile differs from that of the CFA group containing *Alcaligenes faecalis*, *A. xylosoxidans* subsp. *xylosoxidans*, *A. xylosoxidans* subsp. *denitrificans* group 1, and *Bordetella bronchiseptica* by the presence of smaller amounts of 16:1ω7c (2% versus >18%) and larger amounts of 17:0cyc (30% versus 14%). *B. avium* and *B. holmesii* (NO-2) differ from *B. parapertussis* by the absence of 3—OH-10:0, the presence of 2—OH-12:0 (2–4%) and 2—OH-14:0 (3–4%); smaller amounts of 14:0 (1% versus 6%) and 16:1ω7c (2% versus 6%); and larger amounts of 3—OH-14:0 (4–5% versus 1%). Seven of nine *B. holmesii* strains contain small amounts (0.7–1%) of 19:0cyc[9–10] which is absent in *B. avium*.

Literature

1. Dees, S.B., S. Thanabalasundrum, C.W. Moss, D.G. Hollis, and R.E. Weaver. 1980. Cellular fatty acid composition of group IVe, a nonsaccharolytic organism from clinical sources. J. Clin. Microbiol. *11:* 664–668.

2. Moore, C.J., H. Mawhinney, and P.J. Blackall. 1987. Differentiation of *Bordetella avium* and related species by cellular fatty acid analysis. J. Clin. Microbiol. *25:* 1059–1062.

3. Weyant, R.S., D.J. Brenner, R.E. Weaver, M.F.M. Amin, D.G. Hollis, S.P. O'Connor, M.I. Daneshvar, and C.W. Moss. 1995. *Bordetella holmesii* sp. nov., a new gram-negative species associated with septicemia. J. Clin. Microbiol. *33:* 1–7.

Table 4.16. Cellular fatty acid composition of *Bordetella avium* and *Bordetella holmesii* (NO-2).

Fatty Acid	Organism	
	Bordetella avium	*Bordetella holmesii* (NO-2)
	%	%
12:0	T[a]	-
2-OH-12:0	2	4
14:0	1	1
15:0	T	-
2-OH-14:0	3	4
3-OH-14:0	4	5
16:1ω7c	2	2
16:0	40	41
17:0cyc	34	30
17:0	2	-
18:2	-	2
18:1ω9c	-	1
18:1ω7c	1	-
18:0	6	7
i-19:0	T	-
19:0cyc$^{11-12}$	T	-
19:0cyc$^{9-10}$	-	1

[a]T = Trace.

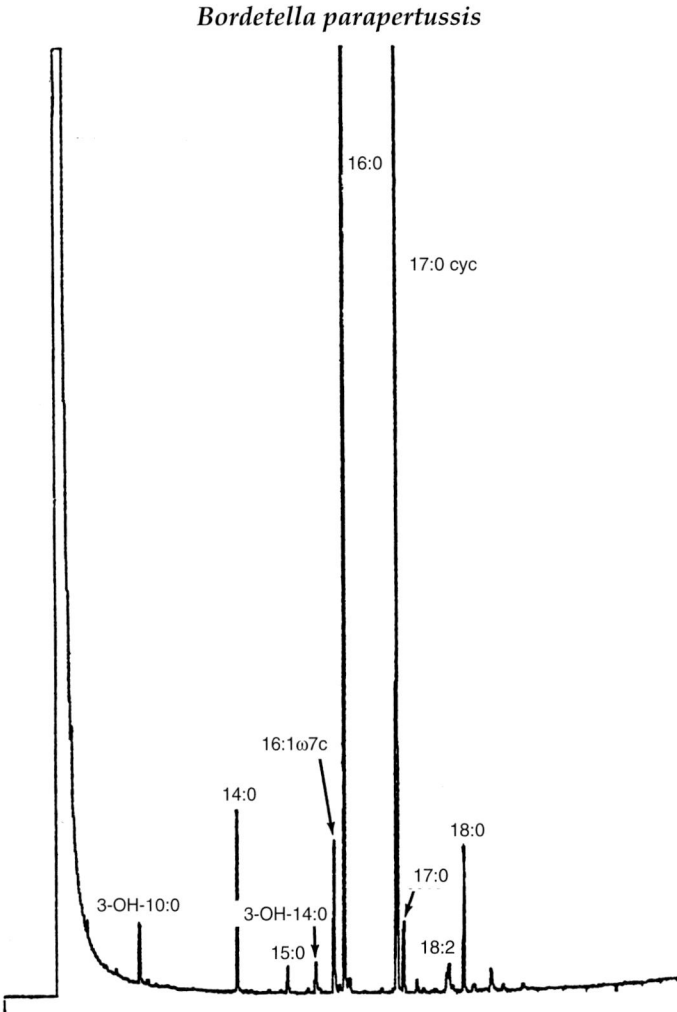

Figure 4.22. Gas chromatogram of the cellular fatty acids of *Bordetella parapertussis*.

Comments

Bordetella parapertussis has a unique CFA profile that is characterized by large amounts of 16:0 (40%) and 17:0cyc (35%), small amounts (1–2%) of 3—OH-10:0 and 3—OH-14:0, and **no** 18-carbon monounsaturated acids (18:1ω9c or 18:1ω7c). It is similar to the CFA group containing *Alcaligenes piechaudii, Comamonas acidovorans, C. testosteroni, Pseudomonas alcaligenes,* and *P. pseudoalcaligenes* except for the absence of 12:0 (0% versus 2%) and 18:1ω7c (0% versus 10%). *B. parapertussis* differs from the CFA group containing *A. faecalis, A. xylosoxidans* subsp. *xylosoxidans, A. xylosoxidans* subsp. *denitrificans* group 1, and *Bordetella bronchiseptica* by the presence of 3—OH-10:0 (2% versus 0%), the absence of 12:0 (0% versus >1%) and 2—OH-12:0 (0% versus >2%), and smaller amounts of 16:1ω7c (6% versus >18%).

Literature

1. Weyant, R.S., D.J. Brenner, R.E. Weaver, M.F.M. Amin, D.G. Hollis, S.P. O'Connor, M.I. Daneshvar, and C.W. Moss. 1995. *Bordetella holmesii* sp. nov., a new gram-negative species associated with septicemia. J. Clin. Microbiol. 33: 1–7.

Table 4.17. Cellular fatty acid composition of *Bordetella parapertussis*.

Fatty Acid	Organism *Bordetella parapertussis* %
3-OH-10:0	2
14:0	6
15:0	1
3-OH-14:0	1
16:1ω7c	6
16:0	40
17:0cyc	35
17:0	3
18:2	1
18:0	5

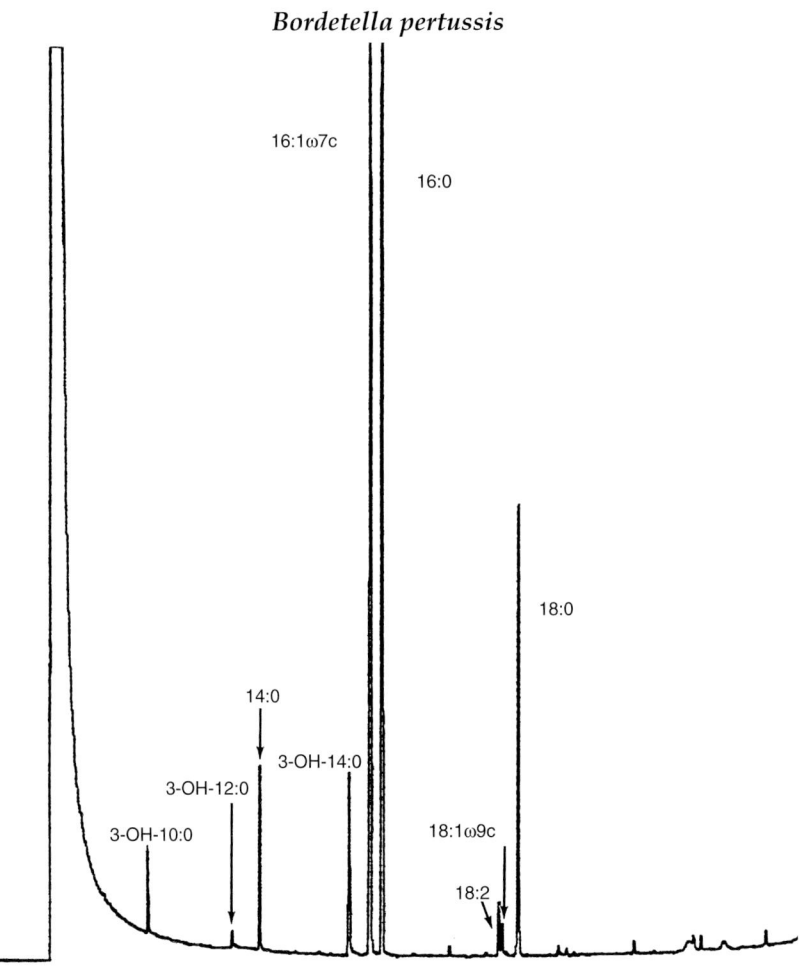

Figure 4.23. Gas chromatogram of the cellular fatty acids of *Bordetella pertussis*.

Comments

Bordetella pertussis has a unique CFA profile that is characterized by large amounts (32–40%) of 16:1ω7c and 16:0; moderate amounts (4–9%) of 3—OH-10:0, 14:0, 3—OH-14:0, and 18:0; and small amounts (1%) of 3—OH-12:0, 18:2, and 18:1ω9c. A significant difference in the CFA of *B. pertussis* compared to other *Bordetella* species is the absence of 17:0cyc. This fatty acid is present in moderate to large amounts (14–35%) in all other *Bordetella* species. *B. pertussis* also differs from *B. avium* and *B. holmesii* (NO-2) by the presence of 3—OH-10:0 and 3—OH-12:0; the absence of 2—OH-12:0 and 2—OH-14:0; and larger amounts of 14:0 (5% versus 1%), 3—OH-14:0 (9% versus 4–5%), and 16:1ω7c (40% versus 2%). The CFA profile of *B. pertussis* also differs from that of *B. parapertussis* by the presence of 1% each of 3—OH-12:0 and 18:1ω9c, the absence of 15:0 and 17:0, and increased amounts of 3—OH-14:0 (9% versus 1%) and 16:1ω7c (40% versus 6%). *B. pertussis* differs from the CFA group containing *B. bronchiseptica*, *Alcaligenes faecalis*, *A. xylosoxidans* subsp. *xylosoxidans*, and *A. xylosoxidans* subsp. *denitrificans* group 1 by the presence of 3—OH-10:0, 3—OH-12:0, and 18:1ω9c; the absence of 12:0, 2—OH-12:0, and 18:1ω7c; and increased amounts of 16:1ω7c (40% versus 18–27%) and 18:0 (8% versus 0–2%).

Literature

1. Weyant, R.S., D.J. Brenner, R.E. Weaver, M.F.M. Amin, D.G. Hollis, S.P. O'Connor, M.I. Daneshvar, and C.W. Moss. 1995. *Bordetella holmesii* sp. nov., a new gram-negative species associated with septicemia. J. Clin. Microbiol. 33: 1–7.

Table 4.18. Cellular fatty acid composition of *Bordetella pertussis*.

Fatty Acid	Organism *Bordetella pertussis* %
3-OH-10:0	4
3-OH-12:0	1
14:0	5
3-OH-14:0	9
16:1ω7c	40
16:0	32
18:2	1
18:1ω9c	1
18:0	8

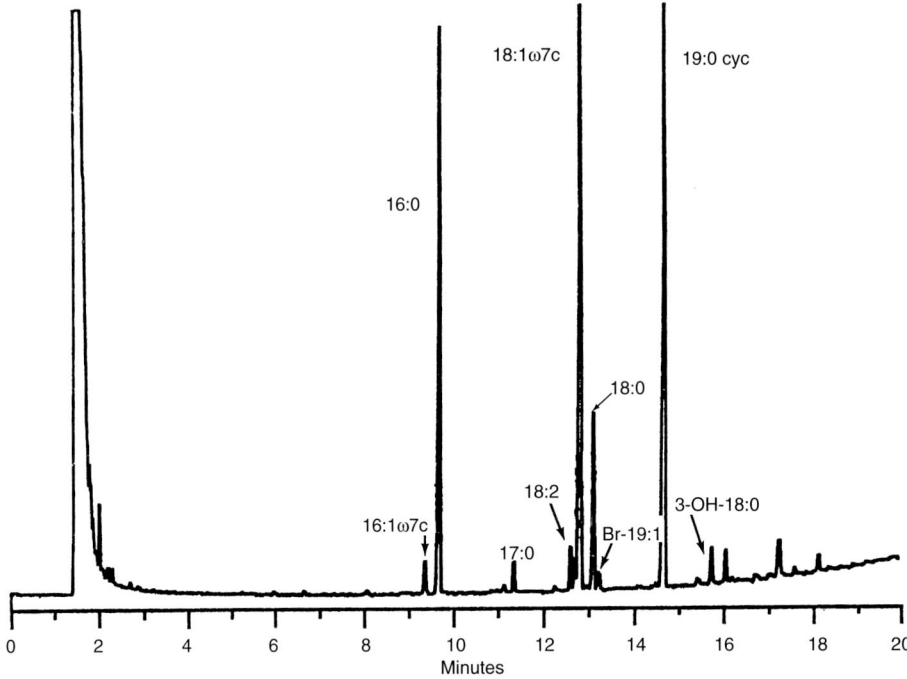

Figure 4.24. Gas chromatogram of the cellular fatty acids of *Brucella abortus*, *Brucella melitensis*, *Brucella ovis*, and *Brucella suis*.

Comments

The CFA profile of *Brucella abortus*, *B. melitensis*, *B. ovis*, and *B. suis* is recognized by the presence of 19:0cyc$^{11-12}$ as the major acid (37–46%); large amounts of 18:1ω7c (18–32%) and 18:0 (5–10%); Br-19:1 (1–4%); and no hydroxy acids (except 1% of 3—OH-18:0 in *B. abortus* and *B. melitensis*). These three species differ from *Brucella canis* by the presence of 19:0cyc$^{11-12}$ (37–46% versus 0%) and much less 18:1ω7c (32% versus 80%). These three species differ from *Ochrobactrum anthropi*, *Roseomonas cervicalis*, *R. gilardii*, and *Roseomonas* genomospecies 4 and 5 by the absence of 2—OH-18:1 and 2—OH-19:0cyc, and from *Bartonella* by the presence of 19:0cyc$^{11-12}$ (>37% versus 0%) and less 18:1ω7c (32% versus 59%).

Literature

1. Dees, S.B., D.G. Hollis, R.E. Weaver, and C.W. Moss. 1981. Cellular fatty acids of *Brucella canis* and *Brucella suis*. J. Clin. Microbiol. *14*: 111–112.

2. Dees, S.B., J. Powell, C.W. Moss, D.G. Hollis, and R.E. Weaver. 1981. Cellular fatty acid composition of organisms frequently associated with human infections resulting from dog bites: *Pasteurella multocida* and groups EF-4, IIj, M-5, and DF-2. J. Clin. Microbiol. *14*: 612–616.

3. Cole, P.J., A.J. Sinclair, J.F. Slattery, and D. Burke. 1984. Differentiation of *Brucella ovis* and *Brucella abortus* by gas-liquid chromatographic analysis of cellular fatty acids. J. Clin. Microbiol. *19*: 896–898.

4. Wallace, P.L., D.G. Hollis, R.E. Weaver, and C.W. Moss. 1990. Biochemical and chemical characterization of pink-pigmented oxidative bacteria. J. Clin. Microbiol. *28*: 689–693.

Table 4.19. Cellular fatty acid composition of *Brucella abortus*, *Brucella melitensis*, *Brucella ovis*, and *Brucella suis*.

Fatty Acid	Organism			
	Brucella abortus %	*Brucella melitensis* %	*Brucella ovis* %	*Brucella suis* %
15:0	–	–	2	–
16:1ω7c	1	1	1	1
16:0	13	15	14	9
17:1ω6c	–	–	T[a]	1
17:0	3	1	8	3
18:2	1	1	–	1
18:1ω9c	T	1	T	T
18:1ω7c	22	22	18	32
18:0	10	9	9	5
Br-19:1	3	4	1	1
19:0cyc$^{11-12}$	43	37	43	46
3-OH-18:0	1	1	–	–
UNK[b] 19.735	3	3	1	1

[a]T = Trace.
[b]UNK = Unknown.

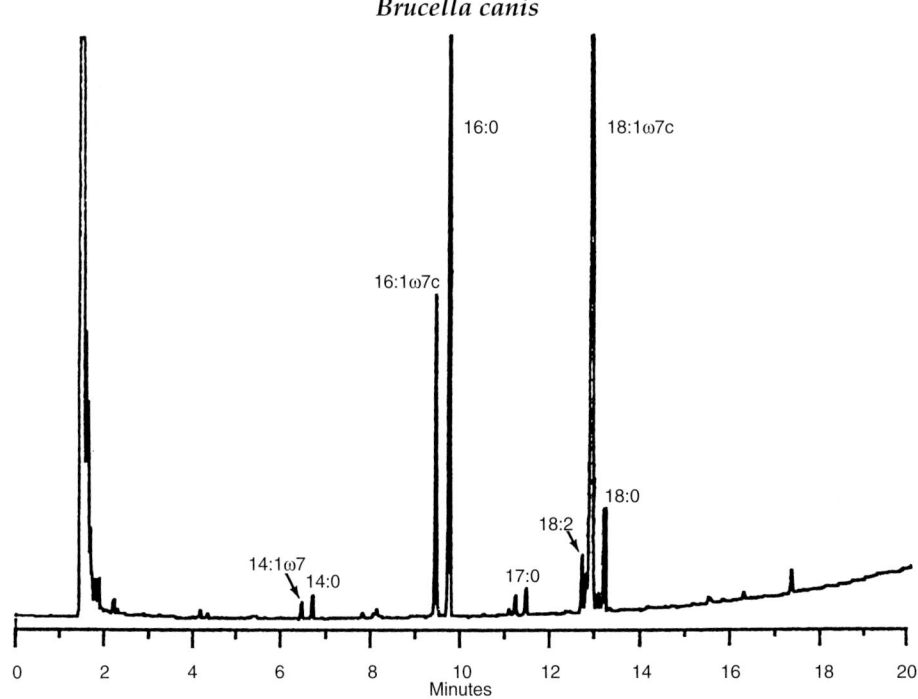

Figure 4.25. Gas chromatogram of the cellular fatty acids of *Brucella canis*.

Comments

This organism contains large amounts of 18:1ω7c (80%); moderate amounts of 16:0 (12%); small amounts (2–4%) of 16:1ω7c and 18:0; and trace amounts of 17:0, 18:1ω9c, and 18:2. *Brucella canis* differs from other *Brucella* species by the absence of 19:0cyc$^{11-12}$ (0% versus 37–46%) and larger amounts of 18:1ω7c (80% versus 32%). *Brucella canis* differs from *Afipia* species by the absence of 17:0cyc (0% versus 1%), Br-19:1 (0% versus 10%), and 19:0cyc$^{11-12}$ (0% versus 2%); and the presence of larger amounts of 18:0 (6% versus 2%), and trace amounts of 17:0 and 18:1ω9c. This organism differs from EO-3 by the presence of larger amounts of 16:1ω7c (3% versus 1%); and the absence of 19:0cyc$^{11-12}$ (0% versus 3%), 2—OH-18:1 (0% versus 1%), and 20:1ω9t (0% versus 1%). *Brucella canis* differs from the CFA group containing *Roseomonas fauriae*, *Roseomonas* genomospecies 6, *Methylobacterium extorquens*, *M. mesophilicum*, and *M. zatmanii* by the absence of 3—OH-14:0 (0% versus 2%).

Literature

1. Dees, S.B., D.G. Hollis, R.E. Weaver, and C.W. Moss. 1981. Cellular fatty acids of *Brucella canis* and *Brucella suis*. J. Clin. Microbiol. *14:* 111–112.

2. Wallace, P.L., D.G. Hollis, R.E. Weaver, and C.W. Moss. 1990. Biochemical and chemical characterization of pink-pigmented oxidative bacteria. J. Clin. Microbiol. *28:* 689–693.

3. Rihs, J.D., D.J. Brenner, R.E. Weaver, A.G. Steigerwalt, D.G. Hollis, and V.L. Yu. 1993. *Roseomonas*, a new genus associated with bacteremia and other human infections. J. Clin. Microbiol. *31:* 3275–3283.

Table 4.20. Cellular fatty acid composition of *Brucella canis*.

	Organism
	Brucella canis
Fatty Acid	%
16:1ω7c	3
16:0	12
17:0	T[a]
18:2	1
18:1ω9c	T
18:1ω7c	80
18:0	2

[a]T = Trace.

SECTION 4: CELLULAR FATTY ACID ANALYSIS

Campylobacter coli, Campylobacter helveticus, Campylobacter jejuni, Campylobacter lari, Campylobacter upsaliensis, Oligella ureolytica, **and** *Oligella urethralis*

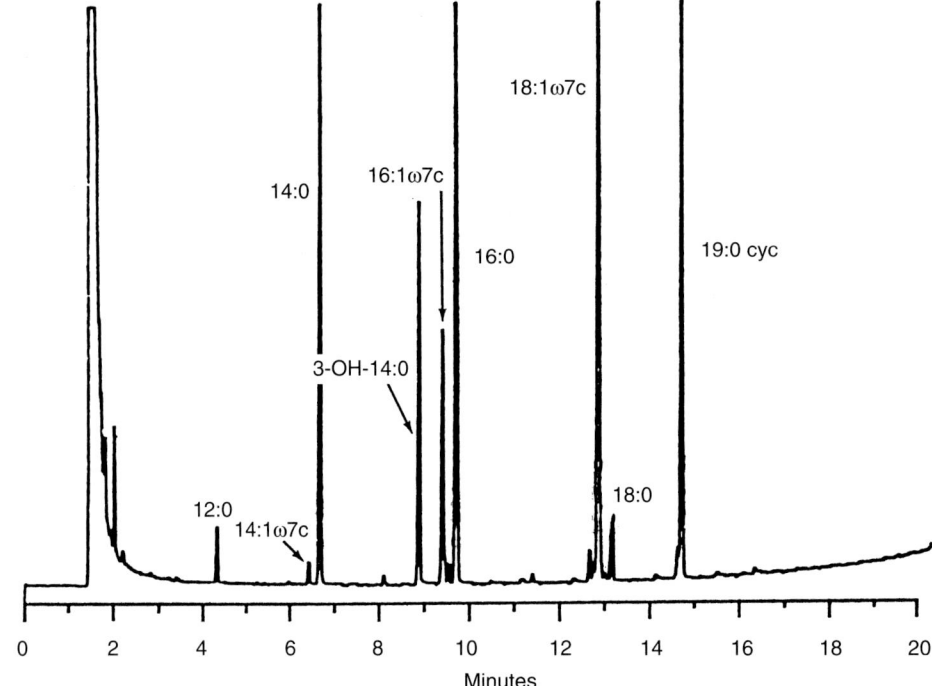

Figure 4.26. Gas chromatogram of the cellular fatty acids of *Campylobacter coli, Campylobacter helveticus, Campylobacter jejuni, Campylobacter lari, Campylobacter upsaliensis, Oligella ureolytica,* and *Oligella urethralis*.

Comments

Organisms in this CFA group contain major amounts of 18:1ω7c (28–47%) and 16:0 (30–44%) with moderate amounts of 14:0 and 16:1ω7c, and small amounts of 18:0. *Oligella ureolytica* and *O. urethralis* contain 3—OH-14:0 and small amounts of 3—OH-16:0 while the *Campylobacter* species contain only 3—OH-14:0. More than 95% of *C. jejuni* strains and about 80% of *C. coli* contain significant amounts (8–20%) of 19:0cyc[11–12] but this acid is absent in *C. helveticus, C. lari, C. upsaliensis,* and *O. urethralis*. Thus, all 19:0cyc[11–12]-containing strains of *C. jejuni* and *C. coli* are readily distinguished from *C. helveticus, C. lari, C. upsaliensis,* and *O. urethralis*. The CFA profile of *C. helveticus, C. lari,* and *C. upsaliensis* (and 19:0cyc[11–12]-negative *C. coli* and *C. jejuni* strains) is most similar to the CFA group containing *Haemophilus, Pasteurella,* and *Actinobacillus* except for smaller amounts of 14:0 (9% versus >15%) and larger amounts of 18:1ω7c (>20% versus <10%). *C. coli* and *C. jejuni* strains that contain 19:0cyc[11–12] are distinguished from *O. ureolyticus* by the absence of 3—OH-16:0 and Br-19:1.

Literature

1. Blaser, M.J., C.W. Moss, and R.E. Weaver. 1980. Cellular fatty acid composition of *Campylobacter fetus*. J. Clin. Microbiol. *11:* 448–451.

2. Leaper, S., and R.J. Owen. 1981. Identification of catalase producing *Campylobacter* species based on biochemical characteristics and cellular fatty acid composition. Curr. Microbiol. *6:* 31–35.

3. Hebert, G.A., D.G. Hollis, R.E. Weaver, M.A. Lambert, M.J. Blaser, and C.W. Moss. 1982. 30 years of *Campylobacter:* Biochemical characteristics and a biotyping proposal for *Campylobacter jejuni*. J. Clin. Microbiol. *15:* 1065–1073.

4. Curtis, M.A. 1983. Cellular fatty acid profile of campylobacters. Med. Lab. Sci. *40:* 333–348.

5. Moss, C.W., A. Kai, M.A. Lambert, and C.M. Patton. 1984. Isoprenoid quinone content and cellular fatty acid composition of *Campylobacter* species. J. Clin. Microbiol. *19:* 772–776.

Table 4.21. Cellular fatty acid composition of *Campylobacter coli, Campylobacter helveticus, Campylobacter jejuni, Campylobacter lari, Campylobacter upsaliensis, Oligella ureolytica,* and *Oligella urethralis*

Fatty Acid	Organism[a]						
	A	B	C	D	E	F	G
	%	%	%	%	%	%	%
14:0	6	6	9	4	8	9	7
15:0	-	-	1	-	-	T[b]	1
3-OH-14:0	5	6	3	5	6	6	7
16:1ω7c	3	5	5	6	7	3	4
16:0	44	39	36	35	35	31	30
17:0	-	-	T	-	-	-	1
3-OH-16:0	-	-	-	-	-	2	2
18:2	T	-	T	T	-	-	-
18:1ω7c	28	42	30	47	41	39	47
18:0	1	1	1	1	1	1	1
Br-19:1	-	-	-	-	-	1	-
19:0cyc[11-12]	11	-	12	-	-	6	-

[a] A, *Campylobacter coli;* B, *Campylobacter helveticus;* C, *Campylobacter jejuni;* D, *Campylobacter lari;* E, *Campylobacter upsaliensis;* F, *Oligella ureolytica;* G, *Oligella urethralis.*
[b] T = Trace.

Campylobacter concisus, *Campylobacter mucosalis*, and *Campylobacter* group DG

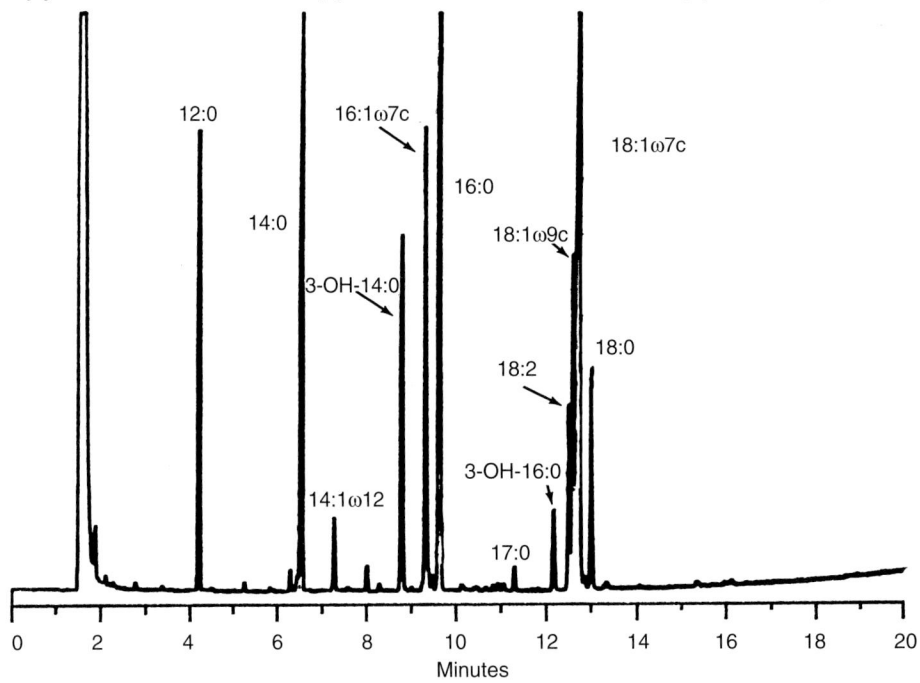

Figure 4.27. Gas chromatogram of the cellular fatty acids of *Campylobacter concisus*, *Campylobacter mucosalis*, and *Campylobacter* group DG.

Comments

The CFA profile of *Campylobacter concisus*, *C. mucosalis*, and *Campylobacter* group DG is characterized by large amounts of 16:0 and 18:1ω7c; moderate amounts of 14:0 and 16:1ω7c; and smaller amounts of 12:0, 3—OH-14:0, 3—OH-16:0, 18:1ω9c, 18:0, and 18:2. The presence of 5–7% of 12:0 distinguishes this CFA group from all other *Campylobacter*. The overall profile of this group is most similar to that of *C. hyointestinalis* but differs by the presence of 12:0 and 3—OH-16:0.

Literature

1. Lambert, M.A., C.M. Patton, T.J. Barrett, and C.W. Moss. 1987. Differentiation of *Campylobacter* and *Campylobacter*-like organisms by cellular fatty acid composition. J. Clin. Microbiol. 25: 706–713.

2. Brondy, I., and I. Olsen. 1991. Multivariate analyses of cellular fatty acids in *Bacteroides*, *Prevotella*, *Porphyromonas*, *Wolinella*, and *Campylobacter* ssp. J. Clin. Microbiol. 29: 183–189.

Table 4.22. Cellular fatty acid composition of *Campylobacter concisus*, *Campylobacter mucosalis*, and *Campylobacter* group DG

Fatty Acid	Organism		
	Campylobacter concisus %	*Campylobacter mucosalis* %	*Campylobacter* group DG %
12:0	7	5	7
14:0	9	10	9
14:1ω12	1	1	1
15:0	1	–	T[a]
3-OH-14:0	7	10	5
16:1ω7c	10	13	8
16:0	26	28	31
17:0	1	–	T
3-OH-16:0	2	6	2
18:2	5	6	5
18:1ω9c	7	4	4
18:1ω7c	21	12	24
18:0	5	5	4

[a]T = Trace.

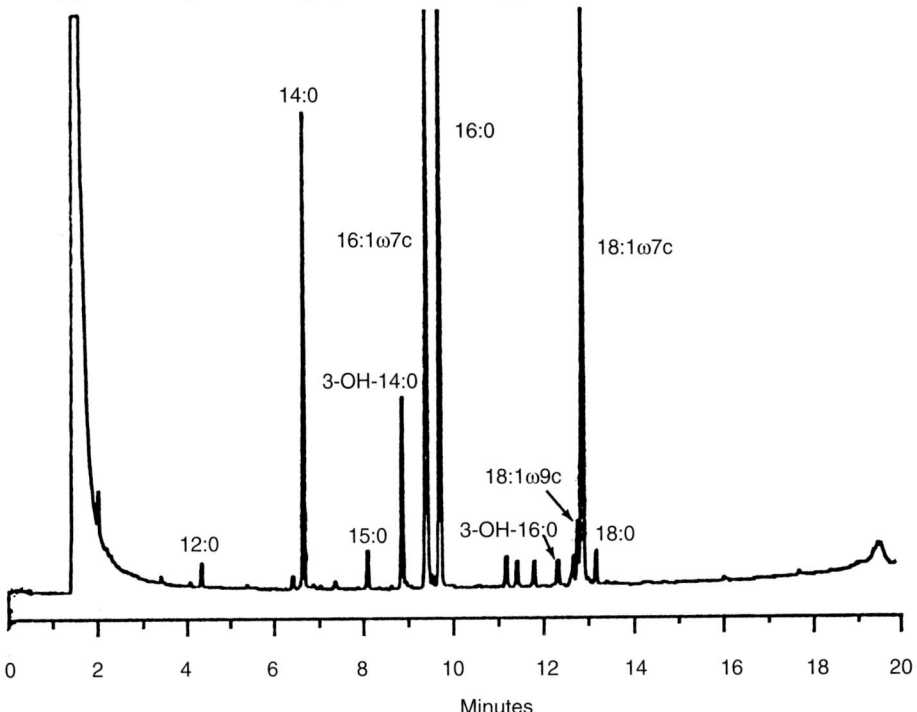

Figure 4.28. Gas chromatogram of the cellular fatty acids of *Campylobacter fetus* subsp. *fetus* and *Campylobacter fetus* subsp. *venerealis*.

Comments

Campylobacter fetus subsp. *fetus* and *C. fetus* subsp. *venerealis* contain major amounts of 18:1ω7c, 16:0, and 16:1ω7c; moderate amounts of 14:0 (8–10%); smaller amounts (1–4%) of 3—OH-14:0, 3—OH-16:0, 15:0, and 18:2; and only 1% amounts of 18:1ω9c and 18:0. This profile differs from *C. hyointestinalis* by the presence of 3—OH-16:0 and smaller amounts of 18:1ω9c and 18:0 (1% versus 2–4%); it differs from the CFA group containing *C. coli*, *C. helveticus*, *C. jejuni*, *C. lari*, and *C. upsaliensis* by the presence of 3—OH-16:0. The *C. fetus* profile is also similar to the CFA group containing *Haemophilus*, *Pasteurella*, and *Actinobacillus* except for the presence of 3—OH-16:0, larger amounts of 18:1ω7c (30% versus 3%), and smaller amounts of 14:0 (<10% versus >10%).

Literature

1. Moss, C.W., A. Kai, M.A. Lambert, and C.M. Patton. 1984. Isoprenoid quinone content and cellular fatty acid composition of *Campylobacter* species. J. Clin. Microbiol. 19: 772–776.

2. Lambert, M.A., C.M. Patton, T.J. Barrett, and C.W. Moss. 1987. Differentiation of *Campylobacter* and *Campylobacter*-like organisms by cellular fatty acid composition. J. Clin. Microbiol. 25: 706–713.

3. Brondy, I., and I. Olsen. 1991. Multivariate analyses of cellular fatty acids in *Bacteroides*, *Prevotella*, *Porphyromonas*, *Wolinella*, and *Campylobacter* ssp. J. Clin. Microbiol. 29: 183–189.

Table 4.23. Cellular fatty acid composition of *Campylobacter fetus* subsp. *fetus* and *Campylobacter fetus* subsp. *venerealis*

Fatty Acid	Organism	
	Campylobacter fetus subsp. *fetus*	*Campylobacter fetus* subsp. *venerealis*
	%	%
14:0	8	10
14:1ω12	T[a]	-
15:0	1	1
3-OH-14:0	4	3
16:1ω7c	19	17
16:0	33	40
17:0	-	1
3-OH-16:0	2	4
18:2	1	1
18:1ω9c	1	1
18:1ω7c	30	20
18:0	1	1

[a]T = Trace.

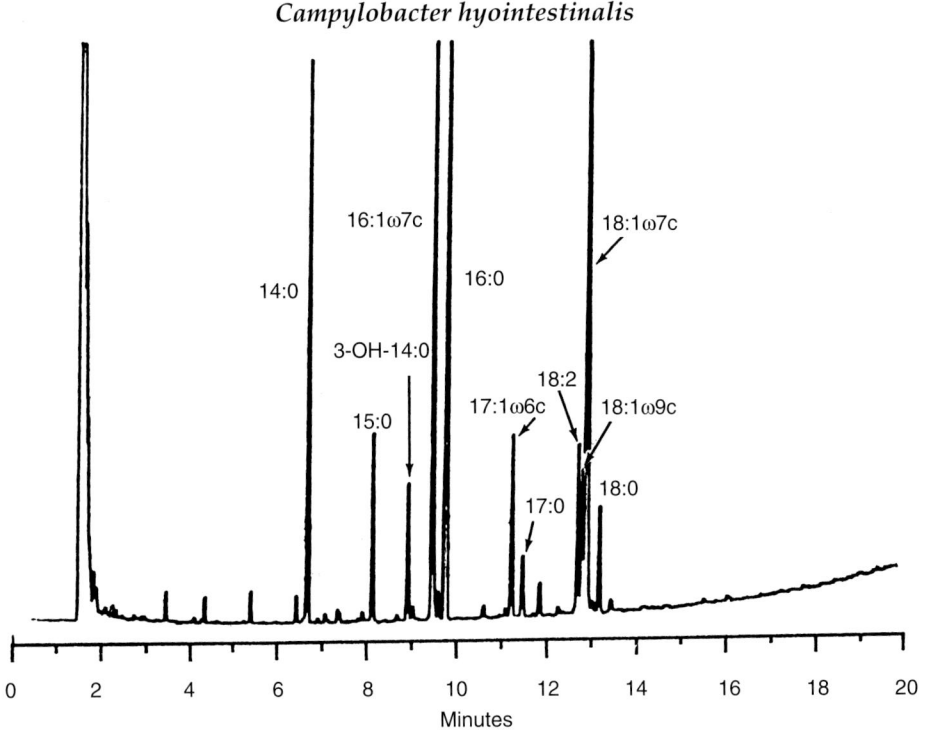

Figure 4.29. Gas chromatogram of the cellular fatty acids of *Campylobacter hyointestinalis*.

Comments

The CFA profile of *Campylobacter hyointestinalis* has large amounts of 16:0 (30%) and 18:1ω7c (25%); moderate amounts (10–13%) of 14:0 and 16:1ω7c; with smaller amounts (1–5%) of 15:0, 17:1ω6c, 17:0, 18:1ω9c, 18:2, and 18:0. This profile is most similar to the CFA group containing *C. coli*, *C. helveticus*, *C. jejuni*, *C. lari*, and *C. upsaliensis* except for the presence of 18:1ω9c (4% versus 0%) and larger amounts of 18:2 (5% versus trace). The profile is also similar to the CFA group containing *Haemophilus*, *Pasteurella*, and *Actinobacillus* except for larger amounts of 18:1ω7c (25% versus <10%).

Literature

1. Moss, C.W., A. Kai, M.A. Lambert, and C.M. Patton. 1984. Isoprenoid quinone content and cellular fatty acid composition of *Campylobacter* species. J. Clin. Microbiol. 19: 772–776.

2. Lambert, M.A., C.M. Patton, T.J. Barrett, and C.W. Moss. 1987. Differentiation of *Campylobacter* and *Campylobacter*-like organisms by cellular fatty acid composition. J. Clin. Microbiol. 25: 706–713.

Table 4.24. Cellular fatty acid composition of *Campylobacter hyointestinalis*

Fatty Acid	*Campylobacter hyointestinalis* %
14:0	10
14:1ω12	1
15:0	2
3-OH-14:0	5
16:1ω7c	13
16:0	30
17:1ω6c	1
17:0	1
18:2	5
18:1ω9c	4
18:1ω7c	25
18:0	2

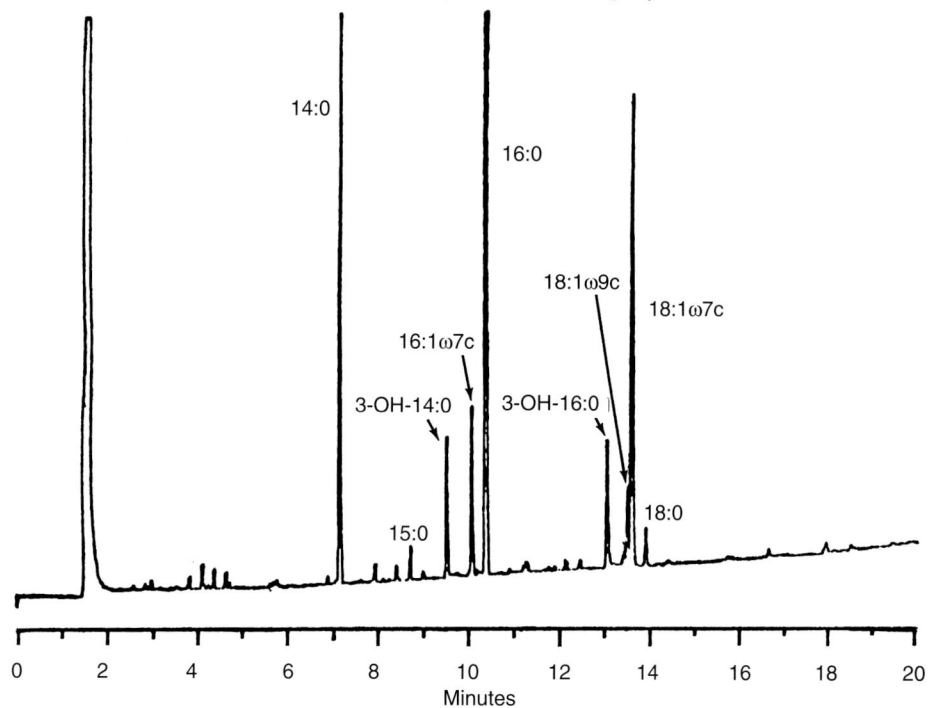

Figure 4.30. Gas chromatogram of the cellular fatty acids of *Campylobacter sputorum* subsp. *bubulus*, *Campylobacter sputorum* subsp. *faecalis*, and *Campylobacter sputorum* subsp. *sputorum*.

Comments

The CFA profile of *Campylobacter sputorum* subsp. *bubulus*, *C. sputorum* subsp. *faecalis*, and *C. sputorum* subsp. *sputorum* contains large amounts of 16:0, 18:1ω7c, and 14:0; moderate amounts (4–7%) of 3—OH-14:0, 3—OH-16:0, and 16:1ω7c; and smaller amounts (1–3%) of 18:2, 18:1ω9c, and 18:0. This profile is most similar to that of the CFA group containing *C. fetus* subsp. *fetus* and *C. fetus* subsp. *venerealis* except for larger amounts of 14:0 (16% versus <10%), higher amounts of 3—OH-16:0 (6% versus 4%), and smaller amounts of 16:1ω7c (6% versus 17%).

Literature

1. Moss, C.W., A. Kai, M.A. Lambert, and C.M. Patton. 1984. Isoprenoid quinone content and cellular fatty acid composition of *Campylobacter* species. J. Clin. Microbiol. *19:* 772–776.
2. Brondy, I., and I. Olsen. 1991. Multivariate analyses of cellular fatty acids in *Bacteroides, Prevotella, Porphyromonas, Wolinella,* and *Campylobacter* ssp. J. Clin. Microbiol. *29:* 183–189.

Table 4.25. Cellular fatty acid composition of *Campylobacter sputorum* subsp. *bubulus*, *Campylobacter sputorum* subsp. *faecalis*, and *Campylobacter sputorum* subsp. *sputorum*

	Organism		
	Campylobacter sputorum subsp. *bubulus*	*Campylobacter sputorum* subsp. *faecalis*	*Campylobacter sputorum* subsp. *sputorum*
Fatty Acid	%	%	%
14:0	18	17	16
14:1ω12	T[a]	T	–
15:0	T	1	1
3-OH-14:0	4	4	5
16:1ω7c	6	6	5
16:0	31	28	37
3-OH-16:0	6	6	7
18:2	1	2	3
18:1ω9c	1	1	2
18:1ω7c	29	31	23
18:0	1	1	1

[a]T = Trace.

Capnocytophaga canimorsus, *Capnocytophaga cynodegmi*, *Capnocytophaga gingivalis*, and *Capnocytophaga ochracea*

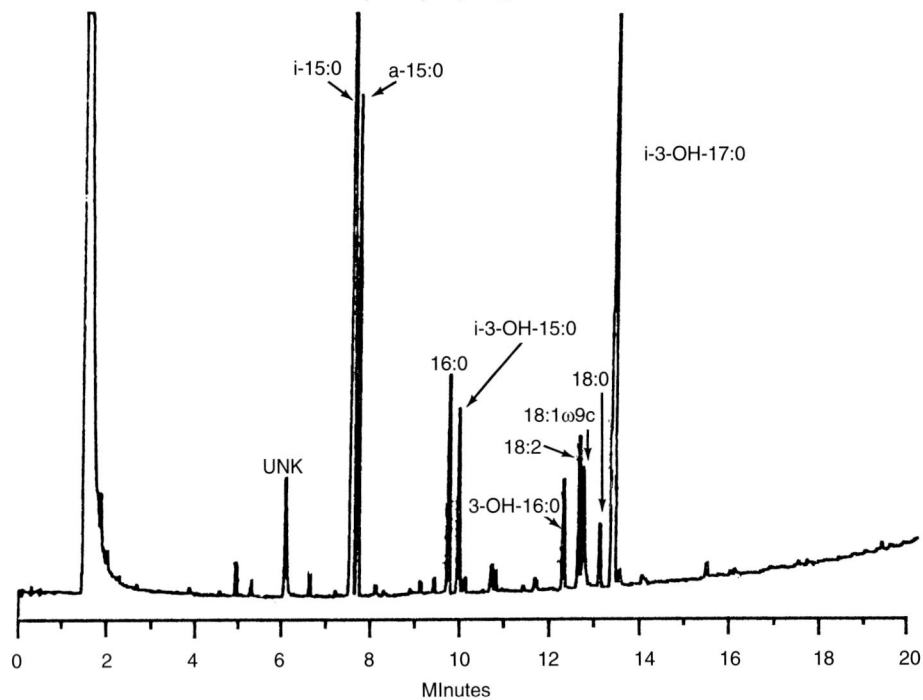

Figure 4.31. Gas chromatogram of the cellular fatty acids of *Capnocytophaga canimorsus*, *Capnocytophaga cynodegmi*, *Capnocytophaga gingivalis*, and *Capnocytophaga ochracea*.

Comments

Capnocytophaga species have a unique CFA profile that is characterized by major amounts (greater than 65%) of i-15:0 and smaller amounts of i-3—OH-15:0, i-3—OH-17:0, 3—OH-16:0, i-13:0, 14:0, 16:0, 18:2, 18:1ω9c, and 18:0. Organisms with generally similar CFA profiles include *Flavobacterium*; *Weeksella*; *Sphingobacterium*; *Cytophaga*; and IIc, IIe, and IIh. *Capnocytophaga* species differ from *Cytophaga johnsonae* by larger amounts of i-15:0 (65% versus 26%) and the absence of 16:1ω7c, 17:1ω6, and 3—OH-14:0 that are present at 22%, 2%, and 2% concentrations, respectively, in *C. johnsonae*. *Capnocytophaga* species differ from the CFA group containing *F. balustinum*, *F. gleum*, *F. indologenes*, *F. meningosepticum*, *F. odoratum*, *Flavobacterium* species (IIb), *W. virosa*, and *W. zoohelcum* by the absence of i-2—OH-15:0, and differs from *F. breve* by the absence of 16:1ω7c and i-17:1ω8c and larger amounts of i-15:0 (>60% versus 30%). *Capnocytophaga* species differ from the CFA group containing *Flavobacterium* (*Sphingobacterium*) *mizutaii*, *Flavobacterium* (*Sphingobacterium*) *multivorum*, *Flavobacterium* (*Sphingobacterium*) *thalpophilum*, *Flavobacterium* (*Sphingobacterium*) *spiritivorum*, and IIi by the absence of 16:1ω7c, i-2—OH-15:0, and i-17:1ω8c; and by larger amounts of i-15:0 (>60% versus 30%).

Literature

1. Dees, S.B., J. Powell, C.W. Moss, D.G. Hollis, and R.E. Weaver. 1981. Cellular fatty acid composition of organisms frequently associated with human infections resulting from dog bite: *Pasteurella multocida* and Groups EF-4, IIj, M-5, and DF-2. J. Clin. Microbiol. *14:* 612–616.

2. Dees, S.B., D.E. Karr, D.G. Hollis, and C.W. Moss. 1982. Cellular fatty acids of *Capnocytophaga* species. J. Clin. Microbiol. *16:* 779–783.

Table 4.26. Cellular fatty acid composition of *Capnocytophaga canimorsus*, *Capnocytophaga cynodegmi*, *Capnocytophaga gingivalis*, **and** *Capnocytophaga ochracea*

Fatty Acid	Organism			
	Capnocytophaga canimorsus %	*Capnocytophaga cynodegmi* %	*Capnocytophaga gingivalis* %	*Capnocytophaga ochracea* %
i-13:0	1	1	3	1
UNK[a] 13.566	1	2	-	-
14:0	1	1	1	2
i-15:0	75	78	70	65
a-15:0	2	2	1	1
15:0	-	-	1	-
16:1ω7c	-	-	-	T[b]
16:0	3	2	3	6
i-3-OH-15:0	3	3	3	3
3-OH-16:0	1	1	3	4
18:2	2	2	2	4
18:1ω9c	2	1	2	5
18:0	1	1	1	2
i-3-OH-17:0	8	4	10	7

[a]UNK = Unknown.
[b]T = Trace.

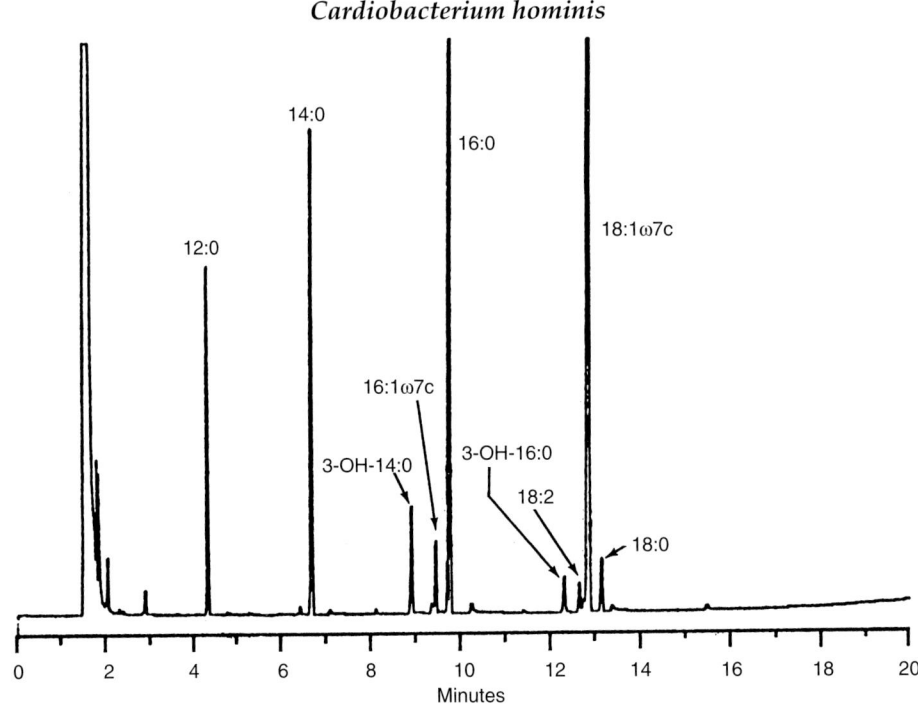

Figure 4.32. Gas chromatogram of the cellular fatty acids of *Cardiobacterium hominis*.

Comments

Cardiobacterium hominis has a unique CFA profile that is characterized by major amounts of 18:1ω7c (41%) and 16:0 (26%); large amounts of 14:0 (14%); with smaller amounts of 12:0 (9%), 16:1ω7c (3%), and 3—OH-14:0 (3%); and trace to 1% amounts of 3—OH-16:0. Does not contain 3—OH-10:0 or 3—OH-12:0. The CFA profile of *C. hominis* differs from that of *Suttonella indologenes* by the presence of 3—OH-16:0, larger amounts of 3—OH-14:0 (3% versus 1%), the absence of 3—OH-12:0 (0% versus 2%), and higher amounts of 18:1ω7c (41% versus 26%).

Literature

1. Wallace, P.L., D.G. Hollis, R.E. Weaver, and C.W. Moss. 1988. Cellular fatty acid composition of *Kingella* species, *Cardiobacterium hominis,* and *Eikenella corrodens*. J. Clin. Microbiol. 26: 1592–1594.

Table 4.27. Cellular fatty acid composition of *Cardiobacterium hominis*

Fatty Acid	Organism *Cardiobacterium hominis* %
12:0	9
14:0	14
3-OH-14:0	3
16:1ω7c	3
16:0	26
3-OH-16:0	1
18:2	1
18:1ω9c	T[a]
18:1ω7c	41
18:0	1

[a]T = Trace.

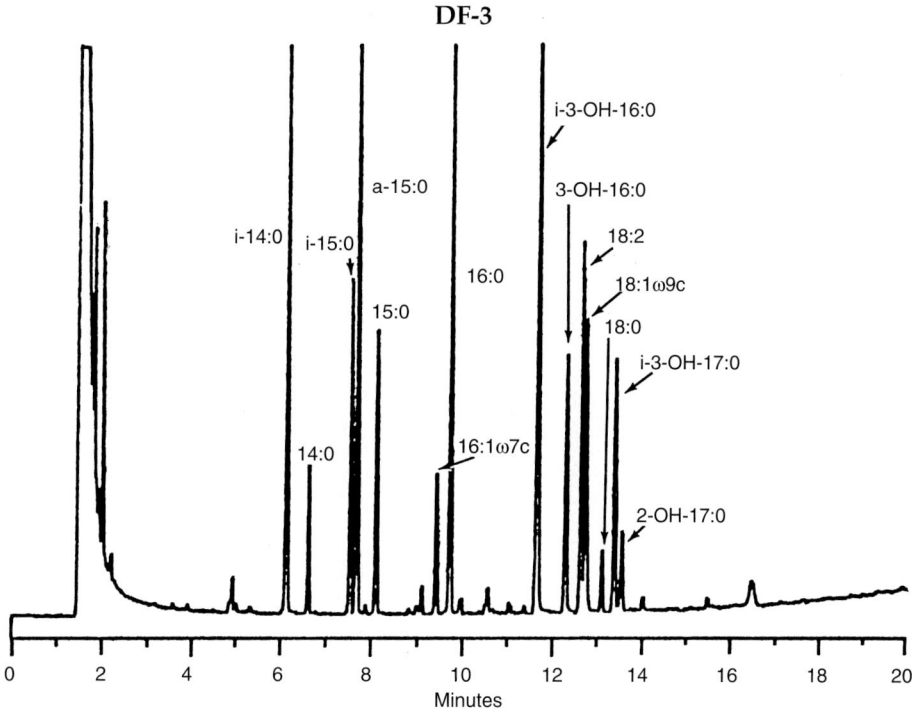

Figure 4.33. Gas chromatogram of the cellular fatty acids of DF-3.

Comments

DF-3 has a unique CFA profile that is characterized by large amounts (25%) of a-15:0; moderate amounts of i-14:0, i-15:0, and 15:0; and small to moderate amounts of both branched- and straight-chain hydroxy acids (i-3—OH-15:0, 3—OH-15:0, i-3—OH-16:0, 3—OH-16:0, i-3—OH-17:0, 2—OH-17:0). Differs from DF-3-like organisms by the presence of i-14:0, 15:0, 3—OH-15:0, and i-3—OH-16:0; and smaller amounts of i-15:0 (12% versus 21% for DF-3-like). Group DF-3 contains a-15:0 as the major fatty acid compared to *Flavobacterium, Capnocytophaga,* and *Cytophaga,* each of which contain i-15:0 as the major acid. In addition to smaller amounts of i-15:0, group DF-3 differs from all *Capnocytophaga* species by the presence of i-14:0, i-3—OH-16:0, and 3—OH-15:0.

Literature

1. Wallace, P.L., D.G. Hollis, R.E. Weaver, and C.W. Moss. 1989. Characterization of CDC group DF-3 by cellular fatty acid analysis. J. Clin. Microbiol. 27: 735–737.

Table 4.28. Cellular fatty acid composition of DF-3

Fatty Acid	Organism DF-3 %
i-13:0	2
a-13:0	1
i-14:0	6
14:0	2
i-15:0	12
a-15:0	25
15:0	3
16:1ω7c	1
16:0	10
i-3-OH-15:0	1
3-OH-15:0	1
i-3-OH-16:0	6
3-OH-16:0	3
18:2	10
18:1ω9c	4
18:0	2
i-3-OH-17:0	6
2-OH-17:0	1

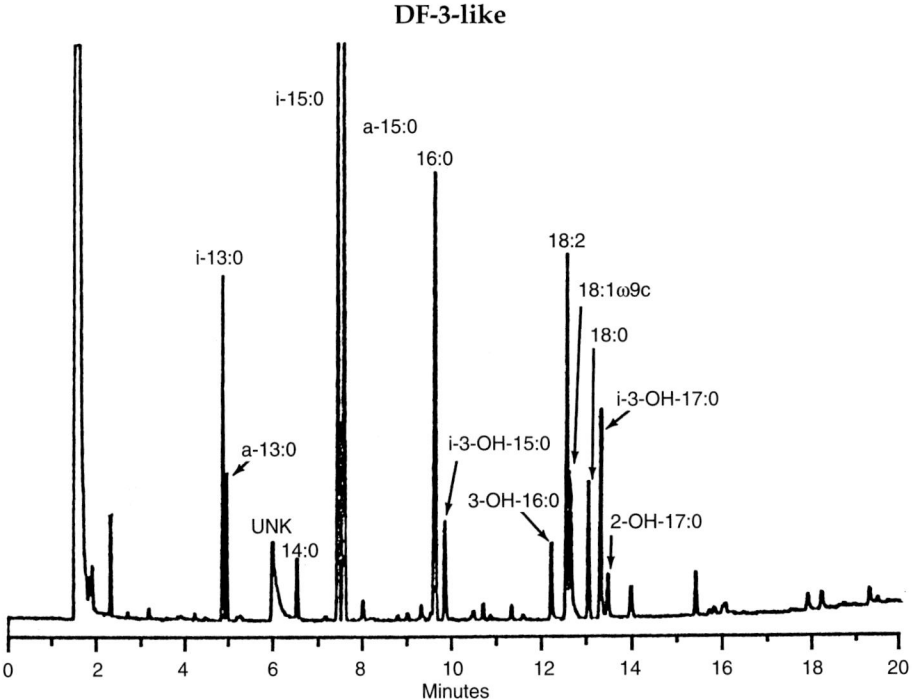

Figure 4.34. Gas chromatogram of the cellular fatty acids of DF-3-like.

Comments

DF-3-like organisms have a distinct CFA profile that is characterized by large amounts (>20%) of both i-15:0 and a-15:0; moderate amounts (3–6%) of i-13:0, a-13:0, and 16:0; and small to moderate amounts of i-3—OH-15:0, 3—OH-16:0, i-3—OH-17:0, and 2—OH-17:0. This CFA profile differs from that of DF-3 organisms by the absence of i-14:0, i-3—OH-16:0, and 15:0; and larger amounts of i-15:0 (21% versus 12%). DF-3-like organisms contain nearly equal amounts of i-15:0 and a-15:0 compared to *Flavobacterium, Capnocytophaga, Cytophaga,* and *Weeksella* species that contain significantly greater amounts of i-15:0 than a-15:0 (i-15:0/a-15:0 greater than 2:1 in all of these species).

Literature

1. Dees, S.B., D.E. Karr, D.G. Hollis, and C.W. Moss. 1982. Cellular fatty acids of *Capnocytophaga* species. J. Clin. Microbiol. *16:* 779–783.

2. Yabuuchi, E., and C.W. Moss. 1982. Cellular fatty acid composition of three species of *Sphingobacterium* gen. nov. and *Cytophaga johnsonae*. FEMS Microbiol. Lett. *13:* 87–91.

3. Daneshvar, M.I., D.G. Hollis, and C.W. Moss. 1991. Chemical characterization of clinical isolates which are similar to CDC group DF-3 bacteria. J. Clin. Microbiol. *29:* 2351–2353.

Table 4.29. Cellular fatty acid composition of DF-3-like

Fatty Acid	Organism DF-3-like %
i-13:0	5
a-13:0	2
14:0	1
i-15:0	21
a-15:0	23
16:0	11
i-3-OH-15:0	2
i-17:0	T[a]
3-OH-16:0	2
18:2	10
18:1ω9c	5
18:0	4
i-3-OH-17:0	10
2-OH-17:0	1

[a]T = Trace.

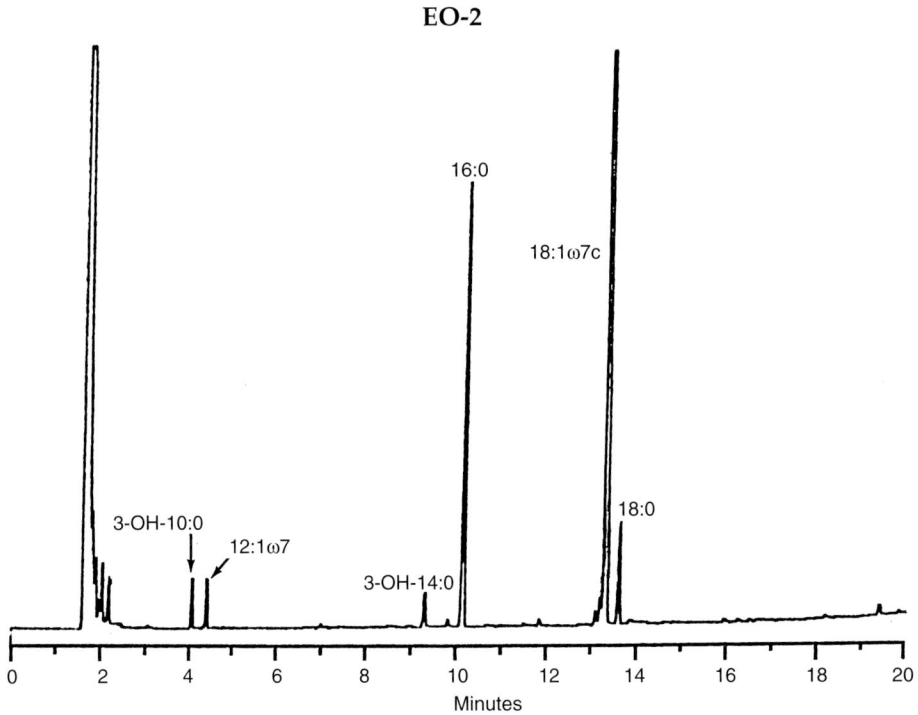

Figure 4.35. Gas chromatogram of the cellular fatty acids of EO-2.

Comments

This CFA group is characterized by very large amounts of 18:1ω7c (70%) and small amounts of 3—OH-10:0; it also contains moderate amounts of 16:0 (16%) and small amounts of 18:2, 18:1ω9c, and 18:0. The overall CFA profile of this group is most like that of the CFA group containing *Methylobacterium* species, *Roseomonas fauriae*, and *Roseomonas* genomospecies 6 except for the presence of 3—OH-10:0 (3% versus 0%). EO-2 has about 3% of 12:1ω7, which is absent in all *Methylobacterium* species and all *Roseomonas* species and genomospecies.

Literature

1. Moss, C.W., P.L. Wallace, D.G. Hollis, and R.E. Weaver. 1988. Cultural and chemical characterization of CDC groups EO-2, M-5, and M-6, *Moraxella (Moraxella)* species, *Oligella urethralis*, *Acinetobacter* species, and *Psychrobacter immobilis*. J. Clin. Microbiol. 26: 484–492.

2. Wallace, P.L., D.G. Hollis, R.E. Weaver, and C.W. Moss. 1990. Biochemical and chemical characterization of pink-pigmented oxidative bacteria. J. Clin. Microbiol. 28: 689–693.

Table 4.30. Cellular fatty acid composition of EO-2

Fatty Acid	Organism EO-2 %
3-OH-10:0	3
12:1ω7c	3
3-OH-14:0	1
16:0	13
17:0	1
18:2	1
18:1ω9c	1
18:1ω7c	71
18:0	5

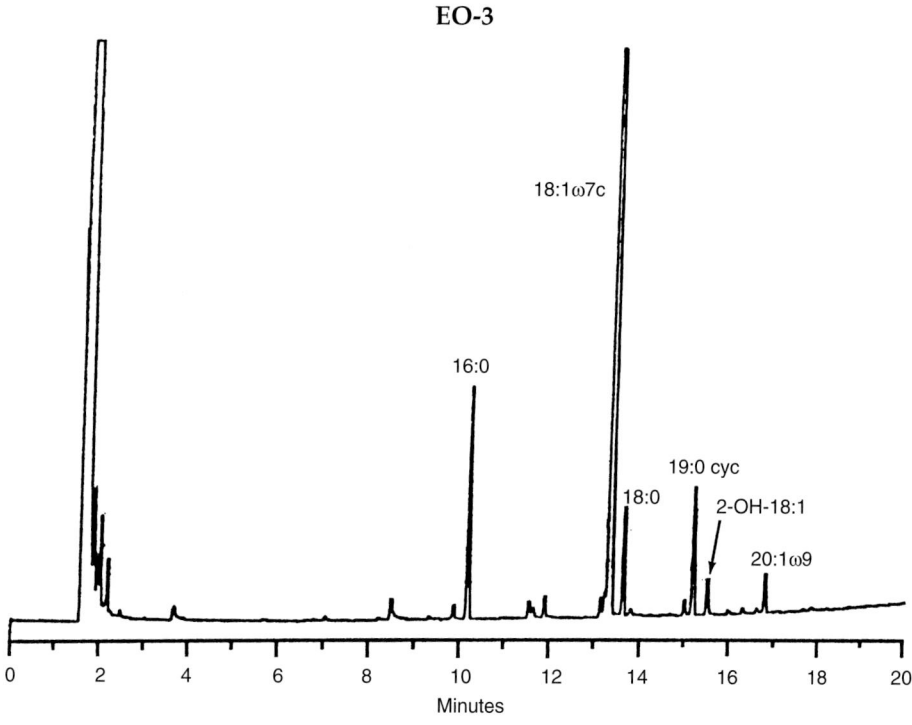

Figure 4.36. Gas chromatogram of the cellular fatty acids of EO-3.

Comments

EO-3 has a distinct CFA profile that is characterized by greater than 80% of 18:1ω7c, 8% 16:0, 3% 19:0cyc$^{11-12}$, and 1% amounts, each of 2—OH-18:1 and 20:1. This profile differs from that of *Brucella canis* by the presence of 19:0cyc$^{11-12}$, 2—OH-18:1, and 20:1, and from *Methylobacterium* species by the presence of 19:0cyc$^{11-12}$ and absence of 3—OH-14:0 and/or 3—OH-18:0 (0% versus 2%) in *Methylobacterium* species. EO-3 differs from the CFA group containing *Roseomonas cervicalis, R. gilardii,* and *Roseomonas* genomospecies 4 and 5, and *Ochrobactrum anthropi* by containing larger amounts of 18:1ω7c (80% versus less than 65%) and no 2—OH-19:0cyc.

Literature

1. Moss, C.W., P.L. Wallace, D.G. Hollis, and R.E. Weaver. 1988. Cultural and chemical characterization of CDC groups EO-2, M-5, and M-6, *Moraxella (Moraxella)* species, *Oligella urethralis, Acinetobacter* species, and *Psychrobacter immobilis*. J. Clin. Microbiol. 26: 484 492.

2. Wallace, P.L., D.G. Hollis, R.E. Weaver, and C.W. Moss. 1990. Biochemical and chemical characterization of pink-pigmented oxidative bacteria. J. Clin. Microbiol. 28: 689–693.

Table 4.31. Cellular fatty acid composition of EO-3.

Fatty Acid	Organism EO-3 %
16:1ω7c	1
16:0	8
17:0	1
18:2	1
18:1ω9c	1
18:1ω7c	81
18:0	3
19:0cyc$^{11-12}$	3
2-OH-18:1	1
20:1ω9t	1

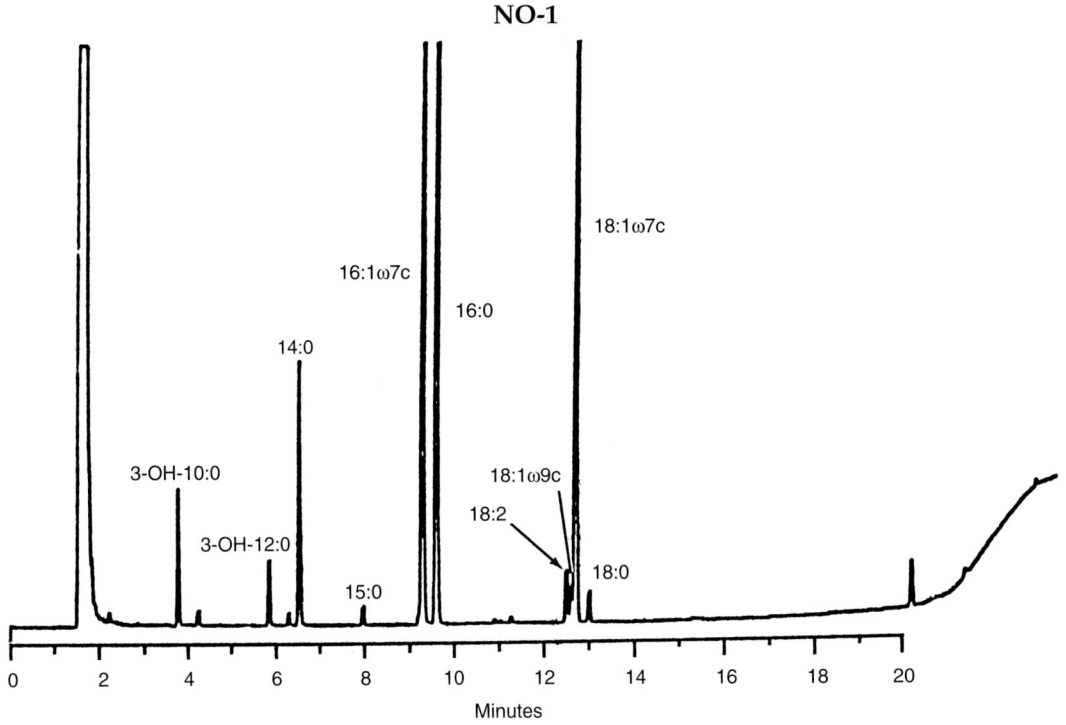

Figure 4.37. Gas chromatogram of the cellular fatty acids of NO-1.

Comments

NO-1 has a unique CFA profile characterized by large amounts (>20%) of 18:1ω7c, 16:0, and 16:1ω7c; moderate amounts (4–8%) of 14:0 and 3—OH-10:0; and small amounts of 18:0, 18:2, and 3—OH-12:0. This profile differs from *Acinetobacter* species which lacks 3—OH-10:0 and contains six acids (12:0, 12:1ω9, 2—OH-12:0, 17:1ω8c, 17:0, 18:1ω9c) that are absent in NO-1. NO-1 differs from the CFA group containing *Chryseomonas luteola*, *Flavimonas oryzihabitans*, and *Pseudomonas syringae* by the absence of 2—OH-12:0, larger amounts of 14:0 (8% versus 1%), and smaller amounts of 12:0 (trace versus 5%).

Literature

1. Hollis, D.G., C.W. Moss, M.I. Daneshvar, L. Meadows, J. Jordan, and B. Hill. 1993. Characterization of Centers for Disease Control group NO-1, a fastidious, nonoxidative, gram-negative organism associated with dog and cat bites. J. Clin. Microbiol. *31:* 746–748.

Table 4.32. Cellular fatty acid composition of NO-1

Fatty Acid	Organism NO-1 %
3-OH-10:0	4
3-OH-12:0	1
14:0	8
16:1ω7c	26
16:0	34
18:2	1
18:1ω7c	24
18:0	1

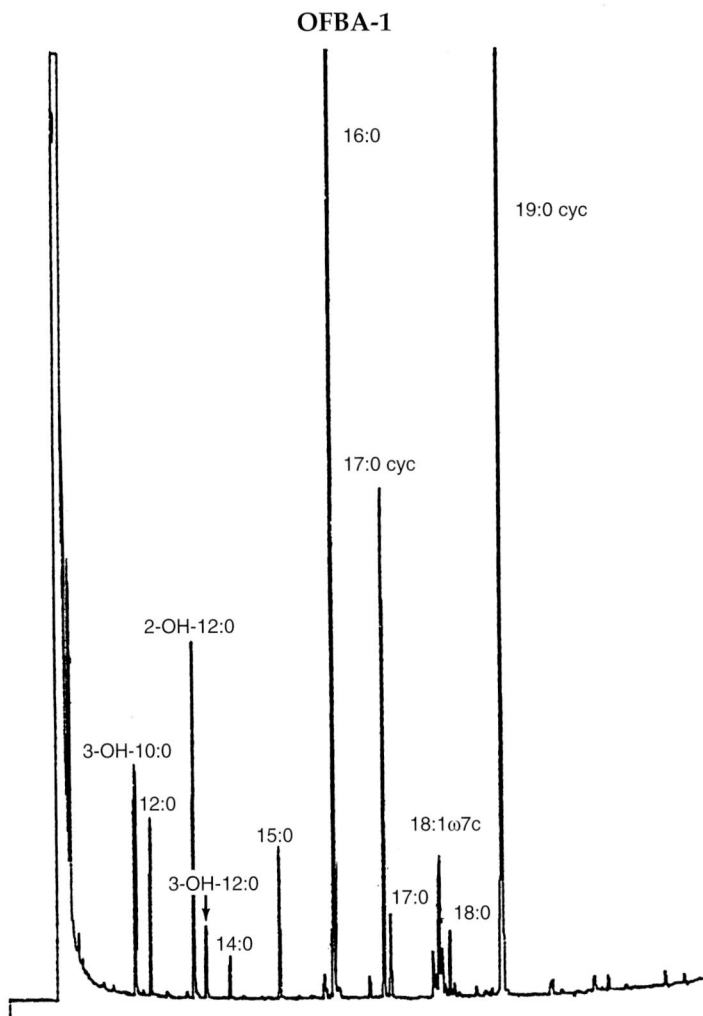

Figure 4.38. Gas chromatogram of the cellular fatty acids of OFBA-1.

Comments

OFBA-1 organisms have a unique CFA profile that consists of major amounts of 16:0 (36%) and 19:0cyc[11–12] (24%); moderate amounts (4–7%) of 3—OH-10:0, 2—OH-12:0, 17:0cyc, and 18:1ω7c; small amounts (1–3%) of 12:0, 3—OH-12:0, 15:0, 16:1ω7c, 17:0, 18:1ω9c, 18:0, and 18:2; and trace amounts of 14:0 and Br-19:1. They do not contain 3—OH-14:0.

Literature

1. von Graevenitz, A., G.E. Pfyffer, M.J. Pickett, R.E. Weaver, and J. Wüst. 1993. Isolation of an unclassified nonfermentative gram-negative rod from a patient on continuous ambulatory peritoneal dialysis. Eur. J. Clin. Microbiol. Infect. Dis. *12:* 568–570.

Table 4.33. Cellular fatty acid composition of OFBA-1

Fatty Acid	Organism OFBA-1 (%)
3-OH-10:0	3
12:0	2
2-OH-12:0	6
3-OH-12:0	2
14:0	1
15:0	2
16:1ω7c	1
16:0	36
17:0cyc	7
17:0	1
18:2	2
18:1ω9c	1
18:1ω7c	6
18:0	2
19:0cyc$^{11-12}$	24

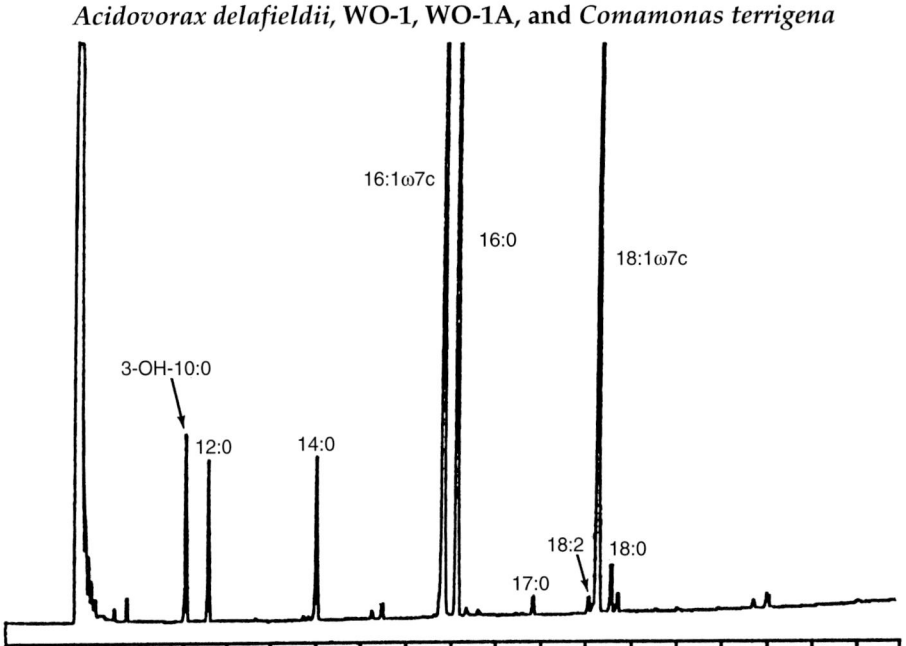

Figure 4.39. Gas chromatogram of the cellular fatty acids of *Acidovorax delafieldii*, WO-1, WO-1A, and *Comamonas terrigena*.

Comments

Species in this CFA group have 16:1ω7c as the major fatty acid (32–42%); large amounts (17–27%) of 16:0; smaller amounts of 18:1ω7c, 12:0, 14:0, 15:0, and 17:0; and no cyclopropane acids. All contain 3–5% of 3—OH-10:0, and *Comamonas terrigena* contains 2% 2—OH-16:0 that is absent in other members of this CFA group. The presence of 3—OH-10:0 distinguishes this CFA group from the CFA group containing *Eikenella corrodens, Suttonella indologenes, Arcobacter nitrofigilis, Neisseria,* and EF-4a and EF-4b that do not contain 3—OH-10:0. The absence of 17:0cyc distinguishes this CFA from the CFA group containing *Alcaligenes piechaudii, C. acidovorans, C. testosteroni, Pseudomonas alcaligenes,* and *P. pseudoalcaligenes* that contain 4–16% of 17:0cyc. WO-1 contains 2% of i-3—OH-15:0 that is absent in other members of the group, and WO-1A differs by the presence of significant amounts of 15:1ω6c (11%), 15:0 (12%), and 17:0 (7%). The overall CFA profile of WO-1A is most similar to that of the type strain of *Acidovorax temperans* that also contains 15:1ω6c, 15:0, and 17:0.

Literature

1. Hollis, D.G., R.E. Weaver, C.W. Moss, M.I. Daneshvar, and P.L. Wallace. 1992. Chemical characterization of CDC group WO-1, a weakly oxidative gram-negative group of organisms isolated from clinical sources. J. Clin. Microbiol. *30:* 291–295.

Table 4.34. Cellular fatty acid composition of *Acidovorax delafieldii*, WO-1, WO-1A, and *Comamonas terrigena*

Fatty Acid	Organism			
	Acidovorax delafieldii %	WO-1 %	WO-1A %	*Comamonas terrigena* %
3-OH-8:0	-	1	-	-
11:0	-	-	1	-
3-OH-10:0	4	4	3	5
12:0	3	7	5	4
13:0	-	-	2	-
14:0	4	2	1	1
15:1ω6c	1	1	12	-
15:0	1	1	13	1
16:1ω7c	38	42	30	40
16:0	27	27	17	27
i-3-OH-15:0	-	2	-	-
17:1ω8c	-	-	1	-
17:1ω6c	-	-	1	-
17:0	1	1	7	1
2-OH-16:0	-	-	-	2
18:2	2	-	-	-
18:1ω9c	1	-	-	-
18:1ω7c	16	11	5	21
18:0	1	T[a]	-	-

[a]T = Trace.

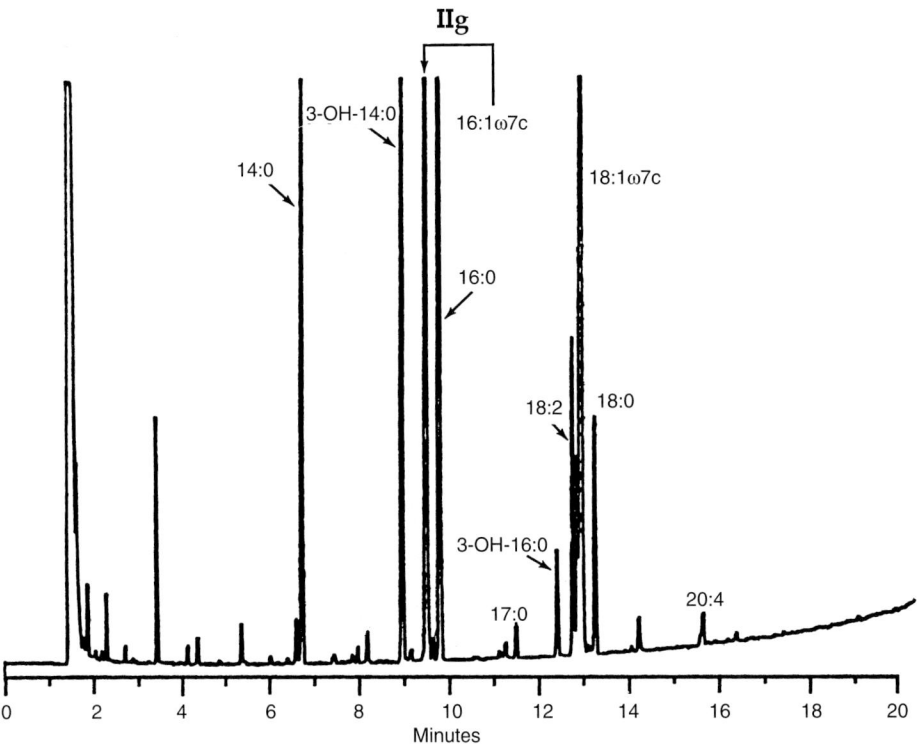

Figure 4.40. Gas chromatogram of the cellular fatty acids of IIg.

Comments

The CFA composition of IIg is differentiated from *Flavobacterium, Sphingobacterium, Weeksella, Capnocytophaga,* or *Cytophaga* species by the absence of branched chain acids. IIg contains 18:1ω7c (29%), 16:0 (27%), and 16:1ω7c (19%) as major acids; with smaller amounts of 3—OH-14:0 (10%), 14:0 (6%), 18:2 (2%), 18:0 (2%), and 3—OH-16:0 (1%). This CFA profile is essentially identical to that of *Campylobacter fetus* but the biochemical characteristics, morphology, and culture requirements of *C. fetus* are completely different from IIg. The CFA profile of IIg differs from the CFA group containing *Haemophilus, Pasteurella,* and *Actinobacillus* by smaller amounts of 14:0 and much larger amounts of 18:1ω7c (29% versus trace).

Literature

1. Dees, S.B., J. Powell, C.W. Moss, D.G. Hollis, and R.E. Weaver. 1981. Cellular fatty acid composition of organisms frequently associated with human infections resulting from dog bite: *Pasteurella multocida* and Groups EF-4, IIj, M-5, and DF-2. J. Clin. Microbiol. *14:* 612–616.

2. Yabuuchi, E., T. Kaneko, I. Yano, C.W. Moss, and N. Miyoshi. 1983. *Sphingobacterium* gen. nov., *Sphingobacterium spiritivorum* comb. nov., *Sphingobacterium multivorum* comb. nov., *Sphingobacterium mizutae* sp. nov., and *Flavobacterium indologenes* sp. nov.: glucose-nonfermenting gram-negative rods in CDC groups IIk-2 and IIb. Int. J. Syst. Bacteriol. *33:* 580–598.

3. Dees, S.B., C.W. Moss, D.G. Hollis, and R.E. Weaver. 1986. Chemical characterization of *Flavobacterium odoratum, Flavobacterium breve,* and *Flavobacterium*-like groups IIc, IIh, and IIf. J. Clin. Microbiol. *23:* 267–273.

Table 4.35. Cellular fatty acid composition of IIg

Fatty Acid	Organism IIg %
14:0	6
15:0	T[a]
3-OH-14:0	10
16:1ω7c	19
16:0	27
17:0	1
3-OH-16:0	1
18:2	2
18:1ω9c	1
18:1ω7c	29
18:0	3

[a]T = Trace.

Vb-3, *Pseudomonas mendocina*, *Pseudomonas pertucinogena*, and *Pseudomonas stutzeri*

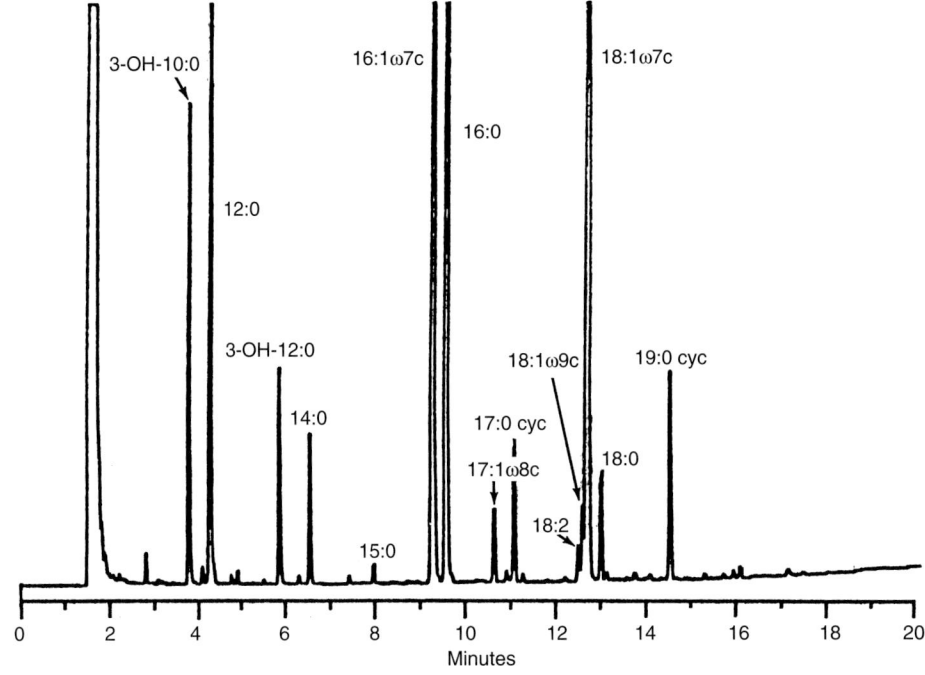

Figure 4.41. Gas chromatogram of the cellular fatty acids of Vb-3, *Pseudomonas mendocina*, *Pseudomonas pertucinogena*, and *Pseudomonas stutzeri*.

Comments

Organisms in this CFA group contain 18:1ω7c, 16:0, and 16:1ω7c as major acids; small to moderate amounts of 17:0cyc and 19:0cyc[11–12]; and small amounts of 3—OH-10:0 and 3—OH-12:0. Do not contain 10:0 or 3—OH-14:0. This CFA profile is most similar to that of *Pseudomonas* sp. CDC Group 1 except for the presence of 19:0cyc[11–12] and the absence of 10:0.

Literature

1. Moss, C.W., and S.B. Dees. 1976. Cellular fatty acids and metabolic products of *Pseudomonas* species obtained from clinical specimens. J. Clin. Microbiol. *4:* 492–502.

2. Oyaizu, H., and K. Komagata. 1983. Grouping of *Pseudomonas* species on the basis of cellular fatty acid composition and the quinone system with special reference to the existence of 3-hydroxy fatty acids. J. Gen. Appl. Microbiol. *29:* 17–40.

3. Veys, A., W. Callewaert, E. Waelkens, and K.V.D. Abbeele. 1989. Application of gas-liquid chromatography to the routine identification of nonfermentative gram-negative bacteria in clinical specimens. J. Clin. Microbiol. *27:* 1538–1542.

4. Osterhout, G.J., V.H. Shull, and J.D. Dick. 1991. Identification of clinical isolates of gram-negative nonfermentative bacteria by an automated cellular fatty acid identification system. J. Clin. Microbiol. *29:* 1822–1830.

Table 4.36. Cellular fatty acid composition of **Vb-3**, *Pseudomonas mendocina*, *Pseudomonas pertucinogena*, and *Pseudomonas stutzeri*

Fatty Acid	Organism			
	Vb-3	Pseudomonas mendocina	Pseudomonas pertucinogena	Pseudomonas stutzeri
	%	%	%	%
3-OH-10:0	3	4	3	4
12:0	12	10	7	9
3-OH-12:0	1	1	2	1
14:0	1	1	3	2
15:0	-	1	1	-
16:1ω7c	20	17	7	23
16:1ω7t	1	-	-	-
16:0	24	24	19	27
i-17:0	1	T[a]	2	T
17:1ω8c	-	1	-	-
17:0cyc	2	2	12	4
17:0	-	1	1	-
18:2	T	-	-	T
18:1ω9c	1	1	-	1
18:1ω7c	29	32	25	23
18:0	1	1	1	1
Br-19:1	-	-	1	-
19:0cyc$^{11-12}$	2	2	14	4

[a]T = Trace.

*Chromobacterium violaceum, Pseudomonas aeruginosa, "Pseudomonas denitrificans," Pseudomonas fluorescens, Pseudomonas-*like Group 2, and *Pseudomonas putida*

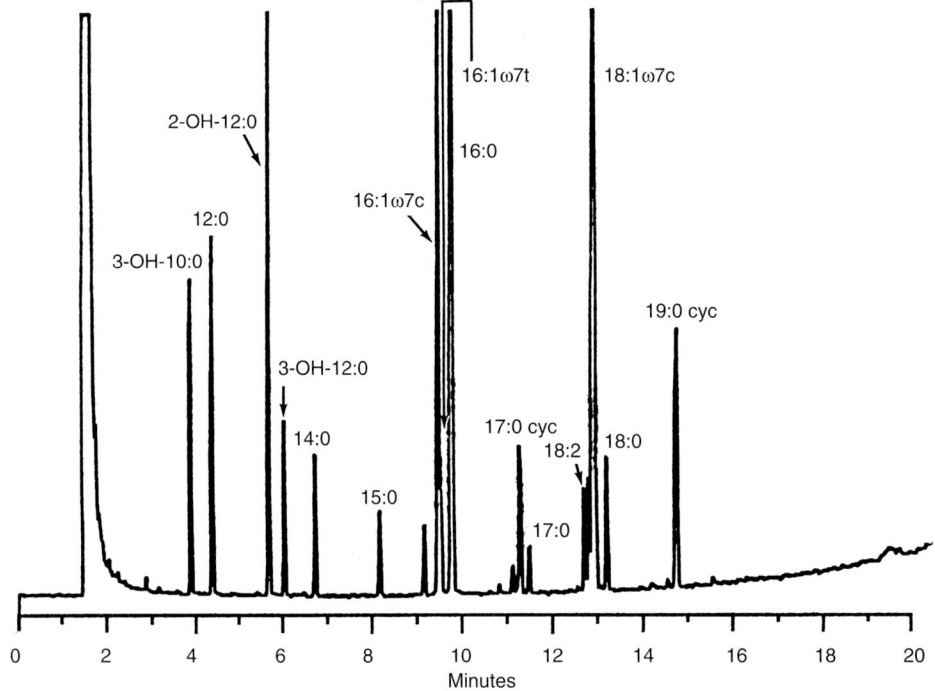

Figure 4.42. Gas chromatogram of the cellular fatty acids of *Chromobacterium violaceum, Pseudomonas aeruginosa, "Pseudomonas denitrificans," Pseudomonas fluorescens, Pseudomonas*-like Group 2, and *Pseudomonas putida*.

Comments

Organisms in this group have a distinct CFA profile that is characterized by small amounts (2%) each of 3—OH-10:0, 2—OH-12:0, and 3—OH-12:0; the absence of 3—OH-14:0; trace to moderate amounts (12%) of 17:0cyc; with larger amounts of 16:1ω7c (10–28%), 16:0 (27–34%), and 18:1ω7c (15–32%). Significant differences in the relative amounts of some acids are observed among strains of each species. For example, some strains contain both 16:1ω7c and 16:1ω7t, and some strains contain small amounts of 2—OH-14:0 and/or 19:0cyc[11–12]. Except for the presence of 17:0cyc, the overall CFA composition is most similar to the CFA group containing *Flavimonas oryzihabitans, Chryseomonas luteola,* and *Pseudomonas syringae*. The CFA profile of *Balneatrix alpica* is also similar to that of the species in this CFA group except for the absence of 12:0 (0% versus 2%) and 2—OH-12:0 (0% versus 1–6%).

Literature

1. Moss, C.W., S.B. Samuels, and R.E. Weaver. 1972. Cellular fatty acid composition of selected *Pseudomonas* species. Appl. Microbiol. *24:* 596–598.

2. Moss, C.W., and S.B. Dees. 1976. Cellular fatty acids and metabolic products of *Pseudomonas* species obtained from clinical specimens. J. Clin. Microbiol. *4:* 492–502.

3. Moss, C.W. 1978. New methodology for identification of nonfermenters: Gas-liquid chromatographic chemotaxonomy. *In* G.L. Gilardi (ed.), Glucose nonfermenting gram-negative bacteria in clinical microbiology, Chapter 9, CRC Press, West Palm Beach, FL, pp. 171–201.

4. Dees, S.B., D.G. Hollis, R.E. Weaver, and C.W. Moss. 1983. Cellular fatty acid compositions of *Pseudomonas marginata* and closely associated bacteria. J. Clin. Microbiol. *18:* 1073–1078.

5. Veys, A., W. Callewaert, E. Waelkens, and K.V.D. Abbeele. 1989. Application of gas-liquid chromatography to the routine identification of nonfermentative gram-negative bacteria in clinical specimens. J. Clin. Microbiol. *27:* 1538–1542.

6. Osterhout, G.J., V.H. Shull, and J.D. Dick. 1991. Identification of clinical isolates of gram-negative nonfermentative bacteria by an automated cellular fatty acid identification system. J. Clin. Microbiol. *29:* 1822–1830.

Table 4.37. Cellular fatty acid composition of *Chromobacterium violaceum, Pseudomonas aeruginosa, "Pseudomonas denitrificans," Pseudomonas fluorescens, Pseudomonas*-like Group 2, *Pseudomonas putida* A, and *Pseudomonas putida* 1

Fatty Acid	Organism[a]						
	A	B	C	D	E	F	G
	%	%	%	%	%	%	%
3-OH-10:0	3	5	3	4	1	4	3
12:0	5	6	3	3	2	3	2
2-OH-12:0	1	6	6	6	2	5	5
3-OH-12:0	2	3	2	2	3	2	1
14:0	3	1	2	1	2	1	1
15:0	2	T[b]	1	-	-	-	1
2-OH-14:0	-	-	-	-	1	-	-
16:1ω7c	35	10	19	27	28	18	21
16:1ω7t	-	2	-	2	1	11	-
16:0	28	29	30	27	29	30	34
17:0cyc	T	2	3	7	7	6	12
17:0	-	-	T	-	T	-	T
18:2	2	-	-	-	2	-	-
18:1ω9c	2	1	T	1	2	-	-
18:1ω7c	13	32	28	18	15	18	15
18:0	1	1	1	1	2	2	1
19:0cyc$^{11\text{-}12}$	-	3	1	1	-	-	1

[a] A, *Chromobacterium violaceum*; B, *Pseudomonas aeruginosa*; C, *"Pseudomonas denitrificans"*; D, *Pseudomonas fluorescens*; E, *Pseudomonas*-like Group 2; F, *Pseudomonas putida* A; G, *Pseudomonas putida* 1.
[b] T = Trace.

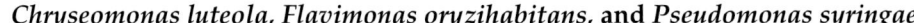

Chryseomonas luteola, Flavimonas oryzihabitans, **and** *Pseudomonas syringae*

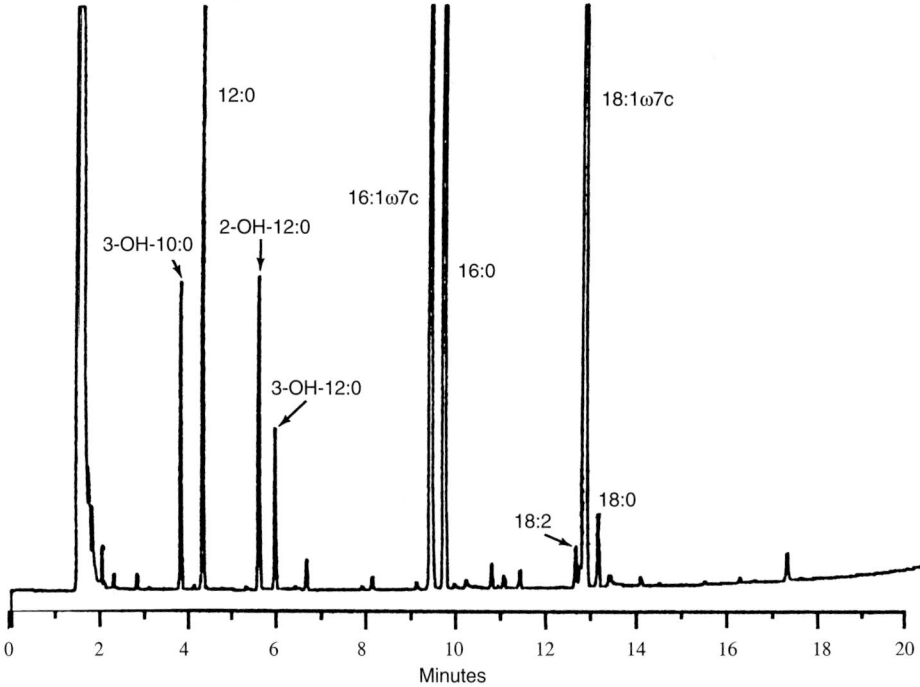

Figure 4.43. Gas chromatogram of the cellular fatty acids of *Chryseomonas luteola, Flavimonas oryzihabitans,* and *Pseudomonas syringae.*

Comments

The CFA profile of the organisms in this group contains 18:1ω7c, 16:1ω7c, and 16:0 as the major acids; smaller amounts of 12:0 (5–8%), and three hydroxy acids (3—OH-10:0, 2—OH-12:0, 3—OH-12:0). Some strains contain trace to small amounts of 19:0cyc$^{11-12}$ and/or Br-19:1. Hydroxy acids with carbon chain lengths greater than 12 carbons are absent. This profile is most similar to that of the CFA group containing *Pseudomonas aeruginosa, "P. denitrificans," P. fluorescens, P. putida,* and *Pseudomonas*-like Group 2 except for the absence of 17:0cyc; the profile differs from that of NO-1 by the presence of 2—OH-12:0 and larger amounts of 12:0 (5% versus trace).

Literature

1. Dees, S.B., C.W. Moss, R.E. Weaver, and D.G. Hollis. 1979. Cellular fatty acid composition of *Pseudomonas paucimobilis* and group IIk-2, Ve-1, and Ve-2. J. Clin. Microbiol. *10:* 206–209.

2. Dees, S.B., C.W. Moss, D.G. Hollis, and R.E. Weaver. 1986. Chemical characterization of *Flavobacterium odoratum, Flavobacterium breve,* and *Flavobacterium*-like groups IIe, IIh, and IIf. J. Clin. Microbiol. *23:* 267–273.

3. Osterhout, G.J., V.H. Shull, and J.D. Dick. 1991. Identification of clinical isolates of gram-negative nonfermentative bacteria by an automated cellular fatty acid identification system. J. Clin. Microbiol. *29:* 1822–1830.

4. Hollis, D.G., C.W. Moss, M.I. Daneshvar, L. Meadows, J. Jordan, and B. Hill. 1993. Characterization of Centers for Disease Control group NO-1, a fastidious, nonoxidative, gram-negative organism associated with dog and cat bites. J. Clin. Microbiol. *31:* 746–748.

Table 4.38. Cellular fatty acid composition of *Chryseomonas luteola, Flavimonas oryzihabitans,* and *Pseudomonas syringae*

Fatty Acid	Organism		
	Chryseomonas luteola %	*Flavimonas oryzihabitans* %	*Pseudomanas syringae* %
3-OH-10:0	4	3	3
12:0	8	7	5
2-OH-12:0	3	2	3
3-OH-12:0	2	1	2
14:0	1	1	T[a]
16:1ω7c	19	19	36
16:0	21	25	26
18:2	1	1	1
18:1ω9c	1	1	1
18:1ω7c	39	36	21
18:0	1	2	1
Br-19:1	-	-	1
19:0cyc$^{11-12}$	1	-	-

[a]T = Trace.

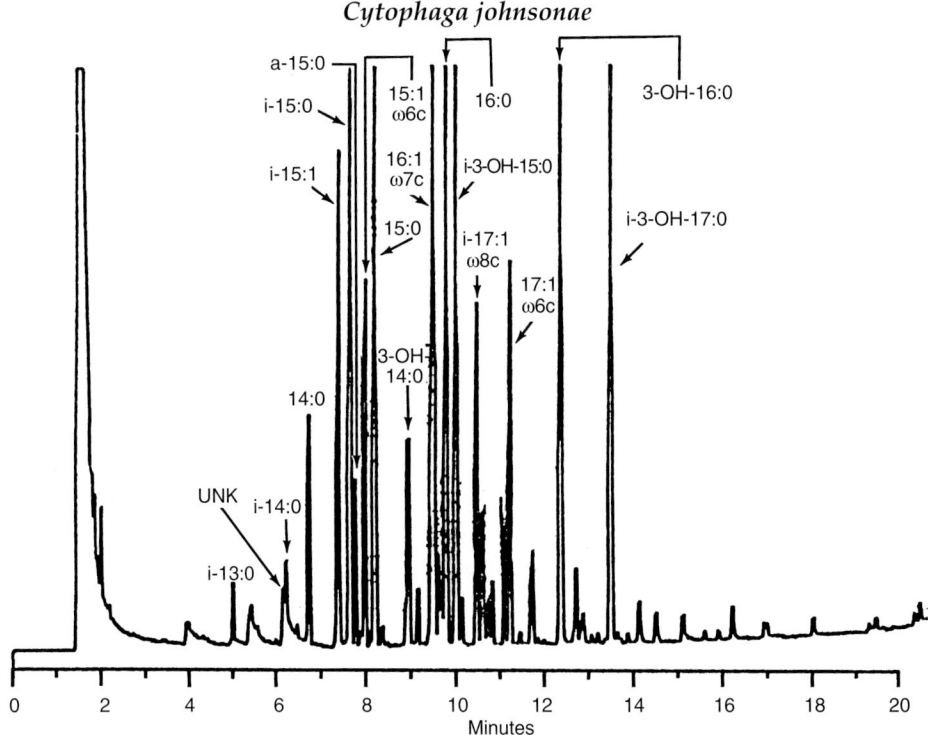

Figure 4.44. Gas chromatogram of the cellular fatty acids of *Cytophaga johnsonae*.

Comments

Cytophaga johnsonae has a distinct CFA profile that contains i-15:0 (26%), 16:1ω7c (22%), 15:0 (9%), and 16:0 (9%) as the four major acids; smaller amounts (2–5%) of four hydroxy acids (3—OH-14:0, i-3—OH-15:0, 3—OH-16:0, i-3—OH-17:0); and small amounts (2%) of i-17:1ω8c and 17:1ω6c. This profile has some similarity to *Flavobacterium, Capnocytophaga,* and *Weeksella,* but differs from these by the presence of larger amounts of 15:0 and 16:0, and the absence of i-2—OH-15:0. The CFA composition of *C. johnsonae* is most like that of *F. breve* [i.e., absence of i-2—OH-15:0, 16:1ω7c (20%), 16:0 (10%)], but differs from *F. breve* by smaller amounts of 16:1ω5 (1% versus 8%), the absence of i-17:1ω12t (0% versus 6%), and larger amounts of 15:0 (9% versus 2%).

Literature

1. Yabuuchi, E., and C.W. Moss. 1982. Cellular fatty acid composition of three species of *Sphingobacterium* gen. nov. and *Cytophaga johnsonae*. FEMS Microbiology Letters. 13: 87–91.

Table 4.39. Cellular fatty acid composition of *Cytophaga johnsonae*

Fatty Acid	Organism *Cytophaga johnsonae* %
UNK[a] 13.566	1
14:0	1
i-15:1G	2
i-15:0	26
a-15:0	2
15:1ω6c	3
15:0	9
3-OH-14:0	2
i-16:0	1
16:1ω7c	22
16:1ω5	1
16:0	9
i-3-OH-15:0	5
i-17:1ω8c	2
i-17:1?[b]	1
3-OH-15:0	1
17:1ω8c	1
17:1ω6c	2
i-3-OH-16:0	1
3-OH-16:0	5
i-3-OH-17:0	4

[a]UNK = Unknown.
[b]?, The position of the double bond is unknown.

Escherichia coli, Salmonella dublin, Salmonella newport, Salmonella oranienburg, Salmonella typhi, **and** *Salmonella zaiman*

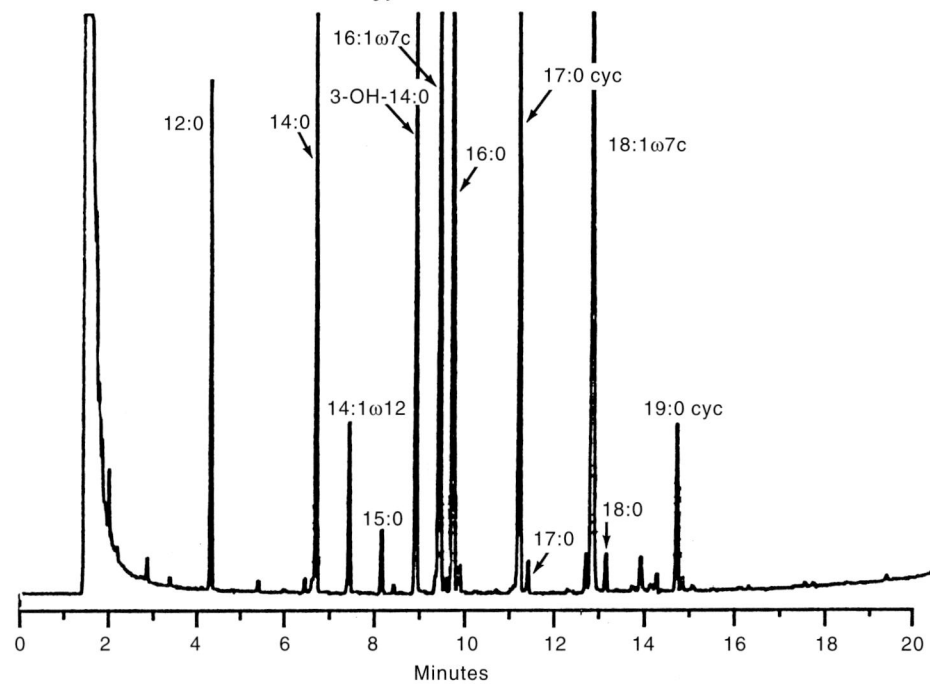

Figure 4.45. Gas chromatogram of the cellular fatty acids of *Escherichia coli, Salmonella dublin, Salmonella newport, Salmonella oranienburg, Salmonella typhi,* and *Salmonella zaiman.*

Comments

Organisms in this CFA group contain 16:0 as the major acid (25–38%) with approximately equal amounts each of 18:1ω7c (11–19%), 17:0cyc (11–18%), and 16:1ω7c (9–14%); and smaller amounts of 19:0cyc$^{11-12}$ (2–4%), 3—OH-14:0 (5–7%), 2—OH-14:0 (trace–1%), 14:1ω12 (trace–1%), 12:0 (4%), 13:0 and 15:0 (trace–7%); and only trace amounts of 18:1ω9c and 18:0. Organisms in this CFA group differ from those in the CFA group containing *Alcaligenes faecalis, A. xylosoxidans* subsp. *xylosoxidans, A. xylosoxidans* subsp. *dentrificans* group 1, and *Bordetella bronchiseptica* by the absence of 2—OH-12:0 (0% versus 4%) and the presence of 14:1ω12 (trace–1% versus 0%); they differ from the CFA group containing *A. piechaudii, Comamonas acidovorans, C. testosteroni, Pseudomonas alcaligenes,* and *P. pseudoalcaligenes* by the absence of 3—OH-10:0 and 2—OH-16:0.

Literature

1. Machtiger, N.A., and W.M. O'Leary. 1973. Fatty acid composition of paracolons: *Arizona, Citrobacter,* and *Providencia.* J. Bacteriol. *114:* 80–85.

2. Boe, B., and J. Gjerde. 1980. Fatty acid patterns in the classification of some representatives of the families *Enterobacteriaceae* and *Vibrionaceae.* J. Gen. Microbiol. *116:* 41–49.

Table 4.40. Cellular fatty acid composition of *Escherichia coli* (A), *Salmonella dublin* (B), *Salmonella newport* (C), *Salmonella oranienburg* (D), *Salmonella typhi* (E), and *Salmonella zaiman* (F)

Fatty Acid	Organism[a]					
	A	B	C	D	E	F
	%	%	%	%	%	%
12:0	3	4	4	4	4	4
13:0	1	1	-	-	-	1
14:0	7	6	6	6	8	7
14:1ω12	1	T[b]	1	1	1	T
15:0	8	7	1	1	T	6
2-OH-14:0	-	1	1	1	1	1
3-OH-14:0	6	5	7	7	7	5
16:1ω7c	13	10	11	10	9	14
16:0	28	25	32	32	38	25
17:1ω8c	1	1	-	-	-	1
17:0cyc	11	14	12	14	18	12
17:0	3	3	1	1	-	3
18:2	-	1	-	-	-	1
18:1ω9c	-	1	-	-	-	1
18:1ω7c	14	16	19	19	11	18
18:0	T	1	T	T	-	1
19:0cyc[11-12]	2	4	2	3	4	2

[a]A, *Escherichia coli*; B, *Salmonella dublin*; C, *Salmonella newport*: D, *Salmonella oranienburg*: E, *Salmonella typhi*; F, *Salmonella zaiman*.
[b]T = Trace.

Flavobacterium balustinum, Flavobacterium gleum, Flavobacterium indologenes, Flavobacterium meningosepticum, Flavobacterium odoratum, Flavobacterium species (IIb), *Weeksella virosa,* and *Weeksella zoohelcum*

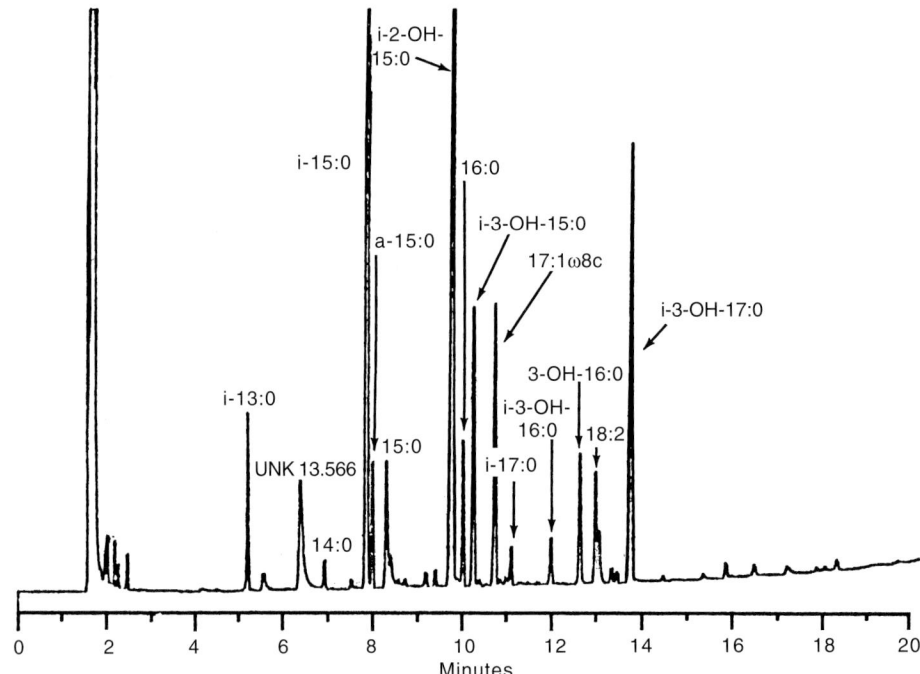

Figure 4.46. Gas chromatogram of the cellular fatty acids of *Flavobacterium balustinum, Flavobacterium gleum, Flavobacterium indologenes, Flavobacterium meningosepticum, Flavobacterium odoratum, Flavobacterium* species (IIb), *Weeksella virosa,* and *Weeksella zoohelcum*.

Comments

Species in this CFA group have a distinct CFA profile that is characterized by large amounts of i-15:0 (29–60%), moderate amounts of i-2—OH-15:0 and/or i-17:1ω8c, small amounts of i-3—OH-15:0 and i-3—OH-17:0, and only trace amounts of 16:1ω7c. The ratio of i-15:0 to i-2—OH-15:0 is ≧2:1; all species have a skewed or tailing peak (designated UNK 13.566) that elutes just before i-14:0 on a nonpolar GLC capillary column. The identity of i-2—OH-15:0 must be confirmed by acetylation, hydrogenation, and/or mass spectrometry since it co-elutes with 16:1ω7t on a 15-meter nonpolar capillary column. Some resolution of i-2—OH-15:0 and 16:1ω7t is possible by using a 50-meter capillary column. Species in this group are distinguished from the CFA group containing *Flavobacterium (Sphingobacterium) multivorum, Flavobacterium (Sphingobacterium) spiritivorum, Flavobacterium (Sphingobacterium) thalpophilum,* IIi, and *Flavobacterium (Sphingobacterium) mizutaii* by the absence of 16:1ω7c (0% versus 3–20%) and the ratio of i-15:0 to i-2—OH-15:0, which is 1:1 in *Flavobacterium (Sphingobacterium)* and IIi compared to 2:1 or greater in the species listed above. Species in this CFA group differ from the CFA group that contains IIc, IIe, and IIh by small amounts of 15:0 (2% versus 8%); they differ from *Cytophaga johnsonae* by the presence of i-2—OH-15:0 and the absence of 16:1ω7c. *Weeksella virosa* differs from other species in this CFA group by the presence of 8% i-17:1ω12t. The possibility of further differentiation of species in this group by CFA analysis requires testing of additional well-documented strains.

Literature

1. Moss, C.W., and S.B. Dees. 1978. Cellular fatty acids of *Flavobacterium meningosepticum* and *Flavobacterium* species group IIb. J. Clin. Microbiol. *8:* 772–774.

2. Dees, S.B., J. Powell, C.W. Moss, D.G. Hollis, and R.E. Weaver. 1981. Cellular fatty acid composition of organisms frequently associated with human infections resulting from dog bite: *Pasteurella multocida* and Groups EF-4, IIj, M-5, and DF-2. J. Clin. Microbiol. *14:* 612–616.

3. Yabuuchi, E., T. Kaneko, I. Yano, C.W. Moss, and N. Miyoshi. 1983. *Sphingobacterium* gen. nov., *Sphingobacterium spiritivorum* comb. nov., *Sphingobacterium multivorum* comb. nov., *Sphingobacterium mizutae* sp. nov., and *Flavobac-*

terium indologenes sp. nov.: glucose-nonfermenting gram-negative rods in CDC groups IIk-2 and IIb. Int. J. Syst. Bacteriol. 33: 580–598.

4. Dees, S.B., C.W. Moss, D.G. Hollis, and R.E. Weaver. 1986. Chemical characterization of *Flavobacterium odoratum*, *Flavobacterium breve*, and *Flavobacterium*-like groups IIc, IIh, and IIf. J. Clin. Microbiol. 23: 267–273.

5. Veys, A., W. Callewaert, E. Waelkens, and K.V.D. Abbeele. 1989. Application of gas-liquid chromatography to the routine identification of nonfermentative gram-negative bacteria in clinical specimens. J. Clin. Microbiol. 27: 1538–1542.

Table 4.41. Cellular fatty acid composition of *Flavobacterium balustinum*, *Flavobacterium gleum*, *Flavobacterium indologenes*, *Flavobacterium meningosepticum*, *Flavobacterium odoratum*, *Flavobacterium* species (**IIb**), *Weeksella virosa*, and *Weeksella zoohelcum*

Fatty Acid	Organism[a]							
	A	B	C	D	E	F	G	H
	%	%	%	%	%	%	%	%
i-13:0	1	T[b]	2	2	3	1	1	2
UNK[c] 13.566	3	3	2	1	1	4	2	-
14:0	T	-	T	T	1	-	T	T
i-15:0	29	39	45	46	55	42	46	60
a-15:0	1	-	-	3	1	-	-	1
15:0	-	-	-	-	2	-	-	-
i-16:0	2	-	1	-	1	-	1	-
i-2-OH-15:0	12	14	14	20	4	11	10	9
16:1ω5	-	-	-	-	-	-	1	-
16:0	2	3	1	2	2	2	4	3
i-3-OH-15:0	4	4	5	3	6	4	5	5
2-OH-15:0	T	-	-	-	-	-	-	-
i-17:1ω8c	28	12	13	5	6	14	5	9
i-17:1ω12t	1	-	-	-	-	-	8	1
i-17:0	1	1	1	1	-	1	3	-
i-3-OH-16:0	1	-	1	1	-	-	-	1
3-OH-16:0	1	1	1	3	4	1	1	-
18:2	2	2	-	1	1	2	2	3
18:1ω9c	1	1	-	1	1	1	1	2
18:0	1	T	-	-	-	-	1	T
i-3-OH-17:0	10	17	11	10	8	14	7	3
20:1ω9	-	-	-	-	-	-	-	1

[a]A, *Flavobacterium balustinum*; B, *Flavobacterium gleum*; C, *Flavobacterium indologenes*; D, *Flavobacterium meningosepticum*; E, *Flavobacterium odoratum*; F, *Flavobacterium* species (IIb); G, *Weeksella virosa*; H, *Weeksella zoohelcum*.
[b]T = Trace.
[c]UNK = Unknown.

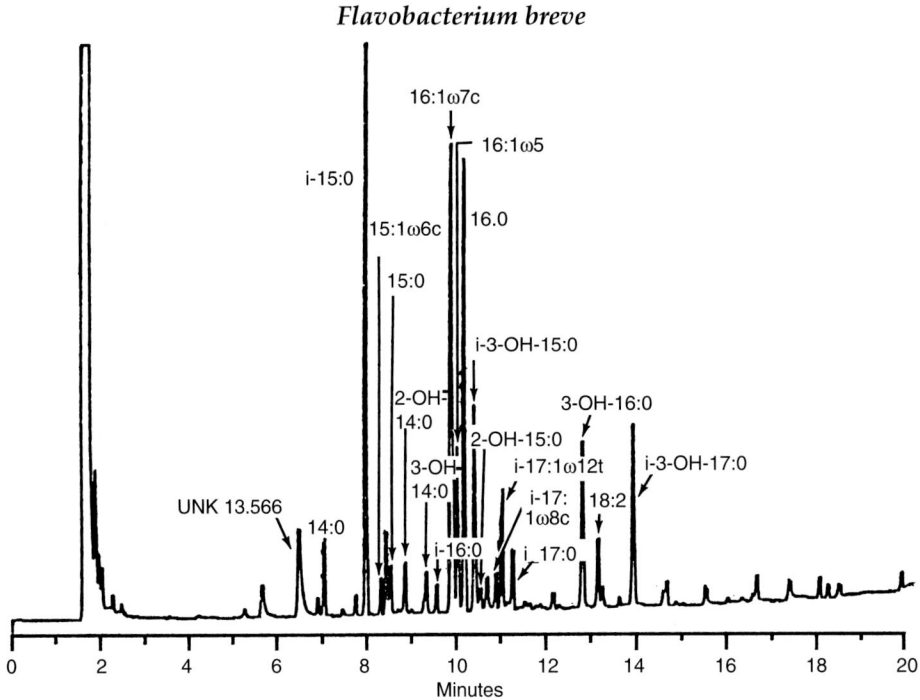

Figure 4.47. Gas chromatogram of the cellular fatty acids of *Flavobacterium breve*.

Comments

Flavobacterium breve has the general CFA profile of other *Flavobacterium* species (UNK 13.566, i-15:0, i-3—OH-15:0, i-17:1ω8c, 3—OH-16:0, i-3—OH-17:0), but differs by the absence of i-2—OH-15:0 and the presence of 16:1ω5 (8%) and i-17:1ω12t (6%). *F. breve* differs from all other *Flavobacterium/Weeksella* organisms by the presence of large amounts (20%) of 16:1ω7c. *Flavobacterium (Sphingobacterium) mizutaii, F. (Sphingobacterium) multivorum*, and *F. (Sphingobacterium) thalpophilum* also have large amounts of 16:1ω7c, but these organisms also contain i-2—OH-15:0 that is absent in *F. breve*. *F. breve* is most similar to *Cytophaga johnsonae* except for the presence of 16:1ω5 and i-17:1ω12t. *F. breve* differs from IIc, IIe, and IIh by the presence of 16:1ω7c, larger amounts of 16:1ω5 and i-17:1ω12t, and the absence of a-15:0 (0% versus >8%).

Literature

1. Moss, C.W., and S.B. Dees. 1978. Cellular fatty acids of *Flavobacterium meningosepticum* and *Flavobacterium* species group IIb. J. Clin. Microbiol. *8*: 772–774.

2. Dees, S.B., C.W. Moss, R.E. Weaver, and D. Hollis. 1979. Cellular fatty acid composition of *Pseudomonas paucimobilis* and groups IIk-2, Ve-1, and Ve-2. J. Clin. Microbiol. *10*: 206–209.

3. Dees, S.B., J. Powell, C.W. Moss, D.G. Hollis, and R.E. Weaver. 1981. Cellular fatty acid composition of organisms frequently associated with human infections resulting from dog bite: *Pasteurella multocida* and Groups EF-4, IIj, M-5, and DF-2. J. Clin. Microbiol. *14*: 612–616.

4. Dees, S.B., C.W. Moss, D.G. Hollis, and R.E. Weaver. 1986. Chemical characterization of *Flavobacterium odoratum*, *Flavobacterium breve*, and *Flavobacterium*-like groups IIe, IIh, IIf. J. Clin. Microbiol. *23*: 267–273.

5. Veys, A., W. Callewaert, E. Waelkens, and K.V.D. Abbeele. 1989. Application of gas-liquid chromatography to the routine identification of nonfermentative gram-negative bacteria in clinical specimens. J. Clin. Microbiol. *27*: 1538–1542.

Table 4.42. Cellular fatty acid composition of *Flavobacterium breve*

Fatty Acid	Organism *Flavobacterium breve* %
UNK[a] 13.566	2
14:0	2
i-15:0	27
15:1ω6c	1
15:0	2
2-OH-14:0	1
i-16:0	1
16:1ω7c	20
16:1ω5	8
16:0	7
i-3-OH-15:0	5
i-17:1ω8c	2
i-17:1ω12t	6
i-17:0	1
3-OH-16:0	4
18:2	1
i-3-OH-17:0	7

[a]UNK = Unknown.

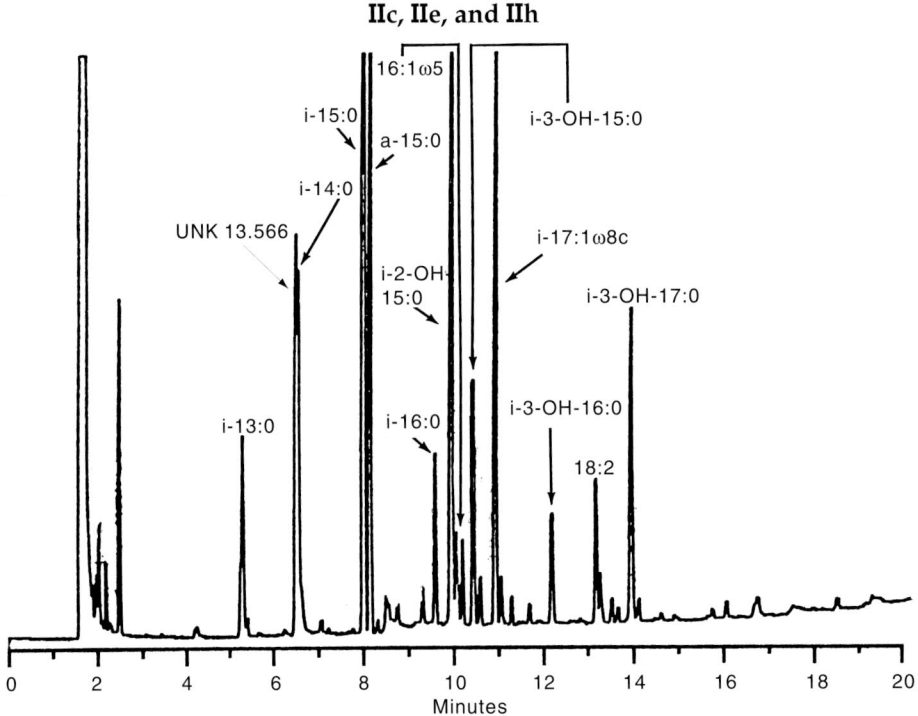

Figure 4.48. Gas chromatogram of the cellular fatty acids of IIc, IIe, and IIh.

Comments

IIc, IIe, and IIh have the general characteristic CFA profile of *Flavobacterium* (UNK 13.566, i-15:0, i-2—OH-15:0, i-3—OH-15:0, i-17:1ω8c, i-3—OH-17:0), but differs by the presence of larger amounts of a-15:0 (8–14% versus <3%). This profile differs from the CFA group containing *Flavobacterium (Sphingobacterium) mizutaii, Flavobacterium (Sphingobacterium) multivorum, Flavobacterium (Sphingobacterium) spiritivorum, Flavobacterium (Sphingobacterium) thalpophilum,* and IIi by the absence of 16:1ω7c and 14:0, and smaller amounts of i-2—OH-15:0 (11% versus 20%). It differs from the *Capnocytophaga* species by the presence of i-2—OH-15:0 and i-17:1ω8c, and smaller amounts of i-15:0 (<40% versus >60%).

Literature

1. Dees, S.B., C.W. Moss, D.G. Hollis, and R.E. Weaver. 1986. Chemical characterization of *Flavobacterium odoratum, Flavobacterium breve,* and *Flavobacterium*-like groups IIe, IIh, IIf. J. Clin. Microbiol. 23: 267–273.

Table 4.43. Cellular fatty acid composition of IIc, IIe, and IIh

Fatty Acid	Organism		
	IIc %	IIe %	IIh %
i-13:0	2	2	5
UNK[a] 13.566	3	4	2
i-14:0	–	1	–
i-15:0	28	29	36
a-15:0	8	14	14
i-16:0	–	1	–
i-2-OH-15:0	11	11	9
16:1ω5	1	1	1
16:0	3	2	3
i-3-OH-15:0	4	4	3
2-OH-15:0	–	2	1
i-17:1ω8c	11	10	7
i-17:1ω12t	1	–	1
i-17:0	1	–	–
i-3-OH-16:0	1	1	1
3-OH-16:0	1	–	1
18:2	2	2	2
18:1ω9c	2	2	2
18:1?[b]	1	–	1
18:0	1	T[c]	–
i-3-OH-17:0	12	9	9
2-OH-17:0	–	–	1

[a]UNK = Unknown.
[b]?, The position of the double bond is unknown.
[c]T = Trace.

IIi, *Flavobacterium (Sphingobacterium) mizutaii, Flavobacterium (Sphingobacterium) multivorum, Flavobacterium (Sphingobacterium) spiritivorum,* and *Flavobacterium (Sphingobacterium) thalpophilum*

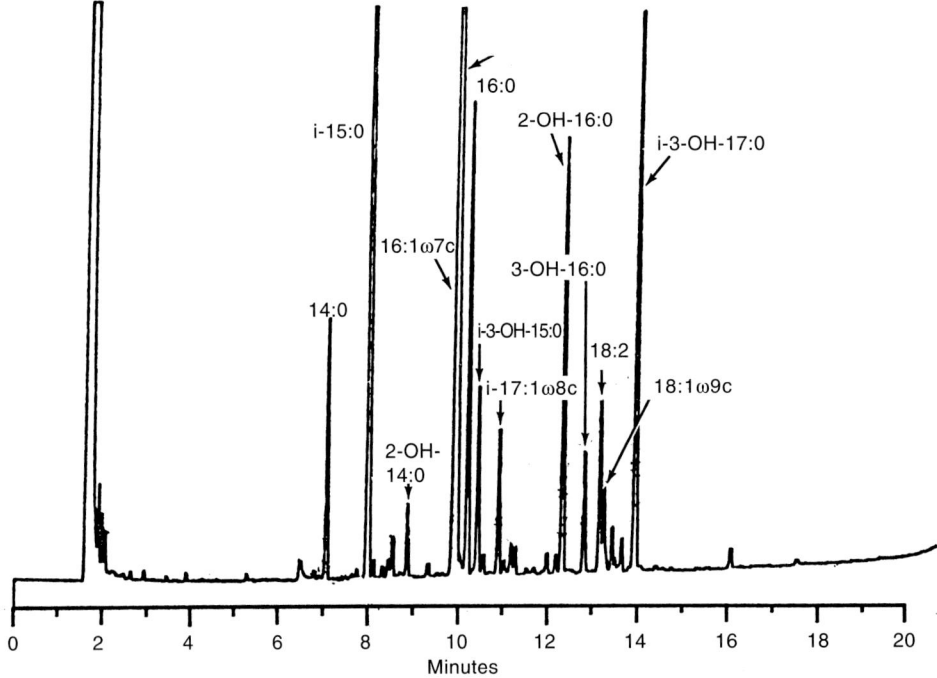

Figure 4.49. Gas chromatogram of the cellular fatty acids of IIi, *Flavobacterium (Sphingobacterium) mizutaii, Flavobacterium (Sphingobacterium) multivorum, Flavobacterium (Sphingobacterium) spiritivorum,* and *Flavobacterium (Sphingobacterium) thalpophilum.*

Comments

The CFA profile of these five species is most like that of the CFA group containing *Flavobacterium balustinum, F. gleum, F. indologenes, F. meningosepticum, F. odoratum, Flavobacterium* species (IIb), *Weeksella virosa,* and *W. zoohelcum,* but differs from this group by much larger amounts of 16:1ω7c (15% versus 0–7%) and the ratio of i-15:0/i-2—OH-15:0 that is 1:1 compared to 2:1 in the *Flavobacterium/Weeksella* CFA group. The CFA profile of these five species differ from *Capnocytophaga* by the presence of 16:1ω7c (16% versus 0%) and i-17:1ω8c (2% versus 0%) and smaller amounts of i-15:0 (30% versus 65%), and differs from *Cytophaga johnsonae* by the presence of i-2—OH-15:0 (25% versus 0%) and smaller amounts of 16:0 (3% versus 10%). This CFA group differs from *F. breve* by the presence of i-2—OH-15:0 and the absence of 16:1ω5 and i-17:1ω12t.

Literature

1. Dees, S.B., C.W. Moss, R.E. Weaver, and D. Hollis. 1979. Cellular fatty acid composition of *Pseudomonas paucimobilis* and group IIk-2, Ve-1 and Ve-2. J. Clin. Microbiol. 10: 206–209.

2. Yabuuchi, E., and C.W. Moss. 1982. Cellular fatty acid composition of three species of *Sphingobacterium* gen. nov. and *Cytophaga johnsonae*, FEMS Microbiol. Lett. 13: 87–91.

3. Yabuuchi, E., T. Kaneko, I. Yano, C.W. Moss, and N. Miyoshi. 1983. *Sphingobacterium* gen. nov., *Sphingobacterium spiritivorum* comb. nov., *Sphingobacterium mizutae* sp. nov., and *Flavobacterium indologenes* sp. nov.: glucose-nonfermenting gram-negative rods in CDC groups IIK-2 and IIb. Int. J. Syst. Bacteriol. 33: 580–598.

4. Dees, S.B., G. Carlone, D. Hollis, R. Weaver, and C.W. Moss. 1985. Chemical and phenotypic characteristics of *Flavobacterium thalpophilum* compared to other *Flavobacterium* and *Sphingobacterium* species. Int. J. Syst. Bact. 35: 16–22.

5. Veys, A., W. Callewaert, E. Waelkens, and K.V.D. Abbeele. 1989. Application of gas-liquid chromatography to the routine identification of nonfermentative gram-negative bacteria in clinical specimens. J. Clin. Microbiol. 27: 1538–1542.

Table 4.44. Cellular fatty acid composition of IIi, *Flavobacterium (Sphingobacterium) mizutaii*, *Flavobacterium (Sphingobacterium) multivorum*, *Flavobacterium (Sphingobacterium) spiritivorum*, and *Flavobacterium (Sphingobacterium) thalpophilum*

Fatty Acid	Organism[a]				
	A	B	C	D	E
	%	%	%	%	%
UNK[b] 13.566	1	-	1	1	1
14:0	1	3	3	1	2
i-15:0	36	29	29	35	32
a-15:0	1	1	-	1	-
15:0	T[c]	T	1	T	T
2-OH-14:0	T	1	1	T	1
16:1ω7c	7	17	15	3	20
i-2-OH-15:0	31	25	27	37	20
16:0	2	6	6	3	4
i-3-OH-15:0	2	3	4	3	3
i-17:1ω8c	2	2	1	1	2
2-OH-16:0	-	T	-	-	2
3-OH-16:0	2	2	2	2	1
18:2	1	1	1	1	2
18:1ω9c	1	1	-	T	1
i-3-OH-17:0	10	6	5	8	6

[a] A, IIi; B, *Flavobacterium (Sphingobacterium) mizutaii*; C, *Flavobacterium (Sphingobacterium) multivorum*; D, *Flavobacterium (Sphingobacterium) spiritivorum*; E, *Flavobacterium (Sphingobacterium) thalpophilum*.
[b] UNK = Unknown.
[c] T = Trace.

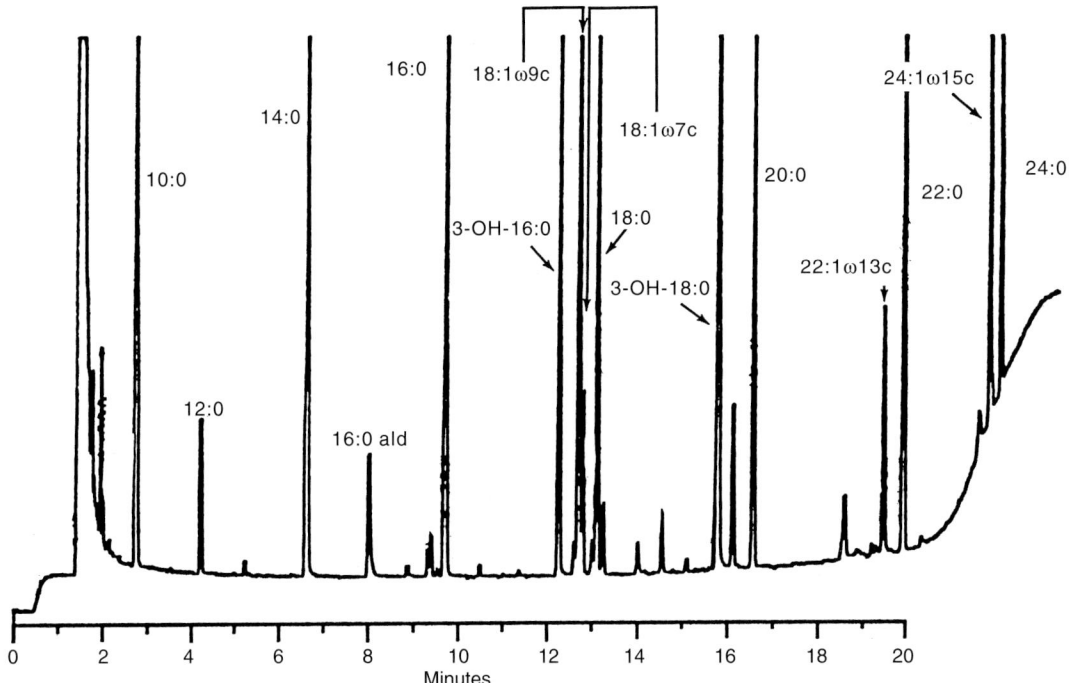

Figure 4.50. Gas chromatogram of the cellular fatty acids of *Francisella philomiragia*, *Francisella tularensis* biogroup novicida, and *Francisella tularensis* biogroup tularensis.

Comments

The CFA composition of *Francisella* species is characterized by the presence of long-chain saturated and monounsaturated 18:0 to 26:0 acids, relatively large amounts of saturated even-chain acids (10:0, 14:0, 16:0), and two long-chain hydroxy acids (3—OH-16:0, 3—OH-18:0). The presence and relative amounts of these acids constitute a fatty acid profile that is unique for *Francisella*. In addition, *Francisella* differs from most other Gram-negative bacteria by the absence of monounsaturated 16-carbon acids in all strains and the absence of 12:0 in most strains.

Literature

1. Jantzen, K., B.P. Berdal, and T. Omland. 1979. Cellular fatty acid composition of *Francisella tularenis*. J. Clin. Microbiol. *10:* 928–930.

2. Hollis, D.G., R.E. Weaver, A.G. Steigerwalt, J.D. Wenger, C.W. Moss, and D.J. Brenner. 1989. *Francisella philomiragia* comb. nov. (formerly *Yersinia philomiragia*) and *Francisella tularensis* biotype novicida (formerly *Francisella novicida*) associated with human disease. J. Clin. Microbiol. *27:* 1601–1608.

Table 4.45. Cellular fatty acid composition of *Francisella philomiragia*, *Francisella tularensis* biogroup novicida, and *Francisella tularensis* biogroup tularensis

Fatty Acid	Francisella philomiragia %	F. tularensis bio. novicida %	F. tularensis bio. tularensis %
10:0	13	10	30
2-OH-10:0	T[a]	5	1
12:0	T	T	T
14:0	16	12	11
16:0 ald	2	2	1
16:0	9	18	10
3-OH-16:0	3	2	3
18:2	2	1	2
18:1ω9c	12	9	7
18:0	9	8	3
3-OH-18:0	9	10	12
20:0	3	2	1
22:1ω13c	1	2	1
22:0	6	6	5
24:1ω15c	9	7	5
24:0	4	3	5

[a]T = Trace.

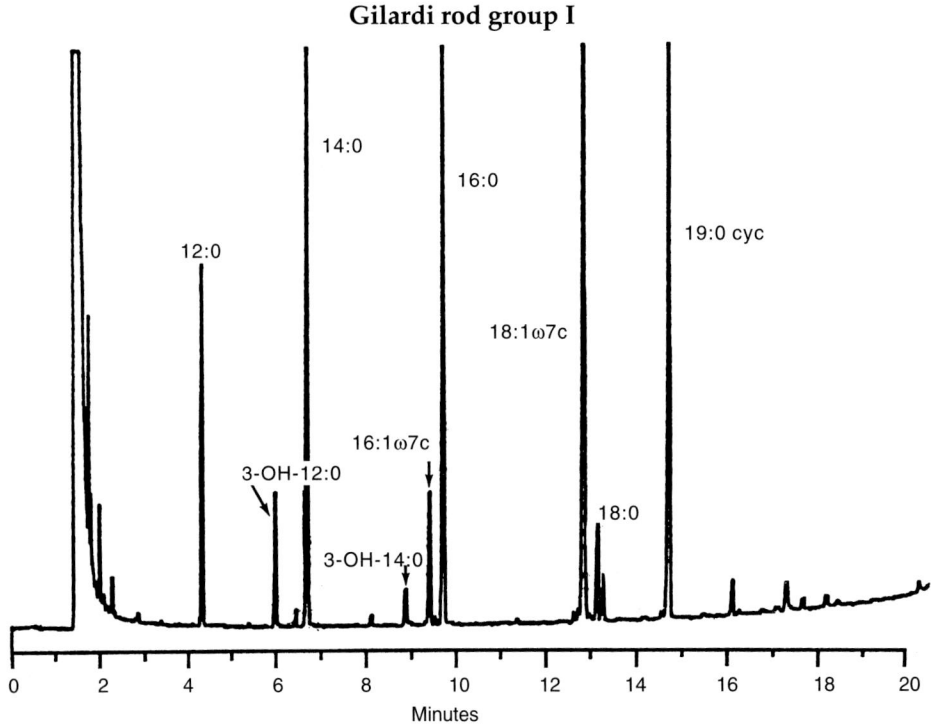

Figure 4.51. Gas chromatogram of the cellular fatty acids of Gilardi rod group I.

Comments

Gilardi rod group 1 has a unique CFA profile that is characterized by large amounts (>15%) of 18:1ω7c, 16:0, 14:0, and 19:0cyc[11–12]; and moderate amounts (3–5%) of 12:0, 3—OH-12:0, and 16:1ω7c. *Kingella, Actinobacillus, Pasteurella,* and *Cardiobacterium hominis* also contain large amounts of 14:0, but unlike Gilardi rod group 1, none of these contain 19:0cyc[11–12].

Literature

1. Moss, C.W., M.I. Daneshvar, and D.G. Hollis. 1993. Biochemical characteristics and fatty acid composition of Gilardi rod group 1 bacteria. J. Clin. Microbiol. *31:* 689–691.

Table 4.46. Cellular fatty acid composition of Gilardi rod group 1

	Organism
	Gilardi rod group 1
Fatty Acid	%
12:0	5
3-OH-12:0	3
14:0	19
15:0	1
3-OH-14:0	T[a]
16:1ω7c	3
16:0	17
18:2	1
18:1ω7c	33
18:0	1
19:0cyc[11-12]	15

[a]T = Trace.

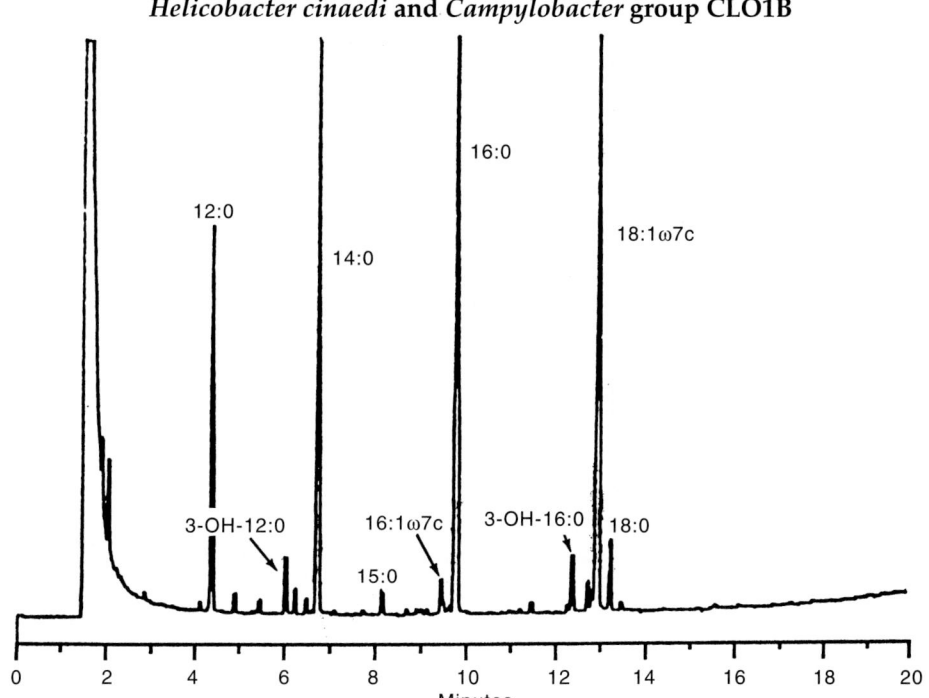

Figure 4.52. Gas chromatogram of the cellular fatty acids of *Helicobacter cinaedi* and *Campylobacter* group CLO1B.

Comments

Helicobacter cinaedi and *Campylobacter* group CLO1B have a unique CFA profile with large amounts of 16:0 (36%), 18:1ω7c (30%), and 14:0 (18%); and small amounts (1–4%) of 12:0, 3—OH-12:0, 15:0, 3—OH-16:0, 18:0, and 18:2. 3—OH-14:0 is **absent** in all strains, and there are only **trace** amounts of 16:1ω7c and 18:1ω9c. All *Campylobacter* and *Arcobacter* species contain greater than 2% concentrations of 16:1ω7c and greater than 4% amounts of 3—OH-14:0. Some *H. cinaedi* strains have moderate amounts (14%) of 12:0 and trace amounts of 3—OH-16:0.

Literature

1. Lambert, M.A., C.M. Patton, T.J. Barrett, and C.W. Moss. 1987. Differentiation of *Campylobacter* and *Campylobacter*-like organisms by cellular fatty acid composition. J. Clin. Microbiol. 25: 706–713.

Table 4.47. Cellular fatty acid composition of *Helicobacter cinaedi* and *Campylobacter* CLO1B

	Organism	
	Helicobacter cinaedi	*Campylobacter* CL01B
Fatty Acid	%	%
12:0	4	2
3-OH-12:0	1	2
14:0	18	14
15:0	1	1
16:0	36	39
17:0	1	1
3-OH-16:0	2	4
18:2	2	2
18:1ω9c	1	1
18:1ω7c	30	30
18:0	3	3

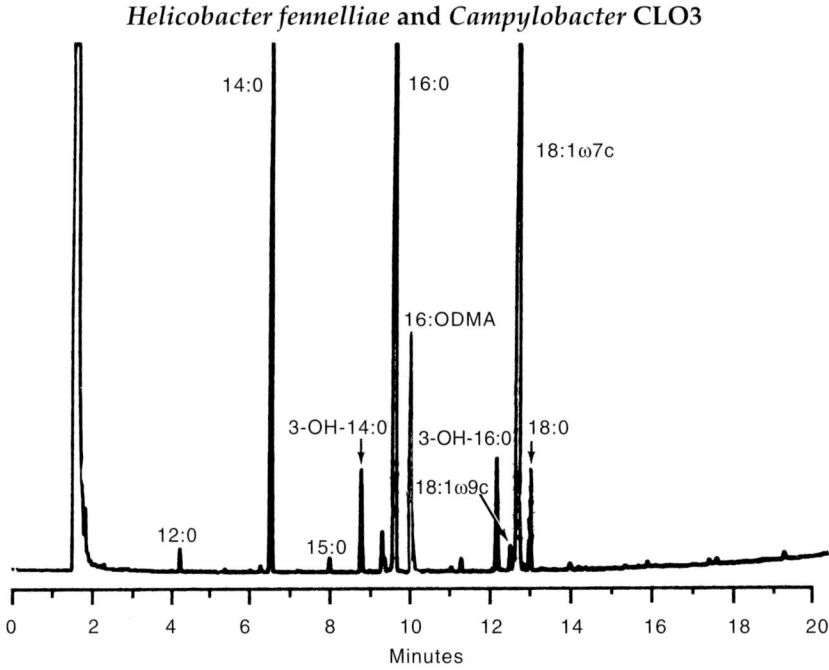

Figure 4.53. Gas chromatogram of the cellular fatty acids of *Helicobacter fennelliae* and *Campylobacter* CLO3.

Comments

Helicobacter fennelliae and *Campylobacter* CLO3 have a unique CFA profile that is characterized by major amounts of 18:1ω7c (35%) and 16:0 (32%); moderate amounts of 14:0 (9%); and small amounts (2–3%) of 12:0, 18:0, 3—OH-14:0, and 3—OH-16:0. In addition, these organisms have trace amounts of aldehydes and 9% of a 16-carbon dimethyl-acetyl (16:0 DMA). *Wolinella curva* and *W. succinogenes* also have significant amounts of 16:0 DMA, but these two species contain greater than 9% of 16:1ω7c and 3% of 18:2 which are both absent in *H. fennelliae*.

Literature

1. Moss, C.W., A. Kai, M.A. Lambert, and C.M. Patton. 1984. Isoprenoid quinone content and cellular fatty acid composition of *Campylobacter* species. J. Clin. Microbiol. *19:* 772–776.

2. Lambert, M.A., C.M. Patton, T.J. Barrett, and C.W. Moss. 1987. Differentiation of *Campylobacter* and *Campylobacter*-like organisms by cellular fatty acid composition. J. Clin. Microbiol. *25:* 706–713.

Table 4.48. Cellular fatty acid composition of *Helicobacter fennelliae* and *Campylobacter* CLO3

	Organism	
	Helicobacter fennelliae	*Campylobacter* CL03
Fatty Acid	%	%
12:0	2	–
13:0	–	1
14:0	9	8
15:0	–	3
3-OH-14:0	3	3
16:0	32	33
16:0 DMA	9	10
17:1ω6c	–	1
17:0	–	4
3-OH-16:0	3	4
18:1ω9c	–	1
18:1ω7c	35	23
18:0	4	9

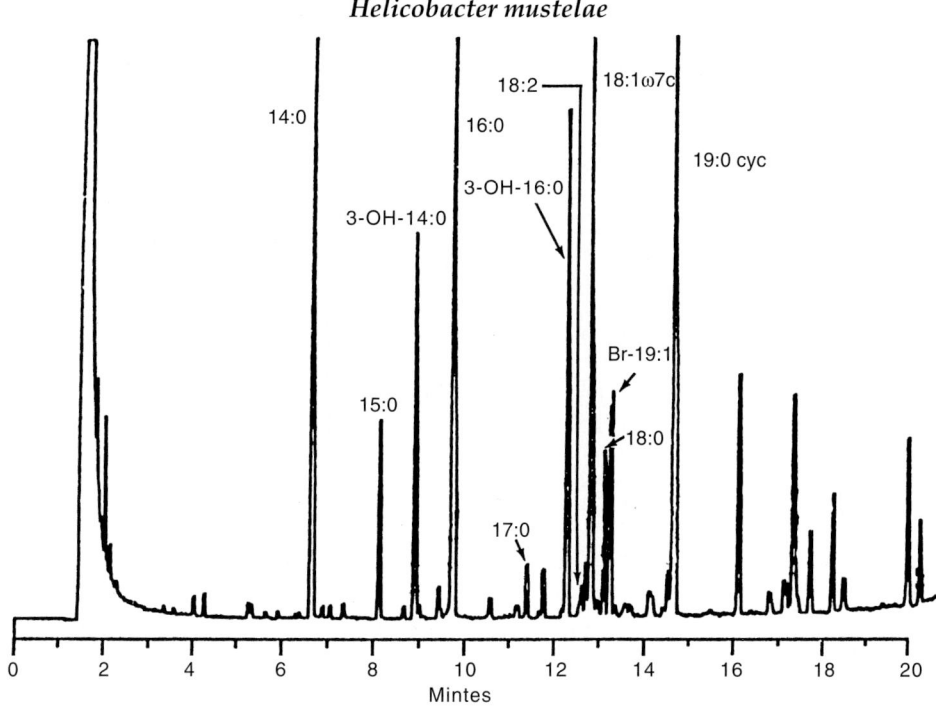

Figure 4.54. Gas chromatogram of the cellular fatty acids of *Helicobacter mustelae*.

Comments

Helicobacter mustelae has a unique CFA profile consisting of major amounts of 16:0 (34%), 19:0cyc$^{11-12}$ (26%), and 14:0 (16%); moderate amounts of 18:1ω7c (9%); and smaller amounts (2–3%) of 15:0, 3—OH-14:0, 3—OH-16:0, and 18:0. This profile is most like that of *H. pylori* except for the presence of 3—OH-14:0, the absence of 3—OH-18:0, smaller amounts of 14:0 (16% versus 34%), and larger amounts of 16:0 (34% versus 5% for *H. pylori*).

Literature

1. Lambert, M.A., C.M. Patton, T.J. Barrett, and C.W. Moss. 1987. Differentiation of *Campylobacter* and *Campylobacter*-like organisms by cellular fatty acid composition. J. Clin. Microbiol. 25: 706–713.

Table 4.49. Cellular fatty acid composition of *Helicobacter mustelae*

Fatty Acid	Organism *Helicobacter mustelae* %
14:0	16
15:0	2
3-OH-14:0	3
16:1ω7c	1
16:0	34
3-OH-16:0	3
18:2	1
18:1ω9c	1
18:1ω7c	9
18:0	1
Br-19:1	1
19:0cyc$^{11-12}$	26
UNK[a] 19.735	1

[a]UNK = Unknown.

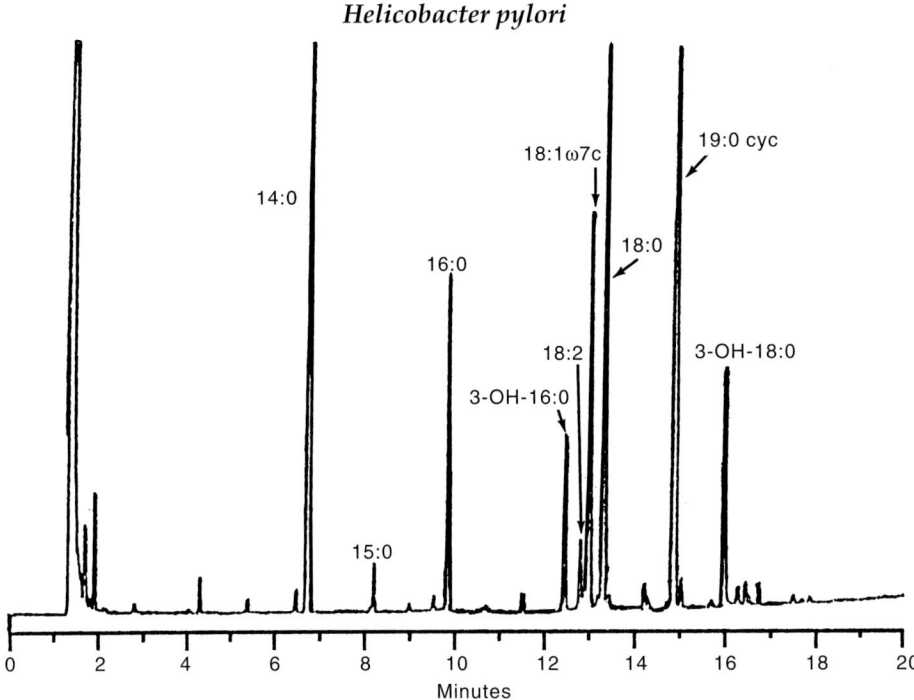

Figure 4.55. Gas chromatogram of the cellular fatty acids of *Helicobacter pylori*.

Comments

The CFA profile of *Helicobacter pylori* is different from other bacteria and is recognized by major amounts of 14:0 (34%) and 19:0cyc[11-12] (28%), together with moderate amounts (3–4%) of 3—OH-16:0 and 3—OH-18:0, low amounts (5%) of 16:0, and the absence of 16:1ω7c and 3—OH-14:0. The CFA group containing *Kingella kingae*, *K. denitrificans*, and *Simonsiella crassa* also contain 14:0 as a major fatty acid, but these organisms do not contain 3—OH-16:0, 3—OH-18:0, and 19:0cyc[11-12] that are present in *H. pylori*.

Literature

1. Goodwin, C.S., R.K. McCulough, J.A. Armstrong, and S.H. Wee. 1985. Unusual cellular fatty acids and distinctive ultra-structure in a new spiral bacterium (*Campylobacter pyloridis*) from the human gastric mucosa. J. Med. Microbiol. *19:* 257–267.

2. Lambert, M.A., C.M. Patton, T.J. Barrett, and C.W. Moss. 1987. Differentiation of *Campylobacter* and *Campylobacter*-like organisms by cellular fatty acid composition. J. Clin. Microbiol. *25:* 706–713.

Table 4.50. Cellular fatty acid composition of *Helicobacter pylori*

	Organism
	Helicobacter pylori
Fatty Acid	%
14:0	34
15:0	1
16:0	5
3-OH-16:0	3
18:2	1
18:1ω9c	1
18:1ω7c	8
18:0	11
19:0cyc[11-12]	28
19:0	1
3-OH-18:0	4

Hydrogenophaga palleronii

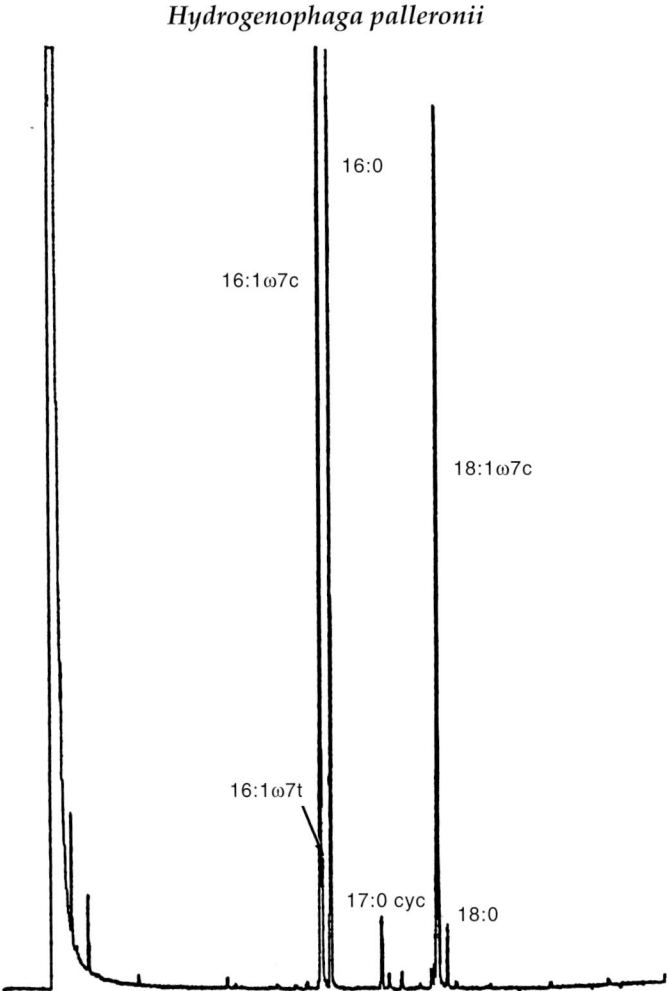

Figure 4.56. Gas chromatogram of the cellular fatty acids of *Hydrogenophaga palleronii*.

Comments

Hydrogenophaga palleronii has a distinct CFA profile consisting of major amounts of 16:1ω7c (43%), 16:0 (25%), and 18:1ω7c (25%); small amounts of 16:1ω7t (2%), 17:0cyc (2%), and 18:0 (1%); and **no** hydroxy acids. The most similar CFA profile is the CFA group containing *Pseudomonas aeruginosa, P. denitrificans, P. fluorescens, P. putida,* and *P.*-like Group 2, but each of these organisms contain one or more hydroxy acids that are absent in *H. palleronii*.

Literature

1. Willems, A., J. Busse, M. Goor, B. Pot, E. Falsen, E. Jantzen, B. Hoste, M. Gillis, K. Kersters, G. Auling, and J. De Ley. 1989. *Hydrogenophaga*, a new genus of hydrogen-oxidizing bacteria that includes *Hydrogenophaga flava* comb. nov. (formerly *Pseudomonas flava*), *Hydrogenophaga palleronii* (formerly *Pseudomonas palleronii*), *Hydrogenophaga pseudoflava* (formerly *Pseudomonas pseudoflava* and *"Pseudomonas carboxydoflava"*), and *Hydrogenophaga taeniospiralis* (formerly *Pseudomonas taeniospiralis*). Int. J. Syst. Bacteriol. 319–333.

Table 4.51. Cellular fatty acid composition of *Hydrogenophaga palleronii*

Fatty Acid	Organism *Hydrogenophaga palleronii* %
16:1ω7c	43
16:1ω7t	2
16:0	25
17:0cyc	2
18:2	T[a]
18:1ω9c	1
18:1ω7c	25
18:0	1

[a]T = Trace.

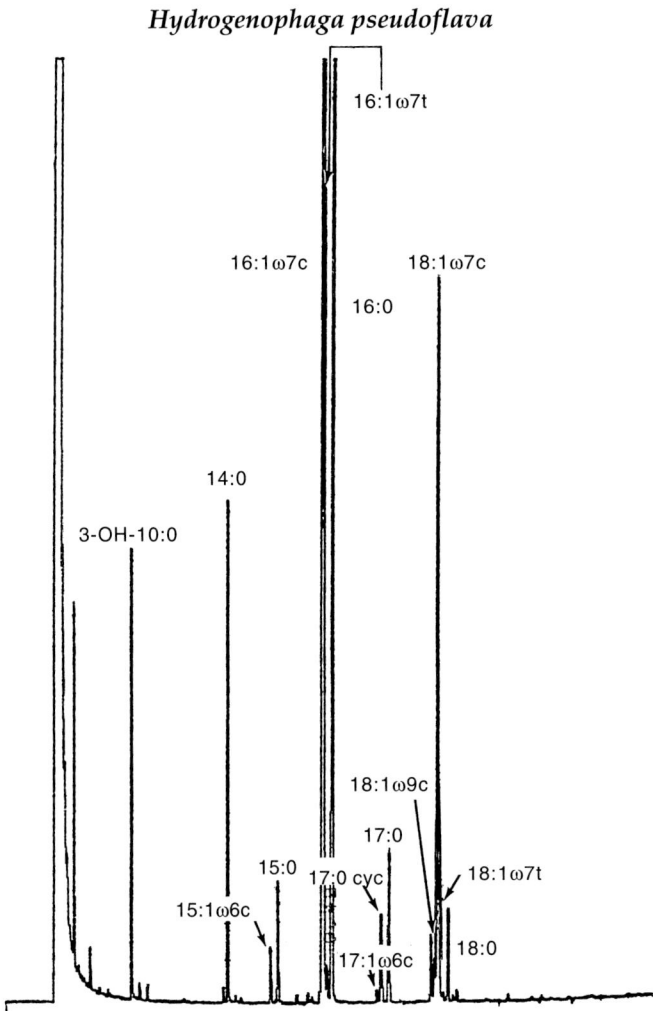

Figure 4.57. Gas chromatogram of the cellular fatty acids of *Hydrogenophaga pseudoflava*.

Comments

Hydrogenophaga pseudoflava has a unique CFA profile that is characterized by major amounts of 16:1ω7c (40%) and 16:0 (26%); moderate amounts (4–12%) of 18:1ω7c, 16:1ω7t, 14:0, and 3—OH-10:0; and small to trace amounts of 15:1ω6c, 15:0, 17:0cyc, 17:0, 18:1ω7t, and 18:0. This species differs from *Hydrogenophaga palleronii* by the presence of 3—OH-10:0, 15:1ω6c, 15:0, 17:0, and 18:1ω7t.

Literature

1. Willems, A., J. Busse, M. Goor, B. Pot, E. Falsen, E. Jantzen, B. Hoste, M. Gillis, K. Kersters, G. Auling, and J. De Ley. 1989. *Hydrogenophaga*, a new genus of hydrogen-oxidizing bacteria that includes *Hydrogenophaga flava* comb. nov. (formerly *Pseudomonas flava*), *Hydrogenophaga palleronii* (formerly *Pseudomonas palleronii*), *Hydrogenophaga pseudoflava* (formerly *Pseudomonas pseudoflava* and "*Pseudomonas carboxydoflava*"), and *Hydrogenophaga taeniospiralis* (formerly *Pseudomonas taeniospiralis*). Int. J. Syst. Bacteriol. 39: 319–333.

Table 4.52. Cellular fatty acid composition of *Hydrogenophaga pseudoflava*

Fatty Acid	Organism *Hydrogenophaga pseudoflava* %
3-OH-8:0	T[a]
3-OH-10:0	4
14:0	5
15:1ω6c	1
15:0	1
16:1ω7c	40
16:1ω7t	12
16:0	26
17:1ω6c	T
17:0cyc	2
17:0	1
18:2	T
18:1ω9c	T
18:1ω7c	5
18:1ω7t	1
18:0	1

[a]T = Trace.

Kingella denitrificans, Kingella kingae, and Simonsiella crassa

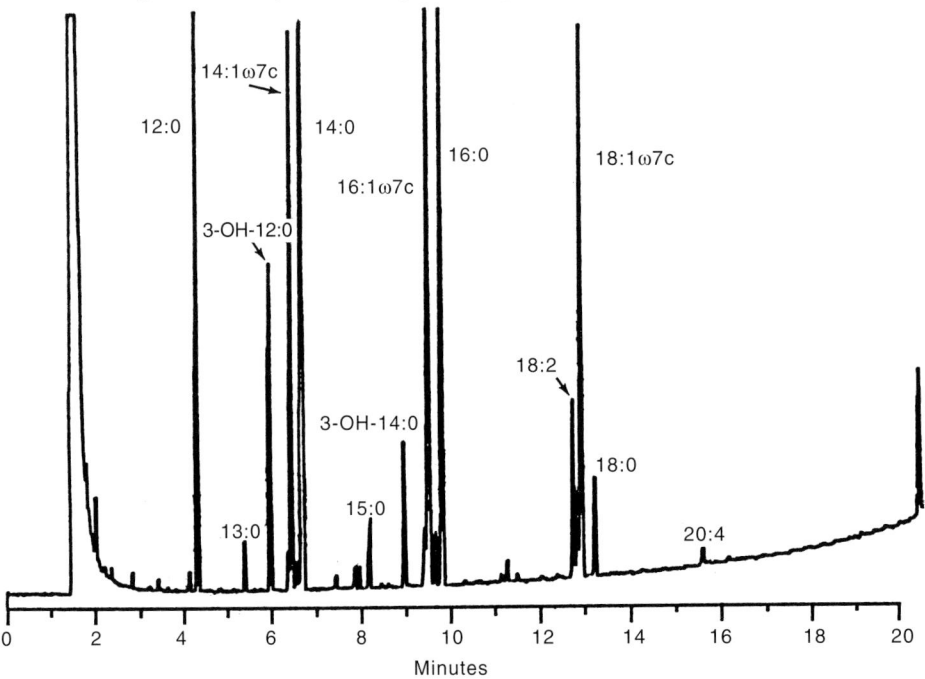

Figure 4.58. Gas chromatogram of the cellular fatty acids of *Kingella denitrificans*, *Kingella kingae*, and *Simonsiella crassa*.

Comments

Organisms in this CFA group are characterized by large amounts of 14:0 (22–34%); moderate to large amounts of 16:0 (10–28%); and small to moderate amounts of 16:1ω7c, 18:2, 18:1ω9c, 3—OH-12:0, 3—OH-14:0, and 18:1ω7c. This CFA group differs from another cellular fatty acid group with large amounts of 14:0 (*Haemophilus, Pasteurella, Actinobacillus*) by the presence of 3—OH-12:0. *Helicobacter pylori* also has large amounts of 14:0 (34%), but this organism has large amounts of 19:0cyc[11–12] and small amounts of 3—OH-16:0 and 3—OH-18:0 that are absent in this CFA group. Additional studies may provide data useful for further differentiation among members of this CFA group. For example, the mean value of 18:1ω7c in *Simonsiella crassa* is 10%, 3% in *Kingella kingae*, and only trace amounts in *K. denitrificans;* however, there is overlap in the relative amounts of 18:1ω7c for one or two strains of each species. This CFA group differs from that of *Cardiobacterium hominis* by smaller amounts of 18:1ω7c (trace–10% versus 41%) and larger amounts of 14:0 (22–34% versus 14%) and from the CFA group containing *Eikenella corrodens* and *Suttonella indologenes* by smaller amounts of 18:1ω7c (trace–10% versus 28– 41%) and larger amounts of 14:0 (22–34% versus 2–14%).

Literature

1. Prefontaine, G., and F.L. Jackson. 1972. Cellular fatty acid profiles as an aid to the classification of "corroding bacilli" and certain other bacteria. Int. J. Syst. Bacteriol. 22: 210–217.

2. Wallace, P.L., D.G. Hollis, R.E. Weaver, and C.W. Moss. 1988. Cellular fatty acid composition of *Kingella* species, *Cardiobacterium hominis*, and *Eikenella corrodens*. J. Clin. Microbiol. 26: 1592–1594.

Table 4.53. Cellular fatty acid composition of *Kingella denitrificans*, *Kingella kingae*, and *Simonsiella crassa*

Fatty Acid	Organism		
	Kingella denitrificans %	Kingella kingae %	Simonsiella crassa %
12:0	10	9	8
3-OH-12:0	3	4	4
14:1ω7c	-	4	4
14:0	22	34	32
15:0	-	1	-
3-OH-14:0	2	2	1
16:1ω7c	5	15	24
16:0	28	13	10
UNK[a]	-	-	1
18:2	13	5	3
18:1ω9c	10	4	2
18:1ω7c	-	3	10
18:0	4	4	1
20:4	1	1	-

[a]UNK = Unknown.

Methylobacterium extorquens, *Methylobacterium mesophilicum*, *Methylobacterium zatmanii*, *Roseomonas fauriae*, and *Roseomonas* genomospecies 6

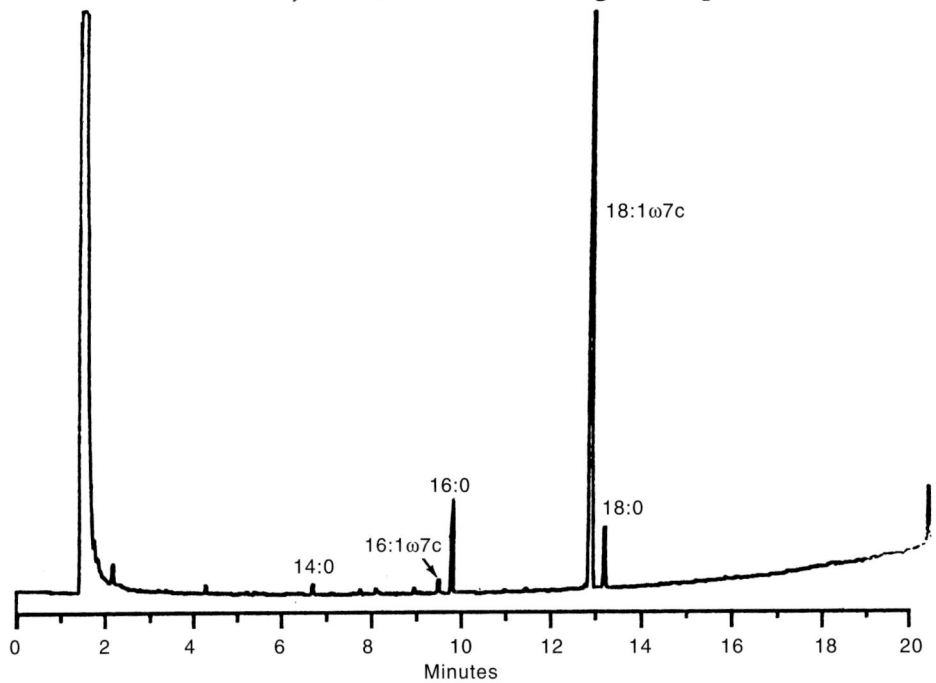

Figure 4.59. Gas chromatogram of the cellular fatty acids of *Methylobacterium extorquens*, *Methylobacterium mesophilicum*, *Methylobacterium zatmanii*, *Roseomonas fauriae*, and *Roseomonas* genomospecies 6.

Comments

Organisms in this CFA group are characterized by very large amounts (63–90%) of 18:1ω7c and small amounts of 3—OH-14:0 (2–6%), 16:1ω7c (3–12%), 18:2, 18:1ω9c, and 18:0. These organisms differ from *Brucella canis* by the presence of 3—OH-14:0 and from CDC group EO-3 by the presence of 3—OH-14:0 and the absence of 2—OH-18:1, 19:0cyc[11–12], and 20:1 (0% versus 1%, 3%, and 1%, respectively, in group EO-3). *Roseomonas fauriae* and *Roseomonas* genomospecies 6 have 4% 2—OH 18:1, 2% 3—OH-16:0, and small to trace amounts of 17:1ω6c that are absent in *Methylobacterium* species, but additional strains of each species/group must be tested to determine if these differences are sufficient for differentiation by CFA data alone.

Literature

1. Urakami, T., and K. Komagata. 1984. *Protomonas*, a new genus of facultative methylotrophic bacteria. Int. J. Syst. Bacteriol. *34:* 188–201.

2. Wallace, P.L., D.G. Hollis, R.E. Weaver, and C.W. Moss. 1990. Biochemical and chemical characterization of pink-pigmented oxidative bacteria. J. Clin. Microbiol. *28:* 689–693.

Table 4.54. Cellular fatty acid composition of *Methylobacterium extorquens*, *Methylobacterium mesophilicum*, *Methylobacterium zatmanii*, *Roseomonas fauriae*, and *Roseomonas* genomospecies 6

Fatty Acid	Organism[a]				
	A	B	C	D	E
	%	%	%	%	%
3-OH-14:0	2	1	2	4	14
16:1ω7c	3	3	-	12	16
16:0	4	4	2	7	6
17:1ω8c	-	-	-	-	T[b]
17:1ω6c	-	-	-	1	1
3-OH-16:0	-	-	-	2	T
18:2	1	-	-	1	1
18:1ω9c	1	-	-	1	1
18:1ω7c	83	89	90	66	52
18:0	5	3	4	1	T
2-OH-18:1	-	-	-	4	6
3-OH-18:0	1	-	2	-	-

[a] A, *Methylobacterium extorquens*; B, *Methylobacterium mesophilicum*; C, *Methylobacterium zatmanii*; D, *Roseomonas fauriae*; E, *Roseomonas* genomospecies 6.
[b] T = Trace.

Moraxella atlantae

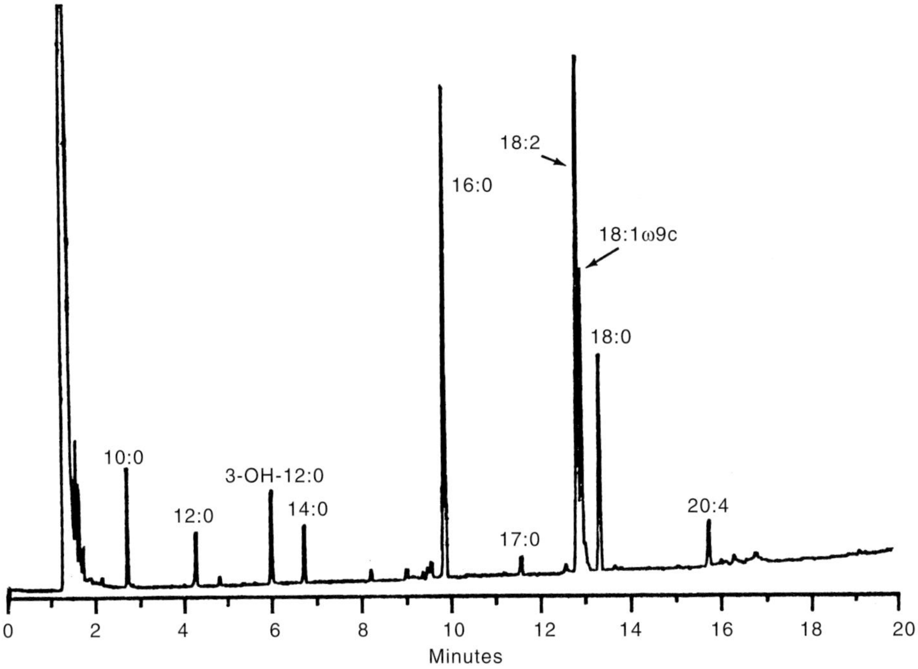

Figure 4.60. Gas chromatogram of the cellular fatty acids of *Moraxella atlantae*.

Comments

Moraxella atlantae has a distinct CFA profile with 18:2, 18:1ω9c, 16:0, and 18:0 as major acids with smaller amounts of 10:0, 12:0, 3—OH-12:0, 14:0, and 3—OH-14:0. The large amount of 18:2 (23%) distinguishes *M. atlantae* from all other *Moraxella* except *M. phenylpyruvica*, but this latter organism contains small amounts (3–4%) of i-11:0 and 11:0 that are absent in *M. atlantae*. The CFA profile of *M. atlantae* differs from that of the CFA group containing *M. bovis*, *M. cuniculi*, *M. ovis*, *M. nonliquefaciens*, *M. lacunata* I, and *M. lacunata* II by larger amounts of 18:2, the absence of 17:1ω8c, and the absence of n-16:0 and n-18:0 alcohols.

Literature

1. Jantzen, E., K. Bryn, T. Bergan, and K. Bøvre. 1974. Gas chromatography of bacterial whole cell methanolysates. V. Fatty acid composition of *Neisseria* and *Moraxella*. Acta. Pathol. Microbiol. Immunol. 82: 767–779.

2. Moss, C.W., P.L. Wallace, D.G. Hollis, and R.E. Weaver. 1988. Cultural and chemical characterization of CDC groups EO-2, M-5, and M-6, *Moraxella (Moraxella)* species, *Oligella urethralis*, *Acinetobacter* species, and *Psychrobacter immobilis*. J. Clin. Microbiol. 26: 484–492.

3. Veys, A., W. Callewaert, E. Waelkens, and K.V.D. Abbeele. 1989. Application of gas-liquid chromatography to the routine identification of nonfermentative gram-negative bacteria in clinical specimens. J. Clin. Microbiol. 27: 1538–1542.

4. Osterhout, G.J., V.H. Shull, and J.D. Dick. 1991. Identification of clinical isolates of gram-negative nonfermentative bacteria by an automated cellular fatty acid identification system. J. Clin. Microbiol. 29: 1822–1830.

Table 4.55. Cellular fatty acid composition of *Moraxella atlantae*

Fatty Acid	Organism *Moraxella atlantae* %
10:0	4
12:0	4
12:1ω9c	1
3-OH-12:0	6
14:0	2
3-OH-14:0	1
16:1ω9c	1
16:1ω7c	1
16:0	23
17:0	1
18:0 alc	1
18:2	23
18:1ω9c	19
18:0	11
20:4	2

Moraxella bovis, Moraxella canis, Moraxella cuniculi, Moraxella lacunata **I**, *Moraxella lacunata* **II**, *Moraxella lincolnii, Moraxella nonliquefaciens,* **and** *Moraxella ovis*

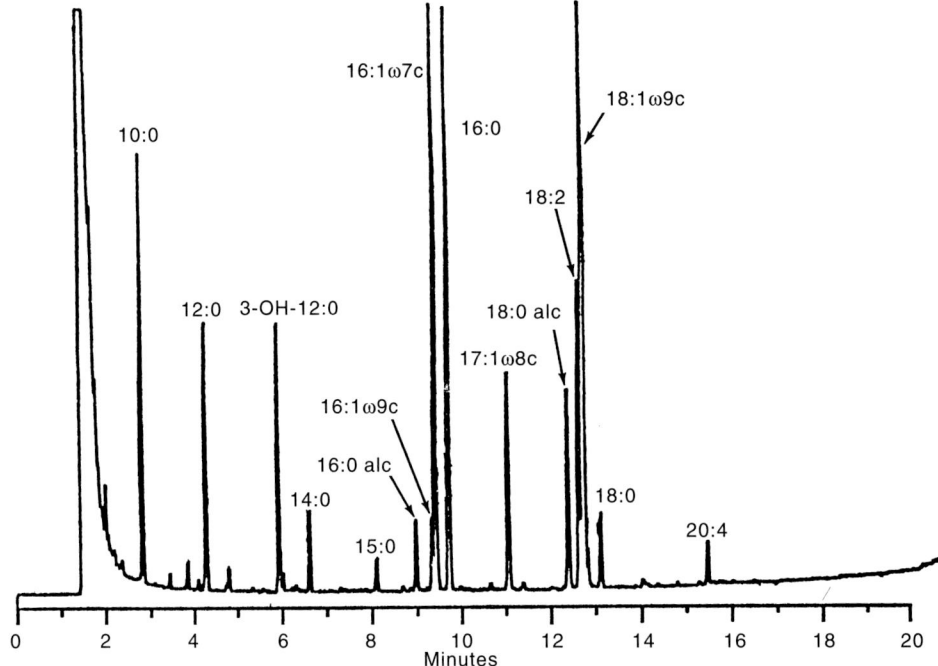

Figure 4.61. Gas chromatogram of the cellular fatty acids of *Moraxella bovis, Moraxella canis, Moraxella cuniculi, Moraxella lacunata* I, *Moraxella lacunata* II, *Moraxella lincolnii, Moraxella nonliquefaciens,* and *Moraxella ovis*.

Comments

Moraxella species in this CFA group are like other *Moraxella* in that they contain large amounts of 18:1ω9c, small amounts of 10:0 and 12:1ω9, and no 18:1ω7c (except 1% in *M. canis* and *M. lincolnii*). This group, however, differs from other *Moraxella* by the presence of small amounts (3–7%) of 17:1ω8c, and two primary alcohols, n-hexadecanol (16:0 alc) and n-octadecanol (18:0 alc). The 16:0 alc elutes from a nonpolar capillary GLC column just before 16:1ω7c, and the 18:0 alc co-elutes with 18:3 (just before 18:2). The profile of this CFA group is most like that of *Psychrobacter immobilis* except for the presence of 16:0 alc and 18:0 alc, and the absence of i-17:0 (0% versus 2% in *P. immobilis*). Most strains of *M. lacunata* II differ from *M. lacunata* I by lower amounts of 17:1ω8c, 16:1ω7c, and 18:1ω9c; higher amounts of 18:0 (16% versus 3%); and higher amounts of 16:0 alc and 18:0 alc. However, there is some overlap in relative amounts of these acids in some strains of both groups that prohibit their differentiation solely on the basis of CFA data. The two primary alcohols (16:0 alc and 18:0 alc) were absent in *M. lincolnii,* but this species is included in this CFA group until additional strains are available for testing.

Literature

1. Jantzen, E., K. Bryn, T. Bergan, and K. Bøvre. 1974. Gas chromatography of bacterial whole cell methanolysates. V. Fatty acid composition of *Neisseria* and *Moraxella*. Acta. Pathol. Microbiol. Immunol. *82:* 767–779.

2. Moss, C.W., P.L. Wallace, D.G. Hollis, and R.E. Weaver. 1988. Cultural and chemical characterization of CDC groups EO-2, M-5, and M-6, *Moraxella (Moraxella)* species, *Oligella urethralis, Acinetobacter* species, and *Psychrobacter immobilis*. J. Clin. Microbiol. *26:* 484–492.

3. Veys, A., W. Callewaert, E. Waelkens, and K.V.D. Abbeele. 1989. Application of gas-liquid chromatography to the routine identification of nonfermentative gram-negative bacteria in clinical specimens. J. Clin. Microbiol. *27:* 1538–1542.

4. Osterhout, G.J., V.H. Shull, and J.D. Dick. 1991. Identification of clinical isolates of gram-negative nonfermentative bacteria by an automated cellular fatty acid identification system. J. Clin. Microbiol. *29:* 1822–1830.

Table 4.56. Cellular fatty acid composition of *Moraxella bovis, Moraxella canis, Moraxella cuniculi, Moraxella lacunata* **I,** *Moraxella lacunata* **II,** *Moraxella lincolnii, Moraxella nonliquefaciens,* and *Moraxella ovis*

Fatty Acid	Organism[a]							
	A %	B %	C %	D %	E %	F %	G %	H %
10:0	3	3	3	3	3	4	3	3
12:0	5	3	3	5	1	3	4	2
12:1ω9c	1	1	1	1	1	T[b]	1	1
3-OH-12:0	5	3	7	4	6	2	5	6
14:0	-	T	T	T	1	T	1	1
15:0	-	-	-	-	-	-	1	-
3-OH-14:0	-	T	1	1	1	-	-	1
16:0 alc	2	-	1	1	5	-	2	2
16:1ω9c	1	1	1	1	-	-	1	1
16:1ω7c	19	5	11	18	1	36	24	11
16:0	9	4	7	8	28	11	9	13
i-17:0	-	T	2	-	-	-	-	-
17:1ω8c	5	T	2	3	-	T	5	2
17:0	-	-	-	-	1	-	-	-
18:0 alc	2	3	2	5	5	-	4	4
18:2	8	1	2	5	16	1	5	1
18:1ω9c	36	67	52	42	12	37	32	46
18:1ω7c	-	-	-	-	-	1	-	-
18:0	2	5	4	3	16	3	2	5
20:4	T	-	-	T	2	-	1	-

[a] A, *Moraxella bovis*; B, *Moraxella canis*; C, *Moraxella cuniculi*; D, *Moraxella lacunata* I; E, *Moraxella lacunata* II; F, *Moraxella lincolnii*; G, *Moraxella nonliquefaciens*; H, *Moraxella ovis*.
[b] T = Trace.

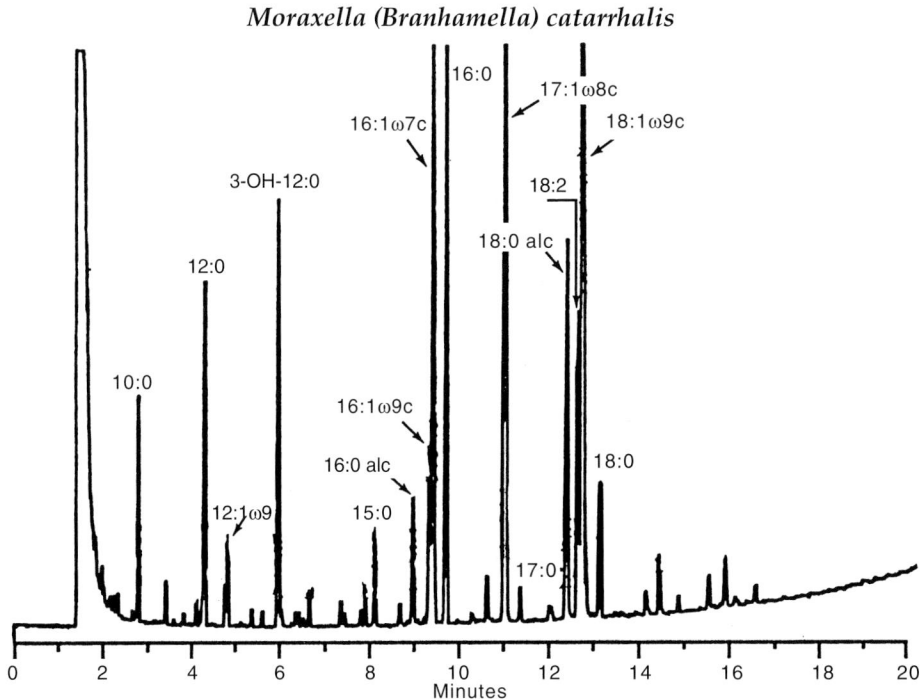

Figure 4.62. Gas chromatogram of the cellular fatty acids of *Moraxella (Branhamella) catarrhalis*.

Comments

Moraxella (Branhamella) catarrhalis has a unique CFA profile that is characterized by major amounts of 18:1ω9c (35%) and 17:1ω8c (25%); and small amounts of 10:0, 11:0, 12:0, 3—OH-12:0, 15:1ω6c, 15:0, 16:1ω7c, 17:0, 18:0, 18:2, 16:0 alc, and 18:0 alc. This profile is most like that of the CFA group containing *M. bovis*, *M. cuniculi*, *M. ovis*, *M. nonliquefaciens*, *M. lacunata* I, and *M. lacunata* II except for the presence of larger amounts of 17:1ω8c (25% versus 7%) and small amounts of 9:0, 11:0, 15:1ω6c, 15:0, and 16:1ω7t. The CFA profile differs from that of *Psychrobacter immobilis* by larger amounts of 17:1ω8c (25% versus 9%); the presence of 9:0, 11:0, 15:1ω6c, 15:0, and 16:1ω7t; and the absence of i-17:0 (0% versus 2% in *P. immobilis*).

Literature

1. Lewis, V.J., R.E. Weaver, and D.G. Hollis. 1968. Fatty acid composition of *Neisseria* species as determined by gas chromatography. J. Bacteriol. *96:* 1–5.

2. Lambert, M.A., D.G. Hollis, C.W. Moss, R.E. Weaver, and M.L. Thomas. 1971. Cellular fatty acids of nonpathogenic *Neisseria*. Can. J. Microbiol. *17:* 1491–1502.

3. Jantzen, E., K. Bryn, T. Bergan, and K. Bøvre. 1974. Gas chromatography of bacterial whole cell methanolysates. V. Fatty acid composition of *Neisseria* and *Moraxella*. Acta. Pathol. Microbiol. Immunol. *82:* 767–779.

4. Moss, C.W., P.L. Wallace, D.G. Hollis, and R.E. Weaver. 1988. Cultural and chemical characterization of CDC groups EO-2, M-5, and M-6, *Moraxella (Moraxella)* species, *Oligella urethralis*, *Acinetobacter* species, and *Psychrobacter immobilis*. J. Clin. Microbiol. *26:* 484–492.

Table 4.57. Cellular fatty acid composition of *Moraxella (Branhamella) catarrhalis*

Fatty Acid	*Moraxella (Branhamella) catarrhalis* %
9:0	1
10:0	4
11:0	1
12:0	3
3-OH-11:0	1
3-OH-12:0	3
14:0	1
15:1ω6c	1
15:0	3
16:1ω9c	1
16:1ω7c	9
16:0	5
17:1ω8c	25
17:0	1
18:0 alc	1
18:2	2
18:1ω9c	35
18:0	2
19:0	1

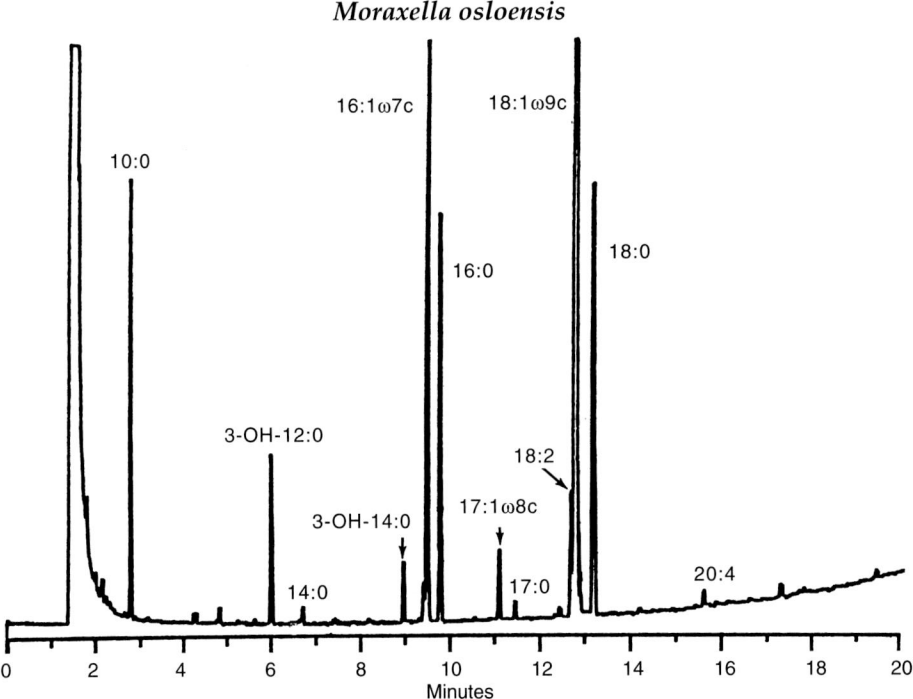

Figure 4.63. Gas chromatogram of the cellular fatty acids of *Moraxella osloensis*.

Comments

Moraxella osloensis has a unique CFA profile that is similar to, but distinct from, other *Moraxella* by the presence of large amounts of 18:1ω9c (range, 46–61% for 12 strains); about equal amounts (7–10%) of 10:0, 16:1ω7c, 16:0, and 18:0; and the absence of 12:0. This CFA profile is most like that of *Psychrobacter immobilis* that also has 10:0 and large amounts (55%) of 18:1ω9c, but *P. immobilis* contains 3% of 12:0, 2% of i-17:0 (that are absent in *M. osloensis*) and larger amounts of 17:1ω8c (7% versus 1%).

Literature

1. Jantzen, E., K. Bryn, T. Bergan, and K. Bøvre. 1974. Gas chromatography of bacterial whole cell methanolysates. V. Fatty acid composition of *Neisseria* and *Moraxella*. Acta. Pathol. Microbiol. Immunol. *82:* 767–779.

2. Moss, C.W., P.L. Wallace, D.G. Hollis, and R.E. Weaver. 1988. Cultural and chemical characterization of CDC groups EO-2, M-5, and M-6, *Moraxella (Moraxella)* species, *Oligella urethralis, Acinetobacter* species, and *Psychrobacter immobilis*. J. Clin. Microbiol. *26:* 484–492.

3. Veys, A., W. Callewaert, E. Waelkens, and K.V.D. Abbeele. 1989. Application of gas-liquid chromatography to the routine identification of nonfermentative gram-negative bacteria in clinical specimens. J. Clin. Microbiol. *27:* 1538–1542.

4. Osterhout, G.J., V.H. Shull, and J.D. Dick. 1991. Identification of clinical isolates of gram-negative nonfermentative bacteria by an automated cellular fatty acid identification system. J. Clin. Microbiol. *29:* 1822–1830.

Table 4.58. Cellular fatty acid composition of *Moraxella osloensis*

Fatty Acid	Organism
	Moraxella osloensis
	%
10:0	7
3-OH-12:0	3
3-OH-14:0	1
16:1ω9c	1
16:1ω7c	10
16:0	6
17:1ω8c	1
18:2	4
18:1ω9c	59
18:0	7

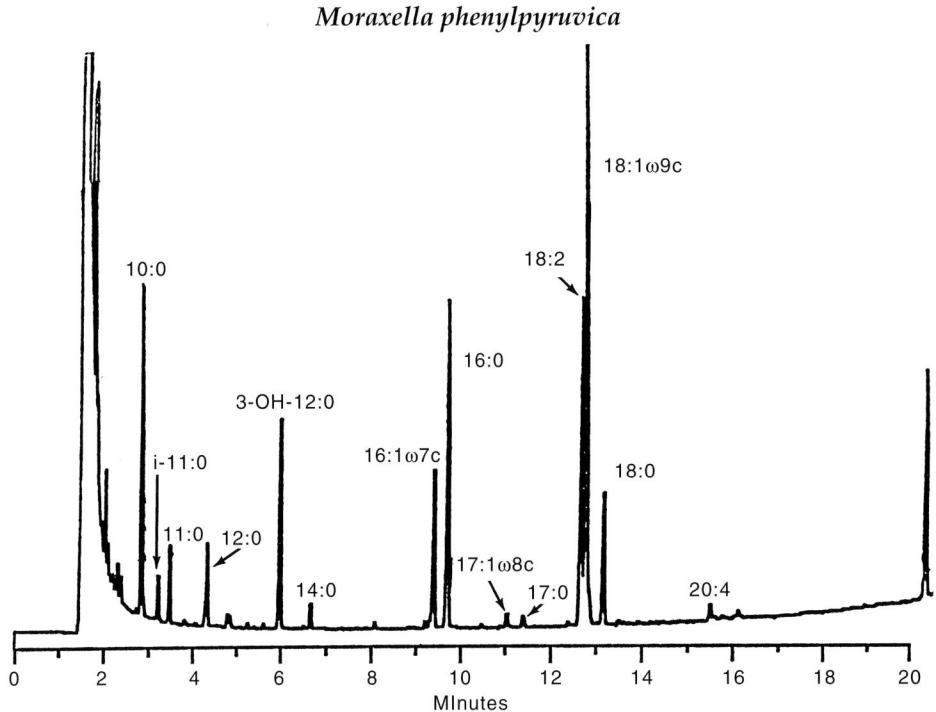

Figure 4.64. Gas chromatogram of the cellular fatty acids of *Moraxella phenylpyruvica*.

Comments

Moraxella phenylpyruvica has a unique CFA profile with 18:2 and 18:1ω9c as major acids; moderate amounts (11%) of 16:1ω7c and 16:0; and smaller amounts of 10:0, 12:0, 3—OH-12:0, 18:0, 20:4, i-11:0, and 11:0. Both *M. phenylpyruvica* and *M. atlantae* differ from other *Moraxella* and biochemically related organisms by the presence of large amounts (>20%) of 18:2; these two organisms are distinguished from each other by the presence of i-11:0 (2%) and 11:0 (4%) in *M. phenylpyruvica*, and larger amounts of 16:1ω7c in *M. phenylpyruvica* (11% versus 1%).

Literature

1. Jantzen, E., K. Bryn, T. Bergan, and K. Bøvre. 1974. Gas chromatography of bacterial whole cell methanolysates. V. Fatty acid composition of *Neisseria* and *Moraxella*. Acta Pathol. Microbiol. Immunol. *82:* 767–779.

2. Moss, C.W., P.L. Wallace, D.G. Hollis, and R.E. Weaver. 1988. Cultural and chemical characterization of CDC groups EO-2, M-5, and M-6, *Moraxella (Moraxella)* species, *Oligella urethralis, Acinetobacter* species, and *Psychrobacter immobilis*. J. Clin. Microbiol. *26:* 484–492.

3. Veys, A., W. Callewaert, E. Waelkens, and K.V.D. Abbeele. 1989. Application of gas-liquid chromatography to the routine identification of nonfermentative gram-negative bacteria in clinical specimens. J. Clin. Microbiol. *27:* 1538–1542.

4. Osterhout, G.J., V.H. Shull, and J.D. Dick. 1991. Identification of clinical isolates of gram-negative nonfermentative bacteria by an automated cellular fatty acid identification system. J. Clin. Microbiol. *29:* 1822–1830.

Table 4.59. Cellular fatty acid composition of *Moraxella phenylpyruvica*

Fatty Acid	Organism *Moraxella phenylpyruvica* %
10:0	5
i-11:0	2
11:0	4
12:0	3
3-OH-11:0	1
12:1ω9c	1
3-OH-12:0	8
14:0	1
16:1ω7c	11
16:0	11
18:2	23
18:1ω9c	22
18:0	4
20:4	2

Ochrobactrum anthropi, Roseomonas cervicalis, Roseomonas gilardii, **and** *Roseomonas* **genomospecies 4 and 5**

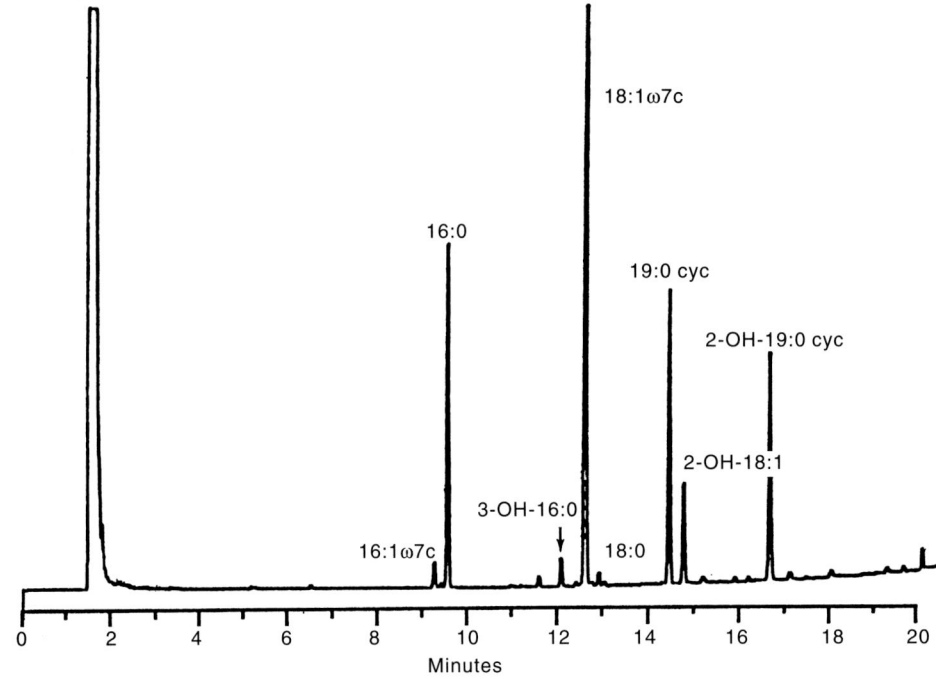

Figure 4.65. Gas chromatogram of the cellular fatty acids of *Ochrobactrum anthropi, Roseomonas cervicalis, Roseomonas gilardii,* and *Roseomonas* genomospecies 4 and 5.

Comments

The profile of the organisms in this CFA group is characterized by large amounts (43–53%) of 18:1ω7c, 19:0cyc$^{11-12}$ (10–25%), 2—OH-18:1 (2–5%), and 2—OH-19:0cyc (3–11%). This profile differs from that of EO-3 by smaller amounts of 18:1ω7c (53% versus 80%), larger amounts of 19:0cyc$^{11-12}$ (10% versus 3%), and the presence of 2—OH-19:0cyc (3% versus 0%). This profile differs from the CFA group containing *Brucella abortus*, *B. melintensis*, and *B. suis* by the presence of 2—OH-19:0cyc and from the CFA group containing *Pseudomonas diminuta* and *P. vesicularis* by the absence of 3—OH-12:0, 14:0, Br-19:1, 15:0, and 17:1ω8c.

Literature

1. Wallace, P.L., D.G. Hollis, R.E. Weaver, and C.W. Moss. 1990. Biochemical and chemical characterization of pink-pigmented oxidative bacteria. J. Clin. Microbiol. *28:* 689–693.

Table 4.60. Cellular fatty acid composition of *Ochrobactrum anthropi*, *Roseomonas cervicalis*, *Roseomonas gilardii*, *Roseomonas* genomospecies 4, and *Roseomonas* genomospecies 5

Fatty Acid	Organism[a]				
	A	B	C	D	E
	%	%	%	%	%
16:1ω7c	1	2	1	6	2
16:1ω5	-	2	-	3	1
16:0	8	15	19	12	13
17:1ω6c	-	2	-	2	1
17:0	1	1	-	-	-
3-OH-16:0	-	T[b]	1	-	16
18:2	1	1	1	1	1
18:1ω9c	T	-	-	T	T
18:1ω7c	47	61	34	61	64
18:0	10	1	1	1	1
19:0cyc$^{11\text{-}12}$	25	3	21	2	6
2-OH-18:1	2	6	3	7	3
2-OH-19:0cyc	3	3	17	2	7

[a]A, *Ochrobactrum anthropi*; B, *Roseomonas cervicalis*; C, *Roseomonas gilardii*; D, *Roseomonas* genomospecies 4; E, *Roseomonas* genomospecies 5.
[b]T = Trace.

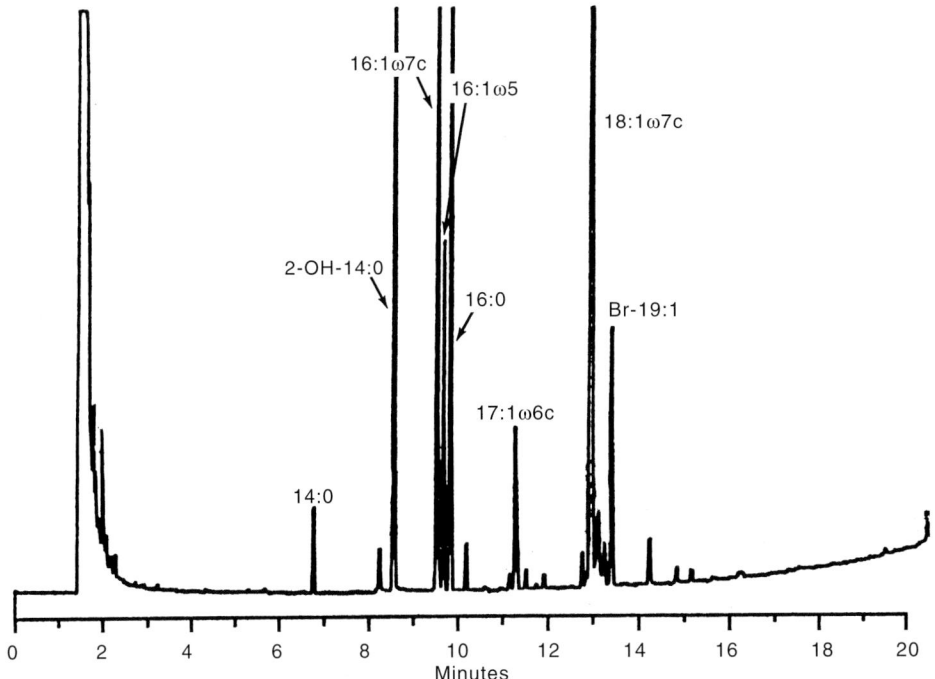

Figure 4.66. Gas chromatogram of the cellular fatty acids of *"Pseudomonas azotocolligans," Sphingomonas capsulata, Sphingomonas* genomospecies 2, *Sphingomonas paucimobilis,* and *Sphingomonas yanoikuyae.*

Comments

Organisms in this CFA group are easily distinguished from other organisms by the presence of large amounts (60%) of 18:1ω7c, 2–7% of 2—OH-14:0 as the only hydroxy acid (except 1% of 2—OH-16:0 in *Sphingomonas yanoikuyae*); small amounts of 16:1ω5 and 17:1ω6c; and **no** cyclopropane acids. This profile is most similar to the CFA group containing *S. adhaesiva* and *S. parapaucimobilis* except for the absence of 17:1ω8c (0% versus 2%) and 17:0 (0% versus 2%), and smaller amounts of 17:1ω6c (3% versus 15%). *Alcaligenes xylosoxidans* subsp. *xylosoxidans*, *A.*-like Group 1, and *Agrobacterium* species also contain significant amounts of 2—OH-14:0, but these organisms contain 3—OH-14:0, 2—OH-16:0, and cyclopropane acids that are absent in this CFA group. Generally, *S. paucimobilis* contains trace to 2% amounts of Br-19:1 that is not detected in *"P. azotocolligans"*.

Literature

1. Yabuuchi, E., E. Tanimura, A. Ohyama, I. Yano, and A. Yamamato. 1979. *Flavobacterium devorans* ATCC 10829: a strain of *Pseudomonas paucimobilis*. J. Gen. Appl. Microbiol. 25: 95–107.

2. Yabuuchi, E., I. Yano, H. Oyaizu, Y. Hashimoto, T. Ezaki, and H. Yamamato. 1990. Proposal of *Sphingomonas paucimobilis* gen. nov. and comb. nov. *Sphingomonas parapaucimobilis* sp. nov., *Sphingomonas yanoikuyae* sp. nov., *Sphingomonas adhaesiva* sp. nov., *Sphingomonas capsulata* comb. nov., and two genospecies of the genus *Sphingomonas*. Microbiol. Immunol. 34: 99–119.

Table 4.61. Cellular fatty acid composition of *"Pseudomonas azotocolligans,"* *Sphingomonas capsulata, Sphingomonas* **genospecies 2**, *Sphingomonas paucimobilis*, **and** *Sphingomonas yanoikuyae*

Fatty Acid	Organism[a]				
	A	B	C	D	E
	%	%	%	%	%
14:0	-	1	1	1	1
15:0	1	1	1	-	-
2-OH-14:0	1	10	6	7	6
16:1ω7c	1	8	6	6	17
16:1ω5	1	2	1	1	2
16:0	20	12	14	13	9
17:1ω6c	4	3	3	3	1
17:0	T[b]	-	T	-	-
2-OH-16:0	-	-	-	-	1
18:2	-	T	T	1	-
18:1ω7c	65	56	63	61	58
18:1?[c]	5	1	1	1	1
18:0	1	4	2	1	2
Br-19:1	-	-	-	2	1

[a]A, *"Pseudomonas azotocolligans"*; B, *Sphingomonas capsulata*; C, *Sphingomonas* genomospecies 2; D, *Sphingomonas paucimobilis*; E, *Sphingomonas yanoikuyae*.
[b]T = Trace.
[c]?, The position of the double bond is unknown.

Pseudomonas cepacia, *Pseudomonas gladioli*, *Pseudomonas mallei*, and *Pseudomonas pseudomallei*

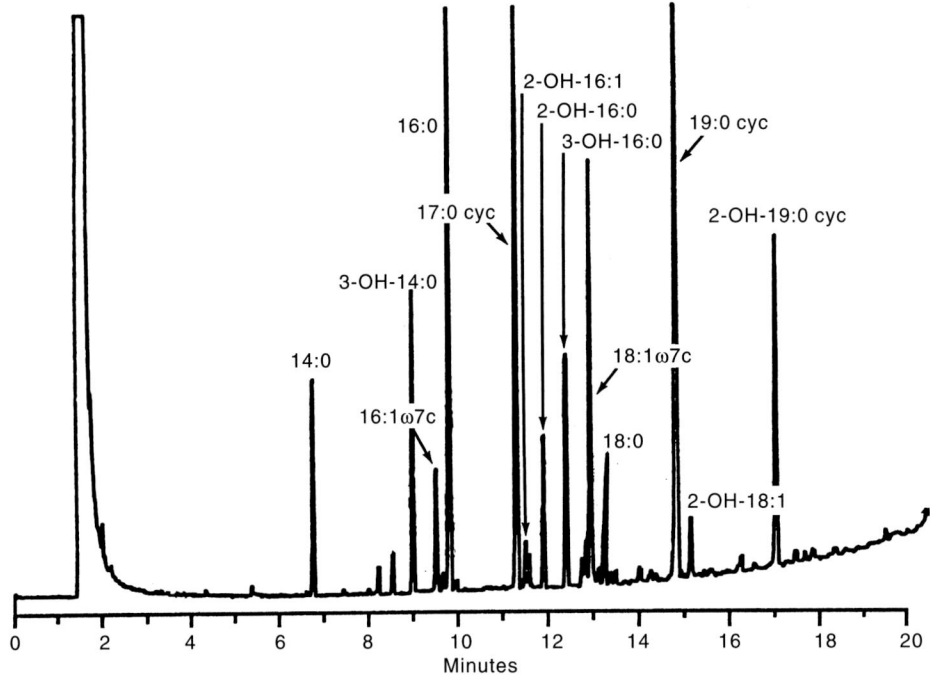

Figure 4.67. Gas chromatogram of the cellular fatty acids of *Pseudomonas cepacia*, *Pseudomonas gladioli*, *Pseudomonas mallei*, and *Pseudomonas pseudomallei*.

Comments

Organisms in this CFA group have a unique profile that is easily recognized by the presence of two cyclopropane acids (17:0cyc, 19:0cyc[11–12]), 16:0 and 18:1ω7c as major acids, and six hydroxy acids (3—OH-14:0, 2—OH-16:1, 2—OH-16:0, 3—OH-16:0, 2—OH-18:1, 2—OH-19:0cyc). The CFA profiles of strains of these four species are essentially identical and cannot be differentiated by CFA data alone.

Literature

1. Samuels, S.B., C.W. Moss, and R.E. Weaver. 1973. The fatty acids of *Pseudomonas multivorans* (*Pseudomonas cepacia*) and *Pseudomonas kingii*. J. Gen. Microbiol. 74: 275–279.

2. Moss, C.W. 1978. New methodology for identification of nonfermenters: gas-liquid chromatographic chemotaxonomy. *In* G.L. Gilardi (Ed.), Glucose nonfermenting gram-negative bacteria in clinical microbiology, Chapter 9, CRC Press, West Palm Beach, FL, pp. 171–201.

3. Oyaizu, H., and K. Komagata. 1983. Grouping of *Pseudomonas* species on the basis of cellular fatty acid composition and the quinone system with special reference to the existence of 3-hydroxy fatty acids. J. Gen. Appl. Microbiol. 29: 17–40.

4. Veys, A., W. Callewaert, E. Waelkens, and K.V.D. Abbeele. 1989. Application of gas-liquid chromatography to the routine identification of nonfermentative gram-negative bacteria in clinical specimens. J. Clin. Microbiol. 27: 1538–1542.

5. Osterhout, G.J., V.H. Shull, and J.D. Dick. 1991. Identification of clinical isolates of gram-negative nonfermentative bacteria by an automated cellular fatty acid identification system. J. Clin. Microbiol. 29: 1822–1830.

Table 4.62. Cellular fatty acid composition of *Pseudomonas cepacia*, *Pseudomonas gladioli*, *Pseudomonas mallei*, and *Pseudomonas pseudomallei*

	Organism			
	Pseudomonas cepacia	*Pseudomonas gladioli*	*Pseudomonas mallei*	*Pseudomonas pseudomallei*
Fatty Acid	%	%	%	%
14:0	4	6	5	3
15:0	T[a]	-	T	1
2-OH-14:0	-	-	-	1
3-OH-14:0	5	6	5	5
16:1ω7c	9	6	7	4
16:0	24	27	24	24
17:0cyc	14	18	8	14
17:0	T	T	T	1
2-OH-16:1	1	1	2	1
2-OH-16:0	2	3	4	3
3-OH-16:0	4	4	4	4
18:2	1	-	1	-
18:1ω9c	1	-	1	1
18:1ω7c	20	10	27	13
18:0	1	1	1	1
19:0cyc[11-12]	10	9	8	14
2-OH-18:1	2	3	1	2
2-OH-19:0cyc	2	3	1	6

[a]T = Trace.

Pseudomonas diminuta and *Pseudomonas vesicularis*

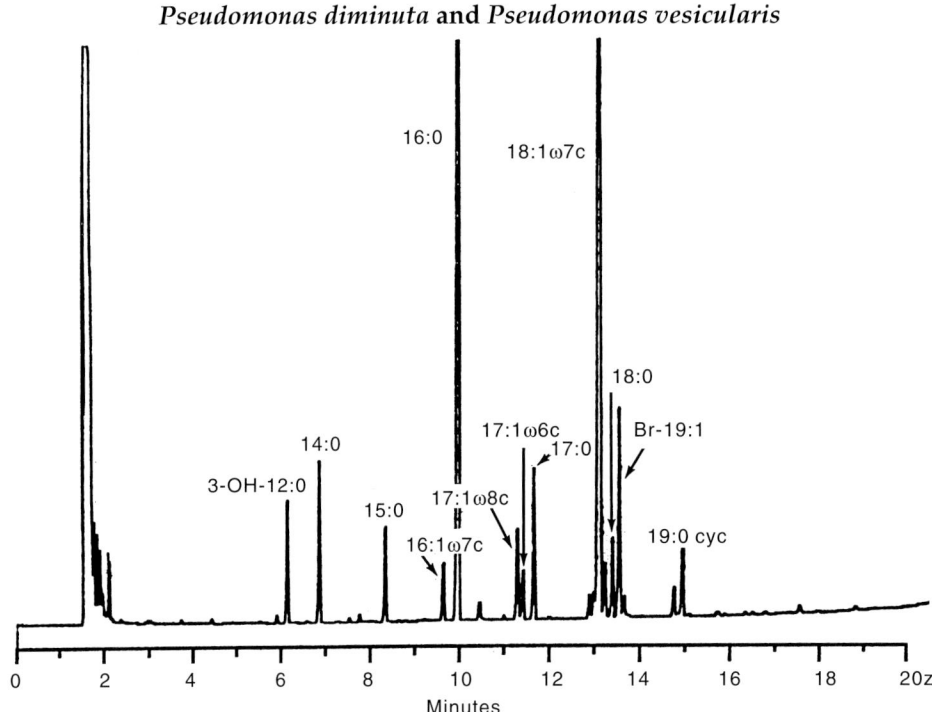

Figure 4.68. Gas chromatogram of the cellular fatty acids of *Pseudomonas diminuta* and *Pseudomonas vesicularis*.

Comments

The CFA profile of these two *Pseudomonas* species is characterized by large amounts of 18:1ω7c (35–54%) and 16:0 (21–33%) and small amounts (1–3%) of 3—OH-12:0, 14:0, 15:0, 17:1ω8c, 17:1ω6c, and Br-19:1. *P. diminuta* generally has more 19:0cyc$^{11-12}$ than does *P. vesicularis* (14% versus 1%), but there are overlaps in the relative amount of this acid in some strains of both species. The overall CFA profile of these two species is most like that of the CFA group containing *Roseomonas cervicalis*, *R. gilardii*, *Roseomonas* genomospecies 4 and 5, and *Ochrobactrum anthropi* except for the presence of 3—OH-12:0, 14:0, 15:0, 17:1ω8c, Br-19:1, and the absence of 2—OH-18:1 and 2—OH-19:0cyc.

Literature

1. Kaltenbach, C.M., C.W. Moss, and R.E. Weaver. 1975. Cultural and biochemical characteristics and fatty acid compositions of *Pseudomonas diminuta* and *Pseudomonas vesiculare*. J. Clin. Microbiol. *1:* 339–344.

2. Moss, C.W., and S.B. Dees. 1976. Cellular fatty acids and metabolic products of *Pseudomonas* species obtained from clinical specimens. J. Clin. Microbiol. *4:* 492–502.

3. Oyaizu, H., and K. Komagata. 1983. Grouping of *Pseudomonas* species on the basis of cellular fatty acid composition and the quinone system with special reference to the existence of 3-hydroxy fatty acids. J. Gen. Appl. Microbiol. *29:* 17–40.

4. Veys, A., W. Callewaert, E. Waelkens, and K.V.D. Abbeele. 1989. Application of gas-liquid chromatography to the routine identification of nonfermentative gram-negative bacteria in clinical specimens. J. Clin. Microbiol. *27:* 1538–1542.

5. Osterhout, G.J., V.H. Shull, and J.D. Dick. 1991. Identification of clinical isolates of gram-negative nonfermentative bacteria by an automated cellular fatty acid identification system. J. Clin. Microbiol. *29:* 1822–1830.

Table 4.63. Cellular fatty acid composition of *Pseudomonas diminuta* and *Pseudomonas vesicularis*

	Organism	
	Pseudomonas diminuta	*Pseudomonas vesicularis*
Fatty Acid	%	%
3-OH-12:0	2	2
14:1ω7c	1	–
14:0	2	3
15:0	2	3
16:1ω7c	2	4
16:0	33	21
17:1ω8c	1	2
17:1ω6c	1	3
17:0	3	2
18:1ω7c	35	54
18:0	1	1
Br-19:1	1	3
19:0cyc$^{11\text{-}12}$	14	1

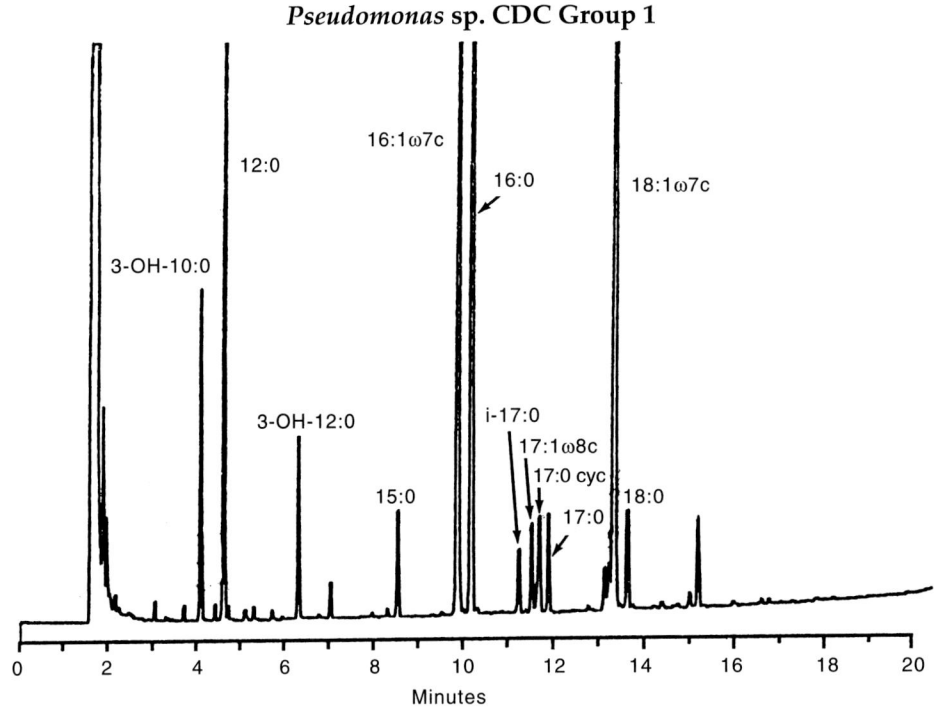

Figure 4.69. Gas chromatogram of the cellular fatty acids of *Pseudomonas* sp. CDC Group 1.

Comments

Pseudomonas sp. CDC Group 1 organisms have a distinct CFA profile that contains 18:1ω7c, 16:0, and 16:1ω7c as major acids with 4% 17:0cyc (but no 19:0cyc[11–12]), 3% 3—OH-10:0, 1% 3—OH-12:0, and 2% 10:0. This CFA profile is most like that of the CFA group containing *P. mendocina, P. pertucinogena, P. stutzeri,* and Vb-3 with the exception that *Pseudomonas* sp. CDC Group 1 does not contain 19:0cyc[11–12].

Literature

1. Ferney, J., W. Hansen, J. Etienne, F. Vandenesch, and J. Fleurette. 1988. Postoperative infant septicemia caused by *Pseudomonas luteola* (CDC group Ve-1) and *Pseudomonas oryzihabitans* (CDC group Ve-2). J. Clin. Microbiol. *26:* 1241–1243.

2. Trotter, J.A., T.L. Kuhls, D.A. Pickett, S. Reyes de la Rocha, and D.F. Welch. 1990. Pneumonia caused by a newly recognized pseudomonad in a child with chronic granulomatous disease. J. Clin. Microbiol. *28:* 1120–1124.

3. Hollis, D.G., R.E. Weaver, C.W. Moss, M.I. Daneshvar, and P.L. Wallace. 1992. Chemical characterization of CDC group WO-1, a weakly oxidative gram-negative group of organisms isolated from clinical sources. J. Clin. Microbiol. *30:* 291–295.

Table 4.64. Cellular fatty acid composition of *Pseudomonas* sp. CDC Group 1

Fatty Acid	*Pseudomonas* sp. CDC Group 1 %
10:0	2
3-OH-10:0	3
12:0	5
3-OH-12:0	1
14:0	3
15:1ω6c	T[a]
15:0	3
16:1ω7c	16
16:0	28
17:1ω8c	1
17:0cyc	4
17:0	1
18:2	1
18:1ω9c	1
18:1ω7c	28
18:0	1

[a]T = Trace.

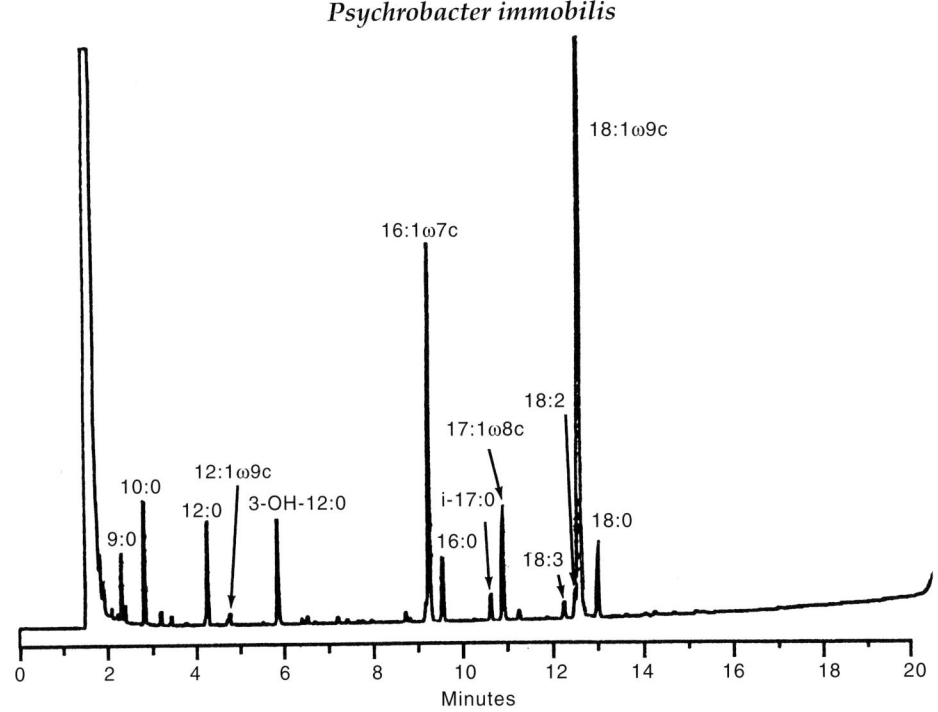

Figure 4.70. Gas chromatogram of the cellular fatty acids of *Psychrobacter immobilis*.

Comments

Psychrobacter immobilis has a unique CFA profile that is characterized by large amounts of 18:1ω9c (54%); moderate amounts (12%) of 16:1ω7c; smaller amounts (3–7%) of 10:0, 12:0, 3:OH-12:0, 16:0, 17:1ω8c, 18:2, and 18:0; with trace to 2% amounts of i-11:0, 12:1ω9, i-17:0, and 17:0. This is most similar to the profile of the CFA group containing *Moraxella bovis, M. cuniculi, M. ovis, M. nonliquefaciens, M. lacunata* groups I and II, except for the presence of i-17:0 (2%) and the absence of n-hexadecanol (16:0 alc) and n-octadecanol (18:0 alc).

Literature

1. Moss, C.W., P.L. Wallace, D.G. Hollis, and R.E. Weaver. 1988. Cultural and chemical characterization of CDC Groups EO-2, M-5, and M-6, *Moraxella (Moraxella)* species, *Oligella urethralis*, *Acinetobacter* species, and *Psychrobacter immobilis*. J. Clin. Microbiol. 26: 484–492.

Table 4.65. Cellular fatty acid composition of *Psychrobacter immobilis*

Fatty Acid	Organism *Psychrobacter immobilis* %
9:0	1
10:0	4
i-11:0	T[a]
12:0	3
12:1ω9c	1
3-OH-12:0	5
16:1ω7c	12
16:0	3
i-17:0	2
17:1ω8c	7
17:0	1
18:2	2
18:1ω9c	54
18:0	3

[a]T = Trace.

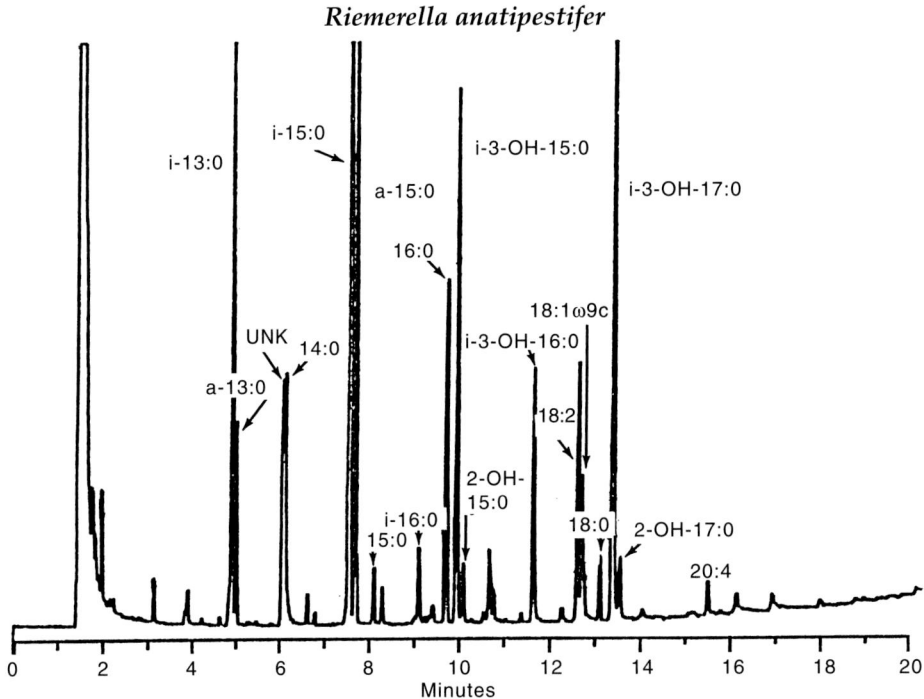

Figure 4.71. Gas chromatogram of the cellular fatty acids of *Riemerella anatipestifer*.

Comments

Riemerella anatipestifer has a unique CFA profile that is characterized by large amounts of i-15:0 (47%) and moderate amounts of i-3—OH-17:0 (10%), a-15:0 (8%), i-13:0 (5%), 16:0 (5%), i-3—OH-15:0 (4%) and 16:1ω7c (4%). This profile has some similarities to *Flavobacterium/Weeksella* organisms except for the absence of i-17:1ω8c, smaller amounts of i-2—OH-15:0, and larger amounts of i-13:0. *R. anatipestifer* differs from the CFA group containing *Flavobacterium (Sphingobacterium) mizutaii, Flavobacterium (Sphingobacterium) multivorum, Flavobacterium (Sphingobacterium) thalpophilum Flavobacterium (Sphingobacterium) spiritivorum,* and IIi by the presence of i-13:0 and smaller amounts of 16:1ω7c (4% versus >15%) and i-2—OH-15:0 (1% versus 20%). *R. anatipestifer* differs from *Capnocytophaga* species by smaller amounts of i-15:0 (47% versus >60%), and from *Cytophaga johnsonae* by the presence of i-13:0, the absence of i-17:1ω8c, and smaller amounts of 16:1ω7c (4% versus >20%).

Literature

1. Lambert, M.A., and C.W. Moss. 1984. Comparison of the cellular fatty acid composition of *Moraxella anatipestifer* and *Legionella pneumophila*. Int. J. Syst. Bacteriol. 34: 490–491.

2. Piechulla, K., S. Pohl, and W. Mannheim. 1986. Phenotypic and genetic relationships of so-called *Moraxella (Pasteurella) anatipestifer* to the *Flavobacterium/Capnocytophaga* group. Vet. Microbiol. 11: 261–270.

3. Segers, P., W. Mannheim, M. Vancanneyt, K. De Brandt, K.-H. Hinz, K. Kersters, and P. Vandamme. 1993. *Riemerella anatipestifer* gen. nov., comb. nov., the causative agent of septicemia anserum exsudativa, and its phylogenetic affiliation within the *Flavobacterium-Cytophaga* rRNA homology group. Int. J. Syst. Bacteriol. 43: 768–776.

Table 4.66. Cellular fatty acid composition of *Riemerella anatipestifer*

Fatty Acid	Organism *Riemerella anatipestifer* %
i-13:0	5
a-13:0	1
i-14:0	1
14:0	2
i-15:0	47
a-15:0	8
15:0	1
16:1ω7c	4
i-2-OH-15:0	1
16:0	5
i-3-OH-15:0	4
i-3-OH-16:0	2
18:2	2
18:1ω9c	1
18:0	1
i-3-OH-17:0	10

Shewanella putrefaciens biotypes 1 and 2

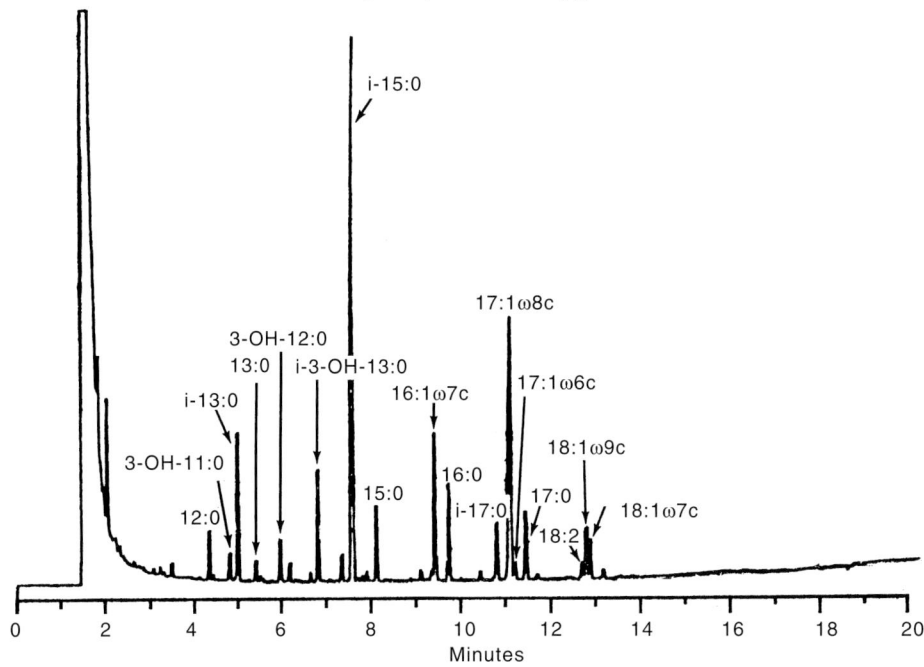

Figure 4.72. Gas chromatogram of the cellular fatty acids of *Shewanella putrefaciens* biotypes 1 and 2.

Comments

The CFA profile of *Shewanella putrefaciens* biotypes 1 and 2 are identical and cannot be differentiated by CFA data alone. This CFA profile is characterized by i-15:0 and 17:1ω8c as the two major acids, with smaller amounts of i-13:0, 15:0, 16:1ω7c, 16:0, i-17:0, 17:1ω6c, 17:0, 18:1ω9c, and 18:1ω7c. i-3—OH-13:0 is the major hydroxy acid (3%), with smaller amounts of 3—OH-12:0 (2%), 3—OH-14:0 (1%), and trace amounts of i-3—OH-15:0. This profile is most similar to that of *Xanthomonas* species except for the presence of 17:1ω8c, 17:1ω6c, and 17:0, and the absence of i-11:0, 12:1ω7, i-3—OH-11:0, a-15:0, and i-17:1ω8c.

Literature

1. Dees, S.B., and C.W. Moss. 1975. Cellular fatty acids of *Alcaligenes* and *Pseudomonas* species isolated from clinical specimens. J. Clin. Microbiol. *1:* 414–419.

2. Moss, C.W. 1978. New methodology for identification of nonfermenters: gas-liquid chromatographic chemotaxonomy. *In* G.L. Gilardi (ed.), Glucose nonfermenting gram-negative bacteria in clinical microbiology, Chapter 9, CRC Press, West Palm Beach, FL, pp. 171–201.

3. Veys, A., W. Callewaert, E. Waelkens, and K.V.D. Abbeele. 1989. Application of gas-liquid chromatography to the routine identification of nonfermentative gram-negative bacteria in clinical specimens. J. Clin. Microbiol. *27:* 1538–1542.

4. Osterhout, G.J., V.H. Shull, and J.D. Dick. 1991. Identification of clinical isolates of gram-negative nonfermentative bacteria by an automated cellular fatty acid identification system. J. Clin. Microbiol. *29:* 1822–1830.

Table 4.67. Cellular fatty acid composition of *Shewanella putrefaciens* biotypes 1 and 2

Fatty Acid	*Shewanella putrefaciens* biotypes 1 and 2 %
UNK[a]	2
12:0	3
3-OH-11:0	1
i-13:0	6
13:0	2
3-OH-12:0	2
i-14:0	1
14:0	1
i-3-OH-13:0	4
i-15:0	22
15:1A	1
15:0	6
3-OH-14:0	1
i-16:0	1
16:1ω9c	1
16:1ω7c	10
16:0	8
i-17:0	2
17:1ω8c	13
17:1ω6c	1
17:0	5
18:1ω9c	2
18:1ω7c	3
18:0	1

[a]UNK = Unknown.

Sphingomonas adhaesiva and *Sphingomonas parapaucimobilis*

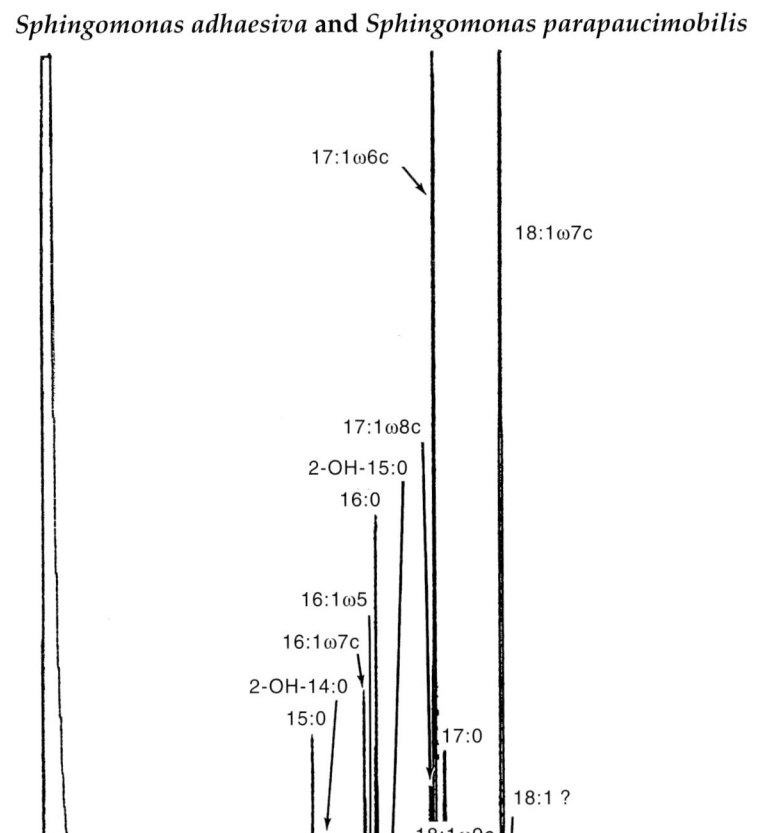

Figure 4.73. Gas chromatogram of the cellular fatty acids of *Sphingomonas adhaesiva* and *Sphingomonas parapaucimobilis*.

Comments

The two species in this CFA group are easily distinguished from other organisms by the presence of relatively large amounts of 18:1ω7c (33%–51%) and 17:1ω6c (15%), 3–9% amounts of 2—OH-14:0, 2% amounts of 17:1ω8c, 1–4% of 2—OH-15:0, and no cyclopropane acids. This CFA profile is most similar to the CFA group containing "*Pseudomonas azotocolligans*", *Sphingomonas capsulata*, *S. paucimobilis*, *S. yanoikuyae*, and *Sphingomonas* genomospecies 2 except for the presence of 17:1ω8c (2% versus 0%), 17:0 (2% versus 0%), and larger amounts of 17:1ω6c (15% versus 1–4%). *S. adhaesiva* contained 1% amounts each of Br-19:1 and 2—OH-18:0, but both of these acids were absent in *S. parapaucimobilis*.

Literature

1. Yabuuchi, E., E. Tanimura, A. Ohyama, I. Yano, and A. Yamamato. 1979. *Flavobacterium devorans* ATCC 10829: A strain of *Pseudomonas paucimobilis*. J. Gen. Appl. Microbiol. 25: 95–107.

2. Yabuuchi, E., I. Yano, H. Oyaizu, Y. Hashimoto, T. Ezaki, and H. Yamamato. 1990. Proposal of *Sphingomonas paucimobilis* gen. nov. and comb. nov. *Sphingomonas parapaucimobilis* sp. nov., *Sphingomonas yanoikuyae* sp. nov., *Sphingomonas adhaesiva* sp. nov., *Sphingomonas capulata* comb. nov., and two genospecies of the genus *Sphingomonas*. Microbiol. Immunol. 34: 99–119.

Table 4.68. Cellular fatty acid composition of *Sphingomonas adhaesiva* and *Sphingomonas parapaucimobilis*

Fatty Acid	Sphingomonas adhaesiva %	Sphingomonas parapaucimobilis %
14:0	1	1
15:0	6	2
2-OH-14:0	9	3
16:1ω7c	4	6
16:1ω5	-	1
16:0	12	10
2-OH-15:0	4	1
17:1ω8c	2	2
17:1ω6c	16	15
17:0	2	2
18:2	T[a]	-
18:1ω9c	1	1
18:1ω7c	33	51
18:1?[b]	-	1
18:0	5	3
Br-19:1	2	-
2-OH-18:0	1	-

[a]T = Trace.
[b]?, The position of the double bond is unknown.

Sporocytophaga myxococcoides

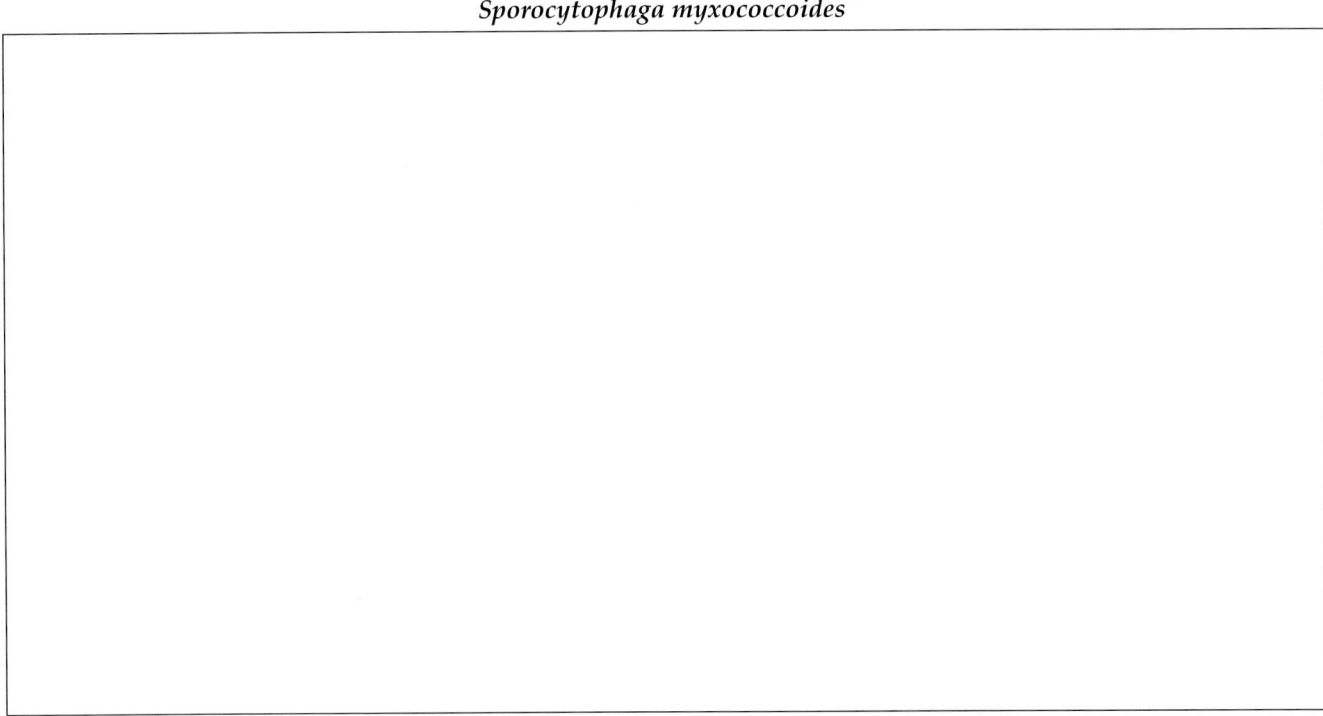

Figure 4.74. Gas chromatogram not available.

Comments

Sporocytophaga myxococcoides has a unique CFA profile with major amounts of branched chain acids and a relatively uncommon monounsaturated 16-carbon acid (16:1ω5). 16:1ω5 is present at a concentration of 15%, i-17:0 at 18%, a-17:0 at 14%, i-15:0 at 9%, a-15:0 at 7%, and i-16:0 at 8%. Only branched-chain hydroxy acids are present (i-3—OH-15:0 at 5% and i-3—OH-17:0 at 2%).

Literature

1. Holt, S.C., G. Forcier, and B.J. Takacs. 1979. Fatty acid composition of gliding bacteria: oral isolates of *Capnocytophaga* compared with *Sporocytophaga*. Infect. Immun. 26: 298–304.

2. McGrath, C.F., C.W. Moss, and R.P. Burchard. 1990. Effect of temperature shifts on gliding mobility adhesion, and fatty acid composition of *Cytophaga* sp. strain U67. J. Bacteriol. 172: 1978–1982.

Table 4.69. Cellular fatty acid composition of *Sporocytophaga myxococcoides*

Fatty Acid	Organism *Sporocytophaga myxococcoides* %
UNK[a] 13.566	1
i-15:0	9
a-15:0	7
i-16:0	8
16:1ω7c	2
16:1ω5	15
16:0	5
i-3-OH-15:0	5
i-17:1?[b]	1
i-17:1G	2
a-17:1A	3
i-17:0	18
a-17:0	14
17:1ω8c	1
17:0	1
18:1ω9c	2
i-3-OH-17:0	2

[a] UNK = Unknown.
[b] ?, The position of the double bond is unknown.

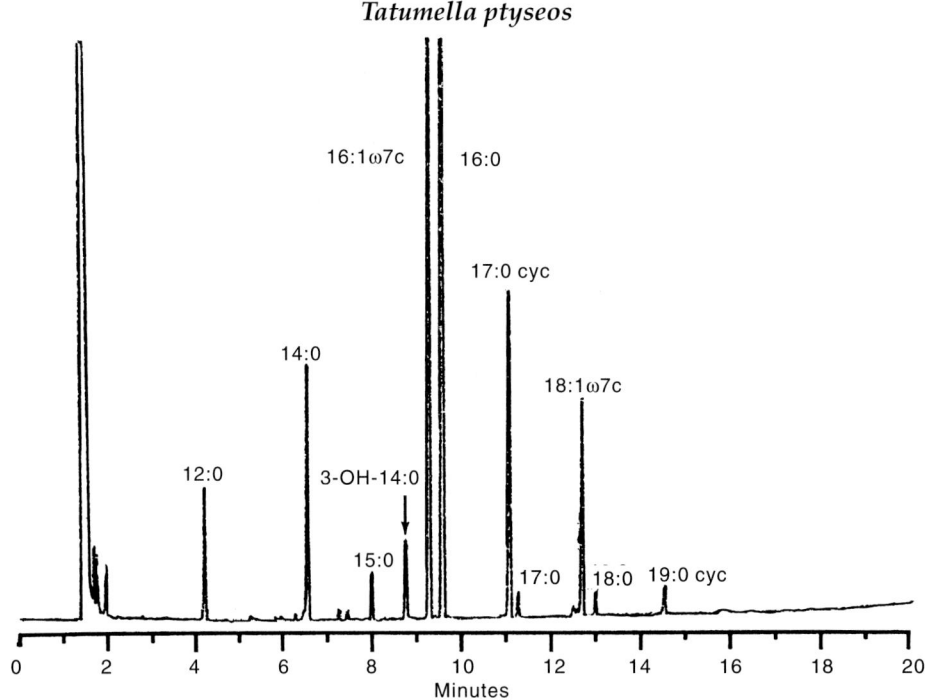

Figure 4.75. Gas chromatogram of the cellular fatty acids of *Tatumella ptyseos*.

Comments

Tatumella ptyseos is characterized by the presence of major amounts of 16:0, 16:1ω7c, 17:0cyc, and 18:1ω7c, with smaller amounts of 12:0, 3—OH-12:0, 14:0, 15:0, 17:0, 18:0, and 19:0cyc[11-12]. This species differs from the CFA group containing *Alcaligenes faecalis*, *A. xylosoxidans* subsp. *xylosoxidans*, *A. xylosoxidans* subsp. *denitrificans* group 1, and *Bordetella bronchiseptica* by the absence of 2—OH-12:0 (0% versus 4%) and the presence of 14:1ω12 (trace–1% versus 0%). *T. ptyseos* differs from the CFA group containing *A.*-like Group 1, *Pseudomonas pickettii*, and IVc-2 by the absence of 2—OH-14:0, 2—OH-16:1, 2—OH-16:0, and 2—OH-18:1 and the presence of 12:0 (3% versus 0%). The overall CFA profile of some strains of *T. ptyseos* is very similar to *Escherichia coli* and *Salmonella* species.

Literature

1. Hollis, D.G., F.W. Hickman, G.R. Fanning, J.J. Farmer III, R.E. Weaver, and D.J. Brenner. 1981. *Tatumella ptyseos* gen. nov., sp. nov., a member of the family *Enterobacteriaceae* found in clinical specimens. J. Clin. Microbiol. *14:* 79–88.

Table 4.70. Cellular fatty acid composition of *Tatumella ptyseos*

Fatty Acid	Organism *Tatumella ptyseos* %
12:0	2
14:0	6
14:1ω12	1
15:0	1
3-OH-14:0	5
16:1ω7c	20
16:0	37
17:0cyc	8
17:0	2
18:1ω9c	1
18:1ω7c	11
18:0	2
19:0cyc[11-12]	3

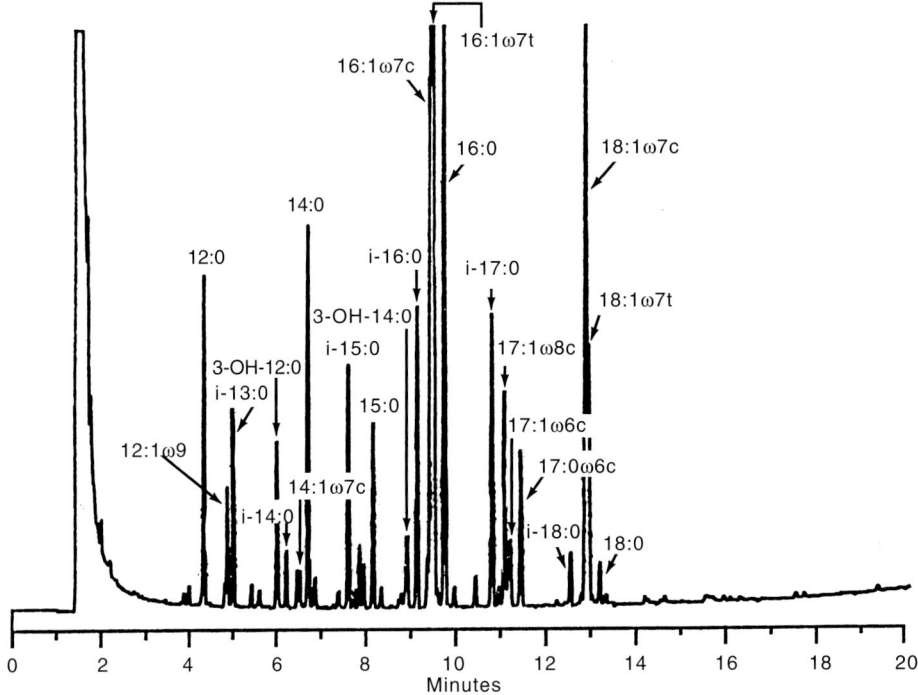

Figure 4.76. Gas chromatogram of the cellular fatty acids of *Vibrio cholerae*, *Vibrio fluvialis*, and *Vibrio parahaemolyticus*.

Comments

The three species in this CFA group contain 16:1ω7c, 16:0, and 18:1ω7c as the major acids with small amounts (1–3%) of 3—OH-12:0 and 3—OH-14:0; 1% each of i-16:0, 17:1ω8c, and/or i-17:0; and **no** cyclopropane acids. In addition to 16:1ω7c, each species contains 4–7% of 16:1ω7t, which is an important chemical marker along with i-16:0, 17:1ω8c, and i-17:0. These three species differ from *Vibrio furnissii* by the presence of 16:1ω7t (4–7% versus 0%). *Vibrio* species differ from the *Enterobacteriaceae* by the presence of i-16:0, 17:1ω8c, and/or i-17:0, and by the ratio of 16:1ω7c to 16:0 which is greater than 1:1.5 compared to 1:0.5 in *Enterobacteriaceae*. These three species differ from *Aeromonas* species by the presence of 3—OH-12:0 and 16:1ω7t and the absence of i-3—OH-15:0, i-16:1ω11c, and/or i-17:1ω8c.

Literature

1. Brian, B.L., and E.W. Gardner. 1968. Fatty acids from *Vibrio cholerae* lipids. J. Infect. Dis. *118:* 47–53.

2. Lambert, M.A., F.W. Hickman-Brenner, J.J. Farmer III, and C.W. Moss. 1983. Differentiation of *Vibrionaceae* by their cellular fatty acid composition. Int. J. Syst. Bacteriol. *33:* 777–792.

3. Urdaci, M.C., M. Marchand, and P.A. Grimont. 1990. Characterization of 22 *Vibrio* species by gas chromatography analysis of their cellular fatty acids. Res. Microbiol. *141:* 437–452.

4. Moss, C.W., and M.I. Daneshvar. 1992. Identification of some uncommon monounsaturated fatty acids of bacteria. J. Clin. Microbiol. *30:* 2511–2512.

Table 4.71. Cellular fatty acid composition of *Vibrio cholerae, Vibrio fluvialis,* and *Vibrio parahaemolyticus*

Fatty Acid	Organism		
	Vibrio cholerae %	*Vibrio fluvialis* %	*Vibrio parahaemolyticus* %
12:0	-	2	2
12:1ω9c	-	T[a]	1
i-13:0	-	1	1
3-OH-12:0	3	2	1
i-14:0	-	1	-
14:0	4	4	3
i-15:0	-	1	T
15:0	-	1	2
3-OH-14:0	2	2	1
i-16:0	2	3	T
16:1ω9c	1	-	-
16:1ω7c	35	31	28
16:1ω7t	6	7	4
16:0	19	14	11
i-17:0	-	1	1
17:1ω8c	1	1	3
17:1ω6c	-	-	1
17:0	-	1	4
i-18:0	-	1	-
18:1ω7c	24	24	32
18:1ω7t	1	-	-
18:0	1	1	2

[a]T = Trace.

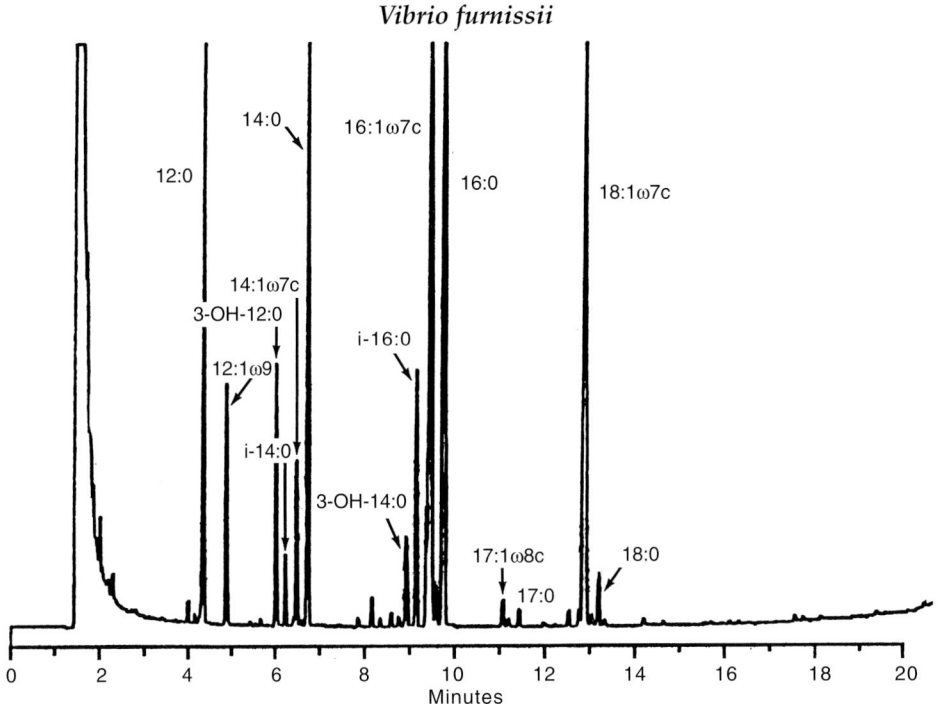

Figure 4.77. Gas chromatogram of the cellular fatty acids of *Vibrio furnissii*.

Comments

The CFA composition of *Vibrio furnissii* is essentially identical to other *Vibrio* species except for the absence of 16:1ω7t. Like all other *Vibrio*, *V. furnissii* contain the important chemical markers of the genus (i-16:0, 17:1ω8c, and/or i-17:0) and are distinguished from *Enterobacteriaceae* and *Aeromonas* species as discussed for the CFA group containing *V. cholerae*, *V. fluvialis*, and *V. parahaemolyticus*.

Literature

1. Lambert, M.A., F.W. Hickman-Brenner, J.J. Farmer III, and C.W. Moss. 1983. Differentiation of *Vibrionaceae* by their cellular fatty acid composition. Int. J. Syst. Bacteriol. *33:* 777–792.

2. Urdaci, M.C., M. Marchand, and P.A. Grimont. 1990. Characterization of 22 *Vibrio* species by gas chromatography analysis of their cellular fatty acids. Res. Microbiol. *141:* 437–452.

Table 4.72. Cellular fatty acid composition of *Vibrio furnissii*

Fatty Acid	Organism
	Vibrio furnissii
	%
12:0	2
i-13:0	1
3-OH-12:0	2
14:1ω7c	1
14:0	4
i-15:0	1
15:0	1
3-OH-14:0	2
i-16:0	2
16:1ω7c	38
16:0	13
i-17:0	2
17:1ω8c	1
17:0	1
i-18:0	T[a]
18:1ω7c	25
18:0	1

[a]T = Trace.

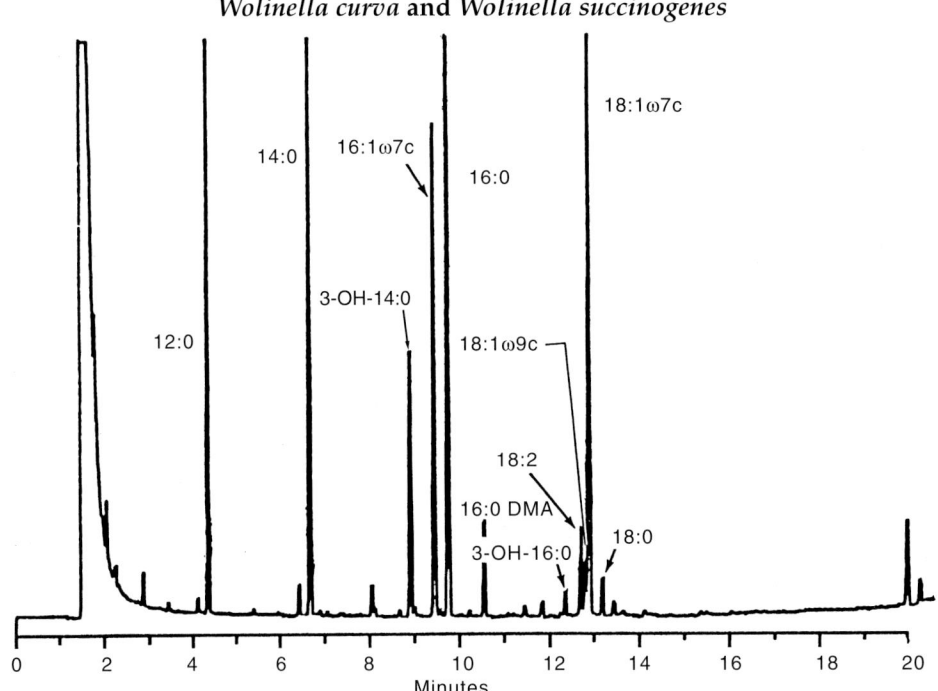

Figure 4.78. Gas chromatogram of the cellular fatty acids of *Wolinella curva* and *Wolinella succinogenes*.

Comments

These two *Wolinella* species have unique CFA profiles that are characterized by the presence of 4–7% of a saturated 16-carbon dimethylacetyl (16:0 DMA) and trace amounts of aldehyde and other DMAs. 18:1ω7c, 16:1ω7c, and 16:0 are the major fatty acids, and each species contains moderate amounts (4–5%) of 3—OH-14:0 and trace to 1% of 3—OH-16:0. There are quantitative differences in the relative amounts of some acids in these two species that would permit their differentiation by CFA data alone; however, these differences must be confirmed with additional strains of each species. *W. curva* and *W. succinogenes* differ from *W. recta* which does not contain aldehydes or DMAs.

Literature

1. Vandamme, P., M.I. Daneshvar, F.E. Dewhirst, B.J. Paster, K. Kersters, H. Goossens, and C.W. Moss. 1995. Chemotaxonomic analyses of *Bacteroides gracilis* and *Bacteroides ureolyticus* and reclassification of *B. gracilis* as *Campylobacter gracilis* comb. nov. Int. J. Syst. Bacteriol. 45: 145–152.

Table 4.73. Cellular fatty acid composition of *Wolinella curva* and *Wolinella succinogenes*

	Organism	
	Wolinella curva	*Wolinella succinogenes*
Fatty Acid	%	%
12:0	7	6
14:0	16	6
15:0	–	1
3-OH-14:0	6	5
16:1ω7c	9	18
16:0	25	23
16:0 DMA	4	7
17:0	–	T[a]
3-OH-16:0	1	1
18:2	3	6
18:1ω9c	2	2
18:1ω7c	22	22
18:0	2	2

[a]T = Trace..

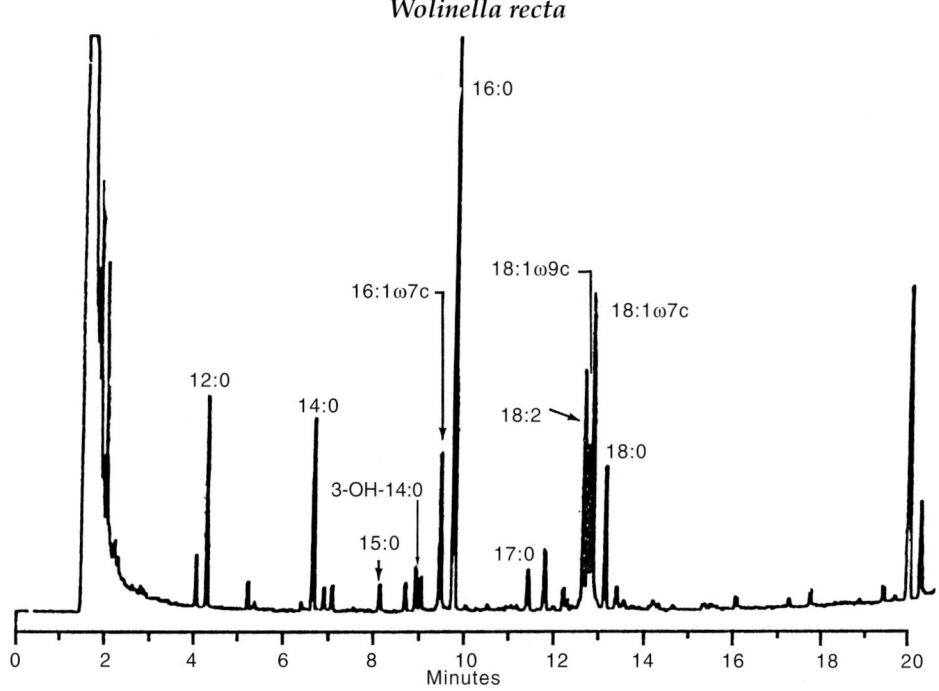

Figure 4.79. Gas chromatogram of the cellular fatty acids of *Wolinella recta*.

Comments

Wolinella recta has a distinct CFA profile with large amounts (31%) of 16:0; smaller amounts of 18:1ω7c (17%) and 12:0 (11%); and moderate amounts of 14:0, 16:1ω7c, 18:1ω9c, 18:2, and 18:0. 3—OH-14:0 is the only hydroxy acid present. The CFA profile of *W. recta* differs from that of *W. curva* and *W. succinogenes* by the absence of dimethylacetyls and/or aldehydes and 3—OH-16:0, and by larger amounts of 12:0 (11% versus 6–7%).

Literature

1. Vandamme, P., M.I. Daneshvar, F.E. Dewhirst, B.J. Paster, K. Kersters, H. Goossens, and C.W. Moss. 1995. Chemotaxonomic analyses of *Bacteroides gracilis* and *Bacteroides ureolyticus* and reclassification of *B. gracilis* as *Campylobacter gracilis* comb. nov. Int. J. Syst. Bacteriol. 45: 145–152.

Table 4.74. Cellular fatty acid composition of *Wolinella recta*

	Organism
	Wolinella recta
Fatty Acid	%
12:0	11
14:0	8
15:0	1
3-OH-14:0	4
16:1ω7c	8
16:0	31
17:0	1
18:2	8
18:1ω9c	5
18:1ω7c	17
18:0	4

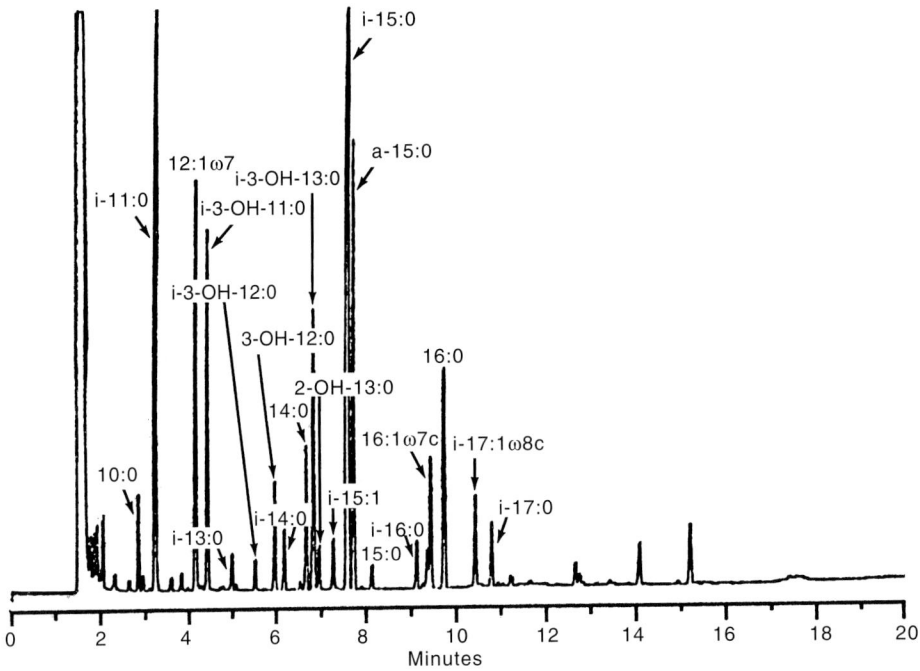

Figure 4.80. Gas chromatogram of the cellular fatty acids of *Xanthomonas beticola*, *Xanthomonas campestris*, *Xanthomonas maltophilia*, *Xanthomonas nakataecorchori*, and *Xanthomonas phaseoli*.

Comments

The *Xanthomonas* species have a unique profile consisting of major amounts of i-15:0 (27%) and a-15:0 (10–20%), and smaller amounts of relatively uncommon branched hydroxy acids (i-3—OH-11:0, i-3—OH-13:0). They also contain small amounts of 12:1ω7, i-11:0, 3—OH-12:0, 2—OH-13:0, 16:1ω9c, and i-17:1ω8c, and only small amounts of 18-carbon or longer carbon-chain acids. The CFA profile of the *Xanthomonas* species is most like that of *Shewanella putrefaciens* biotypes 1 and 2 except for the presence of i-11:0, i-3—OH-11:0, a-15:0, and i-17:1ω8c, and the absence of 17:1ω8c (0% versus 13% in *S. putrefaciens* biotypes 1 and 2).

Literature

1. Moss, C.W., S.B. Samuels, and R.E. Weaver. 1972. Cellular fatty acid composition of selected *Pseudomonas* species. Appl. Microbiol. 24: 596–598.

2. Moss, C.W., S.B. Samuels, J. Liddle, and R.M. McKinney. 1973. Occurrence of branched-chain hydroxy fatty acids of *Pseudomonas maltophilia*. J. Bacteriol. 114: 1018–1024.

3. Moss, C.W., and S.B. Dees. 1975. Identification of microorganisms by gas chromatographic-mass spectrometric analysis of cellular fatty acids. J. Chromatogr. 112: 595–604.

4. Osterhout, G.J., V.H. Shull, and J.D. Dick. 1991. Identification of clinical isolates of gram-negative nonfermentative bacteria by an automated cellular fatty acid identification system. J. Clin. Microbiol. 29: 1822–1830.

Table 4.75. Cellular fatty acid composition of *Xanthomonas beticola*, *Xanthomonas campestris*, *Xanthomonas maltophilia*, *Xanthomonas nakataecorchori*, and *Xanthomonas phaseoli*

Fatty Acid	Organism[a]				
	A	B	C	D	E
	%	%	%	%	%
10:0	1	1	1	T[b]	T
i-11:0	5	4	4	4	3
12:1ω7	2	1	2	1	1
i-3-OH-11:0	3	2	3	1	1
i-13:0	T	-	1	-	-
3-OH-12:0	3	1	2	1	1
i-14:0	T	1	1	T	1
14:0	2	1	3	2	1
i-3-OH-13:0	8	3	3	3	3
2-OH-13:0	1	T	1	1	T
i-15:1F	-	-	1	-	-
i-15:0	34	27	42	34	33
a-15:0	11	16	10	20	20
15:1ω6c	-	-	-	-	1
15:0	1	2	1	1	2
i-16:0	1	3	2	2	3
16:1ω9c	1	1	1	1	1
16:1ω7c	12	12	5	13	12
16:0	5	4	9	5	3
i-17:1ω8c	3	8	2	4	5
i-17:0	4	7	4	4	4
a-17:0	-	1	-	-	1
17:1ω8c	-	1	-	-	1
18:2	T	1	-	1	1
18:1ω9c	2	1	1	1	1
18:0	1	-	-	-	-

[a] A, *Xanthomonas beticola*; B, *Xanthomonas campestris*; C, *Xanthomonas maltophilia*; D, *Xanthomonas nakataecorchori*; E, *Xanthomonas phaseoli*.
[b] T = Trace.

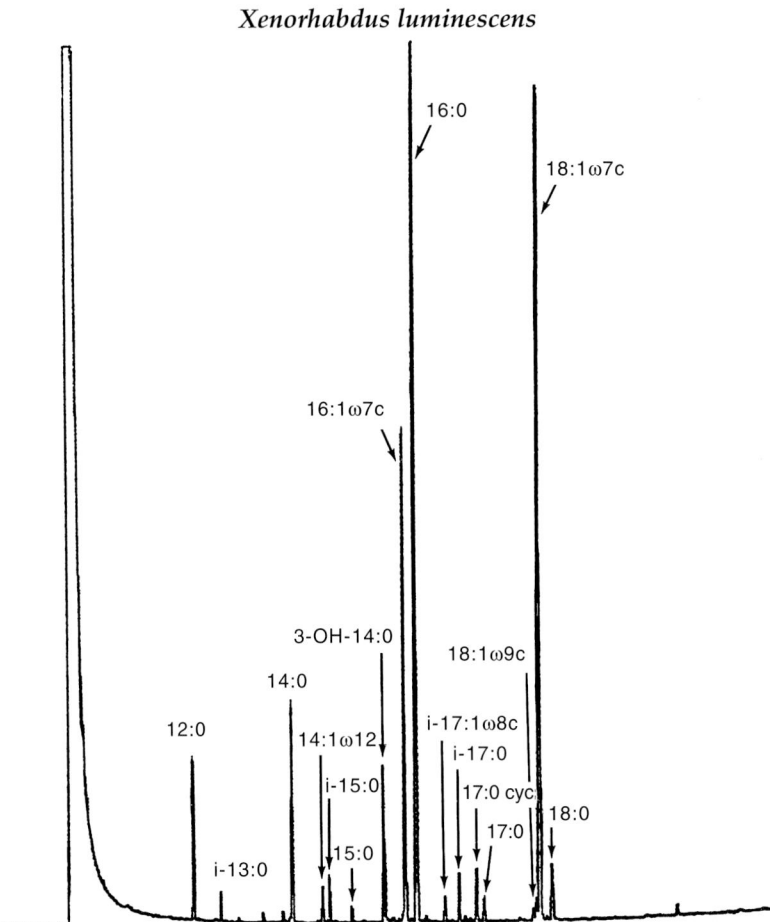

Figure 4.81. Gas chromatogram of the cellular fatty acids of *Xenorhabdus luminescens*.

Comments

The CFA profile of *Xenorhabdus luminescens* differs from that of other organisms by the presence of small amounts (4%) of branched-chain fatty acids and 96% of straight-chain acids. Generally, if branched-chain fatty acids are present in an organism, they will constitute a major part of the total fatty acid content. The presence of small amounts of i-13:0, i-15:0, i-17:1ω8c, i-17:0, 17:0cyc, and 3—OH-14:0 as the only hydroxy acid constitutes a unique CFA profile for *X. luminescens*. The profile has some similarity to that of *Neisseria* species except for the presence of i-13:0, i-15:0, i-17:1ω8c, i-17:0, and 17:0cyc, and the absence of 3—OH-12:0.

Literature

1. Suzuki, T., S. Yamanaka, and Y. Nishimura. 1990. Chemotaxonomic study of *Xenorhabdus*-ssp cellular fatty acids ubiquinone and DNA-DNA hybridization. J. Gen. Appl. Microbiol. 36: 393–402.

Table 4.76. Cellular fatty acid composition of *Xenorhabdus luminescens*

Fatty Acid	Organism *Xenorhabdus luminescens* %
12:0	3
i-13:0	1
14:0	6
14:1ω12	1
i-15:0	1
15:0	1
3-OH-14:0	4
16:1ω7c	18
16:0	31
i-17:1ω8c	1
i-17:0	1
17:0cyc	2
17:0	1
18:2	1
18:1ω9c	1
18:1ω7c	23
18:0	2

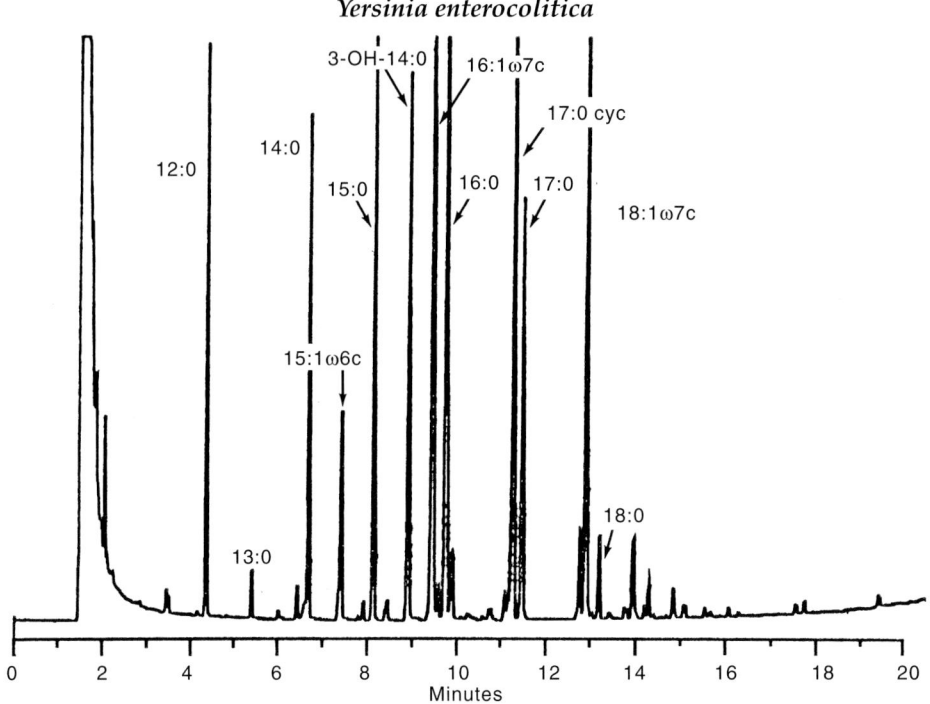

Figure 4.82. Gas chromatogram of the cellular fatty acids of *Yersinia enterocolitica*.

Comments

Yersinia enterocolitica has a unique CFA profile with 16:0, 17:0cyc, and 16:1ω7c as major acids (22%, 21%, and 18%, respectively); moderate amounts (10%) of 15:0; and smaller amounts of 12:0, 13:0, 14:0, 17:0, and 18:1ω7c; and trace to 1% amounts of 11:0, 15:1ω6c, 16:1ω5, 17:1ω8c, 18:2, and 18:0. This profile differs from the CFA group containing *Escherichia coli* and *Salmonella* species by the absence of 19:0cyc[11–12], more 15:0, and the presence of 15:1ω6c and 16:1ω5. The absence of 3—OH-10:0 distinguishes *Y. enterocolitica* from the CFA group containing *Alcaligenes piechaudii, Comamonas acidovorans, C. testosteroni, Pseudomonas alcaligenes, P. pseudoalcaligenes. Y. enterocolitica* differs from *Tatumella ptyseos* by the presence of 15:1ω6c, 16:1ω5, and 17:1ω8c; more 15:0 (10% versus 1%); and the absence of 14:1ω12.

Literature

1. Jantzen, E., and J. Lassen. 1980. Characterization of *Yersinia* species by analysis of whole cell fatty-acids. Int. J. Syst. Bacteriol. *30:* 421–428.
2. Lambert, M.A., F.W. Hickman-Brenner, J.J. Farmer III, and C.W. Moss. 1983. Differentiation of *Vibrionaceae* by their cellular fatty acid composition. Int. J. Syst. Bacteriol. *33:* 777–792.
3. Kanemasa, Y., E. Nagamachi, and S. Shibuya. 1991. The presence of cis-9,10-methylene hexadecanoic acid in *Yersinia enterocolitica*. Microbiol. Immunol. *35:* 77–81.

Table 4.77. Cellular fatty acid composition of *Yersinia enterocolitica*

Fatty Acid	*Yersinia enterocolitica* %
12:0	4
13:0	1
14:0	3
15:1ω6c	1
15:0	10
3-OH-14:0	4
16:1ω7c	18
16:1ω5	1
16:0	22
17:1ω8c	1
17:0cyc	21
17:0	5
18:2	1
18:1ω7c	6
18:0	1

Index

Ib-1 528 (see also *Shewanella putrefaciens*)
Ib-2 528 (see also *Shewanella putrefaciens*)
Ic 48, 83, 85, 550, 551
II (pink coccoid) 514 (see also *Roseomonas*)
IIb 661
IIc 628, 654, 656, 658, 659
IIe 52, 92, 120, 160, 552, 553, 556, 628, 654, 656, 658, 659
IIf 544, 554 (see also *Weeksella virosa*)
IIg 60, 98, 124, 160, 554, 555, 642, 643
IIh 52, 92, 120, 160, 552, 556, 557, 628, 654, 656, 658, 659
IIi 52, 93–96, 120, 160, 558, 559, 628, 654, 658, 660, 661, 700
IIj 546 (see also *Weeksella zoohelcum*)
IIk-2 352 (see also *Flavobacterium multivorum*)
III (pink coccoid) 79, 514 (see also *Roseomonas*)
IV (pink coccoid) 514 (see also *Roseomonas*)
IVc-2 60, 99, 100, 124, 174, 270, 560, 561, 598, 599, 708
IVe 442, 443 (see also *Oligella ureolytica*)
IVf 354 (see also *Flavobacterium odoratum*)
Vb-1 502 (see also *Pseudomonas stutzeri*)
Vb-2 488
Vb-3 48, 84, 86, 118, 198, 502, 562, 563, 644, 645, 696 (see also *Agrobacterium radiobacter* biovar 1)
Vd 438 (see also *Ochrobactrum anthropi*)
Vd-1 438
Vd-2 438
Vd-3 266 (see also *Agrobacterium radiobacter* biovar 1)
Ve-1 318 (see also called *Chryseomonas luteola*)
Ve-2 318 (see also called *Flavimonas oryzihabitans*)
VI 272 (see also *Alcaligenes faecalis*)
2-Ketogluconate reaction 12
   test 12
3-Ketolactonate medium for *Agrobacterium* 12
Acetamide alkalinization slants 1
   alkalinization of 5
Achromobacter 224, 226
   biotypes 1 and 2 438
   biovars A-F 224 (biovars A, C, D are see also biovars of *Ochrobactrum anthropi*)
   group B 44, 86, 87, 116, 182, 224–226, 438, 569, 584, 590
   group E 44, 86, 87, 116, 182, 224, 226, 227, 438, 569, 584, 590
   group F 224, 226
   species biotype 1 438
   species biotype 2 438
   xylosoxidans 224, 226, 278, 280 (see also *Alcaligenes xylosoxidans* subsp. *xylosoxidans*)
      subsp. *denitrificans* 278, 279
      subsp. *xylosoxidans* 280, 281
Acid production
   from carbohydrates 16
   from glucose 16
Acidovorax delafieldii 44, 50, 56, 63, 85, 93, 95, 100, 101, 106–108, 116, 119, 122, 125, 178, 179, 228, 229, 232, 596, 602, 640, 641
   facilis 50, 63, 91–95, 105–108, 119, 125, 178, 179, 230, 231, 233, 234
   temperans 44, 50, 79–82, 91–93, 116, 119, 178, 179, 232, 233, 640

Acinetobacter 7, 20, 21, 42, 54, 62, 75, 76, 97, 103, 115, 121, 124, 170–173, 234, 235, 236, 237–240, 245, 296, 336, 434–437, 585, 636
   anitratis 234
   baumannii 115, 236, 241
   calcoaceticus 21, 115, 236, 241
      var anitratis 234
      var lwoffii 234, 235
   haemolyticus 115, 121, 236, 241
   johnsonii 121, 124, 236, 241
   junii 121, 124, 236, 241
   lwoffii 121, 124, 234, 236, 240, 241
   radioresistens 115, 121, 124, 236, 241
   transformation assay for 20, 21
Actinobacillus 112, 460, 586, 587, 620, 624, 626, 642, 664, 674
   actinomycetemcomitans 10, 36, 38, 70, 72, 73, 112, 113, 140, 242, 243, 587
   equuli 34, 38, 67, 74, 111, 113, 136, 244–246, 587
   hominis 34, 67, 111, 136, 246, 247, 252, 587
   lignieresii 27, 34, 38, 67, 74, 111, 113, 136, 248, 249, 587
   suis 27, 34, 38, 67, 74, 111, 113, 136, 246, 250, 251, 587
   ureae 36, 38, 70, 74, 112, 113, 136, 252, 253, 587
Actinomyces 242
   israelii 242
Aeromonas 67, 111, 588, 710, 712
   hydrophila 2, 588, 589
   salmonicida 588, 589
Afipia 256, 258, 260, 262, 264, 569, 590, 591, 618
   broomeae 63, 106–108, 125, 180, 254, 255, 262, 278, 590, 591
   clevelandensis 63, 104, 106, 125, 180, 256, 257, 590, 591
   felis 56, 63, 99–101, 104, 106–108, 122, 125, 180, 258, 259, 590, 591
   genomospecies 1 63, 106–108, 125, 180, 260, 261, 590, 591 (formerly genospecies)
   genomospecies 2 63, 107, 108, 125, 180, 262, 263, 590, 591 (formerly genospecies)
   genomospecies 3 63, 104, 106–108, 125, 180, 264, 265, 590, 591 (formerly genospecies)
   unnamed genomospecies 1 261 (formerly genospecies)
   unnamed genomospecies 2 254, 263 (formerly genospecies)
Agrobacterium 12, 13, 690
   radiobacter 183, 266, 438,
      biovar 1 13, 44, 78–83, 116, 182, 266, 267, 592, 593 (previously CDC group Vd-3)
      biovar 2 266
   rhizogenes 266
   yellow group 44, 50, 56, 63, 78–81, 91–96, 98–101, 106–108, 116, 119, 122, 125, 184, 268, 269, 534
Alcaligenes 280
   denitrificans 278 (see also *Alcaligenes xylosoxidans* subsp. *denitrificans*)
      subsp. *denitrificans* 278
      subsp. *xylosoxidans* 281
   eutrophus 56, 99, 100, 122, 174, 270, 271, 560

   faecalis 56, 98, 122, 176, 272, 273, 276, 292, 293, 594–596, 610, 612, 614, 652, 708 (formerly CDC group VI; a synonym for *Alcaligenes odorans*)
   odorans 272 (formerly CDC group VI and a synonym for *Alcaligenes faecalis*)
   piechaudii 56, 99, 122, 176, 276, 277, 596–598, 612, 640, 652, 720
   xylosoxidans 102
      subsp. *denitrificans* 56, 102, 122, 176, 274, 278–280 (formerly *Alcaligenes denitrificans* and *Alcaligenes denitrificans* subsp. *denitrificans*)
      subsp. *denitrificans* group 1 594–596, 610, 612, 652, 708
      subsp. *xylosoxidans* 44, 56, 79, 80, 84, 101, 102, 116, 122, 176, 224, 226, 275, 278, 280, 281, 594–596, 610, 612, 614, 652, 690, 708 (formerly *Achromobacter xylosoxidans*)
      xylosoxidans-like group 1 56, 101, 102, 122, 176, 274, 275, 278, 598, 599, 690, 708
Alkaline phosphatase 5
Alteromonas 528
Amylosucrase with 5% sucrose agar 5
Arcobacter 665
   butzleri 600, 601
   cryaerophilus 600, 601
   nitrofigilis 602, 603, 640
   skirrowii 600, 601
Arginine dihydrolase test 7

Bacillus 29
Bacteria
   maintenance of, in motility medium 13
Bacterium anitratum 234
Balneatrix alpica 50, 92, 119, 154, 155, 282, 283, 606, 607, 646
Bartonella 62, 63, 103, 104, 124, 125, 214, 286, 288, 608, 609, 616 (formerly *Rochalimaea*)
   bacilliformis 62, 103, 124, 214, 284, 285, 288, 608, 609
   elizabethae 214, 286, 287, 609
   henselae 214, 286, 287, 609
   quintana 124, 125, 214, 284, 286, 287, 609
   vinsonii 124, 125, 214, 286, 287, 609
Basal medium (Solution B) 20, 21
Benedict's reagent 5
Bergeyella zoohelcum (see *Weeksella zoohelcum*)
Bisgaard's taxon 1, 34, 36, 38, 67, 71, 74, 111–113, 132, 290, 291, 456, 466
Blood agar 5
Blood agar, rabbit 2
Bordet-Gengou agar 6
   agar without peptone 6
Bordetella 614
   avium 56, 98, 122, 174, 272, 276, 292, 610, 611, 614
   bronchiseptica 56, 100, 122, 174, 270, 294, 295, 560, 594–596, 610, 612, 614, 652, 708
   holmesii (NO-2) 54, 62, 97, 103, 121, 124, 170, 436, 437, 610, 611, 614 (originally CDC group NO-2)
   parapertussis 54, 97, 121, 170, 296, 297, 594, 596, 610, 612, 614
   pertussis 6, 18, 63, 105, 125, 212, 213, 298, 299, 490, 594, 614, 615

Bouvert's genomospecies 14   115
Branhamella   64, 105, 125, 164, 220, 394, 395, 682, 683
  catarrhalis   64, 105, 125, 164, 220, 394, 395, 682, 683
Brevundimonas diminuta (see *Pseudomonas diminuta*)
Brevundimonas vesicularis (see *Pseudomonas vesicularis*)
Brown pigment, insoluble   2
Browning   2
Brucella   8, 44, 50, 56, 63, 75, 81, 82, 90, 95, 97, 101–103, 108, 209, 216, 302–305, 566, 590, 618
  abortus   116, 119, 122, 125, 208, 216, 300, 301, 616, 617, 688
  canis   10, 42, 48, 54, 62, 75, 90, 97, 103, 115, 116, 118, 119, 121, 122, 124, 125, 208, 302, 303, 608, 616, 618, 635, 676
    gel formation   10
    dye tolerance test for   8
  melitensis   116, 119, 122, 125, 208, 216, 304, 305, 404, 616, 617, 688
  ovis   616, 617
  suis   116, 119, 208, 216, 306, 307, 616, 617, 688
Burkholderia   474, 482, 486, 492

Campylobacter   56, 63, 99, 106, 122, 125, 566, 586, 620, 622, 665
  coli   592, 620, 621, 624, 626
  concisus   622, 623
  fecalis   586
  fetus   18, 624, 642
    subsp. *fetus*   624, 625, 627
    subsp. *venerealis*   624, 625, 627
    group CLO1B   665
  group CLO3   666
  group DG   622, 623
  helveticus   620, 621, 624, 626
  hyointestinalis   622, 624, 626
  jejuni   592, 620, 621, 624, 626
  lari   592, 620, 621, 624, 626
  mucosalis   622, 623
  sputorum subsp. *bubulus*   627
    subsp. *faecalis*   627
    subsp. *sputorum*   627
  upsaliensis   592, 620, 621, 624, 626
Capnocytophaga   29, 112, 142, 308, 386, 566, 628, 631, 632, 642, 650, 658, 660, 700
  canimorsus   19, 38, 72, 113, 142, 310, 311, 628, 629 (formerly CDC group DF-2)
  cynodegmi   38, 72, 113, 142, 312, 313, 628, 629 (formerly named CDC group DF-2-like)
  gingivalis   112, 308, 628, 629
  granulosa   308
  haemolytica   308
  ochracea   112, 308, 628, 629
  species DF-1   36, 69, 70, 112, 142, 308, 309 (formerly CDC group DF-1)
  sputigena   112, 308
Carbohydrate fermentation   8 (see also Enteric fermentation base)
Cardiobacterium hominis   38, 73, 113, 138, 314, 315, 540, 630, 664, 674
Casein-salt solution   21
Catalase reaction   6
CDC group DF-1   112, 142, 308, 309 (see also *Capnocytophaga* species DF-1)
CDC group DF-2   310 (see also *Capnocytophaga canimorsus*)
CDC group DF-2-like   312 (see also *Capnocytophaga cynodegmi*)

CDC group DF-3   36, 69, 70, 112, 138, 326–328, 631, 632
CDC group DF-3-like   36, 70, 71, 112, 138, 328, 329 631–633
CDC group EF-4   332 (see also EF-4a and EF-4b)
CDC group EF-4a   34, 38, 67, 72, 73, 111, 113, 134, 330–332, 602, 603, 640 (originally designated EF-4)
CDC group EF-4b   44, 50, 79, 91, 116, 119, 148, 332, 333, 406, 602, 603, 640 (originally designated EF-4)
CDC group EO-1   474 (*Pseudomonas cepacia*; other synonyms are *Pseudomonas multivorans* and *Pseudomonas kingii*)
CDC group EO-2   44, 50, 79–82, 94–96, 116, 119, 172, 336–338, 508, 584, 634, 699 (originally included CDC group EO-3)
CDC group EO-3   44, 78–81, 116, 172, 338, 339, 508, 618, 635, 676, 689 (originally included in CDC group EO-2)
CDC group HB-5   452 (see also *Pasteurella bettyae*)
CDC group I (pink coccoid)   514 (see also *Roseomonas*)
CDC group Ib-1   528 (see also *Shewanella putrefaciens*)
CDC group Ib-2   528 (see also *Shewanella putrefaciens*)
CDC group Ic   48, 83, 85, 550, 551
CDC group II (pink coccoid)   514 (see also *Roseomonas*)
CDC group IIb   661
CDC group IIc   628, 654, 656, 658, 659
CDC group IIe   52, 92, 120, 160, 552, 553, 556, 628, 654, 656, 658, 659
CDC group IIf   544, 554 (see also *Weeksella virosa*)
CDC group IIg   60, 98, 124, 160, 554, 555, 642, 643
CDC group IIh   52, 92, 120, 160, 552, 556, 557, 628, 654, 656, 658, 659
CDC group IIi   52, 93–96, 120, 160, 558, 559, 628, 654, 658, 660, 661, 700
CDC group IIj   546 (see also *Weeksella zoohelcum*)
CDC group IIk-2   352 (see also *Flavobacterium multivorum*)
CDC group III (pink coccoid)   79, 514 (see also *Roseomonas*)
CDC group IV (pink coccoid)   514 (see also *Roseomonas*)
CDC group IVc-2   60, 99, 100, 124, 174, 270, 560, 561, 598, 599, 708
CDC group IVe   442, 443 (see also *Oligella ureolytica*)
CDC group IVf   354 (see also *Flavobacterium odoratum*)
CDC group M-3   390 (see also *Moraxella atlantae*)
CDC group M-4f   354 (see also *Flavobacterium odoratum*)
CDC group M-5   432 (see also *Neisseria weaveri*)
CDC group M-6   414 (see also *Neisseria elongata* subsp. *nitroreducens*)
CDC group NO-1   54, 97, 103, 121, 124, 170, 434, 435, 636, 648
CDC group NO-2   54, 62, 97, 103, 121, 124, 170, 436, 437, 610, 611, 614 (see also *Bordetella holmesii* [NO-2])
CDC group O-1   42, 46, 48, 52, 75, 78, 90, 91, 115, 117, 118, 120, 184, 446, 447
CDC group O-2   52, 65, 91–94, 105–107, 120, 126, 184, 448, 449

CDC group OFBA-1   127, 192, 440, 441, 638, 639
CDC group TM-1   382 (see also *Kingella denitrificans*)
CDC group Vb-1   502 (see also *Pseudomonas stutzeri*)
CDC group Vb-2   488
CDC group Vb-3   48, 84, 86, 118, 198, 502, 562, 563, 644, 645, 696 (see also *Agrobacterium radiobacter* biovar 1)
CDC group Vd   438 (see also *Ochrobactrum anthropi*)
CDC group Vd-1   438
CDC group Vd-2   438
CDC group Vd-3   266 (see also *Agrobacterium radiobacter* biovar 1)
CDC group Ve-1   318 (see also *Chryseomonas luteola*)
CDC group Ve-2   318 (see also *Flavimonas oryzihabitans*)
CCDC group VI   272 (see also *Alcaligenes faecalis*)
CDC group WO-1   232, 596, 602, 640, 641
CDC group WO-1A   596, 602, 640, 641
Cellular fatty acid (CFA)
  analysis and bacterial identification   565
  names   583
  protocols   576
Cetrimide agar   6
Chocolate agar   6
Christensen's   22
Chromobacterium violaceum   33, 34, 66–68, 110, 111, 130, 316, 317, 602, 606, 646, 647
Chryseobacterium gleum (see *Flavobacterium gleum*)
Chryseobacterium indologenes (see *Flavobacterium indologenes*)
Chryseobacterium meningosepticum (see *Flavobacterium meningosepticum*)
Chryseomonas luteola   42, 75–77, 115, 186, 318, 319, 340, 636, 646, 648, 649 (formerly CDC group Ve-1; other synonyms are *Chryseomonas polytricha* and *Pseudomonas luteola*)
  polytricha   318 (synonym for *Chryseomonas luteola*)
Citrate   6
Coblentz method   22 (see also Voges-Proskauer [VP] test)
Coccobacilli and cocci, differentiation of   7
Comamonas   56, 99, 100, 122, 204, 320, 322, 323
  acidovorans   56, 100, 122, 204, 320, 321, 323, 596–598, 612, 640, 652, 720 (formerly *Pseudomonas acidovorans*)
  testosteroni   60, 122, 204, 320, 322–324, 596–598, 612, 640, 652, 720 (formerly designated *Pseudomonas testosteroni*)
  terrigena   60, 122, 204, 322–324, 596, 602, 640, 641 (formerly *Variovorax neocistes* and *Variovorax cyclosites*)
Compacted red blood cells   2
Corynebacterium vaginale   366 (see also *Gardnerella vaginalis*)
Crystal violet stain   11
Culture droplet   14 (see also Motility test)
Cysteine agar (*Francisella*)   7
Cytophaga   512, 628, 631, 632, 642
  johnsonae   628, 650, 651, 654, 656, 660, 700

Decarboxylase tests (Moeller's Method)   7
DF-1   112, 142, 308, 309 (see also *Capnocytophaga* species DF-1)
DF-2   310 (see also *Capnocytophaga canimorsus*)
DF-2-like   312 (see also *Capnocytophaga cynodegmi*)

DF-3 36, 69, 70, 112, 138, 326–328, 631, 632
DF-3-like 36, 70, 71, 112, 138, 328, 329, 631–633
Diplobacillus moraxaxenfeld 396 (synonym for *Moraxella lacunata*; another synonym is *Moraxella liquefaciens*)
DNA, preparing crude transforming 21
Dye tolerance test for *Brucella* 8

EF-4 332 (see also EF-4a and EF-4b)
EF-4a 34, 38, 67, 72, 73, 111, 113, 134, 330–332, 602, 603, 640 (originally designated EF-4)
EF-4b 44, 50, 79, 91, 116, 119, 148, 332, 333, 406, 602, 603, 640 (originally designated EF-4)
Eikenella corrodens 63, 104, 125, 164, 334, 335, 602, 603, 640, 674
Empedobacter brevis (see *Flavobacterium breve*)
Enteric fermentation base medium 8
Enterobacteriaceae 17, 33, 66, 110, 710, 712
EO-1 474 (*Pseudomonas cepacia*; other synonyms are *Pseudomonas multivorans* and *Pseudomonas kingii*)
EO-2 44, 50, 79–82, 94–96, 116, 119, 172, 336–338, 508, 584, 634, 699 (originally included CDC group EO-3)
EO-3 44, 78–81, 116, 172, 338, 339, 508, 618, 635, 676, 689 (originally included in CDC group EO-2)
Escherichia coli 652, 653, 708, 720
Esculin agar slant 8

Fatty acid methyl esters (FAME)
  hydrogenation procedure 579
  saponification procedure 576
  trifluroacetylation 578
Flagella broth 8
Flagella stains 9
Flavimonas 608
  oryzihabitans 42, 75–77, 115, 186, 318, 340, 341, 636, 646, 648, 649 (formerly designated CDC group Ve-2 and *Pseudomonas oryzihabitans*)
Flavobacterium 358, 512, 558, 628, 631, 632, 642, 650, 654, 656, 658, 660, 700
  balustinum 628, 654, 655, 660
  breve 44, 89, 116, 156, 342, 343, 628, 650, 656, 657, 660
  breve-like 156
  gleum 119, 156, 157, 344–346, 628, 654, 655, 660
  indologenes 116, 119, 156, 157, 344–346, 628, 654, 655, 660
  meningosepticum 2, 44, 50, 89, 92, 116, 119, 154, 282, 348, 349, 568, 628, 654, 655, 660
  meningosepticum-like 154, 348
  mizutaii 50, 92, 94, 95, 119, 162, 350, 351, 356, 358, 360, 628, 654, 656, 658, 660, 661, 700
  multivorum 44, 81, 116, 162, 350, 352, 353, 356, 358, 360, 628, 654, 656, 658, 660, 661, 700 (formerly CDC group IIk-2)
  odoratum 58, 99, 122, 158, 354, 355, 628, 654, 655, 660 (formerly CDC group M-4f and CDC group IVf)
  species IIb 50, 89, 92–96, 116, 119, 156, 157, 342, 344–346, 348, 628, 654, 655, 660
  spiritivorum 44, 50, 79, 81, 83, 85, 95, 96, 116, 119, 162, 163, 356–358, 360, 628, 654, 658, 660, 661, 700
  thalpophilum 44, 81, 116, 162, 356, 358–360, 628, 654, 656, 658, 660, 661, 700

  yabuuchiae 46, 50, 79, 81, 83, 85, 95, 96, 116, 119, 162, 163, 356, 360, 361
Fluid thioglycollate medium (BBL) 10
Francisella 7, 662
  carbohydrate base for 7
  philomiragia 46, 50, 78, 82, 91, 92, 116, 119, 210, 362, 363, 662, 663 (formerly designated *Yersinia philomiragia*)
  tularensis 7, 11, 48, 62, 90, 103, 118, 124, 362, 364
    biogroup *novicida* 210, 364, 365, 662, 663
    biogroup *palearctica* 210, 364, 365
    biogroup *tularensis* 210, 364, 365, 662, 663

Gardnerella vaginalis 2, 10, 36, 69, 112, 138, 366, 367 (formerly named *Corynebacterium vaginale*, *Haemophilus vaginalis*)
Gas production from fluid thioglycollate medium (BBL) 10
Gel formation 10
Gelatin liquefaction 11
Gilardi rod group 1 58, 98, 122, 168, 368, 369, 664
Glucose cysteine peptone agar 11
Gram stain 11
Greening 2

Haemophilus 5, 17, 18, 19, 24, 218, 242, 372–381, 586, 587, 620, 624, 626, 642, 674
  aegyptius 219, 377 (see also *Haemophilus aegyptius* biogroup *aegyptius*)
  aphrophilus 33, 34, 36, 38, 66, 67, 70, 72, 110–113 140, 370, 371, 587
  aphrophilus-like 370
  canis 218, 374 (see also *Haemophilus haemoglobinophilus*)
  ducreyi 218, 219, 372, 587
  felis 218, 373
  growth factor test for 11
  haemoglobinophilus 218, 219, 374 (also referred to as *Haemophilus canis*)
  haemolyticus 218, 375
  influenzae 218, 219, 376, 377, 587
    biogroup *aegyptius* 219, 377, 587 (see also *Haemophilus aegyptius*)
    biotype I 218
    influenzae biotype II 218
    biotype III 218, 219, 377
  biotype IV 218
    biotype V 218
    biotype VI 218
    biotype VII 218
    biotype VIII 218
  parahaemolyticus 218, 378, 587
  parainfluenzae 218, 379, 587
    biotype I 218
    biotype II 218
    biotype III 218, 378
    biotype IV 218, 219, 379, 381
    biotype V 218, 379 (previously published as *Haemophilus parainfluenzae* biotype IV)
    biotype VI 218, 219
    biotype VII 218
    biotype VIII 218
  paraphrohaemolyticus 379
  paraphrophilus 218, 373, 380, 587
  porphyrin test for 18
  segnis 218, 379, 381, 587
  vaginalis 366 (see also *Gardnerella vaginalis*)
Haemophilus-like 374
HB-5 452 (see also *Pasteurella bettyae*)

Heart infusion agar (Difco) 12
  broth (Difco) 12
  tyrosine (HIT) agar 12
Helicobacter cinaedi 665
  fennelliae 666
  mustelae 667
  pylori 586, 667, 668, 674
Hemolysis 2 (see also Blood agar, rabbit)
  α 2 (see also Blood agar, rabbit)
  β 2 (see also Blood agar, rabbit)
  incomplete 2 (also see Blood agar, rabbit)
Herellea vaginicola 234
Hutner's vitamin-free mineral solution 14
Hydrogenation of fatty acid methyl esters (FAME) procedure 579
Hydrogenophaga
  palleronii 670–672
  pseudoflava 672, 673

Incomplete beta hemolysis 2
Indole production tests 12

King, Elizabeth O. 17, 24, 452
  introduction to identification key 24
Kingella 664
  denitrificans 19, 20, 38, 63, 72, 104, 113, 125, 152, 382, 383, 668, 674, 675 (formerly designated CDC group TM-1)
  indologenes 73, 314, 540, 640 (see also *Suttonella indologenes*)
  kingae 38, 72, 113, 152, 384, 385, 668, 674, 675 (formerly *Moraxella kingae* and *Moraxella kingii*)
Koch-Weeks bacillus 377
Kovacs' reagent 17

Lactate mineral medium (Solution A) 21
Lavender-green 2
Legionella 566
Leifson stain 9
Leptotrichia buccalis 36, 69, 112, 142, 386, 387
Listeria monocytogenes 2
Litmus milk 13
Lysine decarboxylase test 7
Lysis 2
Lysis-inhibition zones 2

MacConkey agar (Difco) 13
M-3 390 (see also *Moraxella atlantae*)
M-4f 354 (see also *Flavobacterium odoratum*)
M-5 432 (see also *Neisseria weaveri*)
M-6 414 (see also *Neisseria elongata* subsp. *nitroreducens*)
Media and methods 4
Metals solution, concentrated 14
Methyl red (MR) test 13
Methylobacterium 46, 50, 58, 64, 78–81, 94, 95, 101, 107, 108, 117, 119, 122, 125, 188, 388, 389, 514, 634, 635, 676
  extorquens 388, 389, 618, 676, 677 (formerly called *Vibrio extorquens*)
  fujisawaense 388
  mesophilicum 388, 389, 618, 676, 677 (formerly called *Pseudomonas mesophilica*)
  organophilum 388
  radiotolerans 389
  rhodesianum 388
  rhodinum 389
  zatmanii 389, 618, 676, 677
Mima polymorpha 234, 402, 444
  var oxidans 444 (many of these strains are see also are also *Moraxella osloensis*; some are see also called *Oligella urethralis*)

INDEX 725

Moeller's Method 7
Moraxella 7, 394, 404, 414, 678, 680, 684, 686
  anatipestifer 512 (see also *Riemerella anatipestifer*)
  atlantae 58, 64, 98, 105, 122, 125, 166, 390, 391, 678, 679, 686 (formerly CDC group M-3)
  bovis 64, 104, 105, 125, 166, 392, 678, 680–682, 698
  canis 680, 681
  catarrhalis 64, 105, 125, 164, 220, 394, 395, 682, 683 (also *Branhamella catarrhalis*; formerly *Neisseria catarrhalis*)
  cuniculi 678, 680–682, 698
  equi 392, 393
  kingae 384 (see also *Kingella kingae*)
  kingii 384 (see also *Kingella kingae*)
  lacunata 64, 105, 125, 164, 396, 397 (synonyms are *Moraxella liquefaciens* and *Diplobacillus moraxaxenfeld*)
    group I 678, 680–682, 698
    group II 678, 680–682, 698
  lincolnii 64, 105, 126, 166, 398, 399, 680, 681
  liquefaciens 396 (synonym for *Moraxella lacunata*; another synonym is *Diplobacillus moraxaxenfeld*)
  nonliquefaciens 64, 105, 126, 164, 400, 401, 678, 680–682, 698
  osloensis 20, 21, 58, 64, 98, 99, 105, 122, 126, 164, 400, 402, 403, 684, 685 (many of these strains were previously *Mima polymorpha*)
    transformation assay for 20, 21
  ovis 678, 680–682, 698
  phenylpyruvica 58, 64, 99, 100, 106, 123, 126, 166, 404, 405, 508, 678, 686, 687
  urethralis 21, 444 (see also called *Oligella urethralis*)
Moraxella-like 404
Motility medium 14
  maintenance of bacteria in 13
Motility test 13

Neisseria 1, 5, 7, 15, 29, 52, 67, 220–222, 395, 585, 602, 640, 718
  canis 46, 58, 64, 79, 91, 99, 105, 117, 119, 123, 126, 144, 220, 406, 407, 603
  catarrhalis 220, 394 (see also *Moraxella [Branhamella] catarrhalis*)
  cinerea 64, 105, 126, 144, 220, 408, 409, 603
  elongata 104, 105, 414
    subsp. *elongata* 58, 64, 98, 104, 123, 126, 148, 410–412, 603
    subsp. *glycolytica* 58, 64, 98, 105, 123, 126, 148, 410, 412, 413, 603
    subsp. *nitroreducens* 46, 52, 58, 64, 79, 91, 99, 104, 117, 119, 123, 126, 148, 410, 412, 414, 415, 603 (formerly CDC group M-6)
  flava 430, 603 (see also *Neisseria subflava*)
  flavescens 58, 64, 98, 105, 123, 126, 146, 221, 416, 417
  gonorrhoeae 5, 6, 15, 19, 21, 52, 91, 119, 144, 221, 222, 382, 408, 418, 419, 603
  gonorrhoeae-like 408
  lactamica 5, 20, 38, 52, 72, 92, 113, 119, 146, 221, 420, 421
  meningitidis 5, 15, 52, 91, 120, 144, 221, 422, 423, 426, 604
  mucosa 34, 40, 58, 64, 67, 73, 101, 105, 111, 113, 123, 126, 146, 222, 424, 425, 604
  nutrient agar (Difco), test for 15
  perflava 430 (see also *Neisseria subflava*)
  polysaccharea 52, 91, 92, 120, 144, 222, 426, 427, 604
  rapid sugar test for 19
  selective medium for 20
  sicca 34, 40, 58, 64, 67, 72, 98, 104, 105, 111, 113, 123, 126, 146, 222, 428, 429, 604
  subflava 34, 40, 58, 65, 67, 72, 98, 104, 105, 111, 113, 123, 126, 146, 222, 428, 430, 431, 567, 581, 582, 604
    biovar perflava 428
  weaveri 58, 65, 98, 105, 123, 126, 148, 368, 432, 433, 604 (formerly CDC group M-5)
Neisseriae 19
Nitrate broth, semiaerobic (Stanier) 14
Nitrate reduction test 14
Nitrite reduction test 15
  test for presence of 15
  test solution #1 15
  test solution #2 15
NO-1 54, 97, 103, 121, 124, 170, 434, 435, 636, 648
NO-2 54, 62, 97, 103, 121, 124, 170, 436, 437, 610, 611, 614 (see also *Bordetella holmesii* [NO-2])
Nutrient agar (Difco) growth test for pathogenic *Neisseria* 15

O-1 42, 46, 48, 52, 75, 78, 90, 91, 115, 117, 118, 120, 184, 446, 447
O-2 52, 65, 91–94, 105–107, 120, 126, 184, 448, 449
Ochrobactrum anthropi 46, 58, 82, 83, 86, 87, 102, 117, 123, 182, 224, 226, 438, 439, 592, 616, 635, 688, 689, 694 (formerly CDC group Vd)
OF (Oxidation/Fermentation) medium 15
  Basal medium 15
    without carbohydrate, table of acid producers 127
OF medium, special (King) 2, 15
OFBA-1 127, 192, 440, 441, 638, 639
Oligella 592
  ureolytica 58, 65, 100, 102, 106, 107, 123, 126, 168, 442, 443, 620, 621 (formerly CDC group IVe)
  urethralis 60, 98, 123, 168, 444, 445, 620, 621 (formerly called *Moraxella urethralis*)
ONPG (O-Nitrophenyl-D-Galactopyranoside) test 16
  medium assay 16
  tablet 16
Ornithine decarboxylase test 7
Oxidase test 16

Pasteurella 17, 112, 386, 450, 456, 586, 587, 620, 624, 626, 642, 664, 674
  aerogenes 34, 67, 68, 111, 132, 450, 451, 587
  anatipestifer 512 (see also *Riemerella anatipestifer*)
  bettyae 33, 34, 36, 40, 66, 67, 71, 73, 110–113, 132, 452, 453, 587 (formerly CDC group HB-5)
  caballi 587
  canis 36, 40, 71, 74, 112, 113, 130, 454, 455, 466, 587
    biotype 1 454
    biotype 2 454
  dagmatis 36, 40, 71, 74, 112, 113, 130, 290, 456, 457, 587 (formerly called Pasteurella gas; *Pasteurella* sp. new species 1; and *Pasteurella pneumotropica* type Henriksen)
  gallinarum 34, 40, 67, 68, 73, 111, 113, 134, 458, 459
  haemolytica 34, 35, 40, 67, 73, 111, 113, 134
    biovar A 460 (see also *Pasteurella haemolytica* sensu stricto)
    sensu stricto 460, 461 (formerly *Pasteurella haemolytica* biovar A)
    biovar T 468 (see also *Pasteurella trehalosi*)
  langaa 587
  multocida 36, 37, 40, 71, 74, 112, 113, 130, 462, 463
    subsp. *gallicida* 131, 462, 463, 587
    subsp. *multocida* 131, 462, 463, 587
    subsp. *septica* 131, 462, 463, 587
  pneumotropica 25, 35, 36, 40, 67, 68, 71, 74, 111–113, 132, 464, 465, 587
    type Henriksen 456 (see also *Pasteurella dagmatis*)
  gas 456 (see also *Pasteurella dagmatis*)
  species new species 1 456 (see also *Pasteurella dagmatis*)
  stomatis 37, 40, 71, 73, 112, 114, 130, 466, 467, 587
  testudinis 586, 587
  trehalosi 11, 67, 111, 134, 468, 469 (formerly biovar T of *Pasteurella haemolytica*)
  ureae 252
Pasteurellaceae 242, 450, 586
Peptic digest of blood 17
Pfeifferella anatipestifer 512 (see also *Reimerella anatipestifer*)
Phenylalanine agar 17
Pigment production by pseudomonads 172
Plesiomonas shigelloides 34, 67, 111
Porphyrin test for *Haemophilus* 18
  substrate solution 18
Potassium hydroxide test for verifying the Gram reaction of bacteria 4, 18
Preservation of bacteria by freezing 18
Pseudomonads, pigment production
Pseudomonas 14, 470, 590, 694
  acidovorans 320 (see also *Comamonas acidovorans*)
  aeruginosa 2, 6, 18, 28, 46, 84, 86–89, 117, 192, 470, 471, 488, 606, 646–648, 670
  alcaligenes 60, 98, 99, 123, 200, 472, 473, 494, 500, 596–598, 612, 640, 652, 720
  azotocolligans 534, 690, 691, 704
  cepacia 42, 46, 76, 77, 79–83, 115, 117, 194, 195, 474, 475, 482, 496, 568, 692, 693 (synonyms are *Pseudomonas multivorans*; *Pseudomonas kingii*; and EO-1)
  denitrificans 46, 84–87, 117, 198, 476, 477, 500, 606, 646–648, 670 (see also *Pseudomonas* species CDC group 1)
  diminuta 46, 60, 78, 80, 98, 99, 117, 123, 202, 478, 479, 592, 688, 694, 695
  fluorescens 28, 46, 88, 89, 117, 192, 480, 481, 498, 606, 646–648, 670
  gladioli 42, 46, 75, 76, 78–82, 115, 117, 194, 482–484, 692, 693 (formerly *Pseudomonas marginata*)
  homology group II species 474, 486
  kingii 474 (*Pseudomonas cepacia*; other synonyms are *Pseudomonas multivorans*; and EO-1)
  luteola 318 (synonym for *Chryseomonas luteola*)
  mallei 42, 46, 48, 52, 75, 76, 83–86, 90–96, 115, 117, 118, 120, 196, 486, 487, 692, 693
  maltophilia 548 (see also *Xanthomonas maltophilia*)

marginata 482, 484 (see also *Pseudomonas gladioli*)
mendocina 46, 84, 86, 117, 192, 488, 489, 644, 645, 696
mesophilica 388 (see also *Methylobacterium mesophilicum*)
multivorans 474 (*Pseudomonas cepacia*; other synonyms are *Pseudomonas kingii* and EO-1)
oryzihabitans 340 (see also *Flavimonas oryzihabitans*)
paucimobilis 534
pertucinogena 46, 78, 117, 212, 490, 491, 644, 645, 696
pickettii 82, 492, 493, 504, 598, 599, 708
  biovar 1 46, 81, 82, 117, 196, 492, 493
  biovar 2 46, 81, 82, 117, 196, 492, 493
pseudoalcaligenes 60, 99, 101, 123, 200, 494, 495, 596–598, 612, 640, 652, 720
pseudomallei 48, 84, 85, 87, 117, 198, 496, 497, 646, 647, 692, 693
putida 28, 48, 83, 85, 88, 117, 192, 498, 499, 606, 646, 648, 670
  1 498, 647
  A 498, 647
  B 498
putrefaciens 528
putrefaciens
  species incertae sedis 528 (see also *Shewanella putrefaciens*)
species CDC group 1 60, 102, 123, 200, 500, 501, 644, 696, 697 (formerly *Pseudomonas denitrificans*)
stutzeri 48, 80, 82, 117, 198, 496, 502, 503, 562, 644, 645, 696 (formerly called CDC group Vb-1)
syringae 636, 646, 648, 649
testosteroni 100, 323, 325 (see also *Comamonas testosteroni*)
thomasii 48, 79–82, 117, 196, 504, 505, 598, 599
vesiculare 506 (see also *Pseudomonas vesicularis*)
vesicularis 48, 52, 60, 65, 78, 91, 94, 98, 100, 104, 106, 107, 117, 120, 123, 126, 202, 506, 507, 592, 688, 694, 695 (formerly called *Pseudomonas* vesiculare)
Pseudomonas-like group 2 46, 81, 82, 85, 86, 117, 194, 484, 485, 606, 646–648, 670, 688
Psychrobacter 21, 98–100, 105, 106, 167, 173, 337, 339, 509, 510
  immobilis 21, 48, 60, 65, 78–81, 83–85, 98–100, 105, 106, 117, 123, 126, 166, 172, 336, 338, 404, 508–510, 680, 682, 684, 698
  transformation assay for 21

Rapid gelatin slant 19
Rapid sugar test for *Neisseria* 19
  reagents 19
Red Phantom 388
Rickettsia 286, 287
Riemerella
  anatipestifer 40, 52, 72, 74, 91, 93, 114, 120, 152, 512, 513, 700, 701 (former names include: *Pfeifferella anatipestifer*; *Pasteurella anatipestifer*; and *Moraxella anatipestifer*)
Rochalimaea 62, 63, 103, 104, 124, 125, 214, 286, 288 (see also *Bartonella*)

Roseomonas 188, 514–516, 518, 520, 522, 524, 526, 634 (formerly CDC pink coccoid groups I-IV)
  cervicalis 60, 98, 99, 101, 123, 188, 515–517, 520, 592, 616, 635, 688, 689, 694
  fauriae 52, 60, 65, 81, 82, 93, 95, 100–102, 107, 108, 117, 120, 123, 126, 190, 515, 518, 519, 524, 618, 634, 676, 677
  genomospecies 4 60, 101, 123, 188, 515, 520, 521, 592, 616, 635, 688, 689, 694
  genomospecies 5 60, 65, 99, 101, 106, 108, 123, 126, 188, 515, 520, 522, 523, 592, 616, 635, 688, 689, 694
  genomospecies 6 60, 65, 100, 107, 123, 126, 190, 515, 524, 525, 618, 634, 676, 677
  gilardii 42, 48, 52, 54, 60, 62, 65, 75, 78, 80, 90–95, 97–101, 103, 105–108, 115, 117, 118, 120, 121, 124, 126, 188, 515, 522, 526, 527, 592, 635, 688, 689, 694
Ryu stain 10

Saline, buffered 20
Salmonella 20, 708, 720
  dublin 652, 653
  newport 652, 653
  oranienburg 652, 653
  typhi 652, 653
  zaiman 652, 653
Salt tolerance test 20
Saponification procedure for cellular fatty acid analysis 576
Selective medium for pathogenic *Neisseria* 20
Shewanella 528
  putrefaciens 528, 529 (formerly CDC groups Ib-1 and Ib-2; designated *Pseudomonas putrefaciens* species incertae sedis in *Bergey's Manual of Systematic Bacteriology*)
  biotype 1 48, 60, 82, 83, 99, 100, 118, 124, 206, 528, 530, 702, 703, 716
  biotype 2 60, 99, 100, 124, 206, 528, 531, 702, 703, 716
  species incertae sedis 528
Shigella 20
Silver stain 10
Simmons citrate 6
Simonsiella 114, 532, 533
  crassa 40, 72, 150, 532, 533, 668, 674, 675
Simonsiella-like 40, 72, 73, 150
  -like species 72, 73
  muelleri 532
  steedae 532
Soluble brown pigments 2
Sphingobacterium 350, 352, 356, 358, 558, 628, 642, 654
  mizutaii 628, 654, 656, 658, 660, 661, 700
  multivorum 628, 654, 656, 658, 660, 661, 700
  spiritivorum 628, 654, 658, 660, 661, 700
  thalpophilum 628, 654, 656, 658, 660, 661, 700
  yabuuchiae
Sphingomonas 534
  adhaesiva 690, 704, 705
  capsulata 690, 691, 704
  genomospecies 2 690, 691, 704
  parapaucimobilis 42, 48, 52, 76, 79, 90, 95, 115, 118, 120, 184, 534–536, 690, 704, 705
  paucimobilis 42, 48, 52, 76, 79, 90, 95, 115, 118, 120, 184, 268, 446, 506, 534, 535, 690, 691, 704
  yanoikuyae 690, 691, 704
Sporocytophaga myxococcoides 706, 707
SS agar (Difco) 20
Stenotrophomonas 548
Streptobacillus 10
  moniliformis 37, 69, 112, 140, 538, 539
Streptococcus 384
Suttonella indologenes 35, 40, 67, 73, 111, 114, 152, 314, 540, 541, 602, 603, 630, 640, 674 (Previously *Kingella indologenes*)

Tatumella ptyseos 598, 708, 720
Taylorella 542
  equigenitalis 60, 65, 98, 105, 124, 126, 158, 542, 543
Tjernberg's genomospecies 14 115
TM-1 382 (see also *Kingella denitrificans*)
Transformation assay 21
  for *Acinetobacter* 20
  for *Moraxella osloensis* 20
  for *Psychrobacter* 22
Trifluoroacetylation of fatty acid methyl esters (FAME) 578
Triple sugar iron (TSI) agar 22
Tryptone glucose yeast extract (TGY) agar 22

Urea agar 22

Variovorax cyclosites 322 (see also *Comamonas terrigena*)
  neocistes 322 (see also *Comamonas terrigena*)
Vibrio 34, 38, 67, 111, 114, 710, 712
  cholerae 710–712
  extorquens 388 (see also *Methylobacterium extorquens*)
  fluvialis 710–712
  furnissii 710, 712, 713
  parahaemolyticus 710–712
Vibrionaceae 72
Voges-Proskauer (VP) test 22

Weeksella 628, 632, 642, 650, 656, 660, 700
  virosa 20, 65, 106, 126, 158, 544, 545, 554, 628, 654, 655, 660 (formerly CDC group IIf)
  zoohelcum 65, 107, 126, 158, 546, 547, 628, 654, 655, 660 (formerly CDC group IIj)
WO-1 232, 596, 602, 640, 641
WO-1A 596, 602, 640, 641
Wolinella 714
  curva 666, 714, 715
  recta 714, 715
  succinogenes 666, 714, 715

Xanthomonas 548, 702, 716
  beticola 716, 717
  campestris 716, 717
  maltophilia 2, 42, 54, 75–77, 97, 115, 121, 186, 548, 549, 568, 716, 717 (previously described as *Pseudomonas maltophilia*)
  nakataecorchori 716, 717
  phaseoli 716, 717
Xenorhabdus luminescens 718, 719

Yersinia enterocolitica 720, 721
Yersinia philomiragia 362 (see also designated *Francisella philomiragia*)